THE MECHANISMS OF ATMOSPHERIC OXIDATION OF AROMATIC HYDROCARBONS

THE MECHANISMS OF ATMOSPHERIC OXIDATION OF AROMATIC HYDROCARBONS

JACK G. CALVERT, ROGER ATKINSON, KARL H. BECKER,
RICHARD M. KAMENS, JOHN H. SEINFELD,
TIMOTHY J. WALLINGTON, AND GREG YARWOOD

OXFORD
UNIVERSITY PRESS

2002

OXFORD
UNIVERSITY PRESS

Oxford New York
Athens Auckland Bangkok Bogotá Buenos Aires Calcutta Cape Town
Chennai Dar es Salaam Delhi Florence Hong Kong Istanbul Karachi
Kolkata Kuala Lumpur Madrid Melbourne Mexico City Mumbai
Nairobi Paris São Paulo Shanghai Singapore Taipei Tokyo Toronto Warsaw

and associated companies in
Berlin Ibadan

Published by Oxford University Press, Inc.
198 Madison Avenue, New York, New York 10016

Library of Congress Cataloging-in-Publication Data
Mechanisms of atmospheric oxidation of aromatic hydrocarbons / by Jack G. Calvert . . . [et al.].
p. cm.
Includes bibliographical references and index.
ISBN 0-19-514628-X
1. Atmospheric chemistry. 2. Aromatic compounds—Oxidation. I. Calvert,
Jack G. (Jack George), 1923–
QC879.7.M42 2001
515.51′1—dc21 00-066948

9 8 7 6 5 4 3 2

Printed in the United States of America
on acid-free paper

Contents

Acknowledgments

The authors, members of the Coordinating Research Council's Review Panel on the Atmospheric Chemistry of the Hydrocarbons (CRC-RPACH), thank several agencies and their members for their support of this research: the Coordinating Research Council (3650 Mansell Road, Suite 140, Alpharetta, GA 30022–8246), the California Air Resources Board (2020 L Street, Sacramento, CA 95812), the National Renewable Energy Laboratory (1617 Cole Boulevard, Golden, CO 80401), and the members of the Coordinating Research Council's Atmospheric Impacts Committee: S. Kent Hoekman (Co-Chair), Chevron Products, San Ramon, CA; Steven M. Japar (Co-Chair), Ford Motor Company, Chemistry Department, Dearborn, MI; David P. Chock, Ford Motor Company, Scientific Research Laboratory, Dearborn, MI; Alan M. Dunker, Chemical & Environmental Sciences, General Motors R&D Center, Warren, MI; Eric Foote, DaimlerChrysler Corporation, Chelsea, MI; Lee Gilmer, Air Quality & Environmental Analytical, Equilon Enterprises, Houston, TX; Miriam Lev-On, ARCO, Los Angeles, CA; Rory MacArthur, Chevron Products, San Ramon, CA; Kenneth D. Rose, Esso Petroleum Company, Ltd., Esso Research Center, Milton Hill, Abingdon, United Kingdom; Charles H. Schleyer, ExxonMobil Research & Engineering Co., Paulsboro Technology Center, Paulsboro, NJ; John Shiller, Ford Motor Company, Fairlane Business Park, Allen Park, MI; Jim J. Simnick, BP Amoco Research Laboratories, Naperville, IL.

The authors also thank the following persons for their significant efforts related to this research: Brent K. Bailey, Timothy C. Belian, Coordinating Research Council (CRC), Inc., Alpharetta, GA; Bart Croes, California Air Resources

Board, Research Division, Sacramento, CA. We thank Lois Mott (CRC) for her careful proofreading of an early draft of this book.

The authors acknowledge their research organizations and educational institutions for support of their participation in this research: National Center for Atmospheric Research (NCAR), Atmospheric Chemistry Division, Boulder, CO (Jack G. Calvert); Air Pollution Research Center (APRC), the Department of Soil and Environmental Sciences, University of California, Riverside, CA (Roger Atkinson); Bergische Universität-Gesamthochschule, Fachbereich Chemie, Wuppertal, Germany (Karl H. Becker); University of North Carolina at Chapel Hill, Department of Environmental Sciences & Engineering, Chapel Hill, NC (Richard M. Kamens); California Institute of Technology, Department of Chemical Engineering, Pasadena, CA (John H. Seinfeld); Ford Motor Company, Dearborn, MI (Tim Wallington); ENVIRON, Novato, CA (Greg Yarwood).

The authors thank their many coworkers who provided important new information to the authors: Björn Klotz, and Ian Barnes of the Bergische Universität-Gesamthochscule, Wuppertal, for their help in providing much new information and spectral data to the authors; Sangdon Lee, University of North Carolina (UNC), Chapel Hill, for literature and vapor pressure work related to the PAHs; Eric Teinemmaa (UNC) for creating and regenerating many of the figures in chapter IX; Michael Tolacka (UNC) for help with the flame combustion literature; Xiaoyu Liu (UNC) for providing unpublished spectra of the unsaturated diones.

THE MECHANISMS OF ATMOSPHERIC OXIDATION OF AROMATIC HYDROCARBONS

I

Importance of Aromatic Hydrocarbons in the Chemistry of Ozone Generation in the Urban Atmosphere

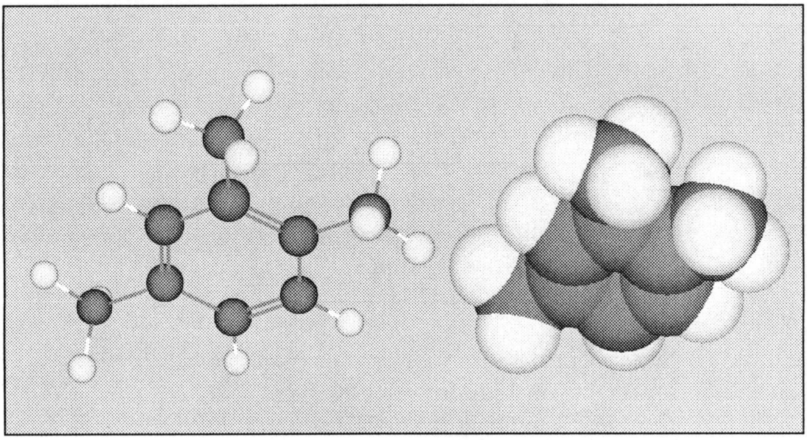

Organic compounds are important trace constituents of the atmosphere that originate from many different natural and man-made sources. The atmospheric reactions of organic compounds influence other atmospheric trace constituents. In particular, reactions of volatile organic compounds (VOCs) with nitrogen oxides (NO_x) in the presence of sunlight can lead to the formation of ozone, other oxidants, and particulate matter (PM), commonly referred to as "smog" (see, e.g., Warneck, 1988; Seinfeld and Pandis, 1998; Finlayson-Pitts and Pitts, 1999). In many urban areas, emissions of VOCs and NO_x, when combined under meteorological conditions that favor smog formation, can lead to the formation of ground-level ozone and PM at harmful levels. In some areas, such as the eastern United States, ozone and PM are a "regional-scale" as well as an "urban-scale" problem because significant atmospheric transport of pollutants occurs between urban and rural areas, or between neighboring urban areas. Similar situations have been observed in Europe.

As mentioned above, formation of ground-level ozone is not the only harmful consequence of the atmospheric reactions of VOCs. Other oxidants present in smog include hydrogen peroxide, peroxy acetyl nitrate (PAN), organic peroxides, organic acids, and oxidizing transient species such as hydroxy (OH), hydroperoxy (HO_2), and organic peroxy (RO_2) radicals. The oxidants produced by smog-forming reactions of VOCs can oxidize atmospheric SO_2 to sulfate and NO_2 to nitrate, and this couples smog formation with other air pollution problems, such as acid deposition, particulate matter formation, and visibility impairment. Also, some air pollution problems result directly from the presence of VOCs in the atmosphere. For example, benzene, 1,3-butadiene, formaldehyde, acetaldehyde, and polycyclic organic matter (POM) are toxic compounds, and in the United States their release into the atmosphere is regulated by the U.S. Environmental Protection Agency (EPA). Some toxic VOCs are emitted directly to the atmosphere (e.g., benzene, 1,3-butadiene, and POM), whereas others are formed in situ from other VOCs by chemical reactions (e.g., formaldehyde and acetaldehyde). Polycyclic aromatic hydrocarbons (PAHs) are an important component of POM and are discussed further in chapter IX. Emissions of VOCs also contribute to PM formation via the formation of semivolatile reaction products that undergo gas-to-particle conversion in the atmosphere (see chapter VIII). Understanding the atmospheric chemistry of VOCs is central to our understanding of the relationships between VOC emissions and these different types of atmospheric pollution.

Aromatic hydrocarbons are an important class of VOCs present in the atmosphere, and chemical oxidation through atmospheric reactions is their main removal pathway from the atmosphere. The main chemical species that initiates reactions of aromatics in the atmosphere is the hydroxyl radical (OH), as discussed in chapter II. Reaction with the nitrate radical (NO_3) is also significant for aromatic hydrocarbons that contain alkene or phenolic groups, as discussed in chapter III, and reactions of aromatic alkenes with ozone are discussed in chapter IV. The reaction of aromatics with ground-state oxygen atoms, $O(^3P)$, is of little importance in the urban atmosphere (see chapter V). Aromatic hydrocarbons generally react rapidly with OH so that the typical atmospheric lifetimes of most of these are in the range of hours to days. Consequently, most aromatic hydrocarbons will tend to react in fairly close proximity to where they are emitted. The least reactive aromatic hydrocarbon, benzene, has an atmospheric lifetime which is in the order of 10 days, and its transport can occur for hundreds or thousands of kilometers.

One of the objectives of this review is to assemble information of use for atmospheric chemistry/transport models. Atmospheric models are valuable tools for integrating our knowledge of the atmospheric chemistry of VOCs with our understanding of other important atmospheric process; for example, emissions, transport, dispersion, and deposition. Various levels of detail are useful in modeling the atmospheric chemistry of VOCs: models with simplified spatial resolution (e.g., box or trajectory models) are capable of incorporating all of the detailed chemical information available for many different species; on the other hand, models with complex physical structure (e.g., three-dimensional grid models) require condensed (i.e., simplified) chemical mechanisms to work within the constraints of limited computational resources. The objective of this review is to assemble and evaluate all the information available in the peer-reviewed literature on the kinetics and mechanisms of the reactions of aromatic hydrocarbons under tropospheric conditions for the aromatic hydrocarbons emitted into the atmosphere and their first-generation products.

I-A. Typical Composition of the Hydrocarbons Present in Urban Air

There are numerous sources of VOCs in urban air, as illustrated by the VOC emissions inventory shown in table I-A-1 (EPA, 1998). This table shows the importance of industrial emissions (particularly solvent usage), as well as transportation emissions—both on- and off-road. The decrease in on-road vehicle emissions from 1990–2000 is due to the impact of emission control measures and improved emissions control technology. Nevertheless, on-road vehicles still tend to be an important source of ambient VOCs in major urban areas because of the high density of roads and vehicle traffic.

To illustrate the relative importance of different VOCs in typical urban air, representative average VOC composition profiles were calculated for air samples taken in recent field studies in the Los Angeles and Boston areas (Fujita et al., 1997). The analysis includes only samples taken during the early morning hours because the composition of VOCs in the atmosphere is expected to resemble fresh emissions most strongly in the morning, when atmospheric mixing and chemical reactions are both relatively slow.

I-A-1. Typical Composition of VOCs in the Atmosphere of Los Angeles, California

A total of 24 samples were collected at eight sites across the Los Angeles urban area from 6 to 7 A.M. or 7 to 8 A.M. on October 10–12, 1995 (Fujita et al., 1997). Samples were analyzed by gas chromatography (GC) for non-methane hydrocarbons (NMHCs) with between 2 and about 11 carbon atoms (C_2–C_{11}). The concentrations of the most abundant identified compounds are shown in table I-A-2. The first column of numbers gives the average concentration across all samples, while the second and third columns give the highest and lowest concentrations observed across all samples in ppbC.* The average concentration for each species was calculated as follows: first, the species concentrations in each sample were converted to percentage of total NMHCs (mass percent); second, the weight

* *Units for ambient air measurements: mass concentrations and volume mixing ratios.* Ambient air concentrations are commonly given either in volume mixing ratios as ppbv (1:10^9) or in mass concentration in $\mu g\, m^3$ (10^{-6} g m^{-3} of air). The unit ppbv or ppb can also be seen as a molar unit. In air chemistry it is common to multiply the unit of ppbv with the number of carbon atoms in a molecule of the particular substance, for example, for benzene, 6, and for toluene, 7. This unit ppbC is now related to the moles of carbon in ambient air and reflects the potential reactivity of all carbon atoms in a molecule. The emission values are given in mass time^{-1} units, usually in Tg yr^{-1} (10^{12} g yr^{-1}) or in tons yr^{-1}. Both the short ton [2000 lb (avoirdupois)] and the metric ton (1000 kg = 2204.6 lb) are in common usage today. Note that 1 Tg yr^{-1} ≡ 1 million metric tons yr^{-1}.

Table I-A-1. Emissions of volatile organic compound in the United States for 1988–1997 (thousands of short tons)

Source Category	1988	1989	1990	1991	1992	1993	1994	1995	1996	1997
1 Fuel combustion	1360	1372	1005	1075	1114	993	989	1073	1079	861
a Electric utilities	37	37	47	44	44	45	45	44	49	51
Coal	27	27	27	27	27	29	29	29	28	29
Oil	7	7	6	5	4	4	4	3	3	3
Gas	2	2	2	2	2	2	2	2	8	8
Internal combustion	1	1	12	10	10	10	10	10	10	10
b Industrial	136	134	182	196	187	186	196	206	208	217
Coal	7	7	7	6	7	6	8	6	6	6
Oil	16	16	12	11	12	12	12	12	12	12
Gas	61	61	58	60	52	51	63	73	73	77
Other	36	36	51	51	49	51	50	50	51	53
Internal combustion	15	15	54	68	66	66	64	65	66	69
c Other	1188	1200	776	835	884	762	748	823	822	593
Residential wood	1155	1169	718	776	822	698	684	759	758	527
Other	33	31	58	59	62	64	63	64	64	9
2 Industrial processes	10,853	10,755	10,000	10,178	10,380	10,578	10,738	10,780	9482	9836
Chemical & allied processing	982	980	634	710	715	701	691	660	436	458
Metals processing	74	74	122	123	124	124	126	125	70	73
Petroleum & related industries	645	639	612	640	632	649	647	642	517	538
Other industrial processes	408	403	401	391	414	442	438	450	439	458
Solvent utilization	5945	5964	5750	5782	5901	6016	6162	6183	6273	6483
Storage & transport	1842	1753	1495	1532	1583	1600	1629	1652	1312	1377
Waste disposal & recycling	959	941	986	999	1,010	1,046	1,046	1,067	433	449
3 Transportation	10,583	9506	8765	8965	8569	8619	8940	8106	7899	7660
On-road vehicles	8290	7192	6313	6499	6072	6103	6401	5701	5502	5230
Non-road sources	2293	2314	2452	2466	2498	2516	2538	2405	2397	2430

4 Miscellaneous	1231	642	1164	845	579	641	798	599	846	858
Other combustion	1230	641	1064	756	485	535	710	511	761	770
Fires	1230	641	1061	753	482	532	707	508	758	767
Other	NA	NA	3	3	3	3	3	3	3	3
Other	1	1	100	89	94	105	88	88	85	87
Total (all sources)	24,027	22,274	20,935	21,063	20,642	20,830	21,465	20,558	19,306	19,214

NA, not available.

Source: EPA (1998).

Table I-A-2. Typical composition of Los Angeles morning air in October 1995 and contribution to reactivity as estimated using OH reactivity and the Carter (1994) maximum incremental reactivity (MIR) factors

	Average ppbC	Min ppbC	Max ppbC	Percent of OH Reactivity	Percent of MIR Reactivity
Methane	2000			1.7	2.3
CO	500			11.6	3.8
Alkanes (42% of NMHC)	211.2			14.6	18.4
2-Methylbutane	28.69	8.01	74.64	2.1	3.0
Propane	27.50	9.18	51.49	1.0	1.0
Ethane	20.23	7.92	47.12	0.2	0.4
n-Butane	16.58	5.01	32.58	1.0	1.3
n-Pentane	15.39	4.58	34.85	1.2	1.3
2-Methylpentane	12.26	3.99	35.78	1.1	1.5
Methylcyclopentane	8.73	2.43	25.96	1.0	1.9
n-Hexane	8.16	2.27	23.67	0.7	0.6
2-Methylpropane	8.08	3.09	12.86	0.4	0.7
2,2,4-Trimethylpentane	7.86	2.11	24.08	0.3	0.6
3-Methylpentane	7.41	2.80	21.28	0.6	0.9
3-Methylhexane	6.72	1.97	16.68	0.7	0.7
Methylcyclohexane	6.28	1.32	13.95	0.9	0.9
n-Heptane	5.02	1.20	11.47	0.5	0.3
2-Methylhexane	4.94	1.45	14.58	0.5	0.4
2,3-Dimethylpentane	4.72	1.23	14.21	0.3	0.6
2,3-Dimethylbutane	3.27	0.91	9.57	0.3	0.3
Cyclohexane	3.14	1.07	8.65	0.4	0.3
2,4-Dimethylpentane	2.81	0.77	8.96	0.2	0.4
2-Methylheptane	2.29	0.70	6.62	0.2	0.2
2,3,4-Trimethylpentane	2.23	0.57	8.67	0.2	0.3
n-Octane	2.09	0.77	6.02	0.2	0.1
Cyclopentane	1.95	0.58	4.80	0.2	0.4
3-Methylheptane	1.94	0.54	6.59	0.2	0.2
n-Nonane	1.62	0.51	4.26	0.2	0.0
2,2-Dimethylbutane	1.24	0.42	3.40	0.0	0.0
Alkenes (7% of NMHC)	34.2		27.3		18.8
Ethene	14.91	4.31	45.67	6.2	7.7
Propene	5.88	1.48	20.22	5.0	4.1
2-Methylpropene and 1-butene	4.97	1.23	20.27	6.2	1.8
2-Methyl-2-butene	1.48	0.38	6.96	2.1	0.9
trans-2-Pentene	1.16	0.00	5.33	1.1	0.8
1-Pentene	0.92	0.14	2.75	0.6	0.4
trans-2-Hexene	0.80	0.13	2.31	0.9	0.5
2-Methyl-1,3-butadiene	0.68	0.14	2.47	1.3	0.4
cis-2-Pentene	0.66	0.13	2.78	0.8	0.5
trans-2-Butene	0.60	0.17	2.48	0.9	0.5
cis-2-Butene	0.56	0.11	2.55	0.8	0.5
Cyclopentene	0.38	0.07	1.77	0.5	0.2
3-Methyl-1-butene	0.35	0.09	1.10	0.2	0.2
2-Methyl-1-pentene	0.32	0.00	1.33	0.3	0.0
4-Methyl-1-pentene	0.29	0.00	0.61	0.2	0.1
cis-2-Hexene	0.23	0.05	1.04	0.2	0.1

Table I-A-2. (*continued*)

	Average ppbC	Min ppbC	Max ppbC	Percent of OH Reactivity	Percent of MIR Reactivity
Aromatics (19% of NMHC)	96.4		17.9		29.9
Toluene	33.02	9.36	91.01	2.7	6.1
m- and *p*-Xylene	19.92	5.77	68.20	4.6	9.4
1,2,4-Trimethylbenzene	12.12	3.90	32.72	4.2	7.1
Benzene	9.73	3.62	30.05	0.2	0.3
o-Xylene	7.14	2.17	23.75	1.2	3.0
Ethylbenzene	5.41	1.76	16.79	0.5	1.0
1,3,5-Trimethylbenzene	3.47	0.83	14.39	2.1	2.3
Styrene	3.06	0.79	10.30	2.2	0.4
n-Propylbenzene	1.60	0.57	4.40	0.1	0.2
iso-Propylbenzene	0.94	0.20	6.10	0.0	0.1
Other (33% of NMHC)	165.2		27.0		26.8
Ethyne	17.35	5.76	54.86	0.7	0.6
Methyl *tert*-butyl ether (MTBE)	20.46	6.80	62.64	1.2	1.1
Other identified	73.86	23.22	222.93	14.6	14.6
Unidentified	53.57	19.38	166.14	10.6	10.6
Total	3007			100	100

Notes

There were 24 samples at eight sites for NMHCs.

Representative values were assumed for CO and methane.

Two pairs of products coeluted and are reported together, *m*- and *p*-xylene, and 2-methylbutene and 1- butene.

Concentrations below detection limit were reported as zero.

"Other identified" and "unidentified" were assumed to have the same OH rate coefficients and MIR reactivities as the average for identified NMHCs.

percentages were averaged over all samples; finally, the average mass percentage for each species was multiplied by the average NMHC concentration to convert back to concentration (parts-per-billion carbon, ppbC). Methane and CO were not quantified in this measurement program, but representative values of 2000 ppb and 500 ppb are included for purposes of the reactivity calculations shown on the right of the table.

On average, about 75% of the carbon detected by the GC system (i.e., C_2–C_{11} NMHCs) is accounted for by the 54 NMHCs reported in table I-A-2, with another 15% accounted for by other identified compounds and 10% by compounds that were quantified, but not identified. The likely contributions of oxygenated compounds other than MTBE and hydrocarbons larger than C_{11} are discussed below.

Since the compounds reported in table I-A-2 differ widely in their reactivity toward atmospheric oxidants and their tendency to form ozone, the average concentration profile was "reactivity weighted" using two different methods. The fourth column shows the average concentrations weighted by OH rate coefficient (Atkinson, 1994), with updated values from this book and renormalized to

100%. This can be thought of as representing the relative rates at which OH would be scavenged by the VOCs in this mixture. The OH rate coefficient weighting is a widely used and easily understood method for reactivity weighting profiles of VOC composition; however, it does not account for reactions with oxidants other than OH or for important differences between VOCs in the yields of reactive products formed during reaction. Carter has developed a useful reactivity scale that accounts approximately for all of these factors based on photochemical model calculations (Carter, 1994). Carter's maximum incremental reactivity (MIR) scale approximates the potential of different compounds to form ozone under conditions where the availability of NO_X does not limit ozone formation. Another approach to the development of a reactivity scale has been taken by Derwent and coworkers (Derwent and Jenkin, 1991; Derwent et al., 1996, 1998). During the transport of polluted air masses, the formation of the ozone increment above a certain area emitting additional ozone precursors has been calculated relative to the ozone increment caused by ethene. Using this procedure, POCP factors (photochemical ozone creation potential) were derived. For example, these reactivity factors have been used in the United Kingdom to calculate POCP-weighted emissions. In the United Kingdom, 39.7% of the POCP-weighted emissions are attributed to the aromatic hydrocarbons, 30.2% to the alkanes, and 14.8% to the alkenes. A comparison of MIR and POCP factors (Derwent et al., 1998) shows reasonable agreement for a number of VOCs. However, in the case of benzene and toluene, the POCP factors are significantly higher than MIR values. In the reactivity estimations made in this chapter, only MIR and OH reactivity factors have been used. In the fifth column in the tables I-A-2 and I-A-3, the average concentrations weighted by MIR factor and renormalized to 100% are shown. The relative contributions of different compounds and groups of compounds to the total carbon and to reactivity are discussed in the following sections.

I-A-1a. Oxygenates

The only oxygenated compound quantified in table I-A-2 is methyl *tert*-butyl ether (MTBE), which is a component of reformulated gasolines. The reason for this is that most carbonyls cannot be quantified using the methodology used for ambient hydrocarbons. However, Grosjean et al. (1996) have reported detailed carbonyl measurements for the Los Angeles area. They collected 4-h integrated samples around the clock at four sites in the Los Angeles urban area on September 8 and 9, 1993. They identified and quantified 23 carbonyl compounds: 14 aliphatic aldehydes (from formaldehyde to tetradecanal), two aromatic aldehydes (benzaldehyde and *m*-tolualdehyde), three ketones (acetone, 2-butanone, and cyclohexanone), one unsaturated aldehyde (crotonaldehyde), and three dicarbonyls (glyoxal, methylglyoxal, and biacetyl). Another 19 carbonyls were tentatively identified, although some may have been artifacts of the measurement technique. On a volume mixing ratio basis (ppbv), the most abundant carbonyls were formaldehyde, acetaldehyde, and acetone, which together account for about half of the total observed carbonyls. Total carbonyl concentrations averaged 22 ppb over all the samples. The morning samples collected between 6 and 10 A.M. are most comparable to the data reported in table I-A-2, and for this time period formaldehyde levels were in the range 3–11 ppb (average about 7 ppb), acetaldehyde levels were in the range 2–9 ppb (average about 5 ppb), and acetone levels were in the range 0–4 ppb (average about 1.5 ppb). The contribution of carbonyls relative to hydrocarbons is expected to be greater later in the day because carbonyls have secondary sources (i.e., chemical reactions) in addition to primary sources.

I-A-1b. Semivolatile Compounds

The standard detection methodology for ambient hydrocarbons, as used for the sampling shown in table I-A-2, does not quantify the contribution of compounds with more than about 10 or 12 carbon atoms because their volatility is too low to allow them to pass quantitatively through the sampling system employed. Zielinska and Fujita (1994) have used a different sampling method designed to capture semivolatile hydrocarbons to look at their contribution to ambient hydrocarbons in Los Angeles. Eight samples were collected in downtown Los Angeles between 6 A.M. and 12 noon on August 21–23, 1991, and hydrocarbons in the range C_2–C_{18} were quantified. It was found that C_{12}–C_{18} compounds contributed 3–8% of the total ppbC and C_{10}–C_{18} compounds contributed 7–15% of the total ppbC. The largest contributors to the semivolatile compounds identified were multisubstituted alkylbenzenes. Other semivolatile compounds identified were polycyclic aromatic hydrocarbons and alkanes.

I-A-2. Typical Composition of VOCs in the Urban Atmosphere of Boston, Massachusetts

A total of 41 samples were collected at five sites in the Boston area from 7 to 8 A.M. on 12 days in August and September of 1995 (Fujita et al., 1997). Samples were analyzed by gas chromatography (GC) for C_2–C_{11} NMHCs. Methane and CO were also quantified in 14 of the samples at four sites. The concentrations of the most abundant identified compounds are shown in table I-A-3. The VOC speciation for the Boston samples was more detailed than for the Los Angeles samples described in the previous section. For some VOCs, identification by mass spectroscopy yielded the carbon number and compound type but not the exact identity, and these are reported in table I-A-3 using names such as "C_{10}-alkane."

On average, about 79% of the carbon detected by the GC system (i.e., C_2–C_{11} NMHCs) is accounted for by the 136 NMHC species reported in table I-A-3, with another 9% accounted for by other identified compounds and 13% by compounds that were quantified, but not identified. The only oxygenated compound quantified was MTBE. As discussed previously for Los Angeles, oxygenated compounds other than MTBE and hydrocarbons larger than C_{11} would also contribute to the total VOCs in Boston air.

Table I-A-3. Typical composition of Boston morning air in August/September 1995 and contribution to reactivity as estimated using OH reactivity and the Carter (1994) maximum incremental reactivity (MIR) factors

	Average ppbC	Min ppbC	Max ppbC	Percent of OH Reactivity	Percent of MIR Reactivity
Methane	1939	1730	2130	0.37	2.97
CO	675	260	1520	3.56	6.70
Alkanes (36% of NMHC)	98.6		11.7		11.2
2-Methylbutane	12.88	1.16	173.75	1.12	1.79
Ethane	8.25	2.10	15.91	0.05	0.22
n-Pentane	6.29	0.57	110.79	0.57	0.67
2-Methylpentane	5.56	0.57	16.36	0.69	0.89
Propane	5.50	1.13	8.51	0.14	0.27
n-Butane	4.64	0.59	23.06	0.26	0.49
2,2,4-Trimethylpentane	4.04	0.36	23.19	0.33	0.38
n-Hexane	3.93	0.42	14.93	0.49	0.40
2,2-Dimethylbutane	3.63	0.31	39.95	0.19	0.30
3-Methylhexane	3.52	0.69	10.94	0.56	0.51
3-Methylpentane	3.43	0.39	8.60	0.43	0.53
2-Methylpropane	2.81	0.26	53.96	0.15	0.33
2,3-Dimethylbutane	2.28	0.00	15.55	0.28	0.23
n-Decane	2.23	0.28	16.84	0.57	0.12
2,3-Dimethylpentane	2.11	0.22	9.31	0.34	0.32
n-Undcane	1.98	0.27	23.03	0.58	0.10
Methylcyclopentane	1.84	0.17	15.22	0.33	0.52
n-Heptane	1.82	0.24	8.97	0.29	0.16
2-Methylhexane	1.77	0.17	5.84	0.27	0.20
2,6-Dimethyloctane	1.62	0.00	11.20	0.31	0.21
Methylcyclohexane	1.47	0.18	7.02	0.33	0.28
2,3,4-Trimethylpentane	1.40	0.10	8.29	0.27	0.23
n-Octane	1.23	0.13	9.62	0.24	0.08
2,4-Dimethylpentane	1.16	0.15	4.53	0.18	0.22
n-Nonane	1.09	0.19	4.78	0.24	0.07
iso-Propylcyclohexane	0.99	0.23	7.25	0.31	0.23
2-Methylheptane	0.96	0.08	6.26	0.18	0.10
2,4-Dimethylhexane	0.93	0.09	5.86	0.18	0.14
3-Methylheptane	0.87	0.15	5.41	0.16	0.09
Cyclohexane	0.80	0.07	15.06	0.13	0.11
Cyclopentane	0.72	0.09	12.40	0.08	0.17
2,5-Dimethylheptane	0.66	0.14	5.36	0.16	0.08
2,3-Dimethylhexane	0.63	0.08	2.83	0.12	0.09
n-Dodecane	0.52	0.00	2.57	0.16	0.02
3-Methyloctane	0.52	0.00	1.71	0.12	0.06
3-Ethylpentane	0.51	0.00	1.01	0.08	0.07
3,3-Dimethylpentane	0.44	0.10	1.13	0.03	0.03
C-10 Alkanes(2)	0.43	0.00	1.21	0.12	0.05
2,5-Dimethylhexane	0.43	0.00	2.70	0.08	0.07
4-Methylheptane	0.36	0.00	4.99	0.07	0.05
1,3-Dimethylcyclopentane	0.31	0.00	0.89	0.06	0.08
2,6-Dimethylheptane	0.29	0.06	1.46	0.07	0.04
C-8 Alkanes(2)	0.26	0.00	1.26	0.05	0.03
3,3-Dimethylheptane	0.25	0.00	1.43	0.06	0.03
3,6-Dimethyloctane	0.24	0.00	1.60	0.06	0.03

Table I-A-3. (*continued*)

	Average ppbC	Min ppbC	Max ppbC	Percent of OH Reactivity	Percent of MIR Reactivity
Alkanes (*continued*)					
C-11 Alkane(1)	0.21	0.00	1.15	0.07	0.03
4,4-Dimethylheptane	0.19	0.00	2.08	0.04	0.02
C-10 Alkane(1)	0.15	0.00	5.35	0.04	0.02
2,2,5-Trimethylhexane	0.13	0.00	1.68	0.02	0.01
C-8 Alkanes(3)	0.10	0.00	1.02	0.02	0.01
C-8 Alkane(1)	0.09	0.00	0.27	0.02	0.01
C-9 Alkanes(2)	0.06	0.00	0.38	0.01	0.00
2-Methyloctane	0.02	0.00	0.10	0.00	0.00
C-9 Alkanes(3)	0.02	0.00	1.45	0.00	0.00
C-9 Alkane(1)	0.00	0.00	0.50	0.00	0.00
C-11Alkanes(2)	0.00	0.00	0.12	0.00	0.00
C-9 Alkanes(4)	0.00	0.00	0.20	0.00	0.00
Alkenes (10% of NMHC)	26.6			24.2	18.0
Ethene	6.86	1.20	14.22	1.27	4.63
Propene	3.09	0.55	8.79	1.77	2.81
2-Methylpropene	2.64	0.42	5.28	2.95	1.29
2-Methyl-1,3-butadiene	1.62	0.21	3.99	3.54	1.37
2-Methyl-2-butene	1.23	0.00	4.61	2.33	1.18
2,4,4-Trimethyl-1-pentene	1.16	0.00	2.36	0.93	0.34
trans-2-Pentene	0.84	0.07	4.47	1.21	0.81
1-Butene	0.69	0.16	1.87	0.47	0.60
1,3-Butadiene	0.64	0.11	1.16	0.93	0.64
2-Methyl-1-butene	0.62	0.00	2.17	0.82	0.28
α-Pinene	0.50	0.00	3.38	0.59	0.16
trans-2-Butene	0.46	0.00	1.78	0.64	0.51
1-Pentene	0.43	0.04	1.56	0.29	0.25
C-9 Alkenes(3)	0.42	0.00	3.54	0.33	0.10
Cyclohexene	0.42	0.00	1.40	0.61	0.19
cis-2-Pentene	0.39	0.00	2.16	0.56	0.37
cis-2-Butene	0.38	0.00	2.29	0.47	0.42
trans-2-Hexene	0.35	0.00	0.73	0.51	0.27
4-Methyl-1-pentene	0.34	0.00	0.84	0.27	0.16
1-Octene	0.33	0.00	11.31	0.27	0.10
C-10 Alkenes(2)	0.32	0.00	2.13	0.46	0.15
C-6 Alkene(1)	0.31	0.00	3.03	0.25	0.14
1-Hexene	0.30	0.00	1.52	0.24	0.14
β-Pinene	0.28	0.00	2.18	0.47	0.04
3-Methyl-1-butene	0.21	0.00	1.07	0.15	0.13
1-Nonene	0.20	0.00	1.13	0.16	0.05
Cyclopentene	0.20	0.00	0.74	0.29	0.14
trans-3-Hexene	0.16	0.00	2.17	0.23	0.12
2-Methyl-1-pentene	0.16	0.00	1.12	0.13	0.08
C-5 Alkene(1)	0.15	0.00	0.85	0.10	0.09
C-8 Alkenes(3)	0.15	0.00	0.46	0.12	0.04
C-9 Alkene(1)	0.14	0.00	0.36	0.11	0.03
1,1-Dimethylcyclohexene	0.13	0.00	0.80	0.18	0.08
cis-2-Hexene	0.12	0.00	0.39	0.18	0.09

(*continued*)

Table I-A-3. (*continued*)

	Average ppbC	Min ppbC	Max ppbC	Percent of OH Reactivity	Percent of MIR Reactivity
Alkenes (*continued*)					
2-Methyl-2-pentene	0.10	0.00	0.44	0.15	0.08
cis-3-Hexene	0.06	0.00	0.15	0.09	0.05
trans-3-Heptene	0.05	0.00	0.79	0.07	0.03
C-7 Alkenes(2)	0.05	0.00	0.13	0.04	0.02
C-8 Alkene(1)	0.03	0.00	0.16	0.02	0.00
C-9 Alkenes(4)	0.03	0.00	0.27	0.02	0.00
C-7 Alkene(1)	0.02	0.00	0.23	0.02	0.00
4-Methylhexene	0.01	0.00	0.35	0.01	0.00
C-8 Alkenes(2)	0.01	0.00	0.10	0.00	0.00
Aromatics (30% of NMHC)	83.1			39.5	41.5
1,2,4-Trimethylbenzene	19.90	2.27	122.97	14.20	15.37
Toluene	18.74	1.94	436.98	2.43	4.52
m- and *p*-Xylene	9.07	1.99	28.75	2.85	5.04
Benzene	4.82	0.74	16.21	0.14	0.22
Ethylbenzene	4.50	0.45	34.19	2.33	3.06
m-Ethyltoluene	3.57	0.64	33.24	1.85	2.16
o-Xylene	3.51	0.84	9.79	1.06	1.91
Styrene	2.78	0.46	22.04	3.50	0.48
1,3,5-Trimethylbenzene	2.29	0.38	23.27	2.89	2.00
p-Ethyltoluene	1.92	0.00	17.01	0.99	1.16
1,2,3-Trimethylbenzene	1.75	0.44	12.30	1.26	1.36
o-Ethyltoluene	1.25	0.20	11.27	0.65	0.75
Diethylbenzenes(2)	1.23	0.31	7.18	0.64	0.67
n-Propylbenzene	1.18	0.18	9.60	0.16	0.22
Naphthalene	0.89	0.00	2.53	0.42	0.07
C-10 Aromatics(2)	0.81	0.13	2.61	0.42	0.44
Indane	0.62	0.12	2.64	0.47	0.05
Indene	0.54	0.12	2.65	0.41	0.04
iso-Propyltoluene	0.53	0.00	1.98	0.28	0.29
1,2,3,5-Tetramethylbenzene	0.46	0.11	1.26	0.58	0.36
1,3-Dimethyl-4-ethylbenzene	0.46	0.00	1.43	0.58	0.36
1,2,4,5-Tetramethylbenzene	0.45	0.00	2.22	0.57	0.35
Diethylbenzenes(3)	0.36	0.00	1.37	0.19	0.20
iso-Propylbenzene	0.31	0.00	1.35	0.04	0.06
2-Methyl-indane	0.30	0.00	1.75	0.22	0.02
C-10 Aromatics(5)	0.27	0.00	1.24	0.14	0.15
C-10 Aromatics(4)	0.18	0.00	0.55	0.09	0.10
n-Butylbenzene	0.16	0.00	2.32	0.02	0.03
1-Methyl-indane	0.11	0.00	0.67	0.08	0.00
Diethylbenzene(1)	0.08	0.00	0.42	0.04	0.05
iso-Butylbenzene	0.05	0.00	0.46	0.00	0.00
C-11 Aromatic(1)	0.04	0.00	0.62	0.02	0.02
C-11 Aromatics(3)	0.02	0.00	0.24	0.01	0.01
C-10 Aromatic(1)	0.00	0.00	0.85	0.00	0.00

Table I-A-3. *(continued)*

	Average ppbC	Min ppbC	Max ppbC	Percent of OH Reactivity	Percent of MIR Reactivity
Other (24% of NMHC)	64.7			20.6	19.5
Ethyne	5.53	0.86	10.86	0.10	0.24
Methyl *tert*-butyl ether	0.97	0.03	3.99	0.06	0.07
Other identified	23.90	1.53	64.74	8.41	7.91
Unidentified	34.25	4.48	187.96	12.05	11.33
Total	2887.0			100.0	100.0

Notes:

There were 41 samples at five sites for NMHCs.

There were 14 samples at four sites for CO and methane. *m* and *p*-Xylene coeluted and are reported together.

Concentrations below detection limit were reported as zero.

Compounds given names like "C-10-alkane" were identified by mass spectroscopy.

"Other identified" and "unidentified" were assumed to have the same OH rate coefficients and MIR reactivities as the average for identified NMHCs.

I-A-3. Potential Differences in Other Areas: Biogenic Emissions

The composition of morning air in Los Angeles and Boston provides a reasonable surrogate for the composition of anthropogenic emissions in other urban areas of North America, although some differences are to be expected between areas due to the influence of particular local emission sources. However, one important difference between Los Angeles and many other areas is the relatively minor contribution of biogenic hydrocarbon emissions to the Los Angeles atmosphere because the area is arid. Isoprene (2-methyl-1,3-butadiene), the important biogenic alkene, accounts for only 0.7% of NMHCs in table I-A-2 on average. Biogenic emissions are important throughout the eastern United States, where warm summer temperatures and humid conditions are conducive to vigorous plant growth and high biogenic emissions of VOCs (Geron et al., 1994). The contribution of isoprene is larger in the Boston samples, 1.6% on average, but table I-A-3 does not fully reflect the importance of biogenic emissions in the eastern United States for the following reasons: the samples were collected in urban areas where anthropogenic emissions dominate, and the samples were collected in the early morning whereas biogenic isoprene emissions peak in the middle of the day.

The difference in composition of atmospheric VOCs in an area dominated by biogenic emissions, as compared with the urban areas described in sections I-A-1 and I-A-2, can be seen from table I-A-4. This table shows average late afternoon VOC concentrations measured at a rural site in Alabama in the summer of 1990 (Goldan et al., 1995). The largest contributors to the total identified VOCs are alkenes and oxygen-containing compounds. Isoprene is the most abundant VOC (in ppbC terms) and the isoprene oxidation products methacrolein and methyl

Table I-A-4. Average late afternoon (16:00) VOC concentrations measured at a rural site in Alabama in the summer of 1990

Compound	ppbC	ppbC % of Identified VOCs
Alkanes		9.4
Propane	3.6	4.2
2-Methylpropane	0.8	0.9
n-Butane	1.2	1.4
2-Methylbutane	1.25	1.5
n-Pentane	0.45	0.5
2,3-Dimethylbutane	0.12	0.1
2-Methylpentane	0.27	0.3
3-Methylpentane	0.132	0.2
n-Hexane	0.12	0.1
Methylcyclohexane	0.056	0.1
Alkenes		42.7
Isoprene	31.5	37.0
α-Pinene	3	3.5
β-Pinene	1.7	2.0
Camphene	0.12	0.1
p-Cymene	0.08	0.1
Aromatics		1.7
Benzene	0.6	0.7
Toluene	0.525	0.6
m- and *p*-Xylene	0.16	0.2
o-Xylene	0.072	0.1
Ethylbenzene	0.08	0.1
Oxygen-containing		46.2
Methanol	11	12.9
Ethanol	2.4	2.8
Acetaldehyde	2.6	3.1
Acetone	12.6	14.8
2-Butanone	1.96	2.3
Methacrolein	6	7.0
Methyl vinyl ketone	2.8	3.3
Sum of Identified	85.2	100.0

Note: Some important VOCs were not quantified, including formaldehyde, ethane, and ethene.

Source: Goldan et al. (1995).

vinyl ketone were also observed. Other abundant oxygen-containing compounds observed, such as methanol, ethanol, and acetone, are also thought to be of biogenic origin (Goldan et al., 1995). Several monoterpenes were also identified at this site, as shown in table I-A-4. Many different biogenic monoterpenes have been observed in the atmosphere, and the monoterpenes observed at any given location will depend upon the plant species present. Table I-A-5 presents a summary of monoterpenes that have been observed and an evaluation of their relative frequency of occurrence in the atmosphere (Guenther et al., 1994). In urban and suburban areas with high biomass, the composition of atmospheric VOCs is expected to resemble a combination of the anthropogenic emission signatures

Table I-A-5. Biogenic monoterpenes that have been observed in the atmosphere and evaluation of their relative frequency of occurrence

Major	Frequent	Occasional
3-Carene	α-Thujene	α-Fenchene
d-Limonene	Tricyclene	β-Fenchene
Myrcrene	Terpinolene	δ-Fenchene
α-Pinene	α-Terpinene	ε-Fenchene
β-Pinene	β-Terpinene	Bornylene
Sabinene	γ-Terpinene	Alloocimeme
Camphene	*p*-Cymene	Methylchavicol
1,8-Cineole	α-Phellandrene	*p*-Cymen-8-ol
β-Phellandrene	*trans*-Ocimene	Linalool
	cis-Ocimene	2-Methyl-6-methylene-1,7-octadiene-3-one
	2-Carene	Pinocarvone
		Verbenone
		Fenchone
		Thujone
		Camphor

Source: Guenther et al. (1994).

shown in tables I-A-2 and I-A-3 and a biogenic emission signature similar to table I-A-4.

It has been reported that plants can emit unsaturated compounds containing an aromatic ring; for example, *p*-cymene (Winer et al., 1992). Studies have also shown that microorganisms can produce benzene (Jüttner, 1988) and toluene (Jüttner and Henatsch, 1986; Dewulf et al., 1998). Recently, it was observed that some plants emit toluene under certain stress situations (Heiden et al., 1999). At present, it is not possible to quantify the emissions of simple aromatic hydrocarbons from the biosphere.

I-A-4. European Ambient Data

Data on the composition of ambient VOCs in Europe provide a perspective on differences within Europe and the United States. Table 1-A-6 summarizes the levels of total non-methane organic compounds (TNMOCs) and the percent contribution of aromatic hydrocarbons reported for various locations in Europe. The TNMOC levels are higher in urban areas than in rural/remote areas, and the aromatic hydrocarbon contribution to TNMOCs is generally higher in urban areas.

Tables 1-A-7a and 1-A-7b show the contribution of identified compounds relative to the total identified hydrocarbons for urban and rural samples in Germany (Kurtenbach, 1999; Kurtenbach et al., 1999, 2000). The urban samples were collected in the city center of Wuppertal and in a nearby traffic tunnel of 1.1 km length. The rural samples were collected in Menz at a distance of about 100 km from the city center of Berlin. Two samples are reported for Menz: one for conditions where the urban plume of Berlin impacted the monitoring location and another that was representative of the local (rural) conditions.

Table I-A-6. Total non-methane organic compounds (TNMOCs) and percent aromatics for various European locations

Location	Characterization	Country	TNMOC	Concentration Aromatic Hydrocarbons	Reference
a Data from several field studies in the late 1980s and early 1990s				mass %	Ciccioli et al., (1999)
Milan	Urban	Italy	490.8($\mu g\, m^{-3}$)	43.7	
Rome	Urban	Italy	583.5	32.9	
Taranto	Urban	Italy	586	34.1	
Montelibretti	Suburban/rural	Italy	360.5	6.0	
Madrid	Suburban	Spain	104.3	34.5	
Castelporzio	Forest/rural	Italy	88.7	2.6	
Castelporzio	Pine–oak forest	Italy	95.5	3.9	
Viols en Laval	Forest maquis	France	36.0	6.0	
Monti Crimini	Pine forest	Italy	96.2	8.7	
Storkow	Pine forest	Germany	91.2	25.8	
Burriana	Orange field	Spain	25.5	5.0	
Apenini Mountains	Forest/rural	Italy	34.0	1.6	
Svalbard Islands	Remote arctic	Norway	48.3	5.2	
b Measurement made in 1997 and 1998 in Germany				ppbC %	
Wuppertal	Urban	Germany	25–184 (ppbC)	44	Kurtenbach et al. (2000)
Wuppertal	Traffic tunnel	Germany	87–5528	47	Kurtenbach et al. (1999)
Menz	Rural, Berlin plume	Germany	23.7	28	Kurtenbach (1999)
Menz	Rural, background	Germany	12	19	Kurtenbach (1999)
c Berlin, Germany. Summer 1996, 4 month average				mass %	
Frohnau	Suburban	Germany	31 ($\mu g\ m^{-3}$)	18	Thijsse et al. (1999)
Nansenstrasse	Urban, roadside	Germany	138	29	Thijsse et al. (1999)
Frankfurter Allee	Urban, roadside	Germany	289	35	Thijsse et al. (1999)
d Munich, Germany. March 15 to April 3, 1996, Urban				ppbC %	
All days	Mean values	Germany	112.4 (ppbC)	34.0	Kern et al. 1997)
Weekdays	Mean values	Germany	117.3	35.3	Kern et al. 1997)
Weekdays 5.00-21.00 hours	Mean values	Germany	134.1	36.3	Kern et al. 1997)

Table I-A-6.(*continued*)

Location	Concentration TNMOC	Concentration Aromatic Hydrocarbons	Reference
e 1996 annual means for 12 cities in the United Kingdom	ppbC	ppbC %	(Derwent et al. (2000)
Birmingham	138	28.9	
Cardiff	132	29.5	
Leeds	138	25.8	
Edinburgh	98	26.0	
Harwell	44	28.2	
Bristol	123	38.4	
London/Eltham	117	30.7	
Middlesbrough	126	22.2	
Belfast	123	26.0	
Southampton	181	35.5	
London/UC	185	35.6	
Liverpool	145	40.5	

Table I-A-7a. Composition of ambient hydrocarbons (percent ppbC of total identified hydrocarbons) for urban and rural locations in Germany

	Location/Characterization			
	Tunnel Wuppertal	Urban Wuppertal	Rural Menz Berlin Plume	Rural Menz
Date:	Weekdays August 1997		July 20, 1998	July 20, 1998
Alkanes	37.0	42.0	63.0	65.0
Ethane	0.5	2.5	10.1	16.3
Propane	0.3	1.6	4.4	4.6
iso-Butane	2.4	6.0	8.8	13.0
n-Butane	5.5	4.9	15.8	15.0
iso-Pentane	12.4	6.9	8.2	5.9
n-Pentane	4.1	4.1	5.0	3.9
2-Methylpentane		7.1		
3-Methylpentane		2.7	2.5	0.7
2,2-Dimethylbutane	0.8			
Cyclopentane	5.1			
2-Methylpentane/cyclopentane			5.7	3.9
n-Hexane	1.4	1.6	1.3	1.3
Methylcyclopentane	1.6			
Cyclohexane	1.1			
n-Heptane	0.9	4.5	1.3	0.7
n-Octane	0.9			
Alkenes/alkynes	16.0	14.0	9.0	16.0
Ethene/ethyne	4.3	1.5	2.2	1.0
Propene	5.4	1.5	3.7	3.5
Isobutene/1-butene	3.4	3.4	2.2	7.5
trans-2-Butene	0.2	0.6		
cis-2-Butene	0.8	0.6		
1,3-Butadiene		4.2		
1-Pentene	1.1			
Isoprene/*trans*-2-pentene		2.0	0.8	2.9
cis-2-Pentene		0.3	0.2	1.1
1-Hexene	0.8			
Aromatics	47.0	44.0	28.0	19.0
Benzene	7.5	8.4	6.4	9.1
Toluene	16.9	21.6	8.7	4.4
Ethylbenzene	9.9	4.4	2.0	1.3
p-Xylene	3.3			
m-Xylene	3.8			
m-Xylene/*p*-xylene		7.0	2.0	1.3
o-Xylene	5.6	2.6	8.7	2.7
Mesitylene			0.3	0.2
Other	0.0[a]	0.0[a]	0.0	0.0
Total NMHC	100.0	100.0	100.0	100.0

[a] The category "Other" was set to zero because of comparison with rural stations. Roughly 3 ppbC % oxygenated aromatic species were measured in the traffic tunnel as well as in the city center shown in table I-A-7b. At the rural site these compounds were not measured.

Sources: Kurtenbach et al. (1999), Kurtenbach et al. (1999, 2000).

Table I-A-7b. Composition of oxygenated aromatic compounds found in the tunnel and urban center of Wuppertal, Germany

Oxygenated Aromatic Compounds	Tunnel (ppbC)	Urban Center (ppbC)
Benzaldehyde	1.4–30.8	0.35–1.4
Phenol	0.3–17.5	0.12–2.16
o-Cresol	$\leq$1.8	$\leq$0.98
m-Cresol	$\leq$2.5	$\leq$1.12
p-Cresol	0.7–7.7	$\leq$0.21

Sources: Kurtenbach et al. (1999, 2000).

I-B. Relative Importance of OH, NO_3, O_3, and $O(^3P)$ as Reactants for Aromatic Hydrocarbons

The pathways for initiating oxidation reactions of VOCs in the lower atmosphere are reaction with O_3, OH, NO_3, and $O(^3P)$, and photolysis at wavelengths longer than about 290 nm. A qualitative comparison of the reaction rates of different classes of VOCs via these channels (Atkinson, 1994) is shown in table I-B-1, from which it is apparent that for aromatic hydrocarbons, reactions with OH are always important and reactions with NO_3 and O_3 are important for specific classes of aromatics.

The relative importance of these reaction pathways for individual aromatics depends upon the concentrations of reaction initiators (O_3, OH, NO_3) and the reaction rate coefficients. The concentrations of reaction initiators are highly variable in time and space because of variations in pollutant concentrations and solar radiation. The reaction rate coefficients may vary somewhat with location due to temperature and/or pressure sensitivities. The relative importance of reactions with O_3, OH, and NO_3 and the unimportance of $O(^3P)$ are illustrated here for conditions representative of a polluted urban atmosphere. The diurnal variations of reactant concentrations calculated using a box model for a polluted urban condition are shown in figure I-B-1. Differences due to the diurnal cycle of

Table I-B-1. Qualitative assessment of the reaction rates for VOCs with the main tropospheric reaction initiators (Blank cells indicate rates that are negligibly slow)

	OH	O_3	NO_3	$O(^3P)$	Photolysis
Alkanes	Slow to medium				
Alkenes	Medium to fast	Fast	Medium to fast	Fast	
Aromatic hydrocarbons	Slow to fast	None to medium[a]	None to fast[b]		
Oxygenated hydrocarbons	Slow to fast		Slow to medium		Slow to fast

[a]Reaction with O_3 is significant only for aromatic alkenes.
[b]Reaction with NO_3 is significant only for aromatic alkenes and phenols.

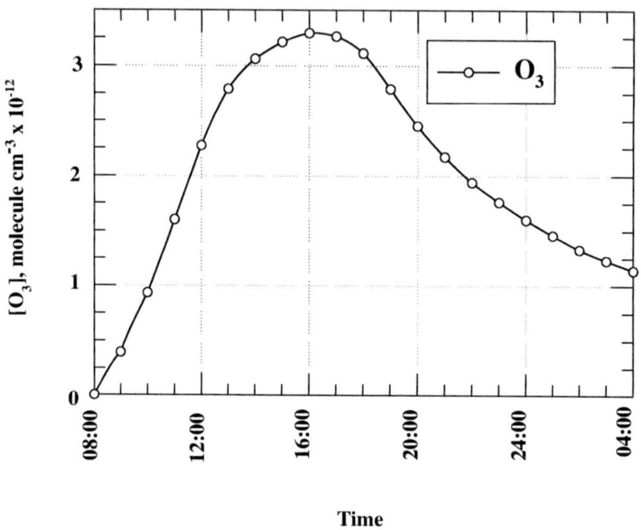

(a)

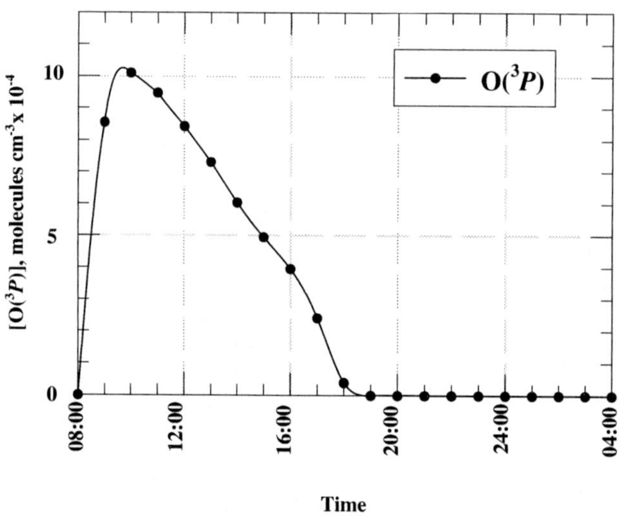

(b)

Figure I-B-1. Diurnal profiles of O_3, OH, NO_3, and $O(^3P)$ concentrations calculated using a photochemical box model for conditions representative of a polluted urban atmosphere.

sunlight and the photolysis reactions are apparent, including the importance of OH, the unimportance of $O(^3P)$ during the day, and the importance of NO_3 at night. The daytime and nighttime average mixing ratios for this urban scenario are summarized in table I-B-2.

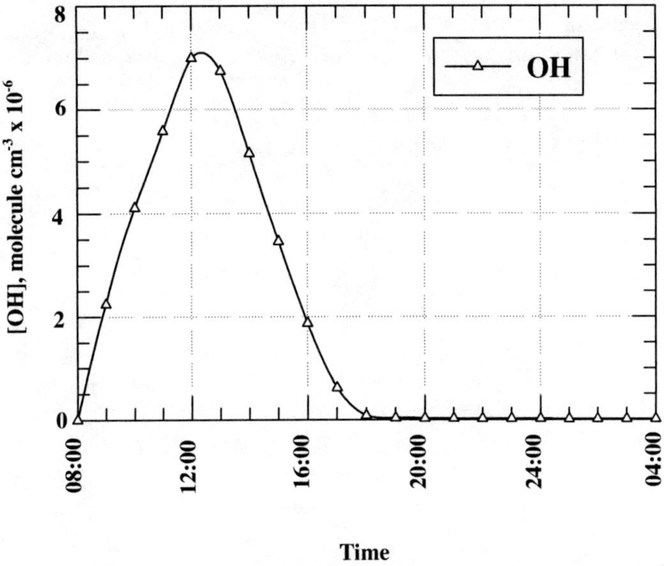

(c)

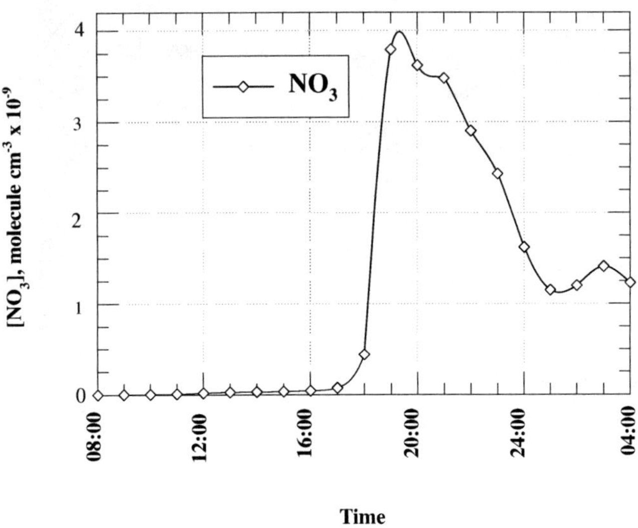

(d)

Figure 1-B-1. (*continued*)

Table I-B-2. Daytime and nighttime average OH, NO_3, O_3, and $O(^3P)$ mixing ratios and concentrations for conditions representative of a polluted urban atmosphere

	Daytime Average		Nighttime Average	
Species	Volume Mixing Ratios	(molecule cm^{-3})	Volume Mixing Ratios	(molecule cm^{-3})
OH	0.16 pptv	3.9×10^6	0.0007 pptv	1.7×10^4
O_3	110 ppbv	2.7×10^{12}	80 ppbv	2.0×10^{12}
NO_3	3 pptv	7×10^7	100 pptv	2.5×10^9
$O(^3P)$	0.003 pptv	7×10^4	0	0

The relative importance of reactions with OH, NO_3, and O_3 at the daytime and nighttime average mixing ratios shown in table I-B-2 are compared for three representative aromatics compounds *m*-xylene, styrene, and *o*-cresol. These aromatics were chosen because they are relatively abundant in the atmosphere and they have different tendencies to react with atmospheric oxidants. For all compounds, reactions with all three potential oxidants were included at the rate coefficients recommended in the later chapters in this book. As shown in figure I-B-2, reaction with OH is the only reaction pathway for *m*-xylene during the day and the dominant (60%) reaction pathway for styrene but only a minor (14%) reaction pathway for *o*-cresol. Styrene is an aromatic

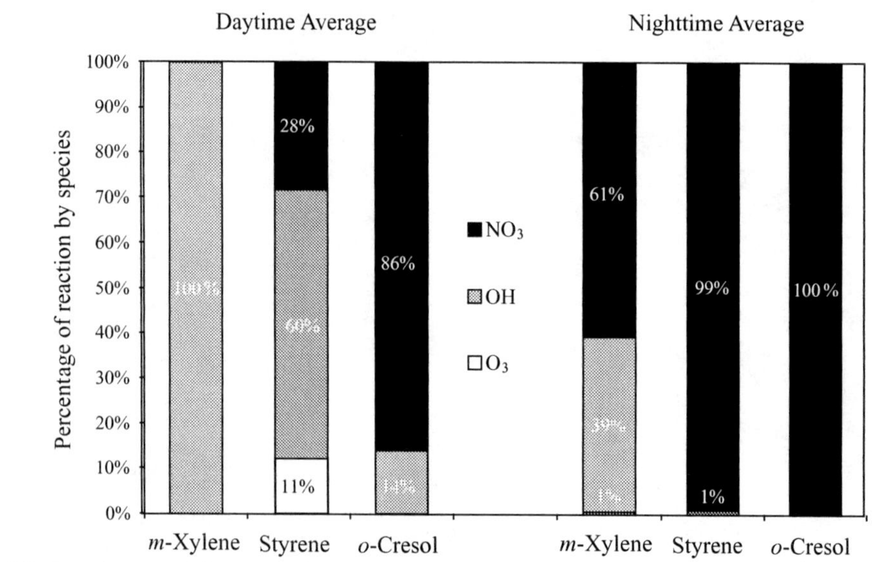

Figure I-B-2. Comparison of relative removal for *m*-xylene, styrene, and *o*-cresol by reaction with OH, NO_3, and O_3 under representative daytime and nighttime average conditions in a polluted urban atmosphere; the contribution of $O(^3P)$ atoms to the reactions is too small to be detected in the graph.

alkene, and it reacts with NO_3 and O_3. The NO_3 reaction is estimated to account for 28% of daytime reaction of styrene whereas O_3 reaction accounts for 11% of styrene removal. The daytime reaction of styrene with NO_3 occurs mainly at low sun angles when NO_3 photolysis becomes slow enough to allow NO_3 to build up from reaction between O_3 and NO_2 ($O_3 + NO_2 \rightarrow NO_3 + O_2$). For *o*-cresol, the NO_3 reaction is rapid and estimated to dominate the daytime removal, although the balance between OH reaction and NO_3 reaction depends strongly on the sun angle.

At night, reaction with NO_3 is the only significant removal pathway for both styrene and *o*-cresol under these conditions. For *m*-xylene, reactions with both NO_3 and OH radicals are important and the balance between these removal pathways will be sensitive to atmospheric conditions. For example, because the NO_3 radical is close to being in equilibrium with NO_2 and N_2O_5 in the lower atmosphere at around 298 K, the NO_3 radical concentration does not scale linearly with the NO_2 concentration. Thus, for example, Platt et al., 1984) observed NO_3 radical concentrations in Death Valley as high as those seen in Los Angeles; in Death Valley the relative humidity was less than 40% and NO_2 mixing ratios < 0.3 ppbv).

Reaction with $O(^3P)$ is a minor pathway accounting for less than 1% of reaction for the three aromatics being compared here. However, $O(^3P)$ reactions are included in this review because they may be important in smog chamber studies of aromatic reactions in which higher NO_2 concentrations are used than are encountered in the atmosphere, and the match between the observed and modeled product concentration profiles is used as an important check on the mechanism choices made in the modeling.

In summary, these results show that OH is the dominant reaction partner for most aromatic hydrocarbons during the day whereas NO_3 and OH dominate at night. The NO_3 reaction is also important during the day (at low sun angles) for aromatic alkenes and phenol-type compounds. The $O(^3P)$ reactions are unimportant in the troposphere.

I-C. Importance of Aromatic Hydrocarbons Relative to Other Volatile Organic Compounds as Reactants for OH, NO_3, and O_3

Section I-A presented information on the composition of VOCs in urban air. As discussed previously, limitations in measurement methodologies underlying tables I-A-2 and I-A-3 mean that these tables do not reflect the contributions of oxygen-containing compounds and semivolatile compounds. The importance of these are discussed for Los Angeles in section I-A-1. Also, not all of the compounds that were detected could be identified. However, these tables do reflect the relative contributions of the most important members of three groups of NMHCs (alkanes, alkenes, and aromatic hydrocarbons) to early morning ambient VOC concentrations. Tables I-A-2 and I-A-3 provide three different ways of evaluating these relative contributions: percent of NMHCs, percent of OH reactivity, and

Table I-C-1. Summary of the relative contribution (ppbC %) of alkanes, alkenes, and aromatics to total NMHCs, OH reactivity, and MIR reactivity presented in tables 1-A-2 (Los Angeles) and 1-A-3 (Boston)

	Alkanes	Alkenes	Aromatics
Los Angeles			
Percent NMHCs	42	7	19
Percent OH reactivity	15	27	18
Percent MIR reactivity	18	19	30
Boston			
Percent NMHCs	36	10	30
Percent OH reactivity	12	24	40
Percent MIR reactivity	11	18	41

percent of MIR reactivity (the derivation and interpretation of the reactivity measures is discussed in section I-A).

Table I-C-1 summarizes the relative contribution of alkanes, alkenes, and aromatics to total NMHCs, OH reactivity, and MIR reactivity derived in section I-B for Los Angeles and Boston. Table I-C-2 gives a similar comparison for data collected in two urban areas in Germany. Alkanes are the largest contributor to total NMHCs but the smallest contributor to both OH and MIR reactivity in both Los Angeles and Boston. In Wuppertal and Munich the aromatic hydrocarbon concentrations are high and the reactivity due to these species is also high. This illustrates the importance of alkanes due to their high contribution to the mass of VOC emissions, but also reflects the lower reactivity of alkanes relative to alkenes and aromatics. Aromatic hydrocarbons are the second largest contributor to total NMHCs in Los Angeles, Boston, and Munich, but the largest class in Wuppertal. The aromatic hydrocarbons are the largest contributor to MIR reactivity in Boston, Los Angeles, and Munich, and the largest contributor to OH reactivity in Boston. Aromatic hydrocarbons are much more important in terms of MIR reactivity than mass because they tend to have high reaction rates and high yields of radicals and reactive products that contribute to ozone formation.

The MIR reactivity scale does not boost the relative importance of alkenes as much as the OH reactivity scale because it takes into account the reactivity of reaction products as well as the initial reaction rate. Furthermore, the MIR scale effectively uses the "fraction reacted" over a 1-day time period, so that, for example, a VOC with a lifetime (τ) of 10 min is effectively no more reactive than one with $\tau = 50$ min. On an MIR reactivity basis, aromatic hydrocarbons are the most important group in cities. Aromatic hydrocarbons are relatively more important in Boston than Los Angeles by all three measures. This may be an artifact of the location of the sampling sites in Boston, which appears to have been strongly influenced by diesel exhaust emissions (Fujita et al., 1997). Overall, considering the relative importance of alkanes, alkenes, and aromatic hydrocarbons from all three perspectives, it is clear that all three classes of compound are important in urban atmospheres.

Table I-C-2. Summary of the relative contributions (ppbC %) of alkanes, alkenes, and aromatics to the total NMHCs, OH reactivity, and MIR reactivity for some cities in Germany: Munich and Wuppertal

	Alkanes	Alkenes	Aromatic Hydrocarbons
Munich			
Percent NMHCs	47	18	35
Percent OH reactivity	15	54	31
Percent MIR reactivity	15	35	51
Wuppertal City			
Percent NMHCs	42	14	44
Percent OH reactivity	11	65	24
Wuppertal Traffic Tunnel			
Percent NMHCs	37	16	47
Percent OH reactivity	13	62	25

Sources: Munich (Kern et al., 1997); Wuppertal (Kurtenbach et al., 1999, 2000).

I-D. Relative Importance of Individual Aromatic Hydrocarbons

In the ambient air samples for Los Angeles and Boston shown in tables I-A-2 and I-A-3, the most abundant aromatics (in ppbC terms) are alkylbenzenes with 7–10 carbon atoms, such as toluene, xylenes, trimethylbenzenes, and their analogues (e.g., ethylbenzene and diethylbenzenes). The abundance of aromatics with more than 10 carbon atoms generally decreases with increasing size. The main impact of factoring in reactivity (either OH-reaction rate or MIR basis) is to reduce the contributions of the very lightest aromatic hydrocarbons: namely, benzene and toluene. In Germany, toluene is the most abundant aromatic species in gasoline (Dietzel and Kägler, 1994). The reason for the importance of C_7–C_{10} alkylbenzenes can be understood by considering the composition of gasoline-powered engine exhaust, illustrated in table I-D-1. Engine exhaust composition is sensitive to both the fuel type and vehicle technology, so the information given in table I-D-1 should be extrapolated with caution; however, this table illustrates the main points that need to be considered to understand the ambient data. Many of the aromatics in the exhaust come directly from unburned fuel, and therefore the exhaust composition strongly reflects fuel aromatic composition (Leppard et al., 1993, 1995; Zervas et al., 1999), as measurements in traffic tunnels and city centers show; see tables I-A-7a and I-A-7b. For the lighter aromatics (particularly benzene and toluene), cracking of higher aromatics during combustion is also a significant source. These factors also influence the exhaust composition for other combustion sources, such as diesel engines or turbines. Since diesel fuel contains more heavy aromatic hydrocarbons, in urban areas diesel exhaust can be a significant source of aromatic compounds with more than about 10 carbon atoms.

In light of the relationship between vehicle emissions composition and the aromatic hydrocarbons content of fuels, regulations have been developed to control the amounts in fuels of both specific aromatic compounds (i.e., ben-

Table I-D-1. Detailed composition of tailpipe exhaust from catalyst-equipped light-duty gasoline vehicles fueled with California Phase 2 reformulated gasoline

	ppbC % Contribution to		
	Total NMHCs	OH Reactivity	MIR Reactivity
Methane	21.8	0.08	0.47
Alkanes	33.5	11.5	9.0
2,2,4-Trimethylpentane	5.42	1.13	0.91
2-Methylbutane	3.97	1.40	1.55
Ethane	3.04	0.19	0.56
2,3-Dimethylpentane	2.13	0.99	0.65
2-Methylpentane	2.11	0.89	0.79
2,4-Dimethylpentane	1.28	0.57	0.48
n-Butane	1.25	0.36	0.46
2,3,4-Trimethylpentane	1.25	0.61	0.35
3-Methylpentane	1.13	0.49	0.41
2,3,3-Trimethylpentane	1.11	0.23	0.19
2,3-Dimethylbutane	0.96	0.40	0.23
2,2-Dimethylbutane	0.79	0.14	0.16
3-Methylhexane	0.74	0.34	0.22
Hexane	0.65	0.27	0.16
2,4-Dimethylhexane	0.64	0.31	0.17
Pentane	0.61	0.22	0.18
2-Methylhexane	0.60	0.26	0.13
Heptane	0.49	0.23	0.09
2,2,5-Trimethylhexane	0.46	0.14	0.07
2,3-Dimethylhexane	0.42	0.21	0.10
Propane	0.39	0.07	0.09
2,5-Dimethylhexane	0.38	0.18	0.08
3-Methylheptane	0.37	0.18	0.07
cis-1,2-Dimethylcyclohexane	0.28	0.22	0.11
2-Methylheptane	0.28	0.13	0.05
2-Methylpropane	0.27	0.07	0.11
n-Octane	0.26	0.13	0.03
Methylcyclopentane	0.22	0.14	0.15
n-Dodecane	0.17	0.09	0.00
2,4-Dimethyloctane	0.16	0.09	0.02
Cyclopentane	0.13	0.06	0.09
3,3-Dimethylpentane	0.12	0.02	0.02
4-Methylheptane	0.12	0.06	0.03
3-Methyloctane	0.11	0.06	0.02
3-Ethylpentane	0.11	0.05	0.03
Methylcyclohexane	0.10	0.07	0.04
2,2,3-Trimethylbutane	0.09	0.02	0.02
2-Methyloctane	0.08	0.04	0.02
2,2-Dimethyloctane	0.08	0.04	0.01
2,2-Dimethylpentane	0.07	0.01	0.01
2,4-Dimethylheptane	0.07	0.03	0.01
3,5-Dimethylheptane	0.06	0.03	0.01
n-Decane	0.06	0.03	0.00
2,3,5-Trimethylhexane	0.06	0.03	0.01
3,3-Dimethylhexane	0.06	0.01	0.00
n-Nonane	0.06	0.03	0.00
cis-1,3-Dimethylcyclopentane	0.06	0.03	0.03

Table I-D-1. (*continued*)

	ppbC % Contribution to		
	Total NMHCs	OH Reactivity	MIR Reactivity
Alkanes (*continued*)			
n-Undecane	0.05	0.03	0.00
Cyclohexane	0.04	0.02	0.01
1-*cis*,2-*trans*,3-Trimethylcyclopentane	0.04	0.02	0.01
2,2-Dimethylhexane	0.04	0.00	0.00
Ethylcyclohexane	0.03	0.02	0.01
2,3-Dimethylheptane	0.02	0.01	0.00
cis-1,3-Dimethylcyclohexane	0.02	0.01	0.00
2,2-Dimethylpropane	0.02	0.00	0.00
trans-1,3-Dimethylcyclohexane	0.02	0.01	0.00
trans-1,2-Dimethylcyclopentane	0.01	0.00	0.00
trans-1,4-Dimethylcyclohexane	0.00	0.00	0.00
Alkenes	13.5	56.8	43.2
Ethene	3.68	6.96	17.41
2-Methylpropene	2.95	16.79	5.06
Propene	2.64	10.25	11.21
Styrene	0.42	1.33	0.13
1-Butene	0.38	1.31	1.15
1,3-Butadiene	0.36	2.63	1.25
2-Methyl-2-butene	0.35	2.69	0.94
trans-2-Butene	0.27	1.89	1.03
2-Methyl-1-butene	0.24	1.88	0.66
cis-2-Butene	0.24	1.49	0.92
2-Methyl-1,3-butadiene	0.19	1.66	0.44
trans-2-Pentene	0.17	1.01	0.46
1-Nonene	0.16	0.29	0.06
Propadiene	0.15	1.15	0.35
1-Pentene	0.10	0.28	0.17
cis-2-Pentene	0.09	0.54	0.25
Cyclopentadiene	0.09	0.51	0.23
3-Methyl-1-butene	0.08	0.23	0.14
4-Methyl-*trans*-2-pentene	0.08	0.38	0.14
3-Methyl-*trans*-2-pentene	0.07	0.36	0.13
3-Methyl-*cis*-2-pentene	0.06	0.30	0.11
trans-2-Hexene	0.05	0.26	0.10
2-Methyl-2-pentene	0.05	0.33	0.10
1-Hexene	0.05	0.14	0.06
Cyclopentene	0.05	0.28	0.10
4-Methyl-1-pentene	0.05	0.13	0.05
2-Methyl-1-pentene	0.04	0.18	0.04
2-Methyl-2-hexene	0.04	0.16	0.06
3-Methyl-1-pentene	0.03	0.09	0.04
trans-3-Hexene	0.03	0.15	0.06
4-Methyl-*cis*-2-pentene	0.03	0.13	0.05
cis-2-Hexene	0.02	0.12	0.04
1-Octene	0.02	0.05	0.01
trans-3-Heptene	0.02	0.10	0.03
cis-2-Heptene	0.02	0.08	0.03
trans-2-Heptene	0.02	0.08	0.03

(*continued*)

Table I-D-1. (*continued*)

	ppbC % Contribution to		
	Total NMHCs	OH Reactivity	MIR Reactivity
Alkenes (*continued*)			
3,4-Dimethyl-1-pentene	0.02	0.04	0.01
2,4,4-Trimethyl-1-pentene	0.02	0.03	0.00
trans-4-Octene	0.01	0.05	0.02
2,4,4-Trimethyl-2-pentene	0.01	0.05	0.02
1-Heptene	0.01	0.03	0.01
Cyclohexene	0.01	0.07	0.01
3,3-Dimethyl-1-butene	0.01	0.04	0.01
cis-3-Hexene	0.01	0.06	0.02
3-Methyl-1-hexene	0.01	0.03	0.00
trans-2-Octene	0.01	0.04	0.01
cis-2-Octene	0.01	0.04	0.01
3-Methylcyclopentene	0.00	0.03	0.00
3-Methyl-*trans*-3-hexene	0.00	0.03	0.00
3-Ethyl-*cis*-2-pentene	0.00	0.03	0.00
2,3-Dimethyl-2-pentene	0.00	0.02	0.00
1-Methylcyclopentene	0.00	0.00	0.00
Alkynes	2.34	0.6	1.1
Ethyne	2.12	0.37	0.64
Propyne	0.19	0.17	0.35
2-Butyne	0.01	0.04	0.04
1-Butyne	0.01	0.04	0.03
Aromatics	20.9	16.7	13.9
Toluene	5.32	2.01	2.57
m- and *p*-Xylene	3.20	2.56	3.12
Benzene	3.05	0.29	0.32
Ethylbenzene	1.40	0.56	0.59
1,2,4-Trimethylbenzene	1.23	1.99	1.48
o-Xylene	1.07	0.82	1.02
1-Methyl-3-ethylbenzene	0.94	1.11	0.89
1,2,3,5-Tetramethylbenzene	0.47	1.20	0.51
1-Methyl-4-ethylbenzene	0.43	0.51	0.41
1,3,5-Trimethylbenzene	0.41	1.17	0.56
1-Methyl-4-*n*-propylbenzene	0.34	0.36	0.26
n-Propylbenzene	0.34	0.10	0.10
Naphthalene	0.29	0.28	0.03
1-Ethyl-2-methylbenzene	0.29	0.34	0.27
1,2,4,5-Tetramethylbenzene	0.28	0.71	0.30
1,2,3-Trimethylbenzene	0.24	0.39	0.29
1-Methyl-3-*n*-propylbenzene	0.16	0.17	0.12
1,2-Dimethyl-4-ethylbenzene	0.16	0.40	0.17
1,4-Dimethyl-2-ethylbenzene	0.12	0.32	0.14
n-Pentylbenzene	0.12	0.03	0.02
1,3-Dimethyl-4-ethylbenzene	0.11	0.28	0.12
1,4-Diethylbenzene	0.10	0.10	0.07
1,3-Diethylbenzene	0.09	0.09	0.07
2-Methylbutylbenzene	0.08	0.08	0.05
1-Methyl-2-*n*-propylbenzene	0.08	0.09	0.06

Table 1-D-1. (*continued*)

	ppbC % Contribution to		
	Total NMHCs	OH Reactivity	MIR Reactivity
Aromatics (*continued*)			
1-Methyl-2-*iso*-propylbenzene	0.07	0.07	0.05
Indan	0.06	0.11	0.00
1,2,3,4-Tetramethylbenzene	0.06	0.16	0.07
1,2-Diethylbenzene	0.06	0.06	0.05
1-Methyl-3-*iso*-propylbenzene	0.05	0.06	0.04
1,3-Dimethyl-2-ethylbenzene	0.05	0.13	0.06
1,2-Dimethyl-3-ethylbenzene	0.04	0.09	0.04
iso-Propylbenzene	0.03	0.01	0.01
1-Methyl-4-*iso*-propylbenzene	0.03	0.03	0.02
tert-1-Butyl-2-methylbenzene	0.02	0.02	0.01
sec-Butylbenzene	0.02	0.00	0.00
tert-1-Butyl-3,5-dimethylbenzene	0.01	0.03	0.01
iso-Butylbenzene	0.00	0.00	0.0
Oxygen-containing organics			
Methyl *tert*-butyl ether	1.67	0.42	0.34
Formaldehyde	1.26	5.50	21.62
Acetaldehyde	0.48	1.70	2.75
Acetone	0.46	0.02	0.12
Benzaldehyde	0.29	0.24	−0.01
p-Tolualdehyde	0.13	0.10	−0.00
Acrolein	0.07	0.21	0.28
Methacrolein	0.06	0.24	0.00
Butanone	0.03	0.00	0.01
n-Butyraldehyde	0.02	0.05	0.05
Propionaldehyde	0.02	0.05	0.07
Crotonaldehyde	0.02	0.03	0.03
Pentanal	0.00	0.00	0.00
Methanol	0.00	0.00	0.00
Ethanol	0.00	0.00	0.00
Hexanal	0.00	0.00	0.00
Unidentified	3.5	5.7	7.1
Sum	100.0	100.0	100.0

Source: AQIRP (1995).

zene) and the total aromatics. In the United States, the specifications for such "reformulated gasolines" (RFG) are determined by the California Air Resources Board (CARB) within California and by the U.S. EPA everywhere else. In Europe, the European Commission regulates the benzene and total aromatic hydrocarbon contents of gasoline (started in January, 2000). Summaries of the RFG expected properties developed by these authorities are given in table 1-D-2. The values for the fuel properties given in table 1-D-2 are generally upper limits. Oil refiners must manufacture gasolines that meet these "environmental performance"–based specifications along with

Table I-D-2. Expected properties for U.S. and European reformulated gasolines (RFG)

	Jurisdiction						
	U.S. EPA RFG			California ARB (CARB) RFG		European Commission	
Program:	Phase I	Phase II	Tier 2	Phase 2	Phase 3		
Implementation Date:	Jan. 1995	Jan. 2000	June 2005	June 1996	Jan. 2003	Jan. 2000	Jan. 2005
Fuel properties							
RVP, psi	7.0	6.7	6.6	6.8	6.9	60.0 kPa	60.0 kPa
Oxygen, wt%	2	2	2.1	2	2	2.7	2.7
Benzene, vol%	0.8	0.8	0.8	0.8	0.5	1	1
Aromatics, vol%	27	25	18	22	22	42	35
Olefins, vol%	8.5	8.5	13	4	4	18.0	18.0
Sulfur, ppm	130	130	30	30	20	150	50
Distillation temperatures							
T_{50}, °F	210	207	196	200	200		
T_{90}, °F	329	321	325	290	290		

Notes:
CARB Phase 3 RFG specifications were in draft form at time of writing.
RVP is the Reid vapor pressure.

Table I-D-3. Flat limit specifications for California RFG

Fuel Properties	Phase 2	Phase 3
RVP, psi	7.0	7.0
Oxygen, wt.%	1.8-2.2	1.8-2.2
Benzene, vol%	1.0	0.8
Aromatics, vol%	25	25
Olefins, vol%	6	6
Sulfur, ppm	40	20
T_{50}, °F	210	213
T_{90}, °F	300	305

numerous other vehicle performance–based requirements. Thus, the composition of individual batches of gasoline produced by different oil refiners will vary within the limits shown in table 1-D-2, depending upon the properties of the crude oil feedstock and the economics of the refining process.

The limits on total aromatics content shown in table 1-D-2 are higher for European than for U.S. gasolines. This reflects a general need for higher octane fuels in Europe than in the United States because European cars traditionally have been designed with smaller displacement engines operating at higher compression ratios. In the past, high-octane fuels were produced in part by the addition of lead compounds. As lead has been phased out, an increase in the aromatic hydrocarbon content of fuels has occurred, largely because aromatics have a high octane rating. Oxygenates such as MTBE and ethanol can also be added to gasolines to boost the octane rating of the fuel, and oxygenates have become a standard component of reformulated gasolines with relatively low aromatic hydrocarbon content sold in the United States. However, the state of California decided to phase out the addition of MTBE to gasolines by January 1, 2003 because of concerns over contamination of groundwater with MTBE. Phase 3 RFG sold in California after January 2003 may contain oxygen in the form of ethanol, and permission has been sought to make the gasoline oxygen-free at the discretion of refiners.

Expected properties for reformulated gasolines listed in table I-D-2 contain only a few actual specifications. Values shown for CARB RFG are average limits, not strictly specifications. Table I-D-3 gives the "flat limit" specifications for California RFG.

I-E. Other Sources of Atmospheric NMHCs

I-E-1. Solvent Use

Solvent use and evaporation of fuels from handling are also important sources of non-methane volatile organic compounds (NMVOCs) and, therefore, of aromatic hydrocarbon emission. These emissions are difficult to estimate because they originate from many different sources. Because of more stringent reduction of NMVOC emissions from cars, in some of the industrialized countries the sources of NMVOCs from solvent use presently exceed those from traffic. In Germany, the emissions from solvent use in 1997 accounted for 56% of the total anthropogenic NMVOC emissions, with 27% from road traffic, not involving fuel distribution; in 1991 both sources were equal (UBA, 1999). More recently it has been calculated that in Germany 56% of the NMVOC emissions from solvent use are released as aromatic hydrocarbons (Friedrich and Obermeier, 1999); this study included industrial varnishes, printing activities, use of adhesives and cleaning activities. In the United States for 1997, 34% of the total anthropogenic NMVOC emissions were attributed to solvent utilization, still slightly under that of 40% from traffic (EPA, 1998). However, when solvent use is taken to include a wider range of industrial activities (storage, etc.) the contribution increases to 51%. In the Lombardia, a highly industrialized area around Milan, for the late 1980s a contribution of 29% of total NMVOC emissions from solvent use was calculated (Bolzacchini et al., 1998). According to this study, 38% of the emissions from solvent use were released as aromatic hydrocarbons. From the total emissions of aromatics in this area, 56% originated from solvent use. In the United Kingdom, 36% of anthropogenic NMVOC emissions were attributed to solvent use in the early 1990s, and 38% to traffic emissions (Derwent et al., 1996). From the NMVOC emissions from solvent use, 25% are considered to be released as aromatic hydrocarbons in the United Kingdom (Derwent et al., 1996). In Germany, the apparently large contribution of NMVOC emissions from solvent use has not yet been observed in the ambient air of city centers. Either the emission calculations are wrong or the solvent emissions are distributed over a large variety of oxygenated VOCs, which are difficult to analyze in the atmosphere.

I-E-2. Biomass Burning

Another source of aromatic hydrocarbons is biomass burning. According to Ehhalt (1999), 25 $Tg\,yr^{-1}$ of benzene and toluene emissions occur globally, with 20% from biomass burning and 80% from the use of fossil fuels.

I-E-3. Global Emissions

The estimation of global emissions of aromatic hydrocarbons ranges from 18.7 $Tg\,yr^{-1}$ [17.1% of the total anthropogenic NMVOC emissions (Piccot et al., 1992)] to 25 $Tg\,yr^{-1}$, with benzene and toluene alone constituting 25% of the total anthropogenic NMVOC emissions (Ehhalt, 1999). Even if the aromatic content of fuel is lowered by further regulations in the industrialized countries to about 20 mass percent, the increasing global oil demand of 3182 million metric tons in 1990 to 4407 million metric tons in 2010 (European Commission, 1999) will keep the global emissions of aromatic hydrocarbons from fossil fuel combustion at a high level.

The aromatic hydrocarbons represent an important class of NMVOC emissions that have a significant impact on atmospheric oxidation processes on a regional as well as on a global scale. For example, independent of the results from model calculations, simulation experiments in an outdoor smog chamber, with authentic exhaust gases taken from a gasoline-fueled engine and mixed with air, have shown that exhaust from aromatic-rich fuel leads to a significant increase of the ozone formation compared with a paraffin-rich fuel (Thüner et al., 1998).

II

Reactions of Aromatic Compounds with OH Radicals

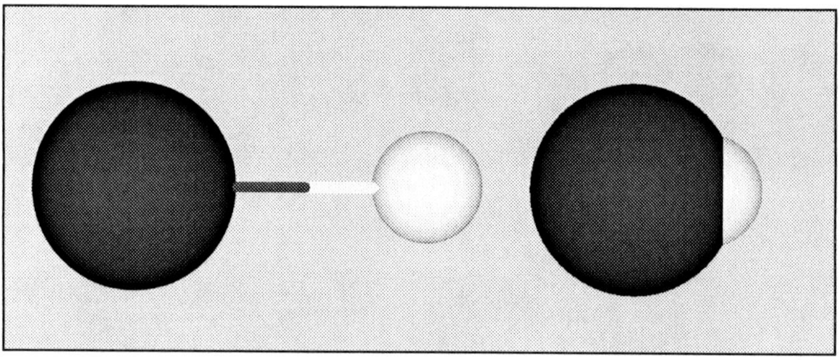

II-A. Mechanisms of the OH–Aromatic Compound Reactions

As discussed in chapter I, the aromatic hydrocarbons are important contributors to the chemical processes in urban air masses. Aromatic hydrocarbons not containing unsaturated $>C=C<$ bond(s) in the substituent groups react only slowly with O_3 and with NO_3 radicals (Atkinson and Carter, 1984; Atkinson, 1991b). However, aromatic compounds react with OH radicals, with a rate coefficient that depends on the substituent groups attached to the aromatic ring (see section II-C; Zetzsch,1982a; and Atkinson, 1991a). As a generalization, in the atmosphere aromatic compounds are more reactive than alkanes but less reactive than alkenes.

The simplest aromatic hydrocarbon, benzene (C_6H_6), is the least reactive of the aromatic hydrocarbons. In view of the fact that benzene has multiple double bonds involved in its structure, at first sight it is surprising that its reactivity is much lower that that of the alkenes. This can be seen in figure II-A-1, where the reactivity of a series of four C_6-ring hydrocarbons that contain an increasing number of carbon–carbon double bonds can be compared. Cyclohexene (C_6H_{10}), which contains one double bond, shows an increase of a factor of about 9 in the rate coefficient for reaction with OH radical (at 298 K) over that for the analogous saturated hydrocarbon, cyclohexane (C_6H_{12}), which contains no double bonds. As one progresses further to the C_6-ring compound with two conjugated carbon–carbon double bonds (1,3-cyclohexadiene), a further increase

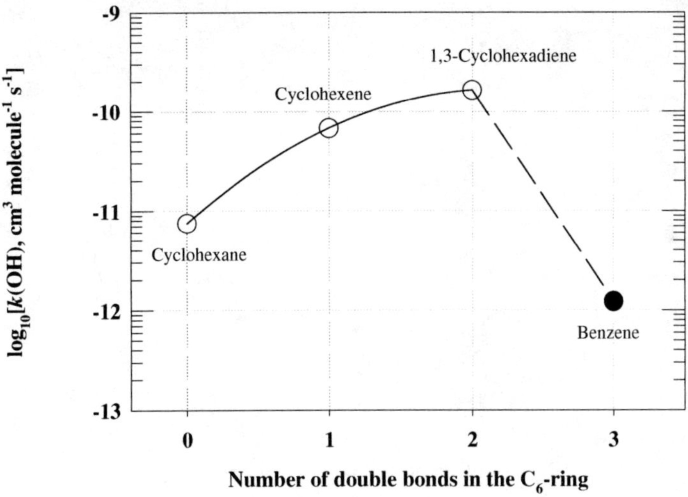

Figure II-A-1. Plot of the logarithm of the rate coefficient for cyclic-C_6 compounds with increasing number of conjugated double bonds.

(factor of 2.4) in OH-radical rate coefficient occurs. However, a sharp decline in rate coefficient occurs as a third conjugated double bond is introduced into the six-membered ring to form the simplest aromatic hydrocarbon, benzene (C_6H_6). In fact the OH rate coefficient for reaction with benzene is a factor of ~6 less than that for the hydrocarbon with no carbon–carbon double bonds, cyclohexane.

The relatively unreactive nature of benzene is attributed to the very special properties associated with the three conjugated double bonds in the six-membered ring. The C–C bonds in benzene are identical; there are no double or single bonds present. The C–C bond distances in benzene (1.40 Å) are somewhat shorter than a C–C single bond (1.54 Å) but longer than a C=C double bond (1.34 Å).

Although we often use the simplified representations of benzene and other aromatic species, with alternating single/double bonds (such as *A* and *B* below), a more accurate representation is structure *C*, which attempts to picture the delocalization of the three pairs of π-bonding electrons present in the molecule.

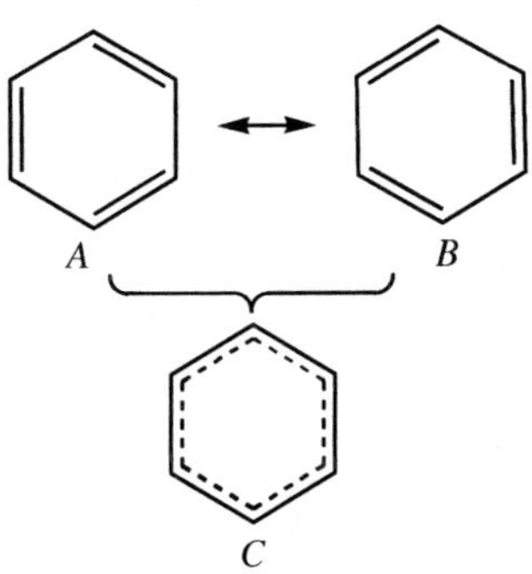

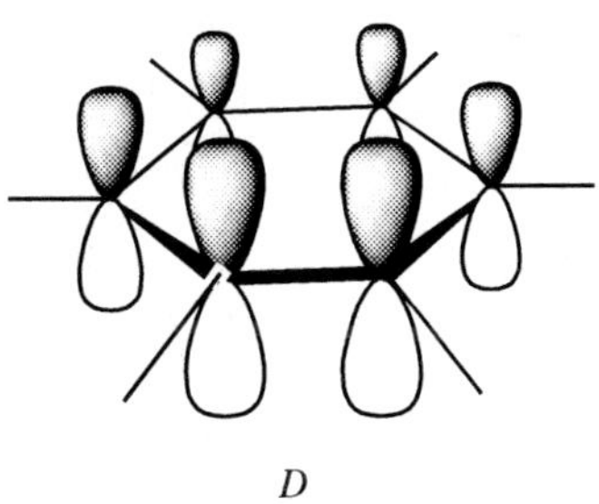

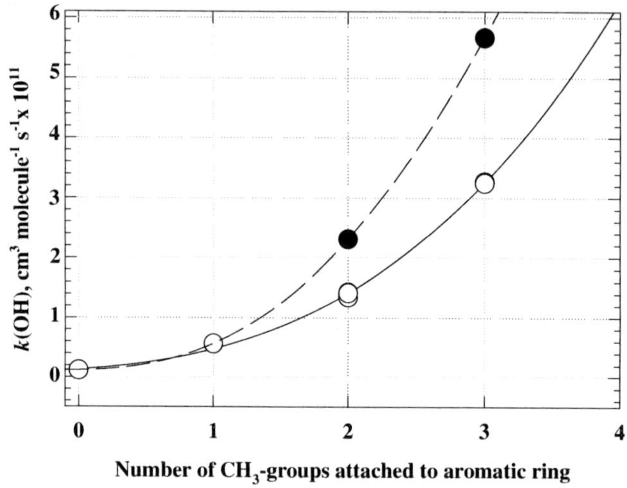

Figure II-A-2. Plot of the rate coefficient for OH radical reaction (298 K) with benzene and methylbenzenes versus the number of methyl groups attached to the benzene ring. The open circles are data for benzene, toluene, *o*- and *p*-xylene, 1,2,3- and 1,2,4-trimethylbenzene, while the filled circles are data for the *m*-xylene, and 1,3,5-trimethylbenzene.

The bonding molecular orbitals in benzene that determine its planar structure with 120° C–C–C bond angles can be thought of as formed from the combination of the hybridized sp^2 atomic orbitals on the carbon atoms (*D*) with the remaining *p*-atomic orbitals on the six carbon atoms, perpendicular to the ring, leading to π-bonding molecular orbitals. The resulting structure has a special stability that characterizes these aromatic structures.

The substitution of methyl groups on benzene increases the rate coefficient for reaction with OH radicals, as seen in figure II-A-2. Presumably, the electron-donating alkyl groups provide a lowered barrier for attack of the electron-seeking OH radical on the π-system of the aromatic molecule. The data in figure II-A-2 follow two different trend curves: the solid curve is determined by the data for benzene, toluene, *o*- and *p*-xylene, 1,2,3- and 1,2,4-trimethylbenzene (open circles), while the dashed curve (determined by the filled circles) follows data for *m*-xylene and 1,3,5-trimethylbenzene. The somewhat higher reactivity of the compounds with the *m*-substituents is consistent with expectations of the influence of these groups as given by their electrophilic substituent, σ^+, values (Brown and Okamoto, 1958); see section II-C. The substitution of other alkyl groups on benzene results in a similar increase in OH rate coefficient over that for benzene; with substitution of a benzene H-atom by CH_3, C_2H_5, *n*-C_3H_7, *iso*-C_3H_7, and *tert*-C_4H_9, the rate coefficient increases by factors of 4.6, 8.2, 7.0, 5.2, and 3.7, respectively. In contrast, the substitution of electron-withdrawing groups on benzene results in a decrease in reactivity toward OH radicals; for example, nitrobenzene and chlorobenzene react at about 0.1 and 0.7 of the rate, respectively, with which benzene reacts.

The reactions of the OH radical with aromatic compounds proceed by three reaction pathways: (1) reaction at the aromatic ring involving initial OH-radical addition to the carbon atoms of the aromatic ring; (2) reaction at the substituent group(s) attached to the aromatic ring, involving H-atom abstraction from the C–H or O–H bonds of the substituent group(s); and (3) OH-radical addition to any >C=C< bond(s) present in these substituent group(s). As an example, these pathways are shown for the reaction of the OH radical with cinnamaldehyde ($C_6H_5CH{=}CHCHO$) (Smith et al., 1996).

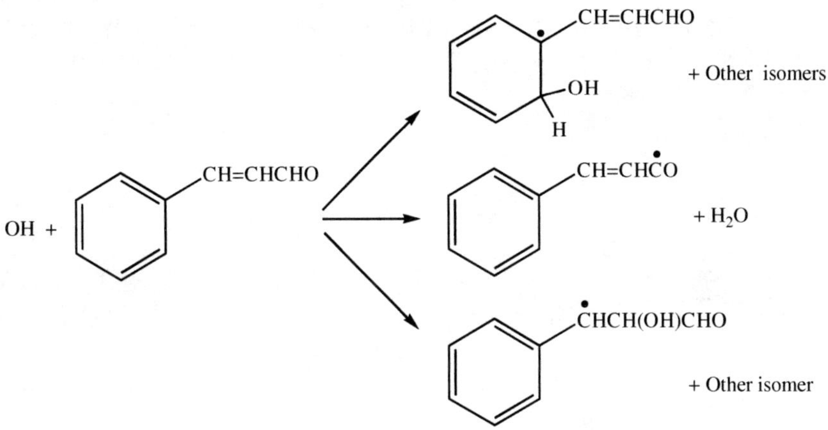

For benzene and the methyl- and ethyl-substituted benzenes, the pathway involving OH-radical addition to the carbon atoms of the aromatic ring dominates, accounting for typically ≥90% of the overall OH-radical reaction at room temperature. The key question then becomes: what is the fate under atmospheric conditions of the hydroxycyclohexadienyl-type radicals (often termed an OH–aromatic adduct for convenience) formed from this OH-radical addition pathway? As discussed in section II-E, the OH–aromatic adducts have been shown to react with O_2 and NO_2, but not with NO (Knispel et al., 1990). For the OH–benzene, OH–toluene, and OH–xylene adducts, the O_2 reactions are slow (with rate coefficients at room temperature in the 10^{-16} cm^3 molecules^{-1} s^{-1} range), while the NO_2 reactions are rapid (with rate coefficients at room temperature of $\sim 3 \times 10^{-11}$ cm^3 molecules^{-1} s^{-1}). Combination of these rate coefficients with atmospheric concentrations of O_2 and NO_2 shows that the dominant reaction of these OH–aromatic adducts in the atmosphere is with O_2, but that in laboratory experiments conducted at elevated NO_X concentrations (at NO_2 mixing ratios of several hundred parts-per-billion or higher), the OH–aromatic adducts will react to a significant (if not dominant) extent with NO_2, and thereby lead to products that may not be representative of those formed for atmospheric conditions.

In section II-B, the initial steps in the atmospheric oxidation of the monocyclic aromatic compounds (both hydrocarbons and their ring-retaining oxidation products) with OH radical are considered in depth. The kinetic data available for the gas-phase reactions of the OH radical with monocyclic aromatic compounds (both those emitted into the atmosphere and those formed in the atmosphere as reaction products) are discussed and evaluated. In section II-C, methods are discussed for estimating OH-radical reaction rate coefficients for monocyclic aromatic compounds. In section II-D, the major products of the reactions of the OH radical with the monocyclic aromatic hydrocarbons are reviewed and discussed. The nature of the reaction mechanisms that lead to the products are discussed in section II-E. In section II-F, the kinetic data available for the gas-phase reactions of the OH radical with the polycyclic aromatic compounds (PACs) are discussed and evaluated. In section II-G, the products and mechanisms of the gas-phase atmospheric oxidation of the PACs are reviewed. Very few of the rate coefficients for the many PACs have been measured to date, and in section II-H we review and suggest some empirical methods of estimating the rate coefficients for the gas-phase reactions of the OH radical with the polycyclic aromatic compounds that have not yet been studied.

II-B. Rate Coefficients for Reactions of OH Radicals with Monocyclic Aromatic Compounds

Arrhenius plots of the high-pressure rate coefficients for the gas-phase reactions of the OH radical with a number of aromatic compounds (including benzene, toluene, xylenes, trimethylbenzenes, phenol, *o*-cresol, and methoxybenzene) over extended temperature ranges exhibit a complex behavior (Atkinson, 1989). These Arrhenius plots generally consist of two distinct temperature regimes: at temperatures < 325–350 K the rate coefficients generally show little change with decreasing temperature, while at temperatures > 400–450 K the rate coefficients increase rapidly with increasing temperature. Furthermore, the rate coefficients at 400–500 K are lower than those at 298 K, and extrapolation of the rate coefficients in this "high-temperature" regime to 298 K leads to rate coefficients much lower than the measured 298 K rate coefficient. In the intermediate temperature range of ~ 325–450 K (with the precise temperature range depending on the specific aromatic compound and the experimental technique used) the pseudo–first-order decays of the OH-radical concentration, as monitored using absolute rate techniques, often exhibit non-exponential behavior.

These observations are interpreted (Atkinson, 1989) as the OH-radical reaction with the aromatic hydrocarbon proceeding mainly by OH-radical addition to the carbon atoms of the aromatic ring to form a thermally stable (on a time-scale of tens of milliseconds) hydroxycyclohexadienyl radical (OH–aromatic adduct) at temperatures < 325–350 K; the back-decomposition of this adduct to reactants occurs on the time-scale of the measurements in the temperature range ~ 325–350 K, and with thermal decomposition of the OH–aromatic adduct being sufficiently rapid at temperatures ~ 400–450 K that only the H-atom abstraction process is observed at > 400–450 K (unless the OH–aromatic adduct decomposes by other routes not leading to reformation of the OH radical). At atmospheric temperatures, the H-atom abstraction path is generally of minor importance (Atkinson, 1994).

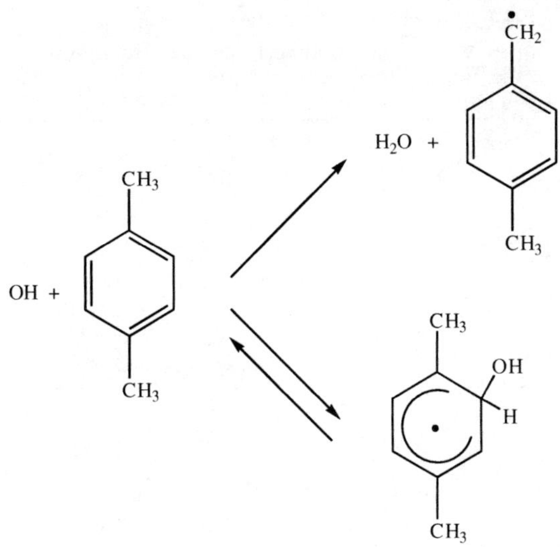

For atmospheric purposes, the rate coefficients measured in the "low" temperature region, < 325–350 K, are most relevant, although the "high" temperature data give valuable information concerning the rate coefficient for the abstraction process(es) and allow (by extrapolation) the relative importance of the H-atom abstraction pathways to be estimated at room temperature and below (Atkinson, 1989, 1994). For benzene and toluene, the rate coefficients at room temperature and below are in the fall-off regime at total pressures < 100 Torr of helium or argon diluents, and the measured rate coefficients k_{obs} can be described using the Troe fall-off expression:

$$k_{obs} = \frac{k_0[M]\,F^Z}{1 + (k_0[M]/k_\infty)}$$

where

$$z = \{1 - [\log_{10}(k_0[M]/k_\infty)]^2\}^{-1}$$

and k_0 is the limiting low-pressure rate coefficient, k_∞ is the limiting high-pressure rate coefficient, F is the broadening factor, and $[M]$ is the concentration of the diluent gas. The temperature dependence of the rate coefficients k_0 and k_∞ is generally given in the form $k(T) = k(298\,\text{K})(T/298)^{-n}$, and the temperature dependence of F is given by $F = \exp(-T/T^*) + \exp(-4T^*/T)$, where T^* is a constant for a given reactant (Atkinson et al., 1999). In the data tables that follow, the temperature-dependent recommendations for the rate coefficients are given in the Arrhenius form $k = A\,e^{-B/T}$.

In the remainder of this section, the measured OH-radical reaction rate coefficients are reviewed and evaluated for each of the atmospherically relevant aromatic hydrocarbons for which experimental data are available. In the tables associated with each reaction, the experimental techniques used are denoted by abbreviations, as given in table II-B-0. Thus, for example, use of a flash photolysis system with resonance fluorescence monitoring of OH radicals is denoted by FP-RF. When relative rate methods were used, the rate coefficient for the reference

Table II-B-0. Abbreviations for the experimental techniques used in the rate coefficient determinations

Abbreviation	Experimental Technique
FP	Flash photolysis
DF	Discharge flow
PLP	Pulsed laser photolysis
RR	Relative rate
PR	Pulse radiolysis
RA	Resonance absorption
RF	Resonance fluorescence
LIF	Laser-induced fluorescence
MS	Mass spectrometry

compound from the most recent review and evaluation is used to reevaluate the rate coefficient for the aromatic compound in question (which therefore may be different from that cited in the original publication).

II-B-1. OH + Benzene

At around room temperature, the rate coefficients for this reaction are pressure dependent below ~50 Torr total pressure of helium or argon diluent (Davis et al., 1975; Lorenz and Zellner, 1983; Wahner and Zetzsch, 1983; Witte et al., 1986; Baulch et al., 1988; Goumri et al., 1991; Lin et al., 1994). Rate coefficients obtained at, or close to, the high-pressure limit at temperatures <360 K are given in table II-B-1. In addition to the rate coefficients given in the table, rate coefficients have been measured at temperatures <325 K, which are in the fall-off region (Davis et al., 1975; Lorenz and Zellner, 1983; Wahner and Zetzsch, 1983; Witte et al., 1986; Baulch et al., 1988; Goumri et al., 1991). The two most extensive investigations of the pressure dependence of the rate coefficient for the reaction of the OH radical with benzene in argon diluent (the diluent gas expected to be closest in its third-body efficiency to N_2 or air) were those of Lorenz and Zellner (1983) and Witte et al. (1986). A limiting low-pressure rate coefficient of $k_0(\mathrm{Ar}) = 3 \times 10^{-29}$ cm^6 molecule^{-2} s^{-1} at 298 K was derived by Witte et al. (1986) using a value of $F = 0.6$, with the data of Lorenz and Zellner (1983) being in good agreement with the resulting calculated fall-off curve (Witte et al., 1986). With a value of $k_\infty \sim 1.2 \times 10^{-12}$ cm^3 molecule^{-1} s^{-1} at 298 K (see table II-B-1 and discussion below), these values of k_0 and F in the Troe fall-off equation result in rate coefficients at 100 Torr and 760 Torr total pressure of Ar (or air, assuming a similar third-body efficiency for Ar and air) being ~12% and ~5–6%, respectively; that is, below the limiting high-pressure rate coefficient k_∞ at 298 K.

The majority of absolute and relative rate measurements at (or close to) the high-pressure limit are in reasonable agreement, and a least-squares analysis of the absolute rate coefficients of Hansen et al. (1975), Perry et al. (1977a), Tully et al. (1981), Lorenz and Zellner (1983), Witte et al. (1986) [judged to supersede the earlier data from the same laboratory of Wahner and Zetzsch (1983) and Rinke

Table II-B-1. OH + benzene → products:

Rate coefficients $k = A\,e^{-B/T}$ at, or close to, the high-pressure limit at $T \leq 365$ K

$10^{12} \times A$ (cm^3 molecule^{-1} s^{-1})	B (K)	$10^{12} \times k$ (cm^3 molecule^{-1} s^{-1})	at T (K)	Technique	Reference	Temperature range covered (K)
		1.59 ± 0.12	298	FP-RF	Davis et al. (1975)	
		≤ 2.6	304 ± 1	RR [relative to k(n-butane) = 2.52×10^{-12}][a]	Doyle et al. (1975)	
		1.24 ± 0.12	298	FP-RF	Hansen et al. (1975)	
		1.20 ± 0.15	297.6	FP-RF	Perry et al. (1977a)	298–422
		1.32 ± 0.30	304.4			
		1.33 ± 0.25	305.8			
		1.66 ± 0.25	322.7			
		0.84	300	RR [relative to k(ethene) = 8.44×10^{-12}][a]	Cox et al. (1980)	
		1.04 ± 0.08	250	FP-RF	Tully et al. (1981)	250–1017
		1.20 ± 0.09	270			
3.1 ± 2.6	270 ± 220	1.24 ± 0.09	298			
		0.93	300	RR [relative to k(ethene) = 8.44×10^{-12}][a]	Barnes et al. (1982)	
		0.76 ± 0.15	244	PLP-RF	Lorenz and Zellner (1983)	244–870
6.3 ± 1.7	500 ± 50	1.15 ± 0.25	298 ± 2			
		0.88 ± 0.04	295	FP-RF	Wahner and Zetzsch (1983)	
		1.02 ± 0.2	296	FP-RF	Rinke and Zetzsch (1984)	
		1.42 ± 0.06	~298	RR [relative to k(n-hexane) = 5.45×10^{-12}][a]	Ohta and Ohyama (1985)	

(continued)

Table II-B-1. *(continued)*

$10^{12} \times A$ (cm^3 molecule^{-1} s^{-1})	B (K)	$10^{12} \times k$ (cm^3 molecule^{-1} s^{-1})	at T (K)	Technique	Reference	Temperature range covered (K)
		1.00 ± 0.04	239	FP-RF	Witte et al. (1986)	239–354
		1.11 ± 0.12	239			
		1.04 ± 0.03	245			
		1.06 ± 0.03	253			
		1.13 ± 0.04	253			
		1.04 ± 0.05	259			
		1.11 ± 0.02	274			
		1.12 ± 0.08	274			
		1.11 ± 0.04	283			
		1.05 ± 0.05	299			
		1.18 ± 0.17	299			
		1.23 ± 0.09	312			
		1.37 ± 0.09	312			
		1.21 ± 0.12	325			
		1.28 ± 0.05	328			
		1.26 ± 0.08	331			
		1.39 ± 0.22	331			
		1.48 ± 0.12	334			
		1.50 ± 0.29	342			
		1.25 ± 0.22	352			
2.3	190 ± 60	1.33 ± 0.36	354			
		1.20 ± 0.44	296	RR [relative to k(propane) $= 1.11 \times 10^{-12}$][a]	Edney et al. (1986)	
		1.40 ± 0.23	234	FP-RF	Wallington et al. (1987)	234–438
		1.30 ± 0.19	263			
		1.29 ± 0.14	296			

	1.11 ± 0.12	298	FP-RF	Knispel et al. (1990)	298–354
	1.04 ± 0.05	315			
	1.06 ± 0.13	333			
	0.90 ± 0.24	354			
	1.25 ± 0.49	297 ± 2	RR [relative to *k* (*n*-hexane) = 5.44×10^{-12}][a]	Sommerlade et al. (1993)	
	1.06 ± 0.07	274	RR [relative to *k*(toluene) = $1.81 \times 10^{-12}\, e^{338/T}$][b]	Semadeni et al. (1995)	274–363
	1.12 ± 0.06	299			
	1.23 ± 0.06	324			
	1.24 ± 0.10	348			
	1.33 ± 0.23	363			
	1.2 ± 0.2	298	PR-RA	Bjergbakke et al. (1996)	

[a] From Atkinson (1997).
[b] This evaluation.

and Zetzsch (1984)], Wallington et al. (1987), and Knispel et al. (1990) at ≤350 K leads to a rate coefficient at ~100 Torr total pressure of Ar diluent of

$$k(\text{benzene}) = 2.33 \times 10^{-12}\,e^{-(193\pm105)/T}\ \text{cm}^3\ \text{molecules}^{-1}\ \text{s}^{-1}$$

over the temperature range 239–342 K, where the indicated error is two standard deviations, with a 298 K rate coefficient of

$$k(\text{benzene}) = 1.22 \times 10^{-12}\ \text{cm}^3\ \text{molecule}^{-1}\ \text{s}^{-1}\ \text{at} \sim 100\,\text{Torr total pressure of Ar},$$

and with an estimated overall uncertainty at 298 K of ±20%. The rate coefficients of Cox et al. (1980), Barnes et al. (1982), Wahner and Zetzsch (1983), Rinke and Zetzsch (1984), Ohta and Ohyama (1985), Edney et al. (1986), Sommerlade et al. (1993), Semadeni et al. (1995), and Bjergbakke et al. (1996) are in reasonably good agreement with the recommendation. In particular, the good agreement of the rate coefficients determined by Semadeni et al. (1995), relative to the rate coefficients for toluene, indicates that our recommendations for the reactions of the OH radical with benzene and toluene are internally consistent. The measurements by Davis et al. (1975) appear to be in error as they obtained a rate coefficient at 100 Torr total pressure of helium diluent significantly higher than the recommended value.

As noted above, the recommended rate coefficient is applicable (to within ~6–7%) to atmospheric pressure of air. The value of k_∞ at 298 K is estimated to be

$$k_\infty(\text{benzene}) = 1.39 \times 10^{-12}\ \text{cm}^3\ \text{molecule}^{-1}\ \text{s}^{-1}\ \text{at 298 K}$$

and

$$k_\infty(\text{benzene}) \sim 3.8 \times 10^{-12}\,e^{-300/T}\ \text{cm}^3\ \text{molecule}^{-1}\ \text{s}^{-1}$$

over the temperature range ~240–350 K.

Extrapolation of the elevated temperature (≥450 K) rate coefficients of Tully et al. (1981), Lorenz and Zellner (1983), and Madronich and Felder (1985) to room temperature indicates that the H-atom abstraction process is of minor importance at room temperature, with $k(\text{benzene}, T \geq 450\,\text{K}) = 4.67 \times 10^{-18}T^2 e^{-(543\pm226)/T}$ cm^3 molecule^{-1} s^{-1} over the temperature range 453–1409 K ($k = 6.7 \times 10^{-14}$ cm^3 molecule^{-1} s^{-1} at 298 K) (Atkinson, 1989). This extrapolated rate coefficient for the H-atom abstraction process indicates that at 298 K the H-atom abstraction pathway accounts for 5–6% of the overall OH-radical reaction at atmospheric pressure of air.

Kinetic studies of the reaction of OH radicals with benzene-d_6 (Tully et al., 1981; Lorenz and Zellner, 1983) show that over the temperature range 250–336 K the rate coefficient for benzene-d_6 is essentially identical to that for benzene (Atkinson, 1989), while at temperatures ≥450 K the rate coefficients for benzene-d_6 are significantly lower than those for benzene (Atkinson, 1989). These data for benzene and benzene-d_6 indicate that at temperatures <350 K OH-radical addition dominates, while at temperatures >450 K the measured rate coefficients are those for the H-atom abstraction process (Atkinson, 1989).

II-B-2. OH + Toluene

At around room temperature, the rate coefficients for this reaction are pressure dependent below ~100 Torr total pressure of helium diluent (Davis et al., 1975; Tully et al., 1981; Bourmada et al., 1988a,1988b). The rate coefficients obtained at, or close to, the high-pressure limit at temperatures < 360 K are given in table II-B-2. In addition to the data given in the table, rate coefficients have been measured at temperatures < 325 K, which are in the fall-off region (Davis et al., 1975; Tully et al., 1981; Bourmada et al., 1988a, 1988b). Based on the low-pressure discharge flow study of Bourmada et al. (1988a, 1988b), the limiting low-pressure rate coefficient at 295 ± 2 K is $k_0(\mathrm{He}) = 4.0 \times 10^{-28}$ cm^6 molecule^{-2} s^{-1}, indicating that with $F \sim 0.6$ and $k_0(\mathrm{Ar}) > k_0(\mathrm{He})$ the measured rate coefficients at 100 Torr total pressure of argon diluent will be within 5% of the high-pressure limiting rate coefficient k_∞.

For temperatures ≤350 K, a least-squares analysis of the absolute rate coefficients of Hansen et al. (1975), Perry et al. (1977a), Tully et al. (1981), and Knispel et al. (1990) (omitting the rate coefficient at 298 K, which has a large associated uncertainty) and the relative rate coefficients of Edney et al. (1986), Atkinson et al. (1983b), Finlayson-Pitts et al. (1993), Semadeni et al. (1995), Anderson and Hites (1996a), and Kramp and Paulson (1998) leads to a rate coefficient applicable to atmospheric pressure of air of

$$k(\text{toluene}) = 1.18 \times 10^{-12}\, e^{(338\pm78)/T}\ \text{cm}^3\ \text{molecule}^{-1}\ \text{s}^{-1}$$

over the temperature range 213–383 K, where the indicated error is two standard deviations, and

$$k(\text{toluene}) = 5.63 \times 10^{-12}\ \text{cm}^3\ \text{molecule}^{-1}\ \text{s}^{-1}$$

at 298 K, with an estimated overall uncertainty at 298 K of ±20%. The rate coefficients of Davis et al. (1975), Cox et al. (1980), Ohta and Ohyama (1985), Markert and Pagsberg (1993), and Sommerlade et al. (1993) are in generally good agreement with the recommendation.

Extrapolation of the elevated temperature (≥450 K) rate coefficients of Perry et al. (1977a) and Tully et al. (1981) to room temperature indicates that H-atom abstraction is of minor importance at room temperature, with $k(\text{toluene}, T \geq 450\ \text{K}) = 7.58 \times 10^{-18}\, T^2\, e^{(11\pm106)/T}$ cm^3 molecule^{-1} s^{-1} over the temperature range 473–1046 K ($k = 7.0 \times 10^{-13}$ cm^3 molecule^{-1} s^{-1} at 298 K) (Atkinson, 1989). This extrapolated rate coefficient indicates that at 298 K the H-atom abstraction pathway accounts for ~12% of the overall OH-radical reaction at atmospheric pressure of air. Zetzsch et al. (1989), using a flash photolysis–resonance fluorescence technique at temperatures of 299, 310, 323, and 345 K, derived Arrhenius expressions for the H-atom abstraction and OH-radical addition pathways of $2 \times 10^{-12}\, e^{-(395\pm200)/T}$ cm^3 molecule^{-1} s^{-1} and $1.9 \times 10^{-13}\, e^{(1040\pm70)/T}$ cm^3 molecule^{-1} s^{-1}, respectively, both over the temperature range 299–345 K. These Arrhenius expressions lead to rate coefficients of 5.3×10^{-13} cm^3 molecule^{-1} s^{-1} at 298 K for H-atom abstraction (in reasonable agreement with the value of 7.0×10^{-13} cm^3 molecule^{-1} s^{-1} obtained by extrapolation of elevated temperature data; see above) and 6.2×10^{-12} cm^3 molecule^{-1} s^{-1} for

Table II-B-2. OH + toluene → products:

Rate coefficients $k = A\,e^{-B/T}$ at, or close to, the high-pressure limit at $T \leq 385\,K$

$10^{12} \times A$ (cm^3 $molecule^{-1}$ s^{-1})	B (K)	$10^{12} \times k$ (cm^3 $molecule^{-1}$ s^{-1})	at T (K)	Technique	Reference	Temperature range covered (K)
		6.11 ± 0.40	298	FP-RF	Davis et al. (1975)	
		3.5 ± 1.6	304 ± 1	RR [relative to k(n-butane) = 2.52×10^{-12}][a]	Doyle et al. (1975)	
		5.78 ± 0.58	298	FP-RF	Hansen et al. (1975)	
		6.40 ± 0.64	297.9	FP-RF	Perry et al. (1977a)	298–473
		4.90 ± 0.6	323.7			
		7.6	300	RR [relative to k(ethene) = 8.44×10^{-12}][a]	Cox et al. (1980)	
		8.20 ± 0.54	213	FP-RF	Tully et al. (1981)	213–1046
		8.73 ± 0.39	231			
		7.97 ± 0.56	250			
		8.53 ± 0.37	260			
		7.44 ± 0.55	270			
3.8 ± 2.5	−180 ± 170	6.36 ± 0.69	298			
		5.89 ± 0.17	~298	RR [relative to k(n-hexane) = 5.45×10^{-12}][a]	Ohta and Ohyama (1985)	
		5.24 ± 0.53	297	RR [relative to k(cyclo-hexane = 7.19×10^{-12}][a]	Edney et al. (1986)	
		5.48 ± 0.16	296 ± 2	RR [relative to k(propene) = 2.66×10^{-11}][a]	Atkinson and Aschmann (1989)	
		7.0 ± 2.3	299	FP-RF	Knispel et al. (1990)	299–340
		6.05 ± 0.61	311			
		5.50 ± 0.36	323			
		4.42 ± 0.99	340			

5.83 ± 0.28	298	RR [relative to k(n-hexane) = 5.45 × 10^{-12}][a]	Finlayson-Pitts et al. (1993)	
6.68 ± 1.57	297 ± 2	RR [relative to k(n-hexane) = 5.44 × 10^{-12}][a]	Sommerlade et al. (1993)	
6.4 ± 0.7	338	PR-RA	Markert and Pagsberg (1993)	
4.98 ± 0.68	258	RR [relative to k(2,3-dimethylbutane) = 1.24 × 10^{-17} T^2 $e^{494/T}$][a]	Semadeni et al. (1995)	258–363
5.55 ± 0.23	264			
5.69 ± 0.46	269			
6.20 ± 0.40	284			
5.61 ± 0.29	299			
5.14 ± 0.12	324			
5.08 ± 0.36	324			
4.63 ± 0.13	353			
4.20 ± 0.71	363			
5.64 ± 0.07	276	RR [relative to k(cyclohexane) = 2.88 × 10^{-17} T^2 $e^{309/T}$][a]	Anderson and Hites (1996b)	276–383
6.09 ± 0.07	285			
5.81 ± 0.14	285			
5.54 ± 0.08	293			
5.81 ± 0.58	296			
5.54 ± 0.22	297			
5.46 ± 0.44	297			
5.39 ± 0.22	297			
5.79 ± 0.08	303			
5.63 ± 0.08	310			
5.55 ± 0.16	323			
5.24 ± 0.16	323			
5.16 ± 0.08	323			
4.93 ± 0.08	323			
5.32 ± 0.16	323			
4.77 ± 0.16	323			
5.00 ± 0.08	323			

(*continued*)

Table II-B-2. *(continued)*

Rate coefficients $k = A\,e^{-B/T}$ at, or close to, the high-pressure limit at $T \leq 385$ K *(continued)*

$10^{12} \times A$ (cm^3 molecule^{-1} s^{-1})	B (K)	$10^{12} \times k$ (cm^3 molecule^{-1} s^{-1})	at T (K)	Technique	Reference	Temperature range covered (K)
		4.71 ± 0.16	324			
		5.34 ± 0.16	324			
		5.18 ± 0.08	324			
		5.02 ± 0.17	344			
		4.80 ± 0.09	363			
		4.62 ± 0.09	363			
		4.53 ± 0.09	363			
		4.62 ± 0.27	363			
		4.53 ± 0.09	363			
		4.45 ± 0.98	363			
		4.45 ± 0.09	363			
		3.37 ± 0.19	371			
		2.94 ± 0.19	383			
		5.55 ± 0.07	296 ± 2	RR [relative to k(1,3-butadiene) $= 6.72 \times 10^{-11}$][a]	Kramp and Paulson (1998)	
		5.15 ± 0.11	296 ± 2	RR [relative to k(1,3-butadiene) $= 6.72 \times 10^{-11}$][a]	Kramp and Paulson (1998)	
		5.77 ± 0.07	296 ± 2	RR [relative to k(propene) $= 2.66 \times 10^{-11}$][a]	Kramp and Paulson (1998)	
		5.55 ± 0.08	296 ± 2	RR [relative to k(cyclo-hexane) $= 7.17 \times 10^{-12}$][a]	Kramp and Paulson (1998)	

[a] From Atkinson (1997).

the OH-radical addition pathway [again in good agreement with our recommendation, although the temperature dependence obtained by Zetzsch et al. (1989) is significantly larger than our recommended value].

Measurements of the rate coefficients for the reaction of the OH radical with toluene-d_3 ($C_6H_5CD_3$) (Tully et al., 1981), toluene-d_5 ($C_6D_5CH_3$) (Tully et al., 1981) and toluene-d_8 ($C_6D_5CD_3$) (Perry et al., 1977a; Tully et al., 1981) show that over the temperature range 250–298 K the rate coefficients for toluene-d_3, toluene-d_5, and toluene-d_8 are essentially identical to that for toluene (Atkinson, 1989), while at temperatures $\geq$450 K the rate coefficients for toluene-d_3 and toluene-d_8 are significantly lower than those for toluene and toluene-d_5 (Atkinson, 1989). These data for toluene and the partially and fully deuterated toluenes therefore indicate that at temperatures $<$350 K OH-radical addition dominates, while at temperatures $>$450 K the measured rate coefficients are those for H-atom abstraction (and mainly from the substituent $-CH_3$ or $-CD_3$ group) (Atkinson, 1989).

II-B-3. OH + Ethylbenzene

The rate coefficients of Lloyd et al. (1976), Ravishankara et al. (1978), and Ohta and Ohyama (1985), all obtained at room temperature, are given in table II-B-3. These rate coefficients are in reasonable agreement, probably within the likely overall experimental uncertainties, and a unit-weighted average leads to the recommendation of

$$k(\text{ethylbenzene}) = 7.0 \times 10^{-12}\ \text{cm}^3\ \text{molecule}^{-1}\ \text{s}^{-1}$$

at 298 K, with an estimated overall uncertainty of 25%. This 298 K rate coefficient is ~25% higher than that for toluene, suggesting that the dominant reaction pathway is OH-radical addition to the aromatic ring. At temperatures $\leq$325 K, the temperature dependence is therefore expected to be close to zero.

II-B-4. OH + *o*-Xylene

Rate coefficients obtained at, or close to, the high-pressure limit are given in table II-B-4. At room temperature the rate coefficient is at, or close to, the high-pressure limit for total pressures $\geq$20 Torr of helium (Ravishankara et al., 1978). In the temperature range $<$350 K, rate coefficients have been obtained over only the small temperature range of 296–320 K, and, as expected because of the narrow temperature range studied, there is no evidence for any significant temperature dependence. Accordingly, a unit-weighted average of the absolute rate coefficients of Hansen et al. (1975), Perry et al. (1977a), Ravishankara et al. (1978), and Nicovich et al. (1981) and the relative rate coefficients of Ohta and Ohyama (1985), Edney et al. (1986), and Atkinson and Aschmann (1989) yields the recommendation

$$k(o\text{-xylene}) = 1.36 \times 10^{-11}\ \text{cm}^3\ \text{molecule}^{-1}\ \text{s}^{-1}$$

independent of temperature over the range 296–320 K, with an estimated overall uncertainty at 298 K of $\pm$25%.

Table II-B-3. OH + ethylbenzene → products:

Rate coefficients *k* at, or close to, the high-pressure limit

$10^{12} \times k$ (cm^3 $molecule^{-1}$ s^{-1})	at *T* (K)	Technique	Reference
6.70 ± 1.34	305 ± 2	RR [relative to *k*(*n*-butane) = 2.53×10^{-12}][a]	Lloyd et al. (1976)
7.95 ± 0.50[b]	298	FP-RF	Ravishankara et al. (1978)
6.32 ± 0.28	~ 298	RR [relative to *k*(*n*-hexane) = 5.45×10^{-12}][a]	Ohta and Ohyama (1985)

[a] From Atkinson (1997).

[b] At 200 Torr total pressure of helium diluent; data also obtained at 3 and 20 Torr total pressure of helium diluent.

Table II-B-4. OH + *o*-xylene → products:

Rate coefficients *k* at, or close to, the high-pressure limit at $T \leq 325$ K

$10^{12} \times k$ (cm^3 $molecule^{-1}$ s^{-1})	at *T* (K)	Technique	Reference	Temperature range covered (K)
10.8 ± 3.3	304 ± 1	RR [relative to *k*(*n*-butane) = 2.52×10^{-12}][a]	Doyle et al. (1975)	
15.3 ± 1.5	298	FP-RF	Hansen et al. (1975)	
14.3 ± 1.5 14.0 ± 2.0	298.5 319.0	FP-RF	Perry et al. (1977a)	298–432
12.4 ± 1.2[b]	298	FP-RF	Ravishankara et al. (1978)	
14.0	300	RR [relative to *k*(ethene) = 8.44×10^{-12}][a]	Cox et al. (1980)	
14.2 ± 1.7 15.8 ± 1.8	298 320	FP-RF	Nicovich et al. (1981)	298–970
12.2 ± 0.6	~ 298	RR [relative to *k*(*n*-hexane) = 5.45×10^{-12}][a]	Ohta and Ohyama (1985)	
13.4	300	RR [relative to *k*(1,3,5-trimethylbenzene) = 5.67×10^{-11}][c]	Klöpffer et al. (1986)	
12.1	297	RR [relative to *k*(cyclo-hexane) = 7.19×10^{-12}][a]	Edney et al. (1986)	
12.2 ± 0.6	296 ± 2	RR [relative to *k*(propene) = 2.66×10^{-11}][a]	Atkinson and Aschmann (1989)	

[a] From Atkinson (1997).

[b] At 200 Torr total pressure of helium diluent; data also obtained at 20 Torr total pressure of helium and argon diluents.

[c] From this evaluation.

Extrapolation of the elevated temperature ($\geq$450 K) rate coefficients of Nicovich et al. (1981) indicates that the H-atom abstraction process is of minor importance at room temperature, with k(o-xylene, $T \geq 450$ K) = $1.75 \times 10^{-17} T^2 e^{-(35 \pm 90)/T}$ cm^3 molecule^{-1} s^{-1} over the temperature range 508–970 K. This leads to a rate coefficient for H-atom abstraction of $k = 1.38 \times 10^{-12}$ cm^3 molecule^{-1} s^{-1} at 298 K, a factor of 2.0 higher than that for toluene and of a similar magnitude to the H-atom abstraction rate coefficients for m- and p-xylene (see the following discussion) (Atkinson, 1989). This extrapolated rate coefficient indicates that H-atom abstraction accounts for ~10% of the overall OH-radical reaction in atmospheric pressure of air at 298 K.

II-B-5. OH + m-Xylene

Rate coefficients obtained at, or close to, the high-pressure limit are given in table II-B-5. At room temperature, the rate coefficient is at, or close to, the high-pressure limit for total pressures $\geq$20 Torr of helium (Ravishankara et al., 1978). In the temperature range <350 K, rate coefficients have been obtained over only a relatively small temperature range of 250–315 K, and there is no evidence for any significant temperature dependence. Accordingly, a unit-weighted average of the absolute rate coefficients of Hansen et al. (1975), Perry et al. (1977a), Ravishankara et al. (1978), and Nicovich et al. (1981) and the relative rate coefficients of Atkinson et al. (1983b), Ohta and Ohyama (1985), Edney et al. (1986), Atkinson and Aschmann (1989), Kramp and Paulson (1998), and Aschmann and Atkinson (1998) yields the recommendation

$$k(m\text{-xylene}) = 2.31 \times 10^{-11} \text{ cm}^3 \text{ molecule}^{-1} \text{ s}^{-1}$$

independent of temperature over the range 250–315 K, with an estimated overall uncertainty at 298 K of ±15%.

Extrapolation of the elevated temperature ($\geq$ 450 K) rate coefficients of Nicovich et al. (1981) to room temperature indicates that the H-atom abstraction process is of minor importance with k(m-xylene, $T \geq 450$ K) = $1.71 \times 10^{-17} T^2 e^{-(127 \pm 235)/T}$ cm^3 molecule^{-1} s^{-1} over the temperature range 508–960 K. This leads to a rate coefficient for H-atom abstraction of $k = 9.9 \times 10^{-13}$ cm^3 molecule^{-1} s^{-1} at 298 K, of similar magnitude to those for o- and p-xylene (Atkinson, 1989). This extrapolated rate coefficient to the H-atom abstraction at 298 K indicates that this pathway accounts for ~4% of the overall OH-radical reaction at atmospheric pressure of air.

II-B-6. OH + p-Xylene

Rate coefficients obtained at, or close to, the high-pressure limit are given in table II-B-6. At room temperature, the rate coefficient is at, or close to, the high-pressure limit for total pressures $\geq$20 Torr of helium (Ravishankara et al., 1978). In the temperature range <350 K rate coefficients have been obtained over only a relatively small temperature range of 296–335 K, and there is no evidence for any significant temperature dependence. However, the measured

Table II-B-5. OH + *m*-xylene → products:
Rate coefficients *k* at, or close to, the high-pressure limit at $T \leq 350$ K

$10^{12} \times k$ (cm^3 $molecule^{-1}$ s^{-1})	at T (K)	Technique	Reference	Temperature range covered (K)
18.9 ± 1.3	304 ± 1	RR [relative to k(n-butane) = 2.52×10^{-12}][a]	Doyle et al. (1975)	
23.6 ± 2.4	297	FP-RF	Hansen et al. (1975)	
18.2 ± 3.7	305 ± 2	RR [relative to k(n-butane) = 2.53×10^{-12}][a]	Lloyd et al. (1976)	
24.0 ± 2.5	298.3	FP-RF	Perry et al. (1977a)	298–427
24.4 ± 3.6	314.5			
20.6 ± 1.3[b]	298	FP-RF	Ravishankara et al. (1978)	
19.6	300	RR [relative to k(ethene) = 8.44×10^{-12}][a]	Cox et al. (1980)	
26.5 ± 2.5	250	FP-RF	Nicovich et al. (1981)	250–960
25.6 ± 4.3	269			
25.4 ± 3.5	298			
20.6 ± 1.4	299 ± 2	RR [relative to k(cyclohexane) = 7.24×10^{-12}][a]	Atkinson et al. (1983b)	
21.6 ± 0.6	~ 299	RR [relative to k(n-hexane) = 5.45×10^{-12}][a]	Ohta and Ohyama (1985)	
22.2	297	RR [relative to k(cyclohexane) = 7.19×10^{-12}][a]	Edney et al. (1986)	
23.0 ± 0.6	296 ± 2	RR [relative to k(propene) = 2.66×10^{-11}][a]	Atkinson and Aschmann (1989)	
20.8 ± 4.0	297 ± 2	RR [relative to k(n-hexane) = 5.44×10^{-12}][a]	Sommerlade et al. (1993)	
21.6 ± 1.8	296 ± 2	RR [relative to k(series of 10 organics)][c]	Kramp and Paulson (1998)	
20.9 ± 1.4	296 ± 2	RR [relative to k(cyclohexane) = 7.17×10^{-12}][a]	Aschmann and Atkinson (1998)	
19.2 ± 4.0	296 ± 2	RR [relative to k(propene) = 2.66×10^{-11}][a]	Alvarado et al. (1998)	

[a] From Atkinson (1997).
[b] At 200 Torr total pressure of helium diluent; data also obtained at total pressures of 3 and 20 Torr of helium diluent and at total pressures of 20 and 200 Torr of argon diluent.
[c] Rate coefficients for the reference organics used were taken from Atkinson (1994, 1997).

rate coefficients exhibit a significant degree of scatter, covering a range of a factor of 1.8. A unit-weighted average of the absolute rate coefficients of Hansen et al. (1975), Perry et al. (1977a), Ravishankara et al. (1978), and Nicovich et al. (1981) and the relative rate coefficients of Ohta and Ohyama (1985), Edney et al. (1986). and Atkinson and Aschmann (1989) yields the recommendation of

$$k(p\text{-xylene}) = 1.43 = 10^{-11}\ \text{cm}^3\ \text{molecule}^{-1}\ \text{s}^{-1}$$

Table II-B-6. OH + *p*-xylene → products:
Rate coefficients k at, or close to, the high-pressure limit at $T \leq 350$ K

$10^{12} \times k$ (cm^3 molecule^{-1} s^{-1})	at T (K)	Technique	Reference	Temperature range covered (K)
10.3 ± 2.3	304 ± 1	RR [relative to k(*n*-butane) = 2.52×10^{-12}][a]	Doyle et al. (1975)	
12.2 ± 1.2	297	FP-RF	Hansen et al. (1975)	
15.3 ± 1.7	298.0	FP-RF	Perry et al. (1977a)	298–428
18.2 ± 2.2	306.3			
18.2 ± 2.2	310.7			
17.3 ± 2.2	313.2			
14.9 ± 2.0	324.2			
10.5 ± 1.0[b]	298	FP-RF	Ravishankara et al. (1978)	
13.5 ± 1.4	298	FP-RF	Nicovich et al. (1981)	298–960
13.8 ± 1.1	320			
12.5 ± 1.3	335			
12.6 ± 0.6	~ 298	RR [relative to k(*n*-hexane) = 5.45×10^{-12}][a]	Ohta and Ohyama (1985)	
13.6	300	RR [relative to k(1,3,5-trimethylbenzene) = 5.67×10^{-11}][a]	Klöpffer et al. (1986)	
13.3	296	RR [relative to k(cyclohexane) = 7.17×10^{-12}][a]	Edney et al. (1986)	
13.6 ± 0.4	298 ± 2	RR [relative to k(cyclohexane) = 7.21×10^{-12}][a]	Becker and Klein (1987)	
13.0 ± 0.5	296 ± 2	RR [relative to k(propene) = 2.66×10^{-11}][a]	Atkinson and Aschmann (1989)	
14.4 ± 3.0	297 ± 2	RR [relative to k(*n*-hexane) = 5.44×10^{-12}][a]	Sommerlade et al. (1993)	

[a] From Atkinson (1997).
[b] At 200 Torr total pressure of helium diluent; data also obtained at a total pressure of 3 Torr of argon diluent and at a total pressure of 20 Torr of argon diluent.

independent of temperature over the range 296–335 K, with an estimated overall uncertainty at 298 K of ±25%.

Extrapolation of the elevated temperature (≥ 450 K) rate coefficients of Nicovich et al. (1981) indicates that the H-atom abstraction process is of minor importance at room temperature, with $k(p\text{-xylene}, T \geq 450\,\text{K}) = 1.74 \times 10^{-17} T^2 e^{-(99 \pm 215)/T}$ cm^3 molecule^{-1} s^{-1} over the temperature range 484–960 K. This leads to a rate coefficient for H-atom abstraction of $k = 1.11 \times 10^{-12}$ cm^3 molecule^{-1} s^{-1} at 298 K, of similar magnitude to those of *o*- and *m*-xylene (Atkinson, 1989). This extrapolated rate coefficient for the H-atom abstraction indicates that at 298 K this pathway accounts for ~7–8% of the overall OH-radical reaction at atmospheric pressure of air.

II-B-7. OH + *n*-Propylbenzene

The rate coefficients of Lloyd et al. (1976), Ravishankara et al. (1978), and Ohta and Ohyama (1985), all obtained at room temperature, are given in table II-B-7. The rate coefficients are in reasonable agreement, probably within the likely overall experimental uncertainties, and a unit-weighted average leads to the recommendation of

$$k(n\text{-propylbenzene}) = 5.8 \times 10^{-12}\ \text{cm}^3\ \text{molecule}^{-1}\ \text{s}^{-1}$$

at 298 K, with an estimated overall uncertainty of ±25%. This 298 K rate coefficient is close to that for toluene, suggesting that the dominant reaction pathway is OH-radical addition to the aromatic ring. At temperatures ≤325 K, the temperature dependence is therefore expected to be close to zero.

Table II-B-7. OH + *n*-propylbenzene → products:
Rate coefficients *k* at, or close to, the high-pressure limit

$10^{12} \times k$ (cm^3 molecule^{-1} s^{-1})	at *T* (K)	Technique	Reference
5.24 ± 1.05	305 ± 2	RR [relative to $k(n\text{-butane}) = 2.53 \times 10^{-12}$][a]	Lloyd et al. (1976)
5.86 ± 0.50[b]	298	FP-RF	Ravishankara et al. (1978)
6.43 ± 0.22	~ 298	RR [relative to $k(n\text{-hexane} = 5.45 \times 10^{-12}$][a]	Ohta and Ohyama (1985)

[a] From Atkinson (1997).
[b] At 200 Torr total pressure of helium diluent; data also obtained at 20 Torr total pressure of helium diluent.

II-B-8. OH + Isopropylbenzene

The rate coefficients of Lloyd et al. (1976), Ravishankara et al. (1978), and Ohta and Ohyama (1985), all obtained at room temperature, are given in table II-B-8.

Table II-B-8. OH + isopropylbenzene → products:
Rate coefficients *k* at, or close to, the high-pressure limit

$10^{12} \times k$ (cm^3 molecule^{-1} s^{-1})	at *T* (K)	Technique	Reference
5.14 ± 1.03	305 ± 2	RR [relative to $k(n\text{-butane}) = 2.53 \times 10^{-12}$][a]	Lloyd et al. (1976)
7.79 ± 0.50	298	FP-RF	Ravishankara et al. (1978)
6.10 ± 0.33	~ 298	RR [relative to $k(n\text{-hexane} = 5.45 \times 10^{-12}$][a]	Ohta and Ohyama (1985)

[a] From Atkinson (1997).

The rate coefficients exhibit significant scatter, covering a range of a factor of 1.5. A unit-weighted average leads to the recommendation of

$$k(\text{isopropylbenzene}) = 6.3 \times 10^{-12}\ \text{cm}^3\ \text{molecule}^{-1}\ \text{s}^{-1}$$

at 298 K, with an estimated overall uncertainty of ±30%. This 298 K rate coefficient is close to that for toluene, suggesting that the dominant reaction pathway is OH-radical addition to the aromatic ring. At temperatures ≤325 K, the temperature dependence is therefore expected to be close to zero.

II-B-9. OH + *o*-Ethyltoluene

The rate coefficients of Lloyd et al. (1976) and Ohta and Ohyama (1985), both obtained at room temperature, are given in table II-B-9. The rate coefficients are in good agreement, and a unit-weighted average leads to the recommendation of

$$k(o\text{-ethyltoluene}) = 1.19 \times 10^{-11}\ \text{cm}^3\ \text{molecule}^{-1}\ \text{s}^{-1}$$

at 298 K, with an estimated overall uncertainty of ±35%. This 298 K rate coefficient is close to that for *o*-xylene, suggesting that the dominant reaction pathway is OH-radical addition to the aromatic ring. At temperatures ≤325 K, the temperature dependence is therefore expected to be close to zero.

Table II-B-9. OH + *o*-ethyltoluene → products: Rate coefficients *k* at, or close to, the high-pressure limit

$10^{12} \times k$ ($\text{cm}^3\ \text{molecule}^{-1}\ \text{s}^{-1}$)	at *T* (K)	Technique	Reference
11.6 ± 2.4	305 ± 2	RR [relative to *k*(*n*-butane) = 2.53×10^{-12}][a]	Lloyd et al. (1976)
12.2 ± 1.2	~ 298	RR [relative to *k*(*n*-hexane = 5.45×10^{-12}][a]	Ohta and Ohyama (1985)

[a] From Atkinson (1997).

II-B-10. OH + *m*-Ethyltoluene

The rate coefficients of Lloyd et al. (1976) and Ohta and Ohyama (1985), both obtained at room temperature, are given in table II-B-10. The rate coefficients are in reasonable agreement and a unit-weighted average leads to the recommendation of

$$k(m\text{-ethyltoluene}) = 1.86 \times 10^{-11}\ \text{cm}^3\ \text{molecule}^{-1}\ \text{s}^{-1}$$

at 298 K, with an estimated overall uncertainty of ±35%. This 298 K rate coefficient is fairly similar to that for *m*-xylene, suggesting that the dominant reaction pathway is OH-radical addition to the aromatic ring. At temperatures ≥325 K, the temperature dependence is therefore expected to be close to zero.

Table II-B-10. OH + *m*-ethyltoluene → products:

Rate coefficients *k* at, or close to, the high-pressure limit

$10^{12} \times k$ (cm^3 molecule^{-1} s^{-1})	at *T* (K)	Technique	Reference
16.4 ± 3.3	305 ± 2	RR [relative to *k*(*n*-butane) = 2.53×10^{-12}][a]	Lloyd et al. (1976)
20.7 ± 1.0	~ 298	RR [relative to *k*(*n*-hexane = 5.45×10^{-12}][a]	Ohta and Ohyama (1985)

[a] From Atkinson (1997).

II-B-11. OH + *p*-Ethyltoluene

The rate coefficients of Lloyd et al. (1976) and Ohta and Ohyama (1985), both obtained at room temperature, are given in table II-B-11. The rate coefficients are in good agreement and a unit-weighted average leads to the recommendation of

$$k(p\text{-ethyltoluene}) = 1.18 \times 10^{-11} \text{ cm}^3 \text{ molecule}^{-1} \text{ s}^{-1}$$

at 298 K, with an estimated overall uncertainty of ±35%. This 298 K rate coefficient is similar to that for *p*-xylene, suggesting that the dominant reaction pathway is OH-radical addition to the aromatic ring. At temperatures ≤325 K, the temperature dependence is therefore expected to be close to zero.

Table II-B-11. OH + *p*-ethyltoluene → products:

Rate coefficients *k* at, or close to, the high-pressure limit

$10^{12} \times k$ (cm^3 molecule^{-1} s^{-1})	at *T* (K)	Technique	Reference
11.0 ± 2.2	305 ± 2	RR [relative to *k*(*n*-butane) = 2.53×10^{-12}][a]	Lloyd et al. (1976)
12.5 ± 1.2	~ 298	RR [relative to *k*(*n*-hexane = 5.45×10^{-12}][a]	Ohta and Ohyama (1985)

[a] From Atkinson (1997).

II-B-12. OH + 1,2,3-Trimethylbenzene

The rate coefficients obtained at, or close to, the high-pressure limit are given in table II-B-12. In the temperature range < 350 K, rate coefficients have been obtained over only a small temperature range of 296–325 K and, as expected because of the narrow temperature range studied, there is no evidence for any significant temperature dependence. The absolute and relative rate coefficients of Hansen et al. (1975), Perry et al. (1977a), Ohta and Ohyama (1985), and Atkinson and Aschmann (1989) are in reasonable agreement. However, because of much larger discrepancies between these four studies for the isomeric 1,3,5-trimethyl-

Table II-B-12. OH + 1,2,3-trimethylbenzene → products:
Rate coefficients k at, or close to, the high-pressure limit at $T \leq 325$ K

$10^{12} \times k$ (cm^3 $molecule^{-1}$ s^{-1})	at T (K)	Technique	Reference	Temperature range covered (K)
19.2 ± 4.1	304 ± 1	RR [relative to k(n-butane) = 2.52×10^{-12}][a]	Doyle et al. (1975)	
26.4 ± 2.6	297	FP-RF	Hansen et al. (1975)	
33.3 ± 4.5	296.9	FP-RF	Perry et al. (1977a)	297–421
24.5 ± 3.7	325.1			
28.9 ± 4.0	~ 298	RR [relative to k(n-hexane) = 5.45×10^{-12}][a]	Ohta and Ohyama (1985)	
32.7 ± 1.9	296 ± 2	RR [relative to k(propene) = 2.66×10^{-11}][a]	Atkinson and Aschmann (1989)	

[a] From Atkinson (1997).

benzene (see section II-B-14), for which the rate coefficients of Hansen et al. (1975) and Ohta and Ohyama (1985) appear to be significantly low, the rate coefficient determined by Atkinson and Aschmann (1989) is recommended:

$$k(\text{1,2,3-trimethylbenzene}) = 3.27 \times 10^{-11}\ \text{cm}^3\ \text{molecule}^{-1}\ \text{s}^{-1}$$

independent of temperature over the range 296–325 K, and with an estimated overall uncertainty at 298 K of ±25%.

Based on the elevated temperature (≥ 450 K) rate coefficients for toluene and o-, m- and p-xylene, a rate coefficient per $-CH_3$ substituent group of k(H-atom abstraction per $-CH_3$ group) = $8.07 \times 10^{-18} T^2 e^{-(38\pm76)T}$ cm^3 $molecule^{-1}$ s^{-1} over the temperature range 470–1046 K can be derived (Atkinson, 1989). Extrapolation of this elevated temperature group rate coefficient to room temperature (k(H-atom abstraction per $-CH_3$ group) = 6.3×10^{-13} cm^3 $molecule^{-1}$ s^{-1} at 298 K) indicates that the H-atom abstraction process in 1,2,3-trimethylbenzene accounts for ~6–7% of the overall OH-radical reaction at atmospheric pressure of air and 298 K.

II-B-13. OH + 1,2,4-Trimethylbenzene

The available rate coefficients obtained at, or close to, the high-pressure limit are given in table II-B-13. In the temperature range < 350 K, rate coefficients have been obtained over only a small temperature range of 296–323 K and, as expected because of the narrow temperature range studied, there is no evidence for any significant temperature dependence. The absolute and relative rate coefficients of Hansen et al. (1975), Perry et al. (1977a), Ohta and Ohyama (1985), and Atkinson and Aschmann (1989) are in fairly good agreement. However, because of much larger discrepancies between these four studies for the isomeric 1,3,5-trimethylbenzene (see section II-B-14), for which the rate coefficients of Hansen et al.

Table II-B-13. OH + 1,2,4-trimethylbenzene → products:

Rate coefficients k at, or close to, the high-pressure limit at $T \leq 325$ K

$10^{12} \times k$ (cm^3 $molecule^{-1}$ s^{-1})	at T (K)	Technique	Reference	Temperature range covered (K)
27.7 ± 5.1	304 ± 1	RR [relative to k(n-butane) = 2.52×10^{-12}][a]	Doyle et al. (1975)	
33.5 ± 3.4	297	FP-RF	Hansen et al. (1975)	
40.0 ± 4.5	298.2	FP-RF	Perry et al. (1977a)	298–430
37.3 ± 4.8	314.3			
32.4 ± 4.8	323.4			
30.8 ± 1.2	~ 298	RR [relative to k(n-hexane) = 5.45×10^{-12}][a]	Ohta and Ohyama (1985)	
32.5 ± 1.1	296 ± 2	RR [relative to k(propene) = 2.66×10^{-11}][a]	Atkinson and Aschmann (1989)	

[a] From Atkinson (1997).

(1975) and Ohta and Ohyama (1985) appear to be significantly low, the rate coefficient determined by Atkinson and Aschmann (1989) is recommended:

$$k(\text{1,2,4-trimethylbenzene}) = 3.25 \times 10^{-11} \text{ cm}^3 \text{ molecule}^{-1} \text{ s}^{-1}$$

independent of temperature over the range 290–325 K, and with an estimated overall uncertainty at 298 K of ±25%.

Based on the elevated temperature (≥ 450 K) rate coefficients for toluene and o-, m-, and p-xylene, a rate coefficient per $-CH_3$ substituent group of k(H-atom abstraction per $-CH_3$ group) = $8.07 \times 10^{-18} T^2 e^{-(38\pm76)/T}$ cm^3 $molecule^{-1}$ s^{-1} over the temperature range 470–1046 K can be derived (Atkinson, 1989). Extrapolation of this elevated temperature group rate coefficient to room temperature [k(H-atom abstraction per $-CH_3$ group) = 6.3×10^{-13} cm^3 $molecule^{-1}$ s^{-1} at 298 K] indicates that the H-atom abstraction process in 1,2,4-trimethylbenzene accounts for ~6% of the overall OH-radical reaction at atmospheric pressure of air and 298 K.

II-B-14. OH + 1,3,5-Trimethylbenzene

The available rate coefficients obtained at, or close to, the high-pressure limit are given in table II-B-14. In the temperature range < 350 K, rate coefficients have been obtained over only a small temperature range of 296–318 K and, as expected because of the narrow temperature range studied, there is no evidence for any significant temperature dependence. The absolute and relative rate coefficient data exhibit a marked degree of scatter, ranging over a factor of 1.65. The two most recent relative rate measurements of Atkinson and Aschmann (1989) and Kramp and Paulson (1998) are in excellent agreement, and an average of the rate coefficients from these two studies leads to the recommendation of

Table II-B-14. OH + 1,3,5-trimethylbenzene → products:

Rate coefficients k at, or close to, the high-pressure limit at $T \leq 325$ K

$10^{12} \times k$ (cm^3 molecule^{-1} s^{-1})	at T (K)	Technique	Reference	Temperature range covered (K)
42.8 ± 5.1	304 ± 1	RR [relative to k(n-butane) = 2.52×10^{-12}][a]	Doyle et al. (1975)	
47.2 ± 4.8	297	FP-RF	Hansen et al. (1975)	
62.4 ± 7.5	298.3	FP-RF	Perry et al. (1977a)	298–420
51.9 ± 6.3	318.4			
37.8 ± 5.2	~ 298	RR [relative to k(n-hexane) = 5.45×10^{-12}][a]	Ohta and Ohyama (1985)	
57.5 ± 3.0	296 ± 2	RR [relative to k(propene) = 2.66×10^{-11}][a]	Atkinson and Aschmann (1989)	
55.9 ± 4.8	296 ± 2	RR [relative to k(series of 10 organics)][b]	Kramp and Paulson (1998)	

[a] From Atkinson (1997).
[b] Rate coefficients for the reference hydrocarbons used were taken from Atkinson (1994, 1997).

$$k(\text{1,3,5-trimethylbenzene}) = 5.67 \times 10^{-11}\ \text{cm}^3\ \text{molecule}^{-1}\ \text{s}^{-1}$$

independent of temperature over the range 290–320 K, and with an estimated overall uncertainty at 298 K of ±20%.

Based on the elevated temperature (≥ 450 K) rate coefficients for toluene and o-, m-, and p-xylene, a rate coefficient per $-CH_3$ substituent group of k(H-atom abstraction per $-CH_3$ group) = $8.07 \times 10^{-18} T^2 e^{-(38\pm76)/T}$ cm^3 molecule^{-1} s^{-1} over the temperature range 470–1046 K can be derived (Atkinson, 1989). Extrapolation of this elevated temperature group rate coefficient to room temperature [k(H-atom abstraction per $-CH_3$ group) = 6.3×10^{-13} cm^3 molecule^{-1} s^{-1} at 298 K] indicates that the H-atom abstraction process in 1,3,5-trimethylbenzene accounts for ~3% of the overall OH-radical reaction at atmospheric pressure of air and 298 K.

II-B-15. OH + *tert*-Butylbenzene

The rate coefficients of Ohta and Ohyama (1985) and Sommerlade et al. (1993), both obtained at room temperature, are given in table II-B-15. The rate coefficients are in good agreement and a unit-weighted average leads to the recommendation of

$$k(\textit{tert}\text{-butylbenzene}) = 4.5 \times 10^{-12}\ \text{cm}^3\ \text{molecule}^{-1}\ \text{s}^{-1}$$

at 298 K, with an estimated overall uncertainty of ±35%. This 298 K rate coefficient is reasonably similar to that for toluene, suggesting that the dominant reaction pathway is OH-radical addition to the aromatic ring. At temperatures ≤ 325 K, the temperature dependence is therefore expected to be close to zero.

Table II-B-15. OH + *tert*-butylbenzene → products:
Rate coefficients k at, or close to, the high-pressure limit

$10^{12} \times k$ (cm^3 molecule^{-1} s^{-1})	at T (K)	Technique	Reference
4.47 ± 0.44	~ 298	RR [relative to k(n-hexane) = 5.45×10^{-12}][a]	Ohta and Ohyama (1985)
4.51 ± 0.94	297 ± 2	RR [relative to k(n-hexane) = 5.44×10^{-12}][a]	Sommerlade et al. (1993)

[a] From Atkinson (1997).

II-B-16. OH + *p*-Cymene (4-Isopropyltoluene)

The rate coefficient of Corchnoy and Atkinson (1990), obtained from a relative rate study at room temperature and atmospheric pressure of air, is given in table II-B-16. This is the only reported study of this reaction.

Table II-B-16. OH + *p*-cymene → products:
Rate coefficients k at, or close to, the high-pressure limit

$10^{12} \times k$ (cm^3 molecule^{-1} s^{-1})	at T (K)	Technique	Reference
14.5 ± 1.5	295 ± 2	RR [relative to k(cyclohexane) = 7.14×10^{-12}][a]	Corchnoy and Atkinson (1990)

[a] From Atkinson (1997).

II-B-17. OH + Indan

The rate coefficients of Baulch et al. (1989) and Kwok et al. (1997b), all obtained at room temperature, are given in table II-B-17. The rate coefficients of Kwok et al. (1997b) at atmospheric pressure of air are a factor of ~2 higher than those measured by Baulch et al. (1989) at 0.32–1.3 Torr total pressure of helium diluent. While Baulch et al. (1989) observed no dependence of the rate coefficient on the

Table II-B-17. OH + indan → products:
Rate coefficients k at, or close to, the high-pressure limit

$10^{12} \times k$ (cm^3 molecule^{-1} s^{-1})	at T (K)	Technique	Reference
9.3 ± 1.4	295	DF-RF	Baulch et al. (1989)
18.5 ± 0.9	297 ± 2	RR [relative to k(n-octane) = 8.69×10^{-12}][a]	Kwok et al. (1997b)
19.8 ± 0.5	297 ± 2	RR [relative to k(cyclohexane) = 7.19×10^{-12}][a]	Kwok et al. (1997b)

[a] From Atkinson (1997).

total pressure over the pressure range studied, it is expected that the major reaction pathway involves OH-radical addition to the aromatic ring, and it is possible that this pathway is still in the fall-off regime at the pressures used by Baulch et al. (1989). An average of the rate coefficients of Kwok et al. (1997b) leads to the recommendation of

$$k(\text{indan}) = 1.9 \times 10^{-11}\ \text{cm}^3\ \text{molecule}^{-1}\ \text{s}^{-1}$$

at 298 K, with an estimated overall uncertainty of ±40%. This 298 K rate coefficient is reasonably similar to that for *o*-xylene, suggesting that the dominant reaction pathway is OH-radical addition to the carbon atoms of the aromatic ring. At temperatures ≤ 325 K, the temperature dependence is therefore expected to be close to zero.

II-B-18. OH + Indene

The rate coefficient of Kwok et al. (1997b), obtained at room temperature and atmospheric pressure of air, is given in table II-B-18. In addition, Baulch et al. (1989) used a discharge flow system to measure absolute rate coefficients over the total pressure range of 0.24–1.81 Torr of helium diluent at 295 K, and they observed that the rate coefficients were in the fall-off regime with a rate coefficient at 1.81 Torr of helium of 5.1×10^{-11} cm^3 molecule^{-1} s^{-1}. The rate coefficients of Baulch et al. (1989) are therefore not inconsistent with the atmospheric pressure rate coefficient of Kwok et al. (1997b). The rate coefficient of Kwok et al. (1997b) is recommended:

$$k(\text{indene}) = 7.8 \times 10^{-11}\ \text{cm}^3\ \text{molecule}^{-1}\ \text{s}^{-1}$$

at 298 K, with an estimated overall uncertainty of ±40%. The magnitude of this rate coefficient, relative to that for indan (see section II-B-17), indicates that the major reaction pathway involves initial OH-radical addition to the carbon atoms of the $>C=C<$ bond in the five-membered ring.

Table II-B-18. OH + indene → products:
Rate coefficients k at, or close to, the high-pressure limit

$10^{12} \times k$ (cm^3 molecule^{-1} s^{-1})	at T (K)	Technique	Reference
77.9 ± 3.3	297 ±2	RR [relative to k(isoprene) = 1.01×10^{-10}][a]	Kwok et al. (1997b)

[a]From Atkinson (1997).

II-B-19. OH + Tetralin

The rate coefficient of Atkinson and Aschmann (1988), obtained from a relative rate study at room temperature and atmospheric pressure of air, is given in table II-B-19. This is the only reported study of this reaction.

Table II-B-19. OH + tetralin → products:

Rate coefficients k at, or close to, the high-pressure limit

$10^{12} \times k$ (cm^3 $molecule^{-1}$ s^{-1})	at T (K)	Technique	Reference
34.3 ±0.6	296 ±2	RR [relative to k(propene) = 2.66×10^{-11}][a]	Atkinson and Aschmann (1988)

[a] From Atkinson (1997).

II-B-20. OH + Styrene

The rate coefficients of Bignozzi et al. (1981) and Atkinson and Aschmann (1988), both obtained from relative rate studies at room temperature and atmospheric pressure of air, are given in table II-B-20. In addition, Baulch et al. (1989) used a discharge flow system to measure absolute rate coefficients over the total pressure range of 0.46–4.50 Torr of helium diluent at 295 K, and they observed that the rate coefficients were in the fall-off regime under these conditions, with a rate coefficient of $(4.31 \pm 0.45) \times 10^{-11}$ cm^3 $molecule^{-1}$ s^{-1} at 4.50 Torr total pressure of helium. The rate coefficient of Atkinson and Aschmann (1988) is recommended:

$$k(\text{styrene}) = 5.8 \times 10^{-11} \text{ cm}^3 \text{ molecule}^{-1} \text{ s}^{-1}$$

at 298 K, with an estimated overall uncertainty of ±20%.

The major products of the reaction of the OH-radical with styrene in the presence of NO are benzaldehyde and HCHO (Bignozzi et al., 1981; Tuazon et al., 1993), indicating that the major initial pathway involves OH-radical addition to the carbon atoms of the $-CH{=}CH_2$ substituent group.

Table II-B-20. OH + styrene → products:

Rate coefficients k at, or close to, the high-pressure limit

$10^{12} \times k$ (cm^3 $molecule^{-1}$ s^{-1})	at T (K)	Technique	Reference
50.7 ± 4.8	298	RR [relative to k(2,2,4-trimethylpentane) = 3.57×10^{-12}][a]	Bignozzi et al. (1981)
58.1 ± 1.5	296 ± 2	RR [relative to k(isoprene) = 1.01×10^{-10}][a]	Atkinson and Aschmann (1988)
43.1 ± 4.5[b]	295	DF-RF	Baulch et al. (1989)

[a] From Atkinson (1997).
[b] At 4.50 Torr total pressure of helium diluent (see text).

II-B-21. OH + α-Methylstyrene

The rate coefficient of Bignozzi et al. (1981), obtained from a relative rate study at room temperature and atmospheric pressure of air, is given in table II-B-21. This is the only reported study of this reaction.

Table II-B-21. OH + α-methylstyrene → products:
Rate coefficients k at, or close to, the high-pressure limit

$10^{12} \times k$ (cm^3 $molecule^{-1}$ s^{-1})	at T (K)	Technique	Reference
50.7 ± 5.8	298	RR [relative to k(2,2,4-trimethylpentane) = 3.57×10^{-12}][a]	Bignozzi et al. (1981)

[a] From Atkinson (1997).

II-B-22. OH + β-Methylstyrene

The rate coefficient of Bignozzi et al. (1981), obtained from a relative rate study at room temperature and atmospheric pressure of air, is given in table II-B-22. This is the only reported study of this reaction.

Table II-B-22. OH + β-methylstyrene → products:
Rate coefficients k at, or close to, the high-pressure limit

$10^{12} \times k$ (cm^3 $molecule^{-1}$ s^{-1})	at T (K)	Technique	Reference
57.4 ± 5.8	298	RR [relative to k(2,2,4-trimethylpentane) = 3.57×10^{-12}][a]	Bignozzi et al. (1981)

[a] From Atkinson (1997).

II-B-23. OH + β,β-Dimethylstyrene

The rate coefficient of Chiorboli et al. (1983), obtained from a relative rate study at room temperature and atmospheric pressure of air, is given in table II-B-23. This is the only reported study of this reaction.

Table II-B-23. OH + β, β-dimethylstyrene → products:
Rate coefficients k at, or close to, the high-pressure limit

$10^{12} \times k$ (cm^3 $molecule^{-1}$ s^{-1})	at T (K)	Technique	Reference
31.6 ± 4.8	298	RR [relative to k(2,2,4-trimethylpentane) = 3.57×10^{-12}][a]	Chiorboli et al. (1983)

[a] From Atkinson (1997).

II-B-24. OH + Benzaldehyde

The rate coefficients of Niki et al. (1978), Kerr and Sheppard (1981), and Semadeni et al. (1995), all obtained from relative rate studies at atmospheric pressure of air, are given in table II-B-24. At room temperature, the rate coeffi-

Table II-B-24. OH + benzaldehyde → products:

Rate coefficients k at, or close to, the high-pressure limit

$10^{12} \times k$ (cm^3 $molecule^{-1}$ s^{-1})	at T (K)	Technique	Reference
14.0 ± 0.9	298 ± 2	RR [relative to k(ethene-d_4) = 8.78×10^{-12}][a]	Niki et al. (1978)
11.8 ± 2.3	298 ± 4	RR [relative to k(ethene) = 8.52×10^{-12}][b]	Kerr and Sheppard (1981)
10.0 ± 2.6	289	RR [relative to k(diethyl ether) = 8.91×10^{-18} T^2 $e^{837/T}$][c]	Semadeni et al. (1995)
12.8 ± 1.0	294		
13.0 ± 0.8	299		
12.4 ± 0.7	307		
12.0 ± 0.9	314		
11.8 ± 1.0	324		
11.7 ± 0.5	334		
11.2 ± 0.5	343		
9.0 ± 0.5	353		
6.3 ± 0.5	373		

[a] From Atkinson (1989).
[b] From Atkinson (1997).
[c] From Atkinson (1994).

cients of Niki et al. (1978), Kerr and Sheppard (1981), and Semadeni et al. (1995) are in good agreement, and an average of the room temperature rate coefficients from these studies is recommended:

$$k(\text{benzaldehyde}) = 1.29 \times 10^{-11}\ \text{cm}^3\ \text{molecule}^{-1}\ \text{s}^{-1}$$

at 298 K, with an estimated overall uncertainty of ±25%. The temperature-dependent study of Semadeni et al. (1995) shows that the rate coefficient is independent of temperature over the restricted temperature range of 289–343 K, although there was a rapid decrease in the measured rate coefficient at >350 K. The recommended room temperature rate coefficient is similar to that for acetaldehyde, implying that the major reaction pathway involves H-atom abstraction from the –CHO group. However, this is not consistent with the observed rapid decrease in rate coefficient at >350 K (Semadeni et al., 1995) if the obvious explanation for this effect is that the OH–benzaldehyde adduct is undergoing rapid back-decomposition at temperatures >350 K (see discussion at the beginning of this section). Until the temperature-dependent data of Semadeni et al. (1995) are confirmed, no recommendation is made concerning the temperature dependence of this rate coefficient.

II-B-25. OH + 2,4-Dimethylbenzaldehyde

The rate coefficient of Tse et al. (1997), obtained from a relative rate study at room temperature and atmospheric pressure of air, is given in table II-B-25. This is the only reported study of this reaction.

Table II-B-25. OH + 2,4-dimethylbenzaldehyde → products: Rate coefficients k at, or close to, the high-pressure limit

$10^{12} \times k$ (cm^3 $molecule^{-1}$ s^{-1})	at T (K)	Technique	Reference
43.2 ± 1.0	298 ± 2	RR [relative to k(1,2,4-trimethylbenzene) = 3.25×10^{-11}][a]	Tse et al. (1997)

[a] This evaluation.

II-B-26. OH + 2,5-Dimethylbenzaldehyde

The rate coefficient of Tse et al. (1997), obtained from a relative rate study at room temperature and atmospheric pressure of air, is given in table II-B-26. This is the only reported study of this reaction.

Table II-B-26. OH + 2,5-dimethylbenzaldehyde → products: Rate coefficients k at, or close to, the high-pressure limit

$10^{12} \times k$ (cm^3 $molecule^{-1}$ s^{-1})	at T (K)	Technique	Reference
43.6 ± 0.7	298 ± 2	RR [relative to k(1,2,4-trimethylbenzene) = 3.25×10^{-11}][a]	Tse et al. (1997)

[a] This evaluation.

II-B-27. OH + 3,4-Dimethylbenzaldehyde

The rate coefficient of Tse et al. (1997), obtained from a relative rate study at room temperature and atmospheric pressure of air, is given in table II-B-27. This is the only reported study of this reaction.

Table II-B-27. OH + 3,4-dimethylbenzaldehyde → products: Rate coefficients k at, or close to, the high-pressure limit

$10^{12} \times k$ (cm^3 $molecule^{-1}$ s^{-1})	at T (K)	Technique	Reference
21.4 ± 1.0	298 ± 2	RR [relative to k(1,2,4-trimethylbenzene) = 3.25×10^{-11}][a]	Tse et al. (1997)

[a] This evaluation.

II-B-28. OH + *trans*-Cinnamaldehyde

The rate coefficient of Smith et al. (1996), obtained from a relative rate study at room temperature and atmospheric pressure of air, is given in table II-B-28. This is the only reported study of this reaction.

Table II-B-28. OH + *trans*-cinnamaldehyde → products:
Rate coefficients k at, or close to, the high-pressure limit

$10^{12} \times k$ (cm^3 molecule^{-1} s^{-1})	at T (K)	Technique	Reference
48.0 ± 4.6	296 ± 2	RR [relative to k(*trans*-2-butene) = 6.48×10^{-11}][a]	Smith et al. (1996)

[a] From Atkinson (1997).

II-B-29. OH + 1,4-Naphthoquinone

The rate coefficient of Atkinson et al. (1989b), obtained from a relative rate study at room temperature and atmospheric pressure of air, is given in table II-B-29. This is the only reported study of this reaction.

Table II-B-29. OH + 1,4-naphthoquinone → products:
Rate coefficients k at, or close to, the high-pressure limit

$10^{12} \times k$ (cm^3 molecule^{-1} s^{-1})	at T (K)	Technique	Reference
3.0 ± 1.1	298 ± 2	RR [relative to k(cyclohexane) = 7.21×10^{-12}][a]	Atkinson et al. (1989b)

[a] From Atkinson (1997).

II-B-30. OH + Phenol

The rate coefficients of Rinke and Zetzsch (1984), Knispel et al. (1990), and Semadeni et al. (1995) are given in table II-B-30. At 296–308 K, the rate coefficients of Rinke and Zetzsch (1984), Knispel et al. (1990), and Semadeni et al. (1995) are in good agreement. A unit-weighted average of the room temperature rate coefficients of Rinke and Zetzsch (1984), Knispel et al. (1990), and that of Semadeni et al. (1995) relative to the rate coefficient for the reaction of the OH-radical with 1,3-butadiene, is recommended:

$$k(\text{phenol}) = 2.7 \times 10^{-11}\ \text{cm}^3\ \text{molecule}^{-1}\ \text{s}^{-1}$$

at 298 K, with an estimated overall uncertainty of ±25%. The temperature dependence of the rate coefficient obtained by Semadeni et al. (1995) (with a value of $B \sim -1400$ K) is significantly more negative than that of Knispel et al. (1990) of $B \sim -845$ K over the same temperature range of 300–374 K. Provisionally, we recommend the temperature dependence obtained in the absolute rate study of Knispel et al. (1990), of $B = -840$ K, to obtain

$$k(\text{phenol}) = 1.73 \times 10^{-12}\, e^{840/T}\ \text{cm}^3\ \text{molecule}^{-1}\ \text{s}^{-1}$$

over the temperature range 290–380 K.

Table II-B-30. OH + phenol → products:

Rate coefficients k at, or close to, the high-pressure limit

$10^{12} \times k$ (cm^3 molecule^{-1} s^{-1})	at T (K)	Technique	Reference	Temperature range covered (K)
28.3 ± 5.7	296 ± 2	FP-RF	Rinke and Zetzsch (1984)	
29 ± 3.1	300	FP-RF	Knispel et al. (1990)	300–374
26 ± 3.1	308			
22.4 ± 2.1	323			
18.6 ± 2.1	339			
17.6 ± 2.1	354			
17.1 ± 2.1	374			
25.6 ± 2.7	301	RR [relative to k(1,3-butadiene) = $1.48 \times 10^{-11} e^{448/T}$][a]	Semadeni et al. (1995)	301–373
21.7 ± 2.5	319			
17.0 ± 4.0	334			
15.0 ± 1.7	348			
11.8 ± 1.6	360			
10.8 ± 1.5	373			
24.6 ± 2.1	301	RR [relative to k(o-cresol) = 4.1×10^{-11}][b]	Semadeni et al. (1995)	301–314

[a] From Atkinson (1997).
[b] This evaluation.

The data of Knispel et al. (1990) show that at ≤ 308 K the reaction proceeds by OH-radical addition, with the H-atom abstraction channel(s) becoming more important as the temperature increases, and accounting for 30 ± 5% of the overall OH-radical reaction at 374 K. The rate coefficient of Semadeni et al. (1995) relative to that for the reaction of the OH-radical with *o*-cresol agrees well with the recommended 298 K value and indicates that our recommendations for the rate coefficients of the phenol and *o*-cresol reactions are internally consistent.

II-B-31. OH + *o*-Cresol

The rate coefficients of Perry et al. (1977b), Atkinson et al. (1978), Atkinson and Aschmann (1990), and Semadeni et al. (1995) are given in table II-B-31. At room temperature, there is a significant degree of scatter between these studies, with the rate coefficient varying by a factor of 1.4 [although the two relative rate studies of Atkinson et al. (1978) and Atkinson and Aschmann (1990) are in good agreement]. A unit-weighted average of the 296–301 K rate coefficients of Perry et al. (1977b), Atkinson et al. (1978), Atkinson and Aschmann (1990), and Semadeni et al. (1995) leads to the recommendation of

$$k(o\text{-cresol}) = 4.1 \times 10^{-11}\ cm^3\ molecule^{-1}\ s^{-1}$$

at 298 K, with an estimated overall uncertainty of ±30%. Both temperature-dependent studies (Perry et al., 1977b; Semadeni et al., 1995) find negative temperature dependencies at around room temperature, although the values of B

Table II-B-31. OH + *o*-cresol → products:

Rate coefficients k at, or close to, the high-pressure limit at $T \leq 375$ K

$10^{12} \times k$ (cm^3 $molecule^{-1}$ s^{-1})	at T (K)	Technique	Reference	Temperature range covered (K)
34.1 ± 6.8	299.4	FP-RF	Perry et al. (1977b)	299–423
29.3 ± 5.9	310.6			
29.8 ± 6.0	322.0			
26.8 ± 5.4	330.7			
25.5 ± 5.1	335.4			
40 ± 4	300 ± 1	RR [relative to {k(*n*-butane) − k(2,2-dimethyl-propane)} = 1.60×10^{-12}][a]	Atkinson et al. (1978)	
42.6 ± 2.2	296 ± 2	RR [relative to k(propene) = 2.66×10^{-11}][a]	Atkinson and Aschmann (1990)	
46.6 ± 4.0	301	RR [relative to k(1,3-butadiene) = 1.48×10^{-11} $e^{448/T}$][a]	Semadeni et al. (1995)	301–373
39.8 ± 3.1	319			
31.1 ± 2.9	334			
26.3 ± 2.2	348			
25.7 ± 2.6	360			
22.6 ± 2.5	373			

[a] From Atkinson (1997).

differ significantly [−720 K for Perry et al. (1977b) and −1165 K for Semadeni et al. (1995)]. Provisionally, we recommend a temperature-dependent rate coefficient of

$$k(o\text{-cresol}) = 1.70 \times 10^{-12}\, e^{950/T}\ \text{cm}^3\ \text{molecule}^{-1}\ \text{s}^{-1}$$

over the temperature range 290–380 K. The rate coefficients obtained by Perry et al. (1977b) at temperatures >400 K indicate that the H-atom abstraction pathway is of minor importance at room temperature (accounting for ~8% of the overall reaction at 298 K).

II-B-32. OH + *m*-Cresol

The rate coefficients of Atkinson et al. (1978), Atkinson and Aschmann (1990), and Semadeni et al. (1995) are given in table II-B-32, all being obtained from relative rate studies at atmospheric pressure of air. In the studies of Atkinson et al. (1978) and Semadeni et al. (1995), rate coefficients for the *m*-cresol reaction were measured relative to the rate coefficient for the reaction of the OH-radical with *o*-cresol, the latter not being a particularly well-established rate coefficient. The rate coefficient ratios k(*m*-cresol)/k(*o*-cresol) measured by Atkinson et al. (1978) (1.42 ± 0.08) and Semadeni et al. (1995) (1.32 ± 0.13) at 300–301 K are in good agreement, and at room temperature the rate coefficients from all three studies are in reasonably good agreement (within 25%). We recommend the rate coefficient obtained by Atkinson and Aschmann (1990) relative to that for the reaction of the OH-radical with propene:

Table II-B-32. OH + *m*-cresol → products:

Rate coefficients *k* at, or close to, the high-pressure limit

$10^{12} \times k$ (cm^3 molecule^{-1} s^{-1})	at *T* (K)	Technique	Reference	Temperature range covered (K)
58.2 ± 3.3	300 ± 1	RR [relative to *k*(*o*-cresol) = 4.1×10^{-11}][a]	Atkinson et al. (1978)	
67.8 ± 4.0	296 ± 2	RR [relative to *k*(propene) = 2.66×10^{-11}][b]	Atkinson and Aschmann (1990)	
54.1 ± 5.4	299	RR [relative to *k*(*o*-cresol) = 4.1×10^{-11}][a]	Semadeni et al. (1995)	299–373

[a] This evaluation.
[b] From Atkinson (1997).

$$k(m\text{-cresol}) = 6.8 \times 10^{-11}\ \text{cm}^3\ \text{molecule}^{-1}\ \text{s}^{-1}$$

at 298 K, with an estimated overall uncertainty of ±35%.

Semadeni et al. (1995) measured the rate coefficient ratio *k*(*m*-cresol)/*k*(*o*-cresol) over the temperature range 299–373 K, with this rate coefficient ratio being essentially independent of temperature over the range 299–363 K and with a value of *k*(*m*-cresol)/*k*(*o*-cresol) = 1.36 ± 0.10 (where the indicated error is two least-squares standard deviations). Hence, the temperature dependencies of the OH-radical reactions with *o*-cresol and *m*-cresol appear to be similar.

II-B-33. OH + *p*-Cresol

The rate coefficients of Atkinson et al. (1978), Atkinson and Aschmann (1990), and Semadeni et al. (1995) are given in table II-B-33, all being obtained from relative rate studies at atmospheric pressure of air. In the studies of Atkinson et al. (1978) and certain of the experiments of Semadeni et al. (1995), rate coefficients for the *m*-cresol reaction were measured relative to the rate coefficient for the reaction of the OH-radical with *o*-cresol, the latter not being a particularly well-established rate coefficient. The rate coefficient ratios *k*(*m*-cresol)/*k*(*o*-cresol) measured by Atkinson et al. (1978) (1.10 ± 0.05) and Semadeni et al. (1995) (1.09 ± 0.17) at 300–301 K are in good agreement, and at room temperature the rate coefficients from all studies are in reasonably good agreement (within 20%). We recommend an average of the rate coefficients obtained by Atkinson and Aschmann (1990) and Semadeni et al. (1995) relative to the rate coefficients for the reactions of the OH-radical with propene and 1,3-butadiene, respectively:

$$k(p\text{-cresol}) = 5.0 \times 10^{-11}\ \text{cm}^3\ \text{molecule}^{-1}\ \text{s}^{-1}$$

at 298 K, with an estimated overall uncertainty of ±30%.

The temperature dependence obtained by Semadeni et al. (1995) over the temperature range 301–373 K corresponds to a value of $B = -1180$ K, quite

Table II-B-33. OH + *p*-cresol → products:

Rate coefficients *k* at, or close to, the high-pressure limit

$10^{12} \times k$ (cm^3 molecule^{-1} s^{-1})	at *T* (K)	Technique	Reference	Temperature range covered (K)
45.1 ± 2.1	300 ± 1	RR [relative to *k*(*o*-cresol) = 4.1×10^{-11}][a]	Atkinson et al. (1978)	
48.4 ± 5.1	296 ± 2	RR [relative to *k*(propene) = 2.66×10^{-11}][b]	Atkinson and Aschmann (1990)	
52.5 ± 5.3	301	RR [relative to *k*(1,3-butadiene) = 1.48×10^{-11} $e^{448/T}$][b]	Semadeni et al. (1995)	301–373
52.5 ± 4.9	319			
39.6 ± 4.6	334			
32.2 ± 3.3	348			
29.3 ± 3.6	360			
27.1 ± 2.5	373			
44.7 ± 7.0	301	RR [relative to *k*(*o*-cresol) = 4.1×10^{-11}][a]	Semadeni et al. (1995)	301–314

[a] This evaluation.
[b] From Atkinson (1997).

similar to the temperature dependencies measured by Semadeni et al. (1995) for *m*- and *p*-cresol.

II-B-34. OH + 2,3-Dimethylphenol

The rate coefficient of Atkinson and Aschmann (1990), obtained from a relative rate study at room temperature and atmospheric pressure of air, is given in table II-B-34. This is the only reported study of this reaction.

Table II-B-34. OH + 2,3-dimethylphenol → products:

Rate coefficients *k* at, or close to, the high-pressure limit

$10^{12} \times k$ (cm^3 molecule^{-1} s^{-1})	at *T* (K)	Technique	Reference
80.2 ± 3.0	296 ± 2	RR [relative to *k*(isoprene) = 1.01×10^{-10}][a]	Atkinson and Aschmann (1990)

[a] From Atkinson (1997).

II-B-35. OH + 2,4-Dimethylphenol

The rate coefficient of Atkinson and Aschmann (1990), obtained from a relative rate study at room temperature and atmospheric pressure of air, is given in table II-B-35. This is the only reported study of this reaction.

Table II-B-35. OH + 2,4-dimethylphenol → products:
Rate coefficients k at, or close to, the high-pressure limit

$10^{12} \times k$ (cm^3 $molecule^{-1}$ s^{-1})	at T (K)	Technique	Reference
71.5 ± 4.1	296 ± 2	RR [relative to k(isoprene) = 1.01×10^{-10}][a]	Atkinson and Aschmann (1990)

[a] From Atkinson (1997).

II-B-36. OH + 2,5-Dimethylphenol

The rate coefficient of Atkinson and Aschmann (1990), obtained from a relative rate study at room temperature and atmospheric pressure of air, is given in table II-B-36. This is the only reported study of this reaction.

Table II-B-36. OH + 2,5-dimethylphenol → products:
Rate coefficients k at, or close to, the high-pressure limit

$10^{12} \times k$ (cm^3 $molecule^{-1}$ s^{-1})	at T (K)	Technique	Reference
80.0 ± 11.0	296 ± 2	RR [relative to k(isoprene) = 1.01×10^{-10}][a]	Atkinson and Aschmann (1990)

[a] From Atkinson (1997).

II-B-37. OH + 2,6-Dimethylphenol

The rate coefficient of Atkinson and Aschmann (1990), obtained from a relative rate study at room temperature and atmospheric pressure of air, is given in table II-B-37. This is the only reported study of this reaction.

Table II-B-37. OH + 2,6-dimethylphenol → products:
Rate coefficients k at, or close to, the high-pressure limit

$10^{12} \times k$ (cm^3 $molecule^{-1}$ s^{-1})	at T (K)	Technique	Reference
65.9 ± 5.0	296 ± 2	RR [relative to k(isoprene) = 1.01×10^{-10}][a]	Atkinson and Aschmann (1990)

[a] From Atkinson (1997).

II-B-38. OH + 3,4-Dimethylphenol

The rate coefficient of Atkinson and Aschmann (1990), obtained from a relative rate study at room temperature and atmospheric pressure of air, is given in table II-B-38. This is the only reported study of this reaction.

Table II-B-38. OH + 3,4-dimethylphenol → products:

Rate coefficients k at, or close to, the high-pressure limit

$10^{12} \times k$ (cm^3 $molecule^{-1}$ s^{-1})	at T (K)	Technique	Reference
81.4 ± 5.8	296 ± 2	RR [relative to k(isoprene) = 1.01×10^{-10}][a]	Atkinson and Aschmann (1990)

[a] From Atkinson (1997).

II-B-39. OH + 3,5-Dimethylphenol

The rate coefficient of Atkinson and Aschmann (1990), obtained from a relative rate study at room temperature and atmospheric pressure of air, is given in table II-B-39. This is the only reported study of this reaction.

Table II-B-39. OH + 3,5-dimethylphenol → products:

Rate coefficients k at, or close to, the high-pressure limit

$10^{12} \times k$ (cm^3 $molecule^{-1}$ s^{-1})	at T (K)	Technique	Reference
113 ± 8	296 ± 2	RR [relative to k(isoprene) = 1.01×10^{-10}][a]	Atkinson and Aschmann (1990)

[a] From Atkinson (1997).

II-B-40. OH + 2,3,5-Trimethylphenol

The rate coefficient of Tse et al. (1997), obtained from a relative rate study at room temperature and atmospheric pressure of air, is given in table II-B-40. This is the only reported study of this reaction.

Table II-B-40. OH + 2,3,5-trimethylphenol → products:

Rate coefficients k at, or close to, the high-pressure limit

$10^{12} \times k$ (cm^3 $molecule^{-1}$ s^{-1})	at T (K)	Technique	Reference
125 ± 3	298 ± 2	RR [relative to k(1,2,4-trimethylbenzene) = 3.25×10^{-11}][a]	Tse et al. (1997)

[a] This evaluation.

II-B-41. OH + 2,3,6-Trimethylphenol

The rate coefficient of Tse et al. (1997), obtained from a relative rate study at room temperature and atmospheric pressure of air, is given in table II-B-41. This is the only reported study of this reaction.

Table II-B-41. OH + 2,3,6-trimethylphenol → products:
Rate coefficients k at, or close to, the high-pressure limit

$10^{12} \times k$ (cm^3 molecule^{-1} s^{-1})	at T (K)	Technique	Reference
118 ± 3	298 ± 2	RR [relative to k(1,2,4-trimethylbenzene) = 3.25×10^{-11}][a]	Tse et al. (1997)

[a] This evaluation.

II-B-42. OH + Nitrobenzene

The rate coefficients of Witte et al. (1986) and Atkinson et al. (1987c) are given in table II-B-42. The upper limits to the rate coefficients of Atkinson et al. (1987c) are consistent with the absolute rate data of Witte et al. (1986). Tentatively, the rate coefficients measured by Witte et al. (1986) are recommended, with

$$k(\text{nitrobenzene}) = 6.2 \times 10^{-13}\, e^{-(444 \pm 136)/T}\ \text{cm}^3\ \text{molecule}^{-1}\ \text{s}^{-1}$$

over the temperature range 259–342 K, where the indicated error is two least-squares standard deviations, and

$$k(\text{nitrobenzene}) = 1.40 \times 10^{-13}\ \text{cm}^3\ \text{molecule}^{-1}\ \text{s}^{-1}$$

at 298 K, with an estimated overall uncertainty of ±35%. Witte et al. (1986) observed non-exponential OH-radical decays throughout the temperature range studied, possibly due to regeneration of OH-radicals from photolysis products (Witte et al., 1986).

II-B-43. OH + *m*-Nitrotoluene

The rate coefficient of Atkinson et al. (1989a), obtained from a relative rate study at room temperature and atmospheric pressure of air, is given in table II-B-43. This is the only reported study of this reaction.

Table II-B-42. OH + nitrobenzene → products:

Rate coefficients $k = A\ e^{-B/T}$ at, or close to, the high-pressure limit at $T \leq 365$ K

$10^{12} \times A$ (cm^3 $molecule^{-1}$ s^{-1})	B (K)	$10^{12} \times k$ (cm^3 $molecule^{-1}$ s^{-1})	at T (K)	Technique	Reference	Temperature range covered (K)
		0.120 ± 0.007	259	FP-RF	Witte et al. (1986)	259–362
		0.110 ± 0.012	265			
		0.119 ± 0.007	274			
		0.126 ± 0.009	283			
		0.132 ± 0.007	288			
		0.137 ± 0.014	299			
		0.146 ± 0.015	312			
		0.169 ± 0.032	316			
		0.137 ± 0.025	316			
		0.181 ± 0.028	331			
		0.154 ± 0.021	339			
		0.169 ± 0.011	342			
		0.163 ± 0.065	352			
		0.136 ± 0.013	358			
		0.133 ± 0.021	358			
0.6	440 ± 80 (259-342 K)	0.158 ± 0.040	362			
		<0.9	296 ± 2	RR [relative to k(dimethyl ether) = 2.82×10^{-12}][a]	Atkinson et al. (1987c)	
		<0.7	296 ± 2	RR [relative to k(benzene) = 1.21×10^{-12}][b]	Atkinson et al. (1987c)	

[a] From Atkinson et al. (1999).
[b] This evaluation.

Table II-B-43. OH + *m*-nitrotoluene → products:

Rate coefficients *k* at, or close to, the high-pressure limit

$10^{12} \times k$ (cm^3 $molecule^{-1}$ s^{-1})	at *T* (K)	Technique	Reference
1.24 ± 0.83	298 ± 2	RR [relative to *k*(benzene) = 1.22×10^{-12}][a]	Atkinson et al. (1989a)

[a] This evaluation.

Table II-B-44. Summary of rate coefficients at 298 K and 760 Torr total presssure of air and Arrhenius parameters ($k = A\,e^{-B/T}$; $T \sim 250$–325 K) for the reaction of OH-radicals with monocyclic aromatic hydrocarbons

Aromatic hydrocarbon	$10^{12} \times k$(298 K) (cm^3 $molecule^{-1}$ s^{-1})	$10^{12} \times A$ (cm^3 $molecule^{-1}$ s^{-1})	*B* (K)	Estimated Uncertainty, *k*(298 K), ±%
Benzene[a]	1.22	2.33	193	20
Toluene	5.63	1.81	−338	20
Ethylbenzene	7.0			25
o-Xylene	13.6	13.6	0	25
m-Xylene	23.1	23.1	0	15
p-Xylene	14.3	14.3	0	25
n-Propylbenzene	5.8			25
iso-Propylbenzene	6.3			30
o-Ethyltoluene	11.9			35
m-Ethyltoluene	18.6			35
p-Ethyltoluene	11.8			35
1,2,3-Trimethylbenzene	32.7	32.7	0	25
1,2,4-Trimethylbenzene	32.5	32.5	0	25
1,3,5-Trimethylbenzene	56.7	56.7	0	20
tert-Butylbenzene	4.5			35
p-Cymene (4-isopropyl-toluene)	14.5			b
Indan	19			40
Indene	78			40
Tetralin	34			b
Styrene	58			20
α-Methylstyrene	51			b
β-Methylstyrene	57			b
β, β-Dimethylstyrene	32			b

[a] The values given in the table for benzene are those for atmospheric pressure; the high-pressure limit values for benzene are $k = 1.39 \times 10^{-12}$ cm^3 $molecule^{-1}$ s^{-1}; $A = 3.8 \times 10^{-12}$ cm^3 $molecule^{-1}$ s^{-1}; $B = 300$ K.

[b] No estimate; single study.

Table II-B-45. Summary of rate coefficients at 298 K and 760 Torr total presssure of air and Arrhenius parameters ($k = A\,e^{-B/T}$; $T \sim$ 250–325 K) for the reaction of OH-radicals with monocyclic aromatic compounds

Aromatic Compound	$10^{12} \times k$(298 K) (cm^3 $molecule^{-1}$ s^{-1})	$10^{12} \times A$ (cm^3 $molecule^{-1}$ s^{-1})	B (K)	Estimated Uncertainty, k(298 K), ±%
Benzaldehyde	12.9			25
2,4-Dimethylbenzaldehyde	43			[a]
2,5-Dimethylbenzaldehyde	43			[a]
3,4-Dimethylbenzaldehyde	21			[a]
trans-Cinnamaldehyde	48.0			[a]
1,4-Naphthaquinone	3.0			[a]
Phenol	27	1.73	−840	25
o-Cresol	41	1.70	−950	30
m-Cresol	68			35
p-Cresol	50			30
2,3-Dimethylphenol	80			[a]
2,4-Dimethylphenol	72			[a]
2,5-Dimethylphenol	80			[a]
2,6-Dimethylphenol	66			[a]
3,4-Dimethylphenol	81			[a]
3,5-Dimethylphenol	113			[a]
2,3,5-Trimethylphenol	125			[a]
2,3,6-Trimethylphenol	118			[a]
Nitrobenzene	0.14	0.62	444	35
m-Nitrotoluene	1.2			[a]

[a] No estimate; single study.

II-C. Estimation of Rate Coefficients for Reactions of OH Radicals with Monocyclic Aromatic Compounds

As evident from section II-B, rate coefficients have been measured for a large number of aromatic compounds, and primarily for alkyl-subsituted benzenes (including the bicyclic hydrocarbons indan, indene, and tetralin). However, given the number of potential aromatic compounds emitted into the atmosphere and their possible atmospheric reaction products, there is a need to be able to estimate rate coefficients reliably for the gas-phase reactions of the OH-radical with other aromatic compounds. Atkinson (1986) and Kwok and Atkinson (1995) have used structure–reactivity relationships to estimate rate coefficients, and this method is briefly described here.

The general approach of this estimation method has been described previously (Atkinson, 1986; Kwok and Atkinson, 1995), and it is based on the observations that gas-phase OH-radical reactions with organic compounds proceed by four reaction pathways that are assumed to be additive: H-atom abstraction from C–H and O–H bonds; OH-radical addition to the carbon atoms of >C=C< and –C≡C– bonds; OH-radical addition to the carbon atoms of aromatic rings; and OH-radical "interaction" with N-, S-, and P-atoms and with more complex structural units such as >P=S, >NC(O)S–, and $>NC(O)O-$ groups. The total rate coefficient is assumed to be the sum of the rate coefficients for these four reaction pathways (Atkinson, 1986); the OH-radical reactions with many organic compounds proceed by more than one of these pathways.

II-C-1. H-Atom Abstraction from C–H and O–H Bonds

Calculation of H-atom abstraction rate coefficients from C–H and O–H bonds is based on the estimation of group rate coefficients for H-atom abstraction from $-CH_3$, $-CH_2$-, $>CH-$, and $-OH$ groups. The rate coefficients for H-atom abstraction from $-CH_3$, $-CH_2-$ and $>CH-$ groups depends on the identity of the substituents attached to these groups, with

$$k(\text{CH}_3\text{-X}) = k_{\text{prim}}\, F(\text{X})$$

$$k(\text{X-CH}_2\text{-Y}) = k_{\text{sec}}\, F(\text{X})\, F(\text{Y})$$

$$k\left(\text{X-CH}\begin{matrix}\text{Y}\\ \text{Z}\end{matrix}\right) = k_{\text{tert}}\, F(\text{X})\, F(\text{Y})\, F(\text{Z})$$

where k_{prim}, k_{sec}, and k_{tert} are the group rate coefficients for H-atom abstraction from $-CH_3$, $-CH_2-$, and $>CH-$ groups, respectively, for a "standard" substituent, and F(X), F(Y), and F(Z) are the substituent factors for the substituent groups X, Y, and Z, respectively. The standard substituent group is chosen to be X (= Y = Z) = $-CH_3$, with $F(-CH_3) = 1.00$ (Atkinson, 1986). It is assumed that the temperature dependence of the substituent factors F(X) can be expressed in the form $F(X) = e^{Ex/T}$, and hence that the pre-exponential factors A (in the Arrhenius expression $k = A\,e^{-B/T}$ or in the alternative three-parameter expression $k = AT^n\,e^{-D/T}$) for H-atom abstraction from C–H bonds are independent of the substituent groups (Atkinson, 1987). Group rate coefficients, k_{prim}, k_{sec}, k_{tert}, and k_{OH} were most recently derived by Kwok and Atkinson (1995) (in $cm^3\ molecule^{-1}\ s^{-1}$ units): $k_{prim} = 4.49 \times 10^{-18}\ T^2\ e^{-320/T}$ (1.36×10^{-13} at 298 K); $k_{sec} = 4.50 \times 10^{-18}\ T^2\ e^{253/T}$ (9.34×10^{-13} at 298 K); $k_{tert} = 2.12 \times 10^{-18}\ T^2\ e^{696/T}$ (1.94×10^{-12} at 298 K); and $k_{OH} = 2.1 \times 10^{-18}\ T^2\ e^{-85/T}$ (1.4×10^{-13} at 298 K). Group substituent factors, F(X), are listed by Kwok and Atkinson (1995) and Atkinson (2000). Ring-strain effects for other than six-membered rings are taken into account through ring-strain factors, F_{ring}, with only those rings containing the $-CH_2-$ or $>CH-$ groups being considered in accounting for ring strain.

Because of the generally greater reactivity of $>C{=}C<$, $-C{\equiv}C-$, and aromatic rings towards OH-radical addition, H-atom abstraction from alkyl or substituted alkyl groups in the alkenes, alkynes, and aromatic compounds is generally of minor importance (as discussed in section II-B), and Kwok and Atkinson (1995) set the substituent group factors $F(>C{=}C<)$, $F(-C{\equiv}C-)$, and $F(-C_6H_5)$ equal to unity.

II-C-2. OH-Radical Addition to >C=C< and –C≡C– Bonds

The calculation of rate coefficients for OH-radical addition to >C=C< and –C≡C– bonds assumes that the rate coefficient for OH-radical addition to these carbon–carbon unsaturated bonds depends on the number, identity, and position of substituent groups around the >C=C< or –C≡C– bond(s). The 298 K rate coefficients for addition to the CH_2=CH–, CH_2=C<, *cis*- and *trans* –CH=CH–, –CH=C<, and >C=C< non-conjugated carbon–carbon double bond units and the various –C≡C– and conjugated >C=C–C=C< units are based on the measured rate coefficients for the alkenes with the appropriate alkyl substitutent groups (Atkinson, 1986), and these are listed by Kwok and Atkinson (1995) and Atkinson (2000). There is no evidence for ring-strain effects on OH-radical addition to cycloalkenes.

For organic compounds containing >C=C< bonds with non-alkyl substituent groups, X, group substituent factors *C*(X) are used (*C*(alkyl) = 1.00). The group factors *C*(X) derived from the available database for haloalkenes and nitrogen- and oxygen-containing organic compounds containing >C=C< bonds are listed by Kwok and Atkinson (1995) and Atkinson (2000).

II-C-3. OH-Radical Addition to Aromatic Rings

The rate coefficients for OH-radical addition to aromatic rings are calculated using the correlation between the OH-radical addition rate coefficient and the sum of the electrophilic substituent constants $\Sigma\sigma^+$ (Zetzsch, 1982a; Atkinson, 1986). As discussed by Zetzsch (1982a), $\Sigma\sigma^+$ is calculated by assuming that (a) steric hindrance can be neglected; (b) $\Sigma\sigma^+$ is the sum of all of the substituent constants of the substituent groups attached to the aromatic ring; (c) in general, the value of $\Sigma\sigma^+$ depends on the position of OH-radical addition to the ring, and the OH-radical adds to the position yielding the most negative value of $\Sigma\sigma^+$; and (d) if all positions on the ring are occupied, the *ipso* position is treated as a *meta* position. Table II-C-1 gives a summary of some σ^+ values for various group substituents attached to the benzene ring (Brown and Okamoto, 1958). The correlation between the OH-radical addition rate coefficients and $\Sigma\sigma^+$ for monocyclic aromatic compounds and biphenyls is given by the relation (Atkinson, 1991a; Kwok and Atkinson, 1995)

$$\log_{10} k_{add}(\text{cm}^3\ \text{molecule}^{-1}\ \text{s}^{-1}) = -11.71 - 1.34\Sigma\sigma^+$$

Note that it is likely that a maximum rate coefficient of 2×10^{-10} cm^3 molecule^{-1} s^{-1} should be used if the calculated value exceeds this number. At the present time, the estimation of rate coefficients for fused-ring polycyclic aromatic hydrocarbons (PAHs) is uncertain; however, see section II-H. For substituted PAHs, the enhancement of the rate coefficient over that for the parent PAH may be approximately estimated using the correlation between the rate coefficient and $\Sigma\sigma^+$ discussed above for the monocyclic aromatic compounds, with the enhancement factor being $e^{1.34\Sigma\sigma^+}$.

Table II-C-1. Electrophilic substituent constants (σ^+)

Group	σ^+ (*ortho, para*)	σ^+ (*meta*)
Methyl ($-CH_3$)	−0.311	−0.066
Ethyl ($-C_2H_5$)	−0.295	−0.064
iso-Propyl ($-CH(CH_3)_2$)	−0.280	−0.060
tert-Butyl ($-C(CH_3)_3$)	−0.256	−0.059
Phenyl ($-C_6H_5$)	−0.179	0.109
Methoxy ($-OCH_3$)	−0.778	0.047
Phenoxy ($-OC_6H_5$)	−0.5	
Hydroxy ($-OH$)	−0.92	
Carboxy ($-C(=O)OH$)	0.421	0.322
Carbomethoxy ($-C(=O)OCH_3$)	0.489	0.368
Carboethoxy ($-C(=O)OC_2H_5$)	0.482	0.366
Formyl ($-CH(=O)$)[a]	0.36	
Nitro ($-NO_2$)	0.790	0.674

[a] From Zetzsch (1982a).

Source: Brown and Okamoto (1958).

II-C-4. Limitations of this Estimation Method

The method proposed by Atkinson (1986) and briefly described in this section requires a reliable rate coefficient database for the various classes of organic compounds to derive the various group rate coefficients and substituent factors. As may be expected, this estimation technique is reasonably reliable when applied within the database used in its derivation (Kwok and Atkinson, 1995). However, *extrapolation* of this estimation method to organic compounds outside of the database used for its development and testing results in a lack of assurance of its reliability, and is therefore not recommended.

Figure II-C-1 shows a log–log plot of the calculated (k_{calc}) versus measured (k_{obs}) rate coefficients for the reactions of the OH-radical with those monocyclic aromatic hydrocarbons evaluated in section II-B and which do not contain unsaturated $>C=C<$ bond(s) in the substituent groups (i.e., excluding indene, the styrenes, and 1,4-naphthoquinone). Of the 36 aromatic compounds included in this analysis (and plotted in figure II-C-1), the calculated 298 K rate coefficients agree with the recommended values (or, for those compounds with only a single study, with that value) to within a factor of 2 for 28 of these compounds. The compounds with disagreements of factors greater than 2 are (with the factor in parentheses) *o*-xylene (2.09), *p*-xylene (2.20), indan (2.29), tetralin (3.11), 2,5-dimethylbenzaldehyde (2.24), 3,5-dimethylphenol (2.01), 2,3,5-trimethylphenol (2.23), and *m*-nitrotoluene (2.13). Hence, for 78% of the 36 aromatic compounds included in figure II-C-1, the calculated values agree with the measured values to within a factor of 2, and for all but one to within a factor of 3.

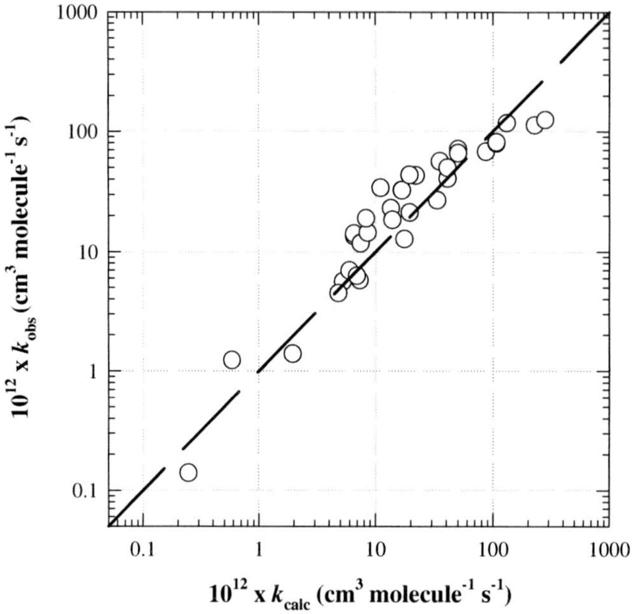

Figure II-C-1. Plot of the measured rate coefficients (k_{obs}) against the calculated rate coefficients (k_{calc}) for those aromatic compounds for which OH-radical reaction rate coefficients have been measured (section II-B), but excluding those which contain >C=C< bond(s) in the substituent groups (indene, styrenes and 1,4-naphthoquinone). The line denotes agreement between measured and calculated rate coefficients.

II-D. Major Products of OH–Aromatic Hydrocarbon Reactions

A number of product studies of the OH-radical–initiated oxidation of aromatic hydrocarbons have been reported. Though many of these studies have focussed on toluene, the most abundant aromatic hydrocarbon found in polluted air, numerous data exist for many other aromatic hydrocarbons, including some polycyclic aromatics; see section II-G.

In this section, we review the products of the OH-initiated oxidation of aromatic hydrocarbons. All reported yields are on a molar basis; that is, molecules of product formed per molecule of reactant consumed. As discussed in sections II-E and VI-B, many of the observed products may be secondary or tertiary in nature, being formed by further oxidation of first- (or second-) generation products. In addition, most of the products are highly reactive themselves; they react with OH and/or NO_3 radicals and also photolyze; see section VI-B and chapter VII. Correction for the influence of these secondary reactions was not made in some of the product studies and, consequently, uncorrected yields are then reported.

One problem encountered in all existing product studies of the OH-initiated oxidation of aromatic hydrocarbons is the deficient carbon balance. Generally, only about 50% of the carbon that reacts with OH-radicals is identified as products. The main reason for this is that all the proposed and identified products of the oxidation are highly reactive themselves, and most of the products react about 1 order of magnitude faster with OH-radicals than the parent aromatic hydrocarbon; see sections II-B and VI-A. This often results in low yields of these products, which makes them difficult to detect and even more difficult to quantify.

Such large differences in the reactivity between products and reactants are not observed in any other class of volatile organic compounds present in the atmosphere. In most cases, the products of the OH-initiated oxidation of VOCs are less reactive than the VOCs themselves; see, for example, Atkinson (1994). The high reactivity of aromatic hydrocarbon photooxidation products can be explained by the stability induced by the three "double bonds" in the benzene ring. In the aromatic benzene ring, these "double bonds" are fully conjugated and consequently very stable; see section II-A. Once the aromaticity is destroyed, however, non-aromatic double bonds are formed, making the products highly reactive.

The discussion in this chapter will be limited to the primary products of the OH-initiated oxidation of aromatic hydrocarbons. No recommendations will be given for the formation yields of secondary products and products for which several formation mechanisms have been proposed. In these cases, yields will depend strongly on experimental conditions.

II-D-1. Benzene

Due to its low reactivity, only limited data are available on the OH-initiated oxidation of benzene. In some studies, detectable product formation could only be achieved by employing very high benzene concentrations, rendering a reliable determination of the benzene consumption, and consequently the product formation yield, impossible; see, for example, Bandow et al. (1985). The available data are listed in table II-D-1.

At ambient temperatures, the reaction of OH-radicals with benzene proceeds almost exclusively by addition of OH to the aromatic ring; see also sections II-B and II-E. The main product observed is phenol, with reported formation yields of around 0.25. The formation of phenol can proceed by several different pathways; see section II-E. With the available database, it is not possible to assess the relative importance of these pathways.

In addition to benzene, nitrobenzene has been observed as a second ring-retaining product. However, under atmospheric conditions where NO_x concentrations are typically 10–50 ppb, formation of nitrobenzene is anticipated to be negligible (Atkinson and Aschmann, 1994).

The only quantified ring-fragmentation product observed to date is glyoxal, with a yield of 0.207 ± 0.019 (Tuazon et al., 1986).

II-D-2. Toluene

Toluene, one of the most abundant of the reactive aromatic hydrocarbons, is the most intensively studied. Its reaction with OH-radicals is about a factor of 5 faster than that of benzene (see section II-B), reducing to some extent the problems of

Table II-D-1. Products of OH-initiated oxidation of benzene

Product	Yield	NO_2 or NO (ppm)[a]	Reference
Phenol	0.236 ± 0.044	NO_2 (0–10)	Atkinson et al. (1989a)
	0.25 ± 0.05	(0–0.34 mbar)[b]	Bjergbakke et al. (1996)
	0.23 ± 0.07	0	Berndt et al. (1999)
p-Benzoquinone	0.055 ± 0.009	0	Berndt et al. (1999)
Formic acid	0.13 ± 0.05	0	Berndt et al. (1999)
Carbonyls (unidentified)	0.59 ± 0.011	0	Berndt et al. (1999)
Nitrobenzene	(0.0336 ± 0.0078) + (3.07 ±0.92) × 10^{-16} [NO_2] cm^3 $molecule^{-1}$;	NO_2 (0–10)	Atkinson et al. (1989a)
	"zero at low NO_x"	NO_2 (0–5)	Atkinson and Aschmann (1994)
Glyoxal	0.207 ± 0.019	$NO_{initial}$ (~5)	Tuazon et al. (1986)
Carbon monoxide	0.30 ± 0.13	0	Berndt et al. (1999)

[a] Range of initial NO_2 concentrations employed in the experiments where given. In some experimental systems, only NO was initially present; in such cases, the initial NO concentration is given. As discussed in the text, it is important to bear in mind that the NO_2 concentration employed in the individual studies can affect the formation yields of many of the products.
[b] Pulse radiolysis experiments.

low reactivity encountered with benzene. Table II-D-2 gives an overview of the available product data.

As discussed in sections II-B-2 and II-E, the reaction of toluene with OH-radicals proceeds by two pathways: abstraction of an H-atom from the methyl substituent, and addition of OH to the aromatic ring. The abstraction channel primarily leads to benzaldehyde, which is apparently formed in constant yield even under a wide variety of experimental conditions. From the studies of Atkinson et al. (1989a), Seuwen and Warneck (1996), Becker et al. (1997), Smith et al. (1998), and Klotz et al. (1998b), an average yield of 0.06 ±0.01 is calculated for benzaldehyde in the OH-initiated oxidation of toluene. Due to their large error limits, the results of Moschonas et al. (1999) have not been included in this evaluation. From these data, it can be concluded that abstraction of a methyl hydrogen atom is not the main pathway in the reaction of OH with toluene, but that addition of OH to the aromatic ring dominates (in agreement with the kinetic data discussed in section II-B).

For *o*-cresol, the product apparently formed in the highest yield on a carbon basis, yields of between $\leq$0.03 and 0.385 have been reported; see table II-D-2. These differences can probably be attributed to problems with the quantitative detection of such a polar and semivolatile compound in the experimental systems employed (Klotz et al., 1998b). A further complication is the fast reaction of cresols with NO_3 radicals (Carter et al., 1981; Klotz et al., 1998b), which affects significantly the measurements of cresol concentrations in aromatic hydrocarbon/ NO_x photooxidation systems.

For the formation of cresols, different pathways have been proposed by different authors (see section II-E). Klotz et al. (1998b) concluded that *o*-cresol is

Table II-D-2. Products of OH-initiated oxidation of toluene

Product	Yield	NO_2/NO(ppm)[a]	Reference
Benzaldehyde	0.12	NO_2 (0.02–0.10)	Atkinson et al. (1980)
	0.073 ± 0.022	$NO_{initial}$ (0.8–1.3)	Atkinson et al. (1983c)
	0.054	$NO_{initial}$ (4.7)	Shepson et al. (1984)
	0.071	NO_2 (0.060–0.245)	Leone et al. (1985)
	0.11 ± 0.01	NO_2 (0.50–1.51)	Bandow et al. (1985)
	0.104 ± 0.029	NO_x (4.5–15.5)[b]	Gery et al. (1985)
	0.0645 ± 0.0080	NO_2 (0–10)	Atkinson et al. (1989a)
	0.053 ± 0.011	[i]	Seuwen and Warneck (1996)
	0.071 ± 0.018	[j]	Becker et al. (1997)
	0.060 ± 0.006	$NO_{initial}$ (0.10-0.42)	Smith et al. (1998)
	0.058 ± 0.008	NO_x (0.003-0.3)	Klotz et al. (1998a)
	0.084 ± 0.05, in air	NO_2 (1–10)	Moschonas et al. (1999)
	0.064 ± 0.03, in N_2	NO_2 (1–10)	Moschonas et al. (1999)
o-Cresol	0.21	NO_2 (0.020–0.102)	Atkinson et al. (1980)
	0.131 ± 0.072	$NO_{initial}$ (0.8–1.3)	Atkinson et al. (1983c)
	0.22	NO_x (4.5–15.5)[b]	Gery et al. (1985)
	0.16	NO_2 (0.060–0.245)	Leone et al. (1985)
	0.204 ± 0.027	NO_2 (0–10)	Atkinson et al. (1989a)
	up to 0.23[c]	[k]	Bierbach et al. (1993)
	= 0.03[d]	[k]	Bierbach et al. (1993)
	0.385 ± 0.083	[i]	Seuwen and Warneck (1996)
	0.123 ± 0.022[e]	NO_x (0)	Atkinson and Aschmann (1994)
	0.145 ± 0.18[f]	NO_2 (1.4)	Atkinson and Aschmann (1994)
	0.160 ± 0.020[g]	NO_2 (7.0)	Atkinson and Aschmann (1994)
	0.123 ± 0.006	$NO_{initial}$ (0.10–0.42)	Smith et al. (1998)
	0.120 ± 0.014	NO_x (0.003–0.3)	Klotz et al. (1998b)
	0.06 ± 0.03, in air	NO_2 (1–10)	Moschonas et al. (1999)
	0.13 ± 0.08, in N_2	NO_2 (1–10)	Moschonas et al. (1999)
m- + *p*-Cresol	0.05	NO_x (4.5–15.5)[b]	Gery et al. (1985)
	0.048 ± 0.009	NO_2 (0–10)	Atkinson et al. (1989a)
	0.056 ± 0.004	$NO_{initial}$ (0.10–0.42)	Smith et al. (1998)
	0.059 ± 0.009	NO_x (0.003–0.3)	Klotz et al. (1998b)
	0.03, in air	NO_2 (1–10)	Moschonas et al. (1999)
	0.05, in N_2	NO_2 (1–10)	Moschonas et al. (1999)
m-Cresol	0.026 ± 0.003	$NO_{initial}$ (0.10–0.42)	Smith et al. (1998)
	0.027 ± 0.007	NO_x (0.003–0.3)	Klotz et al. (1998b)
p-Cresol	0.144 ± 0.031	[i]	Seuwen and Warneck (1996)
	0.030 ± 0.003	$NO_{initial}$ (0.10–0.42)	Smith et al. (1998)
	0.032 ± 0.006	NO_x (0.003–0.3)	Klotz et al. (1998b)
2-Methyl-*p*-benzoquinone	0.050 ± 0.011	[i]	Seuwen and Warneck (1996)
	0.008, in air	NO_2 (1–10)	Moschonas et al. (1999)
	0.07, in N_2	NO_2 (1–10)	Moschonas et al. (1999)
Benzyl alcohol	0.024 ± 0.005	[i]	Seuwen and Warneck (1996)
	0.0015 ± 0.0003	$NO_{initial}$ (0.10–0.42)	Smith et al. (1998)
Benzyl nitrate	0.007 ± 0.004	NO_x (4.5–15.5)[b]	Gery et al. (1985)
	0.0084 ± 0.0017	NO_2 (0–10)	Atkinson et al. (1989a)
o-Nitrotoluene	0.0030 ± 0.0006	$NO_{initial}$ (0.10–0.42)	Smith et al. (1998)

(*continued*)

Table II-D-2. (*continued*)

Product	Yield	NO_2/NO(ppm)[a]	Reference
m-Nitrotoluene	(0.0135 ± 0.0029) + (1.90 ±0.25) × 10^{-16} cm^3 $molecule^{-1}$ [NO_2]	NO_2 (0–10)	Atkinson et al. (1989a)
	"zero at low NO_x"	NO_2 (0–5)	Atkinson and Aschmann (1994)
	0.013 ± 0.0023	$NO_{initial}$ (0.10–0.42)	Smith et al. (1998)
p-Nitrotoluene	0.011 ± 0.002	$NO_{initial}$ (0.10–0.42)	Smith et al. (1998)
Maleic anhydride	0.042 ± 0.011	j	Becker et al. (1997)
α-Angelica lactone	0.029 ± 0.036	$NO_{initial}$ (0.10–0.42)	Smith et al. (1998)
4-Oxo-2-pentenal	0.031 ± 0.004[h]	$NO_{initial}$ (0.10–0.42)	Smith et al. (1998)
Methylglyoxal	0.075	$NO_{initial}$ (4.7)	Shepson et al. (1984)
	0.146 ± 0.014	$NO_{initial}$ (~5)	Tuazon et al. (1984a)
	0.14 ± 0.04	NO_2 (0.50–1.51)	Bandow et al. (1985)
	0.106	NO_x (4.5–15.5)[b]	Gery et al. (1985)
	0.146 ± 0.006	$NO_{initial}$ (~6)	Tuazon et al. (1986)
	0.055 ± 0.018	i	Seuwen and Warneck (1996)
	0.044 ± 0.011	j	Becker et al. (1997)
	0.167 ± 0.018	$NO_{initial}$ (0.10–0.42)	Smith et al. (1998)
Glyoxal	0.08	$NO_{initial}$ (4.7)	Shepson et al. (1984)
	0.111 ± 0.013	$NO_{initial}$ (~5)	Tuazon et al. (1984a)
	0.15 ± 0.04	NO_2 (0.50–1.51)	Bandow et al. (1985)
	0.098	NO_x (4.5–15.5)[b]	Gery et al. (1985)
	0.105 ± 0.019	$NO_{initial}$ (~6)	Tuazon et al. (1986)
	0.041 ± 0.014	i	Seuwen and Warneck (1996)
	0.037 ± 0.009	j	Becker et al. (1997)
	0.238 ± 0.025	$NO_{initial}$ (0.10-0.42)	Smith et al. (1998)

[a] Range of initial NO_2 concentrations employed in the experiments where given. In some experimental systems, only NO was initially present; in such cases, the initial NO concentration is given. As discussed in the text, it is important to bear in mind that the NO_2 concentration employed in the individual studies can affect the formation yields of many of the products.

[b] Continuous stirred tank reactor; NO_x varied between 4.5 and 15.5 ppm; no speciation between NO and NO_2.

[c] OH source: UV-photolysis of H_2O_2; concentrations: (toluene) = (H_2O_2) = 1000 ppmv.

[d] OH source: UV-photolysis of H_2O_2; concentrations: (toluene) = (H_2O_2) = ~1 ppmv.

[e] OH source: alkenes + O_3, NO_x-free.

[f] OH source: CH_3ONO/NO/air/*hν*, 1.4 ppm NO_x.

[g] OH source: CH_3ONO/NO/air/*hν*, 7 ppm NO_x.

[h] Estimated yield based on flame ionization detection (FID) response of similar compound.

[i] NO_x-free in small reactor volumes at very high reactant concentrations with photolysis of H_2O_2 as OH source.

[j] NO_x-free in 1080 L reactor at low reactant concentrations (1–2 ppm) with photolysis of H_2O_2 as OH source.

[k] NO_x-free in 420 L reactor at medium reactant concentrations (2–20 ppm) with photolysis of H_2O_2 as OH source.

probably formed in the reaction of the toluene–OH adduct with O_2 or directly in the reaction of toluene with OH. The latter pathway was proposed for the phenol formation from benzene by Bjergbakke et al. (1996). Formation of *o*-cresol in the reaction of the toluene–OH adduct with ozone, as proposed by Stockwell et al. (1997), was concluded to be insignificant by Klotz et al. (1998b) and cannot be important in reaction systems where O_3 formation (and hence NO_3-radical formation) was suppressed. From the NO_x-free data of Atkinson and Aschmann

(1994) and the low-NO_x data of Smith et al. (1998) and Klotz et al. (1998b), an average for the *o*-cresol yield of 0.12 ± 0.01 is deduced at low NO_x concentration. The recent data of Moschonas et al. (1999) were again not taken into account due to their high degree of uncertainty.

Only limited data exist for the formation of *m*- and *p*-cresol in the OH-initiated oxidation of toluene; see table II-D-2. The existing data of Klotz et al. (1998b) and Smith et al. (1998) show excellent agreement; average formation yields of 0.026 ± 0.005 and 0.031 ± 0.005 are calculated for *m*- and *p*-cresol, respectively.

The nitrogen-containing ring-retaining products observed in numerous laboratory studies are not expected to be formed in any significant yield under atmospheric conditions (Atkinson et al., 1989a; Atkinson and Aschmann, 1994); their formation yields will therefore not be discussed here.

A large number of ring-fragmentation products have been observed in the OH-initiated oxidation of toluene, but only some of these have been determined quantitatively; see table II-D-2. Amongst the identified but not quantified products are highly reactive unsaturated dicarbonyls such as methylated derivatives of 2,4-hexadienedial and 2-butenedial, and other unsaturated carbonyls and epoxides (Dumdei and O'Brien, 1984; O'Brien et al., 1984; Shepson et al., 1984; Yu and Jeffries, 1997; Yu et al., 1997). No statement concerning the importance of these intermediates can be made at the time of this writing.

The best-known ring-fragmentation products of the toluene + OH reaction are glyoxal and methylglyoxal, for which yields have been determined in numerous studies; see table II-D-2. The reported yields for methylglyoxal vary by a factor of 4, while those for glyoxal vary by a factor of 6. This variation may reflect, in part, the fact that these compounds are not exclusively primary products, but can also be formed by the secondary chemistry of other products. Variations could result, in part, from problems with analyses and with variation of product yields with $[NO_2]$, noting that with $[NO_x] = 0$, $RO_2 + NO$ reactions are absent; see section II-E.

II-D-3. *o*-, *m*-, and *p*-Xylene

After toluene, the *o*-, *m*-, and *p*-isomers of xylene are the most intensively studied aromatic hydrocarbons. Tables II-D-3 through II-D-5 give an overview of the existing product data. As for toluene, the OH-radical can react with xylenes both by abstraction of an H-atom from one of the methyl groups and by addition to the aromatic ring. Abstraction, leading to tolualdehyde isomers, is again the minor pathway, with an average yield of about 0.05 for *o*- and *m*-tolualdehyde. A somewhat higher yield of about 0.1 has been observed for the formation of *p*-tolualdehyde in the OH-initiated oxidation of *p*-xylene. The reasons for this difference are not clear.

The yields of the dimethyl phenols from the xylenes appear to be somewhat lower than those of the cresols from toluene, but the database here is very limited. For *p*-xylene, the most-studied xylene isomer, the published 2,5-dimethylphenol yields vary within a factor of 3. At the time of this writing, the determinations of Smith et al. (1999) are probably the most reliable.

Table II-D-3. Products of OH-initiated oxidation of *o*-xylene

Product	Yield	NO_2 or NO (ppm)[a]	Reference
o-Tolualdehyde	0.073 ± 0.036	$NO_{initial}$ (2–10.6)	Takagi et al. (1980)
	0.047	$NO_{initial}$ (4.7)	Shepson et al. (1984)
	0.05 ± 0.01	NO_2 (0.5)	Bandow and Washida (1985a)
	0.172 ± 0.070	NO_x (7.7–8.2)[b]	Gery et al. (1987)
	0.0453 ± 0.0059	NO_2 (0–10)	Atkinson et al. (1991)
2-Methylbenzyl nitrate	0.009 ± 0.002	$NO_{initial}$ (2–10.6)	Takagi et al. (1980)
	~0.012	NO_x (7.7-8.2)[b]	Gery et al. (1987)
	$(0.0135 \pm 0.0051) + (5.5 \pm 4.6) \times 10^{-17}$ cm^3 $molecule^{-1}$ $[NO_2]$	NO_2 (0–10)	Atkinson et al. (1991)
2,3-Dimethylphenol	0.097 ± 0.024	NO_2 (0–10)	Atkinson et al. (1991)
3,4-Dimethylphenol	0.064 ± 0.015	NO_2 (0–10)	Atkinson et al. (1991)
2,3- + 3,4-Dimethylphenol	0.012 ± 0.006	$NO_{initial}$ (2–10.6)	Takagi et al. (1980)
	0.102 ± 0.039	NO_x (7.7–8.2)[b]	Gery et al. (1987)
	0.161 ± 0.039	NO_2 (0–10)	Atkinson et al. (1991)
3-Nitro-*o*-xylene	0.005 ± 0.002	NO_x (7.7–8.2)[b]	Takagi et al. (1980)
	0.0059 ± 0.0018	NO_2 (0–10)	Atkinson et al. (1991)
4-Nitro-*o*-xylene	0.075 ± 0.045	NO_x (7.7–8.2)[b]	Takagi et al. (1980)
	$(0.0111 \pm 0.0029) + (9.9 \pm 2.2) \times 10^{-17}$ cm^3 $molecule^{-1}$ $[NO_2]$	NO_2 (0–10)	Atkinson et al. (1991)
Nitro-*o*-xylenes	0.068 ± 0.019	NO_x (7.7–8.2)[b]	Gery et al. (1987)
Biacetyl	0.18 ± 0.04	NO_x (~0.-5)	Darnall et al. (1979)
	0.260 ± 0.102	NO_x (7.7–8.2)	Takagi et al. (1980)
	0.137 ± 0.016	$NO_{initial}$ (0.8–1.3)	Atkinson et al. (1983c)
	0.085	$NO_{initial}$ (~4.5)	Shepson et al. (1984)
	0.10 ± 0.02	NO_2 (0.5)	Bandow and Washida (1985a)
	0.185[c]	—	Atkinson and Aschmann (1994)
Methylglyoxal	0.116	$NO_{initial}$ (~4.5)	Shepson et al. (1984)
	0.23 ± 0.03	NO_2 (0.5)	Bandow and Washida (1985a)
	0.246 ± 0.020	$NO_{initial}$ (~5)	Tuazon et al. (1986)
Glyoxal	0.034	$NO_{initial}$ (~4.5)	Shepson et al. (1984)
	0.08 ± 0.04	NO_2 (0.5)	Bandow and Washida (1985a)
	0.087 ± 0.012	$NO_{initial}$ (~5)	Tuazon et al. (1986)

[a] Range of initial NO_2 concentrations employed in the experiments where given. In some experimental systems, only NO was initially present; in such cases, the initial NO concentration is given. Where not specifically indicated the systems were NO_x-free. As discussed in the text, it is important to bear in mind that the NO_2 concentration employed in the individual studies can affect the formation yields of many of the products.

[b] Continuous stirred tank reactor; NO_x varied between 7.9 and 8.2 ppm in the inlet; no speciation between NO and NO_2.

[c] Formation yield at low $[NO_x]$ derived by extrapolation of data obtained as a function of NO_2 to zero $[NO_2]$.

Table II-D-4. Products of OH-initiated oxidation of *m*-xylene

Product	Yield	NO_2 or NO (ppm)[a]	Reference
m-Tolualdehyde	0.04 ± 0.01	NO_2 (0.5)	Bandow and Washida (1985a)
	0.122 ± 0.059	NO_x (7.9–8.2)[b]	Gery et al. (1987)
	0.0331 ± 0.0041	NO_2 (0–10)	Atkinson et al. (1991)
	0.0485 ± 0.0036	$NO_{initial}$ (0.157–1.081)	Smith et al. (1999)
3-Methylbenzyl nitrate	~0.010	NO_x (7.9–8.2)[b]	Gery et al. (1987)
	0.0061 ± 0.0027	NO_2 (0-10)	Atkinson et al. (1991)
2,4-Dimethylphenol	0.099 ± 0.023	NO_2 (0-10)	Atkinson et al. (1991)
	0.06 ± 0.004	$NO_{initial}$ (0.157–1.081)	Smith et al. (1999)
2,6-Dimethylphenol	0.111	NO_2 (0-10)	Atkinson et al. (1991)
	0.046 ± 0.003	$NO_{initial}$ (0.157–1.081)	Smith et al. (1999)
2,4- + 2,6-Dimethylphenol	0.178 ± 0.065	NO_x (7.9–8.2)[b]	Gery et al. (1987)
3,5-Dimethylphenol	0.003 ± 0.0003	$NO_{initial}$ (0.157–1.081)	Smith et al. (1999)
4-Nitro-*m*-xylene	0.0018 ± 0.009	NO_2 (0–10)	Atkinson et al. (1991)
5-Nitro-*m*-xylene	$(0.0032 \pm 0.0012) + (1.6 \pm 0.8) \times 10^{-17}$ cm^3 molecule^{-1} [NO_2]	NO_2 (0–10)	Atkinson et al. (1991)
Nitro-*m*-xylenes	0.033 ± 0.025	NO_x (7.9–8.2)[b]	Gery et al. (1987)
3-Methyl-5(2*H*)-furanone	0.0058 ± 0.0006	$NO_{initial}$ (0.157–1.081)	Smith et al. (1999)
4-Oxo-2-pentenal	0.12 ± 0.012[c]	$NO_{initial}$ (0.157–1.081)	Smith et al. (1999)
Methylglyoxal	0.265 ± 0.035	$NO_{initial}$ (~5)	Tuazon et al. (1984a)
	0.42 ± 0.05	NO_2 (0.5)	Bandow and Washida (1985a)
	0.319 ± 0.009	$NO_{initial}$ (~5)	Tuazon et al. (1986)
	0.40 ± 0.018	$NO_{initial}$ (0.157–1.081)	Smith et al. (1999)
Glyoxal	0.104 ± 0.02	$NO_{initial}$ (~5)	Tuazon et al. (1984a)
	0.13 ± 0.03	NO_2 (0.5)	Bandow and Washida (1985a)
	0.086 ± 0.011	$NO_{initial}$ (~5)	Tuazon et al. (1986)
	0.079 ± 0.004	$NO_{initial}$ (0.157–1.081)	Smith et al. (1999)

[a] Range of initial NO_2 concentrations employed in the experiments where given. In some experimental systems, only NO was initially present; in such cases, the initial NO concentration is given. Where not specifically indicated the systems were NO_x-free. As discussed in the text, it is important to bear in mind that the NO_2 concentration employed in the individual studies can affect the formation yields of many of the products.

[b] Continuous stirred tank reactor; NO_x varied between 7.9 and 8.2 ppm in the inlet; no speciation between NO and NO_2.

[c] Estimated yield based on FID response of similar compounds.

Quantitative information is available on a range of ring-fragmentation products that will not be discussed in detail here; see discussion in section II-D-2. Other ring-fragmentation products have been observed frequently but are not likely to be of atmospheric importance and will not be discussed here.

Table II-D-5. Products of OH-initiated oxidation of *p*-xylene

Product	Yield	NO_2 or NO(ppm)[a]	Reference
p-Tolualdehyde	0.08 ± 0.01	NO_2 (0.5)	Bandow and Washida (1985a)
	0.10		Becker and Klein (1987)
	0.0701 ± 0.0103	NO_2 (0-10)	Atkinson et al. (1991)
	0.064 ± 0.015 (FTIR)[f]	[d]	Becker et al. (1997)
	0.112 ± 0.040 (GC-MS)[f]	[e]	Becker et al. (1997)
	0.103 ± 0.016	$NO_{initial}$ (0.157–1.081)	Smith et al. (1999)
4-Methylbenzyl nitrate	0.0082 ± 0.0016	NO_2 (0–10)	Atkinson et al. (1991)
2,5-Dimethylphenol	0.188 ± 0.038	NO_2 (0-10)	Atkinson et al. (1991)
	0.08 (GC-FID)[f]	[d]	Becker et al. (1997)
	0.058 ± 0.020 (GC-MS)[f]	[e]	Becker et al. (1997)
	0.13 ± 0.018	$NO_{initial}$ (0.157–1.081)	Smith et al. (1999)
2-Nitro-*p*-xylene	(0.0120 ± 0.0035) + (2.8 ± 2.6) × 10^{-17} cm^3 $molecule^{-1}$ [NO_2]	NO_2 (0-10)	Atkinson et al. (1991)
	0.012 ± 0.0037	$NO_{initial}$ (0.157–1.081)	Smith et al. (1999)
2,5-Dimethyl-*p*-benzoquinone	0.0094 ± 0.0022	$NO_{initial}$ (0.157–1.081)	Smith et al. (1999)
2,5-Dimethylfuran	0.0074 ± 0.0018	$NO_{initial}$ (0.157–1.081)	Smith et al. (1999)
Z-3-Hexene-2,5-dione	0.083 ± 0.019 (FTIR)[f]	[d]	Becker et al. (1997)
	0.144 ± 0.040 (GC-MS)[f]	[e]	Becker et al. (1997)
	0.176 ± 0.03[b]	$NO_{initial}$ (0.157–1.081)	Smith et al. (1999)
E-3-Hexene-2,5-dione	0.045 ± 0.01[b]	$NO_{initial}$ (0.157–1.081)	Smith et al. (1999)
2-Methylbutenedial	0.071 ± 0.035[c]	$NO_{initial}$ (0.157–1.081)	Smith et al. (1999)
Methylglyoxal	0.111 ± 0.015	$NO_{initial}$ (~5)	Tuazon et al. (1984a)
	0.12 ± 0.02	NO_2 (0.5)	Bandow and Washida (1985a)
	0.105 ± 0.034	$NO_{initial}$ (~5)	Tuazon et al. (1986)
	0.022 ± 0.005	[d]	Becker et al. (1997)
	0.217 ± 0.076	$NO_{initial}$ (0.157–1.081)	Smith et al. (1999)

Glyoxal	0.120 ± 0.020	$NO_{initial}$ (~5)	Tuazon et al. (1984a)
	0.24 ± 0.02	NO_2 (0.5)	Bandow and Washida (1985a)
	0.225 ± 0.039	$NO_{initial}$ (~5)	Tuazon et al. (1986)
	0.040 ± 0.017	[d]	Becker et al. (1997)
	0.394 ± 0.11	$NO_{initial}$ (0.157–1.081)	Smith et al. (1999)

[a] Range of initial NO_2 concentrations employed in the experiments where given. In some experimental systems, only NO was initially present; in such cases, the initial NO concentration is given. As discussed in the text, it is important to bear in mind that the NO_2 concentration employed in the individual studies can affect the formation yields of many of the products.

[b] Estimated yield based on FID response of a similar compound.

[c] Estimated yield.

[d] NO_x-free in 1080 L reactor at low reactant concentrations (1–2 ppm) with photolysis of H_2O_2 as OH source.

[e] NO_x-free in 36.6 L reactor at high reactant concentrations (50–200 ppm) with photolysis of H_2O_2 as OH source.

[f] FTIR (Fourier-transform infrared absorption spectrometry), GC-MS (gas chromatography–mass spectrometry), GC-FID (gas chromatography–flame ionization detection).

II-D-4. Trimethylbenzenes

The database concerning products from the OH-initiated oxidation of the three isomers, 1,2,3-, 1,2,4-, and 1,3,5-trimethylbenzene, is limited; see tables II-D-6 through II-D-8. The only reasonably complete product analysis is that of Smith et al. (1999) , who investigated the 1,2,4- and 1,3,5-isomers. As expected from the high reactivity of the trimethylbenzenes, the abstraction channel, leading to dimethylbenzaldehydes, is of even less importance than in the case of toluene and the xylenes discussed earlier. Smith et al. (1999) found dimethylbenzaldehyde yields of less than 0.04. The combined formation yields of phenolic products, trimethylphenol isomers, is also significantly lower than has been observed for benzene, toluene, and the xylenes. The ring-fragmentation products observed in the OH-initiated oxidation of trimethylbenzenes are methylated derivatives of the compounds detected in the OH-initiated oxidation of toluene and the xylenes; for further discussion, the reader is referred to sections II-D-1 and II-D-2.

II-D-5. Hydroxylated Monocyclic Aromatic Hydrocarbons

The OH-radical–initiated oxidation of phenolic compounds has not been studied extensively. Formation yields are available for only one class of products, nitrophenols. Tables II-D-9 through II-D-12 give an overview of the published data. The formation yields of the nitrophenols are low, and although it is conceivable that they are formed in the OH-radical–initiated oxidation of hydroxylated aromatic hydrocarbons, they may also be artifacts of the fast reaction of phenols with NO_3 radicals. NO_3 radicals can be formed in significant quantities in aromatic hydrocarbon/NO_x photooxidation systems (Carter et al., 1981; Klotz et al., 1998b). However, the experiments of Atkinson et al. (1992) did not involve NO_x–aromatic–air mixtures, but instead involved CH_3ONO–NO–aromatic–air mixtures, and the conditions for which NO_3 became important were easy to identify. Under tropospheric conditions of low NO_x concentrations (in comparison to most laboratory studies), nitrophenols are not expected to be formed in significant yields.

Considering the importance of phenols as products of the OH-radical–initiated oxidation of benzene and its methylated derivatives, more detailed studies of their atmospheric behavior are required.

II-D-6. Styrene and Methyl-Substituted Styrenes

Product data exist for the OH-initiated oxidation reaction of styrene and methyl-substituted styrenes. The OH-radical reaction with styrene produces benzaldehyde and formaldehyde with yields of 0.63 ± 0.06 and 0.72 ± 0.07, respectively (Tuazon et al., 1993). From β-methylstyrene, a formation yield for benzaldehyde of 0.98 ± 0.14 has been reported (Bignozzi et al., 1981). β,β-Dimethylstyrene gave benzaldehyde with a yield of 1.05 ± 0.15 and acetone with a yield of 0.89 ± 0.05 (Chiorboli et al., 1983).

Table II-D-6. Products of OH-initiated oxidation of 1,2,3-trimethylbenzene

Product	Yield	NO_2 or NO (ppm)[a]	Reference
Biacetyl	0.45 ± 0.02	NO_2 (0.5)	Bandow and Washida (1985b)
	0.316 ± 0.036	$NO_{initial}$ (~5)	Tuazon et al. (1986)
	0.444 ± 0.053	NO_2 (0–5)	Atkinson and Aschmann (1994)
Methylglyoxal	0.18 ± 0.01	NO_2 (0.5)	Bandow and Washida (1985b)
	0.152 ± 0.025	$NO_{initial}$ (~5)	Tuazon et al. (1986)
Glyoxal	0.072 ± 0.001	NO_2 (0.5)	Bandow and Washida (1985b)
	0.058 ± 0.008	$NO_{initial}$ (~5)	Tuazon et al. (1986)

[a] Range of initial NO_2 concentrations employed in the experiments where given. In some experimental systems, only NO was initially present; in such cases, the initial NO concentration is given. Where not specifically indicated the systems were NO_x-free. As discussed in the text, it is important to bear in mind that the NO_2 concentration employed in the individual studies can affect the formation yields of many of the products.

Table II-D-7. Products of OH-initiated oxidation of 1,2,4-trimethylbenzene

Product	Yield	NO_2 or NO (ppm)[a]	Reference
2,4-Dimethylbenzaldehyde	0.011 ± 0.0022	$NO_{initial}$ (0.157–1.081)	Smith et al. (1999)
2,5-Dimethylbenzaldehyde	0.010 ± 0.002	$NO_{initial}$ (0.157–1.081)	Smith et al. (1999)
3,4-Dimethylbenzaldehyde	0.015 ± 0.002	$NO_{initial}$ (0.157–1.081)	Smith et al. (1999)
2,4,5-Trimethylphenol[b]	0.0026 ± 0.0004	$NO_{initial}$ (0.157–1.081)	Smith et al. (1999)
2,3,5-Trimethylphenol[b]	0.005 ± 0.002	$NO_{initial}$ (0.157–1.081)	Smith et al. (1999)
2,3,6-Trimethylphenol[b]	0.014 ± 0.002	$NO_{initial}$ (0.157–1.081)	Smith et al. (1999)
Z-3-Hexene-2,5-dione	0.130 ± 0.011[c]	$NO_{initial}$ (0.157–1.081)	Smith et al. (1999)
E-3-Hexene-2,5-dione	0.031 ± 0.003[c]	$NO_{initial}$ (0.157–1.081)	Smith et al. (1999)
3-Methyl-3-hexene-2,5-dione	0.079 ± 0.04[c]	$NO_{initial}$ (0.157–1.081)	Smith et al. (1999)
2-Methyl-2-butenedial	0.045 ± 0.022[d]	$NO_{initial}$ (0.157–1.081)	Smith et al. (1999)
Biacetyl	0.11 ± 0.01	NO_2 (0.5)	Bandow and Washida (1985b)
	0.048 ± 0.009	$NO_{initial}$ (5)	Tuazon et al. (1986)
	0.114 ± 0.024	$NO_{initial}$ (0.157–1.081)	Smith et al. (1999)
Methylglyoxal	0.37 ± 0.01	NO_2 (0.5)	Bandow and Washida (1985b)
	0.357 ± 0.017	$NO_{initial}$ (5)	Tuazon et al. (1986)
	0.44 ± 0.074	$NO_{initial}$ (0.157–1.081)	Smith et al. (1999)
Glyoxal	0.078 ± 0.005	NO_2 (0.5)	Bandow and Washida (1985b)
	0.048 ± 0.005	$NO_{initial}$ (5)	Tuazon et al. (1986)
	0.066 ± 0.011	$NO_{initial}$ (0.157–1.081)	Smith et al. (1999)

[a] Range of initial NO_2 concentrations employed in the experiments where given. In some experimental systems, only NO was initially present; in such cases, the initial NO concentration is given. As discussed in the text, it is important to bear in mind that the NO_2 concentration employed in the individual studies can affect the formation yields of many of the products.

[b] Tentative identification.

[c] Estimated yield based on FID response of similar compounds.

[d] Estimated yield.

Table II-D-8. Products of OH-initiated oxidation of 1,3,5-trimethylbenzene

Product	Yield	NO_2/NO(ppm)[a]	Reference
3,5-Dimethylbenzaldehyde	0.03 ± 0.01	$NO_{initial}$ (0.157–1.081)	Smith et al. (1999)
2,4,6-Trimethylphenol	0.04 ± 0.01[b]	$NO_{initial}$ (0.157–1.081)	Smith et al. (1999)
3,5-Dimethyl-3(2*H*)-2-furanone	0.08 ± 0.02	$NO_{initial}$ (0.157–1.081)	Smith et al. (1999)
3-Dimethyl-5(2*H*)-2-furanone	0.10 ± 0.02	$NO_{initial}$ (0.157–1.081)	Smith et al. (1999)
3,5-Dimethyl-5-methylidene-5(2*H*)-2-furanone	0.10 ± 0.02	$NO_{initial}$ (0.157–1.081)	Smith et al. (1999)
2-Methyl-4-oxo-2-pentenal	0.050 ± 0.015	$NO_{initial}$ (0.157–1.081)	Smith et al. (1999)
Citraconic anhydride[c]	0.03 ± 0.01	$NO_{initial}$ (0.157–1.081)	Smith et al. (1999)
Methylglyoxal	0.64 ± 0.03	NO_2 (0.5)	Bandow and Washida (1985b)
	0.602 ± 0.033	$NO_{initial}$ (5)	Tuazon et al. (1986)
	0.9 ± 0.25	$NO_{initial}$ (0.157–1.081)	Smith et al. (1999)

[a] Range of initial NO_2 concentrations employed in the experiments where given. In some experimental systems, only NO was initially present; in such cases, the initial NO concentration is given. As discussed in the text, it is important to bear in mind that the NO_2 concentration employed in the individual studies can affect the formation yields of many of the products.
[b] Tentative identification.
[c] Methylmaleic anhydride.

Table II-D-9. Products of OH-initiated oxidation of phenol

Product	Yield	Reference
2-Nitrophenol	0.067 ± 0.015	Atkinson et al. (1992)

Table II-D-10. Products of OH-initiated oxidation of *o*-cresol

Product	Yield	Reference
6-Methyl-2-nitrophenol + 6-methyl-4-nitrophenol	0.049–0.114	Grosjean (1985)
6-Methyl-2-nitrophenol	0.051 ± 0.015	Atkinson et al. (1992)

Table II-D-11. Products of OH-initiated oxidation of *m*-cresol

Product	Yield	Reference
3-Methyl-2-nitrophenol	0.016 ± 0.010	Atkinson et al. (1992)
5-Methyl-2-nitrophenol	0.016 ± 0.010[a]	Atkinson et al. (1992)

[a] Estimated yield.

Table II-D-12. Products of OH-initiated oxidation of *p*-cresol

Product	Yield	Reference
4-Methyl-2-nitrophenol	0.10 ± 0.04	Atkinson et al. (1992)

II-D-7. The Aromatic Aldehydes

Product data exist only for the reaction of OH-radicals with benzaldehyde. From structure-reactivity relationships it can be predicted that this reaction will proceed predominantly via abstraction of the aldehydic H-atom. Electrophilic addition of OH is expected to be hindered by the electron-withdrawing nature of the formyl group. The only product, study of the reaction of OH-radicals with benzaldehyde is that of Atkinson and Aschmann (1992), who have observed *o*-nitrophenol as the only detectable product, with a yield of 0.21 ± 0.04. The formation of *o*-nitrophenol, and possibly also *p*-nitrophenol, was also observed in the Cl-initiated oxidation of benzaldehyde; this is expected to proceed via abstraction of the aldehydic H-atom (Niki et al., 1979). Abstraction of the aldehydic H-atom of benzaldehyde gives the benzoyl radical, which adds molecular oxygen to give the benzoylperoxy radical (Fenter et al., 1994; Caralp et al., 1999). Reactions of this radical with other identical radicals, HO_2, and NO_x have been studied (Ohta and Mizoguchi, 1981; Kenley and Hendry, 1982; Kirchner et al., 1992; Nozière et al., 1994; Caralp et al., 1999). These studies showed that formation of benzoylperoxy nitrate, also termed peroxybenzoyl nitrate (PBzN), can be of atmospheric significance. It is a very strong eye irritant (Heuss and Glasson, 1968), and it may be of atmospheric significance as a carrier of NO_x. *E*-Cinnamaldehyde, an analogue to the 2-formyl cinnamaldehyde observed in the OH-radical–initiated oxidation of naphthalene (see section II-D-7), was observed to react to give a 0.90 ± 0.39 yield of benzaldehyde (Smith et al., 1996).

II-D-8. 2,3-Benzofuran

Product studies are available for only a few other monocyclic aromatic hydrocarbons; the reaction of OH-radicals with 2,3-benzofuran was found to give salicylaldehyde (salicylic aldehyde) with a yield of 0.26 ± 0.05 (Atkinson et al., 1992).

II-E. Mechanisms of the Reactions of the Transient First Radical Products of OH-Oxidation of Aromatic Hydrocarbons

In all but a few cases (i.e., styrenes and benzaldehyde), the mechanism of reaction of OH-radicals with aromatic hydrocarbons under atmospheric conditions proceeds predominantly (i.e., $>90\%$) via addition to the aromatic ring. Compelling kinetic and product evidence for this statement is given in sections II-B and II-D. Interestingly, despite many experimental and theoretical investigations, significant uncertainties persist in our understanding of the reactions of the transient first radical products of the OH-oxidation of aromatic hydrocarbons. There are three fundamental problems that inhibit our ability to construct complete quantitative mechanisms for the OH-initiated oxidation of aromatic compounds.

First, as discussed in section II-D, the product data concerning such reactions are incomplete in most cases (in studies of the simulated oxidation of aromatic compounds, the identified products typically account for only 50% of the observed aromatic loss). Second, laboratory experiments often employ NO_x concentrations that are considerably higher than those encountered in the atmosphere. In the presence of high NO_x concentrations, reactions can take place that are not representative of the chemistry occurring in the atmosphere, and care must be exercised in the derivation of atmospheric degradation mechanisms from such laboratory data. Third, many different transient radical species are formed during the OH-initiated oxidation of aromatic compounds, and these radicals have structures and chemical properties that are often different from those of radicals derived from non-aromatic hydrocarbons. There are few convenient clean laboratory sources of these radicals, and consequently there have been few direct laboratory studies of their chemistry.

While our understanding of the reactions occurring during OH-radical–initiated oxidation of aromatic compounds is far from complete, there has been considerable progress in this area during the 1990s. In 1989 it was known that reactions of OH-radicals with aromatic compounds proceeded mainly via addition to form an adduct; however, the nature and the atmospheric fate of the adduct was unclear. At the time of this writing (2000), we have a good understanding of the structure and properties of the adduct. Under ambient conditions, the OH–aromatic adduct has a lifetime of the order of 1 s with respect to thermal decomposition to reform the aromatic and OH; see table II-E-1. As shown in table II-E-2, the OH–aromatic adduct reacts with O_2 and NO_2. Initial reports that the adduct reacts with NO (Zellner et al., 1985) have been shown to be in error (Knispel et al., 1990), probably due to contamination of NO reactant by small amounts of NO_2. It is now clear that under atmospheric conditions the dominant fate of the OH–aromatic adduct is reaction with O_2. The kinetics and mechanisms of the reactions occurring subsequent to reaction of the OH–aromatic adduct with O_2 are less clear and are the focus of considerable research. High-level ab initio quantum mechanical calculations have provided valuable insight during the 1990s and, with ever-increasing computational resources, are likely to play an important role in future advances in this area. New methods under development to facilitate direct laboratory study of the key radical intermediates should provide direct chemical information concerning key steps of the mechanisms.

The current knowledge concerning the reactions of the transient radicals formed during the OH-initiated oxidation of aromatic compounds, and their transformation into primary products, is discussed in the following sections. The kinetics and mechanisms of the atmospheric oxidation of the primary products are discussed in chapter VI. Atmospheric oxidation of the primary products leads to aerosol formation, and chemical mechanisms for aerosol formation are discussed in chapter VIII.

II-E-1. Benzene

Under atmospheric conditions, the reaction of OH with benzene proceeds mainly via addition to the ring to give a σ-bonded hydroxycyclohexadienyl radical (often referred to as an OH–aromatic adduct). In solution-phase studies, evidence has been presented that OH-radical initially forms a π-complex, which either decomposes back to reactants or rearranges to give a σ-bonded hydroxycyclohexadienyl

Table II-E-1. Reported rate parameters for the thermal decomposition of selected OH–aromatic adducts

Aromatic Compound	A-factor (cm^3 molecule^{-1} s^{-1})	E_a/R	k(298K) (s^{-1})	Reference
Benzene	3×10^{13}	9410 ± 1000	0.58	Perry et al. (1977a)
	4×10^{13}	9500 ± 720	0.58	Lorenz and Zellner (1983)
	3×10^{13}	8960 ± 690	2.6	Wahner and Zetzsch (1983)
	3×10^{12}	8180 ± 720	3.6	Witte et al. (1986)
	3×10^{13}	9560	0.35	Wallington et al. (1987)
	9.4×10^{12}	8540 ± 750	3.4	Atkinson (1989)
	$(9.0 \pm 4.2) \times 10^{12}$	8570 ± 200	2.9	Knispel et al. (1990)
Toluene	3×10^{13}	9110 ± 1000	1.6	Perry et al. (1977a)
	$(1.49 \pm 0.74) \times 10^{12}$	7880 ± 200	4.9	Knispel et al. (1990)
o-Xylene	3×10^{13}	9260 ± 1000	1.0	Perry et al. (1977a)
m-Xylene	3×10^{13}	9010 ± 1000	2.2	Perry et al. (1977a)
p-Xylene	3×10^{13}	9410 ± 1000	0.6	Perry et al. (1977a)
1,2,3-Trimethylbenzene	3×10^{13}	9360 ± 1000	0.7	Perry et al. (1977a)
1,2,4-Trimethylbenzene	3×10^{13}	9210 ± 1000	1.1	Perry et al. (1977a)
1,3,5-Trimethylbenzene	3×10^{13}	9110 ± 1000	1.6	Perry et al. (1977a)
Phenol	1×10^{13}	8900 ± 1300	1.1	Witte et al. (1986)
	1.33×10^{14}	9770 ± 700	0.8	Knispel et al. (1990)
Methoxybenzene	3×10^{13}	9110 ± 1000	1.6	Perry et al. (1977b)
o-Cresol	3×10^{13}	9610 ± 1000	0.3	Perry et al. (1977b)

radical (Ashton et al., 1995). However, there is no compelling evidence for the existence of a π-complex in the gas-phase. As discussed in section II-B-1, the kinetic data show that H-atom abstraction is of minor (5–6%) importance under atmospheric conditions and that the bulk of the reaction proceeds to give the hydroxycyclohexadienyl radical. The hydroxycyclohexadienyl radical has been observed directly using its UV absorption in the region 260–350 nm in both gas phases (Fritz et al., 1985; Zellner et al., 1985; Knispel et al., 1990; Bjergbakke et al., 1996) and solution-phases (Dorfman et al., 1962; Mantaka et al., 1971). This UV absorption has provided a convenient tool for kinetic studies of atmospherically relevant reactions of the hydroxycyclohexadienyl radical (Zellner et al., 1985; Knispel et al., 1990; Zetzsch and Bohn, 1999). In addition to experimental investigations, there have been several theoretical studies of the energetics and reaction mechanisms of the hydroxycyclohexadienyl radical (Andino et al., 1996; Lay et al., 1996; Berho et al., 1999; Ghigo and Tonachini, 1998). The combination of experimental and theoretical approaches has provided considerable insight into the chemistry of the hydroxycyclohexadienyl radical. However, there is much that we have yet to learn about the mechanism of the OH-initiated atmospheric oxidation of even the simplest aromatic compound—benzene.

Table II-E-2. Rate coefficients (cm^3 $molecule^{-1}$ s^{-1}) for reactions of selected OH–aromatic adducts with O_2, NO, and NO_2

Radical	$10^{16} \times k(O_2)$	$10^{14} \times k(NO)$	$10^{11} \times k(NO_2)$	Reference
a Benzene–OH				
N_2 (15 Torr), 298 K		100 ± 50	0.85 ± 21	Zellner et al. (1985)
Ar (100 Torr), 299 K	1.88 ± 0.47	$< 1^a$	2.6^b	Knispel et al. (1990)
Ar (100 Torr), 314 K	2.61 ± 0.79			Knispel et al. (1990)
Ar (100 Torr), 333 K	2.94 ± 0.11			Knispel et al. (1990)
Ar (100 Torr), 354 K	3.77 ± 0.43			Knispel et al. (1990)
N_2 (750 Torr), 297 K	20 ± 10	< 1		Zetzsch and Bohn (1999)
b Toluene–OH				
Ar (100 Torr), 353 K	< 10			Perry et al. (1977a)
Ar (100 Torr), 300K	5.4 ± 0.6	$< 3^c$	3.7^d	Knispel et al. (1990)
Ar (100 Torr), 321 K	4.3 ± 0.6			Knispel et al. (1990)
c Phenol–OH				
Ar (98 Torr), 298 K		$< 7^e$	3.2^f	Knispel et al. (1990)
Ar (100 Torr), 330K	300	< 7	3.6	Zetzsch et al. (1997)
d *p*-Xylene–OH				
Ar (98 Torr), 330 K	8	< 10	3.2	Zetzsch et al. (1997)
d *m*-Cresol–OH				
Ar (98 Torr), 330 K	800	< 3	4	Zetzsch et al. (1997)
e Aniline–OH				
Ar (98 Torr), 330 K	10	< 10	5	Zetzsch et al. (1997)

[a]At 319 K; [b]average value 305–349 K; [c]333 K; [d]average value 300–338 K; [e]at 316 –332 K; [f]average value 331–353 K.

The mechanistic complexity is illustrated by the (simplified!) scheme shown in figure II-E-1, which was adapted from Lay et al. (1996), Klotz et al. (1997), and Kwok et al. (1997a). Hydrogen abstraction to give the phenyl radical (A) is of minor (5–6%) importance under atmospheric conditions. The majority of the reaction proceeds via addition of OH to give an energized (denoted by *) hydroxycyclohexadienyl radical (B), which can eliminate an H-atom, or be collisionally stabilized to give a thermalized radical (C). It should be noted that the structures given as (B) and (C) represent one of the three possible resonance structures. There is conflicting experimental evidence concerning the importance of H-atom elimination from the energized adduct (B). Bjergbakke et al. (1996) argue that only the direct displacement reaction $OH + C_6H_6 \rightarrow H + C_6H_5OH$ accounts for their experimental observation of a prompt phenol yield of $25 \pm 5\%$ (Pagsberg, 1997). However, as noted by Koch (1997), this interpretation is inconsistent with the fact that the presence of NO_2 (which scavenges H-atoms and produces OH-radicals) does not impact the value of $k(OH + C_6H_6)$ measured in flash photolysis kinetic studies where the concentration of OH-radicals is monitored. Zetzsch and Bohn (1999) report an upper limit of 5% for the direct displacement reaction. Quantum mechanical calculations also suggest that the direct displacement channel is not important under atmospheric conditions (Lay et al., 1996).

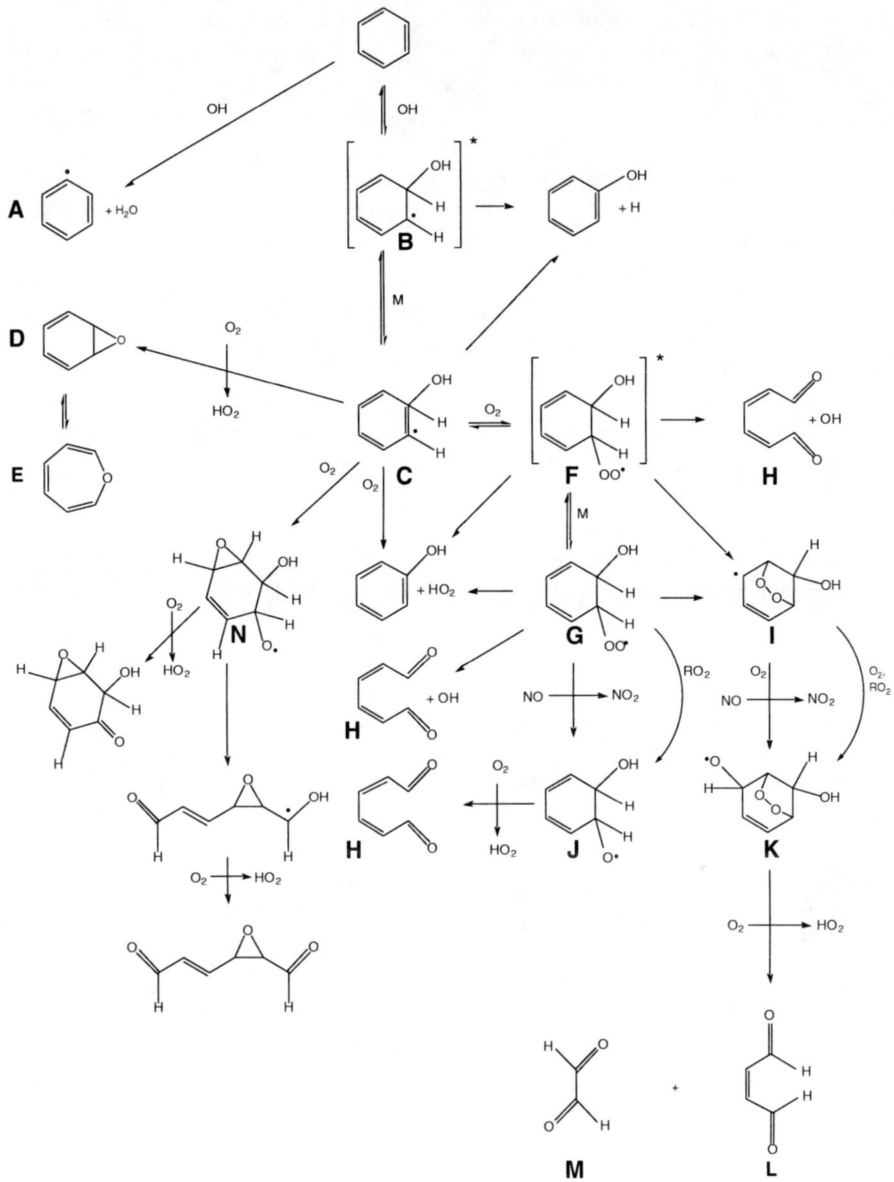

Figure II-E-1. Benzene oxidation mechanism.

The stabilized hydroxycyclohexadienyl radical (C) can undergo unimolecular dissociation to reform the reactants; unimolecular dissociation to give an H-atom and phenol; addition of O_2 to give an energized (denoted by *) peroxy radical (F); bimolecular reaction with O_2 to give phenol and an HO_2 radical; reaction with O_2 to give an epoxide alkoxy radical (N); or reaction with O_2 to give benzene oxide

(D), which is in a rapid equilibrium with oxepin (E) (equilibrium lifetime of 0.2 μs at room temperature (Klotz et al., 1999).

The importance of the arene oxide–forming channel in the reaction of OH with aromatics is uncertain. Klotz et al. (1997) have proposed that this reaction pathway may be significant and, depending on the different fates of the highly reactive benzene oxide/oxepin under different experimental conditions, may explain the considerable variation in the reported phenolic products and prompt HO_2 yields. In contrast, on the basis of a theoretical study, Ghigo and Tonachini (1998) concluded that the benzene oxide–forming channel is not competitive with the other possible fates of the adduct. Furthermore, in a recent experimental study, no evidence was observed for benzene oxide formation (< 1% yield) following reaction of OH-radicals with benzene in 13 and 100 mbar of O_2/He at 298 K (Berndt et al., 1999). Berndt et al. (1999) report a 30% yield of CO and a 13% yield of HC(=O)OH. It is difficult to understand the origin of these products; they may reflect the presence of unwanted secondary reactions that would cause an underestimation of the benzene oxide yield.

Information concerning the rate of decomposition of the adduct (C) to reform the reactants can be derived from the observed non-exponential decay of OH-radicals in flash photolysis kinetic experiments. Kinetic data are listed in table II-E-1 for the hydroxycyclohexadienyl radical and various substituted hydroxycyclohexadienyl radicals. As seen from table II-E-1 the $HO-C_6H_6$ adduct (C) has a lifetime of approximately 0.3 s at 298 K (Knispel et al., 1990) with respect to decomposition to $OH + C_6H_6$. Quantum mechanical calculations suggest that addition of O_2 to the hydroxycyclohexadienyl radical (C) to give a peroxy radical proceeds with a high-pressure limiting rate coefficient of 2×10^{-12} cm^3 $molecule^{-1}$ s^{-1} at 298 K (Lay et al., 1996). This calculated rate coefficient is 3–4 orders of magnitude greater than that derived from experimental studies (Knispel et al., 1990; Zetzsch and Bohn, 1999). The origin of this discrepancy is unclear at the present time.

The energized peroxy radical (F), $(HO-C_6H_6-O_2)^*$, can dissociate back to reactants, decompose to give phenol and HO_2, react via hydrogen transfer and subsequent β-scission to give 2,4-hexadiene-1,6-dial (H, muconaldehyde) + OH, isomerize to give four possible bicyclic peroxy adducts (the most stable of which is shown in figure II-E-1 as (I) (Lay et al., 1996), or undergo collisional deactivation to give the peroxy radical (G), $(HO-C_6H_6-O_2)$. The stablized peroxy radical $(HO-C_6H_6-O_2$ (G) can undergo the same reactions as the energized radical (see figure II-E-1). However, the dominant fate of the $HO-C_6H_6-O_2$ radical is decomposition to the $HO-C_6H_6$ radical and O_2 through the steps (G) → (F) → (C). Unlike aliphatic peroxy radicals, the C–OO bond strength in the $HO-C_6H_6-O_2$ radical (G) is low; a barrier of 11.4 kcal mol^{-1} has been calculated which leads to an estimated decomposition rate of 10^5 s^{-1} at 298 K (Lay et al., 1996).

At this point, it is interesting to contrast the different stabilities and fates of the $OH-C_6H_6$ adduct (C) and the $OH-C_6H_6-O_2$ peroxy radical (G). Thermal decomposition of the $OH-C_6H_6$ (C) adduct occurs at a rate of approximately 3 s^{-1} at 298 K (Knispel et al., 1990). In air at atmospheric pressure, this is 10^4 or 10^7 times slower than the pseudo-first-order loss rate with respect to addition of

O_2, depending on whether the value of Zetzsch and Bohn (1999) or Lay et al. (1996) is adopted for the O_2-addition rate coefficient. The rate coefficients of Zetzsch and coworkers (Knispel et al., 1990; Zetzsch et al., 1997) for O_2 addition to the OH–benzene, OH–toluene, and OH–*p*-xylene adducts are supported by the product study of Atkinson and Aschmann (1994) for *o*-xylene, and hence we recommend the Zetzsch and Bohn (1999) reaction rate coefficient. Under atmospheric conditions, thermal decomposition of the $OH-C_6H_6$ (C) adduct to give OH and C_6H_6 is of negligible importance. Thermal decomposition of the $OH-C_6H_6-O_2$ peroxy radical (G) to give the $OH-C_6H_6$ adduct (C) occurs at a rate estimated to be $10^5\,s^{-1}$ at 298 K (Lay et al., 1996) and is believed to be the dominant fate of the $OH-C_6H_6-O_2$ peroxy radical. Under atmospheric conditions, the two species (C) and (G) interconvert rapidly via the energized species (F). During each pass through the energized species (F) there is a small but significant probability of decomposition to give either phenol and HO_2, or 2,4-hexadiene-1,6-dial (H) and an OH-radical, or isomerization to give the bicyclic peroxy adduct such as structure (I). The repeated formation and dissociation of the benzene–OH–O_2 adduct, C → F → G → F → C, etc., amplifies the importance of decomposition/isomerization of species (F) and enables chemical activation effects to play a significant role in the atmospheric oxidation of benzene. The bicyclic radical (I) can add O_2 to give a peroxy radical (not shown in figure II-E-1) that can react with NO to produce an alkoxy radical [(K) in figure II-E-1], which ring opens by reaction with O_2 to give butenedial (L) and glyoxyl (M). As listed in table II-D-1, two products of the OH-radical–initiated oxidation of benzene have been observed in laboratory experiments conducted using low concentrations of NO_x: phenol ($\approx$25 %) and glyoxal (21%). It is unclear to what extent secondary reactions involving 2,4-hexadiene-1,6-dial (muconaldehyde) or benzene oxide/oxepin contribute to the measured glyoxal yield. Unfortunately, at the time of this writing, it is not possible to assign the relative importance of the various reactions given in figure II-E-1.

It should be noted that in addition to reaction with O_2, it is known that the hydroxycyclohexadienyl radical (C) in figure II-E-1 reacts with NO_2 (Zellner et al., 1985; Knispel et al., 1990; Zetzsch and Bohn, 1999) with a rate coefficient of the order of $(2–3) \times 10^{-11}$ cm^3 molecule^{-1} s^{-1} (see table II-E-2). This reaction may be significant in laboratory experiments where the $[NO_2]/[O_2]$ ratio can be relatively high, but is of little importance in the atmosphere where, even in heavily polluted air with $[NO_x] = 100$ ppb, the $[NO_x]/[O_2]$ ratio is only 5×10^{-7}, and $\sim$8% of the OH-benzene adduct will react with NO_2. With the exception of the results from Berndt et al. (1999), the product data given in table II-D-1 were obtained under conditions for which the OH–benzene reactions with NO_2 were significant. Hence, the product yield data in table II-D-1 may not be applicable to atmospheric conditions.

II-E-2. Toluene

As discussed in section II-D-1, toluene has been the most intensively studied of all the aromatic compounds. The available product data discussed in section II-D-2 show that only 6 ± 1% of the reaction proceeds via H-atom abstraction from the

$-CH_3$ group to give benzaldehyde and (in the presence of NO) benzyl nitrate, with the majority of the reaction proceeding to give an OH–aromatic adduct. The dominant aromatic ring-retaining product is *o*-cresol with a yield of $12 \pm 1\%$, followed by *m*-cresol and *p*-cresol—both of which have yields of approximately 3%. Nitrogen-containing aromatic ring-retaining products such as nitrotoluene and benzylnitrate (the latter probably from $C_6H_5CH_2O_2$ + NO reaction) have been reported as minor products with yields of the order of 1% (Atkinson et al., 1989a; Smith et al., 1998). Nitrotoluene products observed in laboratory experiments are believed to result from reaction of NO_2 with the OH–aromatic adduct (see table II-E-2) (Atkinson et al., 1989a). Hence, nitrotoluenes are only formed under laboratory conditions of high concentration of NO_2 and will not be significant in the atmosphere. Benzyl nitrate, $C_6H_5CH_2ONO_2$, formation is a minor channel in the reaction of the benzylperoxy radical, $C_6H_5CH_2O_2$, with NO, and will be formed in the atmospheric oxidation of toluene.

As shown in table II-D-2, a large number of ring-opened products have been identified as oxidation products following the OH-initiated oxidation of toluene. Glyoxal and methylglyoxal are by far the best characterized of the ring-opened products and have been reported in molar yields of up to 24% and 17%, respectively (Smith et al., 1998).

A proposed mechanism for OH-initiated oxidation of toluene based on the work of Andino et al. (1996), Lay et al. (1996), Klotz et al. (1997), and Kwok et al. (1997a) is given in figure II-E-2. As expected, this oxidation mechanism is very similar to that for benzene (figure II-E-1). Addition of OH to the aromatic ring in the *ortho* position (as shown by structure B) is slightly thermodynamically favored (by 1–2 $kcal\,mol^{-1}$) over addition to the *meta* and *para* positions (not shown in figure II-E-2). While there is no evidence for direct elimination of an H-atom from the adduct to form cresol, this possibility cannot be excluded and hence is included in figure II-E-2, as in the case of benzene (figure II-E-1). Under atmospheric conditions, reaction with O_2 proceeds approximately 2000 times faster than thermal decomposition of the OH–toluene adduct (Knispel et al., 1990). The major atmospheric fate of the OH–toluene adduct is addition of O_2, which proceeds with a pseudo-first order rate of $\sim 3000\,s^{-1}$ in air at atmospheric pressure. Andino et al. (1996) calculated that addition of O_2 to the OH–toluene adduct to give the peroxy radical (G) proceeds with a significant activation energy of approximately 5–10 $kcal\,mol^{-1}$ (depending on the exact reaction site). Such activation energies are consistent with the low rate coefficient of $5.4 \times 10^{-16}\,cm^3\,molecule^{-1}\,s^{-1}$ at room temperature reported by Knispel et al. (1990).

Lay et al. (1996) have presented theoretical evidence that decomposition of the chemically activated OH–benzene–O_2 adduct plays a significant role in the OH–benzene oxidation mechanism. It seems reasonable to suppose that chemical activation of the OH–toluene–O_2 adduct may play a role in the OH-initiated oxidation mechanism for toluene. This possibility and that of the formation of toluene oxide (H) (Klotz et al., 1998a) are included in the mechanism in figure II-E-2. There is some experimental evidence from product studies performed in the absence of NO_x (Barnes et al., 1991; Becker et al., 1992, 1993, 1997; Bierbach et al., 1993) that the ring-opening may proceed as well without the involvement of

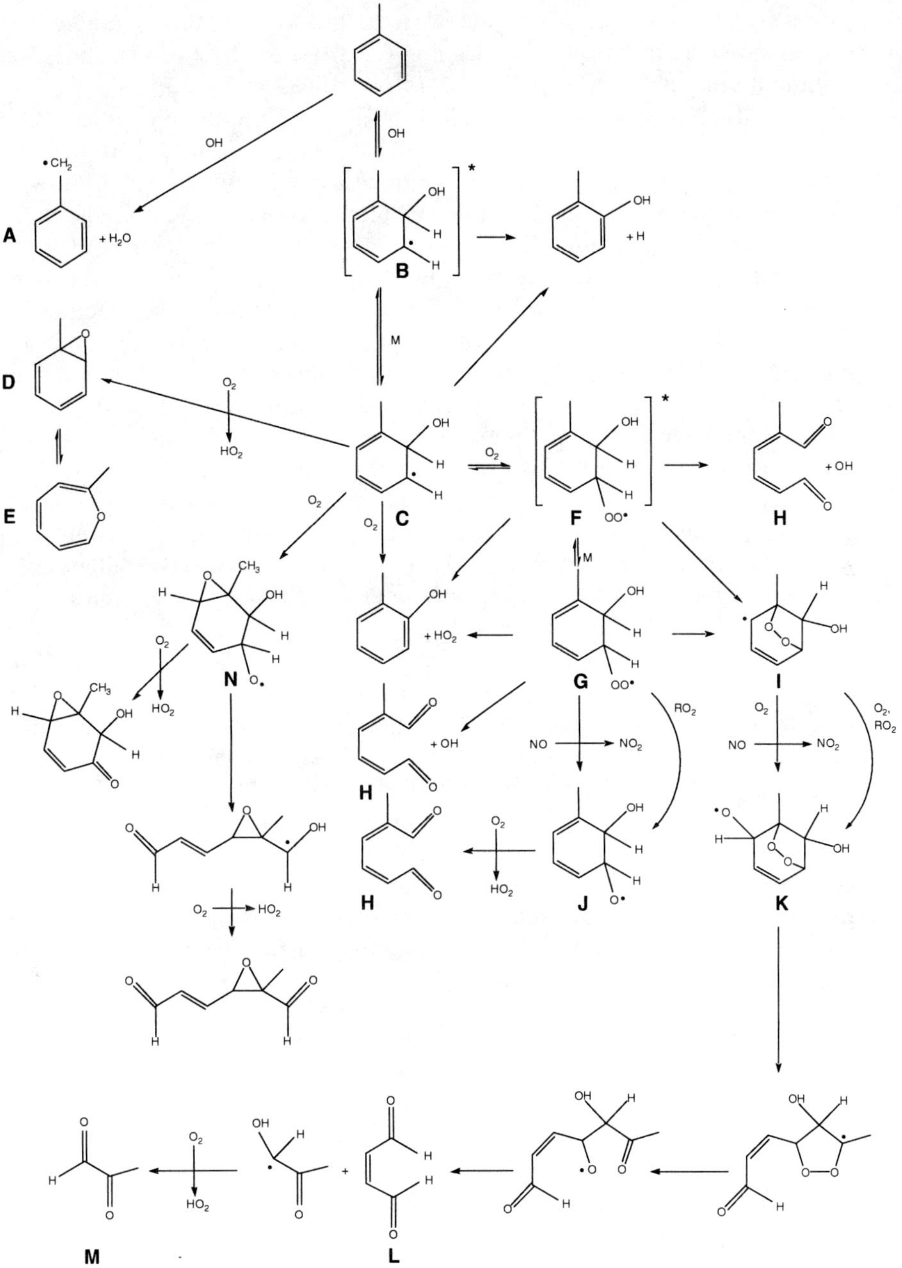

Figure II-E-2. Toluene oxidation mechanism.

NO; however, the extent of this is not clear at the time of this writing. This observation supports the mechanism leading from (C) to (L) and (M) in figure II-E-2 without reaction of NO with peroxy-type radicals.

Andino et al. (1996) considered two possible fates of the peroxy radical (G): cyclization and reaction with NO. For the peroxy radicals derived from toluene (and *m*-xylene, *p*-xylene, 1,2,4-trimethylbenzene, and *m*-ethyltoluene), it was shown that cyclization to form allylically stabilized five-membered bicyclic radicals (e.g., structure I in figure II-E-2) is *exothermic* by approximately 10 kcal mol^{-1}, whereas cyclization to give non-allylically stabilized bicyclic radicals is *endothermic* by 10 kcal mol^{-1} and not favored. In the case of the reaction of the peroxy radicals derived from toluene, their reaction with NO was found to be significantly endothermic and hence unlikely. However, it should be noted that similar reactions of NO with the peroxy radicals derived from *m*-xylene and 1,2,4-trimethylbenzene were substantially exothermic and should be considered in the mechanisms for such aromatics. The question of the relative importance of cyclization versus reaction with NO for peroxy radicals formed in the oxidation of aromatic compounds has to be considered on a case-by-case basis. For completeness, we have included the reaction of NO with the peroxy radical (G) in figure II-E-2. The bicyclic radical (I) adds O_2 rapidly to give a bicyclic peroxy radical (not shown in figure II-E-2), steric hindrance precludes another cyclization, and the bicyclic peroxy radical reacts instead with NO to give a bicyclic oxy radical (K). Fragmentation of the bicyclic oxy radical occurs via two β-scissions to give an acyclic radical, which then fragments via scission of the C–C bond which is not attached to the C=C double bond. In the case of the example in figure II-E-2, such fragmentation produces butenedial (L) and methyl glyoxal (M).

There are several possible bicyclic oxy radicals formed during the OH-radical–initiated oxidation of toluene. These radicals and their fragmentation products are given in figure II-E-3.

As listed in table II-D-2, a large number of products from the OH-initiated oxidation of toluene have been identified. The mechanistic scheme given in figure II-E-2 provides a framework to understand the chemical reactions that lead to the observed products. The mechanism by which a few of the products are formed is clear. For example, benzaldehyde and benzyl nitrate are produced subsequent to H-atom abstraction from the $-CH_3$ group by the following reactions involving the benzyl peroxy radical. The yields of benzyl nitrate from the reaction of benzyl peroxy radicals with NO corresponds to $k_a/(k_a + k_b) \approx 0.10$ (Hoshino et al., 1978; Gery et al., 1985; Atkinson et al., 1989a). In the reactions below, M is a third body that removes the energy liberated by bond formation.

$$C_6H_5CH_2. + O_2 + M \rightarrow C_6H_5CH_2O_2. + M$$

$$C_6H_5CH_2O_2. + NO \rightarrow C_6H_5CH_2O. + NO_2 \quad \text{(a)}$$

$$C_6H_5CH_2O_2. + NO + M \rightarrow C_6H_5CH_2ONO_2 + M \quad \text{(b)}$$

$$C_6H_5CH_2O. + O_2 \rightarrow C_6H_5C(=O)H + HO_2$$

Unfortunately, the observed toluene oxidation products account for only about 50% of the toluene loss, and the exact formation pathway of the majority of the observed products is unclear. For example, as indicated in figure II-E-2, there are

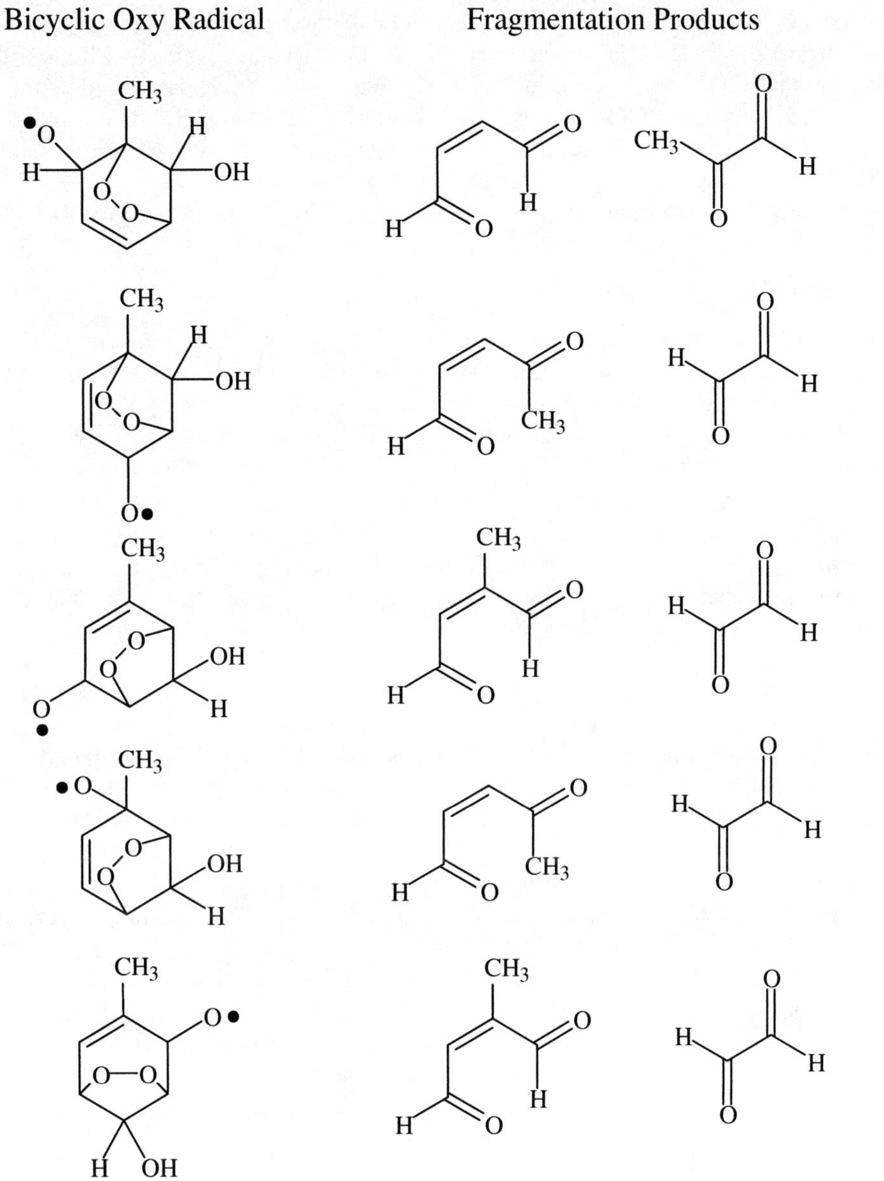

Figure II-E-3. Bicyclic oxy radicals formed in the atmospheric oxidation of toluene and their fragmentation products.

four possible pathways by which cresols can be formed: elimination of a hydrogen atom from the exited OH–toluene adduct (B), bimolecular reaction with O_2 with the stabilized OH–toluene adduct (C), elimination of an HO_2 radical from a chemically activated OH–toluene–O_2 radical (F), or HO_2 elimination from a stabilized OH–toluene–O_2 radical (G). The relative importance of these possible channels is unknown. Further experimental and theoretical work is needed to establish the relative importance of the various pathways given in figure II-E-2.

II-E-3. Xylenes and Trimethylbenzenes

As discussed in sections II-B and II-D, the kinetic and product data for the OH-radical–initiated oxidations of the xylenes and trimethylbenzenes show that such reactions proceed predominantly via addition to the aromatic ring. Kinetic arguments (see section II-B) suggest that hydrogen abstraction from the methyl substituents accounts for approximately 10%, 4%, and 7–8%, of the reactivity of *o*-, *m*-, and *p*-xylene, and 6–7%, 6–7% , and 3–4% for 1,2,3-, 1,2,4-, and 1,3,5-trimethylbenzene, respectively. Hydrogen abstraction from the methyl substituents after O_2 addition leads to the formation of methylbenzyl peroxy radicals, which, like other peroxy radicals, react with NO to give alkoxy radicals and organic nitrates (Wallington et al., 1992). The alkoxy radicals then react with O_2 to give tolualdehydes.

Under atmospheric conditions, the majority ($>90\%$) of the reaction of OH-radicals with the xylenes and trimethylbenzenes proceeds to give OH–aromatic adducts. These adducts react with both O_2 and NO_2. For the *p*-xylene–OH adduct the rate coefficient ratio k_{O2}/k_{NO2} is 2.5×10^{-5} at 330 K (Zetzsch et al., 1997), while for the *o*-xylene–OH adduct the rate coefficient ratio k_{O2}/k_{NO2} is 7.7×10^{-6} at 296 K (Atkinson and Aschmann, 1994). The $[O_2]/[NO_2]$ ratio in urban air is typically $(0.2–2.0) \times 10^7$ (assuming $[NO_2] = 10–100$ ppb), and hence reaction with O_2 is the dominant loss of the OH–aromatic adduct. The reactions occurring subsequent to formation of the OH–aromatic–O_2 peroxy radical are analogous to those described previously for the oxidation of benzene and toluene (see figures II-E-1 and II-E-2). As for benzene and toluene, our knowledge of the relative importance of the possible reaction pathways in the atmospheric oxidation of the xylenes and trimethylbenzenes is far from complete.

II-E-4. Styrene, α-Methylstyrene, and β-Methylstyrene

Reaction proceeds via addition of the OH-radical to the alkene double bond. No evidence of products following addition of OH to the aromatic ring of these compounds has been reported. In the case of styrene, addition of the OH-radical produces two isomeric hydroxy-substituted alkyl radicals, the sole fate of which is addition of O_2 to produce peroxy radicals. The addition of O_2 to alkyl radicals takes place with a rate coefficient that is typically of the order of $10^{-12}–10^{-11}$ cm^3 molecule^{-1} s^{-1}; and hence in the presence of air at atmospheric pressure, the lifetime of the alkyl radicals is of the order of $(0.2–2.0) \times 10^{-7}$ s. As for other alkyl peroxy radicals, in urban air, the two peroxy radicals formed following OH addition will react with NO and NO_2. Reaction with NO_2 gives

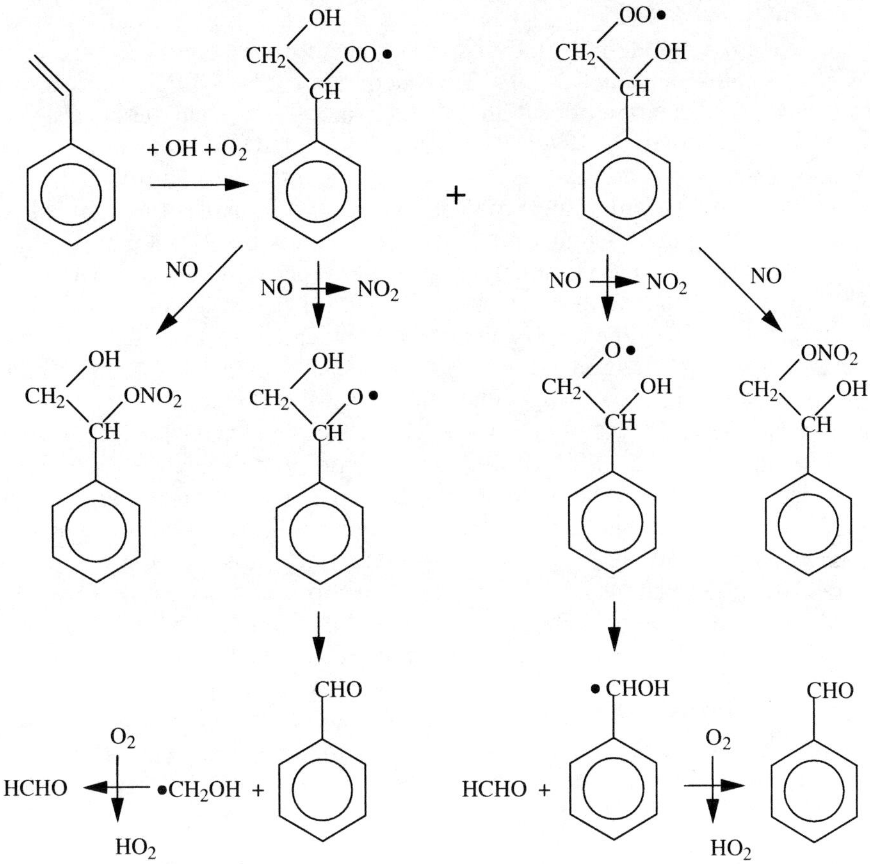

Figure II-E-4. Styrene oxidation mechanism.

a thermally unstable peroxynitrate that decomposes to reform the peroxy radical and NO_2 within a few seconds. Organic peroxy radicals react with NO via two channels to give alkoxy radicals and NO_2 as dominant products and organic nitrates as minor products. Evidence for the formation of organic nitrates in the oxidation of styrene was reported by Tuazon et al. (1993). The mechanism of the oxidation of styrene is given in figure II-E-4.

II-E-5. Benzaldehyde

As discussed above, the reaction of OH-radicals with benzaldehyde proceeds via abstraction of the aldehydic hydrogen atom. Under atmospheric conditions, the resulting C_6H_5CO radical has a lifetime of 3×10^{-8} s with respect to addition of O_2 to give the peroxybenzoyl radical, $C_6H_5C(=O)O_2$ (McDade et al., 1982). In polluted air masses, the fate of $C_6H_5C(=O)O_2$ radicals is reaction with NO_x (NO and NO_2). Typical NO_x concentrations in moderately polluted urban air are 1–10 ppbv. Using $k(C_6H_5C(=O)O_2 + NO) = 1.1 \times 10^{-11}$ and $k(C_6H_5C(=O)O_2 + NO_2)$

$= 1.6 \times 10^{-11}$ cm^3 molecule^{-1} s^{-1} (Caralp et al., 1999), it follows that in typical urban air masses the lifetime of $C_6H_5C(=O)O_2$ radicals with respect to reaction with NO_x is approximately 0.2–4.0 s. The reaction of $C_6H_5C(=O)O_2$ radicals with NO leads to the formation of CO_2 and, by inference, C_6H_5 (phenyl) radicals, with 100% yield (Caralp et al., 1999). Reaction of $C_6H_5C(=O)O_2$ radicals with NO_2 produces a peroxynitrate that decomposes with a rate coefficient of 2.1×10^{16} $e^{-13,600/T}$ s^{-1} (lifetime with respect to thermal decomposition of ~1 hour at 298 K, ~2 days at 273 K, and ~16 days at 263 K) (Caralp et al., 1999; Kirchner et al., 1999). Phenyl radicals add O_2 rapidly to give phenyl peroxy radicals (Yu and Lin, 1994) which are believed to react with NO to give phenoxy radicals (C_6H_5O) (Caralp et al., 1999). Phenoxy radicals react extremely slowly, if at all, with molecular oxygen (Platz et al., 1998), but they react rapidly with NO (Berho et al., 1998; Platz et al., 1998), NO_2 (Platz et al., 1998), and O_3 (Tao and Li, 1999). Using $k(NO) = 1.65 \times 10^{-12}$, $k(NO_2) = 2.1 \times 10^{-12}$, and $k(O_3) = 2.86 \times 10^{-13}$ cm^3 molecule^{-1} s^{-1}, and with NO_x and O_3 concentrations in moderately polluted urban air of 1–10 ppbv and 0–100 ppb, respectively, it follows that these reactions are competitive loss mechanisms for phenoxy radicals. The reactions of phenoxy radicals with NO and NO_2 leads to the formation of nitroso- and nitro-phenols (Niki et al., 1979; Platz et al., 1998). These reactions remove both NO_x and a radical species and are probably responsible for the well-known negative incremental reactivity of benzaldehyde (i.e., addition of benzaldehyde to a hydrocarbon–NO_x–air mixture decreases ozone formation) (Kuntz et al., 1973; Carter and Atkinson, 1987). The mechanism of the reaction of phenoxy radicals with ozone is unknown. The mechanism of benzaldehyde oxidation is given in figure II-E-5.

II-F. Rate Coefficients for Gas-Phase Reactions of Polycyclic Aromatic Hydrocarbons (PAHs) with OH Radicals

The polycyclic aromatic hydrocarbon compounds (PAHs) are characterized by the presence of more than one benzene-like ring structure fused into the molecule. Certain of these molecules partition between the gas and particle phases, and different mechanisms dominate the reactions of PAHs in these two phases. The gas-phase reactions of the OH-radical can be important during the daytime for the more volatile PAHs; namely, naphthalene, acenaphthylene, acenaphthene, fluorene, phenanthrene, anthracene, fluoranthene, and pyrene (Atkinson et al., 1987a, 1989a, 1990a; Atkinson and Arey, 1994). The structures of the PAHs are described in chapter IX and in appendix I-B. In this section, we evaluate the rate coefficients for OH reaction with the limited number of PAH compounds that have been studied at the time of this writing. In section II-G the products and mechanisms of some of these gas-phase reactions are discussed, and in section II-H we discuss some possible empirical methods of estimation of rate coefficients for the PAHs as yet unmeasured. In chapter IX the reactions of the PAHs sorbed on aerosols are described.

Studies of reactions of gas-phase PAHs with OH, NO_3, and O_3 have been reported by Atkinson et al. (1984a, 1987a, 1990a, 1990b), Atkinson and

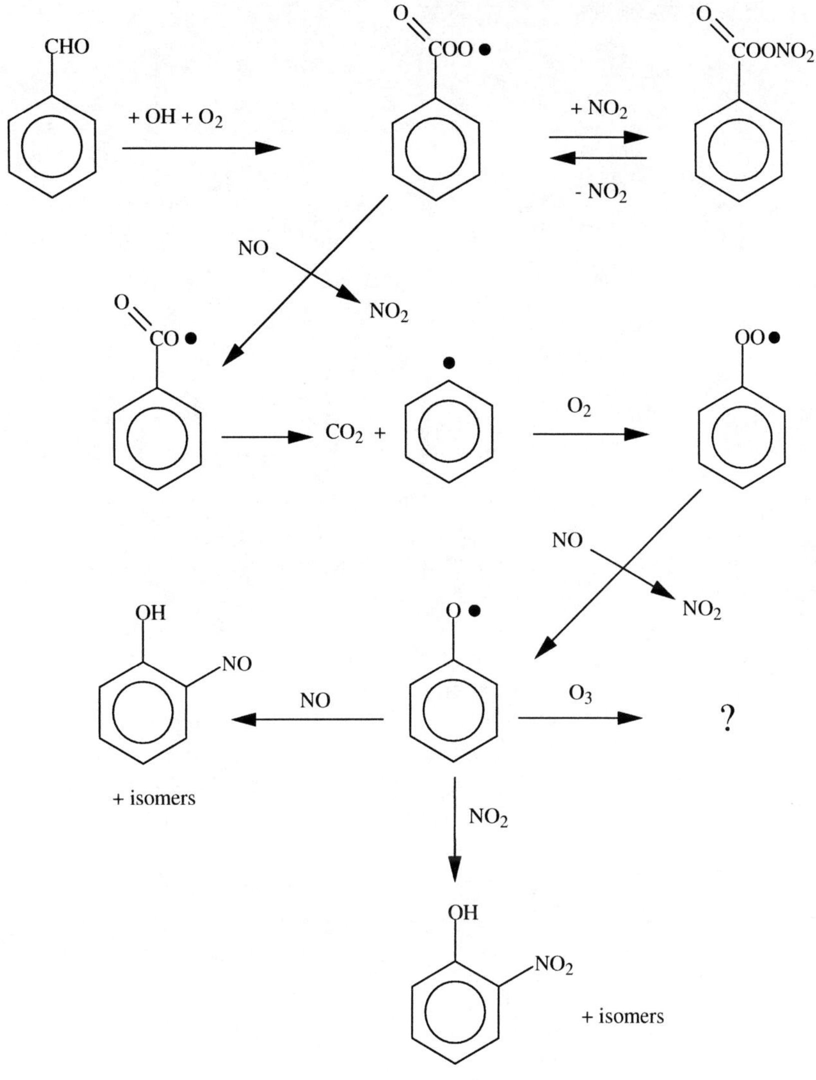

Figure II-E-5. Benzaldehyde oxidation mechanism.

Aschmann (1986, 1988), Arey et al. (1989a, 1989b, 1990), Atkinson (1989, 1991b), Helmig et al. (1992a), Atkinson and Arey (1994), Bunce et al. (1997), and Sasaki et al. (1997). Attack of OH-radical on PAHs parallels the OH-radical reactions with benzene or toluene (Doyle et al., 1975; Perry et al., 1977a; Kenley et al., 1978, 1981; Atkinson et al., 1980, 1987a, 1989a; Dumdei and O'Brien, 1984).

Naphthalene is the first member of the series of fused-ring structures that constitute the PAHs compounds. Structures of many of the PAHs are shown in chapter IX and appendix I-B, where the reactions of these compounds sorbed on aerosols are discussed. In this section, only the gas-phase reactions are considered.

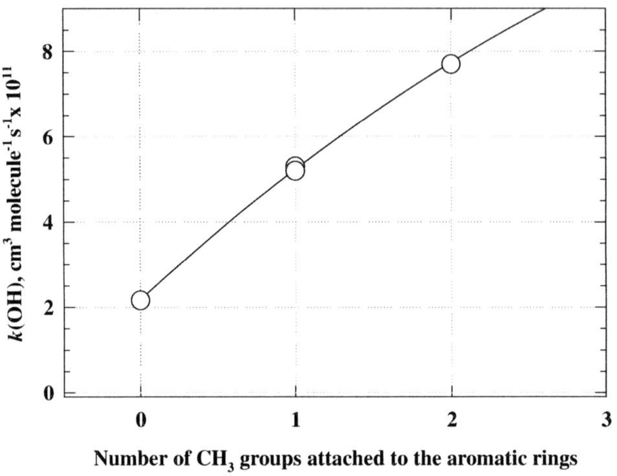

Figure II-F-1. Plot of the rate coefficient for OH-radical reaction with methyl-substituted naphthalenes as a function of the number of CH_3-groups attached to the aromatic rings; data are for naphthalene, 1-, and 2-methylnaphthalene, and 2,3-dimethylnaphthalene.

As with benzene and the alkyl-substituted benzenes described in sections II-A to II-E, the major first step in the OH-radical reaction with the PAHs is the addition to the aromatic rings. For alkyl-substituted naphthalenes, OH-radical attack on the aromatic ring is more important than hydrogen abstraction of an alkyl hydrogen (Atkinson, 1989; Atkinson and Arey, 1994). As the extent of methyl-substitution on naphthalene is increased, an increase in rate coefficient is observed, an effect similar to that observed with benzene (figure II-A-2). This is illustrated for the methyl-substituted naphthalenes in figure II-F-1. Inductive effects of the alkyl groups provide a higher electron density on the *ortho*- and *para*-carbons, which most probably leads to increased rates of reaction of alkyl naphthalenes over unsubstituted naphthalene (Atkinson and Arey, 1994).

Exceptions to the dominance of the mechanism of OH addition to the aromatic ring occur when an olefinic double bond is attached to the PAHs rings. The initial attack of the OH-radical in these cases is on the $>C{=}C<$ group. This is observed in the reactivity of two PAHs, acenaphthene (I) and acenaphthylene (II):

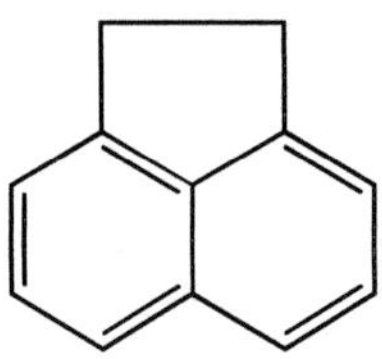

(I)

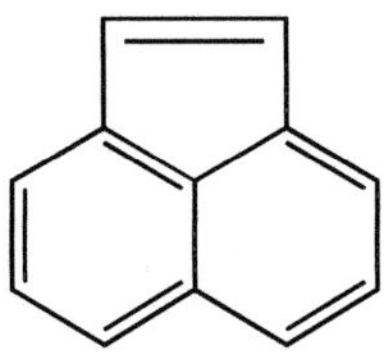

(II)

The alkyl two-carbon alkane-like bridge fused to the naphthalene structure changes the rate coefficient of acenaphthene (I) (6–10 × 10^{-11} cm^3 $molecule^{-1}$ s^{-1}) from that of naphthalene (2.3 × 10^{-11} cm^3 $molecule^{-1}$ s^{-1}). As with the simple alkyl-substituted benzenes, the attack of OH on the aromatic ring system of the PAHs is probably more important that the abstraction of H-atoms from alkyl group substituents. Acenaphthylene (II), however, has a fused alkene-like double-bonded carbon unit that increases its reactivity toward OH significantly (1.1×10^{-10} cm^3 molecule^{1-} s^{-1}) over that for the unsubstituted naphthalene. In this case, the reaction of the OH-radical probably also occurs by addition to the $-CH{=}CH-$ group attached to the naphthalene-like structure.

In the remaining portion of this section, the rate coefficients for the existing data for the OH-radical reactions with the PAHs are evaluated. The format of the review is the same as that used in section II-B. Abbreviations given to describe the techniques employed in the individual measurements are given in table II-B-0.

II-F-1. OH + Naphthalene

The rate coefficients of Lorenz and Zellner (1983), Atkinson et al. (1984a), Biermann et al. (1985), Atkinson and Aschmann (1986), Klöpffer et al. (1986), and Brubaker and Hites (1998b) are given in table II-F-1. The room temperature rate coefficients measured by Atkinson et al. (1984a), Biermann et al. (1985), Atkinson and Aschmann (1986), Klöpffer et al. (1986), and Brubaker and Hites (1998b) are in excellent agreement. However, at elevated temperatures the rate coefficients of Lorenz and Zellner (1983) and Brubaker and Hites (1998b) disagree significantly, with the disagreement increasing with increasing temperature (those of Lorenz and Zellner (1983) show a negative temperature dependence and those of Brubaker and Hites (1998b) are essentially independent of temperature). A least-squares analysis of the rate coefficients of Atkinson et al. (1984a), Biermann et al. (1985), Atkinson and Aschmann (1986), and Brubaker and Hites (1998b) leads to the recommendation of

$$k(\text{naphthalene}) = 1.56 \times 10^{-11}\, e^{(117\pm127)/T}\ \text{cm}^3\ \text{molecule}^{-1}\ \text{s}^{-1}$$

over the temperature range 294–366 K, where the indicated errors are two least-squares standard deviations, and

$$k(\text{naphthalene}) = 2.30 \times 10^{-11}\ \text{cm}^3\ \text{molecule}^{-1}\ \text{s}^{-1}$$

at 298 K, with an estimated overall uncertainty at 298 K of ±25%.

II-F-2. OH + 1-Methylnaphthalene

The rate coefficient of Atkinson and Aschmann (1987), obtained from a relative rate study at room temperature and atmospheric pressure of air, is given in table II-F-2. This is the only reported study of this reaction.

II-F-3. OH + 2-Methylnaphthalene

The rate coefficient of Atkinson and Aschmann (1986), obtained from a relative rate study at room temperature and atmospheric pressure of air, is given in table II-F-3. This is the only reported study of this reaction.

Table II-F-1. $OH + naphthalene \rightarrow products$:

Rate coefficients k at, or close to, the high-pressure limit at $T < 410$ K.

$10^{12} \times A$ (cm^3 molecule^{-1} s^{-1})	B (K)	$10^{12} \times k$ (cm^3 molecule^{-1} s^{-1})	at T (K)	Technique	Reference	Temperature Range Covered (K)
		14.6 ± 5.0	337	FP-RF	Lorenz and Zellner (1983)	337–525
		11.0 ± 4.4	358			
		10.1 ± 4.0	378 ± 2			
2.2	−440 ± 50	10.5 ± 4.0	407			
		22.4 ± 1.5	294 ± 1	RR [relative to k(n-nonane) = 9.92×10^{-12}][a]	Atkinson et al. (1984a)	
		23.5 ± 0.6	298 ± 1	RR [relative to k(propene) = 2.63×10^{-11}][a]	Biermann et al. (1985)	
		25.9 ± 2.4	295 ± 1	RR [relative to k(isoprene) = 1.02×10^{-10}][a]	Atkinson and Aschmann (1986)	
		21.6	300	RR [relative to k(ethene) = 8.44×10^{-12}][a]	Klöpffer et al. (1986)	
		22.1 ± 0.3	306	RR [relative to k(cyclohexane) = $2.88 \times 10^{-17} T^2 e^{309/T}$][a]	Brubaker and Hites (1998b)	306–366
		21.1 ± 0.4	306			
		22.4 ± 0.2	307			
		21.7 ± 0.1	336			
		21.9 ± 0.1	336			
		23.3 ± 0.3	366			
		22.4 ± 0.4	366			
		23.8 ± 0.3	306	RR [relative to k(2-propanol) = $4.06 \times 10^{-18} T^2 e^{788/T}$][b]	Brubaker and Hites (1998b)	306–366
		21.7 ± 0.9	306			
		23.8 ± 0.1	336			
		20.3 ± 0.1	366			
		19.8 ± 0.5	366			

[a] From Atkinson (1997).
[b] From Atkinson et al.(1999).

Table II-F-2. OH + 1-methylnaphthalene → products:
Rate coefficients k at, or close to, the high-pressure limit

$10^{12} \times k$ (cm^3 $molecule^{-1}$ s^{-1})	at T (K)	Technique	Reference
53.0 ± 4.8	298 ± 2	RR [relative to k(isoprene) = 1.01×10^{-10}][a]	Atkinson and Aschmann (1987)

[a] From Atkinson (1997).

Table II-F-3. OH + 2-methylnaphthalene → products:
Rate coefficients k at, or close to, the high-pressure limit.

$10^{12} \times k$ (cm^3 $molecule^{-1}$ s^{-1})	at T (K)	Technique	Reference
52.3 ± 4.2	295 ± 2	RR [relative to k(isoprene) = 1.02×10^{-10}][a]	Atkinson and Aschmann (1986)

[a] From Atkinson (1997).

II-F-4. OH + 2,3-Dimethylnaphthalene

The rate coefficient of Atkinson and Aschmann (1986), obtained from a relative rate study at room temperature and atmospheric pressure of air, is given in table II-F-4. This is the only reported study of this reaction.

Table II-F-4. OH + 2,3-dimethylnaphthalene → products:
Rate coefficients k at, or close to, the high-pressure limit.

$10^{12} \times k$ (cm^3 $molecule^{-1}$ s^{-1})	at T (K)	Technique	Reference
76.8 ± 4.8	295 ± 2	RR [relative to k(isoprene) = 1.02×10^{-10}][a]	Atkinson and Aschmann (1986)

[a] From Atkinson (1997).

II-F-5. OH + Biphenyl

The rate coefficients of Zetzsch (1982b), Atkinson et al. (1984a), Atkinson and Aschmann (1985), Klöpffer et al. (1986), and Anderson and Hites (1996a) are given in table II-F-5. The room temperature rate coefficients measured by Zetzsch (1982b), Atkinson et al. (1984a), Atkinson and Aschmann (1985), Klöpffer et al. (1986), and Anderson and Hites (1996a) are in reasonable agreement. Based on the rate coefficients of Zetzsch (1982b), Atkinson et al. (1984a), and Atkinson and Aschmann (1985), we recommend that

$$k(\text{biphenyl}) = 7.1 \times 10^{-12}\ \text{cm}^3\ \text{molecule}^{-1}\ \text{s}^{-1}$$

Table II-F-5. OH + biphenyl → products:

Rate coefficients k at, or close to, the high-pressure limit.

$10^{12} \times k$ (cm^3 $molecule^{-1}$ s^{-1})	at T (K)	Technique	Reference	Temperature Range Covered (K)
5.8 ± 0.8	296	FP-RF	Zetzsch (1982b)	
7.47 ± 0.66	294 ± 1	RR [relative to k(n-nonane) = 9.92×10^{-12}][a]	Atkinson et al. (1984a)	
8.00 ± 0.72	295 ± 1	RR [relative to k(cyclohexane) = 7.14×10^{-12}][a]	Atkinson and Aschmann (1985)	
8.0	300	RR [relative to k(ethene) = 8.44×10^{-12}][a]	Klöpffer et al. (1986)	
6.41 ± 0.32	323	RR [relative to k(cyclohexane) = 2.88×10^{-17} T^2 $e^{309/T}$][a]	Anderson and Hites (1996a)	323–356
6.67 ± 2.05	324			
6.52 ± 0.32	324			
5.83 ± 0.83	338			
6.81 ± 0.50	338			
7.55 ± 0.33	338			
7.21 ± 0.34	348			
6.69 ± 0.35	356			
6.26 ± 0.18	356			
6.43 ± 0.35	356			
7.47 ± 0.53	356			
6.90 ± 0.35	323	RR [relative to k(2-methylpropane) = 1.16×10^{-17} T^2 $e^{225/T}$][a]	Anderson and Hites (1996a)	323–343
6.22 ± 0.25	323			
6.23 ± 0.64	343			
6.29 ± 0.74	343			
6.10 ± 0.40	343			
6.44 ± 0.43	343			
6.39 ± 1.00	343			

[a] From Atkinson (1997).

at 298 K, with an estimated overall uncertainty of ±30%. The rate coefficients measured by Anderson and Hites (1996a) over the restricted temperature range of 323–356 K do not show a significant temperature dependence (a unit-weighted least-squares analysis yields a value of B in $k = A\,e^{-B/T}$ of $B = 69 \pm 323$ K, where the indicated error is two least-squares standard deviations). A unit-weighted average of all of the Anderson and Hites (1996a) rate coefficients leads to $k(\text{biphenyl}) = 6.6 \times 10^{-12}$ cm^3 $molecule^{-1}$ s^{-1} over the 323–356 K temperature range, consistent with our 298 K recommendation. We therefore tentatively recommend a temperature-independent rate coefficient of

$$k(\text{biphenyl}) = 7.1 \times 10^{-12}\ \text{cm}^3\ \text{molecule}^{-1}\ \text{s}^{-1}$$

over the temperature range 294–356 K, and with an estimated overall uncertainty of ±30% over this temperature range.

II-F-6. OH + Fluorene

The rate coefficients of Klöpffer et al. (1986), Kwok et al. (1997b), and Brubaker and Hites (1998b) are given in table II-F-6. Over the restricted temperature range of 326–366 K the rate coefficients measured by Brubaker and Hites (1998b) are independent of temperature, and they are in good agreement with the room temperature rate coefficients of Klöpffer et al. (1986) and Kwok et al. (1997b). Accordingly, a unit-weighted average of the rate coefficients of Klöpffer et al. (1986), Kwok et al. (1997b), and Brubaker and Hites (1998b) leads to the recommendation of

$$k(\text{fluorene}) = 1.4 \times 10^{-11}\ \text{cm}^3\ \text{molecule}^{-1}\ \text{s}^{-1}$$

independent of temperature over the range 297–366 K, and with an estimated overall uncertainty of a factor of 1.5 over this temperature range.

Table II-F-6. OH + fluorene → products: Rate coefficients k at, or close to, the high-pressure limit

$10^{12} \times k$ (cm^3 molecule^{-1} s^{-1})	at T (K)	Technique	Reference	Temperature Range Covered (K)
13.0	300	RR [relative to k(ethene) = 8.44×10^{-12}][a]	Klöpffer et al. (1986)	
15.2 ± 1.8	297 ± 2	RR [relative to k(cyclohexane) = 7.19×10^{-12}][a]	Kwok et al. (1997b)	
13.7 ± 0.3	326	RR [relative to k(cyclohexane) = 2.88×10^{-17} T^2 $e^{309/T}$][a]	Brubaker and Hites (1998b)	326–366
13.5 ± 0.3	326			
12.2 ± 0.2	326			
13.7 ± 0.1	346			
13.5 ± 0.1	346			
13.9 ± 0.1	366			
14.2 ± 0.1	366			

[a] From Atkinson (1997).

II-F-7. OH + Acenaphthene

The rate coefficients of Klöpffer et al. (1986), Atkinson and Aschmann (1988), and Brubaker and Hites (1998b) are given in table II-F-7. Clearly, the results of these three studies are not in agreement. Note also that Klöpffer et al. (1986) indicated that their rate coefficient was subject to significant uncertainties because of wall losses and scatter of the analytical data. In the Brubaker and Hites (1998b) study the reference compounds used (cyclohexane and toluene) were observed to be a factor of 7–13 less reactive than acenaphthene, resulting in very disparate fractions of the acenaphthene and reference compound being consumed during the experiments and possibly leading to larger uncertainties in the rate coefficient ratio than reported. In the absence of confirmatory data, we

Table II-F-7. OH + acenaphthene → products:

Rate coefficients k at, or close to, the high-pressure limit

$10^{12} \times k$ (cm^3 $molecule^{-1}$ s^{-1})	at T (K)	Technique	Reference	Temperature Range Covered (K)
58.4	300	RR [relative to k(ethene) $= 8.44 \times 10^{-12}$][a]	Klöpffer et al. (1986)	
102 ± 12	296 ± 2	RR [relative to k(2,3-dimethyl-2-butene) $= 1.10 \times 10^{-10}$][a]	Atkinson and Aschmann (1988)	
59.8 ± 2.4	325	RR [relative to k(cyclohexane) $= 2.88 \times 10^{-17}\ T^2\ e^{309/T}$][a]	Brubaker and Hites (1998b)	325–366
60.6 ± 1.6	325			
60.4 ± 1.7	345			
59.8 ± 1.7	346			
60.9 ± 0.7	365			
69.2 ± 0.7	366			
56.8 ± 6.2	325	RR [relative to k(toluene) $= 1.81 \times 10^{-12}\ e^{338/T}$][b]	Brubaker and Hites (1998b)	325–365
55.9 ± 1.0	345			
58.5 ± 1.0	365			

[a] From Atkinson (1997).
[b] This evaluation.

use the average of the rate coefficient measured by Atkinson and Aschmann (1988) and the average 325–366 K rate coefficient of Brubaker and Hites (1998b) to recommend that

$$k(\text{acenaphthene}) = 8 \times 10^{-11}\ \text{cm}^3\ \text{molecule}^{-1}\ \text{s}^{-1}$$

at 298 K, with an estimated overall uncertainty of a factor of 1.5.

II-F-8. OH + Acenaphthylene

The rate coefficient of Atkinson and Aschmann (1988), obtained from a relative rate study at room temperature and atmospheric pressure of air, is given in table II-F-8. This is the only reported study of this reaction.

Table II-F-8. OH + acenaphthylene → products:

Rate coefficients k at, or close to, the high-pressure limit

$10^{12} \times k$ (cm^3 $molecule^{-1}$ s^{-1})	at T (K)	Technique	Reference
109 ± 11	296 ± 2	RR [relative to k(2,3-dimethyl-2-butene) $= 1.10 \times 10^{-10}$][a]	Atkinson and Aschmann (1988)

[a] From Atkinson (1997).

II-F-9. OH + Phenanthrene

The rate coefficients of Biermann et al. (1985), Kwok et al. (1994c), and Brubaker and Hites (1998b) are given in table II-F-9. Clearly, the results of these three studies are not in agreement; note that significant wall losses were observed in the Biermann et al. (1985) study, and these were taken into account in the data analysis. The rate coefficients measured by Brubaker and Hites (1998b) over the restricted temperature range of 346–386 K are essentially independent of temperature; a unit-weighted least-squares analysis yields a value of B in $k = A\,e^{-B/T}$ of $B = -169 \pm 331$ K, where the indicated error is two least-squares standard deviations. We take an average of the 346–386 K rate coefficients of Brubaker and Hites (1998b) and the room temperature rate coefficient of Kwok et al. (1994c) to tentatively recommend that

$$k(\text{phenanthrene}) = 1.8 \times 10^{-11}\ \text{cm}^3\ \text{molecule}^{-1}\ \text{s}^{-1}$$

independent of temperature over the range 297–386 K, and with an estimated overall uncertainty of a factor of 2 over this temperature range.

Table II-F-9. OH + phenanthrene → products:

Rate coefficients k at, or close to, the high-pressure limit

$10^{12} \times k$ (cm^3 molecule^{-1} s^{-1})	at T (K)	Technique	Reference	Temperature Range Covered (K)
33.9 ± 11.6	298 ± 1	RR [relative to k(propene) $= 4.85 \times 10^{-12}\,e^{504/T}$][a]	Biermann et al. (1985)	298–319
28.2 ± 5.5	319 ± 1			
12.7 ± 1.3	296 ± 2	RR [relative to k(propene) $= 2.66 \times 10^{-11}$][a]	Kwok et al. (1994c)	
24.3 ± 0.6	346	RR [relative to k(cyclohexane) $= 2.88 \times 10^{-17}\,T^2\,e^{309/T}$][a]	Brubaker and Hites (1998b)	346–386
26.1 ± 0.9	346			
24.0 ± 0.8	346			
23.2 ± 0.6	365			
22.9 ± 0.4	365			
21.3 ± 0.2	366			
21.9 ± 0.2	366			
25.0 ± 0.4	366			
23.0 ± 0.2	366			
25.9 ± 0.2	375			
25.2 ± 0.3	375			
23.5 ± 0.3	375			
23.9 ± 0.4	376			
23.6 ± 0.3	376			
24.3 ± 0.2	384			
25.1 ± 0.2	385			
20.4 ± 0.4	385			
24.6 ± 0.2	386			
21.0 ± 0.4	386			

[a] From Atkinson (1997).

II-F-10. OH + Anthracene

The rate coefficients of Biermann et al. (1985), Kwok et al. (1994c, 1997b), and Brubaker and Hites (1998b) are given in table II-F-10. Clearly, the results of these three studies are not in agreement. It should be noted that (a) significant wall losses were observed in the Biermann et al. (1985) study, and these were taken into account in the data analysis; (b) the Kwok et al. (1994c) and (1997b) studies were both very limited and consisted of two and three data points, respectively; and (c) the Brubaker and Hites (1998b) study employed acenaphthene as the reference compound, and there are significant uncertainties associated with the rate coefficient for the reaction of the OH-radical with acenaphthene (see section II-F-7). Use of the rate coefficient measured by Brubaker and Hites (1998b) would result in a rate coefficient for the anthracene reaction at 346–365 K of (1.6–1.7) $\times 10^{-10}$ cm^3 molecule^{-1} s^{-1}. Given the order of magnitude discrepancy in the reported rate coefficients, no recommendation is made.

Table II-F-10. OH + anthracene → products:
Rate coefficients k at, or close to, the high-pressure limit

$10^{12} \times k$ (cm^3 molecule^{-1} s^{-1})	at T (K)	Technique	Reference	Temperature Range Covered (K)
112 ± 9	325 ± 1	RR [relative to k(propene) = 2.29×10^{-11}][a]	Biermann et al. (1985)	298–319
18 ± 9	296 ± 2	RR [relative to k(phenanthrene) = 1.8×10^{-11}][b]	Kwok et al. (1994c)	
18.6 ± 3.5	297 ± 2	RR [relative to k(propene) = 2.65×10^{-11}][a]	Kwok et al. (1997b)	
231 ± 43[c]	346–365	RR [relative to k(acenaphthene) = 8×10^{-11}][b]	Brubaker and Hites (1998b)	346–365

[a] From Atkinson (1997).
[b] This evaluation.
[c] Average value, indicated errors are two standard deviations. See text.

II-F-11. OH + Fluoranthene

The rate coefficients of Brubaker and Hites (1998b) are given in table II-F-11. This is the sole reported study of this reaction.

II-F-12. OH + Dibenzo-*p*-dioxin

The rate coefficients of Kwok et al. (1994a) and Brubaker and Hites (1998b) are given in table II-F-12. The rate coefficients of Brubaker and Hites (1998b) are

Table II-F-11. OH + fluoranthene → products: Rate coefficients k at, or close to, the high-pressure limit

$10^{12} \times k$ (cm^3 molecule^{-1} s^{-1})	at T (K)	Technique	Reference	Temperature Range Covered (K)
17.2 ± 0.5	356	RR [relative to k(cyclohexane) = $2.88 \times 10^{-17}\ T^2\ e^{309/T}$][a]	Brubaker and Hites (1998b)	356–386
17.4 ± 0.5	366			
19.7 ± 0.3	366			
20.3 ± 0.7	376			
24.7 ± 0.3	386			
16.3 ± 0.2	376	RR [relative to k(2-propanol) = $4.06 \times 10^{-18}\ T^2 e^{788/T}$][b]	Brubaker and Hites (1998b)	376–386
13.8 ± 0.2	376			
17.2 ± 0.3	386			
19.1 ± 1.0	386			
18.0 ± 0.4	356	RR [relative to k(toluene) = $1.81 \times 10^{-12}\ e^{338/T}$][c]	Brubaker and Hites (1998b)	356–366
18.3 ± 0.4	366			

[a] From Atkinson (1997).
[b] From Atkinson et al.(1999).
[c] This evaluation.

Table II-F-12. OH + dibenzo-*p*-dioxin → products: Rate coefficients k at, or close to, the high-pressure limit

$10^{12} \times k$ (cm^3 molecule^{-1} s^{-1})	at T (K)	Technique	Reference	Temperature Range Covered (K)
13.7 ± 2.5	297 ± 2	RR [relative to k(cyclohexane) = 7.19×10^{-12}][a]	Kwok et al. (1994a)	
15.1 ± 1.6	297 ± 2	RR [relative to k(propene) = 2.65×10^{-11}][a]	Kwok et al. (1994a)	
11.9 ± 0.1	345	RR [relative to k(cyclohexane) = $2.88 \times 10^{-17}\ T^2\ e^{309/T}$][a]	Brubaker and Hites (1998b)	345–385
11.9 ± 0.1	345			
11.9 ± 0.1	345			
12.2 ± 0.1	345			
11.5 ± 0.1	365			
11.6 ± 0.1	365			
12.2 ± 0.1	365			
12.2 ± 0.1	365			
11.6 ± 0.1	385			
12.4 ± 0.1	385			
11.9 ± 0.1	385			
12.2 ± 0.1	385			

[a] From Atkinson (1997).

independent of temperature over the restricted temperature range of 345–385 K, and agree reasonably well with the room temperature rate coefficients of Kwok et al. (1994a). A mean of the two measurements of Kwok et al. (1994a) and the averaged 345–385 K rate coefficient of Brubaker and Hites (1998b) leads to the recommendation of

$$k(\text{dibenzo-}p\text{-dioxin}) = 1.4 \times 10^{-11}\ \text{cm}^3\ \text{molecule}^{-1}\ \text{s}^{-1}$$

independent of temperature over the range 297–385 K, and with an estimated overall uncertainty of a factor of 1.5 over this temperature range.

II-F-13. OH + Dibenzofuran

The rate coefficients of Kwok et al. (1994a) and Brubaker and Hites (1998b) are given in table II-F-13. The rate coefficients of Brubaker and Hites (1998b) increase with increasing temperature, with a unit-weighted least-squares analysis leading to $k = 2.33 \times 10^{-11}\ e^{-(572\pm269)/T}$ cm^3 molecule^{-1} s^{-1} over the restricted temperature range 345–386 K and with $k = 3.4 \times 10^{-12}$ cm^3 molecule^{-1} s^{-1} at 298 K; this in good agreement with the room temperature rate coefficients of Kwok et al. (1994a). Accordingly, a unit-weighted least-squares analysis of the rate coefficients of Kwok et al. (1994a) and Brubaker and Hites (1998b) leads to the recommendation of

$$k(\text{dibenzofuran}) = 1.48 \times 10^{-11}\ e^{-(404\pm126)/T}\ \text{cm}^3\ \text{molecule}^{-1}\ \text{s}^{-1}$$

Table II-F-13. OH + dibenzofuran → products:
Rate coefficients k at, or close to, the high-pressure limit

$10^{12} \times k$ (cm^3 molecule^{-1} s^{-1})	at T (K)	Technique	Reference	Temperature Range Covered (K)
3.90 ± 0.82	297 ± 2	RR [relative to k(cyclo-hexane) = 7.19×10^{-12}][a]	Kwok et al. (1994a)	
3.87 ± 0.56	297 ± 2	RR [relative to k(propene) = 2.65×10^{-11}][a]	Kwok et al. (1994a)	
4.11 ± 0.09	345	RR [relative to k(cyclo-hexane) = $2.88 \times 10^{-17}\ T^2\ e^{309/T}$][a]	Brubaker and Hites (1998b)	345–385
4.28 ± 0.09	345			
4.70 ± 0.09	345			
4.78 ± 0.09	345			
4.75 ± 0.09	366			
5.02 ± 0.09	366			
4.75 ± 0.09	366			
4.84 ± 0.09	366			
4.86 ± 0.10	385			
5.54 ± 0.10	386			
5.16 ± 0.10	386			
5.73 ± 0.10	386			

[a] From Atkinson (1997).

over the temperature range 297–386 K, where the indicated errors are two least-squares standard deviations, and

$$k(\text{dibenzofuran}) = 3.8 \times 10^{-12}\ \text{cm}^3\ \text{molecule}^{-1}\ \text{s}^{-1}$$

at 298 K, with an estimated overall uncertainty at 298 K of ±30%.

II-F-14. OH + 1-Nitronaphthalene

The rate coefficient of Atkinson et al. (1989b), obtained from a relative rate study at room temperature and atmospheric pressure of air, is given in table II-F-14. This is the only reported study of this reaction.

Table II-F-14. OH + 1-nitronaphthalene → products: Rate coefficients k at, or close to, the high-pressure limit

$10^{12} \times k$ (cm^3 molecule^{-1} s^{-1})	at T (K)	Technique	Reference
5.2 ± 1.8	298 ± 2	RR [relative to k(cyclohexane) = 7.21×10^{-12}][a]	Atkinson et al. (1989b)

[a] From Atkinson (1997).

II-F-15. OH + 2-Nitronaphthalene

The rate coefficient of Atkinson et al. (1989b), obtained from a relative rate study at room temperature and atmospheric pressure of air, is given in table II-F-15. This is the only reported study of this reaction.

Table II-F-15. OH + 2-nitronaphthalene → products: Rate coefficients k at, or close to, the high-pressure limit

$10^{12} \times k$ (cm^3 molecule^{-1} s^{-1})	at T (K)	Technique	Reference
5.4 ± 0.9	298 ± 2	RR [relative to k(cyclohexane) = 7.21×10^{-12}][a]	Atkinson et al. (1989b)

[a] From Atkinson (1997).

II-F-16. OH + 2-Methyl-1-nitronaphthalene

The upper limit to the rate coefficient of Arey et al. (1990), obtained from a relative rate study at room temperature and atmospheric pressure of air, is given in table II-F-16. This is the only reported study of this reaction.

Table II-F-16. OH + 2-methyl-1-nitronaphthalene → products: Rate coefficients k at, or close to, the high-pressure limit

$10^{12} \times k$ (cm^3 molecule^{-1} s^{-1})	at T (K)	Technique	Reference
< 8.0	298 ± 2	RR [relative to k(cyclohexane) = 7.21×10^{-12}][a]	Arey et al. (1990)

[a] From Atkinson (1997).

Table II-F-17. Summary of rate coefficients for the reactions of OH with polycyclic aromatic compounds

Polycyclic Compound	10^{12} x k(298 K) (cm^3 molecule^{-1} s^{-1})	10^{12} x A (cm^3 molecule^{-1} s^{-1})	B (K)	Estimated Uncertainty
Naphthalene	23.0	15.6	117 ± 127	25%
1-Methylnaphthalene	53.0			a
2-Methylnaphthalene	52.3			a
2,3-Dimethylnaphthalene	76.8			a
Biphenyl	71			30%
Fluorene	14			Factor of 1.5
Acenaphthene	100			Factor of 2
Acenaphthalene	109			a
Phenanthrene	18			Factor of 2
Anthracene	18–289			b
Fluoranthene	~ 18			a,c
Dibenzo-*p*-dioxin	14			Factor of 1.5
Dibenzofuran	3.8	14.8	404 ± 126	30%
1-Nitronaphthalene	5.2			a
2-Nitronaphthalene	5.4			a
2-Methyl-1-nitro-naphthalene	< 8.0			a

[a] No estimate; determination of a single laboratory.
[b] No recommendation; very large range of values measured; see table II-F-10.
[c] Average value at 356–386 K.

II-G. Mechanisms and Products of OH-Radical Reactions with PAHs

The OH-radical reactions with the gas-phase PAHs are analogous to the corresponding reactions of OH-radicals with monocyclic aromatics discussed in sections II-B and II-E in that the reactions can proceed by three pathways: initial OH-radical addition to the carbon atoms of the aromatic ring; H-atom abstraction from C–H or O–H bonds in the substituent group(s); and/or OH-radical addition to any >C=C< bonds in the substituent group(s).

Based on the kinetic data of Lorenz and Zellner (1983), the lifetime of the OH–naphthalene adduct with respect to thermal back-decomposition to reactants is $\sim 10^3$ s at 298 K (Atkinson, 1989). Assuming that the data of Lorenz and Zellner

(1983) are correct and that the OH–naphthalene adduct is representative of other OH–PAH adducts, then thermal decomposition of the OH–PAH adducts is slower than those of OH–monocyclic aromatic adducts by a factor of $\sim 10^3$. As for the OH–aromatic adducts discussed in section II-E, the OH–PAH adducts are expected to react with O_2 and NO_2 [reactions (d) and (c), respectively, in the reaction scheme shown below for naphthalene], and the relatively long thermal decomposition lifetime of the OH–PAH adduct means that reactions with O_2 and NO_2 will dominate over back-decomposition.

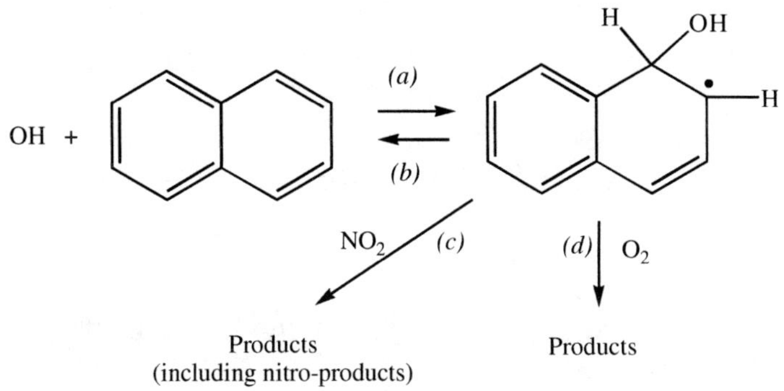

No published data are available concerning the rate coefficients k_c and k_d for reactions of OH–PAH adducts with NO_2 and O_2. Observations of the nitro-PAHs formed from gas-phase OH-radical–initiated reactions of PAHs (see the following discussion) in the ambient atmosphere indicates that reactions of the OH–PAH adducts with NO_2 must be competitive with reaction of the OH–PAH adducts with O_2. Moreover, measurements of nitro-PAHs in the atmosphere are consistent with laboratory nitro-PAHs yields (and isomeric distributions, see below) obtained at NO_2 mixing ratios >1 ppm (see, e.g., Arey et al., 1987, 1989a, 1989b, 1990; Gupta et al., 1996). Because the formation yield of nitro-PAH, $Y_{NPAH} = \alpha k_c[NO_2]/(k_c[NO_2] + k_d[O_2])$, where α is the nitro-PAH formation yield from reaction (c), the rate coefficient ratio k_d/k_c for the OH–PAH adducts must be significantly lower than those measured for the OH–benzene, OH–toluene, and OH–xylene adducts of $k_d/k_c \sim 10^{-5}$; for a ratio of this magnitude, one would predict that at ambient NO_2 levels of a few tens of parts-per-billions the nitro-PAH formation yields would be approximately a factor of 10^2 lower than that measured in laboratory studies (and hence $<0.1\%$). As noted in section III-A, for the NO_3–naphthalene adduct, Atkinson et al. (1994) obtained an upper limit to the rate coefficient ratio k_d/k_c of $<4 \times 10^{-7}$ at 298 ± 2 K, corresponding to reaction of the NO_3-PAH adduct with NO_2 dominating over its reaction with O_2 for NO_2 mixing ratios >80 ppb (and, of course, possibly lower). The observation of nitro-PAHs in the atmosphere under conditions where OH-radical reactions dominate, and with NO_2 mixing ratios typically <100 ppb, therefore suggests that reactions of OH–PAH adducts with NO_2 and O_2 are more similar to those of the NO_3–naphthalene adduct than to the OH–monocyclic aromatic adducts discussed in section II-E.

II-G-1. Naphthalene

Of the PAHs, the most complete reaction product yields have been described for naphthalene + OH in the presence of NO_x in air by Bunce et al. (1997) and Sasaki et al. (1997). The major products observed in the previous studies of the HO + naphthalene reaction are summarized in table II-G-1.

To account for some of these products, both the mechanisms of Sasaki et al. (1997) and Bunce et al. (1997) start with OH attack on either the 1- or 2-carbon of naphthalene.

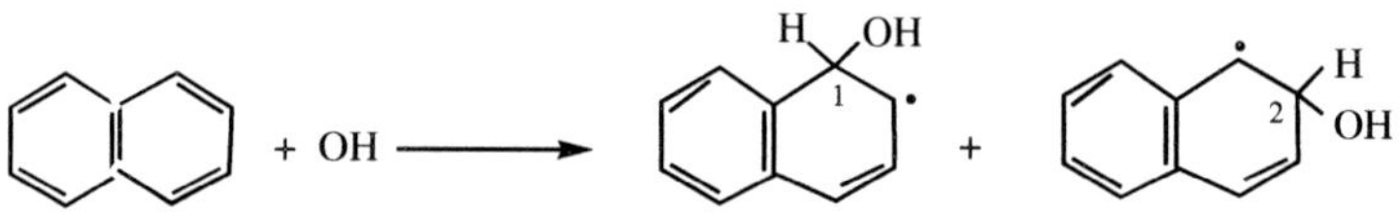

Either O_2 or NO_2 can add to the radical site and yield nitronaphthalenes, naphthols, and 2-formylcinnamaldehyde. Sasaki et al. (1997) illustrated this for OH attack (figure II-G-1) on the 1-carbon atom. Analogous compounds would be observed for attack on the 2-carbon atom.

Bunce et al. (1997) also proposed a scheme to form 1,4-naphthoquinone from the naphthalene–OH adduct (OH attack on the 1-carbon) or from 1-naphthol. Naphthols can also form from the initial OH–naphthalene adduct via loss of a

Table II-G-1. Products observed in the OH-initiated oxidation of naphthalene and their reported yields

Product	Yield	Reference
2-Formylcinnamaldehyde	$0.35^{+0.10}_{-0.05}$	Sasaki et al. (1997)[a]
	0.16	Bunce et al. (1997)
2-Formylbenzaldehyde	ca. 0.027[b]	Sasaki et al. (1997)
1-Naphthol	0.071 ± 0.043	Atkinson et al. (1987a)
	0.029 ± 0.015	Sasaki et al. (1997)
2-Naphthol	0.042 ± 0.025	Atkinson et al. (1987a)
	0.038 ± 0.011	Sasaki et al. (1997)
1-Nitronaphthalene	0.0032	Atkinson et al. (1987a)
	0.012 ± 0.009	Sasaki et al. (1997)
	0.07	Bunce et al. (1997)
2-Nitronaphthalene	0.0027	Atkinson et al. (1987a)
	0.013 ± 0.011	Sasaki et al. (1997)
1-Hydroxy-2-nitronaphthalene	0.011 ± 0.011	Sasaki et al. (1997)
2-Hydroxy-1-nitronaphthalene	Observed	Sasaki et al. (1997)
1,4-Naphthoquinone	0.010 ± 0.003	Sasaki et al. (1997)
	0.06	Bunce et al. (1997)
Phthalic anhydride	ca. 0.03[b]	Sasaki et al. (1997)
Compounds with molecular weight 174 and 176	ca. 0.018	Bunce et al. (1997)

[a] Data of Sasaki et al. (1997) were corrected for secondary reactions of observed products.
[b] Possibly second-generation products formed from 2-formylcinnamaldehyde.

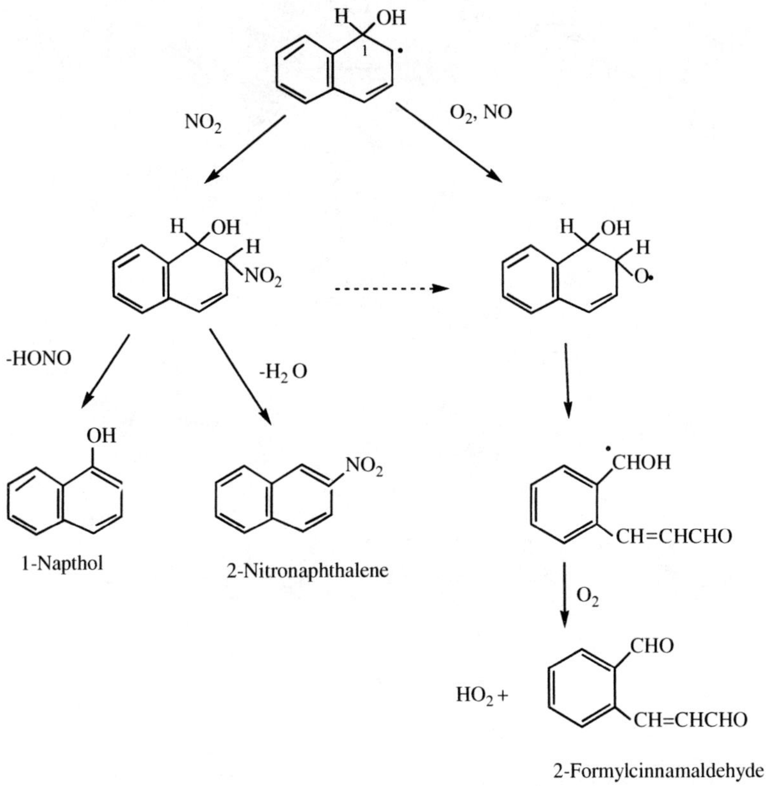

Figure II-G-1. A reaction scheme after Sasaki et al. (1997) for naphthalene + OH.

hydrogen atom, but this channel is not considered important (Atkinson et al., 1987a).

II-G-2. Fluorene

Helmig and coworkers studied the products of OH attack on fluorene in the presence of NO_x (Helmig et al., 1992a). Based on the higher yields of the 1- and 3-nitrofluorene products, compared with the 3- and 4-nitro-substituted isomers, the addition of OH by electrophilic attack on aromatic rings favors the site of the highest electron density (Nielsen, 1984; Pitts et al., 1985c; Arey et al., 1986; Atkinson et al., 1987a). In the case of fluorene, this is the 2-carbon atom (figure II-G-2). The total yield of all the nitrofluorene products in the Helmig et al. (1992a) study was 2–2.7% of the fluorene that reacted. 9-Fluorenone accounted for another 9% of the reacted fluorene and results from OH abstraction of a hydrogen attached to the 9-carbon atom. Other observed products were hydroxyfluorenes, hydroxynitrofluorene, and nitrofluorenones (Helmig et al., 1992a). All of these can be rationalized from the reaction mechanism shown in figure II-G-2.

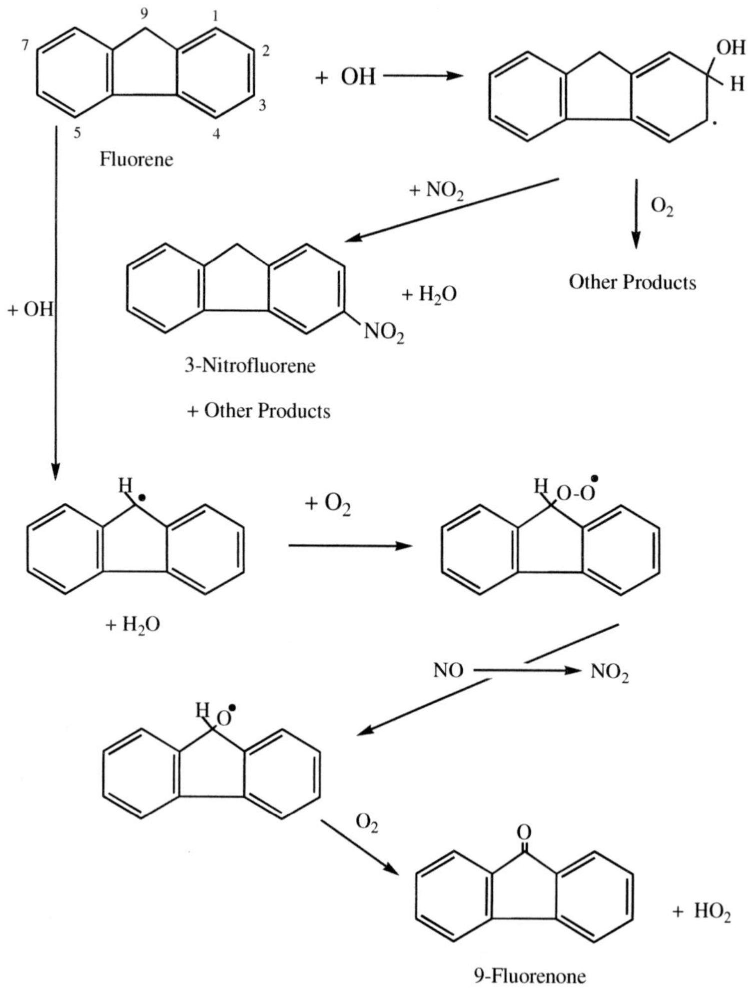

Figure II-G-2. Reaction of gas-phase fluorene with the OH-radical.

II-G-3. Phenanthrene

For phenanthrene, very low yields of nitro-PAHs are observed for reaction with OH in the presence of NO_x (Kwok et al., 1994c). Of additional atmospheric interest is the phenanthrene + OH reaction in the presence of O_3, NO_2, and NO_3, leading to formation of 2- and 4-nitrodibenzopyranones (Helmig et al., 1992b, 1992c; Kwok et al., 1994c). The 2-nitro derivative is very mutagenic in *Salmonella typhimurium* TA98 (-S9) bacterial mutagenicity assays, and can account for a significant portion of the mutagenicity in ambient samples from Claremont, Long Beach and Riverside, CA (Helmig et al., 1992c). In addition, 5-methyl derivatives of these compounds appear to be present in particle samples taken from Riverside, CA (Helmig et al., 1992b). Chamber experiments gave

0.004–0.04% yields for the 2-nitrodibenzopyranone derivative, while estimates from ambient air suggested 0.5–1% yields (Kwok et al., 1994c). This difference could result from either rapid photolysis or the formation of these compounds in additional atmospheric pathways.

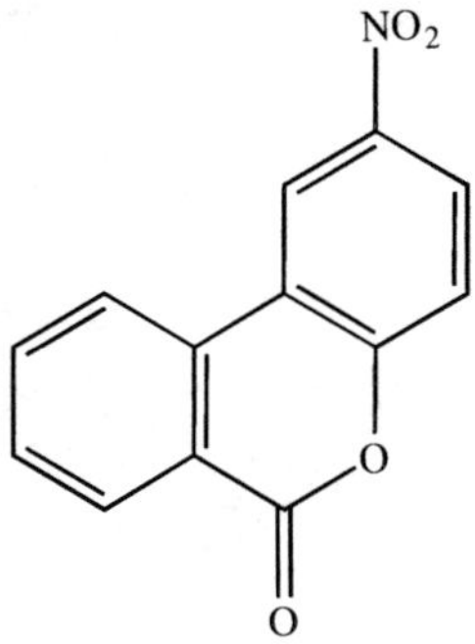

2-Nitro-6*H*-dibenzo[*b*,*d*]pyran-6-one

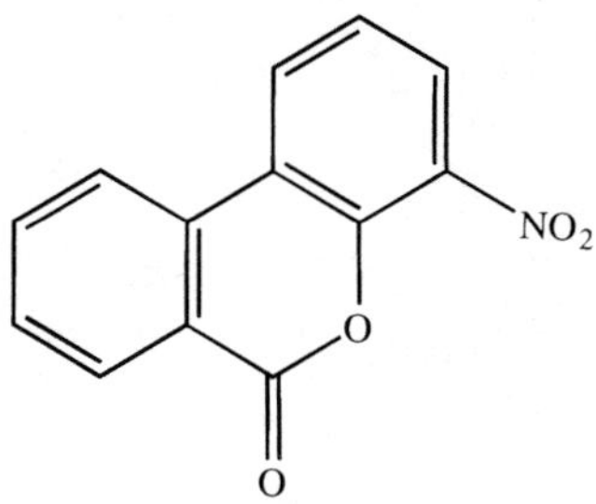

4-Nitro-6*H*-dibenzo[*b*,*d*]pyran-6-one

II-G-4. Fluroanthene and Pyrene

A reaction scheme (figure II-G-3), similar to that of naphthalene and fluorene, can be written for the initial OH attack on fluoranthene and pyrene, and can be used to explain observed nitrated products (Arey et al., 1986; Atkinson et al., 1990a; Atkinson and Arey, 1994). Gas-phase fluoranthene–OH and pyrene–OH studies gave total nitro-PAHs yields that account for ~5 and 0.05%, respectively, of the reacted PAHs (Arey et al., 1989b; Atkinson et al., 1990a; Atkinson and Arey, 1994). Given the low vapor pressures of these compounds, wall losses of product nitro-PAHs were possible in this study (Atkinson and Arey, 1994). In other chamber studies, which used natural sunlight and diesel soot exhaust, the resulting gas-phase 2-nitropyrene and nitrofluoranthenes, given their low vapor pressures, migrated quickly to the diesel soot particles. In these studies, a 10 ± 3% product yield of 2-nitrofluoranthene was observed (Fan et al., 1995, 1996b).

II-G-5. Biphenyl

The reaction of OH-radicals with biphenyl in the presence of NO_x gave 0.205 ± 0.078 of 2-hydroxybiphenyl, and 0.047 ± 0.014 of 3-nitrobiphenyl (Atkinson et al., 1987a). Brubaker and Hites (1998a) observed benzoic acid (with a molar yield of <0.01) in the absence of NO_x.

II-G-6. Yields of Nitro-PAHs

Product yields of nitro-PAHs from the reactions of selected PAHs with OH in the presence of NO_2 are given in table II-G-2. Observed products account for only a fraction of the initial fluoranthene and pyrene (Atkinson and Arey, 1994). Based

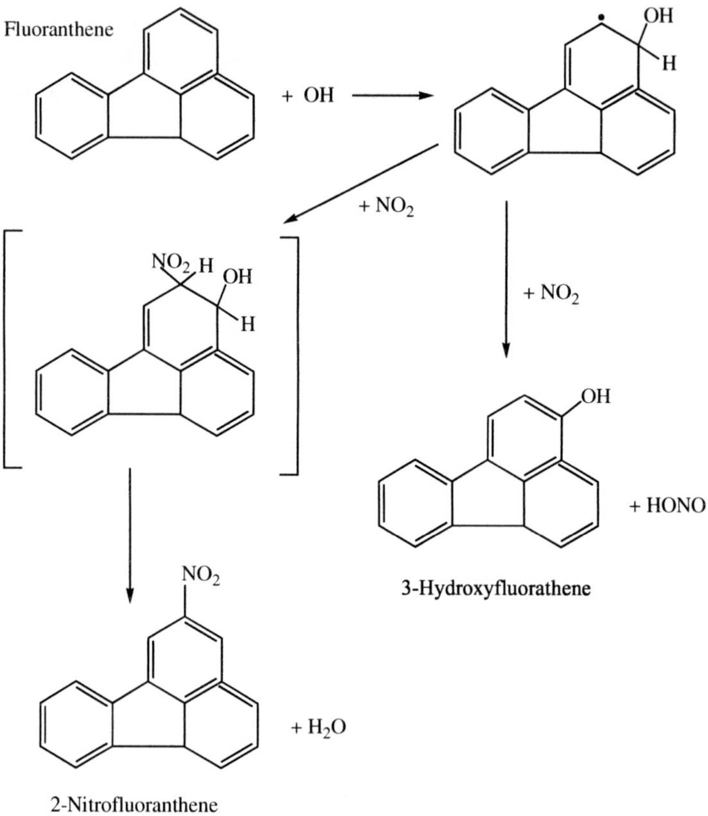

Figure II-G-3. Reactions of gas-phase fluoranthene with the OH-radical.

on the naphthalene studies (Bunce et al., 1997; Sasaki et al., 1997), one would expect quinones, hydroxy compounds, and possibly ring-cleavage products to account for significant portions of the presently unaccounted-for PAH products from OH-initiated reactions.

II-H. Estimation of Rate Coefficients for OH-Radical Reactions with PAHs

In the work of Biermann et al. (1985), a correlation was observed between the measured rate coefficients and the ionization potential of the PAHs. The rate coefficients for OH-radical reaction with the substituted benzenes (as reviewed herein) also show a similar reasonable correlation with the ionization potential of the compounds (filled circles in figure II-H-1). The rate coefficients for PAHs (table II-F-14) do not appear to be well correlated with the ionization potential, probably reflecting uncertainties in k(OH) and the fact that many of the PAHs for which data are available are not part of the same structural sequence. Those for

Table II-G-2. Nitroarene products formed from the gas-phase reactions of polycyclic aromatic hydrocarbons known to be present in ambient air with hydroxyl radicals (in the presence of NO_x) and their yields as listed by Atkinson and Arey (1994)

PAH[a]	Nitroarene Product Yields
Naphthalene	1-Nitronaphthalene (0.3%), (0.7%)[c] 2-Nitronaphthalene (0.3%), (0.05%)[c]
1-Methylnaphthalene	All 1-Methylnitronaphthalene isomers except 1-methyl-2-nitro-naphthalene (≈0.4%)
2-Methylnaphthalene	All 2-Methylnitronaphthalene isomers except 2-methyl-1- and 2-methyl-3-nitronaphthalene (≈0.2%)
Acenaphthene	5-Nitroacenaphthene 3-Nitroacenaphthene (≈0.2%) 4-Nitroacenaphthene
Acenaphthalene	4-Nitroacenaphthylene (2%)
Fluorene	3-Nitrofluorene (≈1.4%) 1-Nitrofluorene (≈0.6%) 4-Nitrofluorene (≈0.3%) 2-Nitrofluorene (≈0.1%)
Phenanthrene	Two nitroisomers (not 9-nitrophenanthrene) in trace yields
Anthracene[b]	1-Nitroanthracene, 2-Nitroanthracene, low yield
Pyrene	2-Nitropyrene (≈0.5%) 4-Nitropyrene (≈0.06%)
Fluoranthene	2-Nitrofluoranthene (≈3%), (10 ± 3%)[d] 7-Nitrofluoranthene (≈1%) 8-Nitrofluoranthene (≈0.3%)
Acephenanthrylene	Two nitroarene isomers (≈0.1%)
Biphenyl	3-Nitrobiphenyl (5%)

[a] PAH, polycyclic aromatic hydrocarbon.
[b] 9-Nitroanthracene was observed in both the OH and NO_3 radical reactions, but may not be a product of these reactions because it is also formed from exposure to NO_2/HNO_3 (Atkinson and Arey, 1994).
[c] Fielberg et al. (1999).
[d] Fan et al. (1995, 1996b).

benzene, naphthalene, and anthracene have a regular increase in the number of fused rings along one line when the higher values estimated for anthracene by Brubaker and Hites (1998a) and Biermann et al. (1985) are used (long-dashed line). Data points for these three compounds are shown as small filled circles in open squares. The data for phenanthrene (shown as small open circles in open squares) may be part of a new sequence of PAHs that have a similar angular attachment of the rings to naphthalene. In this case, the sequence benzene, naphthalene, phenanthrene follows a quadratic function shown by the dash–dot–dot–dash line in figure II-H-1. Until more rate measurements allow further tests of the correlations with the ionization potential, only a qualitative use of such correlations can be made.

Other molecular parameters that appear useful in correlations with rate coefficients for the PAHs are the Clar aromatic index (Clar, 1964), the $T_1 - S_0$ energy separation, the half-wave oxidation potential, and the theoretical "hardness" of

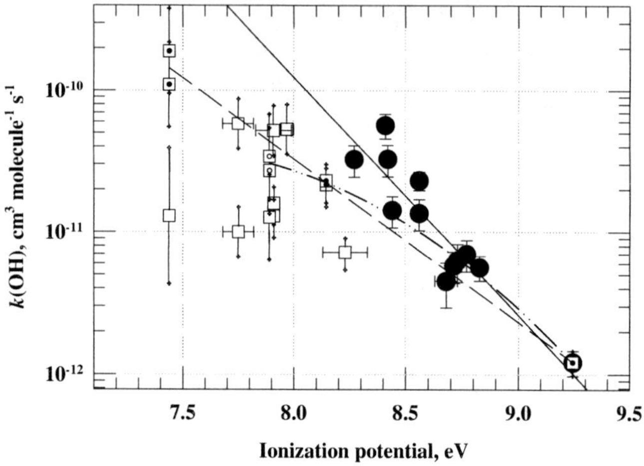

Figure II-H-1. Plot of the measured rate coefficients for alkyl-substituted benzenes (filled circles) and PAHs (open squares) as evaluated in this study versus the ionization potential of each compound as given by Lide (1999). See the text for the meaning of the lines and other features.

the molecule (Zhou and Parr, 1989). The latter term, "hardness" (η) has been defined by the relation

$$\eta = (\varepsilon_{\text{LUMO}} - \varepsilon_{\text{HOMO}})/2$$

where the difference $\varepsilon_{\text{LUMO}} - \varepsilon_{\text{HOMO}}$ is the energy gap between the lowest unoccupied molecular orbital (LUMO) and the highest occupied molecular orbital (HOMO) in the molecule. This quantity is also related theoretically to the ionization potential minus the electron affinity of the molecule. Zhou and Parr (1989) have calculated "Hückel hardness" values for 216 different cyclic conjugated molecules. "Hardness" appears to be a useful quantity that can be used to estimate OH–PAH rate coefficients. For example, note in figure II-H-2 that there is a reasonably good correlation between the measured rate coefficients for OH-radical reactions and the "Hückel hardness" for the only five different PAHs for which data for both "hardness" and OH rate coefficients are now available (benzene, biphenyl, naphthalene, phenanthrene, and anthracene). However, the very large scatter in the anthracene rate coefficient measurements leads to a great uncertainty in the upper portion of the curve for the forced linear least-squares fit. This relationship is given by equation (II-H-1):

$$\log_{10}[k(\text{OH})] = -8.99 - 2.85 \times (\text{hardness}) \qquad \text{(II-H-1)}$$

where the units for k are cm^3 molecule^{-1} s^{-1} and hardness is in measured $-\beta$ units.

In view of the apparent relationship between the rate coefficient for OH-radical reaction with the "Hückel hardness" factor, it is interesting to observe the variation in these factors for series of structurally related molecules; that is, the number and alignment of fused benzene rings in PAHs. Some of these are shown in figures II-H-3a, 3b, 3c, and 3d. Here, the "hardness" is plotted versus the number of benzene rings fused in a linear array (figure II-H-3a); the number of benzene rings fused in a zigzag array (figure II-H-3b); the number of benzene rings multiply fused to the phenanthrene structure (figure II-H-3c); or the number of benzene rings multiply fused to a central benzene ring (figure II-H-3d). In each of the figures II-H-3a, 3b, 3c, and 3d, the logarithmic scale for the corresponding estimated *k*(OH) values is shown on the right axis of the plot. Note that these scales, based on the application of equation (II-H-1), are gross approximations that can, at best, provide an idea of the magnitude of the *k*(OH) for each species.

Assuming that the OH rate coefficients for all of the reactions of OH-radical with PAHs and the "Hückel hardness" follow the trend exhibited by the limited data presented in figure II-H-2, then some interesting features related to reactivity of the PAHs toward OH-radical can be observed. In figure II-H-3a, it can be seen that as more benzene rings are attached to the benzene structure, maintaining the linear alignment of the rings, the hardness decreases markedly and the estimated *k*(OH) markedly increases. The limit to the *k*(OH) value that is approached at the large number of fused benzene rings is near the collision number. The same trend is seen in figure II-H-3b, where additional benzene rings are fused to anthracene to maintain a zigzag pattern of rings; however, in this case, the effect is much less than that with the linear ring alignment. As the structure progresses from three rings to six rings, a 24% decrease in hardness results, with about a factor of 3 increase in the estimated rate coefficient; while in the linear ring case, increasing the number of rings from three to six causes about a 60% decrease in hardness and about a factor of 7.6 increase in the estimated *k*(OH).

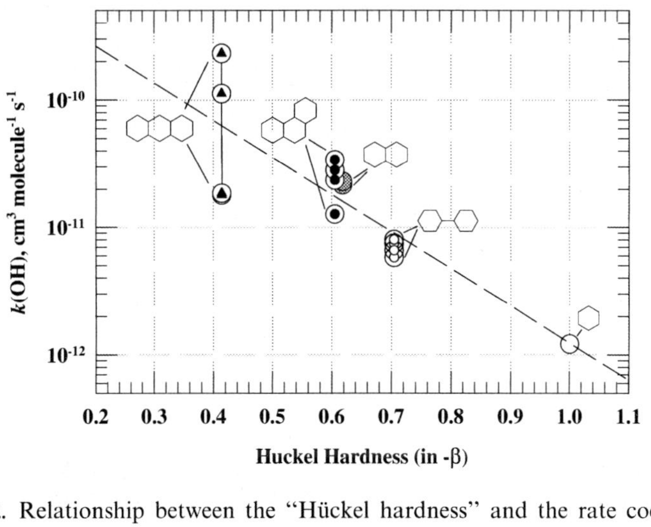

Figure II-H-2. Relationship between the "Hückel hardness" and the rate coefficients for OH-radical reactions with some PAH molecules. The calculated hardness values are from Zhou and Parr (1989).

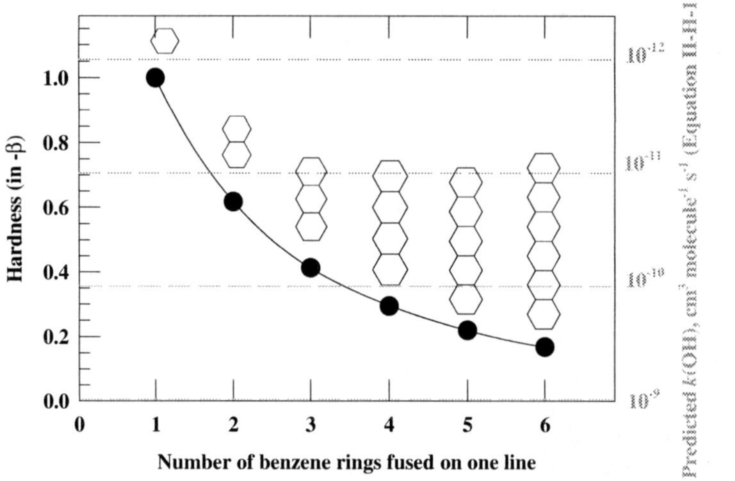

Figure II-H-3a. Relationship of Hückel hardness to a series of compounds formed by fusing successive benzene rings to benzene (in line).

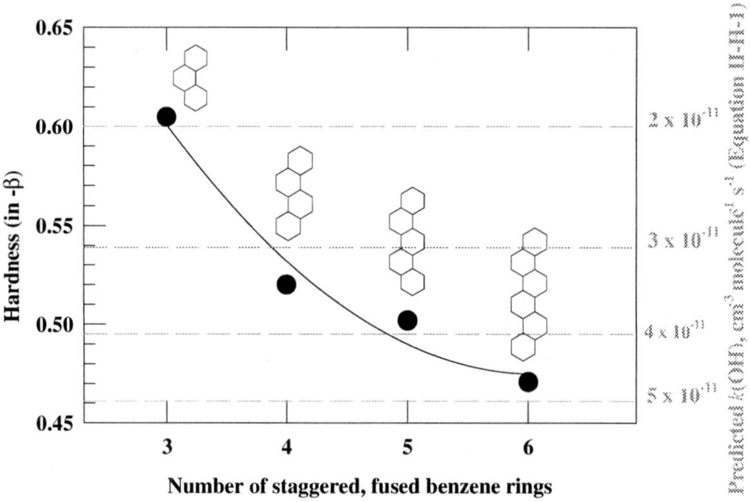

Figure II-H-3b. Relationship of Hückel hardness of compounds versus the number of benzene rings fused in a zigzag pattern to form the structures.

The effect of fusing more rings into a two-dimensional array of rings is seen in figure II-H-3c. Again, as an increasing number of rings are added to the array, the hardness decreases, and the reactivity toward OH increases. The effect in going from three rings to six rings in the fused molecule decreases the hardness by about 54% and increases the estimated k(OH) by about a factor of 13. An exception to

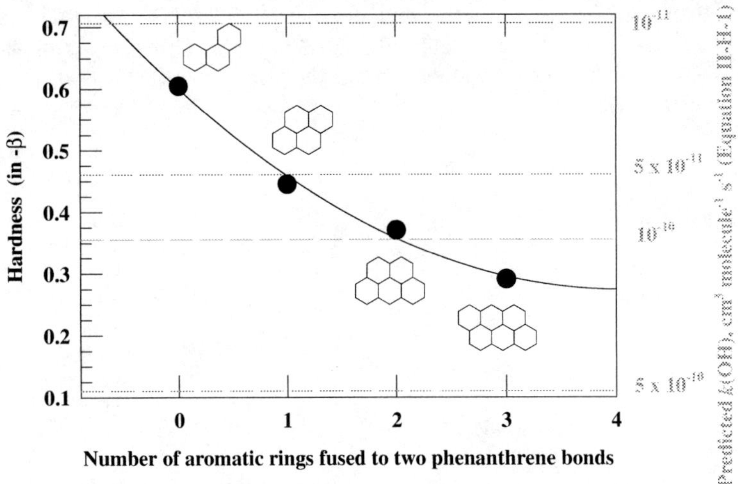

Figure II-H-3c. Relationship of Hückel hardness to compounds structurally related to phenanthrene versus the number of benzene rings fused to two or more rings of the phenanthrene structure.

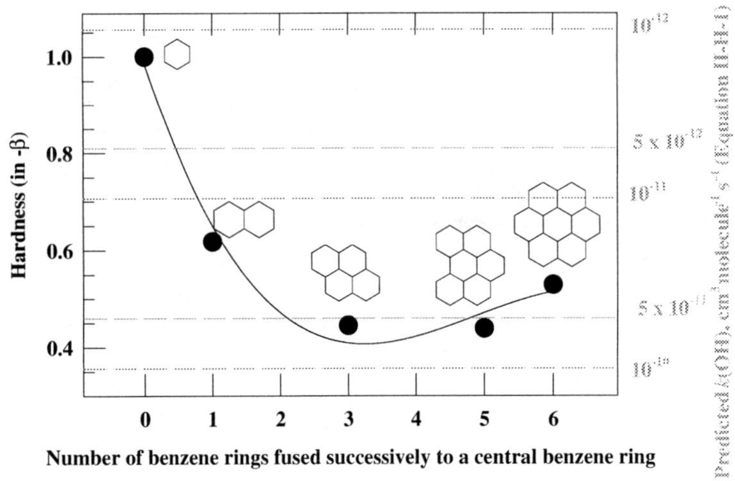

Figure II-H-3d. Relationship of Hückel hardness to the number of benzene rings that are fused consecutively about a central benzene molecule.

this generalization of reactivity increase with number of rings appears in figure II-H-3d, where the structural sequence from benzene to coronene shows a reversal of the effect as one progresses from the fusion of three benzene rings about a central benzene structure to five and six rings.

Obviously, a great deal more rate coefficient data for the PAHs will be required to test adequately how well the ln[k(OH)] relationship with hardness observed in

Table II-H-1. Estimated rate coefficients (cm^3 $molecule^{-1}$ s^{-1}) for OH-radical reactions with PAHs (near 298 K) using the calculated Hückel hardness factors of Zhou and Parr (1989) and the empirical relation relating these to the k(OH) rate coefficients [figure II-H-2; equation (II-H-1)]

PAHs	Hardness	Estimated k(OH)	Experimental k(OH)
Triphenylene	0.684	1.2×10^{-11}	None
Benzo[*c*]phenanthrene	0.568	2.5×10^{-11}	None
Coronene	0.539	3.0×10^{-11}	None
Chrysene	0.520	3.4×10^{-11}	None
Dibenz[*a,c*]anthracene	0.499	3.9×10^{-11}	None
Benzo[*e*]pyrene	0.497	4.0×10^{-11}	None
Dibenz[*a,h*]anthracene	0.473	4.6×10^{-11}	None
Benz[*a*]anthracene	0.452	5.3×10^{-11}	None
Pyrene	0.445	5.6×10^{-11}	None
Benzo[*b*]chrysene	0.405	7.2×10^{-11}	None
Benzo[*a*]pyrene	0.371	9.0×10^{-11}	None
Perylene	0.347	1.1×10^{-10}	None
Tetracene	0.295	1.5×10^{-10}	None
Pentacene	0.220	2.4×10^{-10}	None
Benzene	1.000	1.5×10^{-12}	1.2×10^{-12}
Biphenyl	0.704	1.0×10^{-11}	7.2×10^{-12}
Naphthalene	0.618	1.8×10^{-11}	2.3×10^{-11}
Phenanthrene	0.605	1.9×10^{-11}	1.8×10^{-11}
Anthracene	0.414	6.8×10^{-11}	$(1.8\text{-}29) \times 10^{-11}$

figure II-H-2 [equation (II-H-1)] applies to other PAHs. Until such data are available, very cautious use may be made of the theoretically determined "hardness" values to estimate approximate k(OH) values for other PAHs for which reliable experimental measurements have not yet been made. A few such estimates are given in table II-H-1 for some PAHs in which the approximate relationship observed in figure II-H-2 [equation (II-H-1)] has been applied. Results obtained by this procedure must be considered very approximate. In particular, an accurate measurement of the rate coefficient for anthracene and other very reactive PAHs will be important in improving the relation of k to hardness. It is probable that as more extensive data become available, each structurally related series of compounds will be found to follow its own somewhat different unique relationship, which probably will be somewhat shifted from the linear relation drawn in figure II-H-2.

III

Reactions of Monocyclic and Polycyclic Aromatic Compounds with NO_3 Radicals

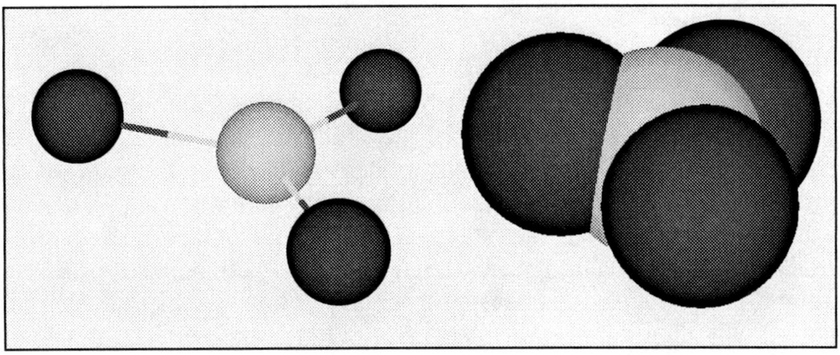

III-A. Mechanisms of NO_3-Radical Reactions with Aromatic Compounds

The gas-phase reactions of aromatic compounds with the NO_3 radical are only significant as a tropospheric loss process for compounds containing unsaturated $>C=C<$ bonds in the substituent groups (e.g., styrene and acenaphthylene and related compounds) and for phenolic compounds (e.g., phenol and cresols). Analogous to the OH-radical reactions (see chapter II), the gas-phase reactions of the NO_3 radical with organic compounds appear to proceed by three pathways: (1) initial addition of the NO_3 radical to the carbon atoms of the aromatic ring to form an "NO_3–aromatic adduct"; (2) H-atom abstraction from the C–H or O–H bonds in the substituent group(s); or (3) NO_3-radical addition to the carbon atoms of $>C=C<$ bonds in the substituent groups (Atkinson, 1991b). The data presently available indicate that the NO_3-radical addition pathway to the aromatic ring is reversible, with back-decomposition of the NO_3-aromatic adduct dominating the atmospheric fate of the adduct (Atkinson, 1991b).

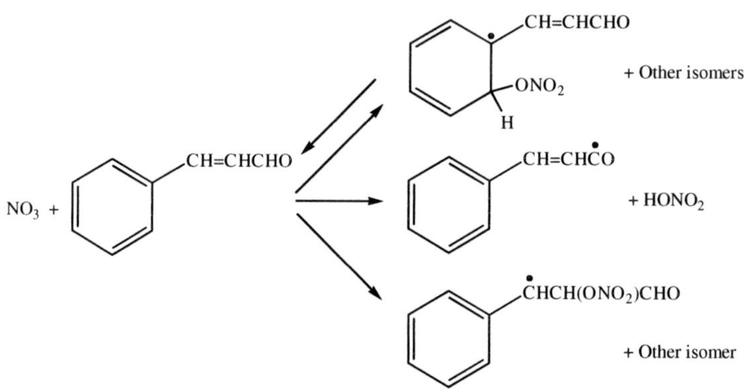

No reaction with the NO_3 radical has been observed for benzene or biphenyl (Atkinson, 1991b). Reactions of the NO_3 radical are observed with toluene and *o*- and *p*-xylene, and the measured rate coefficients exhibit a significant deuterium isotope effect (Atkinson and Aschmann, 1988; Rindone et al., 1991; Chiodini et al., 1993), indicating that the rate-determining steps in these reactions involve H- (or D-) atom abstraction from the C–H (or C–D) bonds of the substituent groups. This conclusion is consistent with the observations that *m*-xylene reacts more slowly than do *o*- or *p*-xylene, and that 1,3,5-trimethylbenzene is the least reactive of the trimethylbenzenes, in contrast to the OH-radical reactions where *m*-xylene and 1,3,5-trimethylbenzene are the most reactive of the xylenes and the trimethylbenzenes, respectively.

For naphthalene and the alkyl-substituted naphthalenes, reaction is observed with the measured rate coefficient increasing with increasing concentration of NO_2 (Atkinson and Aschmann, 1986, 1987, 1988; Atkinson et al., 1994), kinetically equivalent to a reaction with N_2O_5 (Atkinson and Aschmann, 1986, 1987, 1988). Atkinson et al. (1990b) showed, by conducting experiments at 272 K (where the lifetime of N_2O_5 due to thermal decomposition is relatively long, $\sim$10 min) in the presence of 2,3-dimethyl-2-butene (to scavenge NO_3 radicals), that naphthalene reacts with NO_3 radicals but not with N_2O_5. By analogy with the corresponding OH-radical reactions (see chapter II), the NO_3–alkylnaphthalene adducts are also expected to react with O_2, and the expected overall reaction of NO_3 with naphthalene is given in the following scheme.

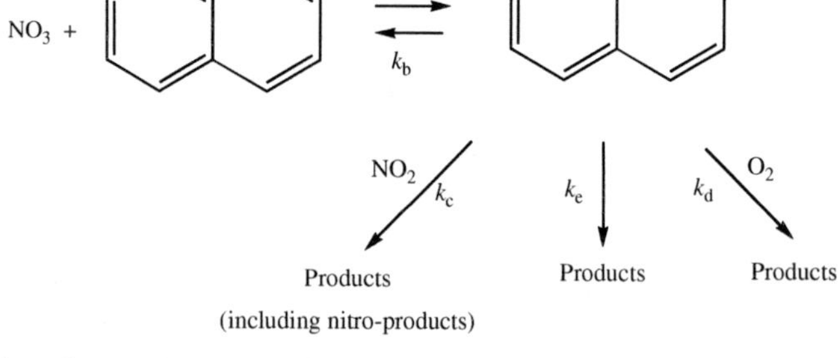

Hence, the observed rate coefficient k_{obs} is given by

$$k_{obs} = k_a(k_c[NO_2] + k_d[O_2] + k_e)/(k_b + k_c[NO_2] + k_d[O_2] + k_e) \qquad (I)$$

where k_e is the reaction rate for reactions of the NO_3–aromatic adduct other than with NO_2 or O_2. Under experimental conditions such that $k_b \gg (k_c[NO_2] + k_d[O_2] + k_e)$ and $k_c[NO]_2 \gg (k_d[O_2] + k_e)$, then equation (I) simplifies to

$$k_{obs} = k_a k_c[NO_2]/k_b \qquad (II)$$

and the measured rate coefficient k_{obs} is linearly dependent on the NO_2 concentration.

Using the kinetic data for the reaction of the NO_3 radical with naphthalene and with assumptions concerning the magnitude of reaction pathways (a) and (c), Atkinson (1991b) concluded that the back-decomposition rate coefficient of the NO_3–naphthalene adduct(s) is $k_b \sim 1 \times 10^{12}\,e^{-4090/T}\,s^{-1}$ ($\sim 1 \times 10^6\,s^{-1}$ at 298 K). Use of thermochemical data for the NO_3–aromatic and OH–aromatic adducts formed from naphthalene and benzene allowed Atkinson (1991b) to derive a rate coefficient for the back-decomposition of the NO_3–monocyclic aromatic adducts of $k_b \sim 3 \times 10^{12}\,e^{-2600/T}\,s^{-1}$ ($\sim 5 \times 10^8\,s^{-1}$ at 298 K). Hence, for the monocyclic aromatic compounds, the pathway involving NO_3-radical addition to the aromatic ring is of no atmospheric importance. The reactions of the NO_3 radical with phenolic compounds are rapid, with room temperature rate coefficients of $\geq 10^{-12}\,cm^3\,molecule^{-1}\,s^{-1}$, and this magnitude of the rate coefficients is not consistent with a simple H-atom abstraction from the O–H bonds in the substituent hydroxyl groups (Carter et al., 1981; Atkinson et al., 1992; Atkinson, 1991b); rather, it has been proposed that the reactions involve initial NO_3-radical addition to the ring at the position to which the –OH group is attached, with the H-atom abstraction from the –OH group proceeding via a six-membered transition state. An alternative explanation is that NO_3 radicals add to the aromatic ring to give a π-adduct where the NO_3 moiety is free to move around the ring. Decomposition of the π-adduct could occur via elimination of HNO_3 to give the phenoxy radical.

III-B. Rate Coefficients for Reactions of NO_3 Radicals with Aromatic Compounds

Rate coefficients are available for a number of aromatic compounds; all data except for some of those for naphthalene are at room temperature. In many cases, the reported rate coefficients are only upper limits. The rate data show that the reactions of the NO_3 radical with monocyclic aromatic compounds containing saturated non-phenolic substituent groups are slow, typically with room temperature rate coefficients of $< 10^{-15}\,cm^3\,molecule^{-1}\,s^{-1}$. For the monocyclic aromatic compounds, rate coefficients are given in the tables as values of k, corresponding to reaction with the substituent groups and with the measured rate coefficients not showing any dependence on the NO_2 concentration (where this has been examined). For the polycyclic aromatic compounds such as naphthalene, values of the NO_2-independent rate coefficient k (anticipated to correspond to H-atom abstraction from C−H or O−H bonds and/or NO_3 radical addition to unsaturated >C=C< bonds) and values of $k_a k_c / k_b$ are given.

To date, all of the rate coefficients reported for the NO_3-radical reactions with monocyclic aromatic compounds have been measured using relative rate methods (RR). In most cases, rate coefficients have been obtained relative to that for the reaction of the NO_3 radical with another organic compound. When this is the case, the rate coefficient used in the tables for the reference organic is taken from a recent evaluation. The rate coefficient for reaction of the NO_3 radical with the aromatic compound of interest has therefore been reevaluated, and it may be different from that cited in the original publication.

In a number of cases, rate coefficients for reaction of NO_3 radical with aromatic compound have been measured relative to the equilibrium constant, K, for the reactions

$$NO_3 + NO_2 \leftrightarrow N_2O_5$$

This equilibrium constant is not accurately known, and at 298 K it has an uncertainty of a factor of ~1.5 [see, e.g., Wängberg et al. (1997) and references therein]. In this section, an equilibrium constant of

$$K(NO_3 + NO_2 \leftrightarrow N_2O_5) = 2.80 \times 10^{-27}\, e^{11000/T}\ \text{cm}^3\ \text{molecule}^{-1}$$

is used, the value recommended in the 1997 IUPAC evaluation (Atkinson et al., 1997a). The large temperature dependence of K means that this equilibrium constant changes rapidly with temperature, by ~10% K^{-1}, and, given the typical uncertainty in the temperature of the experiments of ±1–2 K, this introduces additional uncertainties into the calculation of the rate coefficient for the aromatic compound. In general, rate coefficients measured relative to the $NO_3 + NO_2 \leftrightarrow N_2O_5$ reactions are given less weight in the evaluations than are rate coefficients measured relative to the reaction of the NO_3 radical with another organic compound.

III-B-1. NO_3 + Benzene

The upper limits to the rate coefficients obtained by Japar and Niki (1975) and Atkinson et al. (1984b, 1984c) are given in table III-B-1. Based on these data, an upper limit to the rate coefficient of

$$k(\text{benzene}) < 3 \times 10^{-17}\ \text{cm}^3\ \text{molecule}^{-1}\ \text{s}^{-1}$$

at 298 K is recommended.

III-B-2. NO_3 + Toluene

The rate coefficients, or upper limits thereof, of Japar and Niki (1975), Carter et al. (1981), Atkinson et al. (1984b, 1984c), and Atkinson and Aschmann (1988) are given in table III-B-2. The rate coefficients of Atkinson et al. (1984b, 1984c) and Atkinson and Aschmann (1988), all obtained from relative rate studies (and relative to the reaction of the NO_3 radical with organic compounds and not to the equilibrium constant for the $N_2O_5 \leftrightarrow NO_3 + NO_2$ reactions), are in good

Table III-B-1. NO_3 + benzene → products: Rate coefficients k

k (cm^3 molecule^{-1} s^{-1})	at T (K)	Technique	Reference
$< 5.5 \times 10^{-17}$	300	RR [relative to $K(NO_3 + NO_2 \leftrightarrow N_2O_5) = 2.35 \times 10^{-11}$][a]	Japar and Niki (1975)
$\leq 2.7 \times 10^{-17}$	298 ± 1	RR [relative to k(ethene) = 2.05×10^{-16}][b]	Atkinson et al. (1984c)
$< 6.0 \times 10^{-17}$	298 ± 1	RR [relative to k(n-heptane) = 1.5×10^{-16}][b]	Atkinson et al. (1984b)

[a]From the IUPAC evaluation (Atkinson et al., 1997a).
[b]From Atkinson (1997).

Table III-B-2. NO_3 + toluene → products: Rate coefficients k

k (cm^3 molecule^{-1} s^{-1})	at T (K)	Technique	Reference
$< 5.5 \times 10^{-17}$	300	RR [relative to $K(NO_3 + NO_2 \leftrightarrow N_2O_5) = 2.35 \times 10^{-11}$][a]	Japar and Niki (1975)
$\leq 3.7 \times 10^{-15}$	300 ± 1	RR [relative to k(propene) = 9.73×10^{-15}][b]	Carter et al. (1981)
$(6.56 \pm 2.46) \times 10^{-17}$	298 ± 1	RR [relative to k(ethene) = 2.05×10^{-16}][b]	Atkinson et al. (1984c)
$(6.75 \pm 1.65) \times 10^{-17}$	298 ± 1	RR [relative to k(n-heptane) = 1.5×10^{-16}][b]	Atkinson et al. (1984b)
$(7.81 \pm 0.83) \times 10^{-17}$	296 ± 2	RR [relative to k(n-heptane) = 1.37×10^{-16}][c]	Atkinson and Aschmann (1988)

[a]From the IUPAC evaluation (Atkinson et al., 1997b).
[b]From Atkinson (1997).
[c]From Atkinson (1991b).

agreement. A unit-weighted average of the rate coefficients of Atkinson et al. (1984b) and (1984c), and Atkinson and Aschmann (1988) leads to the recommendation of

$$k(\text{toluene}) = 7.0 \times 10^{-17}\ \text{cm}^3\ \text{molecule}^{-1}\ \text{s}^{-1}$$

at 298 K, with an estimated overall uncertainty of a factor of 2.

Atkinson and Aschmann (1988) also measured rate coefficients at 296 ± 2 K for the reactions of the NO_3 radical with toluene-α,α,α-d_3 ($C_6H_5CD_3$) and toluene-d_8 ($C_6D_5CD_3$) of $(3.84 \pm 0.69) \times 10^{-17}$ cm^3 molecule^{-1} s^{-1} and $(3.43 \pm 1.79) \times 10^{-17}$ cm^3 molecule^{-1} s^{-1}, respectively. The significant deuterium isotope effects indicate that the rate-determining step involves H- (or D-) atom abstraction from the C–H (or C–D) bonds of the $-CH_3$ (or $-CD_3$) substituent group:

$$NO_3 + C_6H_5CD_3 \rightarrow DNO_3 + C_6H_5CD_2$$

III-B-3. NO_3 + Ethylbenzene

The upper limit to the rate coefficient obtained by Atkinson et al. (1987b) is given in table III-B-3. This is the only reported study of this reaction.

Table III-B-3. NO_3 + ethylbenzene → products: Rate coefficients k

k (cm^3 $molecule^{-1}$ s^{-1})	at T (K)	Technique	Reference
$\leq 5.7 \times 10^{-16}$	298 ± 2	RR [relative to k(propene) = 9.49×10^{-15}][a]	Atkinson et al. (1987b)

[a]From Atkinson (1997).

III-B-4. NO_3 + *o*-Xylene

The rate coefficient obtained by Atkinson et al. (1984b) is given in table III-B-4. This is the only reported measurement of this reaction rate coefficient. Rindone et al. (1991) and Chiodini et al. (1993) have also measured the rate coefficient for the gas-phase reaction of *o*-xylene relative to that for *o*-xylene-d_{10}. The reactants were analyzed by GC-MS and in situ FTIR before and after reaction, thereby allowing a rate coefficient ratio k(*o*-xylene)/k(*o*-xylene-d_{10}) = k_H/k_D to be determined. Values of $k_H/k_D = 1.79$ (Rindone et al., 1991) and 1.67 (Chiodini et al., 1993) were measured at room temperature using GC-MS and in situ FTIR analyses, respectively. These kinetic isotope effects indicate that the rate determining step in the gas-phase reaction of the NO_3 radical with *o*-xylene involves H- (or D-) atom abstraction from the $-CH_3$ (or $-CD_3$) substituent groups [consistent with the data and conclusions of Atkinson and Aschmann (1988) for the toluene reaction].

Table III-B-4. NO_3 + *o*-xylene → products: Rate coefficients k

k (cm^3 $molecule^{-1}$ s^{-1})	at T (K)	Technique	Reference
$(4.13 \pm 0.57) \times 10^{-16}$	298 ± 1	RR [relative to k(*n*-heptane) = 1.5×10^{-16}][a]	Atkinson et al. (1984b)

[a]From Atkinson (1997).

III-B-5. NO_3 + *m*-Xylene

The rate coefficients of Atkinson et al. (1984b, 1984c) are given in table III-B-5. The three rate coefficients obtained by Atkinson et al. (1984b, 1984c) are in reasonable agreement, considering the uncertainties in the equilibrium constant for the $N_2O_5 \leftrightarrow NO_3 + NO_2$ reactions. An average of the rate coefficients of

Atkinson et al. (1984b) and (1984c) measured relative to the reactions of the NO_3 radical with *n*-heptane and ethene, respectively, is used to recommend that

$$k(m\text{-xylene}) = 2.6 \times 10^{-16}\ \text{cm}^3\ \text{molecule}^{-1}\ \text{s}^{-1}$$

at 298 K, with an estimated overall uncertainty of a factor of 2.

Table III-B-5. NO_3 + *m*-xylene → products:
Rate coefficients *k*

k (cm^3 molecule^{-1} s^{-1})	at *T* (K)	Technique	Reference
$(1.85 \pm 0.33) \times 10^{-16}$	298 ± 1	RR [relative to $K(NO_3 + NO_2 \leftrightarrow N_2O_5) = 3.01 \times 10^{-11}$][a]	Atkinson et al. (1984c)
$(2.54 \pm 0.50) \times 10^{-16}$	298 ± 1	RR [relative to *k*(ethene) = 2.05×10^{-16}][b]	Atkinson et al. (1984c)
$(2.58 \pm 0.47) \times 10^{-16}$	298 ± 1	RR [relative to *k*(*n*-heptane) = 1.5×10^{-16}][b]	Atkinson et al. (1984b)

[a]From the IUPAC evaluation (Atkinson et al., 1997a).
[b]From Atkinson (1997).

III-B-6. NO_3 + *p*-Xylene

The rate coefficient obtained by Atkinson et al. (1984b) is given in table III-B-6. This is the only reported measurement of this reaction rate coefficient. Rindone et al. (1991) and Chiodini et al. (1993) have also measured the rate coefficient for the gas-phase reaction of *p*-xylene relative to those for *p*-xylene-d_6 and *p*-xylene-d_{10}. The reactants were analyzed by GC-MS and in situ FTIR before and after reaction, thereby allowing rate coefficient ratios $k(p\text{-xylene})/k(p\text{-xylene-}d_x) = k_H/k_D$ to be determined. Values of $k_H/k_D = 1.48$ (Rindone et al., 1991) and 1.40 (Chiodini et al., 1993) were measured at room temperature for $k(p\text{-xylene})/k(p\text{-xylene-}d_6)$ using GC-MS and in situ FTIR analyses, respectively, and corresponding values of $k_H/k_D = 1.87$ (Rindone et al., 1991) and 1.50 (Chiodini et al., 1993) were obtained for $k(p\text{-xylene})/k(p\text{-xylene-}d_{10})$. These kinetic isotope effects indicate that the rate-determining step in the gas-phase reaction of the NO_3 radical with *p*-xylene involves H- (or D-) atom abstraction from the $-CH_3$ (or $-CD_3$) substituent groups (consistent with the corresponding reactions of toluene and *o*-xylene discussed previously).

Table III-B-6. NO_3 + *p*-xylene → products:
Rate coefficients *k*

k (cm^3 molecule^{-1} s^{-1})	at *T* (K)	Technique	Reference
$(4.97 \pm 0.36) \times 10^{-16}$	298 ± 1	RR [relative to *k*(*n*-heptane) = 1.5×10^{-16}][a]	Atkinson et al. (1984b)

[a]From Atkinson (1997).

III-B-7. NO_3 + 4-Ethyltoluene

The rate coefficient obtained by Bolzacchini et al. (1999) is given in table II-B-7. This is the only reported study of this reaction.

Table III-B-7. NO_3 + 4-ethyltoluene → products: Rate coefficients k

k (cm^3 $molecule^{-1}$ s^{-1})	at T (K)	Technique	Reference
$(8.58 \pm 1.23) \times 10^{-16}$	298	RR [relative to k(1,3,5-trimethylbenzene) = 8.76×10^{-16}][a]	Bolzacchini et al. (1999)

[a]From Atkinson et al. (1984b); this evaluation.

III-B-8. NO_3 + 1,2,3-Trimethylbenzene

The rate coefficient obtained by Atkinson et al. (1984b) is given in table III-B-8. This is the only reported study of this reaction.

Table III-B-8. NO_3 + 1,2,3-trimethylbenzene → products: Rate coefficients k

k (cm^3 $molecule^{-1}$ s^{-1})	at T (K)	Technique	Reference
$(1.86 \pm 0.22) \times 10^{-15}$	294	RR [relative to k(n-heptane) = 1.37×10^{-16}][a]	Atkinson et al. (1984b)

[a]From Atkinson (1991b).

III-B-9. NO_3 + 1,2,4-Trimethylbenzene

The rate coefficient obtained by Atkinson et al. (1984b) is given in table III-B-9. This is the only reported study of this reaction.

Table III-B-9. NO_3 + 1,2,4-trimethylbenzene → products: Rate coefficients k

k (cm^3 $molecule^{-1}$ s^{-1})	at T (K)	Technique	Reference
$(1.81 \pm 0.21) \times 10^{-15}$	294	RR [relative to k(n-heptane) = 1.37×10^{-16}][a]	Atkinson et al. (1984b)

[a]From Atkinson (1991b).

III-B-10. NO_3 + 1,3,5-Trimethylbenzene

The rate coefficient obtained by Atkinson et al. (1984b) is given in table III-B-10. This is the only reported study of this reaction.

Table III-B-10. NO_3 + 1,3,5-trimethylbenzene → products: Rate coefficients k

k (cm^3 molecule^{-1} s^{-1})	at T (K)	Technique	Reference
$(8.76 \pm 0.68) \times 10^{-16}$	298 ± 1	RR [relative to k(n-heptane) = 1.5×10^{-16}][a]	Atkinson et al. (1984b)

[a]From Atkinson (1997).

III-B-11. NO_3 + p-Cymene (4-Isopropyltoluene)

The rate coefficients obtained by Corchnoy and Atkinson (1990) and Bolzacchini et al. (1999) are given in table III-B-11. The rate coefficient for the reference compound used by Bolzacchini et al. (1999) is not well established (Atkinson, 1991b) and hence the rate coefficient cited for p-cymene from the Bolzacchini et al. (1999) study is subject to large uncertainties. The rate coefficient of Corchnoy and Atkinson (1990) is used to recommend that

$$k(p\text{-cymene} = 1.0 \times 10^{-15}\ \text{cm}^3\ \text{molecule}^{-1}\ \text{s}^{-1}$$

at 298 K, with an estimated overall uncertainty of a factor of 2.

Table III-B-11. NO_3 + p-cymene → products: Rate coefficients k

k (cm^3 molecule^{-1} s^{-1})	at T (K)	Technique	Reference
$(1.00 \pm 0.07) \times 10^{-15}$	295 ± 2	RR [relative to k(2,3-dimethylbutane) = 4.08×10^{-16}][a]	Corchnoy and Atkinson (1990)
$(6.76 \pm 0.11) \times 10^{-16}$	298	RR [relative to k(methoxybenzene) = 2.30×10^{-16}][a,b]	Bolzacchini et al. (1999)

[a]From Atkinson (1991b).
[b]Rate coefficient was measured at 294 K, and has been increased by ~10% to be applicable to 298 K [see Atkinson (1997) for similar extrapolations for alkanes].

III-B-12. NO_3 + 4-*tert*-Butyltoluene

The rate coefficient obtained by Bolzacchini et al. (1999) is given in table III-B-12. This is the only reported study of this reaction.

Table III-B-12. NO_3 + 4-*tert*-butyltoluene → products:
Rate coefficients k

k (cm^3 $molecule^{-1}$ s^{-1})	at T (K)	Technique	Reference
$(4.09 \pm 0.31) \times 10^{-16}$	298	RR [relative to k(methoxybenzene) $= 2.30 \times 10^{-16}$][a]	Bolzacchini et al. (1999)

[a]From Atkinson (1991b). The rate coefficient for methoxybenzene was measured at 294 K, and has been increased by ~10% to be applicable to 298 K [see Atkinson (1997) for similar extrapolations for alkanes].

III-B-13. NO_3 + Indan

The rate coefficient obtained by Kwok et al. (1997b) is given in table III-B-13. This is the only reported study of this reaction.

Table III-B-13. NO_3 + indan → products:
Rate coefficients k

k (cm^3 $molecule^{-1}$ s^{-1})	at T (K)	Technique	Reference
$(7.35 \pm 0.40) \times 10^{-15}$	297 ± 2	RR [relative to k(1-butene) $= 1.33 \times 10^{-14}$][a]	Kwok et al. (1997b)

[a]From Atkinson (1997).

III-B-14. NO_3 + Tetralin

The rate coefficients of Atkinson et al. (1987b) and Atkinson and Aschmann (1988) are given in table III-B-14. These two rate coefficients, both relative to that for the reaction of the NO_3 radical with propene, agree within the combined experimental uncertainties. The more precise rate coefficient of Atkinson and Aschmann (1988) is used to recommend that

$$k(\text{tetralin}) = 8.5 \times 10^{-15}\ \text{cm}^3\ \text{molecule}^{-1}\ \text{s}^{-1}$$

at 298 K, with an estimated overall uncertainty of a factor of 1.5.

Table III-B-14. NO_3 + tetralin → products:
Rate coefficients k

k (cm^3 $molecule^{-1}$ s^{-1})	at T (K)	Technique	Reference
$(1.11 \pm 0.23) \times 10^{-14}$	298 ± 2	RR [relative to k(propene) $= 9.49 \times 10^{-15}$][a]	Atkinson et al. (1987b)
$(8.45 \pm 0.58) \times 10^{-15}$	296 ± 2	RR [relative to k(propene) $= 9.24 \times 10^{-15}$][a]	Atkinson and Aschmann (1988)

[a]From Atkinson (1997).

III-B-15. NO_3 + Indene

The rate coefficient obtained by Kwok et al. (1997b) is given in table III-B-15. This is the only reported study of this reaction. By comparison with the rate coefficients for the reactions of the NO_3 radical with indan and tetralin (see sections III-B-13 and III-B-14), the NO_3-radical reaction with indene must occur mainly by initial addition to the $>C{=}C<$ bond in the five-membered ring.

Table III-B-15. NO_3 + indene → products: Rate coefficients k

k (cm^3 $molecule^{-1}$ s^{-1})	at T (K)	Technique	Reference
$(4.08 \pm 0.26) \times 10^{-12}$	297 ± 2	RR [relative to k(2-methyl-2-butene) $= 9.37 \times 10^{-12}$][a]	Kwok et al. (1997b)

[a]From Atkinson (1997).

III-B-16. NO_3 + Styrene

The rate coefficient obtained by Atkinson and Aschmann (1988) is given in table III-B-16. This is the only reported study of this reaction. By comparison with the rate coefficients for the reactions of the NO_3 radical with indan and tetralin (see sections III-B-13 and III-B-14), the NO_3-radical reaction with styrene must occur mainly by initial addition to the $>C{=}C<$ bond in the $-CH{=}CH_2$ substituent group.

Table III-B-16. NO_3 + styrene → products: Rate coefficients k

k (cm^3 $molecule^{-1}$ s^{-1})	at T (K)	Technique	Reference
$(1.51 \pm 0.04) \times 10^{-12}$	296 ± 2	RR [relative to k(*trans*-2-butene) $= 3.89 \times 10^{-13}$][a]	Atkinson and Aschmann (1988)

[a] From Atkinson (1997).

III-B-17. NO_3 + Benzaldehyde

The rate coefficients, or upper limits thereof, of Carter et al. (1981) and Atkinson et al. (1984b) are given in table III-B-17. The upper limit to the rate coefficient of Carter et al. (1981) is consistent with the measured rate coefficient of Atkinson et al. (1984b). The rate coefficient of Atkinson et al. (1984b) is used to recommend that

$$k(\text{benzaldehyde}) = 2.4 \times 10^{-15}\ \text{cm}^3\ \text{molecule}^{-1}\ \text{s}^{-1}$$

at 298 K, with an estimated overall uncertainty of a factor of 2.

Table III-B-17. NO_3 + benzaldehyde → products: Rate coefficients k

k (cm^3 $molecule^{-1}$ s^{-1})	at T (K)	Technique	Reference
$\leq 9.9 \times 10^{-15}$	300 ± 1	RR [relative to k(propene) = 9.73×10^{-15}][a]	Carter et al. (1981)
$(2.43 \pm 0.08) \times 10^{-15}$	294	RR [relative to k(propene) = 9.00×10^{-15}][a]	Atkinson et al. (1984b)

[a]From Atkinson (1997).

III-B-18. NO_3 + *trans*-Cinnamaldehyde

The rate coefficient obtained by Smith et al. (1996) is given in table III-B-18. This is the only reported study of this reaction. By comparison with the rate coefficients for the reactions of the NO_3-radical with benzaldehyde, indan, and tetralin (sections III-B-17, III-B-13, and II-B-14, respectively), the NO_3-radical reaction with *trans*-cinnamaldehyde must occur mainly by initial addition to the $>C{=}C<$ bond in the $-CH{=}CHCHO$ substituent group.

Table III-B-18. NO_3 + *trans*-cinnamaldehyde → products: Rate coefficients k

k (cm^3 $molecule^{-1}$ s^{-1})	at T (K)	Technique	Reference
$(1.91 \pm 0.17) \times 10^{-14}$	296 ± 2	RR [relative to k(propene) = 9.24×10^{-15}][a]	Smith et al. (1996)

[a]From Atkinson (1997).

III-B-19. NO_3 + 1,4-Naphthoquinone

The upper limit to the rate coefficient obtained by Atkinson et al. (1989b) is given in table III-B-19. This is the only reported study of this reaction.

Table III-B-19. NO_3 + 1,4-naphthoquinone → products: Rate coefficients k

k (cm^3 $molecule^{-1}$ s^{-1})	at T (K)	Technique	Reference
$< 9.5 \times 10^{-16}$	298 ± 2	RR [relative to k(propene) = 9.49×10^{-15}][a]	Atkinson et al. (1989b)

[a]From Atkinson (1997).

III-B-20. NO_3 + Phenol

The rate coefficients of Carter et al. (1981) and Atkinson et al. (1984b, 1992) are given in table III-B-20. These rate coefficients range over a factor of 2, with that of Carter et al. (1981) relative to the rate coefficient for the reaction of the NO_3 radical being fairly uncertain, at least in part due to the factor of 7 higher reactivity of the reference compound *cis*-2-butene, which thus makes accurate measurement of the rate coefficient ratio k(phenol)/k(*cis*-2-butene) difficult. The rate coefficients of Atkinson et al. (1984b) and (1992) are in good agreement, and an average of these two rate coefficients leads to the recommendation of

$$k(\text{phenol}) = 3.8 \times 10^{-12}\ \text{cm}^3\ \text{molecule}^{-1}\ \text{s}^{-1}$$

at 298 K, with an estimated overall uncertainty of a factor of 1.5.

Table III-B-20. NO_3 + phenol → products:
Rate coefficients k

k (cm^3 molecule^{-1} s^{-1})	at T (K)	Technique	Reference
$(2.59 \pm 0.52) \times 10^{-12}$	300 ± 1	RR [relative to k(*cis*-2-butene) = 3.50×10^{-13}][a]	Carter et al. (1981)
$(3.64 \pm 0.14) \times 10^{-12}$	294	RR [relative to k(2-methyl-2-butene) = 9.37×10^{-12}][a]	Atkinson et al. (1984b)
$(5.23 \pm 0.57) \times 10^{-12}$	298 ± 1	RR [relative to $K(NO_3 + NO_2 \leftrightarrow N_2O_5)$ = 3.01×10^{-11}][b]	Atkinson et al. (1984b)
$(3.92 \pm 0.25) \times 10^{-12}$	296 ± 2	RR [relative to k(2-methyl-2-butene) = 9.37×10^{-12}][a]	Atkinson et al. (1992)

[a] From Atkinson (1997).
[b] From the IUPAC evaluation (Atkinson et al., 1997a).

III-B-21. NO_3 + *o*-Cresol

The rate coefficients of Carter et al. (1981) and Atkinson et al. (1984b, 1992) are given in table III-B-21. These rate coefficients are in surprisingly excellent agreement, given the uncertainty in the rate coefficient for the reaction of the NO_3 radical with phenol (see section III-B-20). The most recent measurement of Atkinson et al. (1992) is used to recommend

$$k(o\text{-cresol}) = 1.4 \times 10^{-11}\ \text{cm}^3\ \text{molecule}^{-1}\ \text{s}^{-1}$$

at 298 K, with an estimated overall uncertainty of a factor of 1.5.

III-B-22. NO_3 + *m*-Cresol

The rate coefficients of Carter et al. (1981) and Atkinson et al. (1984b, 1992) are given in table III-B-22. These rate coefficients range over a factor of 2, with those of Carter et al. (1981) and Atkinson et al. (1992) relative to the rate coefficient for

Table III-B-21. NO_3 + *o*-cresol → products: Rate coefficients *k*

k (cm^3 $molecule^{-1}$ s^{-1})	at *T* (K)	Technique	Reference
$(1.39 \pm 0.24) \times 10^{-11}$	300 ± 1	RR [relative to *k*(2-methyl-2-butene) = 9.37×10^{-12}][a]	Carter et al. (1981)
$(1.43 \pm 0.16) \times 10^{-11}$	298 ± 1	RR [relative to *k*(*m*-cresol) = 1.1×10^{-11}][b]	Atkinson et al. (1984b)
$(1.37 \pm 0.09) \times 10^{-11}$	296 ± 2	RR [relative to *k*(2-methyl-2-butene) = 9.37×10^{-12}][a]	Atkinson et al. (1992)

[a]From Atkinson (1997).
[b]This evaluation.

the reaction of the NO_3 radical with 2-methyl-2-butene (a fairly reliably known rate coefficient) being in reasonable agreement. The most recent rate coefficient of Atkinson et al. (1992) is used to recommend that

$$k(m\text{-cresol}) = 1.1 \times 10^{-11}\ \text{cm}^3\ \text{molecule}^{-1}\ \text{s}^{-1}$$

at 298 K, with an estimated overall uncertainty of a factor of 1.5.

Table III-B-22. NO_3 + *m*-cresol → products: Rate coefficients *k*

k (cm^3 $molecule^{-1}$ s^{-1})	at *T* (K)	Technique	Reference
$(8.10 \pm 1.16) \times 10^{-12}$	300 ± 1	RR [relative to *k*(2-methyl-2-butene) = 9.37×10^{-12}][a]	Carter et al. (1981)
$(1.66 \pm 0.18) \times 10^{-11}$	298 ± 1	RR [relative to *k*(phenol) = 3.8×10^{-12}][b]	Atkinson et al. (1984b)
$\sim 1.5 \times 10^{-11}$	298 ± 1	RR [relative to *K*($NO_3 + NO_2 \leftrightarrow N_2O_5$) = 3.01×10^{-11}][c]	Atkinson et al. (1984b)
$(1.07 \pm 0.10) \times 10^{-11}$	296 ± 2	RR [relative to *k*(2-methyl-2-butene) = 9.37×10^{-12}][a]	Atkinson et al. (1992)

[a]From Atkinson (1997).
[b]This evaluation.
[c]From the IUPAC evaluation (Atkinson et al., 1997a).

III-B-23. NO_3 + *p*-Cresol

The rate coefficients of Carter et al. (1981) and Atkinson et al. (1984b, 1992) are given in table III-B-23. These rate coefficients are in reasonably good agreement. The most recent measurement of Atkinson et al. (1992) is used to recommend

$$k(p\text{-cresol}) = 1.1 \times 10^{-11}\ \text{cm}^3\ \text{molecule}^{-1}\ \text{s}^{-1}$$

at 298 K, with an estimated overall uncertainty of a factor of 2.

Table III-B-23. NO_3 + *p*-cresol → products: Rate coefficients *k*

k (cm^3 molecule^{-1} s^{-1})	at *T* (K)	Technique	Reference
$(1.50 \pm 0.24) \times 10^{-11}$	300 ± 1	RR [relative to *k*(2-methyl-2-butene) $= 9.37 \times 10^{-12}$][a]	Carter et al. (1981)
$(1.52 \pm 0.17) \times 10^{-11}$	298 ± 1	RR [relative to *k*(*m*-cresol) $= 1.1 \times 10^{-11}$][b]	Atkinson et al. (1984b)
$(1.07 \pm 0.10) \times 10^{-11}$	296 ± 2	RR [relative to *k*(2-methyl-2-butene) $= 9.37 \times 10^{-12}$][a]	Atkinson et al. (1992)

[a]From Atkinson (1997).
[b]This evaluation.

III-B-24. NO_3 + 2-Nitrophenol

The upper limit to the rate coefficient obtained by Atkinson et al. (1992) is given in table III-B-24. This is the only reported study of this reaction.

Table III-B-24. NO_3 + 2-nitrophenol → products: Rate coefficients *k*

k (cm^3 molecule^{-1} s^{-1})	at *T* (K)	Technique	Reference
$< 1.2 \times 10^{-14}$	296 ± 2	RR [relative to *k*(*trans*-2-butene) $= 3.89 \times 10^{-13}$][a]	Atkinson et al. (1992)

[a]From Atkinson (1997).

III-B-25. NO_3 + Naphthalene

The rate coefficient ratios $k_a k_c/k_b$ obtained from the experiments of Pitts et al. (1985a), Atkinson et al. (1987a, 1990b, 1994), and Atkinson and Aschmann (1988) are given in table II-B-25. The approximate rate coefficient ratio derived from the data of Pitts et al. (1985a) arises from fitting the measured time–concentration profiles of naphthalene and N_2O_5 in a single experiment using an assumed six-step reaction scheme. At room temperature, the rate coefficient ratios of Atkinson et al. (1987a, 1990b, 1994), and Atkinson and Aschmann (1988), are in reasonably good agreement, and a unit-weighted least-squares analysis of the data from these studies (Atkinson et al., 1987a, 1990b, 1994; Atkinson and Aschmann, 1988) leads to the recommendation of

$$k_a k_c/k_b(\text{naphthalene}) = 1.55 \times 10^{-33}\, e^{(3686 \pm 580)/T}\ \text{cm}^6\ \text{molecule}^{-2}\ \text{s}^{-1}$$

over the temperature range 272–298 K, where the indicated errors are two least-squares standard deviations, and

$$k_a k_c/k_b(\text{naphthalene}) = 3.65 \times 10^{-28}\ \text{cm}^6\ \text{molecule}^{-2}\ \text{s}^{-1}$$

at 298 K, with an estimated overall uncertainty at 298 K of a factor of 1.5.

The data of Atkinson et al. (1990b) show that the rate coefficient ratio $k_a k_c/k_b$ for naphthalene-d_8 is a factor of 1.22 ± 0.10 higher than that for naphthalene-h_8 at 296 ± 2 K, consistent with the rate-determining step being an addition process. The kinetic and mechanistic study of Atkinson et al. (1994) further allowed an upper limit for the rate coefficient ratio k_d/k_c of $< 4 \times 10^{-7}$ at 298 ± 2 K to be derived from the effects of changing the O_2 concentration (by a factor of 4) on the kinetics of the reaction.

Table III-B-25. NO_3 + naphthalene → products: Rate coefficient ratios $k_a k_c/k_b$

$k_a k_c/k_b$ (cm^6 molecule^{-2} s^{-1})	at T (K)	Technique	Reference
~(6–9) × 10^{-28}	298 ± 2	RR [relative to $K(NO_3 + NO_2 \leftrightarrow N_2O_5) = 3.01 \times 10^{-11}$][a]	Pitts et al. (1985a)
(4.21 ± 0.61) × 10^{-28}	298 ± 2	RR [relative to $K(NO_3 + NO_2 \leftrightarrow N_2O_5) = 3.01 \times 10^{-11}$][a]	Atkinson et al. (1987a)
3.23 × 10^{-28}	296 ± 2	RR [relative to k(propene) = 9.24×10^{-15}][b]	Atkinson and Aschmann (1988)
(1.28 ± 0.20) × 10^{-27}	272.0 ± 0.6	RR [relative to k(thiophene) = 3.93×10^{-14}][c]	Atkinson et al. (1990b)
(1.10 ± 0.27) × 10^{-27}	275.2 ± 0.6		
(1.32 ± 0.25) × 10^{-27}	275.2 ± 0.6		
(6.96 ± 1.73) × 10^{-28}	281.9 ± 0.6		
(5.31 ± 1.18) × 10^{-28}	289.1 ± 0.6		
(3.64 ± 1.11) × 10^{-28}	296.3 ± 0.6		
(3.66 ± 1.26) × 10^{-28}	296.9 ± 0.6		
(1.12 ± 0.10) × 10^{-27}	272.0 ± 0.6	RR [relative to $K(NO_3 + NO_2 \leftrightarrow N_2O_5) = 2.80 \times 10^{-27} e^{11,000/T}$][a]	Atkinson et al. (1990b)
(1.03 ± 0.04) × 10^{-27}	275.2 ± 0.6		
(7.23 ± 0.26) × 10^{-28}	275.2 ± 0.6		
(7.50 ± 0.33) × 10^{-28}	281.9 ± 0.6		
(6.12 ± 0.18) × 10^{-28}	289.1 ± 0.6		
(3.67 ± 0.15) × 10^{-28}	296.3 ± 0.6		
(3.72 ± 0.11) × 10^{-28}	296.9 ± 0.6		
(4.4 ± 2.1) × 10^{-28}	298 ± 2	RR [relative to k(propene) = 9.49×10^{-15}][b]	Atkinson et al. (1994)

[a]From the IUPAC evaluation (Atkinson et al.,1997a)
[b]From Atkinson (1997).
[c]From Atkinson (1991b).

III-B-26. NO_3 + Biphenyl

The upper limit to the rate coefficient ratio $k_a k_c/k_b$ obtained from the study of Atkinson et al. (1987a) is given in table III-B-26. This is the only reported study of this reaction.

Table III-B-26. NO_3 + biphenyl → products:
Rate coefficient ratios $k_a k_c / k_b$

$k_a k_c / k_b$ (cm^6 $molecule^{-2}$ s^{-1})	at T (K)	Technique	Reference
$< 6 \times 10^{-30}$	298 ± 2	RR [relative to $k_a k_c / k_b$(naphthalene) $= 3.65 \times 10^{-28}$][a]	Atkinson et al. (1987a)

[a]This evaluation.

III-B-27. NO_3 + 1-Methylnaphthalene

The rate coefficient ratios $k_a k_c / k_b$ obtained from the studies of Atkinson and Aschmann (1987, 1988) are given in table III-B-27, and are seen to be in reasonable agreement. An average of the rate coefficient ratios of Atkinson and Aschmann (1987) and (1988) leads to the recommendation of

$$k_a k_c / k_b(\text{1-methylnaphthalene}) = 7.8 \times 10^{-28}\ \text{cm}^6\ \text{molecule}^{-2}\ \text{s}^{-1}$$

at 298 K, with an estimated overall uncertainty of a factor of 2.

Table III-B-27. NO_3 + 1-methylnaphthalene → products:
Rate coefficient ratios $k_a k_c / k_b$

$k_a k_c / k_b$ (cm^6 $molecule^{-2}$ s^{-1})	at T (K)	Technique	Reference
$(8.50 \pm 1.03) \times 10^{-28}$	298 ± 2	RR [relative to $k_a k_c / k_b$(naphthalene) $= 3.65 \times 10^{-28}$][a]	Atkinson and Aschmann (1987)
7.00×10^{-28}	296 ± 2	RR [relative to k(*trans*-2-butene) $= 3.89 \times 10^{-13}$][b]	Atkinson and Aschmann (1988)

[a]This evaluation.
[b]From Atkinson (1997).

III-B-28. NO_3 + 2-Methylnaphthalene

The rate coefficient ratios $k_a k_c / k_b$ obtained from the studies of Atkinson and Aschmann (1987, 1988) are given in table III-B-28, and are seen to be in excellent agreement. An average of the rate coefficient ratios of Atkinson and Aschmann (1987) and (1988) leads to the recommendation of

$$k_a k_c / k_b(\text{2-methylnaphthalene}) = 1.1 \times 10^{-27}\ \text{cm}^6\ \text{molecule}^{-2}\ \text{s}^{-1}$$

at 298 K, with an estimated overall uncertainty of a factor of 2.

Table III-B-28. NO_3 + 2-methylnaphthalene → products:
Rate coefficient ratios $k_a k_c/k_b$

$k_a k_c/k_b$ (cm^6 molecule^{-2} s^{-1})	at T (K)	Technique	Reference
$(1.08 \pm 0.16) \times 10^{-27}$	298 ± 2	RR [relative to $k_a k_c/k_b$(naphthalene) $= 3.65 \times 10^{-28}$][a]	Atkinson and Aschmann (1987)
1.06×10^{-27}	296 ± 2	RR [relative to k(propene) $= 9.24 \times 10^{-15}$][b]	Atkinson and Aschmann (1988)

[a]This evaluation.
[b]From Atkinson (1997).

III-B-29. NO_3 + 2,3-Dimethylnaphthalene

The rate coefficient ratios $k_a k_c/k_b$ obtained from the studies of Atkinson and Aschmann (1987, 1988) are given in table III-B-29, and are seen to be in excellent agreement. An average of the rate coefficient ratios of Atkinson and Aschmann (1987) and (1988) leads to the recommendation of

$$k_a k_c/k_b(2,3\text{-dimethylnaphthalene}) = 1.5 \times 10^{-27}\ \text{cm}^6\ \text{molecule}^{-2}\ \text{s}^{-1}$$

at 298 K, with an estimated overall uncertainty of a factor of 2.

Table III-B-29. NO_3 + 2,3-dimethylnaphthalene → products:
Rate coefficient ratios $k_a k_c/k_b$

$k_a k_c/k_b$ (cm^6 molecule^{-2} s^{-1})	at T (K)	Technique	Reference
$(1.50 \pm 0.44) \times 10^{-27}$	298 ± 2	RR [relative to $k_a k_c/k_b$(naphthalene) $= 3.65 \times 10^{-28}$][a]	Atkinson and Aschmann (1987)
1.57×10^{-27}	296 ± 2	RR [relative to k(propene) $= 9.24 \times 10^{-15}$][b]	Atkinson and Aschmann (1988)

[a]This evaluation.
[b]From Atkinson (1997).

III-B-30. NO_3 + Acenaphthene

The rate coefficient k and rate coefficient ratio $k_a k_c/k_b$ obtained from the study of Atkinson and Aschmann (1988) are given in table III-B-30; they were derived from a plot of k_{obs} against the NO_2 concentration, with the intercept yielding the NO_2-independent rate coefficient k (interpreted as involving H-atom abstraction from the C–H bonds of the two CH_2 groups in the five-membered penta-fused ring) and the slope yielding the rate coefficent ratio $k_a k_c/k_b$. This is the only reported study of this reaction.

Table III-B-30. NO_3 + acenaphthene → products:

Rate coefficients k and rate coefficient ratios $k_a k_c / k_b$

k (cm^3 molecule^{-1} s^{-1})	$k_a k_c / k_b$ (cm^6 molecule^{-2} s^{-1})	at T (K)	Technique	Reference
$(4.59 \pm 2.46) \times 10^{-13}$	$(1.65 \pm 0.38) \times 10^{-27}$	296 ± 2	RR [relative to k(*trans*-2-butene) = 3.89×10^{-13}][a]	Atkinson and Aschmann (1988)

[a]From Atkinson (1997).

III-B-31. NO_3 + Acenaphthylene

The rate coefficient k obtained from the study of Atkinson and Aschmann (1988) is given in table III-B-31. No dependence of the rate coefficient on the NO_2 concentration was observed, indicating that the reaction involves initial NO_3-radical addition to the unsaturated $-CH{=}CH-$ bond in the five-membered penta-fused ring (with this $-CH{=}CH-$ moiety behaving as does an unsaturated $>C{=}C<$ bond in an alkene). This is the only reported study of this reaction.

Table III-B-31. NO_3 + acenaphthylene → products:

Rate coefficients k

k (cm^3 molecule^{-1} s^{-1})	at T (K)	Technique	Reference
$(5.45 \pm 0.47) \times 10^{-12}$	296 ± 2	RR [relative to k(*trans*-2-butene) = 3.89×10^{-13}][a]	Atkinson and Aschmann (1988)

[a]From Atkinson (1997).

III-B-32. NO_3 + Fluorene

The rate coefficient k obtained from the study of Kwok et al. (1997b) is given in table III-B-32. No dependence of the rate coefficient on the NO_2 concentration was observed, suggesting that the reaction involves H-atom abstraction from the C−H bonds of the $-CH_2-$ group in the five-membered ring. This is the only reported study of this reaction.

Table III-B-32. NO_3 + fluorene → products:

Rate coefficients k

k (cm^3 molecule^{-1} s^{-1})	at T (K)	Technique	Reference
$(3.95 \pm 0.51) \times 10^{-14}$	297 ± 2	RR [relative to k(1-butene) = 1.33×10^{-14}][a]	Kwok et al. (1997b)

[a]From Atkinson (1997).

III-B-33. NO_3 + Phenanthrene

The rate coefficient k and rate coefficient ratio $k_a k_c/k_b$ obtained by Kwok et al. (1994c) are given in table III-B-33. The rate parameters were derived from a plot of k_{obs} against the NO_2 concentration, with the intercept yielding the NO_2-independent rate coefficient k and the slope yielding the rate coefficient ratio $k_a k_c/k_b$. This is the only reported study of this reaction. The NO_2-independent rate coefficient k can be interpreted (Kwok et al., 1994c) as involving initial NO_3-radical addition to the 9,10-bond (assuming that this –CH=CH– bond behaves as does a >C=C< bond in an alkene in that the resulting NO_3-adduct is thermally "stable") and/or that it is due to occurrence of reaction pathway (d) (i.e., that involving reaction of the NO_3–phenanthrene adduct with O_2).

Table III-B-33. NO_3 + phenanthrene → products:
Rate coefficients k and rate coefficient ratios $k_a k_c/k_b$

k (cm^3 molecule^{-1} s^{-1})	$k_a k_c/k_b$ (cm^6 molecule^{-2} s^{-1})	at T (K)	Technique	Reference
$(1.31 \pm 0.07) \times 10^{-13}$	$(7.77 \pm 0.53) \times 10^{-28}$	296 ± 2	RR [relative to k(1-butene) = 1.32×10^{-14}][a]	Kwok et al. (1994c)

[a]From Atkinson (1997).

III-B-34. NO_3 + Fluoranthene-d_{10}

The rate coefficient ratio $k_a k_c/k_b$ obtained by Atkinson et al. (1990a) is given in table III-B-34 (and is expected to be also applicable to fluoranthene-h_{10}). This is the only reported study of this reaction.

Table III-B-34. NO_3 + fluoranthene-d_{10} → products:
Rate coefficient ratios $k_a k_c/k_b$

$k_a k_c/k_b$ (cm^6 molecule^{-2} s^{-1})	at T (K)	Technique	Reference
$\sim 5 \times 10^{-28}$[a]	296 ± 2	RR [relative to $k_a k_c/k_b$(naphthalene) = 3.97×10^{-28}][b]	Atkinson et al. (1990a)

[a]With an estimated uncertainty of a factor of ~2.
[b]This evaluation.

III-B-35. NO_3 + Pyrene

The rate coefficient ratio $k_a k_c/k_b$ obtained by Atkinson et al. (1990a) is given in table III-B-35. This is the only reported study of this reaction.

Table III-B-35. NO_3 + pyrene → products:
Rate coefficient ratios $k_a k_c/k_b$

$k_a k_c/k_b$ (cm^6 molecule^{-2} s^{-1})	at T (K)	Technique	Reference
$\sim 1.6 \times 10^{-27}$[a]	296 ± 2	RR [relative to $k_a k_c/k_b$(naphthalene) = 3.97×10^{-28}][b]	Atkinson et al. (1990a)

[a]With an estimated uncertainty of a factor of ~2.
[b]This evaluation.

III-B-36. NO_3 + Dibenzofuran

The upper limits to the rate coefficient k and rate coefficient ratio $k_a k_c/k_b$ obtained by Kwok et al. (1994b) are given in table III-B-36. This is the only reported study of this reaction.

Table III-B-36. NO_3 + dibenzofuran → products:
Rate coefficients k and rate coefficient ratios $k_a k_c/k_b$

k (cm^3 molecule^{-1} s^{-1})	$k_a k_c/k_b$ (cm^6 molecule^{-2} s^{-1})	at T (K)	Technique	Reference
$< 1.4 \times 10^{-15}$	$< 5.6 \times 10^{-30}$	297 ± 2	RR [relative to k(1-butene) = 1.33×10^{-14}][a]	Kwok et al. (1994b)

[a]From Atkinson (1997).

III-B-37. NO_3 + Dibenzo-*p*-dioxin

The upper limits to the rate coefficient k and the rate coefficient ratio $k_a k_c/k_b$ obtained by Kwok et al. (1994b) are given in table III-B-37. This is the only reported study of this reaction. The rate coefficient was observed to increase with increasing NO_2 concentration, but reached a value that was essentially independent of the NO_2 concentration for NO_2 concentrations in the range (4.8–24) $\times 10^{-13}$ molecule cm^{-3}. This behavior suggests that for these NO_2 concentrations the NO_3–dibenzo-*p*-dioxin adduct reacts with NO_2 much faster than the back-decomposition rate, and hence that for these NO_2 concentrations the measured rate coefficient is that for the initial NO_3-radical addition reaction (k_a). Values of ($k_a = 7.6 \times 10^{-14}$ cm^3 molecule^{-1} s^{-1} and $k_a k_c/k_b = 4.4 \times 10^{-27}$ cm^6 molecule^{-2} s^{-1} were derived from an analysis of the data using equation (I) with neglect of the O_2 reaction pathway k_c; both are uncertain by a factor of ~2. The rate coefficient k_a obtained from the study of Kwok et al. (1994c) seems too low by 1–2 orders of magnitude.

Table III-B-37. NO_3 + dibenzo-*p*-dioxin → products:
Rate coefficients k and rate coefficient ratios $k_a k_c/k_b$

k (cm^3 molecule^{-1} s^{-1})	$k_a k_c/k_b$ (cm^6 molecule^{-2} s^{-1})	at T (K)	Technique	Reference
$< 6.7 \times 10^{-15}$	4.4×10^{-27} [b]	297 ± 2	RR [relative to k(1-butene) = 1.33×10^{-14}][a]	Kwok et al. (1994b)

[a]From Atkinson (1997).
[b]Estimated to have an uncertainty of a factor of ~ 2 (Kwok et al., 1994b).

III-B-38. NO_3 + 1-Nitronaphthalene

The rate coefficient k and rate coefficient ratio $k_a k_c/k_b$ obtained by Atkinson et al. (1989b) are given in table III-B-38; they were derived from a plot of k_{obs} against the NO_2 concentration, with the intercept yielding an upper limit to the NO_2-independent rate coefficient k and the slope yielding the rate coefficient ratio $k_a k_c/k_b$. This is the only reported study of this reaction.

Table III-B-38. NO_3 + 1-nitronaphthalene → products:
Rate coefficients k and rate coefficient ratios $k_a k_c/k_b$

k (cm^3 molecule^{-1} s^{-1})	$k_a k_c/k_b$ (cm^6 molecule^{-2} s^{-1})	at T (K)	Technique	Reference
$< 7.3 \times 10^{-15}$	$(3.04 \pm 1.43) \times 10^{-29}$	298 ± 2	RR [relative to k(propene) = 9.49×10^{-15}][a]	Atkinson et al. (1989b)

[a]From Atkinson (1997).

III-B-39. NO_3 + 2-Nitronaphthalene

The rate coefficient k and rate coefficient ratio $k_a k_c/k_b$ obtained by Atkinson et al. (1989b) are given in table III-B-39; they were derived from a plot of k_{obs} against the NO_2 concentration, with the intercept yielding an upper limit to the NO_2-independent rate coefficient k and the slope yielding the rate coefficient ratio $k_a k_c/k_b$. This is the only reported study of this reaction.

III-B-40. NO_3 + 2-Methyl-1-nitronaphthalene

The rate coefficient k and rate coefficient ratio $k_a k_c/k_b$ obtained by Arey et al. (1990) are given in table III-B-40; they were derived from a plot of k_{obs} against the NO_2 concentration, with the intercept yielding the NO_2-independent rate coefficient k and the slope yielding the rate coefficient ratio $k_a k_c/k_b$. This is the only reported study of this reaction.

Table III-B-39. NO_3 + 2-nitronaphthalene → products: Rate coefficients k and rate coefficient ratios $k_a k_c/k_b$

k (cm^3 $molecule^{-1}$ s^{-1})	$k_a k_c/k_b$ (cm^6 $molecule^{-2}$ s^{-1})	at T (K)	Technique	Reference
$< 7.4 \times 10^{-15}$	$(2.75 \pm 1.14) \times 10^{-29}$	298 ± 2	RR [relative to k(propene) $= 9.49 \times 10^{-15}$][a]	Atkinson et al. (1989b)

[a]From Atkinson (1997).

Table III-B-40. NO_3 + 2-methyl-1-nitronaphthalene → products: Rate coefficients k and rate coefficient ratios $k_a k_c/k_b$

k (cm^3 $molecule^{-1}$ s^{-1})	$k_a k_c/k_b$ (cm^6 $molecule^{-2}$ s^{-1})	at T (K)	Technique	Reference
$(1.14 \pm 0.57) \times 10^{-14}$	$(2.75 \pm 1.81) \times 10^{-29}$	298 ± 2	RR [relative to k(propene) $= 9.49 \times 10^{-15}$][a]	Arey et al. (1990)

[a]From Atkinson (1997).

Table III-B-41. Summary of rate coefficients (298 K and 760 Torr of air) for the reactions of NO_3 with monocyclic aromatic hydrocarbons

Aromatic Hydrocarbon	k(298 K) (cm^3 $molecule^{-1}$ s^{-1})	Estimated Uncertainty k(298 K)
Benzene	$< 3 \times 10^{-17}$	[a]
Toluene	7.0×10^{-17}	Factor of 2
Ethylbenzene	$\leq 5.7 \times 10^{-16}$	[a]
o-Xylene	4.1×10^{-16}	[b]
m-Xylene	2.6×10^{-16}	Factor of 2
p-Xylene	5.0×10^{-16}	[b]
1,2,3-Trimethylbenzene	1.9×10^{-15}	[b]
1,2,4-Trimethylbenzene	1.8×10^{-15}	[b]
1,3,5-Trimethylbenzene	8.8×10^{-16}	[b]
p-Cymene	1.0×10^{-15}	[b]
Indane	7.4×10^{-15}	[b]
Tetralin	8.5×10^{-15}	Factor of 1.5
Indene	4.1×10^{-12}	[b]
Styrene	1.5×10^{-12}	[b]

[a]Upper limit.
[b]No estimate, since a single study only.

Table III-B-42. Summary of rate coefficients (298 K and 760 Torr of air) for reactions of NO_3 with monocyclic aromatic compounds

Aromatic Compound	k(298K) (cm^3 molecule^{-1} s^{-1})	Estimated Uncertainty k(298 K)
Benzaldehyde	2.4×10^{-15}	Factor of 2
trans-Cinnamaldehyde	1.9×10^{-14}	a
1,4-Naphthaquinone	$<9.5 \times 10^{-16}$	b
Phenol	3.8×10^{-12}	Factor of 1.5
o-Cresol	1.4×10^{-11}	Factor of 1.5
m-Cresol	1.1×10^{-11}	Factor of 1.5
p-Cresol	1.1×10^{-11}	Factor of 2
2-Nitrophenol	$<1.2 \times 10^{-14}$	b

[a]No estimate, since a single study only.
[b]Upper limit.

Table III-B-43. Summary of rate coefficients (298 K and 760 Torr of air) for reactions of NO_3 with polycyclic aromatic compounds

Polycyclic Aromatic Compound	k(~298 K) (cm^3 molecule^{-1} s^{-1})	$k_a k_c / k_b$ (cm^6 molecule^{-2} s^{-1})	Estimated Uncertainly k(298 K)
Naphthalene		3.7×10^{-28}	Factor of 1.5
Biphenyl		$<6 \times 10^{-30}$	a
1-Methylnaphthalene		7.8×10^{-28}	Factor of 2
2-Methylnaphthalene		1.1×10^{-27}	Factor of 2
2,3-Dimethylnaphthalene		1.5×10^{-27}	Factor of 2
Acenaphthene	4.0×10^{-13}	1.7×10^{-27}	a
Acenaphthalene	5.5×10^{-12}		a
Fluorene	4.0×10^{-14}		a
Phenanthrene	1.3×10^{-13}	7.8×10^{-28}	a
Fluoranthene-d_{10}		$\sim 5 \times 10^{-28}$	a
Pyrene		$\sim 1.6 \times 10^{-27}$	a
Dibenzofuran	$<1.4 \times 10^{-15}$	$<5.6 \times 10^{-30}$	a
Dibenzo-*p*-dioxin	$<6.7 \times 10^{-15}$	4.4×10^{-27}	a
1-Nitronaphthalene	$<7.3 \times 10^{-15}$	3.0×10^{-29}	a
2-Nitronaphthalene	$<7.4 \times 10^{-15}$	2.8×10^{-29}	a
2-Methyl-1-nitronaphthalene	1.1×10^{-14}	2.8×10^{-29}	a

[a]No estimate; measurement from a single study.

III-C. Estimation of Rate Coefficients for Reactions of NO_3 Radicals with Monocyclic Aromatic Compounds

As discussed in sections III-A and III-B, the gas-phase reactions of the NO_3 radical with monocyclic aromatic compounds occur at significant rates (with room temperature rate coefficients of $>10^{-15}$ cm^3 molecule^{-1} s^{-1}) only if the substituent groups(s) contain a $>C{=}C<$ bond or the aromatic compound contains a substituent $-OH$ group (i.e., it is phenolic). Otherwise, for benzene and alkyl-substituted benzenes, the 298 K rate coefficients are $\leq 10^{-15}$ cm^3 $molecule^{-1} s^{-1}$. For aromatic compounds containing >C=C< bond(s) in the substituent group(s), it may be anticipated that the rate coefficient can be approximately estimated in a manner analogous to those for the alkenes (Atkinson, 1991b; Calvert et al., 2000). While the database for aromatic compounds containing >C=C< bonds in the substituent group(s) is too small to draw any conclusions concerning estimation methods for the NO_3-radical reaction rate coefficients, in the alkenes the rate coefficient for the NO_3-radical reaction is related to the number and position of alkyl substituents around the >C=C< bond (Atkinson, 1991b; Calvert et al., 2000). Based on the simple alkenes, NO_3-radical reaction rate coefficients at 298 K (in units of cm^3 $molecule^{-1}$ s^{-1}) for the structural units are $-CH{=}CH_2$, 1.0×10^{-14}; $>C{=}CH_2$, 3.3×10^{-13}; *cis*- $-CH{=}CH-$, 3.5×10^{-13}; *trans*- $-CH{=}CH-$, 3.9×10^{-13}; $-CH{=}C<$, 9.4×10^{-12}; and $>C{=}C<$, 5.7×10^{-11}. Assuming that the phenyl group behaves similarly to an alkyl group (which may or may not be the case), then the 298 K rate coefficient for styrene is calculated to be identical to that for propene (1.0×10^{-14} cm^3 $molecule^{-1}$ s^{-1}), a factor of 100 lower than the measured rate coefficient. The calculated rate coefficient for indene would be the same as that for *cis*-2-butene (3.5×10^{-13} cm^3 $molecule^{-1}$ s^{-1}), while the measured rate coefficient is 4.1×10^{-12} cm^3 $molecule^{-1}$ s^{-1} (Kwok et al., 1997b), a factor of 10 higher.

For hetero-atom substituents around a >C=C< bond, their effect could be taken into account using a group substituent factor C(X) (as for the OH-radical reactions), but no such group factors have been tabulated to date and hence no quantitative estimation method is available for such compounds. However, analogous compounds could be used to estimate the reaction rate coefficients; for example, crotonaldehyde (*trans*-$CH_3CH{=}CHCHO$) can be used as a surrogate for *trans*-cinnamaldehyde. The literature rate coefficient for the reaction of the NO_3 radical with crotonaldehyde [5.1×10^{-15} cm^3 $molecule^{-1}$ s^{-1} at 298 K (Atkinson, 1991b)] is lower, by a factor of 3, than the sole measurement of Smith et al. (1996) for *trans*-cinnamaldehyde of 1.9×10^{-14} cm^3 $molecule^{-1}$ s^{-1}. These examples indicate that, at present, there is no reliable method to quantitatively estimate rate coefficients for the reactions of the NO_3 radical with aromatic compounds.

III-D. Products of Reactions of NO_3 Radicals with Monocyclic Aromatic Hydrocarbons

There is very limited information concerning the products of NO_3-radical–initiated atmospheric oxidation of aromatic compounds, in part because most of these reactions are slow and not important in the atmosphere. The few studies that have been performed employed high concentrations of NO_2 (typically 1–10 ppm), and may not provide an accurate picture of the products formed under real atmospheric conditions where NO_2 levels are much lower (typically 10–100 ppb or less).

III-D-1. Phenol, *o*-Cresol, *m*-Cresol and *p*-Cresol

There have been two studies of the products formed following reaction of NO_3 radicals with phenol and the cresols in 1 atmosphere of dry air at ambient temperature (Grosjean, 1985; Atkinson et al., 1992). As seen from table III-D.1,

Table III-D-1. Molar product yields from the reaction of NO_3 radicals with phenolic compounds

Reactant	Product	Yield	Reference
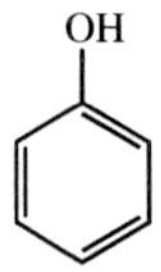	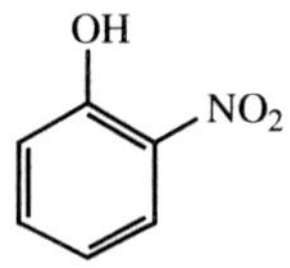	0.251 ± 0.051	Atkinson et al. (1992)
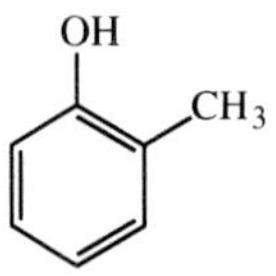	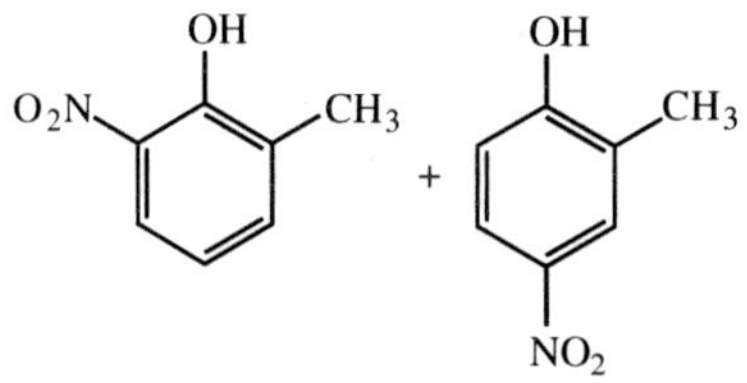	0.024–0.23	Grosjean (1985)
	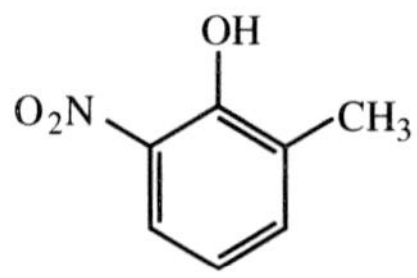	0.128 ± 0.028	Atkinson et al. (1992)
	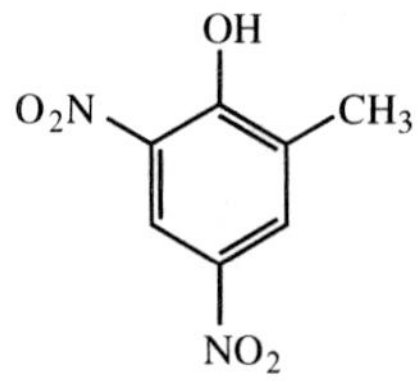	Detected	Grosjean (1985)
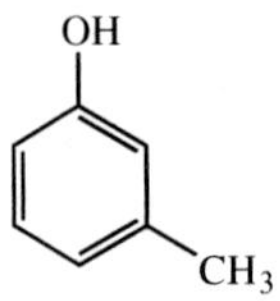	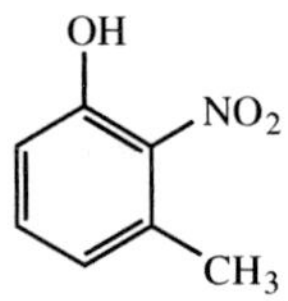	0.168 ± 0.029	Atkinson et al. (1992)
	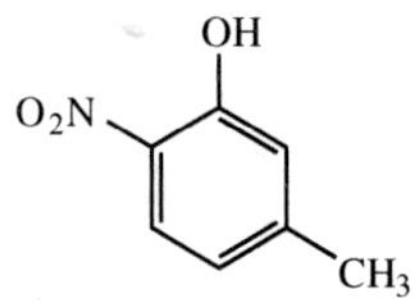	0.196 ± 0.036	Atkinson et al. (1992)
	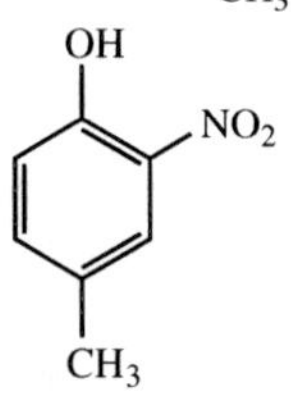	0.74 ± 0.16	Atkinson et al. (1992)

where comparison is possible, there are significant differences in results of the two studies. The experimental approach used in the two studies was similar, although not identical.

Grosjean (1985) introduced 1.1–1.4 ppm of *o*-cresol and then 0.28–0.58 ppm of NO_2 into a large (4 m^3) Teflon chamber containing air with 0.17–0.6 ppm of O_3. The NO_2 reacts with O_3 with a rate constant of 3.2×10^{-17} cm^3 $molecule^{-1}s^{-1}$ at room temperature (DeMore et al., 1997) to produce NO_3 radicals. In the presence of 0.4 ppm of O_3, the lifetime of NO_2 with respect to reaction with O_3 is 53 min. The NO_3 radicals reacted with *o*-cresol to give the observed reaction products. Gas-phase samples were collected on impingers containing 10 mL of 0.005 N aqueous KOH approximately every hour for 3–4 h and analyzed using ion chromatography with UV detection. Grosjean (1985) observed 2-methyl-6-nitrophenol and 2-methyl-4-nitrophenol as products in a combined yield that varied over the range 22.7–2.4%. Grosjean (1985) did not provide any information concerning the relative yields of the two methyl-nitrophenol isomers. The combined product yield of the two isomers decreased with increasing reaction time (and hence conversion of *o*-cresol), indicating a significant loss process for the methyl-nitrophenols in the chamber. Grosjean (1985) detected, but did not quantify, a dinitrocresol (2-methyl-4,6-dinitrophenol) in the aerosol formed in the system. Grosjean (1985) also observed 2-methyl-6-nitrophenol and 2-methyl-4-nitrophenol in the aerosol. While phase distribution results showed that the two methyl-nitrophenols reside at >95% in the gas-phase, this observation does not preclude a significant loss of such compounds via adsorption to the aerosol followed by particle deposition at the chamber walls.

Atkinson et al. (1992) prepared mixtures of 0.7–1.3 ppm of aromatic reactant (phenol or cresol) and 5–10 ppm of NO_2 in 1 atmosphere of air in a 6.4 m^3 Teflon chamber. Thermal decomposition of N_2O_5 was used as a source of NO_3 radicals. At 298 K in air at atmospheric pressure, N_2O_5 has a lifetime of 14 s with respect to thermal decomposition to give NO_3 radicals and NO_2. Three or four separate additions of 0.1–0.7 ppm of N_2O_5 were made to the chamber; gas samples were analyzed using gas chromatography with flame ionization detection and using gas chromatography–mass spectrometric (GC-MS) techniques. The experimental technique used by Atkinson et al. (1992) was considerably more rapid than that employed by Grosjean (1985); there was no evidence for any significant loss of products, and the results from Atkinson et al. (1992) are preferred. It is unclear why Grosjean (1985) found and reported two nitrophenol products while Atkinson et al. (1992) using GC-MS techniques found only one.

From table III-D-1 it can be seen that, with the exception of *p*-cresol, the observed products account for only a small fraction of the loss of the parent aromatic compound. The majority of the reaction products have yet to be identified.

As discussed in section III-A, reaction of NO_3 radicals with phenolic compounds is believed to proceed via an overall H-atom abstraction mechanism that occurs after the NO_3 radicals add to the aromatic ring:

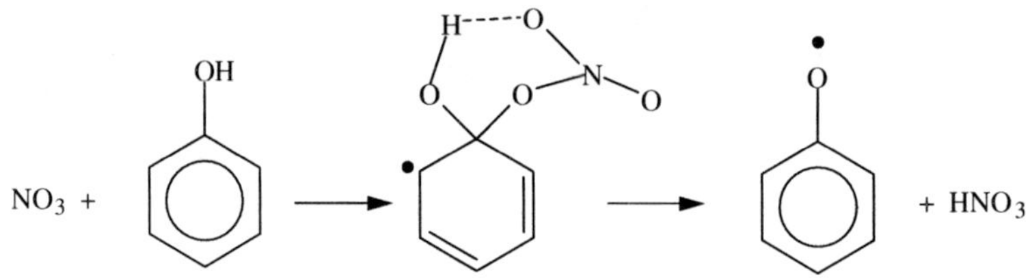

The electronic structure of the $C_6H_5O\cdot$ radical can be represented as an admixture of three principal resonance structures:

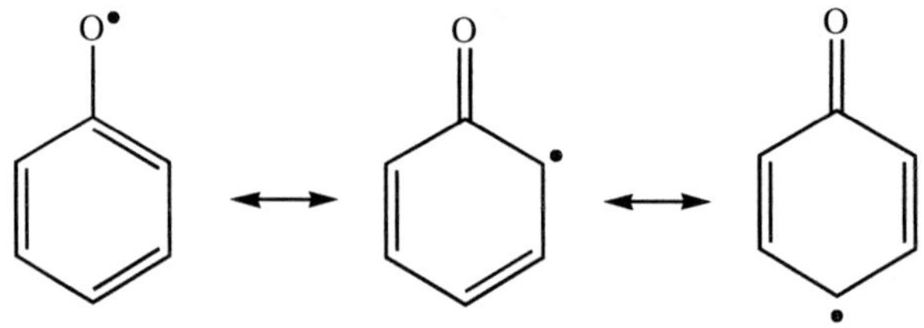

While intuitively the first resonance structure might be expected to dominate, in fact density functional calculations predict Mulliken spin densities of 0.40, 0.27, and 0.36 electrons at the phenoxy oxygen, *ortho*- and *para*-carbon positions, respectively. In contrast, the *meta*-position has a Mulliken spin density of only −0.10 electrons (Platz et al., 1998). Hence, one would expect to observe formation of phenyl nitrate, *o*-nitrophenol, and *p*-nitrophenol. The $C_6H_5O{-}NO$ bond dissociation energy in phenyl nitrite is only 21 kcal mol^{-1} (Berho et al., 1998), and it seems reasonable to expect that phenyl nitrate will also be thermally unstable. It is therefore not surprising that only *o*- and *p*-nitrophenolic compounds are observed as significant products. As discussed by Atkinson et al. (1992), from the simple mechanistic picture presented above one would expect to observe nitrophenolic products in a combined yield of 100%. As seen from table III-D-1, with the exception of the product study of *p*-cresol, this is not observed. Clearly, there are one or more significant reaction channels (perhaps involving opening of the ring) that are unknown at present. Further product studies are clearly needed.

III-D-2. Alkyltoluenes and Alkoxytoluenes

Carbonyl compounds, benzyl alcohols, nitro-derivatives, formaldehyde, methanol, and ethanol have been reported as products of the NO_3-initiated oxidation of a series of 4-alkyltoluenes and 4-methoxytoluenes (Bolzacchini et al., 1999). The product yields reported by Bolzacchini et al. (1999) are inconsistent with the kinetic database discussed in section III-D-1. For example, following reaction of NO_3 radicals with 4-ethyltoluene, the following products (and molar yields) were reported: 4-ethylbenzaldehyde (0.16 ± 0.02) and 4-methylstyrene (0.04 ± 0.01) (Bolzacchini et al., 1999). Product yields were not corrected for secondary loss via reaction with NO_3 radicals. It seems reasonable to assume that methyl- and ethylbenzaldehydes are more reactive toward NO_3 radicals than benzaldehyde and that methylstyrene is more reactive than styrene. Using

$k(NO_3 + \text{4-ethyltoluene}) = 8.6 \times 10^{-16} cm^3 molecule\, s^{-1}$ (table III-B-7) and $k(NO_3 + \text{benzaldehyde}) = 2.4 \times 10^{-15}$ (table III-B-17), we calculate that the yield of 4-ethylbenzaldehyde needs to be revised upward by a factor of at least 4.2. This correction leads to a carbon balance substantially above 100%, a physically unreasonable result. The NO_3 radicals react at least 2000 times ($1.5 \times 10^{-12}/7.7 \times 10^{-16}$) more rapidly with 4-methylstyrene than with 4-ethyltoluene. The observation of $4 \pm 1\%$ yield of 4-methylstyrene by Bolzacchini et al. (1999) is not reasonable, and that work is not considered further.

III-D-3. Styrene

The reaction of NO_3 radicals with styrene in air at atmospheric pressure has been reported to give CH_2O and C_6H_5CHO in yields of approximated 11% each, with the remaining unidentified products consisting of transient and stable species containing $-ONO_2$, $-OONO_2$, and $>C=O$ functional groups (Tuazon et al., 1993).

III-E. Mechanisms of Reactions of NO_3 Radicals with Polycyclic Aromatic Compounds

Initial interpretations of kinetic studies of naphthalene and methylnaphthalenes suggested that the reactions involved N_2O_5 rather than the NO_3 radical (Atkinson and Aschmann, 1988). In a subsequent study, 2,3-dimethyl-2-butene was used as an NO_3-radical scavenger at low enough temperatures that N_2O_5 was present while NO_3 radicals were removed rapidly by reaction. Under these conditions, the rate of reaction of naphthalene was dramatically reduced and the reactive species was therefore shown to be the NO_3 radical (Atkinson et al., 1990b). This is illustrated by the mechanism for naphthalene given in section III-A and shown in figure III-E-1. The NO_3-radical–initiated reactions therefore proceed by a number of potential pathways, involving:

- Addition of NO_3 radical to the carbon atoms of the aromatic rings to form an NO_3–PAH adduct [reaction (a) in figure III-E-1], with the NO_3–PAH adduct undergoing back-decomposition to reactants [reaction (b)] in competition with bimolecular reactions with NO_2 [reaction (c)] and O_2 [reaction (d)], as well as (possibly) unimolecular decomposition reactions to form products other than the reactants [reaction (e)].
- Abstraction of an H-atom from the O−H and/or C−H bonds of substituent groups (e.g., H-atom abstraction from the C−H bonds of the two $-CH_2-$ groups in the five-membered cyclopenta-fused ring in acenaphthene).
- Addition of NO_3 radical to the carbon atoms of any unsaturated $>C=C<$ bonds in substituent groups (an example being NO_3-radical addition to the $-CH=CH-$ moiety in the unsaturated five-membered cyclopenta-fused ring in acenaphthylene).

The available database indicates that under laboratory conditions the rate of NO_3-radical addition to the aromatic ring pathway (through the formation of

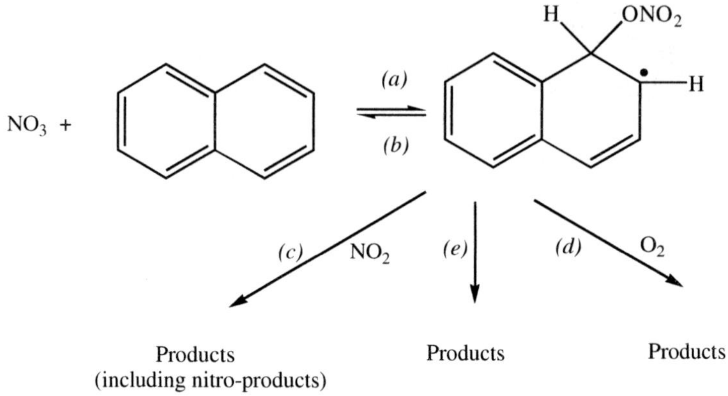

Figure III-E-1. The mechanism of NO_3-radical reaction with naphthalene.

an NO_3–PAH adduct) increases with the NO_2 concentration, with (see section III-A) an observed rate coefficient of

$$k_{obs} = k_a(k_c[NO_2] + k_d[O_2] + k_e)/(k_b + k_c[NO_2] + k_d[O_2] + k_e) \qquad (II)$$

Considering also the H-atom abstraction and NO_3-radical addition reactions with the substituent groups, the measured rate coefficient will then be

$$k_{obs} = k + k_a(k_c[NO_2] + k_d[O_2] + k_e)/(k_b + k_c[NO_2] + k_d[O_2] + k_e) \qquad (III)$$

where k is the rate coefficient for the H-atom abstraction pathway from C–H or O–H bonds of the substituent group(s) and/or NO_3-radical addition to the $>C{=}C<$ bond(s) of the substituent group(s).

Two cases therefore arise for atmospheric conditions:

- For PAHs not containing substituent groups (and apparently for the methyl-substituted naphthalenes), the rate coefficient depends on the NO_2 concentration and the products can be formed by reaction pathways (c), (d), and (e). A knowledge of the rate coefficients (c), (d), and (e) (or a knowledge of their relative importance) is therefore necessary to assess the kinetics, products, and mechanism under atmospheric conditions.
- For PAHs with substituent groups such that H-atom abstraction from C–H and/or O–H bonds and/or NO_3-radical addition to $>C{=}C<$ bonds in the substituent groups is sufficiently rapid that these reaction pathways dominate over the overall process initiated by NO_3-radical addition to the carbons atoms of the PAH aromatic rings under atmospheric conditions {i.e., $k > k_a(k_c[NO_2] + k_d[O_2] + k_e)/(k_b + k_c[NO_2] + k_d[O_2] + k_e)$}, then the kinetics, products, and mechanism of the reaction are those for H-atom abstraction from C–H and/or O–H bonds and/or NO_3-radical addition to $>C{=}C<$ bonds in the substituent groups (and the NO_3-radical addition process to the aromatic rings is of no consequence). The reactions are then analogous to those for the NO_3-radical reactions with the monocyclic aromatic compounds. Examples are the NO_3-radical reactions with acenaphthene and acenaphthylene, with the reaction with acenaphthene under laboratory conditions employed by Atkinson and Aschmann (1988) and Arey et al. (1989b) showing both an NO_2-dependent reaction involving initial NO_3-radical addition

to the aromatic rings and an NO_2-independent reaction involving H-atom abstraction from the C–H bonds of the two $-CH_2-$ groups in the five-membered cyclopenta-fused ring.

As may be expected from the above discussion, only for acenaphthene and acenaphthylene (and their homologs) are the NO_3-radical reactions important as an atmospheric loss process; this may also be the case for phenanthrene, if the data of Kwok et al. (1994c) are confirmed. Furthermore, for acenaphthene and acenaphthylene, nitro-products arising after the addition of NO_3 radicals to the carbon atoms of the aromatic rings will not be formed to any extent under atmospheric conditions (although nitroacenaphthenes are observed in laboratory experiments at elevated NO_2 concentrations) because the dominant reactions under atmospheric conditions involve H-atom abstraction (acenaphthene) and NO_3-radical addition to the $>C{=}C<$ bond of the cyclopenta-fused ring (which behaves like an alkene $>C{=}C<$ bond system).

Table III-E-1. Nitroarene products formed from the gas-phase reactions of polycyclic aromatic hydrocarbons known to be present in ambient air with nitrate radicals (both in the presence of NO_x) and their molar yields as recommended by Atkinson and Arey (1994)

PAH[a]	Nitroarene Product Yields
Naphthalene	1-Nitronaphthalene (0.17; 0.24[b]) 2-Nitronaphthalene (0.07; 0.11[b])
1-Methylnaphthalene	All 1-Methylnitronaphthalene isomers ($\approx$0.03)
2-Methylnaphthalene	All 2-Methylnitronaphthalene isomers ($\approx$0.03)
Acenaphthene	4-Nitroacenaphthene (0.040)[c] 3-Nitroacenaphthene ($\approx$0.02)[c] 5-Nitroacenaphthene ($\approx$0.015)[c]
Acenaphthylene	No nitro-isomers formed
Phenanthrene	Four nitro-isomers (including 9-nitrophenanthrene) in trace yields
Anthracene[d]	1-Nitroanthracene (0.0003[e]) 2-Nitroanthracene (0.00002[e]) 3-Nitroanthracene (0.00002[e]) 4-Nitroanthracene (0.00004[e]) 9-Nitroanthracene (0.001[e])
Pyrene	4-Nitropyrene ($\approx$0.0006)[f]
Fluoranthene	2-Nitrofluoranthene ($\approx$0.24)[f]
Acephenanthrylene	None observed
Biphenyl	No reaction observed

[a]PAH, polycyclic aromatic hydrocarbon.
[b]Sasaki et al. (1997).
[c]Yields for the NO_3-radical addition pathway to the fused aromatic rings. Reaction expected to proceed by H-atom abstraction from the C–H bonds of the cyclopenta-fused ring under atmospheric conditions.
[d]9-Nitroanthracene was observed in both the OH- and NO_3-radical reactions, but may not be a product of these reactions because it is also formed from exposure to NO_2/HNO_3.
[e]Kwok et al. (1994c).
[f]Fan et al. (1995,1996b)

For naphthalene and the methyl-substituted naphthalenes (and possibly also for pyrene and fluoranthene), the question then concerns the rate coefficients k_c, k_d, and k_e (or their rate coefficient ratios) and hence the dominant reactions of the NO_3–PAH adducts under atmospheric conditions. The sole available data concerning these rate coefficients arise from the study of naphthalene by Atkinson et al. (1994), where a rate coefficient ratio of $k_d/k_c < 4 \times 10^{-7}$ at 298 ± 2 K was obtained, corresponding to the reaction of the NO_3–naphthalene adduct with NO_2 dominating over its reaction with O_2 for NO_2 mixing ratios >80 ppb (and, of course, possibly lower).

The observation of nitro-PAHs in the atmosphere (Arey et al., 1990), where NO_2 mixing ratios are typically <100 ppb, suggests that the NO_3–PAH, and possibly also the OH–PAH adducts, react with NO_2 to a significant extent under ambient atmospheric conditions.

Reactions of the PAHs with NO_3 lead to reported nitro-PAH yields that range from <1% for phenanthrene to 25–40% for naphthalene, acenaphthene, and fluoranthene (Arey et al., 1989b; Atkinson and Arey, 1994). These yields are tabulated in table III-E-1. For phenanthrene, five nitrophenanthrenes result, including the 1-, 2-, 3-, 4-, and 9-isomers (Kwok et al., 1994c). Although the combined nitrophenanthrene yield was less than 0.2% of the reacted phenanthrene, it is not surprising that the 9-nitro isomer dominated the reactivity of the nitrophenanthrenes, given the double character of the 9–10 carbons of phenanthrene (Kwok et al., 1994c).

Except for acenaphthylene (alkene-like PAH), the gas-phase reactions of PAHs with NO_3 are not a significant loss pathway from the atmosphere (Atkinson and Arey, 1994). On the other hand, the very high yields of nitro-PAHs from PAH reactions with NO_3 mean that these reactions can be important in the formation of nitro-PAHs in the atmosphere at night (Arey et al., 1989b; Gupta et al., 1996).

IV

Reactions of Ozone with Aromatic Compounds

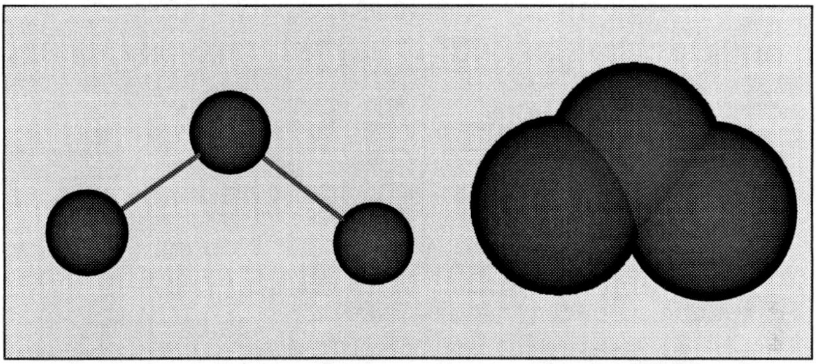

IV-A. Mechanisms of Ozone–Aromatic Compound Reactions

The gas-phase reactions of aromatic compounds with O_3 are generally very slow and hence difficult to measure accurately because of the possible presence of impurities and the occurrence of secondary reactions. These reactions are only significant for those compounds containing unsaturated $>C{=}C<$ bonds in the substituent groups. The reactions of O_3 with monocyclic aromatic compounds not containing $>C{=}C<$ bonds in the substituent group(s) either show no reaction or react only slowly (with rate coefficients of $\leq 10^{-20}$ cm^3 molecule^{-1} s^{-1} at 298 K). This behavior is expected, because the gas-phase reactions of O_3 with organic compounds generally involve initial addition of O_3 across a $>C{=}C<$ bond to form a primary ozonide, which then rapidly decomposes to (in general) two sets of carbonyl plus biradical (Atkinson and Carter, 1984). For example, for styrene the initial reaction sequence is

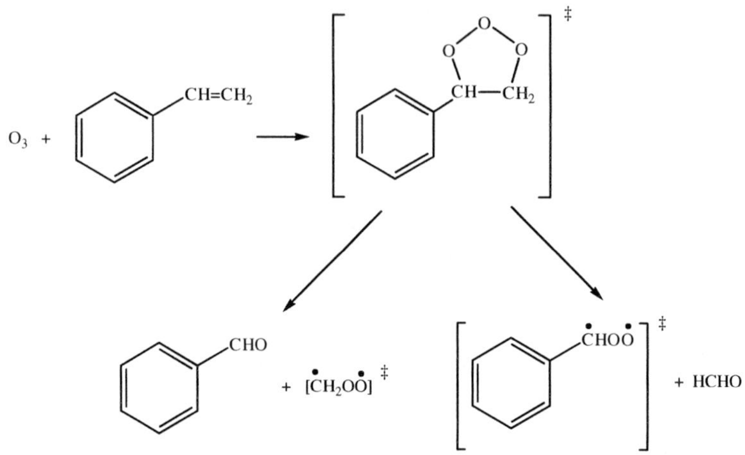

‡ Indicates an energy-rich product.

The reaction of ethene with O_3 to form HCHO + $\cdot CH_2OO\cdot$ is exothermic by ~42 kcal mol^{-1} (Olzmann et al., 1997). The aromatic ring resonance energy in benzene of ~36 kcal mol^{-1} (Benson, 1976) markedly decreases the exothermicity for O_3 to add across the aromatic ring C–C bonds, and the slowness of these O_3 reactions with the aromatic ring is therefore not surprising.

IV-B. Rate Coefficients for Reactions of Ozone with Aromatic Compounds

Rate coefficients are available for a number of aromatic compounds, mostly only at room temperature, and in many cases the reported rate coefficients are upper limits. The rate data show that the reactions of O_3 with aromatic compounds containing saturated substituent groups (i.e., with substituent groups not containing >C=C< bonds) are slow, typically with room temperature rate coefficients $< 10^{-20}$ cm^3 $molecule^{-1}$ s^{-1}. While rate coefficients (and not upper limits) have been measured for benzene, toluene, the xylenes, and 1,3,5-trimethylbenzene, it is not totally clear whether these are rate coefficients for the elementary reactions of O_3 with the aromatic hydrocarbons, or whether they are "artifacts" caused by secondary chain reactions or [less likely, given the general agreement between the studies of Pate et al. (1976), Toby et al. (1985), and Kramp and Paulson (1998) to within a factor of ~5] the presence of impurities in the aromatic hydrocarbon samples used. Note that a similar situation exists for the gas-phase reactions of O_3 with alkanes (Atkinson and Carter, 1984). Slow reactions of O_3 with the cresols and phenanthrene have also been reported (Atkinson et al., 1978, 1982; Kwok et al., 1994c), and again it is not clear whether or not the rate coefficients measured refer to the elementary gas-phase reaction with O_3, or are due to secondary reactions.

In the data tables given in the remainder of this section, the experimental techniques used are denoted by abbreviations; for example, S-CL to denote a static reaction system with the O_3 concentration being monitored by chemiluminescence. These abbreviations used throughout this section are summarized in table IV-B-0.

Table IV-B-0. Key to abbreviations used to designate the experimental techniques employed in the determination of rate coefficients for the ozone reactions discussed in this section

Abbreviation	Technique
CA	Wet chemical analysis of O_3
CL	Chemiluminescence analysis of O_3
F	Flow reactor
GC	Gas chromatographic analysis of the aromatic compound
S	Static reaction system
RR	Relative rate measurement
UV	Ultraviolet absorption analysis of O_3

For rate coefficients obtained using relative rate methods, the reference rate coefficients are taken from recent evaluations. The rate coefficient for the reaction of O_3 with the aromatic compound may therefore be different from that cited in the original publication.

IV-B-1. O_3 + Benzene

The rate coefficients of Pate et al. (1976) and Toby et al. (1985) are given in table IV-B-1. The reaction is slow, with the measured rate coefficients at room temperature being $\sim 10^{-22}$ cm^3 molecule^{-1} s^{-1}. Such a small reaction rate coefficient is difficult to measure accurately because of the possible presence of reactive impurities and the occurrence of secondary reactions [as observed by Toby et al. (1985)]. Therefore, we recommend a conservative upper limit of

$$k(\text{benzene}) < 1 \times 10^{-21} \text{ cm}^3 \text{ molecule}^{-1} \text{ s}^{-1}$$

at 298 K.

IV-B-2. O_3 + Toluene

The rate coefficients of Stedman and Niki (1973), Pate et al. (1976), and Toby et al. (1985) are given in Table IV-B-2. The rate coefficient reported by Stedman and Niki (1973) is a factor of 100 higher than those of Pate et al. (1976) and Toby et al. (1985), and Stedman and Niki (1973) observed no reaction for an isomeric xylene mixture (with $k \leq 5 \times 10^{-21}$ cm^3 molecule^{-1} s^{-1} at 298 ± 2 K). The reported rate coefficient of Stedman and Niki (1973) for toluene is therefore discounted. Based on the data of Pate et al. (1976) and Toby et al. (1985), the reaction is slow, with the measured rate coefficients at room temperature being in the 10^{-22} cm^3 molecule^{-1} s^{-1} range. Such a small reaction rate coefficient is difficult to measure accurately because of the possible presence of reactive impurities and the occur-

Table IV-B-1. O_3 + benzene → products:
Rate coefficients $k = A\ e^{-B/T}$

A (cm^3 $molecule^{-1}$ s^{-1})	B(K)	k (cm^3 $molecule^{-1}$ s^{-1})	at T(K)	Technique	Reference	Temperature Range Covered (K)
		$(7 \pm 4) \times 10^{-23}$	297 ± 2	S-CL	Pate et al. (1976)	
1.05×10^{-11}	7398 ± 353	1.7×10^{-22}[a]	298	S-UV	Toby et al. (1985)	298–423

[a]Calculated from the cited Arrhenius expression.

Table IV-B-2. O_3 + toluene → products:
Rate coefficients $k = A\,e^{-B/T}$

A (cm^3 molecule^{-1} s^{-1})	B(K)	k (cm^3 molecule^{-1} s^{-1})	at T(K)	Technique	Reference	Temperature Range Covered (K)
		$(1.2 \pm 0.6) \times 10^{-20}$	298 ± 2	S-CL	Stedman and Niki (1973)	
		$(1.5 \pm 0.8) \times 10^{-22}$	297 ± 2	S-CL	Pate et al. (1976)	
2.34×10^{-12}	6694 ± 403	4.1×10^{22} [a]	298	S-UV	Toby et al. (1985)	298–398

[a]Calculated from the cited Arrhenius expression.

rence of secondary reactions [as observed by Toby et al. (1985)]. Therefore, we recommend a conservative upper limit of

$$k(\text{toluene}) < 1 \times 10^{-21} \text{ cm}^3 \text{ molecule}^{-1} \text{ s}^{-1}$$

at 298 K.

IV-B-3. O_3 + *o*-Xylene

The rate coefficients (or upper limits thereof) of Pate et al. (1976), Atkinson et al. (1982), and Toby et al. (1985) are given in table IV-B-3. The upper limits reported by Stedman and Niki (1973) ($k \leq 5 \times 10^{-21}$ cm^3 molecule^{-1} s^{-1} at 298 ± 2 K for an isomeric mixture of xylenes) and Atkinson et al. (1982) are consistent with the data of Pate et al. (1976) and Toby et al. (1985). The reaction is slow, with the measured rate coefficients at room temperature being $\sim 10^{-21}$ cm^3 molecule^{-1} s^{-1} (Pate et al., 1976; Toby et al., 1985). Such a small reaction rate coefficient is difficult to measure accurately because of the possible presence of reactive impurities and the occurrence of secondary reactions. Therefore, we recommend a conservative upper limit of

$$k(o\text{-xylene}) < 1 \times 10^{-20} \text{ cm}^3 \text{ molecule}^{-1} \text{ s}^{-1}$$

at 298 K.

IV-B-4. O_3 + *m*-Xylene

The rate coefficients (or upper limits thereof) of Pate et al. (1976), Atkinson et al. (1982), Toby et al. (1985), and Kramp and Paulson (1998) are given in table IV-B-4. The upper limits of Stedman and Niki (1973) ($k \leq 5 \times 10^{-21}$ cm^3 molecule^{-1} s^{-1} at 298 ± 2 K for an isomeric mixture of xylenes), Atkinson et al. (1982), and Kramp and Paulson (1998) are consistent with the data of Pate et al. (1976) and Toby et al. (1985). The reaction is slow, with the measured rate coefficients at room temperature being in the 10^{-22} cm^3 molecule^{-1} s^{-1} range (Pate et al., 1976; Toby et al., 1985). Such a small reaction rate coefficient is difficult to measure accurately because of the possible presence of reactive impurities and the occurrence of secondary reactions. Therefore, we recommend a conservative upper limit of

$$k(m\text{-xylene}) < 3 \times 10^{-21} \text{ cm}^3 \text{ molecule}^{-1} \text{ s}^{-1}$$

at 298 K.

IV-B-5. O_3 + *p*-Xylene

The rate coefficients of Pate et al. (1976) and Toby et al. (1985) are given in table IV-B-5. In addition, Stedman and Niki (1973) reported an upper limit of $k \leq 5 \times 10^{-21}$ cm^3 molecule^{-1} s^{-1} at 298 ± 2 K for an isomeric mixture of xylenes. The reaction is slow, with the measured rate coefficients at room temperature being in the 10^{-22} cm^3 molecule^{-1} s^{-1} to 10^{-21} cm^3 molecule^{-1} s^{-1} range (Pate et al., 1976; Toby et al., 1985). Such a small reaction rate coefficient is

Table IV-B-3. O_3 + *o*-xylene → products:

Rate coefficients $k = A\,e^{-B/T}$

A (cm^3 $molecule^{-1}$ s^{-1})	B(K)	k (cm^3 $molecule^{-1}$ s^{-1})	at T(K)	Technique	Reference	Temperature Range Covered (K)
		$(7 \pm 2) \times 10^{-22}$	297 ± 2	S-CL	Pate et al. (1976)	
		$< 1 \times 10^{-20}$	296 ± 2	S-CL	Atkinson et al. (1982)	
2.40×10^{-13}	5586 ± 453	1.7×10^{21} [a]	298	S-UV	Toby et al. (1985)	298–423

[a]Calculated from the cited Arrhenius expression.

Table IV-B-4. O_3 + *m*-xylene → products:
Rate coefficients $k = A\,e^{-B/T}$

A (cm^3 $molecule^{-1}$ s^{-1})	B(K)	k (cm^3 $molecule^{-1}$ s^{-1})	at T(K)	Technique	Reference	Temperature Range Covered (K)
		$(6 \pm 2) \times 10^{-22}$	297 ± 2	S-CL	Pate et al. (1976)	
		$< 5 \times 10^{-21}$	296 ± 2	S-CL	Atkinson et al. (1982)	
5.37×10^{-13}	6039 ± 202	8.5×10^{22}[a]	298	S-UV	Toby et al. (1985)	298–423
		$< 2.9 \times 10^{-21}$	296 ± 2	S-UV/GC	Kramp and Paulson (1998)	

[a]Calculated from the cited Arrhenius expression.

Table IV-B-5. O_3 + *p*-xylene → products:
Rate coefficients $k = A\ e^{-B/T}$

A (cm^3 molecule^{-1} s^{-1})	B(K)	k (cm^3 molecule^{-1} s^{-1})	at T(K)	Technique	Reference	Temperature Range Covered (K)
		$(4 \pm 2) \times 10^{-22}$	297 ± 2	S-CL	Pate et al. (1976)	
1.91×10^{-13}	5586 ± 302	1.4×10^{-21} [a]	298	S-UV	Toby et al. (1985)	298–423

[a]Calculated from the cited Arrhenius expression.

difficult to measure accurately because of the possible presence of reactive impurities and the occurrence of secondary reactions. Therefore, we recommend a conservative upper limit of

$$k(p\text{-xylene}) < 1 \times 10^{-20}\ \text{cm}^3\ \text{molecule}^{-1}\ \text{s}^{-1}$$

at 298 K.

IV-B-6. O_3 + 1,2,3-Trimethylbenzene

The rate coefficient of Pate et al. (1976) is given in Table IV-B-6. This is the only reported study of this reaction. The reaction is slow, and the rate coefficient is difficult to measure accurately because of the possible presence of reactive impurities and the occurrence of secondary reactions. Therefore, we recommend a conservative upper limit of

$$k(1,2,3\text{-trimethylbenze}) < 1 \times 10^{-20}\ \text{cm}^3\ \text{molecule}^{-1}\ \text{s}^{-1}$$

at 298 K.

Table IV-B-6. O_3 + 1,2,3-trimethylbenzene → products: Rate coefficients k

k (cm^3 molecule^{-1} s^{-1})	at T(K)	Technique	Reference
$(1.6 \pm 0.4) \times 10^{-21}$	297 ± 2	S-CL	Pate et al. (1976)

IV-B-7. O_3 + 1,2,4-Trimethylbenzene

The rate coefficient of Pate et al. (1976) is given in table IV-B-7. This is the only reported study of this reaction. The reaction is slow, and the rate coefficient is difficult to measure accurately because of the possible presence of reactive impurities and the occurrence of secondary reactions. Therefore, we recommend a conservative upper limit of

$$k(1,2,4\text{-trimethylbenzene}) < 1 \times 10^{-20}\ \text{cm}^3\ \text{molecule}^{-1}\ \text{s}^{-1}$$

at 298 K.

Table IV-B-7. O_3 + 1,2,4-trimethylbenzene → products: Rate coefficients k

k (cm^3 molecule^{-1} s^{-1})	at T(K)	Technique	Reference
$(1.3 \pm 0.3) \times 10^{-21}$	297 ± 2	S-CL	Pate et al. (1976)

IV-B-8. O_3 + 1,3,5-Trimethylbenzene

The rate coefficients (or upper limits thereof) of Bufalini and Altshuller (1965), Pate et al. (1976), and Kramp and Paulson (1998) are given in table IV-B-8. The reaction is slow, and the rate coefficient is difficult to measure accurately because of the possible presence of reactive impurities and the occurrence of secondary reactions (Kramp and Paulson, 1998). Therefore, we recommend a conservative upper limit of

$$k(\text{1,3,5-trimethylbenze}) < 1 \times 10^{-20}\ \text{cm}^3\ \text{molecule}^{-1}\ \text{s}^{-1}$$

at 298 K.

Table IV-B-8. O_3 + 1,3,5-trimethylbenzene → products: Rate coefficients k

k (cm^3 $molecule^{-1}$ s^{-1})	at T(K)	Technique	Reference
$< 1 \times 10^{-19}$	303	F-CA	Bufalini and Altshuller (1965)
$(2.2 \pm 0.6) \times 10^{-21}$	297 ± 2	S-CL	Pate et al. (1976)
$(2.9 \pm 1.9) \times 10^{-21}$	296 ± 2	S-UV/GC	Kramp and Paulson (1998)

IV-B-9. O_3 + *p*-Cymene (4-Isopropyltoluene)

The upper limits to the rate coefficients of Atkinson et al. (1990a) are given in table IV-B-9. The reaction is slow, and the rate coefficient is difficult to measure accurately because of the possible presence of reactive impurities and the occurrence of secondary reactions. We recommend the upper limit obtained in the absolute measurement of Atkinson et al. (1990a) of

$$k(p\text{-cymene}) < 5 \times 10^{-20}\ \text{cm}^3\ \text{molecule}^{-1}\ \text{s}^{-1}$$

at 298K. This upper limit to the rate coefficient is consistent with those for the methyl-substituted benzenes (see above).

Table IV-B-9. O_3 + *p*-cymene → products: Rate coefficients k

k (cm^3 $molecule^{-1}$ s^{-1})	at T(K)	Technique	Reference
$< 5 \times 10^{-20}$	296 ± 2	S-CL	Atkinson et al. (1990a)
$\leq 2.1 \times 10^{-18}$	296 ± 2	RR [relative to $k(\alpha\text{-pinene}) = 8.52 \times 10^{-17}$][a]	Atkinson et al. (1990a)

[a]From Atkinson (1997).

IV-B-10. O_3 + Indan

The upper limit to the rate coefficient of Kwok et al. (1997b) is given in table IV-B-10. This is the only reported study of this reaction. The reaction is slow, and the rate coefficient is difficult to measure accurately because of the possible presence of reactive impurities and the occurrence of secondary reactions. We recommend the upper limit obtained by Kwok et al. (1997b) of

$$k(\text{indan}) < 3 \times 10^{-19} \text{ cm}^3 \text{ molecule}^{-1} \text{ s}^{-1}$$

at 298 K. This upper limit to the rate coefficient is consistent with those for the alkyl-substituted benzenes; see the discussion in the previous sections.

Table IV-B-10. O_3 + indan → products: Rate coefficients k

k (cm^3 $molecule^{-1}$ s^{-1})	at T(K)	Technique	Reference
$< 3 \times 10^{-191}$	297 ± 2	S-UV/GC	Kwok et al. (1997b)

IV-B-11. O_3 + Indene

The rate coefficient of Kwok et al. (1997b) is given in table IV-B-11. This is the only reported study of this reaction. By comparison with indan and other alkyl-substituted benzenes, the reaction obviously proceeds at the >C=C< bond in the five-membered ring.

Table IV-B-11. O_3 + indene → products: Rate coefficients k

k (cm^3 $molecule^{-1}$ s^{-1})	at T(K)	Technique	Reference
$(1.66 \pm 0.05) \times 10^{-16}$	297 ± 2	RR [relative to k(*cis*-2-butene) = 1.24×10^{-16}][a]	Kwok et al. (1997b)

[a] From Atkinson (1997).

IV-B-12. O_3 + Styrene

The rate coefficients of Bufalini and Altshuller (1965), Atkinson et al. (1982), Tuazon et al. (1993), and Zhang et al. (1994) are given in table IV-B-12. The rate coefficient reported by Zhang et al. (1994) from a single experiment with $[\text{styrene}]_{\text{initial}} > 10[O_3]_{\text{initial}}$ is a factor of 2–5 higher than the other reported rate coefficients (Bufalini and Altshuller, 1965; Atkinson et al., 1982; Tuazon et al., 1993), for reasons that are not known; Zhang et al. (1994) also obtained a rate coefficient for the reaction of O_3 with limonene that was a factor of 2 higher than the most recent literature data (Atkinson, 1997). We recommend the rate coefficient measured in the more extensive study of Tuazon et al. (1993) of

$$k(\text{styrene}) = 1.7 \times 10^{-17}\ \text{cm}^3\ \text{molecule}^{-1}\ \text{s}^{-1}$$

at 298 K, with an estimated overall uncertainty of ±30%.

Table IV-B-12. O_3 + styrene → products: Rate coefficients k

k (cm^3 $molecule^{-1}$ s^{-1})	at T(K)	Technique	Reference
3.0×10^{-17}	303	F-CA	Bufalini and Altshuller (1965)
$(2.16 \pm 0.46) \times 10^{-17}$	296 ± 2	S-CL	Atkinson et al. (1982)
$(1.71 \pm 0.18) \times 10^{-17}$	296 ± 2	S-CL/GC	Tuazon et al. (1993)
7.9×10^{-17}	297.4	S-CL/GC	Zhang et al. (1994)

IV-B-13. O_3 + *trans*-Cinnamaldehyde

The rate coefficient of Smith et al. (1996) is given in table IV-B-13. This is the only reported study of this reaction. By comparison with other alkyl-substituted benzenes and with aliphatic aldehydes (Atkinson and Carter, 1984), the reaction obviously proceeds by addition of O_3 at the >C=C< bond in the –CH=CHCHO substitutent group.

Table IV-B-13. O_3 + *trans*-cinnamaldehyde → products: Rate coefficients k

k (cm^3 $molecule^{-1}$ s^{-1})	at T(K)	Technique	Reference
$(2.2 \pm 1.8) \times 10^{-18}$	296 ± 2	RR [relative to k(propene) $= 9.68 \times 10^{-18}$][a]	Smith et al. (1996)

IV-B-14. O_3 + 1,4-Naphthoquinone

The upper limit to the rate coefficient of Atkinson et al. (1989b) is given in table IV-B-14. This is the only reported study of this reaction.

Table IV-B-14. O_3 + 1,4-naphthoquinone → products: Rate coefficients k

k (cm^3 $molecule^{-1}$ s^{-1})	at T(K)	Technique	Reference
$< 2 \times 10^{-19}$	298 ± 2	S-UV/GC	Atkinson et al. (1989b)

IV-B-15. O_3 + *o*-Cresol

The rate coefficients of Atkinson et al. (1978) and (1982) are given in table IV-B-15. The reaction is slow and the rate coefficient is difficult to measure accurately because of the possible presence of impurities and the occurrence of secondary reactions. Based on the more extensive study of Atkinson et al. (1982), we recommend that

$$k(o\text{-cresol}) = 3 \times 10^{-19} \text{ cm}^3 \text{ molecule}^{-1} \text{ s}^{-1}$$

at 298 K, with an estimated overall uncertainty of a factor of 2.

Table IV-B-15. O_3 + *o*-cresol → products: Rate coefficients *k*

k (cm^3 $molecule^{-1}$ s^{-1})	at *T*(K)	Technique	Reference
$\sim 6 \times 10^{-19}$[a]	300 ± 1	S-UV/GC	Atkinson et al. (1978)
$(2.55 \pm 0.39) \times 10^{-19}$	296 ± 2	S-CL	Atkinson et al. (1982)

[a]With an uncertainty of a factor of ~ 2 (Atkinson et al., 1978).

IV-B-16. O_3 + *m*-Cresol

The rate coefficients of Atkinson et al. (1978) and (1982) are given in table IV-B-16. The reaction is slow and the rate coefficient is difficult to measure accurately because of the possible presence of impurities and the occurrence of secondary reactions. Based on the more extensive study of Atkinson et al. (1982), we recommend

$$k(m\text{-cresol}) = 2 \times 10^{-19} \text{ cm}^3 \text{ molecule}^{-1} \text{ s}^{-1}$$

at 298 K, with an estimated overall uncertainty of a factor of 2.

Table IV-B-16. O_3 + *m*-cresol → products: Rate coefficients *k*

k (cm^3 $molecule^{-1}$ s^{-1})	at *T*(K)	Technique	Reference
$\sim 6 \times 10^{-19}$[a]	300 ± 1	S-UV/GC	Atkinson et al. (1978)
$(1.94 \pm 0.35) \times 10^{-19}$	296 ± 2	S-CL	Atkinson et al. (1982)

[a]With an uncertainty of a factor of ~ 2 (Atkinson et al., 1978).

IV-B-17. O_3 + *p*-Cresol

The rate coefficients of Atkinson et al. (1978) and (1982) are given in table IV-B-17. The reaction is slow and the rate coefficient is difficult to measure accurately because of the possible presence of impurities and the occurrence of secondary reactions. Based on the more extensive study of Atkinson et al. (1982), we recommend that

$$k(p\text{-cresol}) = 5 \times 10^{-19} \text{ cm}^3 \text{ molecule}^{-1} \text{ s}^{-1}$$

at 298 K, with an estimated overall uncertainty of a factor of 2.

Table IV-B-17. O_3 + *p*-cresol → products: Rate coefficients *k*

k (cm^3 molecule^{-1} s^{-1})	at *T*(K)	Technique	Reference
$\sim 6 \times 10^{-19}$[a]	300 ± 1	S-UV/GC	Atkinson et al. (1978)
$(4.71 \pm 0.66) \times 10^{-19}$	296 ± 2	S-CL	Atkinson et al. (1982)

[a]With an uncertainty of a factor of ~ 2 (Atkinson et al., 1978).

IV-B-18. O_3 + Nitrobenzene

The upper limit to the rate coefficient of Atkinson et al. (1987c) is given in table IV-B-18. This is the only reported study of this reaction.

Table IV-B-18. O_3 + nitrobenzene → products: Rate coefficients *k*

k (cm^3 molecule^{-1} s^{-1})	at *T*(K)	Technique	Reference
$< 7 \times 10^{-21}$	296 ± 2	S-UV/GC	Atkinson et al. (1987c)

IV-B-19. O_3 + Biphenyl

The upper limit to the rate coefficient of Atkinson et al. (1984a) is given in table IV-B-19. This is the only reported study of this reaction.

Table IV-B-19. O_3 + biphenyl → products: Rate coefficients *k*

k (cm^3 molecule^{-1} s^{-1})	at *T*(K)	Technique	Reference
$< 2 \times 10^{-19}$	294 ± 1	S-GC	Atkinson et al. (1984a)

IV-B-20. O_3 + Naphthalene

The upper limits to the rate coefficients of Atkinson et al. (1984a) and Atkinson and Aschmann (1986) are given in table IV-B-20. Based on these studies of

Atkinson et al. (1984a) and Atkinson and Aschmann (1986), we recommend an upper limit to the rate coefficient of

$$k(\text{naphthalene}) < 2 \times 10^{-19} \text{ cm}^3 \text{ molecule}^{-1} \text{ s}^{-1}$$

at 298 K.

Table IV-B-20. O_3 + naphthalene → products: Rate coefficients k

k (cm^3 $molecule^{-1}$ s^{-1})	at T(K)	Technique	Reference
$< 2 \times 10^{-19}$	294 ± 1	S-GC	Atkinson et al. (1984a)
$< 3 \times 10^{-19}$	295 ± 1	S-GC	Atkinson and Aschmann (1986)

IV-B-21. O_3 + 1-Methylnaphthalene

The upper limit to the rate coefficient of Atkinson and Aschmann (1987) is given in table IV-B-21. This is the only reported study of this reaction.

Table IV-B-21. O_3 + 1-methylnaphthalene → products: Rate coefficients k

k (cm^3 $molecule^{-1}$ s^{-1})	at T(K)	Technique	Reference
$< 1.3 \times 10^{-19}$	298 ± 2	S-GC	Atkinson and Aschmann (1987)

IV-B-22. O_3 + 2-Methylnaphthalene

The upper limit to the rate coefficient of Atkinson and Aschmann (1986) is given in table IV-B-22. This is the only reported study of this reaction.

Table IV-B-22. O_3 + 2-methylnaphthalene → products: Rate coefficients k

k (cm^3 $molecule^{-1}$ s^{-1})	at T(K)	Technique	Reference
$< 4 \times 10^{-19}$	295 ± 1	S-GC	Atkinson and Aschman (1986)

IV-B-23. O_3 + 2,3-Dimethylnaphthalene

The upper limit to the rate coefficient of Atkinson and Aschmann (1986) is given in table IV-B-23. This is the only reported study of this reaction.

Table IV-B-23. O_3 + 2.3-dimethylnaphthalene → products: Rate coefficients k

k (cm^3 $molecule^{-1}$ s^{-1})	at T(K)	Technique	Reference
$< 4 \times 10^{-19}$	295 ± 1	S-GC	Atkinson and Aschmann (1986)

IV-B-24. O_3 + Acenaphthene

The upper limit to the rate coefficient of Atkinson and Aschmann (1988) is given in table IV-B-24. This is the only reported study of this reaction.

Table IV-B-24. O_3 + acenaphthene → products: Rate coefficients k

k (cm^3 $molecule^{-1}$ s^{-1})	at T(K)	Technique	Reference
$< 5 \times 10^{-19}$	296 ± 2	S-GC	Atkinson and Aschmann (1988)

IV-B-25. O_3 + Acenaphthylene

The rate coefficient measured by Atkinson and Aschmann (1988) is given in table IV-B-25. The decay rate of acenaphthylene was measured in the presence of O_3 under approximately second-order kinetic conditions in the absence of an OH-radical scavenger, and a 1:1 stoichiometry (i.e., $\Delta[O_3]/\Delta$[acenaphthylene] = 1.0) was assumed. Because of the short lifetime of acenaphthylene under the conditions employed, relatively few data points were obtained, thereby leading to only an approximate value of the rate coefficient. This is the only reported study of this reaction.

Table IV-B-25. O_3 + acenaphthalene → products: Rate coefficients k

k (cm^3 $molecule^{-1}$ s^{-1})	at T(K)	Technique	Reference
$\sim 5.5 \times 10^{-16}$[a]	296 ± 2	S-GC	Atkinson and Aschmann (1988)

[a]With an estimated uncertainty of a factor of 2 (Atkinson and Aschmann, 1988).

IV-B-26. O_3 + Fluorene

The upper limit to the rate coefficient of Kwok et al. (1997b) is given in table IV-B-26. This is the only reported study of this reaction.

Table IV-B-26. O_3 + fluorene → products:
Rate coefficients k

k (cm^3 $molecule^{-1}$ s^{-1})	at T(K)	Technique	Reference
$< 2 \times 10^{-19}$	297 ± 2	S-GC	Kwok et al. (1997b)

IV-B-27. O_3 + Phenanthrene

The rate coefficient measured by Kwok et al. (1994c) is given in table IV-B-27. The reaction is slow and the rate coefficient is difficult to measure accurately because of the possible presence of impurities and the occurrence of secondary reactions. This is the only reported study of this reaction

Table IV-B-27. O_3 + phenanthrene → products:
Rate coefficients k

k (cm^3 $molecule^{-1}$ s^{-1})	at T(K)	Technique	Reference
$(4.0 \pm 1.0) \times 10^{-19}$	296 ± 2	S-GC	Kwok et al. (1994c)

IV-B-28. O_3 + Dibenzo-*p*-dioxin

The upper limit to the rate coefficient of Kwok et al. (1994a) is given in table IV-B-28. This is the only reported study of this reaction.

Table IV-B-28. O_3 + dibenzo-*p*-dioxin → products:
Rate coefficients k

k (cm^3 $molecule^{-1}$ s^{-1})	at T(K)	Technique	Reference
$< 5 \times 10^{-20}$	297 ± 2	S-GC	Kwok et al. (1994a)

IV-B-29. O_3 + Dibenzofuran

The upper limit to the rate coefficient of Kwok et al. (1994a) is given in table IV-B-29. This is the only reported study of this reaction.

Table IV-B-29. O_3 + dibenzofuran → products:
Rate coefficients k

k (cm^3 $molecule^{-1}$ s^{-1})	at T(K)	Technique	Reference
$< 8 \times 10^{-20}$	297 ± 2	S-GC	Kwok et al. (1994a)

IV-B-30. O_3 + 1-Nitronaphthalene

The upper limit to the rate coefficient of Atkinson et al. (1989b) is given in table IV-B-30. This is the only reported study of this reaction.

Table IV-B-30. O_3 + 1-nitronaphthalene → products:
Rate coefficients k

k (cm^3 $molecule^{-1}$ s^{-1})	at T(K)	Technique	Reference
$< 6 \times 10^{-19}$	298 ± 2	S-GC	Atkinson et al. (1989b)

IV-B-31. O_3 + 2-Nitronaphthalene

The upper limit to the rate coefficient of Atkinson et al. (1989b) is given in table IV-B-31. This is the only reported study of this reaction to date.

Table IV-B-31. O_3 + 2-nitronaphthalene → products:
Rate coefficients k

k (cm^3 $molecule^{-1}$ s^{-1})	at T(K)	Technique	Reference
$< 6 \times 10^{-19}$	298 ± 2	S-GC	Atkinson et al. (1989b)

IV-B-32. O_3 + 2-Methyl-1-nitronaphthalene

The upper limit to the rate coefficient of Arey et al. (1990) is given in table IV-B-32. This is the only reported study of this reaction.

Table IV-B-32. O_3 + 2-methyl-1-nitronaphthalene → products:
Rate coefficients k

k (cm^3 $molecule^{-1}$ s^{-1})	at T(K)	Technique	Reference
$< 3 \times 10^{-19}$	298 ± 2	S-GC	Arey et al. (1990)

IV-C. Estimation of Rate Coefficients for Reactions of Aromatic Compounds with Ozone

As discussed above in sections IV-A. and IV-B, the gas-phase reactions of O_3 with aromatic compounds only occur at significant rates (with room temperature rate coefficients of 10^{-18} cm^3 $molecule^{-1}$ s^{-1}) if the substituent group(s) contain(s) a $>C{=}C<$ bond. Otherwise, for benzene and alkyl-substituted benzenes, the 298 K rate coefficients are $< 10^{-20}$ cm^3 $molecule^{-1}$ s^{-1}. The cresols are apparently somewhat different in that they react slowly with O_3. For mono-

Table IV-B-33. Summary of the rate coefficients (298 K and 1 atm of air) for reactions of O_3 with monocyclic aromatic hydrocarbons

Aromatic Hydrocarbon	k(298 K) (cm^3 $molecule^{-1}$ s^{-1})	Estimated Uncertainty, k(298), ±%
Benzene	$< 1 \times 10^{-21}$	a
Toluene	$< 1 \times 10^{-21}$	a
o-Xylene	$< 1 \times 10^{-20}$	a
m-Xylene	$< 3 \times 10^{-21}$	a
p-Xylene	$< 1 \times 10^{-20}$	a
1,2,3-Trimethylbenzene	$< 1 \times 10^{-20}$	a
1,2,4-Trimethylbenzene	$< 1 \times 10^{-20}$	a
1,3,5-Trimethylbenzene	$< 1 \times 10^{-20}$	a
p-Cymene	$< 5 \times 10^{-20}$	a
Indan	$< 3 \times 10^{-19}$	a
Indene	1.7×10^{-16}	b
Styrene	1.7×10^{-17}	30

[a]Upper limit.
[b]No estimate; single study only.

Table IV-B-34. Summary of the rate coefficients (298 K and 1 atm of air) for reactions of O_3 with some monocyclic aromatic compounds

Aromatic Compound	k(298 K) (cm^3 $molecule^{-1}$ s^{-1})	Uncertainty Factor for k(298)
trans-Cinnamaldehyde	2.2×10^{-18}	a
o-Cresol	3×10^{-19}	Factor of 2
m-Cresol	2×10^{-19}	Factor of 2
p-Cresol	5×10^{-19}	Factor of 2
Nitrobenzene	$< 7 \times 10^{-21}$	b

[a]No estimate; single study only.
[b]Upper limit.

Table IV-B-35. Summary of the rate coefficients (~298 K and 1 atm of air) for reactions of O_3 with polycyclic aromatic compounds

Polycyclic Aromatic Hydrocarbon	k(298 K) (cm^3 $molecule^{-1}$ s^{-1})	Estimated Uncertainty in k(298)
Biphenyl	$< 2 \times 10^{-19}$	a
Naphthalene	$< 2 \times 10^{-19}$	a
1-Methylnaphthalene	$< 1.3 \times 10^{-19}$	a
2-Methylnaphthalene	$< 4 \times 10^{-19}$	a
2.3-Dimethylnaphthalene	$< 4 \times 10^{-19}$	a
Acenaphthene	$< 5 \times 10^{-19}$	a
Acenaphthylene	$\sim 5.5 \times 10^{-16}$	b
Fluorene	$< 2 \times 10^{-19}$	a
Phenanthrene	4.0×10^{-19}	b
Dibenzo-*p*-dioxin	$< 5 \times 10^{-20}$	a
Dibenzofuran	$< 8 \times 10^{-20}$	a
1,4-Naphthaquinone	$< 2 \times 10^{-19}$	a
1,-Nitronaphthalene	$< 6 \times 10^{-19}$	a
2-Nitronaphthalene	$< 6 \times 10^{-19}$	a
2-Methyl-1-nitronaphthalene	$< 3 \times 10^{-19}$	a

[a]Upper limit.
[b]No estimate; single study only.

cyclic aromatic compounds containing >C=C< bond(s) in the substituent group(s), the rate coefficient can be estimated approximately in a manner analogous to that for the alkenes (Atkinson and Carter, 1984; Calvert et al., 2000). The database for aromatic compounds containing >C=C< bonds in the substituent group(s) is too small to draw any conclusions concerning estimation methods for O_3 reaction rate coefficients (see section IV-B). In the alkenes, the rate coefficient for the O_3 reaction is related to the number and position of alkyl substituents around the >C=C< bond (Atkinson and Carter, 1984; Calvert et al., 2000). Based on the simple alkenes, O_3 reaction rate coefficients at 298 K (in units of cm^3 $molecule^{-1}$ s^{-1}) for the structural units are $-CH=CH_2$, 1.01×10^{-17}; $>C=CH_2$, 9.6×10^{-18}; *cis*- $-CH=CH-$, 1.25×10^{-16}; *trans*- $-CH=CH-$, 1.90×10^{-16}; $-CH=C<$, 4.0×10^{-16}; and $>C=C<$, 1.13×10^{-15}. Assuming that the phenyl group behaves similarly to an alkyl group (which may or may not be the case), then the 298 K rate coefficient for styrene is calculated to be identical to that for propene (1.0×10^{-17} cm^3 $molecule^{-1}$ s^{-1}) and is in reasonable agreement with our recommendation of 1.7×10^{-17} cm^3 $molecule^{-1}$ s^{-1} at 298 K. For >C=C< bond systems in bicyclic compounds (e.g., indene), the effects of ring strain on the O_3 reaction rate coefficients need to be considered (Atkinson and Carter, 1984). For example, neglecting ring-strain effects the calculated rate coefficient for indene would be the same as that for *cis*-2-butene (1.25×10^{-16} cm^3 $molecule^{-1}$ s^{-1}), while the measured 298 K rate coefficient for cyclopentene is 5.7×10^{-16} cm^3 $molecule^{-1}$ s^{-1} (Atkinson, 1997). The only measured rate coefficient for the reaction of O_3 with indene is that of Kwok et al. (1997b) of 1.66×10^{-16} cm^3 $molecule^{-1}$ s^{-1}, which lies between the rate coefficients for *cis*-2-butene and cyclopentene.

The effect of hetero-atom substituents around a >C=C< bond is taken into account using a group substituent factor C(X) (Atkinson and Carter, 1984), but no such group factors have been tabulated to date and hence no quantitative estimation method is available for such compounds (Atkinson and Carter, 1984). However, analogous compounds can be used to estimate the reaction rate coefficients; for example, crotonaldehyde [*trans*-$CH_3CH=CHCHO$] can be used as a surrogate for *trans*-cinnamaldehyde, and the literature rate coefficient for the reaction of O_3 with crotonaldehyde, 9.0×10^{-19} cm^3 $molecule^{-1}$ s^{-1} at 298 K (Atkinson, 1994), is in reasonable agreement with the sole measurement of Smith et al. (1996) for *trans*-cinnamaldehyde of $(2.2 \pm 1.8) \times 10^{-18}$ cm^3 $molecule^{-1}$ s^{-1}.

IV-D. Products and Mechanisms of Reactions of Ozone with Aromatic Hydrocarbons

IV-D-1. Reactions with the Monocyclic Aromatic Hydrocarbons

As discussed previously, ozone does not react at any significant rate with aromatic rings or with saturated aliphatic substituents. Reaction with ozone is significant only for aromatic compounds with substituent groups containing unsaturated

Table IV-D-1. Molar product yields from the reaction of O_3 with styrene

Product	Yield	Reference
HCHO	0.37 ± 0.05	Tuazon et al. (1993)
HCHO	0.34 ± 0.05	Grosjean and Grojean (1996)
C_6H_5CHO	0.41 ± 0.05	Tuazon et al. (1993)
C_6H_5CHO	0.64 ± 0.07	Grosjean and Grojean (1996)
CO	0.07 ± 0.02	Tuazon et al. (1993)
CO_2	0.04 ± 0.02	Tuazon et al. (1993)
HC(O)OH	0.015 ± 0.005	Tuazon et al. (1993)
$C_6H_5C(O)OH$	< 0.01	Tuazon et al. (1993)

$>C{=}C<$ bonds. Styrene is the only such compound for which product data are available. The products formed following reaction of ozone with styrene in 1 atm of air at room temperature have been studied by Tuazon et al. (1993) and Grosjean and Grosjean (1996), and the results are given in table IV-D-1.

Both studies report the formation of HCHO in a yield of approximately 0.35; however, there is a significant difference in the reported C_6H_5CHO yields. As discussed at the beginning of this chapter, the reaction mechanism involves O_3 addition across the $>C{=}C<$ double bond to form an ozonide that decomposes via two pathways, giving two different sets of carbonyl and excited biradical products: C_6H_5CHO and $[\cdot CH_2OO\cdot]^{\ddagger}$ or HCHO and $[C_6H_5C\cdot HOO\cdot]^{\ddagger}$ (see illustration in section IV-A). The excited biradical was first proposed by Criegee (1957) and is often referred to as a "Criegee biradical," a "Criegee radical," or simply a "Criegee intermediate." The product data from Tuazon et al. (1993) and Grosjean and Grosjean (1996) provide a reasonably consistent picture of the reaction mechanism. The fact that the combined yields of C_6H_5CHO and HCHO reported in the studies of both products does not exceed 100% shows that the excited Criegee biradicals are not converted into either C_6H_5CHO or HCHO in any significant yield. It is well established that a significant yield of OH radicals can be produced in the reaction of O_3 with alkenes (Calvert et al., 2000); these OH radicals can lead to complications in product studies. In the study of Grosjean and Grosjean (1996), cyclohexane was added as an OH scavenger. No OH scavenger was used by Tuazon et al. (1993), and we prefer the product data reported by Grosjean and Grosjean (1996). Based on the data of Grosjean and Grosjean (1996), we recommend a branching ratio of 0.65:0.35 for the primary C_6H_5CHO- and HCHO-producing channels. While there is no information available concerning the atmospheric fates of the $[\cdot CH_2OO\cdot]^{\ddagger}$ and $[C_6H_5C\cdot HOO\cdot]^{\ddagger}$ radicals produced in the reaction of O_3 with styrene, there is a substantial database concerning Criegee biradicals formed in reactions of ozone with alkenes (Calvert et al., 2000). The fate of these excited biradicals is either collisional relaxation to give stabilized radicals, or decomposition to give a variety of products. Thus, for the $[\cdot CH_2OO\cdot]^{\ddagger}$ radical:

$$[\cdot CH_2OO\cdot]^{\ddagger} + M \rightarrow \cdot CH_2OO\cdot + M$$
$$[\cdot CH_2OO\cdot]^{\ddagger} \rightarrow CO + H_2O$$

$[\cdot CH_2OO\cdot]^{\ddagger} \rightarrow CO_2 + H_2$
$[\cdot CH_2OO\cdot]^{\ddagger} \rightarrow HCO + OH$

The atmospheric fate of stabilized Criegee biradicals is believed to be reaction with water vapor to give the corresponding acid, and in the case of $\cdot CH_2OO\cdot$, hydroxymethylhydroperoxide (and H_2O_2), presumably from decomposition of $HOCH_2OOH$:

$\cdot CH_2OO\cdot + H_2O \rightarrow HC(O)OH + H_2O$
$\cdot CH_2OO\cdot + H_2O \rightarrow HOCH_2OOH \rightarrow (CH_2O + H_2O_2)$
$C_6H_5C\cdot HOO\cdot + H_2O \rightarrow C_6H_5C(O)OH + H_2O$

IV-D-2. PAH Reactions with O_3

Gas-phase reactions of fused aromatic ring PAHs with O_3 are generally very slow, with lower limits to the atmospheric lifetimes ranging from 30 to 100 days in the presence of an average global background of 0.029 ppm of O_3 (Logan, 1985). For example, in the presence of 0.029 ppm of O_3, phenanthrene has an atmospheric lifetime with respect to reaction with O_3 of 40 days. Polycyclic aromatic hydrocarbons with fused cyclopenteno rings that contain a carbon–carbon double bond, such as acenaphthene, like all other alkenes react relatively quickly with O_3 (Zielinska et al., 1988; Atkinson and Arey, 1994) and have an atmospheric lifetime of ~40 min in the presence of 0.029 ppm O_3.

V

Reactions of $O(^3P)$ Atoms with Aromatic Compounds

V-A. Mechanisms of $O(^3P)$–Aromatic Compound Reactions

Reactions of $O(^3P)$ atoms with aromatic compounds are of little importance in the ambient atmosphere, but may be significant in laboratory irradiations of NO_x–aromatic–air mixtures and possibly also in plumes containing elevated concentrations of NO_2. As with most other classes of atmospheric reactions, there is a substantial database concerning the kinetics of the reactions of $O(^3P)$ atoms with aromatic compounds, but relatively few product studies have been performed.

Analogous to the OH- and NO_3-radical reactions reviewed in previous chapters, reactions of the $O(^3P)$ atom with aromatic compounds proceed mainly by initial addition of the $O(^3P)$ atom to the carbon atoms of the aromatic ring (Sloane, 1977; Sibener et al., 1980). The measured room temperature rate coefficients increase with the number of alkyl-substituent groups attached to the aromatic ring (see table V-B-1); the positions of the substituent groups also have an effect on the rate coefficients (with *m*-xylene and 1,3,5-trimethylbenzene being the most reactive xylene and trimethylbenzene, respectively). This behavior is very similar to that observed for the OH-radical reactions (chapter II). Grovenstein and Mosher (1970) showed that formation yields of the phenolic products (formed by rearrangement of the initially formed biradical) are consistent with electrophilic addition, with alkyl-substituent groups activating the *ortho*- and *para*- positions.

As discussed by Frerichs et al. (1990b), at temperatures < 870 K the rate coefficients for reactions of $O(^3P)$ atoms with toluene, ethylbenzene, and isopropylbenzene (all monoalkylbenzenes) are very similar, with 298 K rate coefficients of (6.0–7.4) × 10^{-14} cm^3 molecule^{-1} s^{-1} and values of B (in $k = A\,e^{-B/T}$) in the range 1800–1910 K (see table V-B-1). The similarity of the rate coefficients for these three monoalkylbenzenes indicates that H-atom abstraction from C-H bonds of the alkyl-substituent groups is of little importance and that the dominant reaction pathway involves O(^3P)-atom addition to the carbon atoms of the aromatic ring.

Rate coefficients for the reactions of O(^3P) atoms and OH radicals with the 15 aromatic hydrocarbons for which data are available are plotted in figure V-A-1, where the correlation between the two series of measurements suggests strongly that there is a common mechanism operative in the two series; namely, dominance of addition of the attacking species to the aromatic ring.

V-B. Rate Coefficients for Reactions of O(^3P) Atoms with Aromatic Compounds

Absolute rate coefficients for the gas-phase reactions of O(^3P) atoms with aromatic compounds at around room temperature have been measured by Mani and Sauer (1968), Bonanno et al. (1972), Atkinson and Pitts (1974, 1975a, 1975b, 1979), Colussi et al. (1975), Nicovich et al. (1982a, 1982b), Schliephake et al. (1986), Frerichs et al. (1989a, 1989b, 1990a, 1990b, 1991), Tappe et al. (1989), and Buchta et al. (1993). The rate coefficients reported by Mani and Sauer (1968) are systematically higher than those from all of the other studies. The experiments of Mani and Sauer (1968) were conducted using gas mixtures containing 15–100 atm pressure of Ar and CO_2, Ar and N_2, or CO_2, with 0.009–0.11 atm of aromatic reactant. Kinetic data were obtained by monitoring the rate of increase

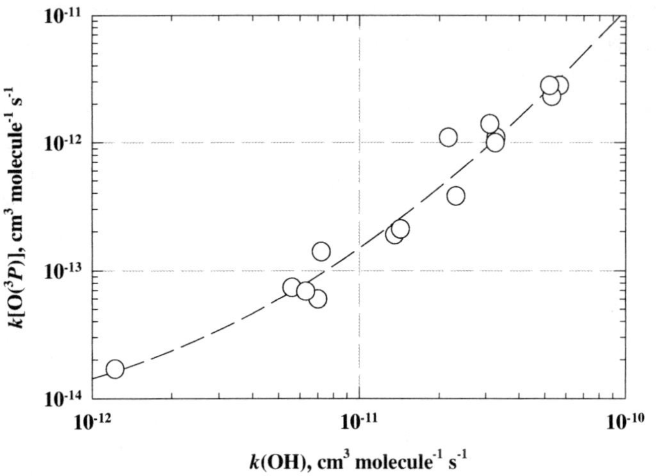

Figure V-A-1. Rate coefficients for reactions of aromatic hydrocarbons with O(^3P)-atoms versus those for OH radicals for the 15 different aromatic hydrocarbons for which data are available. The dashed curve is the least-squares second-order fit to the data.

in absorption at 275–285 nm attributed to formation of an O(^3P)–aromatic adduct. It is now known that for the conditions employed by Mani and Sauer (1968), the effective second-order rate coefficient for reaction of O(^3P) atoms with O_2 is of the order of 10^{-13} to 10^{-12} cm^3 molecule^{-1} s^{-1} (Atkinson et al., 1997a), comparable to that for the reactions of the O(^3P) atom with aromatic compounds. The presence of a small (0.001%) O_2 impurity in the diluent gas would then be sufficient to explain the high rate coefficients reported by Mani and Sauer (1968) . This potential problem was not recognized by Mani and Sauer (1968), and their results are therefore not considered here in the evaluation of the O(^3P)-atom reaction rate coefficients.

The rate coefficients measured by Bonanno et al. (1972) and Schliephake et al. (1986) for the reaction of O(^3P) atoms with benzene are a factor of approximately 2 (or more) higher than those measured in more recent studies. The study of Schliephake et al. (1986) was repeated by the same research group (Tappe et al., 1989), with the more recent data of Tappe et al. (1989) being in good agreement with other studies as shown in table V-B-1. Hence, we assume in our evaluation that the rate coefficients of Schliephake et al. (1986) are superceded by those of Tappe et al. (1989). In the study of Bonanno et al. (1972), rate coefficients were derived from two sets of experiments in which either the loss of O(^3P) atoms was monitored (by electron paramagnetic resonance, EPR) in the presence of excess concentrations of benzene, or the loss of benzene was monitored (by mass spectrometry, MS) in the presence of excess concentrations of O(^3P) atoms. Rate coefficients obtained by monitoring the loss of O(^3P) atoms were an order of magnitude higher than those derived from the benzene loss rates, presumably due to secondary reactions of O(^3P) atoms under the conditions employed.

Table V-B-1 gives the Arrhenius parameters and room temperature rate coefficients determined by Atkinson and Pitts (1974, 1975a, 1975b, 1979), Colussi et al. (1975), Nicovich et al. (1982a, 1982b), Schliephake et al. (1986), Frerichs et al. (1989a, 1989b, 1990a, 1990b, 1991), Tappe et al. (1989), and Buchta et al. (1993) for the reactions of the O(^3P) atom with aromatic compounds (including a number of polycyclic aromatic hydrocarbons). In general, the agreement is reasonable for those aromatic compounds for which more than a single study has been carried out (for benzene, toluene, the xylenes, and ethylbenzene, but not for isopropylbenzene), although the temperature dependencies, B, measured by Atkinson and Pitts (1974, 1975b, 1979) are consistently lower (at ~200–300 K) than those measured in the other studies cited in table V-B-1 for benzene, toluene, and the xylenes.

Recommended values of the Arrhenius parameters A and B (in $k = A\,e^{-B/T}$) and of the 298 K rate coefficients are listed in table V-B-2; the recommended parameters were derived from the data given in table V-B-1. Except for the benzene, toluene, xylene, ethylbenzene, and isopropylbenzene reactions (see above), only single studies have been carried out, and hence the rate data for these reactions may contain systematic errors; for example, as discussed above, the temperature dependencies B given for 1,2,3-, 1,2,4-, and 1,3,5-trimethylbenzene may well be too low since they are based on the data of Atkinson and Pitts (1974, 1975b).

Table V-B-1. Rate coefficients, $k = A\,e^{-B/T}$, for the reactions of the $O(^3P)$ atom with aromatic compounds

Aromatic Compound	$10^{11} \times A$ (cm^3 $molecule^{-1}$ s^{-1})	B (K)	$10^{14} \times k$ (298 K) (cm^3 $molecule^{-1} s^{-1}$)	References
Benzene	1.84	2003 ± 202	2.39 ± 0.34	Atkinson and Pitts (1974, 1975b)
	1.81	2114 ± 217	1.54 ± 0.08	Colussi et al. (1975)
	1.68	2011 ± 101	2.00 ± 0.20	Atkinson and Pitts (1979)
	4.62	2396 ± 76[a]	1.58 ± 0.39	Nicovich et al. (1982a)
	4.8	2140 ± 230	3.66[b]	Schliephake et al. (1986)
	3.5	2250 ± 230	1.83[b]	Tappe et al. (1989)[c]
Toluene	1.36	1560 ± 151	7.47 ± 0.75	Atkinson and Pitts (1974, 1975b)
	3.82	1943 ± 21	5.68 ± 0.25	Colussi et al. (1975)
	1.64	1535 ± 101	9.62 ± 0.96	Atkinson and Pitts (1979)
	4.26	1873 ± 101[a]	7.14 ± 0.89	Nicovich et al. (1982a)
	2.8	1770 ± 180	7.4[b]	Tappe et al. (1989)
Ethylbenzene	3.0	1840 ± 180	6.2[b]	Tappe et al. (1989)
	3.65	1912 ± 191	6.0[b]	Frerichs et al. (1991)[c]
o-Xylene	1.04	1223 ± 151	17.4 ± 1.9	Atkinson and Pitts (1974, 1975b)
	3.90	1540 ± 80	22.0 ± 2.0	Nicovich et al. (1982b)
	3.3	1550 ± 160	18.3[b]	Frerichs et al. (1989b)
m-Xylene	1.28	1082 ± 151	35.2 ± 3.5	Atkinson and Pitts (1974, 1975b)
	3.78	1350 ± 90	39.3 ± 4.3	Nicovich et al. (1982b)
	3.8	1360 ± 140	39.8[b]	Frerichs et al. (1989a)
p-Xylene	1.31	1278 ± 151	18.1 ± 1.9	Atkinson and Pitts (1974, 1975b)
	1.91	1540 ± 90	21.9 ± 2.2	Nicovich et al. (1982b)
	4.3	1540 ± 150	24.6[b]	Frerichs et al. (1989b)
Isopropylbenzene	3.3	1616 ± 162	14.7[b]	Frerichs et al. (1990a)
	3.3	1839 ± 184	6.9[b]	Frerichs et al. (1991)[c]
1,2,3-Trimethylbenzene	1.71	805 ± 151	115 ± 12	Atkinson and Pitts (1974, 1975b)
1,2,4-Trimethylbenzene	1.55	830 ± 151	100 ± 10	Atkinson and Pitts (1974, 1975b)
1,3,5-Trimethylbenzene	1.00	388 ± 151	279 ± 34	Atkinson and Pitts (1974, 1975b)

(continued)

Table V-B-1. *(continued)*

Aromatic Compound	$10^{11} \times A$ (cm^3 $molecule^{-1}$ s^{-1})	B (K)	$10^{14} \times k$ (298 K) (cm^3 $molecule^{-1}$ s^{-1})	References
Styrene	0.23[d]	-198 ± 20[d]	450[b]	Buchta et al. (1993)
Phenol	2.1	1455 ± 138	16[b,e]	Frerichs et al. (1990a)
o-Cresol			58 ± 14	Atkinson and Pitts (1975a)
Biphenyl	2.0	1485 ± 149	13.7[b]	Frerichs et al. (1990b)
Naphthalene	2.3	902 ± 90	113[b,c]	Frerichs et al. (1989a, 1990b)
1-Methylnaphthalene	2.0	647 ± 65	227[b,c]	Frerichs et al. (1990b)
2-Methylnaphthalene	2.2	606 ± 61	283[b,c]	Frerichs et al. (1990b)
Phenanthrene	3.2	921 ± 92	143[b,c]	Frerichs et al. (1990b)

[a]There are discrepancies in this publication concerning the Arrhenius activation energy. The value cited is from Table II of Nicovich et al. (1982a).
[b]Calculated from the cited Arrhenius expression.
[c]Supercedes earlier work from the same laboratory.
[d]The cited Arrhenius expression is applicable only for temperatures ≤ 400 K, where the overall reaction proceeds mainly by addition of the $O(^3P)$ atom to the $>C{=}C<$ bond of the $-C{=}CH_2$ substituent group. At 298 K, $O(^3P)$-atom addition to the aromatic ring was estimated to have a rate coefficient of $\sim 2.8 \times 10^{-14}$ cm^3 $molecule^{-1}$ s^{-1} (Buchta et al., 1993).
[e]Rate coefficients measured at temperatures 342–872 K for phenol, 362–773 K for naphthalene, ~ 335–875 K for 1-methylnaphthalene, ~ 335–575 K for 2-methylnaphthalene, and ~ 420–675 K for phenanthrene.

Table V-B-2. Recommended rate coefficients at 298 K and temperature-dependent parameters, $k = A\,e^{-B/T}$, for reactions of $O(^3P)$ atoms with aromatic compounds

Aromatic Compound	A (cm^3 $molecule^{-1}$ s^{-1})	B (K)	k (298 K) (cm^3 $molecule^{-1}$ s^{-1})	Based on Reference(s)
Benzene	2.7×10^{-11}	2200	1.7×10^{-14}	Colussi et al. (1975); Atkinson and Pitts (1979); Nicovich et al. (1982a); Tappe et al. (1989)
Toluene	3.1×10^{-11}	1800	7.4×10^{-14}	Colussi et al. (1975); Atkinson and Pitts (1979); Nicovich et al. (1982a); Tappe et al. (1989)
Ethylbenzene	3.6×10^{-11}	1910	6.0×10^{-14}	Frerichs et al. (1991)
o-Xylene	2.9×10^{-11}	1500	1.9×10^{-13}	Atkinson and Pitts (1974, 1975b); Nicovich et al. (1982b); Frerichs et al. (1989b)
m-Xylene	3.0×10^{-11}	1300	3.8×10^{-13}	Atkinson and Pitts (1974, 1975b); Nicovich et al. 1982b); Frerichs et al. (1989b)
p-Xylene	3.2×10^{-11}	1500	2.1×10^{-13}	Atkinson and Pitts (1974, 1975b); Nicovich et al. (1982b); Frerichs et al. (1989b)
Isopropylbenzene	3.3×10^{-11}	1840	6.9×10^{-14}	Frerichs et al. (1991)
1,2,3-Trimethylbenzene	1.7×10^{-11}	810	1.1×10^{-12}	Atkinson and Pitts (1974, 1975b)
1,2,4-Trimethylbenzene	1.6×10^{-11}	830	1.0×10^{-12}	Atkinson and Pitts (1974, 1975b)
1,3,5-Trimethylbenzene	1.0×10^{-11}	390	2.8×10^{-12}	Atkinson and Pitts (1974, 1975b)
Styrene[a]	2.3×10^{-12}	−200	4.5×10^{-12}	Buchta et al. (1993)
Phenol	2.1×10^{-11}	1450	1.6×10^{-13}	Frerichs et al. (1990a)
o-Cresol			5.8×10^{-13}	Atkinson and Pitts (1975a)
Biphenyl	2.0×10^{-11}	1485	1.4×10^{-13}	Frerichs et al. (1990b)

(continued)

Table V-B-2. (*continued*)

Aromatic Compound	A (cm^3 $molecule^{-1}$ s^{-1})	B (K)	k (298 K) (cm^3 $molecule^{-1}$ s^{-1})	Based on Reference(s)
Naphthalene	2.3×10^{-11}	902	1.1×10^{-12}	Frerichs et al. (1989a, 1990b)
1-Methylnaphthalene	2.0×10^{-11}	647	2.3×10^{-12}	Frerichs et al. (1990b)
2-Methylnaphthalene	2.2×10^{-11}	606	2.8×10^{-12}	Frerichs et al. (1990b)
Phenanthrene	3.2×10^{-11}	921	1.4×10^{-12}	Frerichs et al. (1990b)

[a]The cited Arrhenius expression is applicable for temperatures ≤ 400 K, where the overall reaction proceeds mainly by addition of the $O(^3P)$ atom to the $>C{=}C<$ bond of the $-CH{=}CH_2$ substituent group. At 298 K, $O(^3P)$-atom addition to the aromatic ring was estimated to have a rate coefficient of $\sim 2.8 \times 10^{-14}$ cm^3 $molecule^{-1}$ s^{-1} (Buchta et al., 1993).

V-C. Products and Mechanisms for Reactions of O(^3P) Atoms with Aromatic Hydrocarbons

The product database for the reactions of O(^3P) atoms with aromatic hydrocarbons is sparce. With the exception of the study by Gaffney et al. (1976), all product studies have been performed at room temperature. Product data are available from the studies of Boocock and Cvetanovic (1961), Jones and Cvetanovic (1961), Mani and Sauer (1968), Grovenstein and Mosher (1970), Bonanno et al. (1972), Gaffney et al. (1976), Sloane (1977), Sibener et al. (1980), and Nicovich et al. (1982a). The data from these product studies and the kinetic data discussed in section V-B show that (with the exception of styrene) the reaction mechanism is electrophilic addition to the aromatic ring to give a triplet biradical species that undergoes rapid isomerization to give phenolic compounds and H-atom elimination to give phenoxy radicals.

Boocock and Cvetanovic (1961) conducted a product study of the reaction with benzene and reported the formation of phenol in 13% yield, with the major product being a yellow-colored low-volatility polymeric material. Jones and Cvetanovic (1961) studied the reaction with toluene and reported the formation of *o*-cresol and *p*-cresol in a combined yield of 15–20% with no observable *m*-cresol, a 5–10% yield of CO, and a reddish-yellow-colored low-volatility polymeric material that accounted for the majority of the reaction. As with the study of the benzene reaction, there was a total absence of products that could be attributed to H-atom abstraction. Based upon these results, a reaction mechanism was proposed in which the O-atom adds to the aromatic ring to form an adduct, which either isomerizes to give phenolic compounds or undergoes ring cleavage/contraction to give olefinic compounds. It was proposed that subsequent reaction of the reactive olefinic compounds with O-atoms produced the low-volatility polymeric material.

In the pulse radiolysis study of the reaction of O(^3P) atoms with benzene, toluene, ethylbenzene, *o*-xylene, *m*-xylene, *p*-xylene, fluorobenzene, and chlorobenzene by Mani and Sauer (1968), a transient species with strong absorption in the region 260–320 nm was observed that was attributed to the formation of the O(^3P)–aromatic species. Mani and Sauer (1968) used GC techniques in end-product analysis for the O(^3P) + benzene reaction and reported phenol as a product.

Grovenstein and Mosher (1970) conducted a product study of the reaction of O(^3P) atoms with 13 different aromatic compounds (benzene, benzene-d_6, toluene, ethylbenzene, *tert*-butylbenzene, *o*-, *m*-, and *p*-xylene, 1,2,3-, and 1,3,5-trimethylbenzene, fluorobenzene, trifluoromethylbenzene, and methoxybenzene). In all cases there were two major products: phenolic compounds and polymeric tars. In 12 out of the 13 compounds studied, phenolic compounds accounted for >95% of the volatile products. For ethylbenzene, the phenolic products accounted for 82% of the volatile products. The observed substitution pattern was consistent with that expected from electrophilic attack by the O(^3P) atom.

Bonanno et al. (1972) studied the reaction of O(^3P) atoms with benzene and reported a 10% yield of phenol together with "copious" quantities of a yellow polymeric material. Mass spectrometric analysis of the polymeric material gave many mass peaks, with $m/e = 185$ being the most intense.

Gaffney et al. (1976) studied the products following reaction of $O(^3P)$ atoms with toluene at 373 K. Reaction products were CO, *o*-cresol, *m*-cresol, *p*-cresol, phenol, an unknown species, and a reddish-yellow tar. The volatile addition products accounted for 15–25% of the $O(^3P)$ atoms consumed. Nuclear magnetic resonance (NMR) analysis of the tar showed no evidence of aromatic protons. While the results of Jones and Cvetanovic (1961), Grovenstein and Mosher (1970), and Gaffney et al. (1976) are in agreement with respect to the identity of the products, there were significant differences in the relative yields of the individual products.

Sloane (1977) and Sibener et al. (1980) conducted crossed molecular beam studies of the reaction of $O(^3P)$ atoms with benzene. Both studies concluded that the reaction proceeded via two pathways and that one pathway leads to the formation of phenol. However, there was disagreement as to the nature of the second pathway. Sloane (1977) reported CO and a C_5H_6 hydrocarbon (probably 3-penten-1-yne) as products. In contrast, Sibener et al. (1980) did not find CO to be a major reaction channel, and showed that the CO observed by Sloane was not a primary reaction product but instead was a fragment formed in the mass spectrometer. Sibener et al. (1980) argued that opening of the aromatic ring is unlikely because of the high energy threshold involved. Sibener et al. (1980) concluded that the reaction of $O(^3P)$ with benzene proceeds via formation of a triplet biradical, C_6H_6O, which undergoes either H-atom elimination to give a phenoxy radical, or isomerization to give, most likely, phenol.

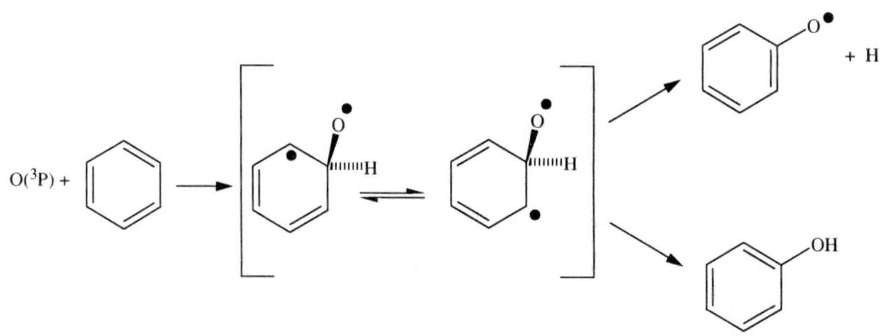

Support for this mechanism comes from the flash photolysis study of Nicovich et al. (1982a), where resonance fluorescence was used to search for CO formation following reaction of $O(^3P)$ with benzene; none ($< 5\%$) was observed. Based upon the work of Sibener et al. (1980) and Nicovich et al. (1982a), it seems likely that the observed CO product in the previous end-product studies was formed in secondary reactions and is not a primary product.

It has been shown in more recent studies (Berho and Lesclaux, 1997; Platz et al., 1998) that phenoxy radicals undergo rapid self-reaction to give 4-phenoxy-phenol, which has a molecular weight of 186 and may explain the prominent $m/e = 185$ feature reported by Bonanno et al. (1972). While the reaction of $O(^3P)$ atoms with 4-phenoxy-phenol could explain the formation of the polymeric tar reported in the end-product studies, it is difficult to understand why such tars would not show evidence of aromatic protons. An alternative explanation is to postulate that the O(^3P)–aromatic adduct forms benzene oxide. Benzene oxide reacts rapidly with OH radicals (Klotz et al., 1997; see section VI-A-17a) and it seems likely that reaction with O(^3P) would also be rapid. The tars could then be ascribed to products of the reaction of O(^3P) with benzene oxide rather than to products of reaction of O(^3P) atoms with olefins produced following ring opening/contraction of the O(^3P)–aromatic adduct (Sibener et al., 1980). Further product work is needed to clarify this matter.

From the available product and kinetic data, we draw the following conclusions with regard to the products and mechanism of the reaction of O(^3P) atoms with aromatics. First, the kinetic data and product distributions show that the rate-determining step is electrophilic addition of the O(^3P) atom to the aromatic ring to form an adduct. Second, the fate of the O(^3P)–aromatic adduct is either isomerization to phenolic compounds (in a 10–20% yield), conversion to benzene oxide, or H-atom elimination to give phenoxy radicals. The atmospheric fate of phenolic compounds and benzene oxide are discussed in chapters II, VI, and VII. Under atmospheric conditions, the fate of the phenoxy radicals will likely be reaction with NO and NO_2 to give nitroso- and nitro-phenolic compounds (Platz et al., 1998) and reaction with ozone to give unknown products (Tao and Li, 1999).

VI

Reactions of Non-Aromatic Products of Aromatic Compound Oxidation

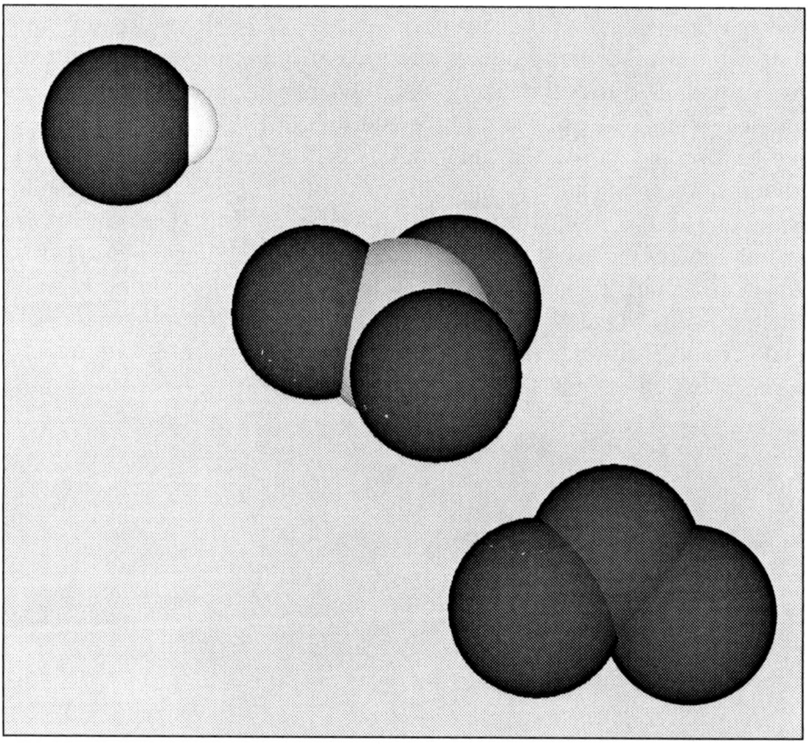

The first-generation products formed from the atmospheric reactions of aromatic compounds, discussed in sections II-D, III-D, and IV-D, can also react in the atmosphere with OH radicals, NO_3 radicals, and O_3, and may photolyze. The available rate coefficient data for ring-retaining products are reviewed and evaluated in sections II-B, III-B, and IV-B and photolysis is considered in chapter VII. The available rate coefficient data concerning the reactions of non-aromatic reaction products with OH radicals, NO_3 radicals, and O_3 are dealt with in this chapter.

VI-A. Kinetic Data for Reactions of OH Radicals, NO_3 Radicals, and Ozone with Non-Aromatic Products of Aromatic Compound Oxidation

The format of this section is similar to that of sections II-B, III-B, and IV-B, and the same abbreviations (see tables II-B-0, IV-B-0) are used in the tables below for the experimental techniques. The available kinetic data for the OH radical, NO_3 radical, and O_3 reactions are all presented here.

VI-A-1. Kinetic Data for Glyoxal, HC(=O)C(=O)H

VI-A-1a. OH-Radical Reaction

The rate coefficient of Plum et al. (1983) is given in table VI-A-1a. This is the only reported study of this reaction. Niki et al. (1985a) investigated the reactions of the HC(C=O)CO radical formed after H-atom abstraction from glyoxal, and showed that the subsequent reactions are

$$HC(=O)CO \rightarrow HCO + CO \quad (a)$$

$$HC(=O)CO + O_2 \rightarrow HC(=O)C(=O)OO \quad (b)$$

$$HC(=O)CO + O_2 \rightarrow CO + CO + HO_2 \quad (c)$$

with $k_b \sim k_c$ and $k_a/k_b = 3.5 \times 10^{18}$ molecule cm^{-3} at 298 K and 700 Torr total pressure.

Table VI-A-1a. OH + glyoxal → products:
Rate coefficients k

$10^{12} \times k$ (cm^3 molecule^{-1} s^{-1})	at T(K)	Technique	Reference
11.0 ± 0.4	298 ± 2	RR [relative to k(cyclohexane) $= 7.21 \times 10^{-12}$][a]	Plum et al. (1983)

VI-A-1b. O_3 Reaction

The upper limit to the rate coefficient of Plum et al. (1983) is given in table VI-A-1b. This is the only reported study of this reaction.

Table VI-A-1b. O_3 + glyoxal → products:
Rate coefficients k

k (cm^3 molecule^{-1} s^{-1})	at T(K)	Technique	Reference
$< 3 \times 10^{-21}$	298 ± 2	S-CL	Plum et al. (1983)

VI-A-2. Kinetic Data for Methylglyoxal, $CH_3C(=O)C(=O)H$

VI-A-2a. OH-Radical Reaction

The rate coefficients of Kleindienst et al. (1982), Plum et al. (1983), and Tyndall et al. (1995) are given in table VI-A-2a. The rate coefficient of Kleindienst et al. (1982) is significantly lower than those of Plum et al. (1983) and Tyndall et al. (1995), possibly because of the presence of non-reactive impurities (such as CO_2) in the methylglyoxal sample used. The room temperature rate coefficients of Plum et al. (1983) were determined by a relative rate method and are not influenced by potential impurities. These results and those of Tyndall et al. (1995) are in reasonable agreement. An average of the 298 K rate coefficients of Plum et al. (1983) and Tyndall et al. (1995), combined with a least-squares analysis of the temperature-dependent rate coefficients of Tyndall et al. (1995), leads to the recommendation of

$$k(\text{OH} + \text{methylglyoxal}) = 9.2 \times 10^{-13}\, e^{(828 \pm 151)/T}\ \text{cm}^3\ \text{molecule}^{-1}\ \text{s}^{-1}$$

over the temperature range 260–333 K, where the indicated errors are two least-squares standard deviations, and $k(\text{OH} + \text{methylglyoxal}) = 1.49 \times 10^{-11}$ cm^3 molecule^{-1} s^{-1} at 298 K, with an estimated overall uncertainty at 298 K of $\pm30\%$.

Green et al. (1990) investigated the subsequent reactions of the $CH_3C(=O)CO$ radical (formed by H-atom abstraction from the –CHO group) under atmospheric conditions, and concluded that the dominant fate of this radical is decomposition to form $CH_3CO + CO$:

$$CH_3C(=O)CO \rightarrow CH_3CO + CO$$

rather than addition of O_2 to form the $CH_3C(=O)C(=O)OO\cdot$ radical.

VI-A-2b. O_3 Reaction

The rate coefficients (or upper limits thereof) of Pate et al. (1976), Atkinson et al. (1981), and Plum et al. (1983) are given in table VI-A-2b. The reaction is slow, and hence the rate coefficient is difficult to measure accurately because of the potential for secondary reactions and the possible presence of impurities. We recommend a conservative upper limit of

$$k(O_3 + \text{methylglyoxal}) < 1 \times 10^{-20}\ \text{cm}^3\ \text{molecule}^{-1}\ \text{s}^{-1}$$

at 298 K.

VI-A-3. Kinetic Data for 2,3-Butanedione, $CH_3C(=O)C(=O)CH_3$; OH-radical Reaction

The absolute rate coefficients of Darnall et al. (1979) and Dagaut et al. (1988) are given in table VI-A-3. At room temperature the rate coefficients from these studies are in excellent agreement. A unit-weighted least-squares analysis of the rate coefficients of Darnall et al. (1979) and Dagaut et al. (1988), using the three-

Table VI-A-2a. OH + methylglyoxal → products:

Rate coefficients k and temperature-dependent parameters $k = A\,e^{-B/T}$

$10^{12} \times A$ (cm^3 $molecule^{-1}$ s^{-1})	B(K)	$10^{12} \times k$ (cm^3 $molecule^{-1}$ s^{-1})	at T(K)	Technique	Reference
		7.1 ± 1.6	297	FP-RF	Kleindienst et al. (1982)
		16.5 ± 12	298 ± 2	RR [relative to k(cyclohexane) $= 7.21 \times 10^{-12}$][a]	Plum et al. (1983)
		19.6 ± 2.5	260	DF-LIF	Tyndall et al. (1995)
		18.4 ± 2.5	273		
		13.2 ± 1.5	298		
0.84 ± 0.12	−830 ± 50	10.1 ± 1.5	333		

[a]From Atkinson (1997).

Table VI-A-2b. O_3 + methylglyoxal → products:
Rate coefficients k

k (cm^3 $molecule^{-1}$ s^{-1})	at T(K)	Technique	Reference
$(1.1 \pm 0.5) \times 10^{-21}$	297 ± 1	S-CL	Pate et al. (1976)
$< 7 \times 10^{-20}$	296 ± 2	S-CL	Atkinson et al. (1981)
$< 6 \times 10^{-21}$	298 ± 2	S-CL	Plum et al. (1983)

parameter expression $k = CT^2 e^{-D/T}$, leads to the recommendation of Atkinson (1989):

$$k(\mathrm{OH} + 2{,}3\text{-butanedione}) = 1.40 \times 10^{-18}\ T^2\ e^{(194 \pm 112)/T}\ \mathrm{cm^3\ molecule^{-1}\ s^{-1}}$$

over the temperature range 240–440 K, where the indicated errors are two least-squares standard deviations, and k(OH + 2,3-butanedione) = 2.4×10^{-13} cm^3 $molecule^{-1}$ s^{-1} at 298 K, with an estimated overall uncertainty at 298 K of ±35%.

VI-A-4. Kinetic Data for Glycidaldehyde (2,3-Epoxypropanal); OH-Radical Reaction

The rate coefficient of Ma et al. (1998) is given in table VI-A-4. This is the only reported study of this reaction. The magnitude of the 298 K rate coefficient, which is very similar to the rate coefficients for the reactions of the OH radical with acetaldehyde and propanal (Atkinson, 1994), indicates that the reaction proceeds mainly, if not exclusively, by H-atom abstraction from the –CHO group (Ma et al., 1998).

VI-A-5. Kinetic Data for *cis*-Butenedial, HC(= O)CH=CHC(=O)H; OH-Radical Reaction

The rate coefficient of Bierbach et al. (1994) is given in table VI-A-5. This is the only reported study of this reaction.

VI-A-6. Kinetic Data for *trans*-Butenedial, HC(=O)CH=CHC(=O)H; OH-Radical Reaction

The rate coefficient of Bierbach et al. (1994) is given in table VI-A-6. This is the only reported study of this reaction.

VI-A-7. Kinetic Data for *cis-/trans*-4-Oxo-2-pentenal, $CH_3C(=O)CH=CHC(=O)H$; OH-Radical Reaction

The rate coefficient of Bierbach et al. (1994) for the *cis-/trans*-mixture of $CH_3C(=O)CH=CHC(=O)H$ is given in table VI-A-7; only data for the *cis-/trans-*

Table VI-A-3. OH + 2.3-butanedione → products:

Rate coefficients k and temperature-dependent parameters $k = A\ e^{-B/T}$

$10^{12} \times A$ (cm^3 $molecule^{-1}$ s^{-1})	B(K)	$10^{12} \times k$ (cm^3 $molecule^{-1}$ s^{-1})	at T(K)	Technique	Reference
		$0.24^{+0.08}_{-0.06}$	298	FP-RF	Darnall et al. (1979)
		0.19 ± 0.02	240	FP-RF	Daguat et al. (1988)
		0.23 ± 0.02	298		
		0.26 ± 0.02	350		
		0.39 ± 0.02	400		
1.12 ± 0.65	450 ± 90	0.44 ± 0.02	440		

Table VI-A-4. OH + glycidaldehyde → products:
Rate coefficients k

$10^{12} \times k$ (cm^3 $molecule^{-1}$ s^{-1})	at T(K)	Technique	Reference
16.9 ± 0.4	298 ± 2	RR [relative to k(ethene) $= 8.52 \times 10^{-12}$][a]	Ma et al. (1998)

[a]From Atkinson (1997).

Table VI-A-5. OH + *cis*-butenedial → products:
Rate coefficients k

$10^{12} \times k$ (cm^3 $molecule^{-1}$ s^{-1})	at T(K)	Technique	Reference
52.9 ± 1.1	296 ± 2	RR [relative to k(propene) $= 2.66 \times 10^{-11}$][a]	Bierbach et al. (1994)

[a]From Atkinson (1997).

Table VI-A-6. OH + *trans*-butenedial → products:
Rate coefficients k

$10^{12} \times k$ (cm^3 $molecule^{-1}$ s^{-1})	at T(K)	Technique	Reference
≥ 24.5	296 ± 2	RR [relative to k(propene) $= 2.66 \times 10^{-11}$][a]	Bierbach et al. (1994)

[a]From Atkinson (1997).

Table VI-A-7. OH + *cis-/trans*-4-oxo-2-pentenal → products:
Rate coefficients k

$10^{12} \times k$ (cm^3 $molecule^{-1}$ s^{-1})	at T(K)	Technique	Reference
56.7 ± 2.2	296 ± 2	RR [relative to k(propene) $= 2.66 \times 10^{-11}$][a]	Bierbach et al. (1994)

[a]From Atkinson (1997).

mixture could be obtained because of rapid *cis-/trans*-photoisomerization. This is the only reported study of this reaction.

VI-A-8. Kinetic Data for *cis*-3-Hexene-2,5-dione, $CH_3C(=O)CH=CHC(=O)CH_3$

VI-A-8a. OH-Radical Reaction

The rate coefficients of Tuazon et al. (1985) and Bierbach et al. (1994) are given in table VI-A-8a, both obtained at room temperature using relative rate methods. The rate coefficients of Tuazon et al. (1985) and Bierbach et al. (1994) are in good agreement, although that of Bierbach et al. (1994) has large associated uncertainties (a single standard deviation of ±30%). The more precise rate coefficient of Tuazon et al. (1985) is recommended:

$$k(\textit{cis}\text{-3-hexene-2,5-dione}) = 6.3 \times 10^{-11}\ \text{cm}^3\ \text{molecule}^{-1}\ \text{s}^{-1}$$

at 298 K, with an estimated overall uncertainty of ±30%.

Table VI-A-8a. OH + *cis*-3-hexene-2,5-dione → products: Rate coefficients *k*

$10^{12} \times k$ (cm^3 molecule^{-1} s^{-1})	at *T*(K)	Technique	Reference
63.1 ± 6.1	298 ± 2	RR [relative to *k*(propene) = 2.63×10^{-11}][a]	Tuazon et al. (1985)
68.1 ± 20.8	296 ± 2	RR [relative to *k*(propene) = 2.66×10^{-11}][a]	Bierbach et al. (1994)

[a]From Atkinson (1997).

VI-A-8b. O_3 Reaction

The rate coefficient of Tuazon et al. (1985) is given in table VI-A 8b. This is the only reported study of this reaction.

Table VI-A-8b. O_3 + *cis*-3-hexene-2,5-dione → products: Rate coefficients *k*

k (cm^3 molecule^{-1} s^{-1})	at *T*(K)	Technique	Reference
$(1.8 \pm 0.2) \times 10^{-18}$	298 ± 2	S-CL	Tuazon et al. (1985)

VI-A-9. Kinetic Data for *trans*-3-Hexene-2,5-Dione, $CH_3C(=O)CH=CHC(=O)CH_3$

VI-A-9a. OH-Radical Reaction

The rate coefficients of Tuazon et al. (1985) and Bierbach et al. (1994) are given in table VI-A-9a, both obtained at room temperature using relative rate methods. For this isomer, the rate coefficient of Tuazon et al. (1985) is 35% higher than that of Bierbach et al. (1994). A unit-weighted average of the rate coefficients of Tuazon et al. (1985) and Bierbach et al. (1994) leads to the recommendation of

$$k(\textit{trans}\text{-3-hexene-2,5-dione}) = 4.6 \times 10^{-11}\ \text{cm}^3\ \text{molecule}^{-1}\ \text{s}^{-1}$$

at 298 K, with an estimated overall uncertainty of a factor of 1.5.

Table VI-A-9a. OH + *trans*-3-hexene-2,5-dione → products:
Rate coefficients k

$10^{12} \times k$ (cm^3 $molecule^{-1}$ s^{-1})	at T(K)	Technique	Reference
53.1 ± 2.4	298 ± 2	RR [relative to k(propene) $= 2.63 \times 10^{-11}$][a]	Tuazon et al. (1985)
39.4 ± 4.0	296 ± 2	RR [relative to k(propene) $= 2.66 \times 10^{-11}$][a]	Bierbach et al. (1994)

[a]From Atkinson (1997).

VI-A-9b. O_3 Reaction

The rate coefficient of Tuazon et al. (1985) is given in table VI-A-9b. This is the only reported study of this reaction.

Table VI-A-9b. O_3 + *cis*-3-hexene-2,5-dione → products:
Rate coefficients k

k (cm^3 $molecule^{-1}$ s^{-1})	at T(K)	Technique	Reference
$(8.3 \pm 1.2) \times 10^{-18}$	298 ± 2	S-CL	Tuazon et al. (1985)

VI-A-10. Kinetic Data for Maleic Anhydride; OH-Radical Reaction

The rate coefficient of Bierbach et al. (1994) is given in table VI-A-10. This is the only reported study of this reaction.

Table VI-A-10. OH + maleic anhydride → products:
Rate coefficients k

$10^{12} \times k$ (cm^3 $molecule^{-1}$ s^{-1})	at T(K)	Technique	Reference
1.40 ± 0.08	296 ± 2	RR [relative to k(n-butane) $= 2.42 \times 10^{-12}$][a]	Bierbach et al. (1994)

VI-A-11. Kinetic Data for 3*H*-furan-2-one

VI-A-11a. OH-Radical Reaction

The rate coefficient of Bierbach et al. (1994) is given in table VI-A-11a. This is the only reported study of this reaction.

Table VI-A-11a. OH + 3*H*-furan-2-one → products: Rate coefficients *k*

$10^{12} \times k$ (cm^3 $molecule^{-1}$ s^{-1})	at *T*(K)	Technique	Reference
44.7 ± 2.7	296 ± 2	RR [relative to *k*(propene) $= 2.66 \times 10^{-11}$][a]	Bierbach et al. (1994)

[a]From Atkinson (1997).

VI-A-11b. NO_3-Radical Reaction

The rate coefficient of Berndt et al. (1997) is given in table VI-A-11b. This is the only reported study of this reaction.

Table VI-A-11b. NO_3 + 3*H*-furan-2-one → products: Rate coefficients *k*

k (cm^3 $molecule^{-1}$ s^{-1})	at *T*(K)	Technique	Reference
$(1.76 \pm 0.24) \times 10^{-13}$	298	RR [relative to *k*(*trans*-2-butene) $= 3.90 \times 10^{-13}$][a]	Berndt et al. (1997)

[a]From Atkinson (1997).

VI-A-12. Kinetic Data for 5-Methyl-3*H*-furan-2-one; OH-Radical Reaction

The rate coefficient of Bierbach et al. (1994) is given in table VI-A-12. This is the only reported study of this reaction.

Table VI-A-12. OH + 5-methyl-3*H*-furan-2-one → products: Rate coefficients *k*

$10^{12} \times k$ (cm^3 $molecule^{-1}$ s^{-1})	at *T*(K)	Technique	Reference
68.7 ± 4.6	296 ± 2	RR [relative to *k*(*trans*-2-butene) $= 6.48 \times 10^{-11}$][a]	Bierbach et al. (1994)

[a]From Atkinson (1997).

VI-A-13. Kinetic Data for *trans,trans*-2,4-Hexadienedial, HC(=O)CH=CHCH=CHC(=O)H

VI-A-13a. OH-Radical Reaction

The rate coefficients of Klotz et al. (1995) and (1999) are given in table VI-A-13a. The agreement is reasonable and an average of

$$k(\textit{trans,trans}\text{-2,4-hexadenedial}) = 8.2 \times 10^{-11}\ \text{cm}^3\ \text{molecule}^{-1}\ \text{s}^{-1}$$

at 298 K is recommended, with an estimated overall uncertainty of ±30%. However, Klotz et al. (1999) judge that the more recent determination is more accurate.

Table VI-A-13a. OH + *trans,trans*-2,4-hexadiendial → products: Rate coefficients *k*

$10^{12} \times k$ (cm^3 $molecule^{-1}$ s^{-1})	at T(K)	Technique	Reference
87.7 ± 3.2	298 ± 2	RR [relative to k(*trans*-2-butene) = 6.40×10^{-11}][a]	Klotz et al. (1995)
76 ± 8	298 ± 3	RR [relative to k(*trans*-2-butene) = 6.40×10^{-11}][a]	Klotz et al. (1999)

[a]From Atkinson (1997).

VI-A-13b. O_3 Reaction

The rate coefficient of Klotz et al. (1995) is given in table VI-A-13b. This is the only reported study of this reaction, and no details were provided (the study was reported to be preliminary). Because a relative rate method was used without an OH-radical scavenger (with ethene and propene as the reference compounds), Klotz et al. (1995) stated that the rate coefficient is strictly an upper limit.

Table VI-A-13b. O_3 + *trans,trans*-2,4-hexadiendial → products: Rate coefficients *k*

k (cm^3 $molecule^{-1}$ s^{-1})	at T(K)	Technique	Reference
$\leq 1 \pm 10^{-17}$	298 ± 2	RR [relative to k(ethene) and k(propene)][a]	Klotz et al. (1995)

[a]No details given.

VI-A-13c. NO_3-Radical Reaction

The rate coefficient of Klotz et al. (1995) is given in table VI-A-13c. This is the only reported study of this reaction.

Table VI-A-13c. NO_3 + *trans,trans*-2,4-hexadiendial → products: Rate coefficients *k*

k (cm^3 $molecule^{-1}$ s^{-1})	at T(K)	Technique	Reference
$(5.34 \pm 0.12) \times 10^{-15}$	298 ± 2	RR [relative to k(propene) = 9.49×10^{-15}][a]	Klotz et al. (1995)

VI-A-14. Kinetic Data for *cis,trans*-2,4-Hexadienedial, HC(=O)CH=CHCH=CHC(=O)H

VI-A-14a. OH-Radical Reaction

The rate coefficients of Klotz et al. (1995) and Klotz et al. (1999) are given in table VI-A-14a. There is a significant discrepancy between the two studies; an average of

$$k(cis,\ trans\text{-2,4-hexadenedial}) = 9.2 \times 10^{-11}\ \text{cm}^3\ \text{molecule}^{-1}\ \text{s}^{-1}$$

at 298 K is recommended, with an estimated overall uncertainty of a factor of 1.5. However, Klotz et al. (1999) judge that the more recent determination is more accurate.

Table VI-A-14a. OH + *cis,trans*-2,4-hexadiendial → products: Rate coefficients *k*

$10^{12} \times k$ (cm^3 molecule^{-1} s^{-1})	at *T*(K)	Technique	Reference
109 ± 11	298 ± 2	RR [relative to *k*(*trans*-2-butene) = 6.40×10^{-11}][a]	Klotz et al. (1995)
74 ± 19	298 ± 3	RR [relative to *k*(*trans*-2-butene) = 6.40×10^{-11}][a]	Klotz et al. (1999)

[a]From Atkinson (1997).

VI-A-14b. O_3 Reaction

The rate coefficient of Klotz et al. (1995) is given in table VI-A-14b. This is the only reported study of this reaction, and no details were provided (the study was reported to be preliminary). Because a relative rate method was used without an OH-radical scavenger (with ethene and propene as the reference compounds), Klotz et al. (1995) stated that the rate coefficient is strictly an upper limit.

Table VI-A-14b. O_3 + *cis,trans*-2,4-hexadiendial → products: Rate coefficients *k*

k (cm^3 molecule^{-1} s^{-1})	at *T*(K)	Technique	Reference
$\leq 1 \times 10^{-17}$	298 ± 2	RR [relative to *k*(ethene) and *k*(propene)][a]	Klotz et al. (1995)

[a]No details given.

VI-A-14c. NO_3-Radical Reaction

The rate coefficient of Klotz et al. (1995) is given in table VI-A-14c. This is the only reported study of this reaction.

Table VI-A-14c. NO_3 + *cis,trans*-2,4-hexadiendial → products: Rate coefficients k

k (cm^3 $molecule^{-1}$ s^{-1})	at T(K)	Technique	Reference
$(5.26 \pm 0.18) \times 10^{-15}$	298 ± 2	RR [relative to k(propene) $= 9.49 \times 10^{-15}$][a]	Klotz et al. (1995)

[a]From Atkinson (1997).

VI-A-15. Kinetic Data for *trans,trans*-2-Methyl-2,4-Hexadienedial, $HC(=O)C(CH_3)=CHCH=CHC(=O)H$

VI-A-15a. OH-Radical Reaction

The rate coefficient of Klotz et al. (1995) is given in table VI-A-15a. This is the only reported study of this reaction.

Table VI-A-15a. OH + *trans,trans*-2-methyl-2,4-hexadiendial → products: Rate coefficients k

$10^{12} \times k$ (cm^3 $molecule^{-1}$ s^{-1})	at T(K)	Technique	Reference
118 ± 2	298 ± 2	RR [relative to k(*trans*-2-butene) $= 6.40 \times 10^{-11}$][a]	Klotz et al. (1995)

[a]From Atkinson (1997).

VI-A-15b. O_3 Reaction

The rate coefficient of Klotz et al. (1995) is given in table VI-A-15b. This is the only reported study of this reaction, and no details were provided (the study was reported to be preliminary). Because a relative rate method was used without an OH-radical scavenger, the rate coefficient is strictly an upper limit.

Table VI-A-15b. O_3 + *trans,trans*-2-methyl-2,4-hexadiendial → products: Rate coefficients k

k (cm^3 $molecule^{-1}$ s^{-1})	at T(K)	Technique	Reference
$\leq 2 \pm 10^{-17}$	298 ± 2	RR [relative to k(ethene) and k(propene)][a]	Klotz et al. (1995)

[a]No details given.

VI-A-15c. NO_3-Radical Reaction

The rate coefficient of Klotz et al. (1995) is given in table VI-A-15c. This is the only reported study of this reaction.

Table VI-A-15c. NO_3 + *trans,trans*-2-methyl-2,4-hexadiendial → products: Rate coefficients *k*

k (cm^3 molecule^{-1} s^{-1})	at *T*(K)	Technique	Reference
$(1.02 \pm 0.03) \times 10^{-14}$	298 ± 2	RR [relative to *k*(propene) $= 9.49 \times 10^{-15}$][a]	Klotz et al. (1995)

[a]From Atkinson (1997).

VI-A-16. Kinetic Data for 3,4-Dihydroxy-3-hexene-2,5-dione, $CH_3C(=O)C(OH)=C(OH)C(=O)CH_3$; OH-Radical Reactions

The rate coefficients of Wiesen et al. (1995) are given in table VI-A-16. This is the only reported study of this reaction. An average of the two independently measured rate coefficients leads to the recommendation of

$$k(\text{3,4-dihydroxy-3-hexene-2,5-dione}) = 3.0 \times 10^{-11}\ \text{cm}^3\ \text{molecule}^{-1}\ \text{s}^{-1}$$

at 298 K, with an estimated overall uncertainty of a factor of 1.5.

Table VI-A-16. OH + 3,4-dihydroxy-3-hexene-2,5-dione → products: Rate coefficients *k*

$10^{12} \times k$ (cm^3 molecule^{-1} s^{-1})	at *T*(K)	Technique	Reference
22.5 ± 6.2	296 ± 2	RR [relative to *k*(propene) $= 2.66 \times 10^{-11}$][a]	Wiesen et al. (1995)
29 ± 7	296 ± 2	RR [relative to *k*(isoprene) $= 1.01 \times 10^{-10}$][a]	Wiesen et al. (1995)

[a]From Atkinson (1997).

VI-A-17. Kinetic Data for Benzene Oxide/Oxepin

VI-A-17a. OH-Radical Reaction

The rate coefficient of Klotz et al. (1997) is given in table VI-A-17a. This is the only reported study of this reaction.

Table VI-A-17a. OH + benzene oxide/oxepin → products: Rate coefficients k

$10^{12} \times k$ (cm^3 molecule^{-1} s^{-1})	at T(K)	Technique	Reference
100 ± 4	298 ± 3	RR [relative to k(*trans*-2-butene) $= 6.40 \times 10^{-11}$][a]	Klotz et al. (1997)

[a]From Atkinson (1997).

VI-A-17b. NO_3-Radical Reaction

The rate coefficient of Klotz et al. (1997) is given in table VI-A-17b. This is the only reported study of this reaction.

Table VI-A-17b. NO_3 + benzene oxide/oxepin → products: Rate coefficients k

k (cm^3 molecule^{-1} s^{-1})	at T(K)	Technique	Reference
$(9.27 \pm 0.31) \times 10^{-12}$	298 ± 3	RR [relative to k(2-methyl-2-butene) $= 9.37 \times 10^{-12}$][a]	Klotz et al. (1995)

[a]From Atkinson (1997).

VI-A-18. Kinetic Data for Furan

VI-A-18a. OH-Radical Reaction

The rate coefficients of Lee and Tang (1982), Atkinson et al. (1983a), Tuazon et al. (1984b), Wine and Thompson (1984), and Bierbach et al. (1992) are given in table VI-A-18a. The room temperature rate coefficient of Lee and Tang (1982) is a factor of ~2.5 higher than the other rate coefficients and is hence not used in the evaluation. Note that the rate coefficient given by Lee and Tang (1982) for thiophene is a factor of ~5 higher than other literature data (Atkinson et al., 1989c). A least-squares analysis of the rate coefficients of Atkinson et al. (1983a), Tuazon et al. (1984b), Wine and Thompson (1984), and Bierbach et al. (1992), which are in excellent agreement at room temperature, leads to the recommendation of

$$k(\text{OH} + \text{furan}) = 1.32 \times 10^{-11}\, e^{334\pm67)/T}\ \text{cm}^3\ \text{molecule}^{-1}\ \text{s}^{-1}$$

over the temperature range 254–424 K, where the indicated errors are two least-squares standard deviations, and $k(\text{OH} + \text{furan}) = 4.05 \times 10^{-11}$ cm^3 molecule^{-1} s^{-1} at 298 K, with an estimated overall uncertainty at 298 K of ±15%.

Table VI-A-18a. OH + furan → products:

Rate coefficients k and temperature-dependent parameters $k = A\ e^{-B/T}$

$10^{12} \times A$ (cm^3 $molecule^{-1}$ s^{-1})	B(K)	$10^{12} \times k$ (cm^3 $molecule^{-1}$ s^{-1})	at T(K)	Technique	Reference
		105 ± 8	295 ± 1	DF-RF	Lee and Tang (1982)
		38.4 ± 2.8	298 ± 2	RR [relative to k(n-hexane) = 5.45×10^{-12}][a]	Atkinson et al. (1983a)
		49.6 ± 3.3	254	FP-RF	Wine and Thompson (1984)
		40.8 ± 1.8	297		
		43.1 ± 1.2	297		
		38.7 ± 2.2	299		
		41.6 ± 3.5	299		
		38.3 ± 4.5	300		
		31.9 ± 1.6	365		
13.2 ± 2.9	−333 ± 67	29.9 ± 2.0	424		
		42.3 ± 3.2	295 ± 2	RR [relative to k(isoprene) = 1.02×10^{-10}][a]	Tuazon et al. (1984b)
		41.9 ± 2.1	300 ± 2	RR [relative to k(propene) = 2.60×10^{-11}][a]	Bierbach et al. (1992)

[a]From Atkinson (1997).

VI-A-18b. O_3 Reaction

The rate coefficient of Atkinson et al. (1983a) is given in table VI-A-18b. This is the only reported study of this reaction.

Table VI-A-18b. O_3 + furan → products:
Rate coefficients k

k (cm^3 $molecule^{-1}$ s^{-1})	at T(K)	Technique	Reference
$(2.42 \pm 0.28) \times 10^{-18}$	298 ± 2	S-CL/GC	Atkinson et al. (1983a)

VI-A-18c. NO_3-Radical Reaction

The rate coefficients of Atkinson et al. (1985) and Kind et al. (1996) are given in table VI-A-18c. The rate coefficients from these two studies (Atkinson et al., 1985; Kind et al., 1996) disagree by a factor of 1.45. A unit-weighted average of the rate coefficients of Atkinson et al. (1985) and Kind et al. (1996) leads to the recommendation of

$$k(\text{furan}) = 1.2 \times 10^{-12}\ \text{cm}^3\ \text{molecule}^{-1}\ \text{s}^{-1}$$

at 298 K, with an estimated overall uncertainty of a factor of 1.5.

Table VI-A-18c. NO_3 + furan → products:
Rate coefficients k

k (cm^3 $molecule^{-1}$ s^{-1})	at T(K)	Technique	Reference
$(1.44 \pm 0.02) \times 10^{-12}$	295 ± 1	RR [relative to k(*trans*-2-butene) $= 3.88 \times 10^{-13}$][a]	Atkinson et al. (1985)
$(9.93 \pm 0.62) \times 10^{-13}$	295 ± 2	RR [relative to k(*trans*-2-butene) $= 3.88 \times 10^{-13}$][a]	Kind et al. (1996)

[a]From Atkinson (1997).

VI-A-19. Kinetic Data for 2-Methylfuran

VI-A-19a. OH-Radical Reaction

The rate coefficient of Bierbach et al. (1992) is given in table VI-A-19a. This is the only reported study of this reaction.

VI-A-19b. NO_3-Radical Reaction

The rate coefficient of Kind et al. (1996) is given in table VI-A-19b. This is the only reported study of this reaction.

Table VI-A-19a. OH + 2-methylfuran → products:
Rate coefficients k

$10^{12} \times k$ (cm^3 $molecule^{-1}$ s^{-1})	at T(K)	Technique	Reference
61.9 ± 3.2	300 ± 2	RR [relative to k(propene) $= 2.60 \times 10^{-11}$][a]	Bierbach et al. (1992)

[a]From Atkinson (1997).

Table VI-A-19b. NO_3 + 2-methylfuran → products:
Rate coefficients k

k (cm^3 $molecule^{-1}$ s^{-1})	at T(K)	Technique	Reference
$(2.57 \pm 0.18) \times 10^{-18}$	295 ± 2	RR [relative to k(2.3-dimethyl-2-butene) $= 5.72 \times 10^{-11}$][a]	Kind et al. (1996)

[a]From Atkinson (1997).

VI-A-20. Kinetic Data for 3-Methylfuran

VI-A-20a. OH-Radical Reaction

The rate coefficient of Atkinson et al. (1989c) is given in table VI-A-20a. This is the only reported study of this reaction.

Table VI-A-20a. OH + 3-methylfuran → products:
Rate coefficients k

$10^{12} \times k$ (cm^3 $molecule^{-1}$ s^{-1})	at T(K)	Technique	Reference
92.6 ± 2.4	296 ± 2	RR [relative to k(2,3-dimethyl-2-butene) $= 1.10 \times 10^{-10}$][a]	Atkinson et al. (1989c)

[a]From Atkinson (1997).

VI-A-20b. O_3 Reaction

The rate coefficient of Alvarado et al. (1996) is given in table VI-A-20b. This is the only reported study of this reaction.

Table VI-A-20b. O_3 + 3-methylfuran → products:
Rate coefficients k

k (cm^3 $molecule^{-1}$ s^{-1})	at T(K)	Technique	Reference
$(2.05 \pm 0.08) \times 10^{-17}$	296 ± 2	RR [relative to k(propene) $= 9.68 \times 10^{-18}$][a]	Alvarado et al. (1996)

[a]From Atkinson (1997).

VI-A-20c. NO_3-Radical Reaction

The rate coefficients of Alvarado et al. (1996) and Kind et al. (1996) are given in table VI-A-20c. The rate coefficients from these two studies (Alvarado et al., 1996; Kind et al., 1996) disagree by a factor of 2. A unit-weighted average of the rate coefficients of Alvarado et al. (1996) and Kind et al. (1996) leads to the tentative recommendation of

$$k(\text{3-methylfuran}) = 2.1 \times 10^{-11} \text{ cm}^3 \text{ molecule}^{-1} \text{ s}^{-1}$$

at 298 K, with an estimated overall uncertainty of a factor of 2.

Table VI-A-20c. NO_3 + 3-methylfuran → products:
Rate coefficients k

k (cm^3 molecule^{-1} s^{-1})	at T(K)	Technique	Reference
$(2.86 \pm 0.06) \times 10^{-11}$	295 ± 2	RR [relative to k(2,3-dimethyl-2-butene) $= 5.72 \times 10^{-11}$][a]	Kind et al. (1996)
$(1.31 \pm 0.02) \times 10^{-11}$	296 ± 2	RR [relative to k(2-methyl-2-butene) $= 9.37 \times 10^{-12}$][a]	Alvarado et al. (1996)

[a]From Atkinson (1997).

VI-A-21. Kinetic Data for 2-Ethylfuran; OH-Radical Reaction

The rate coefficient of Bierbach et al. (1992) is given in table VI-A-21. This is the only reported study of this reaction.

Table VI-A-21. OH + 2-ethylfuran → products:
Rate coefficients k

$10^{12} \times k$ (cm^3 molecule^{-1} s^{-1})	at T(K)	Technique	Reference
108 ± 22	300 ± 2	RR [relative to k(propene) $= 2.60 \times 10^{-11}$][a]	Bierbach et al. (1992)

[a]From Atkinson (1997).

VI-A-22. Kinetic Data for 2,5-Dimethylfuran

VI-A-22a. OH-Radical Reaction

The rate coefficient of Bierbach et al. (1992) is given in table VI-A-22a. This is the only reported study of this reaction.

Table VI-A-22a. OH + 2,5-dimethylfuran → products:
Rate coefficients k

$10^{12} \times k$ (cm^3 $molecule^{-1}$ s^{-1})	at T(K)	Technique	Reference
132 ± 10	300 ± 2	RR [relative to k(*trans*-2-butene) $= 6.32 \times 10^{-11}$][a]	Bierbach et al. (1992)

[a]From Atkinson (1997).

VI-A-22b. NO_3-Radical Reaction

The rate coefficient of Kind et al. (1996) is given in table VI-A-22b. This is the only reported study of this reaction.

Table VI-A-22b. NO_3 + 2,5-dimethylfuran → products:
Rate coefficients k

k (cm^3 $molecule^{-1}$ s^{-1})	at T(K)	Technique	Reference
$(5.78 \pm 0.35) \times 10^{-11}$	295 ± 2	RR [relative to k(2,3-dimethyl-2-butene) $= 5.72 \times 10^{-11}$][a]	Kind et al. (1996)

[a]From Atkinson (1997).

VI-A-23. Kinetic Data for 2,3-Dimethylfuran; NO_3-Radical Reaction

The rate coefficient of Kind et al. (1996) is given in table VI-A-23. This is the only reported study of this reaction.

Table VI-A-23. NO_3 + 2,3-dimethylfuran → products:
Rate coefficients k

k (cm^3 $molecule^{-1}$ s^{-1})	at T(K)	Technique	Reference
$(5.83 \pm 0.46) \times 10^{-11}$	295 ± 2	RR [relative to k(2,3-dimethyl-2-butene) $= 5.72 \times 10^{-11}$][a]	Kind et al. (1996)

[a]From Atkinson (1997).

VI-A-24. Kinetic Data for 2,4-Dimethylfuran; NO_3-Radical Reaction

The rate coefficient of Kind et al. (1996) is given in table VI-A-24. This is the only reported study of this reaction.

Table VI-A-24. NO_3 + 2,4-dimethylfuran → products: Rate coefficients k

k (cm^3 $molecule^{-1}$ s^{-1})	at T(K)	Technique	Reference
$(5.72 \pm 0.46) \times 10^{-11}$	295 ± 2	RR [relative to k(2,3-dimethyl-2-butene) $= 5.72 \times 10^{-11}$][a]	Kind et al. (1996)

[a]From Atkinson (1997).

VI-A-25. Kinetic Data for Tetramethylfuran; NO_3-Radical Reaction

The rate coefficient of Kind et al. (1996) is given in table VI-A-25. This is the only reported study of this reaction.

Table VI-A-25. NO_3 + tetramethylfuran → products: Rate coefficients k

k (cm^3 $molecule^{-1}$ s^{-1})	at T(K)	Technique	Reference
$(1.18 \pm 0.16) \times 10^{-10}$	295 ± 2	RR [relative to k(2,3-dimethyl-2-butene) $= 5.72 \times 10^{-11}$][a]	Kind et al. (1996)

[a]From Atkinson (1997).

VI-A-26. Kinetic Data for Furan-2-aldehyde; OH-Radical Reaction

The rate coefficient of Bierbach et al. (1995) is given in the table VI-A-26. This is the only reported study of this reaction.

Table VI-A-26. OH + furan-2-aldehyde → products: Rate coefficients k

$10^{12} \times k$ (cm^3 $molecule^{-1}$ s^{-1})	at T(K)	Technique	Reference
34.8 ± 1.3	300 ± 2	RR [relative to k(*trans*-2-butene) $= 6.32 \times 10^{-11}$][a]	Bierbach et al. (1995)

[a]From Atkinson (1997).

VI-A-27. Kinetic Data for Furan-3-aldehyde; OH-Radical Reaction

The rate coefficient of Bierbach et al. (1995) is given in table VI-A-27. This is the only reported study of this reaction.

Table VI-A-27. OH + furan-3-aldehyde → products: Rate coefficients k

$10^{12} \times k$ (cm^3 $molecule^{-1}$ s^{-1})	at T(K)	Technique	Reference
48.7 ± 1.9	300 ± 2	RR [relative to k(*trans*-2-butene) $= 6.32 \times 10^{-11}$][a]	Bierbach et al. (1995)

[a]From Atkinson (1997).

VI-A-28. Kinetic Data for 5-Methylfurfural; OH-Radical Reaction

The rate coefficient of Bierbach et al. (1995) is given in table VI-A-28. This is the only reported study of this reaction.

Table VI-A-28. OH + 5-methylfuran → products: Rate coefficients k

$10^{12} \times k$ (cm^3 $molecule^{-1}$ s^{-1})	at T(K)	Technique	Reference
51.2 ± 1.9	300 ± 2	RR [relative to k(*trans*-2-butene) $= 6.32 \times 10^{-11}$][a]	Bierbach et al. (1995)

[a]From Atkinson (1997).

VI-B. Products and Mechanisms of Reactions of OH Radical with Some Major Oxidation Products of Aromatic Hydrocarbons

In this section, the mechanisms of the further oxidation of selected products occurring in the OH-initiated oxidation of aromatic hydrocarbons are described. Only the non-aromatic hydrocarbon products will be described; the ring-retaining (aromatic) products have been discussed in chapters II through V of this book.

As shown in section II-E, different authors have proposed a variety of compounds that may occur as intermediates in the OH-initiated oxidation, even in simple aromatic hydrocarbons such as benzene and methyl-substituted benzenes. Evidence for the actual formation of some of these compounds in some aromatic hydrocarbon oxidation systems exists, though the formation yields become successively less certain as the molecular weight increases; see section II-D. The atmospheric chemistry of some derivatives of the compounds proposed as intermediates has been studied in the laboratory; the results are summarized in this section.

VI-B-1. Arene Oxides

Arene oxides are putative intermediates in the OH-initiated oxidation of aromatic hydrocarbons; their formation was proposed to explain some ill-understood

Table VI-A-29. Summary of rate coefficients (298 K and 760 Torr of air) for reactions of OH with some non-aromatic, first-generation oxidation products of aromatic hydrocarbons

Organic Hydrocarbon Oxidation Product	k(298 K) (cm^3 molecule^{-1} s^{-1})	Estimated Uncertainty k(298)
Glyoxal	1.1×10^{-11}	[a]
Methylglyoxal	1.5×10^{-11}	30%
2,3-Butanedione	2.4×10^{-13}	35%
Glycidaldehyde	1.7×10^{-11}	[a]
cis-Butenedial	5.3×10^{-11}	[a]
trans-Butenedial	$\geq 2.5 \times 10^{-11}$	[a]
cis,trans-4-Oxo-2-pentenal	5.7×10^{-11}	[a]
cis-3-Hexene-2,5-dione	6.3×10^{-11}	30%
trans-3-Hexene-2,5-dione	4.6×10^{-11}	Factor of 1.5
Maleic anhydride	1.4×10^{-12}	[a]
3*H*-Furan-2-one	4.5×10^{-11}	[a]
5-Methyl-3*H*-furan-2-one	6.9×10^{-11}	[a]
trans,trans-2,4-Hexadiendial	8.8×10^{-11}	[a]
cis,trans-2,4-Hexadienedial	1.1×10^{-10}	[a]
trans,trans-2-Methyl-2,4-hexadiene	1.2×10^{-10}	[a]
3,4-Dihydroxy-3-hexene-2,5-dione	3.0×10^{-11}	Factor of 1.5
Benzene oxide/oxepin	1.0×10^{-10}	[a]
Furan	4.05×10^{-11}	15%
2-Methylfuran	6.2×10^{-11}	[a]
3-Methylfuran	9.3×10^{-11}	[a]
2-Ethylfuran	1.1×10^{-10}	[a]
2,5-Dimethylfuran	1.3×10^{-12}	[a]
Furan-2-aldehyde	3.5×10^{-11}	[a]
Furan-3-aldehyde	4.9×10^{-11}	[a]
5-Methylfurfural	5.1×10^{-11}	[a]

[a]No estimate, since a single study only.

Table VI-A-30. Summary of rate coefficients (298 K and 760 Torr of air) for reactions of NO_3 with some non-aromatic, first-generation oxidation products of aromatic hydrocarbons

Organic Hydrocarbon Oxidation Product	k(298 K) (cm^3 molecule^{-1} s^{-1})	Estimated Uncertainty k(298)
3*H*-Furan-2-one	1.8×10^{-13}	[a]
trans,trans-2,4-Hexadiendial	5.3×10^{-15}	[a]
cis,trans-2,4-Hexadiendial	5.3×10^{-15}	[a]
trans,trans-2-Methyl-2,4-hexadiendial	1.0×10^{-14}	[a]
Benzene oxide/oxepin	9.3×10^{-12}	[a]
Furan	1.2×10^{-12}	Factor of 1.5
2-Methylfuran	2.6×10^{-11}	[a]
3-Methylfuran	2.1×10^{-11}	Factor of 2
2,5-Dimethylfuran	5.8×10^{-11}	[a]
2,3-Dimethylfuran	5.8×10^{-11}	[a]
2,4-Dimethylfuran	5.7×10^{-11}	[a]
Tetramethylfuran	1.2×10^{-10}	[a]

[a]No estimate, since a single study only.

Table VI-A-31. Summary of rate coefficients (298 K and 760 Torr of air) for reactions of O_3 with some non-aromatic, first-generation oxidation products of aromatic hydrocarbons

Organic Hydrocarbon Oxidation Product	k(298 K) (cm^3 $molecule^{-1}$ s^{-1})	Estimated Uncertainty k(298)
Glyoxal	$< 3 \times 10^{-21}$	a
Methylglyoxal	$< 1 \times 10^{-20}$	a
cis-3-Hexene-2,5-dione	1.8×10^{-18}	b
trans-3-Hexene-2,5-dione	8.3×10^{-18}	b
trans,trans-2,4-Hexadiendial	$\leq 1 \times 10^{-17}$	a
cis,trans-2,4-Hexadiendial	$\leq 1 \times 10^{-17}$	a
trans,trans-2-Methyl-2,4-hexadiendial	$\leq 2 \times 10^{-17}$	a
Furan	2.4×10^{-18}	b
3-Methylfuran	2.1×10^{-17}	b

[a]Upper limit.
[b]No estimate, since a single study only.

aspects of this reaction. No experimental evidence for the occurrence of these highly reactive compounds in the gas-phase is available. Most arene oxides undergo an extremely rapid equilibrium reaction between the oxide form and its monocyclic oxepin form (Vogel and Günther, 1967). Due to the rapidity of this equilibrium reaction, the oxide and oxepin forms are generally treated as one compound.

Aspects of the gas-phase photooxidation of several arene oxides have been studied by Nojima et al. (1982) and Nojima and Isogami (1994). These investigations were, however, performed in a very small reaction vessel and at very high reactant concentrations. The results therefore may not be readily applicable to atmospheric conditions. Studies of the atmospheric chemistry of two arene oxides, performed under conditions more representative of the troposphere, have been conducted by Klotz et al. (1997, 2000).

VI-B-2. Benzene Oxide/Oxepin

The reaction of benzene oxide/oxepin with OH radicals has been investigated by Klotz et al. (1997). The reaction was found to lead to E/Z-isomers of 2,4-hexadienedial in almost quantitative yields. The mechanism proposed for the formation of these products is shown in figure VI-B-1. Oxidation of both the oxide and the oxepin forms can lead to the formation of 2,4-hexadienedial isomers. A rapid conversion of Z,Z- and E,Z-2,4-hexadienedial into the more stable E,E-isomer will occur under atmospheric conditions; see the discussion in the following sections and chapter VII.

VI-B-3. Toluene-1,2-oxide/2-Methyloxepin

According to Klotz et al. (2000), the reaction of OH radicals with toluene-1,2-oxide/2-methyloxepin proceeds through a mechanism similar to that for benzene oxide/oxepin. The major products are E/Z-isomers of 6-oxo-2,4-heptadienal.

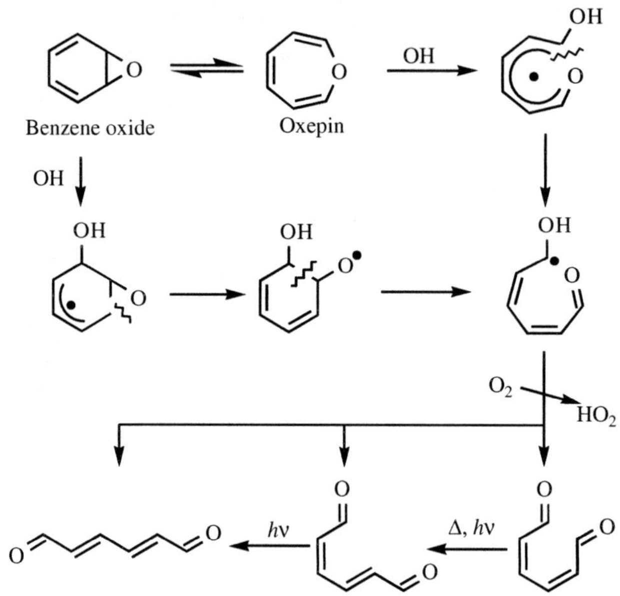

E,E-, *E,Z*- and *Z,Z*-2,4-hexadienedial (left to right)

Figure VI-B-1. Proposed mechanisms for the reaction of benzene oxide/oxepin with OH radicals.

Formation of 2-methyl-2,4-hexadienedial, which is also feasible, was not observed. The mechanism of the OH-initiated oxidation of toluene-1,2-oxide/2-methyloxepin is similar to that shown for the benzene oxide/oxepin mechanism in figure VI-B-1; see also chapter VII.

VI-B-4. Unsaturated Dicarbonyls

The unsaturated dicarbonyls can be grouped into three classes: (1) the unsaturated 1,6-dicarbonyls; (2) the unsaturated 1,4-dicarbonyls; and (3) the 1,2-dicarbonyls or α-dicarbonyls. Amongst the first group of compounds, tentative evidence for the formation of 2,4-hexadienedial (or its methylated derivatives in the OH-initiated oxidation of toluene) has been reported by Hoshino et al. (1978), Dumdei and O'Brien (1984), and Dumdei et al. (1988). Reliable identifications of compounds of this type were also reported by Yu et al. (1997) and Kwok et al. (1997a). Though it can now be regarded as certain that unsaturated 1,6-dicarbonyls are formed in the OH-initiated oxidation of aromatic hydrocarbons, their yields are unknown and consequently their importance remains highly uncertain.

VI-B-4a. E/Z-isomers of 2,4-Hexadienedial

The reaction of *Z,Z*-2,4-hexadienedial with OH has not been investigated due to fast thermal and photolytic isomerization to *E,Z*-2,4-hexadienedial (Klotz et al.,

1995). *E,Z*-2,4-hexadienedial undergoes a rapid photoisomerization to *E,E*-2,4-hexadienedial. The reaction of *E,Z*- and *E,E*-2,4-hexadienedial with OH radicals has been investigated by Klotz et al. (1995); formation of 2-butenedial and glyoxal were the main products observed, but no yields were reported. The degradation mechanism proposed is shown in figure VI-B-2.

VI-B-4b. E/Z-isomers of 2-Methyl-2,4-hexadienedial

According to Klotz et al. (1995), the *E/Z*-isomers of 2-methyl-2,4-hexadienedial show behavior analogous to the *E/Z*-isomers of 2,4-hexadienedial, with the *Z,Z*-isomer rapidly isomerizing to the *E,Z*- and *E,E*-2-methyl-2,4-hexadienedial under atmospheric conditions. Only the reaction of OH radicals with *E,E*-2-methyl-2,4-hexadienedial has been studied, and it was found to give methylglyoxal and butenedial as the main observed products. The formation mechanism of these products is similar to that shown for the formation of butenedial and glyoxal from *E,E*-2,4-hexadienedial in figure VI-B-2.

VI-B-4c. E/Z-isomers of 3-Hexene-2,5-dione

The unsaturated 1,4-dicarbonyl-3-hexene-2,5-dione has been identified as an intermediate in the OH-initiated oxidation of *p*-xylene and 1,2,4-trimethylbenzene (Takagi et al., 1982; Bierbach et al., 1993; Becker et al., 1997; Kwok et al., 1997a; Yu et al., 1997). Smith et al. (1999) estimated formation yields of 16.6 and 10.8% C, respectively, in the OH-initiated oxidation of *p*-xylene and 1,2,4-trimethylbenzene.

The carbon balance reported for the reaction of 3-hexene-2,5-dione with OH radicals is poor, with only about 50% C being accounted for. The major product of this reaction is expected to be methylglyoxal, which was observed in yields of only around 20% C by Bierbach et al. (1994). However, formation of methyl-

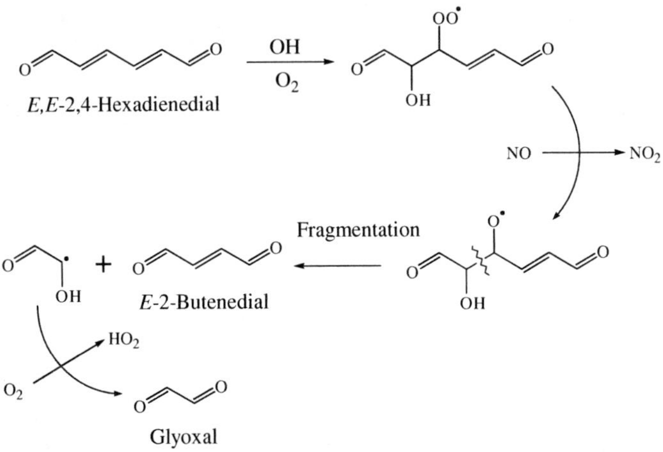

Figure VI-B-2. Proposed mechanism for the reaction of 2,4-hexadienedial with OH radicals.

glyoxal was not observed in an earlier study (Tuazon et al., 1985). It appears to be established that the initial attack of OH on 3-hexene-2,5-dione proceeds predominately via addition to the double bond, followed by addition of O_2 and reaction with NO to give NO_2 and a 4-hydroxy-hexane-2,5-dione-3-yloxyl radical; see figure VI-B-3. This radical then either undergoes a β-scission to yield either (1) two molecules of methylglyoxal or (2) $CH_3C(=O)CH(OH)C(=O)H + CH_3CO$; or (3) it undergoes isomerization to give a 3,4-dihydroxy-hexane-2,5-dione-1-yl radical. The further reactions of this 3,4-dihydroxy-hexane-2,5-dione-1-yl radical are thought to yield polyketones. As a fourth alternative, reaction with O_2 to form HO_2 and 4-hydroxy-hexane-2,3,5-trione has been proposed (Tuazon et al., 1985). Wiesen et al. (1995) have studied the atmospheric chemistry of 3,4-dihydroxy-3-hexene-2,5-dione, the enol form of 4-hydroxy-hexane-2,3,5-trione, and tentatively identified polyketone-type compounds as the major products of the reaction with OH radicals. Similar products were tentatively identified in the OH-initiated oxidation of 3-hexene-2,5-dione by Bierbach et al. (1994).

The fate of *E*- and *Z*-3-hexene-2,5-dione upon reaction with OH radicals is not completely understood, but the seemingly simplest reaction channel, formation of methylglyoxal, does not appear to be dominant. Instead, current experimental

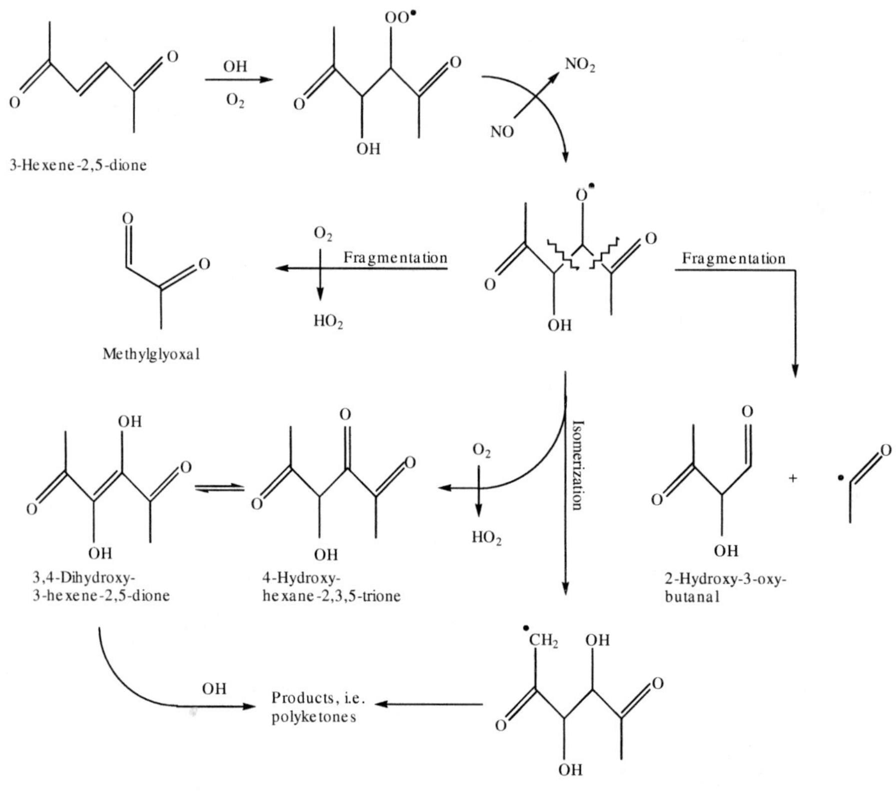

Figure VI-B-3. Proposed mechanisms for the reaction of 3-hexene-2,5-dione with OH radicals.

evidence suggests polyketone-type compounds to be important oxidation products of 3-hexene-2,5-dione.

VI-B-4d. E/Z-isomers of 4-Oxo-2-pentenal

E/Z-Isomers of 4-oxo-2-pentenal have been observed in the OH-initiated photooxidations of toluene and *m*-xylene (O'Brien et al., 1984; Dumdei and O'Brien, 1984; Shepson et al., 1984; Dumdei et al., 1988; Kwok et al., 1997a; Yu et al. 1997; Smith et al., 1998, 1999). Formation yields of 1% C from toluene and 7.5% C from *m*-xylene were estimated by Smith et al. (1998, 1999).

The only study of the atmospheric oxidation of 4-oxo-2-pentenal is that of Bierbach et al. (1994), who found maleic anhydride (40% C) to be the major oxidation product. Glyoxal and methylglyoxal were also observed as major products, but no yields were reported due to uncertainties caused by formation of these from impurities in the oxidation system (Bierbach et al., 1994). In addition, 5-methyl-3*H*-furanone was observed as a minor product (1% C). Proposed formation mechanisms for the observed major products are shown in figure VI-B-4.

VI-B-4e. E/Z-isomers of 2-Butenedial

2-Butenedial has been observed as an intermediate product in the OH-initiated oxidation of toluene (Besemer, 1982; Dumdei and O'Brien, 1984 ; Dumdei et al.,

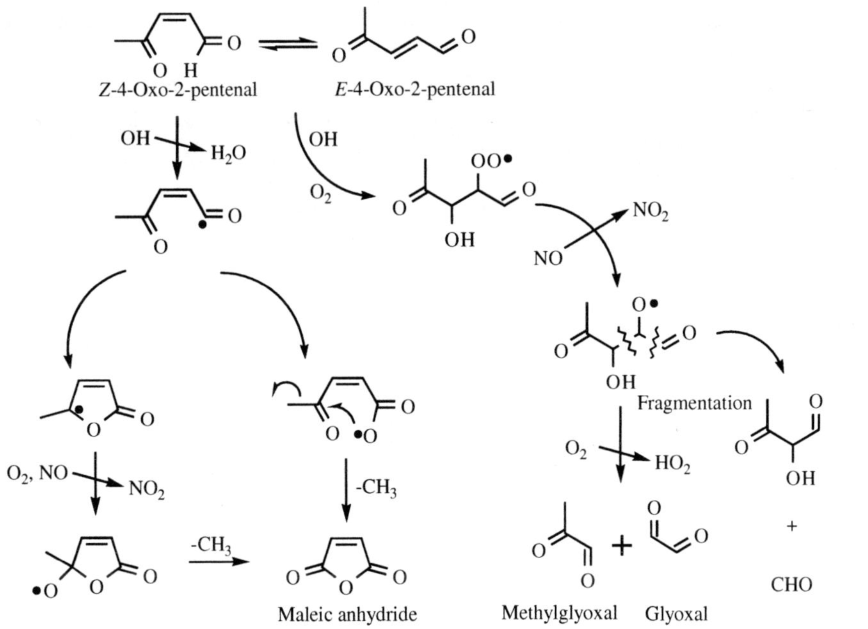

Figure VI-B-4. Proposed mechanisms for the reaction of 4-oxo-2-pentenal with OH radicals.

1988; Yu et al., 1997; Smith et al., 1998 and *o*-xylene (Kwok et al., 1997a; Yu et al., 1997). As was the case for all the other unsaturated 1,4-dicarbonyls, only limited information is available on its reaction with OH radicals. Bierbach et al. (1994) found the major products to be maleic anhydride (50% C) and glyoxal. No yield of glyoxal could be obtained because of interferences by products of impurities in the reaction system. Formation of 3*H*-furan-2-one in low yield (4% C) was also observed. The formation mechanisms of the major products are analogous to those shown in figure VI-B-4.

VI-B-4f. 1,2-Dicarbonyls

α-Dicarbonyls or 1,2-dicarbonyls are the best established ring-fragmentation products to arise from the OH-initiated oxidation of aromatic hydrocarbons. Their formation is reported in virtually all studies of such systems. The earliest determinations were made by Nojima et al. (1974), Darnall et al. (1979), and Takagi et al. (1980). A detailed review of α-dicarbonyl formation yields is given by Atkinson (1994), and more recent determinations by Smith et al. (1998, 1999) are in reasonable agreement with previous results.

VI-B-4g. Glyoxal

No mechanistic data are available on the OH-initiated oxidation of glyoxal, but a study of the Cl-initiated oxidation of glyoxal (Niki et al., 1985a) indicated that after abstraction of an aldehydic H-atom, the resulting carbonyl radical can undergo either fragmentation or reaction with O_2,; see figure VI-B-5. Niki et al. (1985a) determined rate coefficient ratios of $k_a/k_b \approx k_a/k_c \approx 3.5 \times 10^{18}$ molecules cm^{-3}. Under tropospheric conditions (1 atm of air at 298 K), fragmentation will be the dominant process accounting for ~60% of the reaction, while the two O_2-addition channels account for ~20% of the decay of each. The formyl radical (·CHO) will react with molecular oxygen to give HO_2 and CO, while the HC(=O)C(=O)OO· radical is expected to react like other acyl peroxy radicals; see, for example, Atkinson (1994).

VI-B-4h. Methylglyoxal and Biacetyl

No mechanistic data exist on the OH-initiated oxidation of methylglyoxal and biacetyl; however, it can be expected that methylglyoxal will react like glyoxal, with the reaction being initiated by abstraction of the aldehydic H-atom. This will be followed by fragmentation into CO and an acetyl radical (Green et al., 1990). The reaction of biacetyl with OH radicals is initiated by abstraction of an H-atom from one of the methyl groups. Christensen et al. (1998) have shown that the resulting $CH_3C(=O)C(=O)CH_2$· radical adds O_2 rapidly to give a peroxy radical, which reacts with NO to give an alkoxy radical; this radical decomposes to give CH_2O, CO, and a CH_3CO radical.

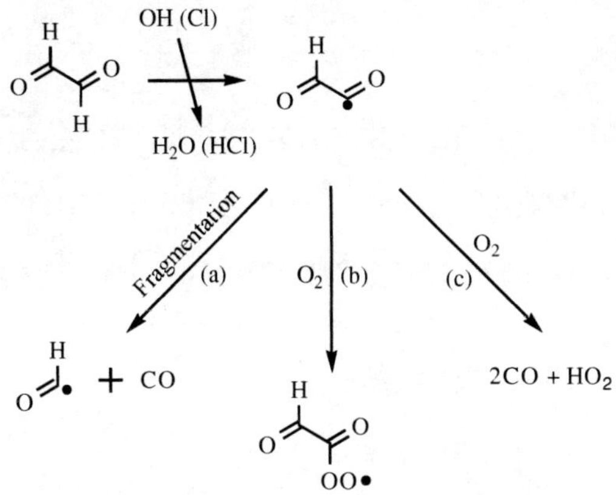

Figure VI-B-5. Proposed mechanisms for the reaction of glyoxal with OH radicals.

VI-B-4i. Maleic Anhydride

Maleic anhydride is, next to the α-dicarbonyls, among the first ring-fragmentation products to be identified in the OH-initiated oxidation of aromatic hydrocarbons (Kobayashi, 1979; Bandow et al., 1985; Bandow and Washida, 1985a; Bierbach et al., 1993; Becker et al., 1997). The only product study of the reaction of maleic anhydride with OH radicals is that of Bierbach et al. (1994). The only significant products identified were CO (20% C) and formic acid (3% C). The major oxidation products were not identified.

VI-B-4j. Low-Molecular-Weight Products

The mechanisms of the further oxidation of low-molecular-weight (C_1 and C_2) products like formaldehyde, formic acid, PAN, and so on, arising from the OH-initiated oxidation of aromatic hydrocarbons, have been extensively reviewed in the existing literature (Atkinson, 1994; Atkinson et al., 1999; Calvert et al., 2000) and no additional recommendations will be given here.

VII

Primary Photochemical Processes of the Aromatic Hydrocarbons and Some of Their Common Oxidation Products

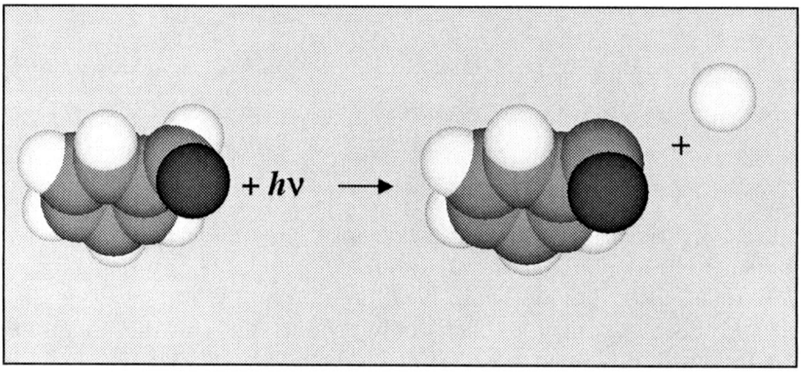

VII-A. Tropospheric Photochemistry of the Aromatic Hydrocarbons

The monocyclic aromatic hydrocarbons, benzene and its alkyl-substituted benzenes, absorb very little of the sunlight present in the troposphere, and as a consequence their photochemistry is unimportant in the urban atmosphere. Substitution of an increasing number of alkyl groups on benzene results in a shift in the absorption of the compounds to longer wavelengths, but even with the trimethyl-substituted benzenes, there is no overlap of the absorption region with the wavelengths of sunlight present in the lower troposphere; see figure VII-A-1. Note that all cross sections (symbolized by σ in this work) are in the conventional units of cm^2 molecule^{-1} with $\sigma = \ln(I_0/I)/d[c]$, where ln is the natural logarithm (base e), I_0 and I are the incident and transmitted intensity of light, respectively, d is the path length in centimeters, and $[c]$ is the concentration of the absorbing species in molecules per cubic centimeter.

However, as each additional aromatic ring fragment is fused to benzene to form the various polycyclic aromatic hydrocarbons (naphthalene, anthracene, etc.), the conjugated π-system grows, the energy to promote an electron to the

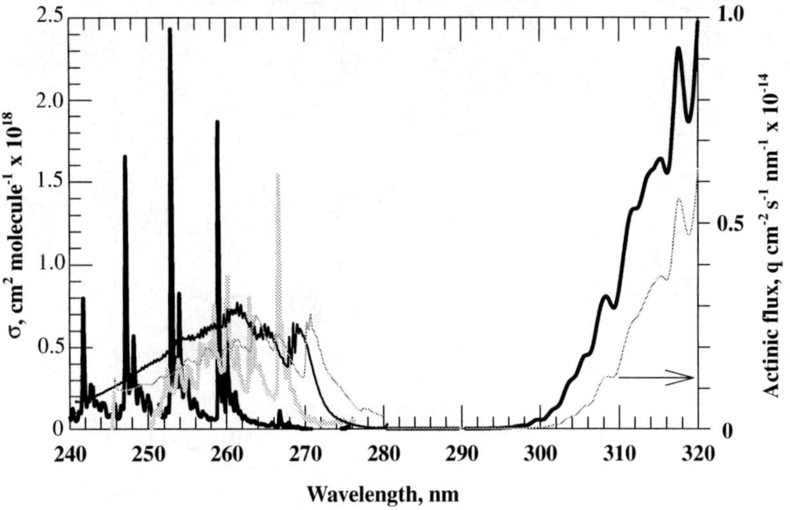

Figure VII-A-1. Wavelength regions for cross sections of vapors of some methyl-substituted benzenes: benzene (heavy black curve), toluene (heavy gray curve), *o*-xylene (medium black curve), and 1,3,5-trimethylbenzene (light gray curve) compared with the wavelength distribution of the actinic flux measured for a cloud-free day (June 19, 1998) near Boulder, CO: black and gray curves to the right in the figure are 10-min averaged data for 08:08 h (gray curve) and 12:08 h (black curve), mountain standard time, respectively. Spectral data are from Etzkorn et al. (1999).

first excited state decreases, the light absorption increases in magnitude, and the absorption band extends further into the longer wavelength region. See figure VII-A-2.

The large cross sections typical of the polycyclic aromatic hydrocarbons result in very large rates of sunlight absorption. In figure VII-A-3, compare the $j_{excitation}$ values (first-order rate coefficients for formation of the first excited singlet state) for naphthalene, anthracene, and naphthacene (tetracene) with the photochemically significant values of $j(NO_2)$ and $j[O(^1D)]$ measured on a cloud-free day (June 19, 1998) at the same location and time (near Boulder, CO). Note that the rates of excitation of anthracene and naphthacene (3- and 4-fused benzene rings, respectively) are much larger than those for $O(^3P)$ atom formation from NO_2 photolysis, while that for naphthalene (2-fused rings) is somewhat smaller than $j(NO_2)$ but larger than that of $j[O(^1D)]$, $O(^1D)$ atom formation from O_3 photolysis. In view of the large rates of excitation of the polycyclic aromatic hydrocarbons that can occur within the lower troposphere, the fate of the excited molecules is of special interest to atmospheric chemists, particularly if significant chemical changes result from the subsequent reactions of the excited singlet species formed in the atmosphere. We consider these reactions in section VII-B.

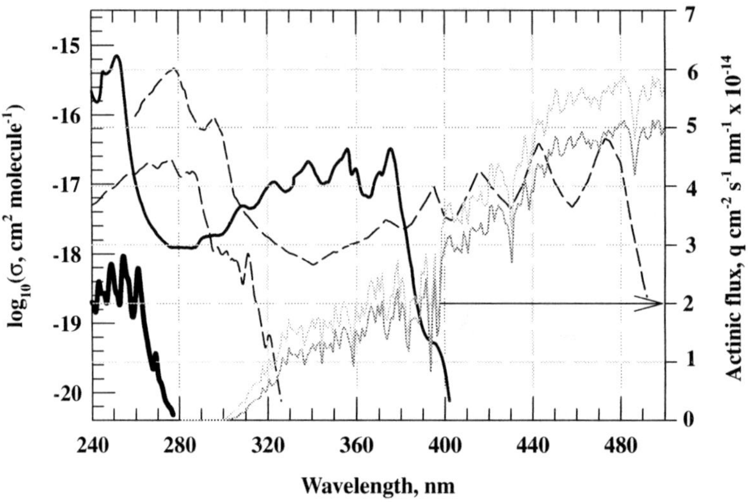

Figure VII-A-2. Absorption cross sections (σ) for a series of polycyclic hydrocarbons in non-polar solvents [data of Láng (1961–1971); Jaffé and Orchin (1962)]: benzene (heavy black curve), naphthalene (medium black dashed curve), anthracene (medium solid black curve) and tetracene (light dashed black curve), compared with the distribution of actinic flux as measured near Boulder, CO (June 19, 1998) at 08:08 h (dark gray curve, right side) and 12:08 h mountain standard time (light gray curve, right side), respectively.

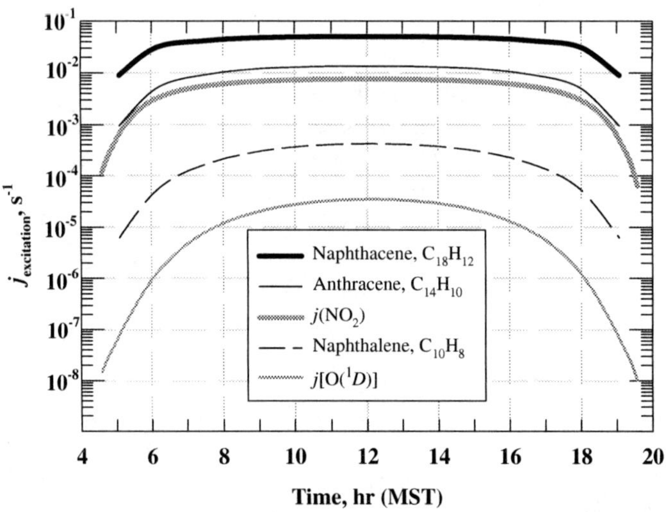

Figure VII-A-3. Values of $j_{\mathrm{excitation}}$, first-order rate coefficients for excited singlet molecule formation in sunlight estimated using the actinic flux data for a cloudless day (June 19, 1998) near Boulder, CO, and cross sections (solution phase) reported by Jaffé and Orchin (1962) and Láng (1961–1971); for comparison, $j(NO_2)$ and $j[O(^1D)]$ values are shown as measured for the same time and location.

VII-B. Photochemistry of Polycyclic Aromatic Hydrocarbons in the Troposphere

The maximum energy of the quantum of light available for absorption in the troposphere is about 412 kJ mol^{-1} ($\lambda = 290$ nm), much less than that necessary to rupture the aromatic carbon atom bonds in benzene and other aromatic ring compounds (> 481 kJ mol^{-1}), the C–H bond on a ring (~464 kJ mol^{-1}), and the C–C bond of an alkyl group attached to a ring (~426 kJ mol^{-1}). As a consequence, the excited states of the polycyclic aromatic species formed by light absorption within the troposphere are stable toward bond rupture. However, it was recognized many years ago (Khan et al., 1967; Kummler and Bortner, 1969; Pitts et al., 1969; Steer et al., 1969) that the excited states of the aromatics compounds are capable of exciting molecular oxygen from its ground electronic state, [$O_2(^3\Sigma_g^-)$], to its low-lying excited singlet states, $O_2(^1\Delta_g)$ and $O_2(^1\Sigma_g^+)$, 94.1 and 156.9 kJ mol^{-1}, respectively, above the ground-state O_2. Since the singlet oxygen states are reasonably long-lived, their participation in the chemistry of the urban atmosphere, through oxidation of aromatic and other organic species, was suggested as potentially important. Many studies have confirmed the notion that electronically excited aromatic molecules can generate singlet oxygen by energy transfer. Thus, benzene in an O_2-containing gas mixture excited at 253.7 nm generated $O_2(^1\Sigma_g^+)$ and $O_2(^1\Delta_g)$ as detected by their characteristic emissions (Snelling, 1968). Similarly, the unambiguous paramagnetic resonance spectra of $O_2(^1\Delta_g)$ was detected in irradiated anthracene–O_2 mixtures (Kearns et al., 1969) and in derivatives of anthracene in O_2 mixtures (Wasserman et al., 1969). Singlet delta oxygen can be formed through the following mechanism, where $PAH(S_0)$, $PAH(S_1)$, and $PAH(T_1)$ represent the ground, first excited singlet, and first excited triplet electronic states, respectively, of the polycyclic aromatic hydrocarbon:

$$PAH(S_0) + h\nu \rightarrow PAH(S_1) \tag{1}$$

$$PAH(S_1) \rightarrow PAH(T_1) \tag{2}$$

$$PAH(T_1) + O_2(^3\Sigma_g^-) \rightarrow O_2(^1\Delta_g) + PAH(S_0) \tag{3}$$

The first excited triplet state is formed by intersystem crossing, reaction (2), in many aromatic species, and during its lifetime the triplet state formed can transfer energy reasonably efficiently to molecular oxygen to form the $O_2(^1\Delta_g)$ in reaction (3).

Demerjian et al. (1972, 1974) and Penzhorn et al. (1974) suggested that the role of singlet oxygen in gas-phase chemistry of the urban atmosphere was unimportant. However, interest in this possible importance was renewed by Ogawa et al., who developed a method believed to be unique to the detection of $O_2(^1\Delta_g)$ in the atmosphere (Ogawa et al., 1991). It employs the reaction of $O_2(^1\Delta_g)$ with α-terpinene that forms a unique product, α-terpinene endoperoxide, not a product of reactions of α-terpinene with OH radicals O_3, or H_2O_2. In their application of this method to the urban atmosphere in Kyoto, Japan, Ogawa and coworkers detected concentrations of $O_2(^1\Delta_g)$, or a species that is similar chemically to singlet oxygen (termed SOLO, "singlet oxygen-like oxidant"), which peaked near noon at concentrations comparable to those of ozone (Ogawa et al.,

1996). Our interest here is in applying the known chemistry of formation and destruction of $O_2(^1\Delta_g)$ to test again for the possible significant involvement of the aromatic compounds, both in formation of singlet oxygen and their possible subsequent reactions with it.

The known sources and sinks of $O_2(^1\Delta_g)$ in the urban atmosphere can be summarized in the following sequence of reactions; rate coefficients shown are from the summary of DeMore et al. (1997); *j*-values are estimates from Demerjian et al. (1974):

$$O_2(^3\Sigma_g^-) + h\nu\,(\lambda \approx 760\,\text{nm})(+M) \rightarrow O_2(^1\Sigma_g^+)(+M); \quad j = 2.50 \times 10^{-10}\,\text{s}^{-1} \tag{4}$$

$$O_2(^3\Sigma_g^-) + h\nu\,(\lambda \approx 1270\,\text{nm})(+M) \rightarrow O_2(^1\Delta_g)(+M); \quad j = 5.06 \times 10^{-11}\,\text{s}^{-1} \tag{5}$$

$$O_2(^1\Sigma_g^+)(+M) \rightarrow O_2(^3\Sigma_g^-) + h\nu(+M); \quad k \geq 8.5 \times 10^{-2}\,\text{s}^{-1} \tag{6}$$

$$O_2(^1\Delta_g)(+M) \rightarrow O_2(^3\Sigma_g^-) + h\nu(+M); \quad k \approx 6.2 \times 10^{-4}\,\text{s}^{-1} \tag{7}$$

$$O_2(^1\Sigma_g^+) + O_2 \rightarrow O_2(^1\Delta_g;{}^3\Sigma_g^-) + O_2; \quad k = 3.9 \times 10^{-17}\,\text{cm}^3\,\text{molecule}^{-1}\,\text{s}^{-1} \tag{8}$$

$$O_2(^1\Sigma_g^+) + H_2O \rightarrow O_2(^1\Delta_g;{}^3\Sigma_g^-) + H_2O; \quad k = 5.4 \times 10^{-12} \tag{9}$$

$$O_2(^1\Sigma_g^+) + N_2 \rightarrow O_2(^1\Delta_g;{}^3\Sigma_g^-) + N_2; \quad k = 2.1 \times 10^{-15} \tag{10}$$

$$O_2(^1\Delta_g) + O_2 \rightarrow O_2(^3\Sigma_g^-) + O_2; \quad k = 1.74 \times 10^{-18} \tag{11}$$

$$O_2(^1\Delta_g) + N_2 \rightarrow O_2(^3\Sigma_g^-) + N_2; \quad k \leq 10^{-20} \tag{12}$$

$$O_2(^1\Delta_g) + H_2O \rightarrow O_2(^3\Sigma_g^-) + H_2O; \quad k = 4.8 \times 10^{-18} \tag{13}$$

$$O_3 + h\nu\,(\lambda \sim 290\text{–}306\,\text{nm}) \rightarrow O_2(^1\Delta_g) + O(^1D) \tag{14}$$

$$O_3 + h\nu\,(\lambda \sim 306\text{–}350\,\text{nm}) \rightarrow O_2(^1\Delta_g) + O(^3P) \tag{15}$$

(14), (15): $j = 7.17 \times 10^{-5}\,\text{s}^{-1}$

$$NO_2 + h\nu\,(\lambda < 400\,\text{nm}) \rightarrow NO_2^* \tag{16}$$

$$NO_2^* + O_2 \rightarrow NO_2 + O_2(^1\Delta_g) \tag{17}$$

$$NO_2^* + N_2\,(\text{or }O_2) \rightarrow NO_2 + N_2\,(\text{or }O_2) \tag{18}$$

(16)–(18): $j = 1.15 \times 10^{-4}\,\text{s}^{-1}$

$$O_2(^1\Delta_g) + \text{PAH} \rightarrow \text{Products} \tag{19}$$

where NO_2^* represents an energy-rich, non-dissociating NO_2 molecule.

Singlet oxygen can be formed by direct light absorption within the O_2 forbidden transitions near 760 and 1270 nm [reactions (4) and (5)], through O_3 photolysis [reactions (14) and (15); Kummler et al., 1969], electronic energy transfer from NO_2 excited at wavelengths greater than its photodissociation threshold [reactions (16)–(18); Jones and Bayes, 1971; Frankiewicz and Berry, 1972], and by energy transfer from electronically excited organic species [reaction (3); Khan et al., 1967; Kummler and Bortner, 1969; Pitts et al., 1969; Steer et al., 1969]. Using the *j*-values estimated by Demerjian et al. (1974) for the reactions (4), (5), (14), (15), and (16), estimated for a solar zenith angle of 40°, and that for reaction

Table VII-B-1. Concentrations of $O_2(^1\Delta_g)$ estimated to be present for varied concentrations of urban pollutants (clear sky, solar zenith angle = 40°)

$[NO_2]$, ppb	$[O_3]$, ppb	$[H_2O]$, % Relative Humidity	$[PAH]$, ppb	$[O_2(^1\Delta_g)]$, ppb[a]
15	100	0	0	0.00728
30	100	0	0	0.00744
15	50	0	0	0.00687
30	50	0	0	0.00703
15	100	0	0	0.00613
15	100	50	0	0.00528
30	100	100	0	0.00626
30	100	50	0	0.00541
15	50	100	0	0.00577
30	50	50	0	0.00512
30	100	100	0.01	0.00670
30	100	50	0.20	0.0154
30	100	50	2.00	0.0980
30	100	50	20.00	0.923
30	100	50	200.00	9.14

[a]Calculated using the rate coefficients given in the text (DeMore et al., 1997), *j*-values for reactions (4), (5), (14), and (15) as given by Demerjian et al. (1974) and that for reaction (1) equal to $0.5\ s^{-1}$. It was assumed that the reactivity of the PAH for $O_2(^1\Delta_g)$ was equal to that for tetramethylethylene ($1.3 \times 10^{-15}\ cm^3\ molecule^{-1}\ s^{-1}$).

(1) equal to $0.5\ s^{-1}$ (a very large PAH), we can estimate the $O_2(^1\Delta_g)$ concentration given in table VII-B-1. These estimates represent the maximum possible values, since it is assumed that ultimately the only singlet oxygen product of the $O_2(^1\Sigma_g^+)$ in reactions (8)–(10) and O_3 photolysis in reactions (14) and (15) is $O_2(^1\Delta_g)$, that the quantum yield of intersystem crossing of the excited $PAH(S_1)$ state to $PAH(T_1)$ is unity, and that $PAH(T_1)$ formed in the urban atmosphere always reacts to form $O_2(^1\Delta_g)$.

Note that with only the sources and sinks of $O_2(^1\Delta_g)$ given in reactions (4)–(18), the $[O_2(^1\Delta_g)]$ is relatively insensitive to the concentration of O_3, NO_2, and H_2O over the range chosen, and the influence of reactions (1)–(3) on the concentration becomes significant at $(PAH] > 0.01$ ppb. It is clear from the data in table VII-B-1 that only if the $(PAH]$ is over 200 ppb does the $[O_2(^1\Delta_g)]$ approach that for the SOLO observed in Kyoto. Typical concentrations of polycyclic aromatic compounds observed and reported by Graedel (1978) are much less that this. Thus, the $\Sigma[PAH]$ in Los Angeles, California in 1989 was ~1–2 ppb. Even if the highest concentrations of *PAH* chosen here were to occur, then the reaction of the $O_2(^1\Delta_g)$ with the aromatic hydrocarbons will still play a very minor role in the removal of PAHs.

However, one may argue that $O_2(^1\Delta_g)$ is indeed the species observed by Ogawa et al. (1996) and that singlet oxygen is formed in some, as yet ill-defined important reactions in addition to those outlined above. For example, some exothermic thermal reactions may generate single oxygen (Pitts et al., 1969). The rapid and important reactions of NO with O_3 are energetically feasible sources of singlet oxygen:

$$NO + O_3 \rightarrow NO_2 + O_2(^3\Sigma_g^-); \qquad \Delta H^\circ_{298} = -198.8\,\text{kJ mol}^{-1} \tag{20a}$$

$$NO + O_3 \rightarrow NO_2 + O_2(^1\Delta_g); \qquad \Delta H^\circ_{298} = -104.6\,\text{kJ mol}^{-1} \tag{20b}$$

$$NO + O_3 \rightarrow NO_2 + O_2(^1\Sigma_g^+); \qquad \Delta H^\circ_{298} = -41.9\,\text{kJ mol}^{-1} \tag{20c}$$

It is well known that most of the excess energy released in reaction (20) appears as electronic excitation in the NO_2 product that emits strongly over a large wavelength range. There is no evidence that reaction channels (20b) and (20c) occur (Michael et al., 1981; Borders and Birks, 1982). Exothermic reactions such as reactions (20) usually deposit the majority of the excess energy in the newly formed bond; certainly, the strong chemiluminescence exhibited by this reaction gives evidence that excited NO_2 molecules are the major, if not exclusive, excited product in which the excess energy appears in reactions (20). One can use the analytical data of Ogawa et al. (1996) to test for the occurrence of reaction (20b) in their atmospheric experiments in which the SOLO was measured. The maximum product of $[O_3](NO]$, 4.8×10^{23} molecule^2 cm^{-6}, occurred near 10 A.M. during their measurements. We estimate from this that the maximum rate of NO_2 and O_2 product generation in reactions (20) was about 8.7×10^9 molecule cm^{-3} s^{-1}. Even if the product channel (20b) that forms $O_2(^1\Delta_g)$ were the dominant channel, this rate of generation is far less than that necessary to rationalize the observed concentration of SOLO; a rate of about 6×10^{13} molecule cm^{-3} s^{-1}, a rate $\sim 10^4$ times that expected from the NO + O_3 reaction, is necessary to maintain a $[O_2(^1\Delta_g)]$ of 20 ppbv when one considers the rate of its loss processes in just the reactions (7)–(13). Obviously, reactions (20) cannot rationalize the high SOLO levels observed by Ogawa et al. (1996).

Singlet oxygen generation is also possible energetically from the molecular channel of the self-reactions of the peroxy radicals; consider, for example, the reactions of the methyl and ethyl peroxy radicals:

$$CH_3O_2 + CH_3O_2 \rightarrow CH_2O + CH_3OH + O_2(^1\Delta_g); \qquad \Delta H^\circ_{298} = -195\,\text{kJ mol}^{-1} \tag{21a}$$

$$CH_3O_2 + CH_3O_2 \rightarrow CH_2O + CH_3OH + O_2(^1\Sigma_g^+); \qquad \Delta H^\circ_{298} = -132\,\text{kJ mol}^{-1} \tag{21b}$$

$$C_2H_5O_2 + C_2H_5O_2 \rightarrow CH_3CHO + C_2H_5OH + O_2(^1\Delta_g); \qquad \Delta H^\circ_{298} = -250\,\text{kJ mol}^{-1} \tag{22a}$$

$$C_2H_5O_2 + C_2H_5O_2 \rightarrow CH_3CHO + C_2H_5OH + O_2(^1\Sigma_g^+); \qquad \Delta H^\circ_{298} = -213\,\text{kJ mol}^{-1} \tag{22b}$$

The HO_2–HO_2 reaction is less exothermic than the RO_2–RO_2 reactions, but in theory it is sufficiently exothermic to form singlet oxygen as well:

$$HO_2 + HO_2 \rightarrow H_2O_2 + O_2(^1\Delta_g); \qquad \Delta H^\circ_{298} = -71\,\text{kJ mol}^{-1} \tag{23a}$$

$$HO_2 + HO_2 \rightarrow H_2O_2 + O_2(^1\Sigma_g^+); \qquad \Delta H^\circ_{298} = -8\,\text{kJ mol}^{-1} \tag{23b}$$

The reactions of the HO_2 radical with the simple alkyl peroxy radicals are even less exothermic, but are sufficiently exothermic to allow formation $O_2(^1\Delta_g)$ in theory:

$$HO_2 + CH_3O_2 \rightarrow CH_3O_2H + O_2(^1\Delta_g); \qquad \Delta H^\circ_{298} = -60\,\text{kJ mol}^{-1} \tag{24a}$$

$$HO_2 + CH_3O_2 \rightarrow CH_3O_2H + O_2(^1\Sigma_g^+); \qquad \Delta H^\circ_{298} = +3\,\text{kJ mol}^{-1} \tag{24b}$$

For reactions in solutions of *tert*-butylbenzene, the occurrence of RO_2–RO_2 reactions like (21a) and (22a) occur with a yield of 3–18% $O_2(^1\Delta_g)$ (Niu and

Mendenhall, 1992). No $O_2(^1\Delta_g)$ was detected from the reaction (23a) (Sugimoto and Sawyer, 1988; Niu and Mendenhall, 1992). There is no experimental evidence related to the reaction (24) available at this writing. It is probable that in the gas phase, the efficiency of generation of $O_2(^1\Delta_g)$ from the alkyl peroxy radical reactions is greater than that in solution. However, even if it occurred with perfect efficiency, this cannot be a major source of singlet oxygen in the highly polluted urban atmosphere, since termination of radical chains occurs largely by the $HO + NO_2 (+M) \rightarrow HONO_2 (+M)$ reaction. Thus, if all of the "$HO_2 + RO_2$" radical signal measured by Cantrell et al. (1992, 1993) in rural atmospheres was from primary and secondary alkyl peroxy radicals (200 pptv maximum), and the yield of $O_2(^1\Delta_g)$ was 100% from reactions equivalent to (22a) and (23a), the rate of singlet delta oxygen generation would be only about 6.8×10^6 molecule cm^{-3} s^{-1}, far short of the 6×10^{13} molecule cm^{-3} s^{-1} necessary to rationalize the SOLO concentrations in terms of $O_2(^1\Delta_g)$. Thus, we conclude that the SOLO observed by Ogawa et al. (1996) cannot be rationalized by the known chemistry of singlet delta oxygen. In any case, the significant generation of $O_2(^1\Delta_g)$ by polycyclic aromatic compounds and PAH removal by reaction with $O_2(^1\Delta_g)$ in the gas phase are probably not important in the urban atmosphere.

However, PAHs absorbed within aerosol solutions may entrap $O_2(^1\Delta_g)$ locally as it is formed in reactions (1)–(3), and reactions with the PAH that generated it or a nearby PAH may be possible before the other possible quenching mechanisms outlined above can occur. For the gas-phase reactions, this "cage" mechanism cannot be operative.

VII-C. Photochemistry of Some Common Ring-Fragmentation Products of Atmospheric Oxidation of Aromatic Hydrocarbons

The photochemistry of simple aldehydes and ketones, and other atmospheric oxidation products of both the alkene and aromatic hydrocarbons, has been reviewed recently (Calvert et al., 2000). Preferred absorption cross sections and quantum yields were given for the following compounds: CH_2O, CH_3CHO, CH_3CH_2CHO, CHOCHO, CH_3COCHO, CH_3COCH_3, CH_3O_2H, and CH_3ONO_2. Data for these and other oxygen- and nitrogen-containing compounds are also presented in Atkinson et al. (1999). Our recommendations concerning these photochemical data have not changed since those reviews, and they will not be discussed further here, except where contrasts or similarities in photochemical behavior are to be noted.

In the following section, the photochemistry of several other of the ring-fragmentation products of the oxidation of aromatic hydrocarbons is reviewed. Unfortunately, the literature related to the photochemistry of many of the aromatic hydrocarbon oxidation products is very limited, and many of the conclusions and recommendations are necessarily also limited. In many cases, the products such as phenols, aromatic nitrates, polyfunctional aldehydes, and ketones, are of rather low vapor pressure, and are very difficult to study in the vapor phase. In these cases, we must resort to the use of absorption spectra for solutions of the compounds (usually in non-polar solvents). In those cases where quantum yield data related to the primary processes of the aromatic hydrocarbon products are unknown or highly uncertain, recommendations must be made based upon an extrapolation of our knowledge of the better known photochemical properties of other related compounds. With many of the less volatile species discussed here, there is a significant uncertainty as to their lifetime in the vapor phase, and in many cases it is likely that the species will agglomerate on existing aerosols, deposit on surfaces, or be incorporated into cloud-water, rain droplets, or groundwater. In such cases, the photochemical and thermal oxidation processes that continue to transform the species may be quite different than those observed in the gas phase. Our discussions in this chapter will focus on gas-phase photochemistry. In chapter IX, the photochemistry of polycyclic aromatic compounds sorbed on surfaces and in aerosols is described.

VII-C-1. Photolysis Model Description

Atmospheric photolysis rate coefficients (photolysis frequencies), or *j*-values, were calculated by numerical summation over wavelength λ,

$$j = \Sigma \langle\sigma\rangle_i \langle\phi\rangle_i \langle F\rangle_i \, \Delta\lambda_i \tag{1}$$

where $\langle\sigma\rangle_i$, $\langle\phi\rangle_i$, and $\langle F\rangle_i$ are, respectively, the absorption cross section, quantum yield, and spectral actinic flux averaged over small wavelength intervals $\Delta\lambda_i$. Cross sections and quantum yields specific to each photodissociation reaction are discussed later in this chapter.

The spectral actinic flux was calculated with the Tropospheric Ultraviolet Visible (TUV) model version 4.0 (Lantz et al., 1996; Madronich and Flocke, 1998). The model code is publicly available and may be downloaded from

http://acd.ucar.edu/UV

The specific conditions assumed for the calculations were as follows (unless otherwise noted): The atmosphere was subdivided from 0 to 80 km in 1-km layers, with the exception of the lowest two layers, which were 500 m each. Vertical profiles of temperature and number densities for air and ozone, representative of annual means for 45° N, were taken from U.S. Standard Atmosphere (1976), with a total ozone column of 350 Dobson units (1 DU = 2.69×10^{16} molecule cm^{-2}). At 500 m above sea level, for which many calculations are given in this section, the temperature was 285 K and air density was 2.430×10^{19} molecule cm^{-3}. No attenuation by other atmospheric gases (e.g., NO_2, SO_2) was considered. An aerosol profile typical for continental rural conditions, from Elterman (1968), was specified with total vertical optical depth of 0.38 and surface attenuation of 0.24 km^{-1} at 340 nm. The optical depth was assumed to scale inversely with wavelength, yielding a surface visible range of ca. 25 km at 550 nm according to the Koshmeider 2% contrast definition. Aerosol scattering was represented by a Henyey–Greenstein function, with the asymmetry factor $g \sim 0.61$, and weak absorption by a single scattering albedo $\omega_0 \sim 0.99$, both independent of wavelength. A 10% Lambertian surface albedo was assumed, independent of wavelength.

Extraterrestrial irradiances are from Van Hoosier et al. (1988) for $\lambda < 350$ nm, and from Neckel and Labs (1984) for longer wavelengths, for annual mean Earth-Sun distance. The O_3 absorption cross sections over 185–350 nm are from Molina and Molina (1986) and include temperature dependence over 240.5–350 nm. Cross sections of O_3 at other wavelengths and those for Rayleigh scattering are from WMO (1986). Absorption of O_2, important at higher altitudes, is taken from WMO (1986), with parameterization of the Schumann–Runge bands according to Kockarts (1994).

At each wavelength interval, the spectral actinic flux was calculated using a pseudo-spherical eight-stream discrete ordinates model (Stamnes et al., 1988; Madronich and Flocke, 1998). The actinic flux included contributions of direct solar light, as well as diffuse radiation from both the up and down directions.

The calculation of atmospheric photolysis rate coefficients via equation (1) requires the computation of average values of $\langle\sigma\rangle_i$, $\langle\phi\rangle_i$, and $\langle F\rangle_i$ over sufficiently small wavelength intervals $\Delta\lambda_i$ to minimize numerical integration errors. Because computational cost increases approximately linearly with the number of intervals, excessively small $\Delta\lambda_i$ are undesirable. Here, uniform $\Delta\lambda_i = 1$-nm intervals centered on integer (nm) values over 200–700 nm were used. Numerical errors induced by the use of coarser resolution have been discussed by Madronich and Weller (1990) and, for HCHO photodissociation, by Cantrell et al. (1990), who showed that acceptable accuracy may be obtained using 5-nm intervals, but this degrades rapidly for wider intervals.

A further cautionary note is required on the use of laboratory data that were obtained at low spectral resolution. These are often reported as average values over broad wavelength intervals that are much wider than those used in the *j*-value calculations ($\Delta\lambda_i$). If the underlying true spectral dependence of the measured quantity (e.g., cross sections and quantum yields) is believed to be reasonably smooth, interpolation between adjacent points is preferable to the use of a constant value over the reported wavelength interval. For example, the quantum yields of HCHO are commonly reported as averaged over 10-nm intervals, although the original data provide much greater spectral detail. When calculating *j*-values with $\Delta\lambda_i = 1$ nm, linear interpolation between adjacent points (the average quantum yields associated with the center of each 10-nm interval) produces errors smaller than about 4% relative to using the original spectrally detailed data. On the other hand, treating the quantum yields as truly constant over the specified 10-nm intervals produces errors significantly larger (10% or more). Such resolution-dependent errors can confound the comparisons among different studies in which different spectral resolutions were used in the measurements or for the reporting of the data.

VII-C-2. Glycolaldehyde [Hydroxyacetaldehyde; $HOCH_2C({=}O)H$]

Few studies of the photochemistry of glycolaldehyde have been made to date, but from the study of Bacher et al. (2001) it is suggested that the primary photodecomposition occurs analogous to that in acetaldehyde:

$$HOCH_2C({=}O)H + h\nu \rightarrow HOCH_2 + CHO \qquad \text{(I)}$$

VII-C-2a. Absorption Cross Sections

The vapor-phase absorption spectrum of glycolaldehyde has been determined by Bacher et al. (2001). It is compared with that for acetaldehyde (Martinez et al., 1992) in figure VII-C-1. The significant shift of the absorption to shorter wavelengths that occurs with the introduction of an HO-group on the methyl group of acetaldehyde decreases the importance of the tropospheric photochemistry of glycolaldehyde compared with that of acetaldehyde. The recommended cross sections are listed in table VII-C-1.

VII-C-2b. Quantum Yields

Limited data are available from the study of Bacher et al. (2001). An average quantum yield of >0.5 for process (I) was estimated for photolysis of glycolaldehyde with light in the wavelength range $265 < \lambda < 305$ nm and in 1 atm of air. However, when the light was limited to wavelengths >300 nm, more representative of those present in the troposphere, the photolysis rate was very low, and that observed could be attributed to wall loss alone; no products were observed. The limited overlap of the lamp radiation with the absorption of the glycolaldehyde in this experiment prevented any realistic estimate of the average quantum yield. However, it is likely that the photodissociation is wavelength dependent, and the quantum yields of process (I) may be much less than 0.5 for $\lambda < 290$ nm. Photolyses in aqueous solution (Beeby et al., 1987) gave, as major products, CH_3OH, CO, CH_2O, and malonaldehyde [HC(=O)CH_2C(=O)H].

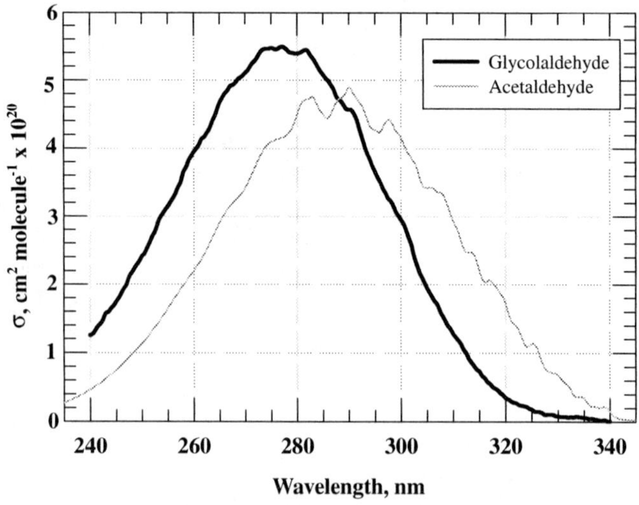

Figure VII-C-1. Comparison of the absorption spectra of glycolaldehyde (Bacher et al., 2001) and acetaldehyde (Martinez et al., 1992) in the gas-phase.

Table VII-C-1. Glycolaldehyde absorption cross sections in the gas phase

λ (nm)	$\sigma \times 10^{20}$	λ (nm)	$\sigma \times 10^{20}$	λ (nm)	$\sigma \times 10^{20}$	λ (nm)	$\sigma \times 10^{20}$
240	1.25	265	4.68	290	4.57	315	0.732
241	1.32	266	4.81	291	4.50	316	0.638
242	1.42	267	4.91	292	4.28	317	0.555
243	1.57	268	4.96	293	4.04	318	0.486
244	1.64	269	5.04	294	3.84	319	0.420
245	1.74	270	5.11	295	3.66	320	0.348
246	1.87	271	5.21	296	3.51	321	0.301
247	2.00	272	5.30	297	3.35	322	0.271
248	2.17	273	5.42	298	3.23	323	0.237
249	2.29	274	5.46	299	3.07	324	0.217
250	2.41	275	5.47	300	2.95	325	0.178
251	2.53	276	5.45	301	2.79	326	0.148
252	2.70	277	5.49	302	2.58	327	0.127
253	2.88	278	5.45	303	2.30	328	0.106
254	3.04	279	5.40	304	2.11	329	0.102
255	3.15	280	5.39	305	1.95	330	0.0779
256	3.27	281	5.42	306	1.81	331	0.0717
257	3.43	282	5.43	307	1.70	332	0.0623
258	3.65	283	5.32	308	1.54	333	0.0649
259	3.82	284	5.19	309	1.4	334	0.0688
260	3.95	285	5.08	310	1.28	335	0.0569
261	4.06	286	4.99	311	1.18	336	0.0412
262	4.17	287	4.87	312	1.05	337	0.0365
263	4.29	288	4.72	313	0.916	338	0.0194
264	4.47	289	4.63	314	0.81	339	0.0225
						340	0.00679

Source: Bacher et al. (2001).
Note: Units on σ are cm^2 molecule^{-1}.

VII-C-2c. Tropospheric Photolysis

Photolysis rate coefficients for glycolaldehyde that are representative of the lower atmosphere were calculated with the TUV model (as described in the introduction to this section) using the measured σ and limited ϕ data given in this section. The data are shown for two assumed average quantum yields: $\phi = 0.5$ and 0.02; see figure VII-C-2.

VII-C-3. Hydroxyacetone ($HOCH_2COCH_3$)

In the study of Orlando et al. (1999), the identified products of photolysis of hydroxyacetone—CH_2O, HCOOH, CO, CO_2, CH_3COOH, CH_3COO_2H, and CH_3OH—were consistent with the occurrence of primary photodecomposition processes similar to that of acetone:

$$HOCH_2COCH_3 + h\nu \rightarrow HOCH_2 + COCH_3 \quad \text{(Ia)}$$

$$HOCH_2COCH_3 + h\nu \rightarrow HOCH_2CO + CH_3 \quad \text{(Ib)}$$

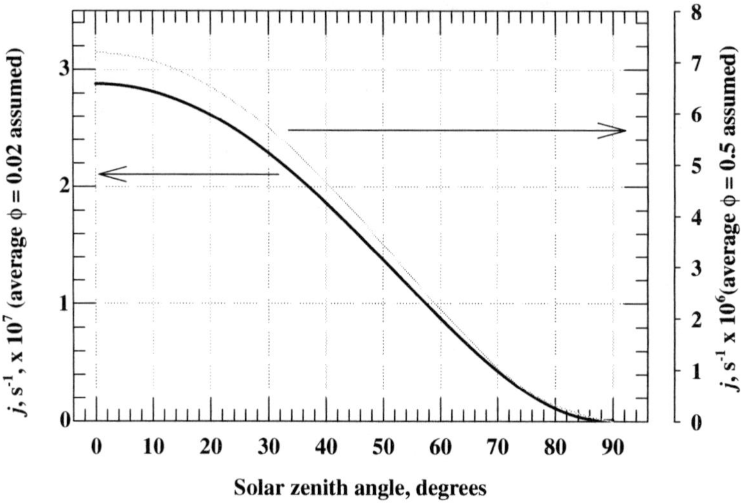

Figure VII-C-2. Representative glycolaldehyde photodissociation coefficients at 0.5 km altitude as a function of solar zenith angle calculated from the σ and limited ϕ data given in this section. Two different values for average ϕ were assumed: $\phi = 0.5$ (gray curve); $\phi = 0.02$ (black curve).

VII-C-3a. Absorption Spectra

The absorption spectra of hydroxyacetone (Orlando et al., 1999) is compared with that of acetone (Martinez et al., 1992) in figure VII-C-3. As observed with hydroxyacetaldehyde and acetaldehyde, a marked shift in the spectrum toward the shorter wavelengths occurs on substitution of an OH-group on the

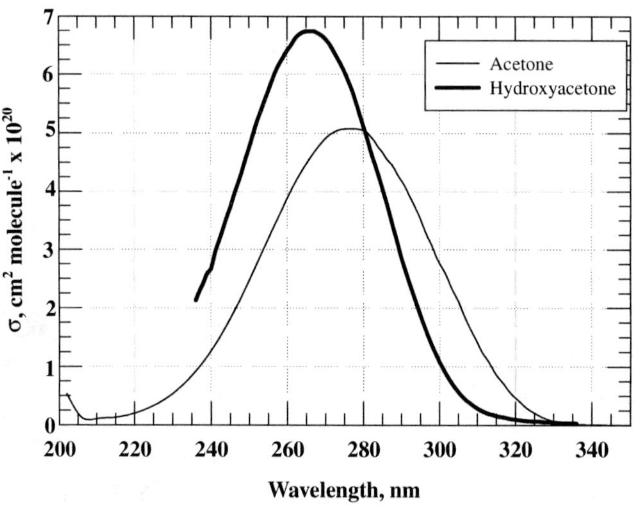

Figure VII-C-3. Comparison of the gas-phase absorption spectra of hydroxyacetone (Orlando et al., 1999) and acetone (Martinez et al., 1992).

methyl group of acetone. This effect acts to diminish the importance of tropospheric photochemistry as a loss mechanism for hydroxyacetone. The cross sections for hydroxyacetone as a function of wavelength are summarized in table VII-C-2.

VII-C-3b. Quantum Yields

The computer simulations of the product yields observed by Orlando et al. (1999) from experiments within the 240–420 nm region indicate that the average quantum yield of hydroxyacetone loss was 0.65 ± 0.25, with about equal importance of reactions (Ia) and (Ib). Experiments by Orlando et al. (1999) employing light of

Table VII-C-2. Hydroxyacetone absorption cross sections in the vapor phase

λ (nm)	$\sigma \times 10^{20}$	λ (nm)	$\sigma \times 10^{20}$	λ (nm)	$\sigma \times 10^{20}$
236	2.13	270	6.61	304	0.652
237	2.29	271	6.53	305	0.569
238	2.43	272	6.42	306	0.493
239	2.59	273	6.3	307	0.431
240	2.678	274	6.18	308	0.379
241	2.94	275	6.05	309	0.331
242	3.14	276	5.9	310	0.287
243	3.34	277	5.73	311	0.249
244	3.55	278	5.54	312	0.219
245	3.74	279	5.33	313	0.192
246	3.95	280	5.12	314	0.177
247	4.14	281	4.91	315	0.157
248	4.34	282	4.69	316	0.142
249	4.56	283	4.48	317	0.133
250	4.76	284	4.27	318	0.117
251	4.98	285	4.05	319	0.104
252	5.18	286	3.82	320	0.0954
253	5.36	287	3.58	321	0.0867
254	5.54	288	3.34	322	0.0776
255	5.72	289	3.11	323	0.0717
256	5.89	290	2.87	324	0.0672
257	6.06	291	2.66	325	0.0629
258	6.19	292	2.45	326	0.0651
259	6.3	293	2.26	327	0.0566
260	6.41	294	2.06	328	0.0507
261	6.5	295	1.87	329	0.0509
262	6.61	296	1.69	330	0.0463
263	6.68	297	1.52	331	0.0406
264	6.72	298	1.36	332	0.0372
265	6.74	299	1.22	333	0.0358
266	6.74	300	1.08	334	0.0374
267	6.74	301	0.961	335	0.0345
268	6.71	302	0.843	336	0.0313
269	6.67	303	0.743		

Source: Orlando et al. (1999).
Note: Units on σ are cm^2 $molecule^{-1}$.

$\lambda < 290$ nm (rough match with the solar spectrum within the troposphere) gave the approximate estimate of the average quantum yield of decomposition of 0.3 ± 0.2. Until more quantitative information becomes available for estimating j-values for hydroxyacetone photodecomposition within the troposphere, it is recommended that $\phi_{Ia} = \phi_{Ib} = 0.15$ be used over the effective tropospheric photolysis wavelengths ($\lambda > 290$ nm).

VII-C-3c. Tropospheric Photolysis

Photolysis rate coefficients for hydroxyacetone that are representative of the lower atmosphere were calculated with the TUV model (as described in the introduction to this section) using the measured σ and limited ϕ data given in this section. See figure VII-C-4.

VII-C-4. Acrolein ($CH_2{=}CHCHO$)

Although acrolein, methylacrolein, and methyl vinyl ketone are not major products of the oxidation of most aromatic hydrocarbons, we discuss their photochemistry here and in sections VII-C-8 and VII-C-9 to allow a comparison between the photochemistry of these compounds that contain the $>C{=}C{-}C({=}O){-}R$ group with those of the important products of the atmospheric oxidation of aromatic hydrocarbons that are structurally related unsaturated dicarbonyls [$HC({=}O)CH{=}CH{-}C({=}O)H$, $CH_3(C{=}O){-}CH{=}CH{-}C({=}O)H$, $CH_3{-}C({=}O){-}CH{=}CH{-}C({=}O){-}CH_3$, $HC({=}O)CH{=}CH{-}CH{=}CH{-}C({=}O)H$].

Previous studies of the photodecomposition of acrolein (Coomber and Pitts, 1969; Ulmstead et al., 1978; Shinohara and Nishi, 1982; Fujimoto et al., 1985; Gardner et al., 1987; Haas et al., 1991; Fahr et al., 1993; Magneron et al., 1999)

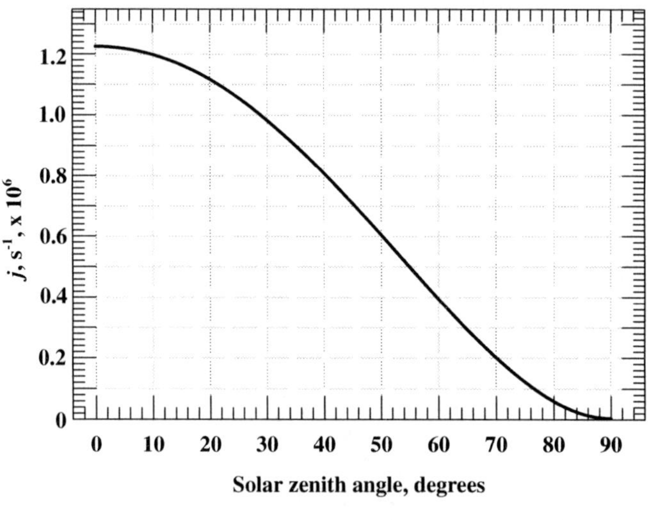

Figure VII-C-4. Representative hydroxyacetone photodissociation rate coefficients ($j_I + j_{II}$) at 0.5 km altitude as a function of solar zenith angle.

suggest that a number of photodecomposition modes are operative, although the extent of each for tropospheric conditions remains uncertain. The experimental conditions (25°C, low pressures of acrolein in air) employed in the study of Gardner et al. (1987) are the most representative of those of the troposphere, but the majority of their experiments were carried out at only one wavelength of light (313 nm). They concluded that the primary photodissociation could be described reasonably well through the occurrence of the primary processes [(I)–(IV)], which involve both the first excited singlet (S_1) and triplet (T_1) states; S_0 is the ground electronic state:

$$CH_2{=}CHCHO + h\nu \rightarrow S_1$$

$$S_1(+M) \rightarrow T_1(+M)$$

$$T_1 + N_2 \rightarrow S_0 + N_2$$

$$T_1 + O_2 \rightarrow S_0 + O_2(^1\Delta_g, {}^1\Sigma_g^+, \text{ and/or } {}^3\Sigma_g^-)$$

$$S_1 \text{ or } T_1 \rightarrow CH_2{=}CH_2 + CO \quad \text{(I)}$$

$$\rightarrow CH_2{=}CH + CHO \quad \text{(II)}$$

$$\rightarrow CH_3CH + CO \quad \text{(III)}$$

$$\rightarrow CH_2{=}CHCO + H \quad \text{(IV)}$$

The acrolein molecule excited by 313-nm light at 25°C will have about 398 kJ mol^{-1} mean total internal energy (382 from the light quantum, 8.0 translational, 8.0 rotational, and 6.7 vibrational). The enthalpy change associated with the ground-state acrolein molecule undergoing processes (I) through (IV), respectively, can be estimated (± 20 kJ mol^{-1}) to be 17, 414, 347 (if the product ethylidene diradical, CH_3CH, is in the singlet state), or 305 (if CH_3CH is in the triplet state in the initial fragment), and 364 kcal mole^{-1}, respectively; Gardner et al. (1987) concluded from their experiments at 313 nm and low pressures of air that the T_1 state appears to be the major source of the reaction channels. For photolyses in 1 atm of air quenching of the excited molecules in the T_1 state becomes important, and they suggest that about 60% of the decomposition occurs from the S_1 state under these conditions. In 1 atm of air the ϕ for decomposition of acrolein is very small (~0.0065 at 313 nm).

VII-C-4a. Absorption Cross Sections

The absorption spectrum of acrolein as determined by Gardner et al. (1987) is given in figure VII-C-5, where it can be compared with the spectra of methacrolein (Meller in Röth et al., 1997) and methyl vinyl ketone (Schneider and Moortgat in Röth et al., 1997). The striking similarity in the spectra of these three compounds reflects their common chromophore: $R{-}C({=}O){-}CR'{=}CH_2$, where R and R$'$ are CH_3-groups or H-atoms. The cross sections as a function of wavelength are summarized in table VII-C-3, as digitized from the graphic data plotted in figure 1 in Gardner et al. (1987).

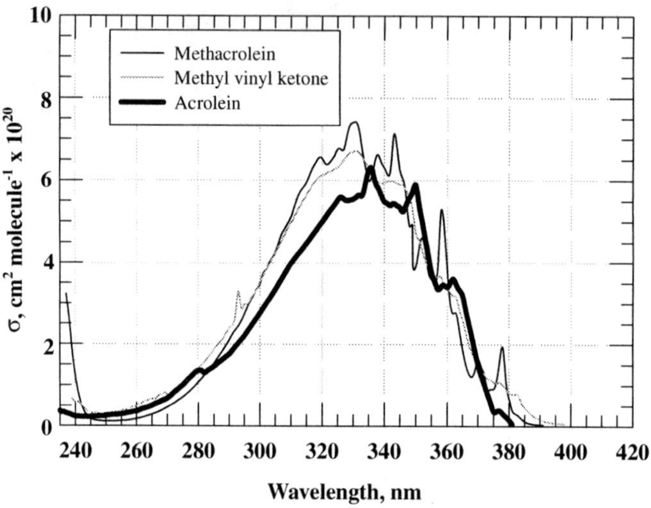

Figure VII-C-5. Gas-phase absorption spectra of acrolein (Gardner et al., 1987), methacrolein (Meller in Röth et al., 1997), and methyl vinyl ketone (Schneider and Moortgat in Röth et al. (1997). Here, the high-resolution spectral data reported for methacrolein and methyl vinyl ketone were averaged over each nanometer.

VII-C-4b. Quantum Yields of the Primary Processes

Detailed information on the quantum yields of primary processes in acrolein were derived by Gardner et al. (1987) from photolyses at ~25°C and 313 nm. The estimated quantum yields of three of the four major primary processes show marked dependence on the pressure of the synthetic air added; see figure VII-C-6. The data in figure VII-C-6 are described by the following equations:

$$\phi_{\mathrm{I}} \approx 0.0018 \text{ (independent of } [M])$$

$$\log_{10}(\phi_{\mathrm{II}}) \approx -2.787 - 4.502 \times 10^{-21}\,[M] - 8.276 \times 10^{-41}\,[M]^2$$

$$\log_{10}(\phi_{\mathrm{III}}) \approx 1.684 - 0.0543 \times 10^{-18}\,[M] + 9.109 \times 10^{-40}\,[M]^2$$

$$\log_{10}(\phi_{\mathrm{IV}}) \approx -2.552 - 0.0142 \times 10^{-18}\,[M] - 2.889 \times 10^{-43}\,[M]^2$$

where $[M]$ is the concentration of air, in units of molecules cm^3.

The fractions of the photodecomposition of acrolein that are estimated to occur by the various primary photochemical processes as a function of air concentration are summarized in figure VII-C-7 (Gardner et al., 1987).

The quenching effect of air on the quantum yield of decomposition of acrolein is shown in figure VII-C-8; data of Gardner et al. (1987).

For the range of $2.58 \times 10^{19} > [M] > 0.082 \times 10^{19}$ the data for the quantum yield of acrolein loss at 313 nm and ~25°C are represented reasonably well by the fit of a second-order least-squares equation to the data in figure VII-C-8:

$$1/(\phi_{\mathrm{decomp}}) = 0.23 + 1.21 \times 10^{-17}[M] - 2.40 \times 10^{-37}[M]^2$$

Table VII-C-3. Absorption cross sections for acrolein vapor

λ (nm)	σ × 10^20	λ (nm)	σ × 10^20	λ (nm)	σ × 10^20	λ (nm)	σ × 10^20
222	7.75	263	0.44	304	3.22	345	5.35
223	7.38	264	0.47	305	3.34	346	5.25
224	7.01	265	0.50	306	3.47	347	5.41
225	6.64	266	0.54	307	3.60	348	5.58
226	6.27	267	0.57	308	3.73	349	5.73
227	5.90	268	0.61	309	3.86	350	5.90
228	4.58	269	0.64	310	3.99	351	5.47
229	3.27	270	0.68	311	4.09	352	5.04
230	1.95	271	0.74	312	4.19	353	4.61
231	1.63	272	0.81	313	4.28	354	4.18
232	1.32	273	0.87	314	4.38	355	3.75
233	1.00	274	0.94	315	4.48	356	3.55
234	0.69	275	1.00	316	4.58	357	3.35
235	0.37	276	1.08	317	4.69	358	3.40
236	0.34	277	1.16	318	4.79	359	3.45
237	0.31	278	1.23	319	4.90	360	3.40
238	0.29	279	1.31	320	5.00	361	3.50
239	0.26	280	1.35	321	5.10	362	3.60
240	0.23	281	1.35	322	5.21	363	3.46
241	0.23	282	1.30	323	5.31	364	3.33
242	0.22	283	1.35	324	5.41	365	3.20
243	0.22	284	1.40	325	5.52	366	2.86
244	0.22	285	1.45	326	5.60	367	2.52
245	0.22	286	1.51	327	5.56	368	2.18
246	0.23	287	1.57	328	5.52	369	1.84
247	0.23	288	1.63	329	5.53	370	1.50
248	0.24	289	1.69	330	5.55	371	1.25
249	0.24	290	1.75	331	5.60	372	1.03
250	0.25	291	1.84	332	5.65	373	0.80
251	0.26	292	1.93	333	5.65	374	0.58
252	0.27	293	2.02	334	5.95	375	0.36
253	0.27	294	2.11	335	6.25	376	0.37
254	0.28	295	2.20	336	6.30	377	0.38
255	0.29	296	2.31	337	5.99	378	0.30
256	0.30	297	2.42	338	5.83	379	0.23
257	0.32	298	2.53	339	5.66	380	0.15
258	0.33	299	2.64	340	5.50	381	0.05
259	0.35	300	2.75	341	5.45		
260	0.36	301	2.87	342	5.40		
261	0.39	302	2.99	343	5.45		
262	0.42	303	3.10	344	5.40		

Source: Gardner et al. (1987).
Note: Units on σ are cm^2 $molecule^{-1}$.

The limited data from photolyses at 334 nm show a somewhat stronger dependence on pressure of added air than the results at 313 nm (squares in figure VII-C-8), but they are too limited to derive a useful relationship between $[M]$ and ϕ. The ratio $(\phi_{decomp})_{334nm}/(\phi_{decomp})_{313nm} = 0.27$ and 0.76 at about 25 and 400 Torr of added air, respectively. In view of the very limited availability of data at wavelengths other than 313 nm, it is recommended that in the calculation of *j*-values the data for 313 nm be used to represent the acrolein quantum yields over the

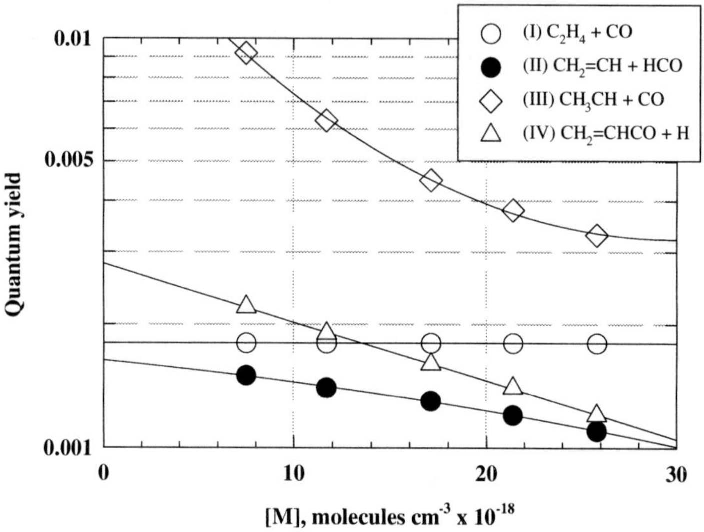

Figure VII-C-6. Estimated quantum yields for the photodecomposition modes of acrolein at $\sim 25^\circ C$ as a function of the concentration of air [*M*] (Gardner et al., 1987).

entire region of tropospheric light absorption; however, the use of this procedure will probably result in an overestimation of the importance of the photodecomposition modes. Magneron et al. (1999) determined an apparent *j*-value for acrolein photodecomposition in the European outdoor photolysis chamber in Valencia to be $< 2 \times 10^{-6}\ s^{-1}$, in reasonable agreement with the calculations presented in figure VII-C-9.

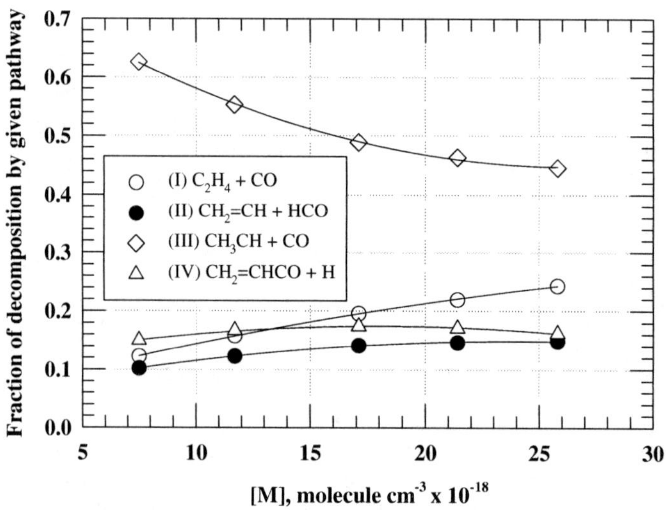

Figure VII-C-7. The fraction of the photodecomposition of acrolein that is estimated to occur by the various primary photodecomposition modes; data of Gardner et al. (1987) in the photolysis at 313 nm and 25°C.

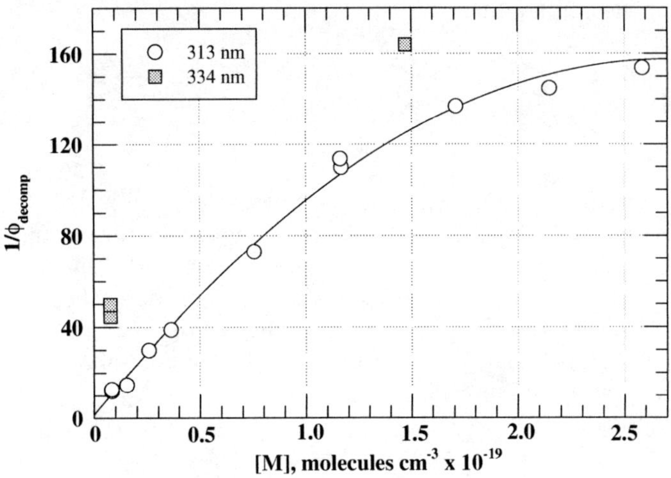

Figure VII-C-8. Stern–Volmer plots of the $1/\phi_{decomp}$ versus $[M]$ data for acrolein vapor for excitation at 313 and 334 nm (Gardner et al., 1987) where $[M]$ = concentration of the synthetic air mixture.

VII-C-4c. Tropospheric Photolysis

Photolysis rate coefficients for acrolein which are representative of the lower atmosphere were calculated with the TUV model (as described in the introduction to this section) using the measured σ and ϕ data given in this section. See figure VII-C-9.

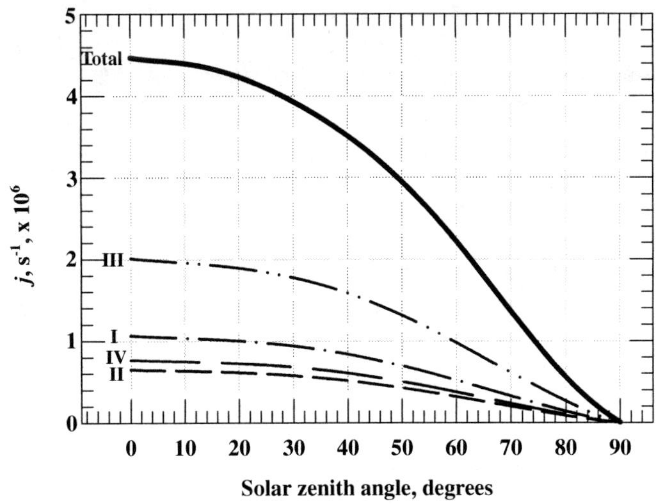

Figure VII-C-9. Acrolein photodissociation coefficients at 0.5 km altitude as a function of solar zenith angle. Quantum yield estimations assume 1 atm, 25°C. Lines I–IV refer to the j-values for primary processes (I)–(IV) defined in this section. Upper curve gives the total j(photodecomposition).

VII-C-5. Glycidaldehyde

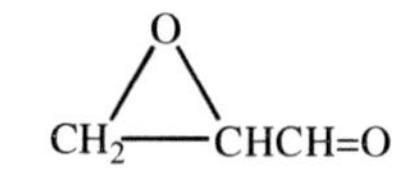

Glycidaldehyde, along with other epoxides, has been identified by means of GC-MS as a product in the OH-radical initiated degradation of benzene and toluene (Yu et al., 1997). Although the yields were not determined and may not be particularly high, the importance of the epoxides in atmospheric chemistry arises principally from their high toxicity. The chemical nature of the primary processes is undetermined, but by analogy with other aldehydes the following primary processes might be expected to occur:

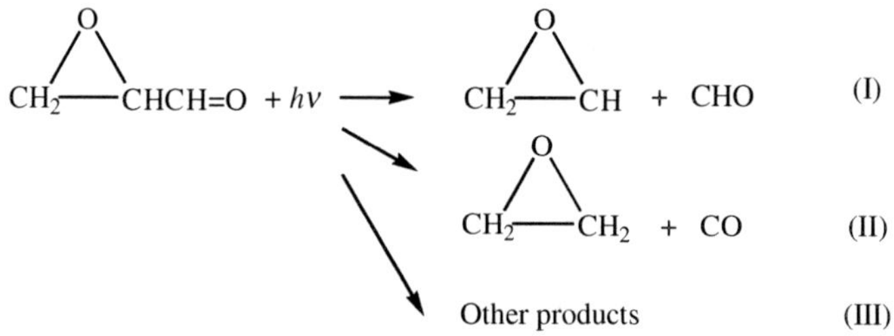

As with acetaldehyde, the dominant mode of photodecomposition in the troposphere may occur by reaction (I), with a rupture of the R–CHO bond.

VII-C-5a. Absorption Cross Sections

The cross sections for glycidaldehyde have been determined by Ma et al. (1998). These high-resolution data are shown in figure VII-C-10 and the data (as averaged over each nanometer) are shown in tabular form in table VII-C-4.

VII-C-5b. Quantum Yields

No information on the wavelength dependence of the quantum yields of glycidaldehyde photolysis have been reported. However, some useful information is available (Ma et al., 1998). The rate of glycidaldehyde photodecomposition (disappearance) in synthetic air has been determined in a large photochemical reactor irradiated using 32 fluorescent lamps (Phillips TL05, 40 W; $\lambda_{max} = 360$ nm) that surrounded the cell. In additional experiments a large excess of propylene was added to the mixture (as a scavenger for OH radicals); the observed rates of loss of glycidaldehyde were the same with or without the added propylene, so it is likely that artifact OH-radical production was unimportant in this system, and that the observed rates of photochemical decay were primarily due to the total rates of the primary photochemical processes. Comparison of rates of photolysis of glycidaldehyde and NO_2 were made in air in separate experiments in the same reactor: j(glycidaldehyde disappearance) $\sim 1.9 \times 10^{-5}\ s^{-1}$; $j(NO_2) \sim 2.7 \times 10^{-3}\ s^{-1}$. Although the action spectrum of these two compounds is different,

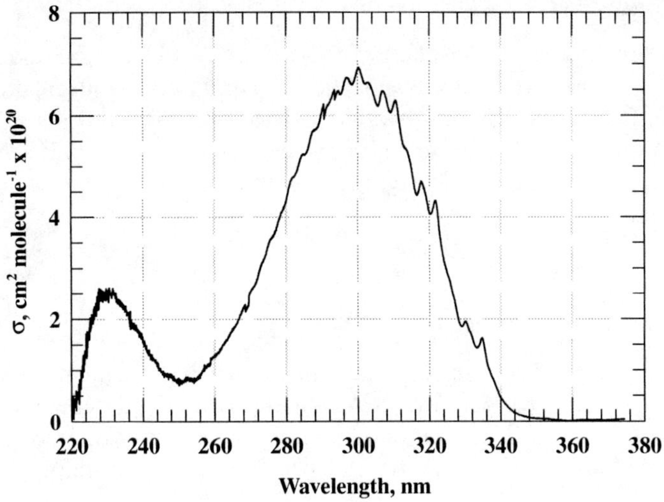

Figure VII-C-10. Wavelength dependence of the absorption cross sections for glycidaldehyde (gas) (Ma et al., 1998).

Table VII-C-4. Cross sections for glycidaldehyde (gas); the high-resolution data of Ma et al. (1998) were averaged over each nanometer and centered on the wavelength shown

λ (nm)	$\sigma \times 10^{20}$	λ (nm)	$\sigma \times 10^{20}$	λ (nm)	$\sigma \times 10^{20}$
290	6.08	318	4.64	346	0.11
291	6.08	319	4.43	347	0.085
292	6.24	320	4.11	348	0.069
293	6.41	321	4.15	349	0.060
294	6.40	322	4.21	350	0.054
295	6.43	323	3.63	351	0.049
296	6.56	324	3.18	352	0.044
297	6.71	325	2.90	353	0.044
298	6.63	326	2.65	354	0.033
299	6.67	327	2.37	355	0.026
300	6.88	328	2.04	356	0.020
301	6.78	329	1.86	357	0.015
302	6.62	330	1.92	358	0.009
303	6.62	331	1.81	359	0.006
304	6.48	332	1.64	360	0.007
305	6.26	333	1.46	361	0.013
306	6.26	334	1.50	362	0.020
307	6.45	335	1.56	363	0.023
308	6.30	336	1.21	364	0.025
309	6.08	337	0.96	365	0.023
310	6.19	338	0.79	366	0.021
311	6.14	339	0.62	367	0.020
312	5.65	340	0.45	368	0.018
313	5.38	341	0.36	369	0.018
314	5.24	342	0.28	379	0.020
315	4.94	343	0.21	371	0.026
316	4.55	344	0.16	372	0.031
317	4.51	345	0.13	373	0.036

Note: Units on σ are cm^2 molecule^{-1}.

these data and an assumed atmospheric $j(NO_2) \sim 8.5 \times 10^{-3}\ s^{-1}$ for noon at 40°N were used to estimate for tropospheric conditions j(glycidaldehyde) $\sim 6.0 \times 10^{-5}\ s^{-1}$. By comparison of this value with that calculated from the observed cross sections and a total quantum yield of decomposition of unity, Ma et al. (1998) estimated that the effective average quantum yield of photodecomposition was approximately 0.6.

VII-C-5c. Tropospheric Photolysis

Photolysis rate coefficients for glycidaldehyde that are representative of the lower atmosphere were calculated with the TUV model (as described in the introduction to this section) using the measured σ values and the estimated value of $\phi_I + \phi_{II} + \phi_{III} = 0.6$, assumed to be independent of wavelength; see figure VII-C-11.

VII-C-6. Pyruvic Acid ($CH_3C(=O)C(=O)OH$)

The tropospheric photochemistry of pyruvic acid has received little attention. However, two studies do provide very useful information: Grosjean (1983) and Yamamoto and Back (1985). Two different primary processes have been suggested to explain the results:

$$CH_3C(=O)C(=O)OH + h\nu \rightarrow CH_3C(=O)H + CO_2 \quad \text{(I)}$$

$$\rightarrow [CH_3C(OH)CO_2] \rightarrow CH_3COH + CO_2 \quad \text{(II)}$$

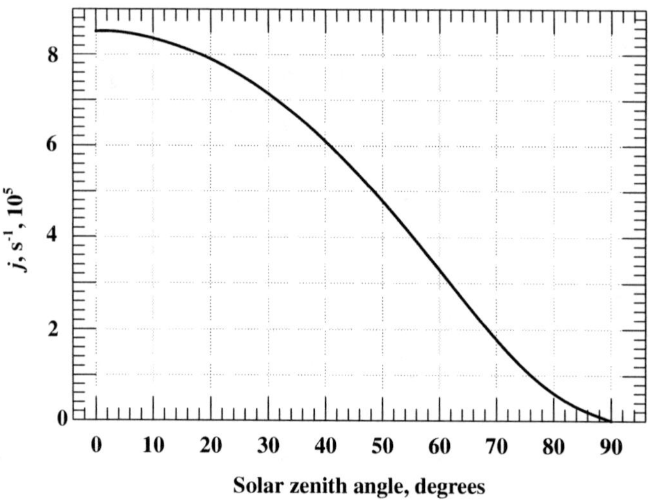

Figure VII-C-11. Approximate glycidaldehyde photodissociation coefficients at 0.5 km altitude as a function of solar zenith angle. A total quantum yield of decomposition of 0.6 was assumed in the calculations.

Acetaldehyde and CO_2 have been observed as products of the sunlight photolysis in air (Grosjean, 1983), and it was suggested that primary process (I) was the dominant source of these products. However, in experiments with radiation at 366, 345, and 320 nm and at a temperature 67°C, Yamamoto and Back (1985) found that the ratio of acetaldehyde to CO_2 in the products, near unity at low pressures of added O_2, decreased dramatically as the pressure of air was increased. They suggested that process (II) was dominant and that acetaldehyde was largely a secondary product formed following intermolecular H-atom abstraction by the carbonyl group followed by decomposition of the intermediate species to the carbene (CH_3COH) and CO_2. Presumably, this reaction occurs from the excited singlet state of pyruvic acid, since the quantum yield of CO_2 was relatively unaffected by O_2 addition, normally an efficient quencher of triplet states.

VII-C-6a. Absorption Cross Sections

Yamamoto and Back (1985) reported the relative absorption spectrum of pyruvic acid determined at 85°C and the approximate absorption cross section at the maximum. Using these data (figure 1 of Yamamoto and Back, 1985), the approximate cross sections were derived and are shown in figure VII-C-12. The approximate cross section data for pyruvic acid, as digitized from the data of figure 1 in Yamamoto and Back (1985) as determined at 85°C, are shown in table VII-C-5.

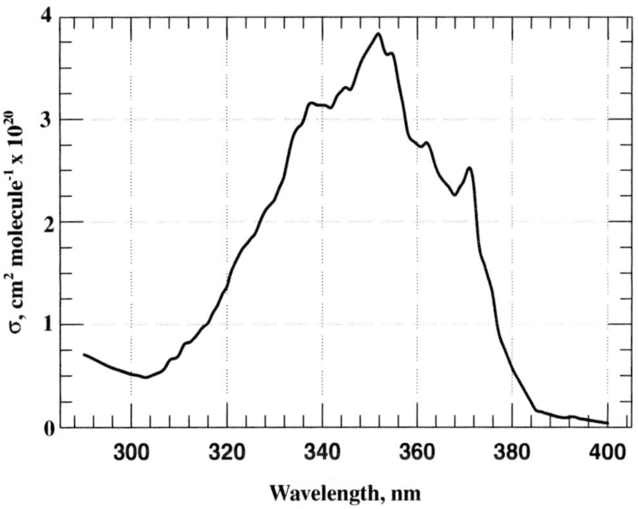

Figure VII-C-12. Approximate wavelength dependence of the absorption cross sections for pyruvic acid vapor at 85°C; derived from the data of Yamamoto and Back (1985).

Table VII-C-5. Approximate absorption cross sections for pyruvic acid vapor at 85°C derived from the data of Yamamoto and Back (1985)

λ (nm)	$\sigma \times 10^{20}$	λ (nm)	$\sigma \times 10^{20}$	λ (nm)	$\sigma \times 10^{20}$	λ (nm)	$\sigma \times 10^{20}$
290	0.71	319	1.30	349	3.63	378	0.80
291	0.69	320	1.38	350	3.71	379	0.69
292	0.66	321	1.53	351	3.78	380	0.57
293	0.64	323	1.72	352	3.82	381	0.49
294	0.62	324	1.78	353	3.67	382	0.40
295	0.60	325	1.83	354	3.63	383	0.32
296	0.58	326	1.89	355	3.61	384	0.24
297	0.56	327	2.01	356	3.36	385	0.16
298	0.55	328	2.10	357	3.13	386	0.15
299	0.53	329	2.16	358	2.87	387	0.13
300	0.52	330	2.22	359	2.79	388	0.12
301	0.51	331	2.33	360	2.75	389	0.11
302	0.50	332	2.44	361	2.73	390	0.10
303	0.49	333	2.67	362	2.77	391	0.09
304	0.50	334	2.85	363	2.67	392	0.10
305	0.52	335	2.92	364	2.52	393	0.10
306	0.53	336	2.98	365	2.43	394	0.08
307	0.57	337	3.12	366	2.37	395	0.08
308	0.65	338	3.15	367	2.31	396	0.07
309	0.67	339	3.13	368	2.25	397	0.06
310	0.71	340	3.13	369	2.33	398	0.05
311	0.80	341	3.12	370	2.41	399	0.05
312	0.82	342	3.11	371	2.52	400	0.04
313	0.86	343	3.21	372	2.33		
314	0.92	344	3.27	373	1.80		
315	0.97	345	3.30	374	1.60		
316	1.01	346	3.29	375	1.45		
317	1.11	347	3.40	376	1.26		
318	1.18	348	3.53	377	0.96		

Note: Units on σ are cm^2 molecule^{-1}.

VII-C-6b. Quantum Yields

The quantum yield of CO_2 has been measured at 366 nm: 0.9 ± 0.1 (67°C) by Yamamoto and Back (1985) and 1.02 ± 0.06 (80–85°C) by Vesly and Leermakers (1964). Evidence suggests strongly that process (II) is the major, if not exclusive, primary process. Until more extensive data are available, it is suggested that modelers assume $\phi_{II} = 1.0$ over the wavelength region 290–400 nm. This assumption leads to calculated *j*-values consistent with the measurements of pyruvic acid loss made by Grosjean (1983) in sunlight: $j \sim 2.7 \times 10^{-4}\ s^{-1}$.

VII-C-6c. Tropospheric Photolysis

Photolysis rate coefficients for pyruvic acid that are representative of the lower atmosphere were calculated with the TUV model (as described in the introduction to this section) using the measured σ and ϕ data given in this section. See figure VII-C-13.

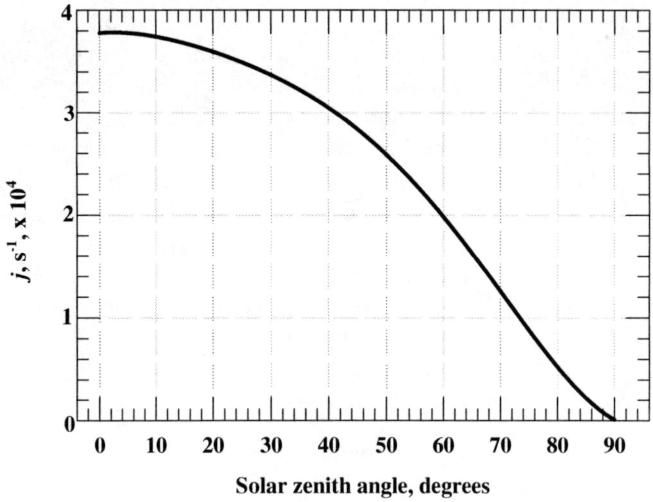

Figure VII-C-13. Approximate pyruvic acid photodissociation coefficients at 0.5 km altitude as a function of solar zenith angle.

VII-C-7. Butenedial [CH(=O)CH=CHC(=O)H]

Butenedial is difficult to prepare in a pure state, and previous workers who have studied its atmospheric chemistry point out that significant impurities were present in the starting material used in photochemical studies. In the studies of Bierbach et al. (1994), infrared spectra of the starting material showed that ~70% of the butenedial was in the *cis*-form and ~30% in the *trans*-form. Liu et al. (1999) found that the purity of the butenedial (liquid phase) used in their studies was >90% as estimated from the NMR spectra. The following primary photodecomposition modes appear to rationalize the products observed in the previous studies:

$$trans\text{-}/cis\text{-}CH(=O)CH=CHC(=O)H + h\nu \rightarrow CH(=O)CH=CHC(=O)H^*$$
$$CH(=O)CH=CHC(=O)H^* \rightarrow cis\text{-}/trans\text{-}CH(=O)CH=CHC(=O)H \quad (I)$$
$$\rightarrow CH_2=CHCHO + CO \quad (II)$$
$$\rightarrow CHOCH=CH + CHO \quad (III)$$
$$\rightarrow CHOCH=CHCO + H \quad (IV)$$

The rapid decay of the butenedial, $\tau_{1/2} \sim 1$ h (7–8 A.M. exposure) in air and rapid oxidation of NO to NO_2 observed in sunlight in chamber experiments with synthetic air–NO_x mixtures (Liu, 1999; Liu et al., 1999), suggest that free radical formation in processes (III) and/or (IV) must occur in addition to processes (I) and (II), since the photolysis of the product acrolein [formed in process (II)] is expected to be relatively slow under tropospheric conditions. In the studies of Bierbach et al. (1994), the major products of the ultraviolet photolysis (254 nm) of butenedial were found to be maleic anhydride and 3-*H*-furan-2-one, and <40% of the primary photochemical reactions occurred by isomerization in the *trans*-compound in process (I). Photolyses with 360–480 nm light gave 3-*H*-furan-2-one and maleic anhydride as major and minor products, respectively, and < 45% of primary photochemical reaction occurred by isomerization in process (I). Presumably, most of these products arise from subsequent reactions of the oxygenated radicals formed following process (IV). 3-*H*-furan-2-one is an exception to this generalization; it is formed by intramolecular rearrangement (Bierbach et al., 1994). The carbonyl products observed by Liu et al. (1999) in daylight chamber experiments—formaldehyde, acrolein, glycolaldehyde, and glyoxal—although minor, suggest the possible occurrence of reaction (II) as well as (IV). The observed malonaldehyde formation cannot be explained in terms of current mechanisms; this might have resulted from impurities or background.

VII-C-7a. Absorption Cross Sections

The existing cross section data for *trans*-butenedial are shown in figure VII-C-14. The solid curve was calculated from the relative absorption data for a water solution of unknown concentration (Liu et al., 1999). The wavelength for the absorption maximum, $\lambda \sim 313\,\text{nm}$, is similar to that observed by Liu and coworkers for the structurally similar species *trans*-3-hexene-2,5-dione ($\lambda \sim 315\,\text{nm}$) in water solution. The nature of the transitions in the two compounds should be very similar. In order to approximate the relation between λ and σ, it was assumed that the cross section in *trans*-butenedial at the maximum is identical to that observed in *trans*-3-hexene-2,5-dione, and the butenedial data were normalized to the value of σ_{max} measured for *trans*-3-hexene-2,5-dione in water solution (Liu et al., 1999). The dashed curve represents the digitized data from a plot of extinction coefficients for *trans*-butenedial (fumaric dialdehyde)

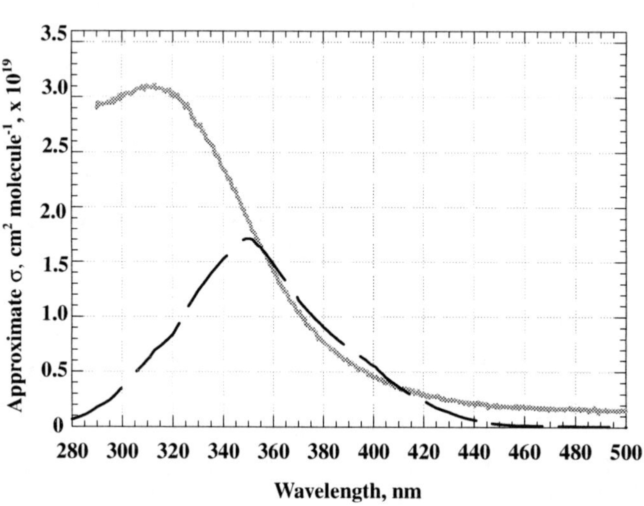

Figure VII-C-14. Approximate cross sections for *trans*-butenedial as determined from relative absorbance data in water solution by (Liu et al. 1999) and normalized to the σ_{max} measured for *trans*-3-hexene-2,5-dione in water solution (Liu et al. 1999) (solid gray curve) and as measured in *iso*-octane solution by Hufford et al. (1952) (dashed black curve).

given by Hufford et al. (1952) as determined in *iso*-octane solution. Estimates of the cross section data for *trans*-butenedial derived from the data of Liu et al. and Hufford et al. are given in tables VII-C-6 and VII-C-7, respectively. There are obvious differences in the spectra: the shift in the maximum and the absolute magnitude of the σ value at a given wavelength. These data may be of qualitative value in estimation of the *j*-values for *trans*-butenedial until more quantitative estimates are available.

VII-C-7b. Quantum Yields

It is likely that the RC(=O)−H bond in the butenedial molecule is somewhat weaker than the R−C(=O)H bond as estimated for acrolein. In view of this and the product data, it appears that process (IV) may be the dominant mode of photolysis within the troposphere. No quantum yield data are available at the time of this writing, but measurements of *j*-values for photodecompostion have been made by Sørensen and Barnes (1998) in sunlight-irradiated chamber experiments. For *cis*-butenedial the *j*-value = $(1.80 \pm 0.01) \times 10^{-3}\ s^{-1}$ for rate of loss in sunlight was obtained in dilute air mixtures with cyclohexane added (but no NO_x) to suppress OH attack on butenedial. Sørensen and Barnes measured $j(NO_2) = (10.1 \pm 0.13) \times 10^{-3}\ s^{-1}$ for these conditions. The apparent *j*-values were ~30% lower in the experiments with the added radical scavenger cyclohexane than in its absence; this suggests that, indeed, in butenedial sunlight photolysis, the free radical generation pathways [presumably largely reaction (IV)] are important. It seems unlikely that photodissociation would be perfectly efficient for the energies available over the entire absorption region, and at the time of writing it is uncertain as to what limit for dissociation should be employed in *j*-value estimates. For calculated *j*-values to match those measured by Sørensen and Barnes (1998), dissociation for wavelengths >340 nm (352 kJ mol^{-1}) must be invoked. See figure VII-C-15.

VII-C-7c. Tropospheric Photolysis

Photolysis rates representative of the lower atmosphere were calculated using the TUV model (as described in the introduction to this section). Both sets of cross-section data presented above were evaluated in combination with the recommended quantum yield of unity for wavelengths <340 nm. The resulting photolysis rates are presented in figure VII-C-15. As described above, Sørensen and Barnes (1998) measured a photolysis rate for *cis*-butenedial of $(1.80 \pm 0.01) \times 10^{-3}\ s^{-1}$ in chamber experiments. There is no information on the impact of *cis*/*trans* configuration on the photolysis of butenedial, so we assume here that the rates are the same for both isomers. For Sørensen and Barnes (1998) chamber conditions, the measured $j(NO_2)$ was $(10.1 \pm 0.13) \times 10^{-3}\ s^{-1}$, which corresponds to a zenith angle of 20° for the calculations shown in figure VII-C-15. Accordingly, the photolysis rate calculated for 20° using the cross-section data of table VII-C-6 ($1.3 \times 10^{-3}\ s^{-1}$) is more consistent with the measured rate than the rate calculated using the data of table VII-C-7 ($5.1 \times 10^{-4}\ s^{-1}$).

Table VII-C-6. Approximate cross sections for *trans*-butenedial in water solution; these were estimated using the measured relative absorbance data for *trans*-butenedial given in figure VII-C-14 (Liu et al., 1999), adjusted so that the σ at λ_{max} for *trans*-butenedial is equal to that measured at λ_{max} for *trans*-3-hexene-2,5-dione, ~ 3.08 × 10^{-19} cm^2 molecule^{-1}

λ (nm)	σ × 10^{19}	λ (nm)	σ × 10^{19}	λ (nm)	σ × 10^{19}	λ (nm)	σ × 10^{19}	λ (nm)	σ × 10^{19}	λ (nm)	σ × 10^{19}
290	2.87	330	2.71	370	1.03	410	0.35	450	0.18	490	0.16
291	2.90	331	2.72	371	1.00	411	0.34	451	0.20	491	0.15
292	2.91	332	2.67	372	0.97	412	0.34	452	0.19	492	0.17
293	2.91	333	2.62	373	0.93	413	0.32	453	0.19	493	0.15
294	2.91	334	2.58	374	0.92	414	0.34	454	0.19	494	0.15
295	2.91	335	2.55	375	0.88	415	0.32	455	0.18	495	0.16
296	2.94	336	2.53	376	0.88	416	0.32	456	0.18	496	0.15
297	2.91	337	2.44	377	0.85	417	0.31	457	0.18	497	0.15
298	2.96	338	2.44	378	0.79	418	0.33	458	0.17	498	0.15
299	2.99	339	2.36	379	0.80	419	0.29	459	0.19	499	0.15
300	2.99	340	2.32	380	0.77	420	0.30	460	0.18	500	0.16
301	2.98	341	2.30	381	0.75	421	0.28	461	0.17	502	0.14
302	3.00	342	2.23	382	0.73	422	0.27	462	0.18	503	0.15
303	3.00	343	2.22	383	0.70	423	0.28	463	0.17	504	0.17
304	2.99	344	2.15	384	0.70	424	0.27	464	0.17	505	0.15
305	3.02	345	2.14	385	0.66	425	0.27	465	0.17	506	0.15
306	3.04	346	2.06	386	0.64	426	0.27	466	0.16	507	0.15
307	3.05	347	2.03	387	0.63	427	0.25	467	0.18	508	0.15
308	3.07	348	1.96	388	0.61	428	0.26	468	0.18	509	0.15
309	3.04	349	1.91	389	0.61	429	0.24	469	0.17	510	0.14
310	3.07	350	1.85	390	0.59	430	0.24	470	0.17	511	0.15
311	3.06	351	1.83	391	0.58	431	0.24	471	0.16	512	0.15
312	3.06	352	1.75	392	0.56	432	0.24	472	0.17	513	0.14
313	3.08	353	1.72	393	0.54	433	0.24	473	0.16	514	0.16
314	3.05	354	1.67	394	0.51	434	0.22	474	0.17	515	0.15
315	3.05	355	1.63	395	0.51	435	0.22	475	0.16	516	0.14
316	3.04	356	1.58	396	0.51	436	0.21	476	0.17	517	0.15
317	3.03	357	1.56	397	0.50	437	0.22	477	0.17	518	0.15
318	3.03	358	1.50	398	0.46	438	0.20	478	0.18	519	0.14
319	3.03	359	1.45	399	0.46	439	0.22	479	0.16	520	0.16
320	3.01	360	1.40	400	0.45	440	0.20	480	0.16	521	0.15
321	2.99	361	1.36	401	0.45	441	0.21	481	0.17	522	0.16
322	2.99	362	1.30	402	0.42	442	0.21	482	0.16	523	0.14
323	2.95	363	1.29	403	0.42	443	0.21	483	0.17	524	0.14
324	2.92	364	1.24	404	0.40	444	0.20	484	0.17	525	0.14
325	2.87	365	1.22	405	0.40	445	0.21	485	0.16	526	0.14
326	2.90	366	1.17	406	0.40	446	0.19	486	0.16	527	0.14
327	2.85	367	1.13	407	0.38	447	0.19	487	0.16	528	0.14
328	2.78	368	1.12	408	0.38	448	0.20	488	0.15	529	0.15
329	2.73	369	1.06	409	0.37	449	0.20	489	0.12	530	0.14

Source: Liu (1999); Liu et al. (1999).
Note: Units on σ are cm^2 molecule^{-1}.

Table VII-C-7. Approximate cross sections for *trans*-butenedial; these were estimated using the extinction coefficient data for *trans*-butenedial given in figure 1 of Hufford et al. (1952) as measured in *iso*-octane solution

λ (nm)	$\sigma \times 10^{19}$	λ (nm)	$\sigma \times 10^{19}$	λ (nm)	$\sigma \times 10^{19}$	λ (nm)	$\sigma \times 10^{19}$	λ (nm)	$\sigma \times 10^{19}$	λ (nm)	$\sigma \times 10^{19}$
290	0.18	330	1.24	370	1.15	410	0.36	450	0.03	490	0.01
291	0.19	331	1.26	371	1.12	411	0.35	451	0.02	491	0.01
292	0.20	332	1.29	372	1.10	412	0.33	452	0.02	492	0.01
293	0.21	333	1.32	373	1.07	413	0.32	453	0.02	493	0.01
294	0.23	334	1.36	374	1.05	414	0.30	454	0.02	494	0.01
295	0.24	335	1.39	375	1.02	415	0.29	455	0.02	495	0.01
296	0.26	336	1.41	376	1.00	416	0.28	456	0.02	496	0.01
297	0.28	337	1.44	377	0.98	417	0.27	457	0.02	497	0.01
298	0.30	338	1.47	378	0.95	418	0.26	458	0.02	498	0.01
299	0.33	339	1.49	379	0.93	419	0.25	459	0.01	499	0.01
300	0.36	340	1.52	380	0.91	420	0.24	460	0.01	500	0.01
301	0.38	341	1.54	381	0.89	421	0.23	461	0.01		
302	0.40	342	1.56	382	0.87	422	0.21	462	0.01		
303	0.43	343	1.59	383	0.85	423	0.20	463	0.01		
304	0.45	344	1.61	384	0.83	424	0.19	464	0.01		
305	0.48	345	1.63	385	0.81	425	0.17	465	0.01		
306	0.50	346	1.65	386	0.79	426	0.17	466	0.01		
307	0.53	347	1.68	387	0.77	427	0.16	467	0.01		
308	0.55	348	1.69	388	0.75	428	0.15	468	0.01		
309	0.58	349	1.71	389	0.73	429	0.14	469	0.01		
310	0.61	350	1.71	390	0.72	430	0.14	470	0.01		
311	0.63	351	1.71	391	0.70	431	0.13	471	0.01		
312	0.66	352	1.69	392	0.68	432	0.12	472	0.01		
313	0.70	353	1.67	393	0.67	433	0.11	473	0.01		
314	0.71	354	1.64	394	0.65	434	0.10	474	0.01		
315	0.73	355	1.63	395	0.64	435	0.09	475	0.01		
316	0.75	356	1.60	396	0.62	436	0.09	476	0.01		
317	0.77	357	1.57	397	0.61	437	0.08	477	0.01		
318	0.79	358	1.54	398	0.58	438	0.07	478	0.01		
319	0.81	359	1.51	399	0.57	439	0.07	479	0.01		
320	0.84	360	1.49	400	0.55	440	0.06	480	0.01		
321	0.88	361	1.45	401	0.54	441	0.06	481	0.01		
322	0.92	362	1.42	402	0.52	442	0.05	482	0.01		
323	0.96	363	1.39	403	0.49	443	0.05	483	0.01		
324	1.00	364	1.36	404	0.47	444	0.04	484	0.01		
325	1.05	365	1.32	405	0.45	445	0.04	485	0.01		
326	1.09	366	1.29	406	0.43	446	0.04	486	0.01		
327	1.12	367	1.26	407	0.41	447	0.03	487	0.01		
328	1.16	368	1.24	408	0.40	448	0.03	488	0.01		
329	1.20	369	1.21	409	0.38	449	0.03	489	0.01		

Note: Units on σ are cm^2 $molecule^{-1}$.

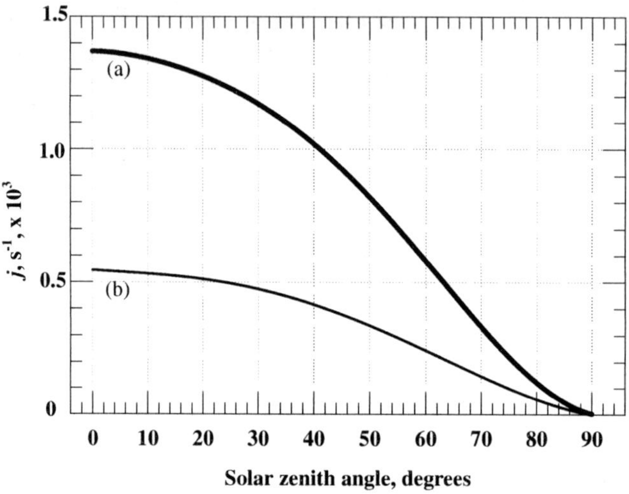

Figure VII-C-15. *trans*-Butenedial photodissociation coefficients at 0.5 km altitude as a function of solar zenith angle. It was assumed that the total quantum yield for pathways (I)–(IV) = 1 at wavelengths < 340 nm, and 0 at wavelengths > 340 nm. In calculating the curves (a) and (b), the cross section data of tables VII-C-7 and VII-C-8, respectively, were used. The simulation (a) appears to match best the Sørensen and Barnes (1998) chamber data.

Recognizing uncertainties in both cross-section and quantum yield data, our recommendation is that cross sections of table VII-C-7 be used with a quantum yield of one for wavelenths < 340 nm and zero for longer wavelengths. Since the cutoff in the quantum yield at 340 nm is based on the estimated RC(=O)–H bond dissociation energy, we recommend that this photolysis estimate be equated to dissociation pathways such as channels (II)–(IV), rather than to isomerization channel (I).

VII-C-8. Methacrolein [$CH_2{=}C(CH_3)C({=}O)H$]

The primary pathways for photodissociation of methacrolein are uncertain, but processes analogous to those for acrolein (I) through (IV) should be considered in addition to process (V), which was suggested from the studies of methacrolein in air by Raber and Moortgat (1995):

$$CH_2{=}C(CH_3)C({=}O)H + h\nu \rightarrow CH_2{=}C(CH_3)C({=}O)H^*$$

$$CH_2{=}C(CH_3)C({=}O)H^* + M \rightarrow CH_2{=}C(CH_3)C({=}O)H + M$$

$$\rightarrow CH_2{=}CHCH_3 + CO \quad \text{(I)}$$

$$\rightarrow CH_2{=}CCH_3 + CHO \quad \text{(II)}$$

$$\rightarrow CH_3CCH_3 \text{ (diradical)} + CO \quad \text{(III)}$$

$$\rightarrow CH_2{=}C(CH_3)CO + H \quad \text{(IV)}$$

$$\rightarrow (CH_3)C{=}C{=}O \quad \text{(V)}$$

VII-C-8a. Cross Sections

Cross sections for methacrolein have been determined by Meller in Röth et al. (1997). These data are presented in figure VII-C-5, where they can be compared with those of the similar compounds acrolein and methyl vinyl ketone. It is obvious that the nature of the electronic transitions involved in these three molecules are very similar, as anticipated from the nature of the chromophore, which is common to all, $CH_2{=}C(R')C({=}O)R$. The cross sections for methacrolein are summarized in table VII-C-8.

VII-C-8b. Quantum Yields

There are no quantum yield data as a function of wavelength, but some apparent overall quantum yields for photodecomposition have been reported (Raber and Moortgat, 1995) using radiation from TL/12 sunlamps (Phillips, 40 W); these lamps emit in a broad continuum from 270 to 380 nm with a peak near 300 nm that is superimposed on spikes near the two strong mercury lines at 313 and 366 nm. The reported effective quantum yields (estimated using chlorine actinometry) vary from 0.03 to 0.08 and show no significant trend with pressure from 50 to 760 Torr of added air. The effective *j*-values for decomposition range from 1.9×10^{-5} to $5.3 \times 10^{-5}\,s^{-1}$, again with no significant trend with pressure of air. The results of Raber and Moortgat (1995) are somewhat surprising in that the analogous compound acrolein shows smaller *j*-values and a strong dependence on the pressure of the air. Since methacrolein contains more atoms per molecule, one might expect its lifetime to be somewhat longer than that of acrolein, its chance for quenching by air to be greater, and its *j*-values smaller. Until more quantitative data for the wavelength dependence and the nature of the primary processes become available, it is recommended that the quantum yield data for acrolein be used to describe the analogous processes (I)–(IV) outlined above.

VII-C-8c. Tropospheric Photolysis

Photolysis rate coefficients representative of the lower atmosphere were calculated with the TUV model (as described in the introduction to this section) using the measured cross sections and assuming that the quantum yield data for methacrolein can be represented by processes analogous to those for acrolein, as given in section VII-C-4. See figure VII-C-16.

VII-C-9. Methyl Vinyl Ketone [$CH_3C({=}O)CH{=}CH_2$]

The nature and the magnitude of the primary pathways are uncertain for photodissociation of methyl vinyl ketone excited by sunlight within the troposphere. Excitation in experiments in the far ultraviolet, 193.3 nm (628 kJ mol^{-1}), gave complete rupture of both C–C bonds attached to the carbonyl in the molecule, with quantum yields of CH_3 and $CH_2{=}CH$ radicals (and, presumably, CO) near unity (Fahr et al., 1993). However, at the wavelengths available within the tropo-

Table VII-C-8. Methacrolein absorption cross sections in the vapor phase; here, the given high-resolution data have been averaged over each nanometer and centered on the wavelength shown

λ (nm)	$\sigma \times 10^{20}$	λ (nm)	$\sigma \times 10^{20}$	λ (nm)	$\sigma \times 10^{20}$	λ (nm)	$\sigma \times 10^{20}$
240	1.26	280	1.07	320	6.54	360	3.89
241	0.82	281	1.15	321	6.42	361	3.03
242	0.53	282	1.24	322	6.41	362	2.75
243	0.35	283	1.32	323	6.50	363	2.75
244	0.24	284	1.41	324	6.61	364	2.33
245	0.18	285	1.52	325	6.76	365	1.89
246	0.15	286	1.63	326	6.76	366	1.52
247	0.13	287	1.74	327	6.76	367	1.29
248	0.12	288	1.85	328	7.08	368	1.21
249	0.11	289	1.97	329	7.35	369	1.36
250	0.11	290	2.09	330	7.41	370	1.64
251	0.11	291	2.23	331	7.42	371	1.46
252	0.11	292	2.34	332	7.19	372	1.28
253	0.11	293	2.44	333	6.70	373	0.93
254	0.12	294	2.57	334	6.32	374	0.90
255	0.12	295	2.73	335	6.01	375	0.91
256	0.13	296	2.91	336	6.10	376	1.12
257	0.13	297	3.06	337	6.48	377	1.62
258	0.15	298	3.19	338	6.63	378	1.92
259	0.16	299	3.36	339	6.45	379	1.10
260	0.17	300	3.46	340	6.32	380	0.67
261	0.19	301	3.73	341	6.16	381	0.38
262	0.21	302	3.82	342	6.28	382	0.30
263	0.23	303	4.00	343	7.10	383	0.25
264	0.26	304	4.12	344	6.82	384	0.18
265	0.28	305	4.27	345	6.26	385	0.12
266	0.31	306	4.48	346	6.03	386	0.08
267	0.34	307	4.70	347	5.59	387	0.06
268	0.38	308	4.83	348	4.90	388	0.05
269	0.42	309	4.97	349	3.98	389	0.04
270	0.46	310	5.11	350	3.91	390	0.03
271	0.50	311	5.33	351	4.35	391	0.03
272	0.55	312	5.56	352	4.61		
273	0.60	313	5.73	353	4.32		
274	0.66	314	5.84	354	3.77		
275	0.71	315	5.96	355	3.59		
276	0.77	316	6.14	356	3.50		
277	0.84	317	6.32	357	3.87		
278	0.91	318	6.45	358	5.21		
279	0.99	319	6.56	359	4.94		

Source: Meller in Röth et al. (1997).
Note: Units on σ are cm^2 molecule^{-1}.

sphere, photodissociation by primary processes (I) and (II) probably occurs but, by analogy with acrolein, probably with low efficiency:

$$CH_3C(=O)CH=CH_2 + h\nu \rightarrow CH_3 + COCH=CH_2 \quad \text{(I)}$$
$$\rightarrow CH_3CO + CH=CH_2 \quad \text{(II)}$$
$$\rightarrow CH_3CH=CH_2 + CO \quad \text{(III)}$$

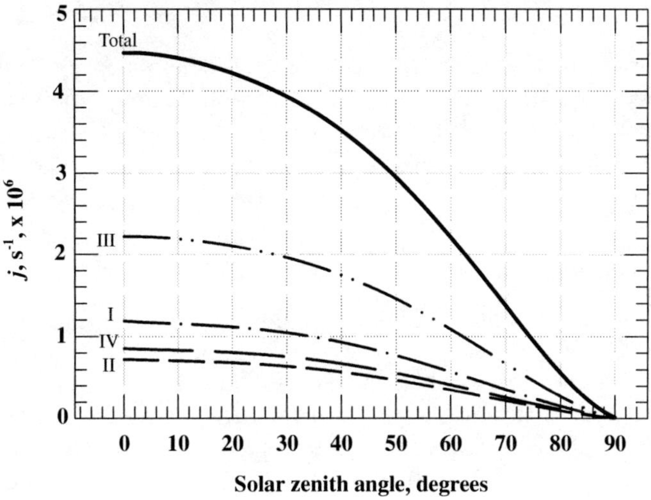

Figure VII-C-16. Estimated methacrolein photodissociation coefficients at 0.5 km altitude as a function of solar zenith angle, calculated using the measured σ values and assuming acrolein quantum yield estimates for channels (I)–(IV) represent those for methacrolein. Products of channels (I)–(IV) are defined in this section. Heavy solid curve represents the total j (photodecomposition).

The additional process (III) has been suggested by Raber and Moortgat (1995) from photolyses in air. The molecular rearrangement of process (III) is unexpected in view of the photolysis of other aliphatic ketones that show either dissociation at the RC(=O)−R bond or, in the case of the ketones with attached alkyl groups with γ-H atoms relative to the carbonyl group, an intramolecular rearrangement to form an enol-ketone and an alkene (McMillan et al., 1964; Nicol and Calvert, 1967). Process (III), the expulsion of CO and formation of a stable R−R′ product, appears to be unique among the photochemistry of alkyl ketones [RC(=O)R′] and its occurrence should be confirmed in further studies.

Process (I) appears to require somewhat less energy (~71 kJ mol^{-1}) than process (II), so one might expect this to be the major primary mode of photodecomposition within the troposphere.

Raber and Moortgat (1995) have reported the results of some photolyses of methyl vinyl ketone in air using the same broad-band radiation source they employed in their methacrolein studies; see section VII-C-8b. The main infrared absorbing products identified were C_3H_6, CH_2O, and CO. Other compounds found were HC(=O)OH, CH_3OH, CO_2, $CH_3C(=O)OH$, and traces of CH_2=CHC(=O)OH. The carbon balance was < 50%. Within the error of the measurements, no significant quenching of the products with increased air pressure (54–760 Torr) was observed.

VII-C-9a. Cross Sections

Cross sections for methyl vinyl ketone have been determined by Schneider and Moortgat and appear in Röth et al. (1997). These data are plotted versus wavelength in figure VII-C-5, where they can be compared with those of the similar compounds acrolein and methacrolein. The cross sections for methyl vinyl ketone are summarized in table VII-C-9.

VII-C-9b. Quantum Yields

There are no published estimates of the quantum yields of the primary processes in methyl vinyl ketone that occur in air exposed to the sunlight within the troposphere. However, there are some effective overall quantum yields of photodecomposition that were determined in broad-band irradiation studies of Raber and Moortgat (1995); these varied from 0.05 to 0.12. Effective *j*-values for photodecomposition varied from 3.2×10^{-5} to $7.8 \times 10^{-5}\ s^{-1}$, and were relatively insensitive to the total pressure of air (54–760 Torr). Until measurements of the wavelength dependence of the quantum yields as a function of wavelength and pressure have been made, it may be reasonable to assume that quantum yields of decomposition by process (I) are equal to those observed in the analogous process in acetone photolysis for a given wavelength and pressure of air; see the tables given in Calvert et al. (2000). The nature of the carbon–carbon single bond adjacent to the carbonyl group ($CH_3{-}COC_2H_3$) is similar to that present in acetone ($CH_3{-}COCH_3$). However, acetone photodecomposition quantum yields may be expected to be somewhat greater than those of methyl vinyl ketone for a given wavelength, temperature, and pressure of air, as a result of the somewhat greater complexity of the methyl vinyl ketone molecule, its somewhat longer lifetime, and the greater probability of quenching of the excited states by air molecules.

VII-C-9c. Tropospheric Photolysis

Photolysis rate coefficients representative of the lower atmosphere were calculated using the TUV model (as described in the introduction to this section) with the measured cross sections and assuming that the quantum yield data for acetone can describe those for methyl vinyl ketone. Note that the effective *j*-values measured for methyl vinyl ketone by Raber and Moortgat (1995) are comparable to those calculated using this procedure. See figure VII-C-17.

VII-C-10. 4-Oxo-2-Pentenal ($CH_3COCH{=}CHCHO$)

The tropospheric photochemistry of 4-oxo-2-pentenal may involve any or all of the following primary processes:

$$\textit{trans-/cis-}CH_3COCH{=}CHCHO + h\nu \rightarrow CH_3COCH{=}CHCHO^*$$

$$CH_3COCH{=}CHCHO^* \rightarrow \textit{cis-/trans-}CH_3COCH{=}CHCHO \quad \text{(I)}$$

$$\rightarrow CH_3 + COCH{=}CHCHO \quad \text{(II)}$$

$$\rightarrow CH_3CO + CH{=}CHCHO \quad \text{(III)}$$

$$\rightarrow CH_3COCH{=}CH + HCO \quad \text{(IV)}$$

Table VII-C-9. Methyl vinyl ketone absorption cross sections in the gas phase; here the high-resolution data have been averaged over each nanaometer and centered on the wavelength shown

λ (nm)	$\sigma \times 10^{20}$	λ (nm)	$\sigma \times 10^{20}$	λ (nm)	$\sigma \times 10^{20}$	λ (nm)	$\sigma \times 10^{20}$
240	0.59	280	1.40	320	6.13	360	3.40
241	0.59	281	1.47	321	6.16	361	3.28
242	0.52	282	1.55	322	6.21	362	3.19
243	0.44	283	1.67	323	6.24	363	3.14
244	0.37	284	1.76	324	6.28	364	2.85
245	0.32	285	1.85	325	6.33	365	2.60
246	0.32	286	1.95	326	6.41	366	2.30
247	0.32	287	2.05	327	6.50	367	2.00
248	0.28	288	2.14	328	6.59	368	1.82
249	0.29	289	2.26	329	6.67	369	1.63
250	0.30	290	2.39	330	6.70	370	1.50
251	0.32	291	2.46	331	6.72	371	1.37
252	0.33	292	2.69	332	6.68	372	1.28
253	0.34	293	3.28	333	6.57	373	1.18
254	0.34	294	2.89	334	6.44	374	1.13
255	0.37	295	2.97	335	6.30	375	1.08
256	0.39	296	2.97	336	6.24	376	1.05
257	0.40	297	3.10	337	6.11	377	1.07
258	0.45	298	3.20	338	5.99	378	0.99
259	0.46	299	3.36	339	5.94	379	0.90
260	0.50	300	3.51	340	5.94	380	0.81
261	0.54	301	3.67	341	5.99	381	0.79
262	0.57	302	3.82	342	5.99	382	0.79
263	0.62	303	3.95	343	5.99	383	0.75
264	0.64	304	4.10	344	5.94	384	0.56
265	0.67	305	4.24	345	5.91	385	0.45
266	0.66	306	4.39	346	5.88	386	0.36
267	0.74	307	4.54	347	5.76	387	0.29
268	0.77	308	4.69	348	5.58	388	0.24
269	0.82	309	4.85	349	5.32	389	0.20
270	0.78	310	4.97	350	4.75	390	0.16
271	0.80	311	5.10	351	4.57	391	0.14
272	0.90	312	5.25	352	4.31	392	0.11
273	0.92	313	5.39	353	4.07	293	0.10
274	0.96	314	5.53	354	3.90	394	0.09
275	1.01	315	5.65	355	3.76	395	0.08
276	1.08	316	5.78	356	3.67	396	0.07
277	1.16	317	5.90	357	3.65	397	0.06
278	1.24	318	6.01	358	3.65	398	0.06
279	1.31	319	6.10	359	3.47		

Source: Schneider and Moortgat in Röth et al. (1997).
Note: Units on σ are cm^2 $molecule^{-1}$.

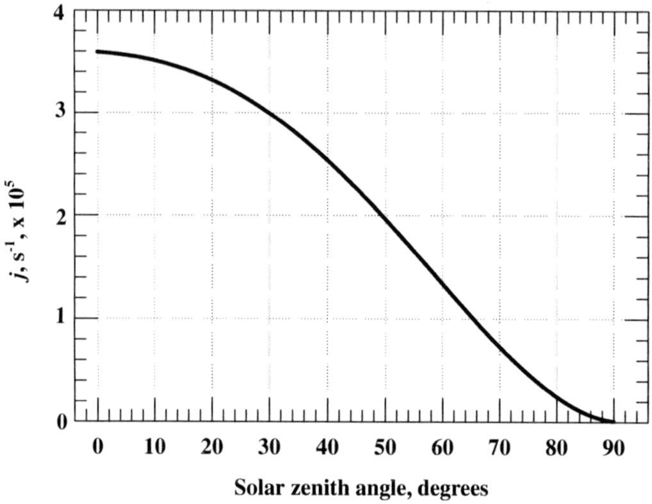

Figure VII-C-17. Methyl vinyl ketone photodissociation coefficients at 0.5 km altitude as a function of solar zenith angle. Quantum yields were assumed to be equivalent to those of acetone (McKeen et al., 1997). Quantum yields were calculated for 1 atm at 25°C.

The rapid photodecomposition of the 4-oxo-2-pentenal in synthetic air, $\tau_{1/2} \sim 1.5\,h$ (6–8 A.M. exposure), and rapid oxidation of added NO to NO_2 observed in sunlight in chamber experiments using a synthetic air–NO_x mixture (Liu et al., 1999) suggests that free radical formation in processes (II), (III), and/or (IV) must occur in addition to process (I). In the studies of Bierbach et al. (1994) the major infrared light–absorbing products of the ultraviolet photolysis (254 nm) of *cis-/trans*-4-oxo-2-pentenal were found to be maleic anhydride, 5-methyl-3-*H*-furan-2-one, CH_2O, CH_3OH, and CH_3O_2H, and <20–30% of the primary photochemical reactions occurred by isomerization in the *trans*-compound in process (I). Photolyses with 360–480 nm light gave 5-methyl-3-*H*-furan-2-one, CH_2O, CH_3OH, and CH_3O_2H as major products and maleic anhydride as a minor product; < 20% of primary photochemical reaction occurred by isomerization in process (I). Presumably, the observed products arise from subsequent reactions of the oxygenated radicals formed following processes (II), (III), and/or (IV). 5-Methyl-3-*H*-furan is an exception that is formed by an intramolecular rearrangement (Bierbach et al., 1994). The carbonyl products observed by Liu et al. (1999) in daylight chamber experiments—formaldehyde, methacrolein, glycolaldehyde, hydroxyacetone, glyoxal, and methylglyoxal—suggest the possible occurrence of free-radical–forming reactions. The formation of the observed malonaldehyde, methacrolein, and hydroxyacetone cannot be explained in terms of current mechanisms; these may have resulted from impurities in the system.

VII-C-10a. Absorption Cross Sections

The absorption cross sections for *cis-/trans*-4-oxo-2-pentenal have been measured in water and acetonitrile solutions (Liu et al., 1999). The low volatility of these molecules makes the measurement of gas-phase spectra very difficult. The solution-phase spectra are given in figure VII-C-18.

The shape of the absorption bands and absolute magnitude of σ in this molecule are very different in the two solvents. Some differences are expected since water may react in part with the carbonyl groups (>C=O) to form gem-diol groups [–C(OH)$_2$–], which would alter the chromophore significantly. Furthermore, the very different solvent polarities can cause some of the wavelength shifts seen in the figure. There is a question as to which of the spectra would be most like that of the gas-phase molecule. In any case, the results will remain uncertain. The use of the acetonitrile spectra in simulations of gas-phase chemistry may be more appropriate, while those spectra for the water solutions may be useful in modeling the solution-phase photochemistry of this molecule. The absorption cross section data for 2-oxo-2-pentenal are summarized in table VII-C-10.

VII-C-10b. Quantum Yields

There are no direct measurements of the quantum yields of the primary processes in the photolysis of 4-oxo-2-pentenal. However the *j*-value for decomposition has been measured in sunlight (in the presence of the radical scavenger, cyclohexane) as j(decomposition) = $(7.99 \pm 0.13) \times 10^{-3}\ s^{-1}$ for atmospheric conditions for which the measured $j(NO_2) = (10.1 \pm 0.13) \times 10^{-3}\ s^{-1}$ (Sørenson and Barnes, 1998). These data seem to be consistent with calculations based on an assumed total ϕ for decomposition of unity at all absorbed wavelengths; see figure VII-C-19. The major rearrangement products of the photolysis of 4-oxo-2-pentenal—maleic-anhydride and 5-methyl-3*H*-furan-2-one—can in theory be derived in reactions following primary processes (II) and (III). The observed methyl-radical

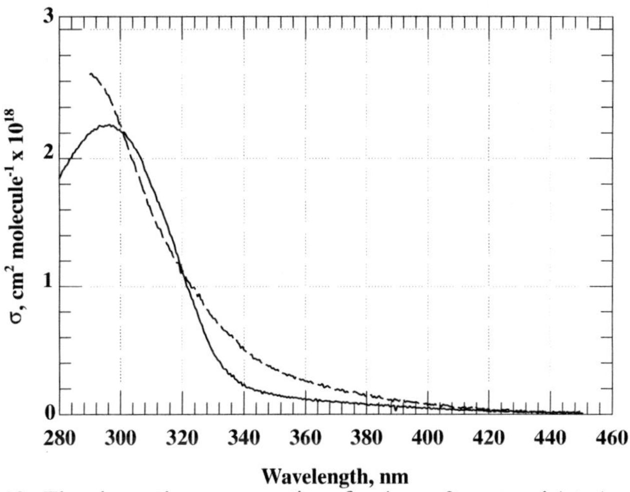

Figure VII-C-18. The absorption cross sections for 4-oxo-2-pentenal (*cis-/trans*-mixture), in water (solid line) and acetonitrile solutions (dashed line); data of Liu et al. (1999); Liu (1999).

Table VII-C-10. Cross sections for 4-oxo-2-pentenal (*cis-/trans*-mixture) as obtained in water and in acetonitrile solutions

(a) Water solution

λ (nm)	$\sigma \times 10^{19}$	λ (nm)	$\sigma \times 10^{19}$	λ (nm)	$\sigma \times 10^{19}$	λ (nm)	$\sigma \times 10^{19}$
290	21.97	330	4.81	370	0.96	410	0.29
291	22.17	331	4.43	371	0.91	411	0.31
292	22.28	332	4.14	372	0.94	412	0.33
293	22.57	333	3.68	373	0.94	413	0.30
294	22.52	334	3.51	374	0.83	414	0.29
295	22.58	335	3.10	375	0.85	415	0.29
296	22.62	336	2.93	376	0.88	416	0.30
297	22.49	337	2.76	377	0.75	417	0.28
298	22.48	338	2.53	378	0.80	418	0.23
299	22.26	339	2.48	379	0.76	419	0.21
300	22.16	340	2.33	380	0.68	420	0.17
301	22.07	341	2.05	381	0.76	421	0.21
302	21.76	342	2.03	382	0.72	422	0.26
303	21.47	343	1.94	383	0.75	423	0.22
304	21.18	344	1.86	384	0.68	424	0.22
305	20.76	345	1.80	385	0.61	425	0.18
306	20.36	346	1.60	386	0.67	426	0.19
307	19.90	347	1.66	387	0.59	427	0.23
308	19.11	348	1.59	388	0.62	428	0.21
309	18.59	349	1.62	389	0.26	429	0.13
310	17.91	350	1.50	390	0.59	430	0.17
311	17.44	351	1.44	391	0.58	431	0.24
312	16.67	352	1.38	392	0.61	432	0.14
313	16.14	353	1.34	393	0.48	433	0.11
314	15.54	354	1.29	394	0.54	434	0.17
315	14.91	355	1.25	395	0.56	435	0.11
316	14.12	356	1.28	396	0.49	436	0.14
317	13.57	357	1.28	397	0.49	437	0.14
318	12.64	358	1.21	398	0.48	438	0.11
319	11.88	359	1.14	399	0.44	439	0.07
320	11.19	360	1.18	400	0.51	440	0.09
321	10.42	361	1.16	401	0.41	441	0.09
322	9.75	362	1.08	402	0.42	442	0.13
323	9.05	363	1.11	403	0.38	443	0.15
324	8.40	364	0.98	404	0.41	444	0.08
325	7.83	365	1.05	405	0.42	445	0.06
326	7.12	366	1.00	406	0.36	446	0.11
327	6.44	367	1.03	407	0.39	447	0.07
328	5.86	368	1.01	408	0.39	448	0.16
329	5.33	369	0.95	409	0.35	449	0.09

Table VII-C-10. *(continued)*

(b) Acetonitrile solution

λ (nm)	$\sigma \times 10^{19}$	λ (nm)	$\sigma \times 10^{19}$	λ (nm)	$\sigma \times 10^{19}$	λ (nm)	$\sigma \times 10^{19}$
290	26.54	330	7.49	370	1.88	410	0.51
291	26.38	331	7.31	371	1.90	411	0.48
292	26.27	332	7.12	372	1.91	412	0.53
293	25.92	333	6.69	373	1.69	413	0.39
294	25.61	334	6.48	374	1.68	414	0.48
295	25.29	335	6.23	375	1.72	415	0.36
296	24.93	336	6.15	376	1.75	416	0.51
297	24.37	337	5.72	377	1.64	417	0.42
298	23.79	338	5.60	378	1.52	418	0.43
299	23.30	339	5.16	379	1.58	419	0.28
300	22.64	340	5.02	380	1.45	420	0.36
301	21.89	341	4.81	381	1.45	421	0.28
302	21.28	342	4.63	382	1.29	422	0.28
303	20.58	343	4.34	383	1.27	423	0.30
304	19.87	344	4.27	384	1.42	424	0.28
305	19.27	345	4.24	385	1.22	425	0.30
306	18.70	346	4.03	386	1.25	426	0.36
307	17.91	347	3.87	387	1.11	427	0.20
308	17.28	348	3.72	388	1.11	428	0.24
309	16.40	349	3.57	389	1.16	429	0.27
310	15.86	350	3.47	390	1.08	430	0.23
311	15.26	351	3.45	391	1.11	431	0.15
312	14.67	352	3.13	392	1.14	432	0.19
313	14.44	353	3.16	393	1.00	433	0.24
314	13.79	354	2.99	394	0.97	434	0.19
315	13.35	355	3.02	395	0.92	435	0.18
316	12.83	356	2.92	396	0.92	436	0.12
317	12.33	357	2.93	397	0.84	437	0.17
318	11.92	358	2.74	398	0.79	438	0.16
319	11.68	359	2.69	399	0.78	439	0.15
320	11.25	360	2.56	400	0.70	440	0.10
321	10.77	361	2.48	401	0.72	441	0.20
322	10.36	362	2.47	402	0.69	442	0.15
323	9.89	363	2.40	403	0.62	443	0.19
324	9.85	364	2.30	404	0.71	444	0.10
325	9.18	365	2.26	405	0.74	445	0.13
326	8.97	366	2.10	406	0.68	446	0.12
327	8.46	367	2.18	407	0.62	447	0.13
328	8.17	368	2.06	408	0.59	448	0.10
329	7.80	369	2.00	409	0.58	449	0.09

Source: Parts (a) and (b), Liu et al. (1999); Liu (1999).
Note: Units on σ are cm^2 molecule^{-1}.

products are consistent with the occurrence of process (II), although the occurrence of process (III) followed by suitable alkoxy-radical decomposition steps can in theory lead to these products as well. Note that maleic anhydride and 5-methyl-3*H*-furan-2-one were both found to be relatively stable photochemically for exposure to the wavelengths of light prevalent within the troposphere. Presumably, the significant ozone-forming potential of the 4-oxo-2-pentenal is derived largely from

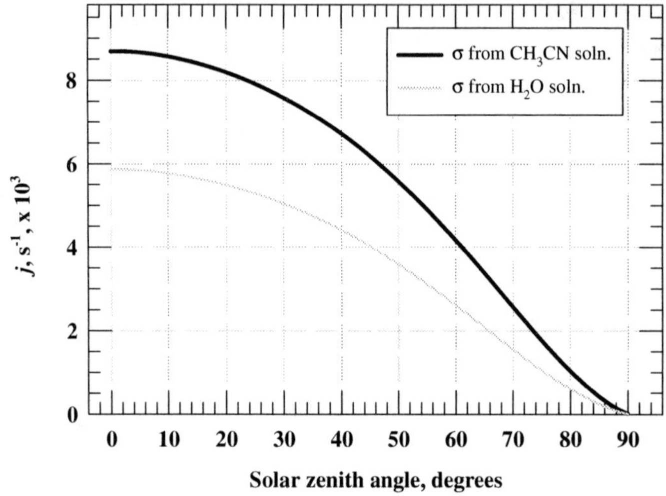

Figure VII-C-19. 4-Oxo-2-pentenal photodissociation coefficients at 0.5 km altitude as a function of solar zenith angle. Total quantum yield of decomposition was assumed to be unity. The upper and lower curves were calculated using the σ data for solutions in acetonitrile (table VII-C-10b) and water solutions (table VII-C-10a), respectively.

the formation of peroxy radicals from the primary photodissociation fragments of the free radical processes (II) and (III).

VII-C-10c. Tropospheric Photolysis

Photolysis rate coefficients representative of the lower atmosphere were calculated using the TUV model (as described in the introduction to this section) with the measured cross sections and an assumed total quantum yield of decomposition of unity at all absorbed wavelengths of light; see figure VII-C-19.

VII-C-11. 3-Hexene-2,5-Dione

By analogy with the photolysis pathways suggested for the similar compounds butendial and 4-oxo-2-pentenal, one might expect the following primary photochemical pathways to occur:

$$\textit{trans-/cis-}CH_3C(=O)CH=CHC(=O)CH_3 + h\nu \rightarrow$$

$$CH_3C(=O)CH=CHC(=O)CH_3^*$$

$$CH_3C(=O)CH=CHC(=O)CH_3^* \rightarrow \textit{cis-/trans-}CH_3C(=O)CH=CHC(=O)CH_3 \quad (I)$$

$$\rightarrow CH_3 + COCH=CHCOCH_3 \quad (II)$$

$$\rightarrow CH_3CO + CH=CHCOCH_3 \quad (III)$$

Tuazon et al. (1985) found that $>80\%$ of the photolysis rate was accounted for by isomeriation in process (I). Bierbach et al. (1994) arrived at a similar conclusion from their photolysis studies using the *trans*-isomer with either 254 nm or 320–480 nm radiation. Isomerization is the major reaction pathway. The photochemical steady-state ratio of [*trans*]/[*cis*] was determined to be ~0.5 in each of these studies and in the more recent work of Becker and Graedler (1999). In chamber studies, Liu et al. (1999) observed the relatively slow disappearance of 3-hexene-2,5-dione; with 0.95 ppm initially present, about half of the compound disappeared during the 3.5-h exposure to the natural sunlight (6:00 to 9:30 A.M.), with the generation of small amounts of several carbonyl products: methylglyoxal, acetaldehyde, and hydroxyacetone. The methylglyoxal was the major carbonyl product observed, and it peaked at a concentration of 0.15 ppm at about 10:30 A.M. These data suggest that perhaps a small fraction of the 3-hexene-2-5-dione excited by sunlight within the troposphere undergoes decomposition to form free radical products in processes (II) or (III). In experiments in the European photochemical reactor EUPHORE, Becker and Graedler (1999) observed a very rapid isomerization of the *cis*-isomer (reaching a photostationary state within 30–35 min) with *j*-values for *cis*-3-hexene-2,5-dione + *trans*-3-hexene-2,5-dione disapearance together with $j(NO_2)$ values; starting with the pure *cis*- or *trans*-isomers, they observed for comparable daylight conditions: $j(NO_2)/j$(*cis*-hexene-2,5-dione) = $(5.9 \pm 2.3) \times 10^{-3}/(1.1 \pm 2.0) \times 10^{-5}$; $j(NO_2)/j$(*trans*-3-hexene-2,5-dione) = $(5.8 \pm 1.0) \times 10^{-3}/(1.3 \pm 1.2) \times 10^{-5}$; $j(NO_2)/j$(*cis*- + *trans*-3-hexene-2,5-dione) = $(5.9 \pm 1.0) \times 10^{-3}/(1.2 \pm 0.9) \times 10^{-5}$.

VII-C-11a. Absorption Cross Sections

The absorption cross sections for *cis*- and *trans*-isomers of 3-hexene-2,5-diones were determined in water and in acetonitrile solutions by Liu et al. (1999). These are shown in figure VII-C-20. The σ data as a function of λ (for each nanometer) are summarized in table VII-C-11.

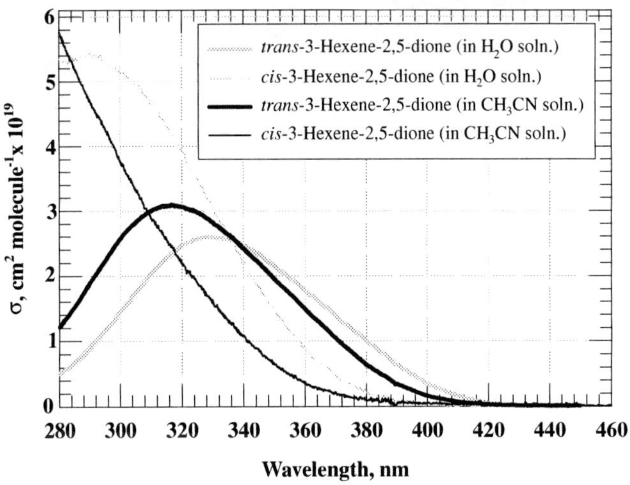

Figure VII-C-20. The absorption cross sections for *trans*- and *cis*-3-hexene-2,5-dione as measured in water and acetonitrile solutions (Liu, 1999; Liu et al., 1999).

Table VII-C-11. Cross sections for *trans*-3-hexene-2,5-dione in solution

(a) *trans*-3-Hexene-2,5-dione in water solution

λ (nm)	$\sigma \times 10^{19}$	λ (nm)	$\sigma \times 10^{19}$	λ (nm)	$\sigma \times 10^{19}$	λ (nm)	$\sigma \times 10^{19}$
290	0.89	330	2.59	370	1.40	410	0.17
291	0.94	331	2.58	371	1.35	411	0.16
292	0.99	332	2.59	372	1.32	412	0.14
293	1.05	333	2.58	373	1.28	413	0.13
294	1.09	334	2.56	374	1.25	414	0.12
295	1.14	335	2.55	375	1.21	415	0.11
296	1.21	336	2.54	376	1.17	416	0.10
297	1.26	337	2.54	377	1.14	417	0.08
298	1.32	338	2.52	378	1.09	418	0.08
299	1.38	339	2.49	379	1.06	419	0.06
300	1.44	340	2.47	380	1.02	420	0.05
301	1.51	341	2.44	381	0.98	421	0.04
302	1.57	342	2.41	382	0.95	422	0.04
303	1.61	343	2.38	383	0.92	423	0.03
304	1.68	344	2.36	384	0.88	424	0.02
305	1.73	345	2.33	385	0.84	425	0.02
306	1.80	346	2.29	386	0.81	426	0.02
307	1.86	347	2.26	387	0.77	427	0.01
308	1.91	348	2.22	388	0.74	428	0.01
309	1.97	349	2.18	389	0.70	429	0.01
310	2.02	350	2.15	390	0.63	430	0.01
311	2.09	351	2.11	391	0.63	431	0.003
312	2.13	352	2.08	392	0.60	432	0.01
313	2.18	353	2.06	393	0.56	433	0.002
314	2.22	354	2.03	394	0.53	434	0.002
315	2.27	355	1.99	395	0.51	435	0.002
316	2.30	356	1.96	396	0.48	436	0.002
317	2.35	357	1.92	397	0.45		
318	2.38	358	1.88	398	0.42		
319	2.41	359	1.84	399	0.39		
320	2.46	360	1.80	400	0.37		
321	2.49	361	1.76	401	0.35		
322	2.51	362	1.72	402	0.32		
323	2.53	363	1.68	403	0.30		
324	2.55	364	1.64	404	0.28		
325	2.56	365	1.60	405	0.26		
326	2.58	366	1.56	406	0.24		
327	2.59	367	1.53	407	0.22		
328	2.60	368	1.49	408	0.20		
329	2.59	369	1.44	409	0.19		

VII-C-11b. Quantum Yields

There are no quantum yield data related to the primary processes in the photolysis of 3-hexene-2,5-dione, although in the European outdoor chamber EUPHORE in Valencia, measurements of the apparent *j*-values for disappearance of the *cis*- and *trans*-isomers have been reported by Becker and Graedler (1999); see section VII-

Table VII-C-11. *(continued)*

(b) trans-3-Hexene-2,5-dione in acetonitrile solution

λ (nm)	$\sigma \times 10^{19}$	λ (nm)	$\sigma \times 10^{19}$	λ (nm)	$\sigma \times 10^{19}$	λ (nm)	$\sigma \times 10^{19}$
290	1.89	330	2.82	370	1.04	410	0.07
291	1.96	331	2.78	371	1.00	411	0.06
292	2.03	332	2.75	372	0.96	412	0.05
293	2.10	333	2.71	373	0.91	413	0.05
294	2.16	334	2.66	374	0.88	414	0.05
295	2.22	335	2.63	375	0.84	415	0.05
296	2.30	336	2.58	376	0.79	416	0.05
297	2.36	337	2.55	377	0.76	417	0.04
298	2.43	338	2.50	378	0.73	418	0.03
299	2.50	339	2.46	379	0.69	419	0.03
300	2.56	340	2.41	380	0.64	420	0.03
301	2.63	341	2.36	381	0.61	421	0.03
302	2.68	342	2.32	382	0.58	422	0.02
303	2.73	343	2.28	383	0.54	423	0.02
304	2.78	344	2.24	384	0.52	424	0.02
305	2.82	345	2.20	385	0.49	425	0.02
306	2.87	346	2.14	386	0.47	426	0.02
307	2.91	347	2.09	387	0.44	427	0.02
308	2.93	348	2.05	388	0.40	428	0.01
309	2.97	349	2.00	389	0.36	429	0.01
310	2.99	350	1.95	390	0.35	430	0.01
311	3.03	351	1.91	391	0.33	431	0.01
312	3.03	352	1.86	392	0.30	432	0.01
313	3.06	353	1.82	393	0.29	433	0.01
314	3.07	354	1.77	394	0.27	434	0.01
315	3.07	355	1.72	395	0.25	435	0.01
316	3.08	356	1.68	396	0.23	436	0.01
317	3.09	357	1.63	397	0.21	437	0.01
318	3.08	358	1.58	398	0.19	438	0.01
319	3.07	359	1.53	399	0.18	439	0.01
320	3.06	360	1.48	400	0.16	440	0.004
321	3.05	361	1.43	401	0.15	441	0.003
322	3.03	362	1.39	402	0.14	442	0.004
323	3.02	363	1.35	403	0.13	443	0.01
324	3.00	364	1.30	404	0.13	444	0.01
325	2.97	365	1.26	405	0.12	445	0.01
326	2.95	366	1.22	406	0.10	446	0.003
327	2.93	367	1.18	407	0.10	447	0.01
328	2.88	368	1.13	408	0.09	448	0.003
329	2.85	369	1.08	409	0.07	449	0.01

(continued)

C-11. It appears that the isomerization, reaction (I), is dominant. The molecules that do not undergo *cis–trans*-isomerization, $<20\%$ of the excited molecules, may take part in some photodecomposition process. The data of Becker and Graedler (1999) suggest that the extent of photodecomposition may be much less than 20%, with *j*-values for photodecomposition about ten times those observed in acrolein; see section VII-C-4.

Table VII-C-11. (*continued*)

(c) *cis*-3-Hexene-2,5-dione in water solution

λ (nm)	$\sigma \times 10^{19}$	λ (nm)	$\sigma \times 10^{19}$	λ (nm)	$\sigma \times 10^{19}$	λ (nm)	$\sigma \times 10^{19}$
290	5.41	330	3.06	370	0.49	410	0.003
291	5.39	331	2.94	371	0.45	411	0.00
292	5.40	332	2.88	372	0.42	412	0.00
293	5.36	333	2.78	373	0.40	413	0.00
294	5.36	334	2.69	374	0.38	414	0.00
295	5.32	335	2.61	375	0.34	415	0.02
296	5.32	336	2.55	376	0.30	416	0.04
297	5.24	337	2.47	377	0.28	417	0.001
298	5.22	338	2.36	378	0.28	418	0.00
299	5.16	339	2.30	379	0.24	419	0.00
300	5.14	340	2.21	380	0.22	420	0.003
301	5.14	341	2.09	381	0.19		
302	5.06	342	2.05	382	0.19		
303	5.03	343	1.97	383	0.18		
304	5.03	344	1.90	384	0.15		
305	4.97	345	1.82	385	0.14		
306	4.95	346	1.73	386	0.16		
307	4.89	347	1.66	387	0.15		
308	4.78	348	1.60	388	0.10		
309	4.73	349	1.54	389	0.10		
310	4.67	350	1.47	390	0.09		
311	4.66	351	1.39	391	0.08		
312	4.54	352	1.31	392	0.08		
313	4.46	353	1.26	393	0.09		
314	4.42	354	1.20	394	0.06		
315	4.34	355	1.15	395	0.05		
316	4.26	356	1.11	396	0.05		
317	4.20	357	1.06	397	0.05		
318	4.06	358	0.98	398	0.04		
319	4.01	359	0.91	399	0.05		
320	3.90	360	0.88	400	0.05		
321	3.85	361	0.84	401	0.02		
322	3.75	362	0.79	402	0.04		
323	3.66	363	0.74	403	0.02		
324	3.62	364	0.69	404	0.02		
325	3.49	365	0.66	405	0.04		
326	3.41	366	0.64	406	0.02		
327	3.32	367	0.61	407	0.02		
328	3.21	368	0.58	408	0.02		
329	3.16	369	0.50	409	0.005		

(*continued*)

VII-C-11c. Tropospheric Photolysis

Photolysis rates representative of the lower atmosphere were calculated using the TUV model (as described in the introduction to this section) with the measured cross sections in two solvents. From the *j*-values related to disappearance of the sum of both *cis*- and *trans*-isomers determined by Becker and Graedler (1999), it appears that less than a few percent of excited molecules decompose. The $j(NO_2)$

Table VII-C-11. (*continued*)

(d) cis-3-Hexene-2,5-dione in acetonitrile solution

λ (nm)	$\sigma \times 10^{19}$	λ (nm)	$\sigma \times 10^{19}$	λ (nm)	$\sigma \times 10^{19}$	λ (nm)	$\sigma \times 10^{19}$
290	4.73	330	1.63	370	0.22	410	0.03
291	4.60	331	1.54	371	0.19	411	0.04
292	4.53	332	1.45	372	0.17	412	0.03
293	4.46	333	1.43	373	0.17	413	0.03
294	4.40	334	1.36	374	0.16	414	0.03
295	4.29	335	1.28	375	0.14	415	0.04
296	4.20	336	1.25	376	0.12	416	0.04
297	4.08	337	1.21	377	0.11	417	0.03
298	4.00	338	1.13	378	0.15	418	0.03
299	3.87	339	1.09	379	0.12	419	0.05
300	3.76	340	1.08	380	0.09	420	0.04
301	3.72	341	0.99	381	0.09	421	0.05
302	3.58	342	0.97	382	0.09	422	0.03
303	3.49	343	0.93	383	0.09	423	0.03
304	3.44	344	0.89	384	0.08	424	0.03
305	3.32	345	0.85	385	0.09	425	0.02
306	3.26	346	0.79	386	0.10	426	0.01
307	3.20	347	0.76	387	0.11	427	0.02
308	3.04	348	0.72	388	0.11	428	0.03
309	2.97	349	0.68	389	0.04	429	0.03
310	2.88	350	0.65	390	0.04	430	0.01
311	2.86	351	0.60	391	0.07	431	0.02
312	2.75	352	0.56	392	0.05	432	0.06
313	2.69	353	0.55	393	0.06	433	0.02
314	2.64	354	0.50	394	0.06	434	0.01
315	2.55	355	0.47	395	0.04	435	0.04
316	2.47	356	0.47	396	0.04	436	0.04
317	2.43	357	0.45	397	0.06	437	0.03
318	2.32	358	0.40	398	0.04	438	0.03
319	2.27	359	0.39	399	0.04	439	0.02
320	2.18	360	0.35	400	0.04	440	0.02
321	2.14	361	0.33	401	0.03	441	0.02
322	2.08	362	0.31	402	0.06	442	0.04
323	2.00	363	0.28	403	0.05	443	0.02
324	1.95	364	0.26	404	0.05	444	0.02
325	1.90	365	0.27	405	0.05	445	0.01
326	1.85	366	0.23	406	0.04	446	0.03
327	1.78	367	0.25	407	0.03	447	0.01
328	1.71	368	0.25	408	0.03	448	0.003
329	1.65	369	0.20	409	0.03	449	0.03

Source: Parts (a)–(d), data from Liu (1999); Liu et al. (1999).
Note: Units on σ are cm^2 $molecule^{-1}$.

values for the experimental condition of Becker and Graedler correspond to a zenith angle of 60° in the photolysis rate calculations performed here. By matching the calculated and measured photolysis rates for the *cis*- and *trans*-isomers, we estimate that the quantum yield is in the range 0.007–0.016. From an average of the three estimates that agree well, we recommend a best estimate quantum yield of photodecomposition of 0.008; this value is used in calculating tropospheric *j*-values given in the data presented in figure VII-C-21.

VII-C-12. *E,Z*-2,4-Hexadienedial

The photolysis of *E,Z*-2,4-hexadienedial has been studied in a laboratory reaction chamber employing visible and ultraviolet lamps (Klotz et al., 1995) and in a large outdoor chamber using sunlight (Klotz et al., 1999). The only significant reaction of the compound was its isomerization to *E,E*-2,4-hexadienedial.

VII-C-12a. Absorption Cross Sections for E,Z-2,4-Hexadienedial

Cross sections for *E,Z*-2,4-hexadienedial are plotted in figure VII-C-22, where they can be compared with those for other related 2,4-hexadienedials. The high-resolution spectral data of Klotz et al. (1995), as averaged over each nanometer (centered on the wavelength shown) are summarized in table VII-C-12. Because of the low volatility of the hexadenedials described in this work, the reported gas-phase cross sections may be significantly in error (by as much as ±50%).

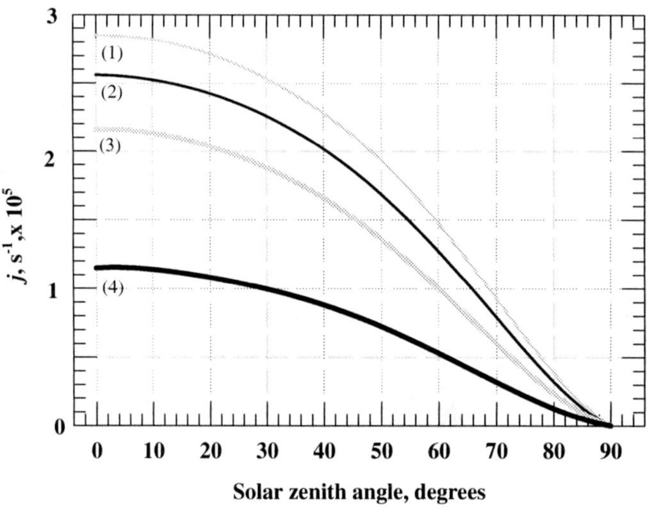

Figure VII-C-21. 3-Hexene-2,5-dione photolysis coefficients (total) at 0.5 km altitude as a function of solar zenith angle. Quantum yield of photodecompostion was assumed to be 0.008 in order to match calculations with measurements. Curves (1) and (2) were calculated using σ data for the *trans*-isomer in H_2O (table VII-C-11a) or acetonitriile solution (table VII-C-11b). Curves (3) and (4) were calculated using the σ data for the *cis*-isomer as measured in H_2O (table VII-C-11c) or acetonitrile (table VII-C-11d) solutions.

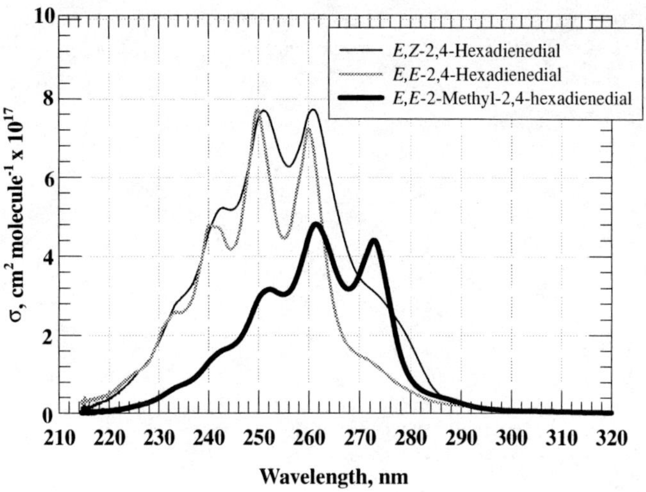

Figure VII-C-22. Gas-phase absorption spectra of some 2,4-hexadienedials as observed in the studies of Klotz et al. (1995).

VII-C-12b. Quantum Yields

There are no direct quantum yield data that have been measured, but *j*-values for the rates of *E,Z*-2,4-hexadienedial isomerization *j*(isomerization) in sunlight and those for $j(NO_2)$ for similar solar zenith angle have been determined in the experiments of Klotz et al. (1999): $j(\text{isomerization})/j(NO_2)$ = $[(1.68 \pm 0.12) \times 10^{-3}\text{s}^{-1}]/[(11.87 \pm 0.80) \times 10^{-3}\text{s}^{-1}]$; $[(0.445 \pm 0.021) \times 10^{-3}\text{s}^{-1}]/[(7.31 \pm 2.39) \times 10^{-3}\text{s}^{-1}]$; $[(1.76 \pm 0.12) \times 10^{-3}]/[(11.22 \pm 1.06) \times 10^{-3}\text{s}^{-1}]$; $[(1.27 \pm 0.07) \times 10^{-3}\,\text{s}^{-1}]/[(4.75 \pm 1.50) \times 10^{-3}\,\text{s}^{-1}]$. The regions of absorption and reaction for the 2,4-hexadienedial do not overlap well with those of NO_2, but at least there is qualitative value to these ratios. From the limited data at hand we assume that the ϕ for isomerization is near unity.

VII-C-12c. Tropospheric Photolysis

Photolysis rate coefficients for isomerization representative of the lower atmosphere were calculated with the TUV model (as described in the introduction to this section) using the measured cross sections and assuming the quantum yield for isomerization to be unity at all absorbed wavelengths of light. See figure VII-C-23.

VII-C-13. *E,E*-2,4-Hexadienedial

The photolysis of *E,E*-2,4-hexadienedial has been studied in the laboratory (Klotz et al., 1995) and in sunlight photolyses in a large photochemical reactor (Klotz et al., 1999). Although the products of the reaction are uncertain, the kinetics of the

Table VII-C-12. Cross sections for *E,Z*-2,4-hexadienedial gas; the high-resolution data of Klotz et al. (1995) have been averaged over each nanometer (centered on the wavelength shown)

λ (nm)	$\sigma \times 10^{19}$	λ (nm)	$\sigma \times 10^{19}$	λ (nm)	$\sigma \times 10^{19}$	λ (nm)	$\sigma \times 10^{19}$
215	12.5	250	749	285	57.2	320	0.293
216	14.7	251	768	286	44.7	321	0.241
217	19.7	252	755	287	35.7	322	0.199
218	25.2	253	718	288	29.1	323	0.146
219	30.0	254	673	289	24.3	324	0.117
220	35.7	255	641	290	20.6	325	0.0915
221	44.0	256	628	291	17.5	326	0.0770
222	53.9	257	641	292	15.0	327	0.0832
223	65.3	258	672	293	12.9	328	0.0734
224	78.4	259	714	294	11.1	329	0.111
225	92.5	260	759	295	9.69	330	0.0943
226	107	261	771	296	8.48	331	0.0775
227	120	262	743	297	7.47	332	0.0671
228	136	263	681	298	6.58	333	0.0532
229	156	264	614	299	5.76	334	0.0375
230	180	265	550	300	5.15	335	0.0445
231	209	266	493	301	4.59	336	0.0430
232	239	267	444	302	4.08	337	0.0520
233	268	268	402	303	3.66	338	0.0595
234	287	269	369	304	3.24	339	0.0794
235	300	270	346	305	2.90	340	0.0888
236	315	271	331	306	2.53	341	0.0951
237	336	272	319	307	2.20	342	0.105
238	369	273	308	308	1.94	343	0.112
239	410	274	295	309	1.65	344	0.129
240	455	275	280	310	1.45	345	0.138
241	490	276	263	311	1.28	346	0.132
242	514	277	245	312	1.11	347	0.124
243	521	278	226	313	0.954	348	0.107
244	518	279	204	314	0.815	349	0.0984
245	519	280	179	315	0.692	350	0.118
246	533	281	152	316	0.589	351	0.0906
247	569	282	124	317	0.501		
248	623	283	98.0	318	0.424		
249	691	284	75.6	319	0.356		

Note: Units on σ are cm^2 $molecule^{-1}$.

loss of the molecules through photochemical processes have been described by the following sequence:

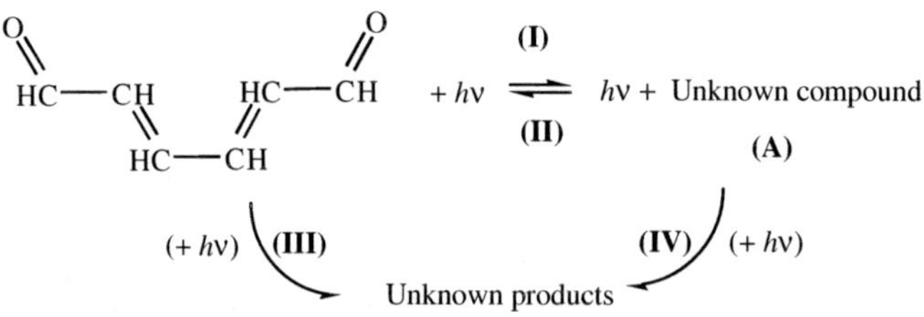

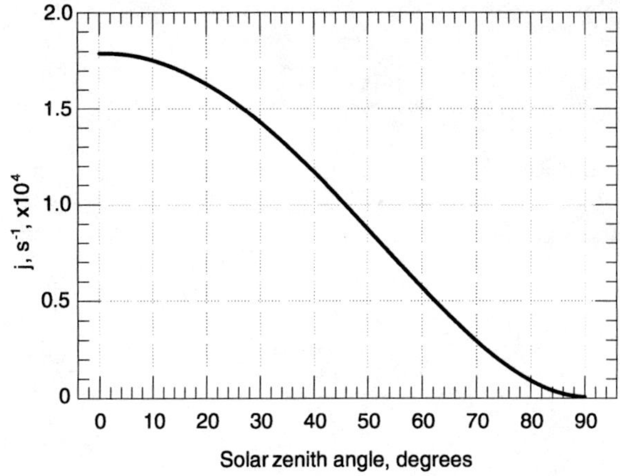

Figure VII-C-23. Estimated *E,Z*-2,4-hexadienedial photoisomerization coefficients at 0.5 km altitude as a function of solar zenith angle using the measured σ data and an assumed ϕ for isomerization of unity.

The apparent *j*-values estimated by Klotz et al. (1999) for the various reactions shown in the diagram are summarized in the table VII-C-13.

The identity of the product A (shown in the reaction sequence), which rapidly achieves a photostationary state with *E,E*-2,4-hexadienedial, is not proven, but *Z*-3,4-diformyl-cyclobutene appears to be a likely candidate:

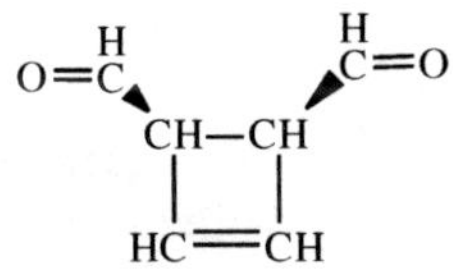

The possible identity of the product as 2-formyl-2*H*-pyran,

Table VII-C-13. Summary of measured *j*-values for daylight exposure of *E,E*-2,4-hexadienedial where the reaction numbers refer to the reactions in the sequence outlined in this section. The $j(NO_2)$ values given were measured toward the end of the experiments in which other *j*-values were measured

$j_I \times 10^3$, s^{-1}	$j_{II} \times 10^3$, s^{-1}	$j_{III} = j_{IV} \times 10^5$, s^{-1}	$j(NO_2) \times 10^3$, s^{-1}
1.3	2.2	26.2 ± 0.66	11.87 ± 0.80
0.36	0.62	13.0 ± 0.76	7.30 ± 2.39
1.3	2.2	15.76 ± 0.27	11.22 ± 1.06
0.9	1.5	1.97 ± 0.10	4.75 ± 1.50
1.17 ± 0.12	2.05 ± 0.21	13.60 ± 0.33	11.71 ± 0.40
0.80 ± 0.08	1.25 ± 0.13	5.4 ± 0.54	4.59 ± 1.52
1.17 ± 0.12	2.1 ± 0.21	14.14 ± 0.63	10.24 ± 0.65

Source: Data of Klotz et al. (1999).

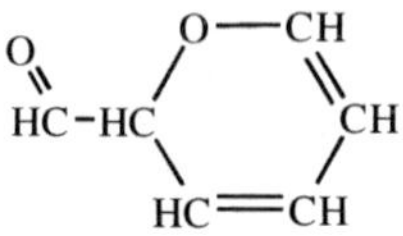

was ruled out from the kinetic data obtained. Although the ultimate products of the reactions are not known, the infrared spectra suggest the presence of a ketene, possibly 2-butenal-4-yl-ketene:

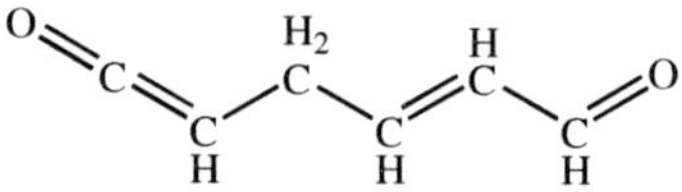

Also, a pronounced generation of aerosol was observed in these photolyses.

VII-C-13a. Absorption Cross Sections

A plot of the wavelength dependence of the absorption cross sections for *E,E*-2,4-hexadienedial is given in figure VII-C-22, where it can be compared with other 2,4-hexadienedials. The absorption cross sections are summarized in table VII-C-14.

VII-C-13b. Quantum Yields

Under typical atmospheric noontime light conditions for a clear sky and low aerosol concentrations on July 1 at a latitude of 40° N, corresponding to the time and location where the Klotz et al. (1999) experiments were performed, a typical $j(NO_2)$ value is $(8.5 \pm 0.5) \times 10^{-3}\ s^{-1}$. Klotz et al. (1999) estimate that for these conditions a lifetime (τ) of *E,E*-2,4-hexadienedial of 17.4 ± 1.7 min [reaction (1)] can be expected while the $\tau_{III} = \tau_{IV} = 160 \pm 60$ min. The nature of the products formed in photolysis reactions (III) and (IV) remains uncertain. However, structural similarities of *E,E*-2,3-hexadienedial to acrolein and the butenedials (RCH=CH(C(=O)H] suggest that some free radical formation may occur through rupture of either the –C(=O)–H or the –CH–CHO bonds in the molecule. The process forming an H-atom plus acyl radical appears to be somewhat favored energetically over that forming HCO plus HCOCH=CHCH=CH radical. Formation of the ketene observed among the products was rationalized following the occurrence of this process, although its formation in a primary photochemical step was suggested as an alternative mechanism (Klotz et al., 1999).

VII-C-13c. Tropospheric Photolysis

Photolysis rate coefficients representative of the lower atmosphere were calculated with the TUV model (as described in the introduction to this section) using the measured cross sections and an assumed quantum yield of photodecomposition of unity (figure VII-C-24).

Table VII-C-14. Cross sections for *E,E*-2,4-hexadienedial; high-resolution data of Klotz et al. (1995) have been averaged over each nanometer (centered on the wavelength shown)

λ (nm)	$\sigma \times 10^{19}$	λ (nm)	$\sigma \times 10^{19}$	λ (nm)	$\sigma \times 10^{19}$	λ (nm)	$\sigma \times 10^{19}$
215	29.6	250	762	285	28.1	320	1.08
216	31.8	251	698	286	25.4	321	0.976
217	35.0	252	620	287	23.3	322	0.858
218	39.4	253	538	288	21.6	323	0.747
219	42.5	254	471	289	20.3	324	0.658
220	47.8	255	447	290	19.2	325	0.596
221	56.8	256	462	291	18.2	326	0.557
222	68.4	257	518	292	17.2	327	0.541
223	80.7	258	590	293	16.2	328	0.549
224	91.8	259	666	294	15.2	329	0.572
225	102	260	716	295	14.3	330	0.538
226	109	261	662	296	13.3	331	0.509
227	119	262	555	297	12.4	332	0.463
228	133	263	438	298	11.4	333	0.405
229	158	264	341	299	10.4	334	0.343
230	189	265	269	300	9.51	335	0.298
231	222	266	220	301	8.67	336	0.259
232	240	267	189	302	7.84	337	0.259
233	255	268	169	303	7.11	338	0.262
234	258	269	157	304	6.41	339	0.282
235	260	270	148	305	5.80	340	0.319
236	274	271	141	306	5.19	341	0.316
237	304	272	139	307	4.63	342	0.330
238	354	273	124	308	4.08	343	0.355
239	414	274	113	309	3.60	344	0.373
240	471	275	103	310	3.20	345	0.378
241	476	276	91.0	311	2.88	346	0.377
242	471	277	80.3	312	2.58	347	0.354
243	450	278	71.3	313	2.31	348	0.279
244	422	279	63.7	314	2.05	349	0.218
245	423	280	56.3	315	1.84	350	0.171
246	452	281	49.2	316	1.64	351	0.114
247	523	282	42.4	317	1.46		
248	615	283	36.4	318	1.31		
249	721	284	31.6	319	1.17		

Note: Units on σ are cm^2 $molecule^{-1}$.

VII-C-14. *E,E*-2-Methyl-2,4-Hexadienedial

In laboratory photoreactor experiments, Klotz et al. (1995) found that the rates of photolysis of *E,E*-2-methyl-2,4-hexadienedial and *E,E*-2,4-hexadienedial were very slow when exposed to radiation in the range $320 < \lambda < 480$ nm (360 maximum); this probably resulted from the poor match of the absorption spectra of these compounds with the lamp emission wavelengths (Klotz et al., 1999). Exposure to 254-nm radiation resulted in significant and near-equal rates of photolysis of *E,E*-2,4-hexadienedial and its related methyl derivative. The infrared spectra of the products of the photolysis showed formation of CO and strong

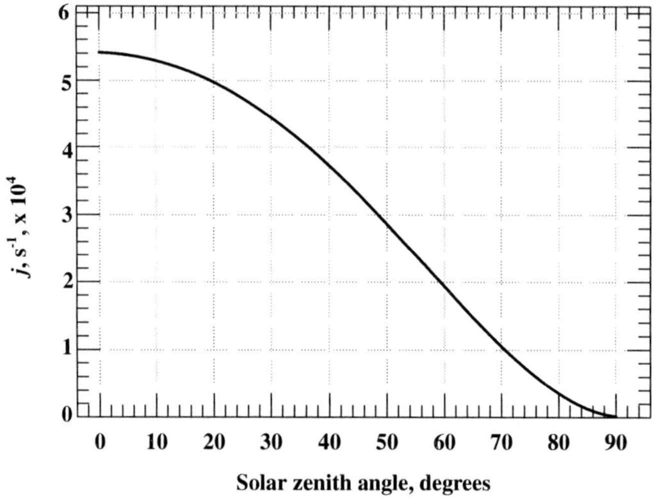

Figure VII-C-24. *E,E*-2-4-hexadienedial photodissociation coefficients at 0.5 km altitude as a function of solar zenith angle. The measured σ values were used and the total quantum yield of photodecomposition was assumed to be unity.

features of a ketene and carbonyl compounds; the latter two bands decayed on further irradiation, with the appearance of a new carbonyl band.

VII-C-14a. Absorption Cross Sections

The wavelength dependence of the light absorption by *E,E*-2-methyl-2,4-hexadienedial is plotted in figure VII-C-22 with other similar compounds. The recommended cross sections for use in *j*-estimates for each nanometer are summarized in table VII-C-15.

VII-C-14b. Quantum Yields

From the laboratory photoreactor studies of Klotz et al. (1995) and the more detailed kinetic studies of *E,E*-2,4-hexadienedial photolysis in sunlight (Klotz et al., 1999), it is suggested that the *j*-values obtained for the unsubstituted *E,E*-2,4-hexadienedial be used to describe the photolysis of *E,E*-2-methyl-2,4-hexadienedial. See figure VII-C-24 in section VII-C-13.

VII-C-14c. Tropospheric Photolysis

Photolysis rates representative of the lower atmosphere were calculated with the TUV model (as described in the introduction to this section) using the measured cross section and assuming the quantum yield for photodecomposition to be unity at all wavelengths of absorbed light (figure VII-C-25).

Table VII-C-15. Cross sections for *E,E*-2-methyl-2,4-hexadienedial vapor; the high-resolution data of Klotz et al. (1995) have been averaged over each nanometer (centered on the wavelength shown)

λ (nm)	$\sigma \times 10^{19}$	λ (nm)	$\sigma \times 10^{19}$	λ (nm)	$\sigma \times 10^{19}$	λ (nm)	$\sigma \times 10^{19}$
215	2.31	250	292	285	44.1	320	1.44
216	1.94	251	309	286	40.4	321	1.30
217	2.69	252	316	287	36.9	322	1.17
218	3.76	253	313	288	33.1	323	1.05
219	4.56	254	307	289	29.3	324	0.936
220	5.34	255	304	290	25.4	325	0.839
221	7.11	256	310	291	21.6	326	0.756
222	8.96	257	329	292	18.3	327	0.687
223	11.2	258	363	293	15.5	328	0.631
224	13.8	259	406	294	13.3	329	0.591
225	17.1	260	449	295	11.5	330	0.543
226	19.9	261	476	296	10.0	331	0.492
227	23.1	262	475	297	8.81	332	0.446
228	27.1	263	452	298	7.82	333	0.397
229	32.2	264	416	299	7.04	334	0.357
230	38.6	265	374	300	6.45	335	0.321
231	46.2	266	342	301	6.01	336	0.292
232	54.3	267	323	302	5.67	337	0.272
233	62.3	268	318	303	5.40	338	0.251
234	68.8	269	329	304	5.16	339	0.243
235	74.2	270	356	305	4.95	340	0.225
236	80.5	271	391	306	4.70	341	0.205
237	88.8	272	424	307	4.41	342	0.187
238	99.9	273	437	308	4.16	343	0.171
239	114	274	402	309	3.86	344	0.164
240	130	275	341	310	3.62	345	0.154
241	144	276	274	311	3.38	346	0.139
242	155	277	206	312	3.13	347	0.125
243	162	278	151	313	2.88	348	0.0977
244	168	279	113	314	2.63	349	0.0810
245	175	280	88.5	315	2.39	350	0.0673
246	186	281	72.2	316	2.16	351	0.0440
247	205	282	61.6	317	1.96		
248	232	283	53.9	318	1.77		
249	263	284	48.5	319	1.60		

Note: Units on σ are cm^2 $molecule^{-1}$.

VII-C-15. Benzene Oxide/Oxepin

In the atmospheric oxidation of benzene, postulated intermediates are benzene oxide (1) and oxepin (2):

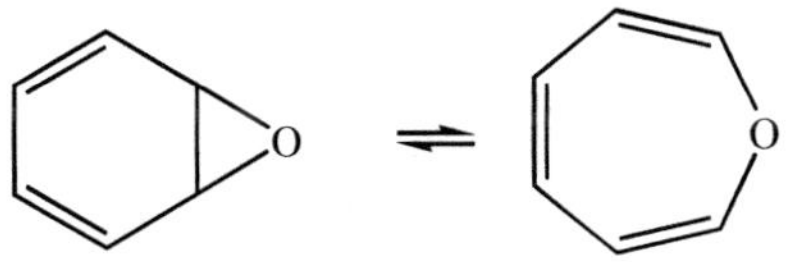

(1) (2)

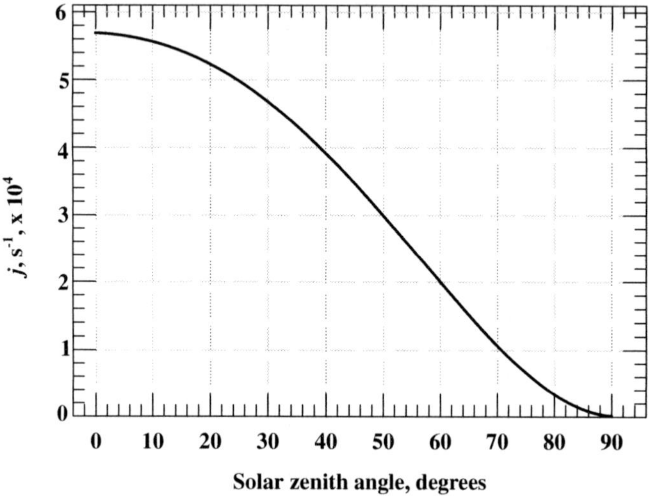

Figure VII-C-25. *E,E*-2-Methyl-2,4-hexadienedial photodissociation coefficients at 0.5 km altitude as a function of solar zenith angle. The total quantum yield of decomposition was assumed to be 1 for all wavelengths of absorbed light.

The equilibrium is expected to be only slightly shifted toward the oxepin form (Klotz et al., 1997). These compounds absorb ultraviolet radiation that is present within the troposphere (see figure VII-C-26), and their photochemistry may be important. The primary process (I) has been suggested to explain the phenol product observed (Klotz et al., 1998a). Process (II) is believed to occur from the oxepin form (Kaubisch et al., 1972):

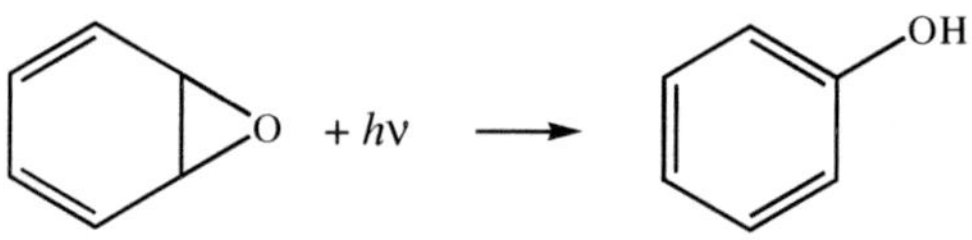

(I)

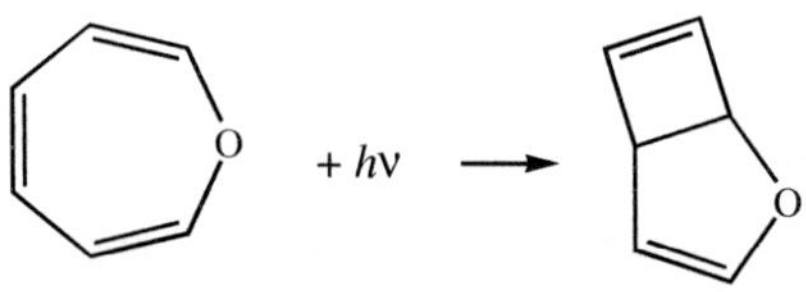

(II)

VII-C-15a. Cross Sections

The absorption cross sections for the benzene oxide/oxepin are plotted in figure VII-C-26 and given in tabular form in table VII-C-16, where the high-resolution data of Klotz et al. (1998a) have been averaged over each nanometer (centered on the wavelengths shown).

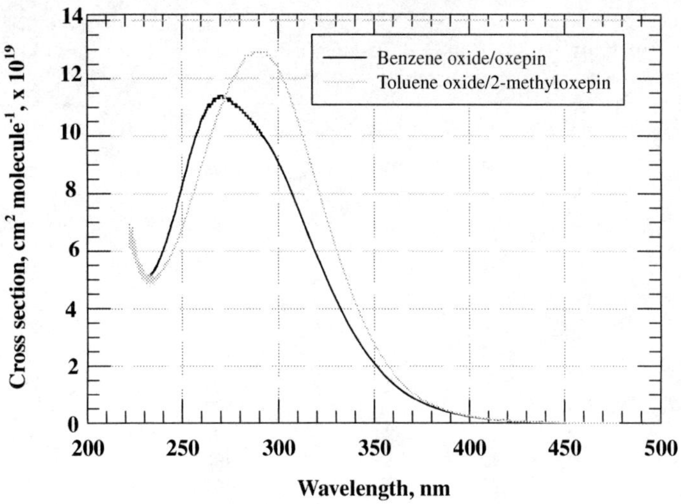

Figure VII-C-26. Cross sections versus wavelength for benzene oxide/oxepin and toluene oxide/2-methyloxepin; data from Klotz et al. (1998a).

VII-C-15b. Quantum Yields

No quantum yield data have been reported for processes (I) or (II), but the apparent *j*-values for benzene oxide/oxepin disappearance have been determined in the European outdoor smog chamber (EUPHORE) in Valencia, Spain (Klotz et al., 1998a). The ratio $j(C_6H_6O)/j(NO_2) = (4.41 \pm 0.44) \times 10^{-2}$, independent of solar zenith angle (or radiation intensity). Thus, for the usual sunlight intensities where $j(NO_2) \sim 8 \times 10^{-3}\ s^{-1}$, the value for disappearance of benzene/oxepin is $j(C_6H_6O) \sim 3.5 \times 10^{-4}\ s^{-1}$. It was further determined that for each benzene oxide/oxepin reacted, 0.432 ± 0.045 moles of phenol formed. So the experimental *j*-value for formation of phenol from the photoisomerization of benene oxide/oxepin is $\sim 1.5 \times 10^{-4}\ s^{-1}$. The extent of process (II) is unknown.

VII-C-15c. Tropospheric Photolysis

Photolysis rate coefficients for benzeneoxide/oxepin photodecomposition that are representative of the lower atmosphere were calculated using the TUV model (as described in the introduction to this section) with the measured cross sections and an assumed effective quantum yield for photodecomposition (disappearance) of 0.12 (figure VII-C-27).

VII-C-16. Toluene oxide/2-Methyloxepin

In the atmospheric oxidation of toluene, similar to the case with benzene, two postulated intermediates (presumably in equilibrium) are toluene oxide (1) and 2-methyloxepin (2). In this case, the equilibrium is believed to be significantly shifted toward the 2-methyloxepin form.

Table VII-C-16. Absorption cross sections for gas-phase benzene oxide/oxepin. The high-resolution data of Klotz et al. (1998a) have been averaged over 1-nm bins, with the center of the bins at the wavelengths shown

λ (nm)	$\sigma \times 10^{19}$	λ (nm)	$\sigma \times 10^{19}$	λ (nm)	$\sigma \times 10^{19}$	λ (nm)	$\sigma \times 10^{19}$	λ (nm)	$\sigma \times 10^{19}$
290	10.18	325	5.03	360	1.37	395	0.29	430	0.05
291	10.07	326	4.87	361	1.32	396	0.28	431	0.05
292	9.98	327	4.72	362	1.26	397	0.26	432	0.04
293	9.89	328	4.57	363	1.21	398	0.24	433	0.04
294	9.79	329	4.43	364	1.16	399	0.24	434	0.04
295	9.69	330	4.30	365	1.11	400	0.23	435	0.04
296	9.58	331	4.16	366	1.06	401	0.22	436	0.04
297	9.47	332	4.03	367	1.02	402	0.21	437	0.03
298	9.34	333	3.89	368	0.97	403	0.20	438	0.03
299	9.20	334	3.75	369	0.93	404	0.19	439	0.03
300	9.07	335	3.62	370	0.89	405	0.18	440	0.03
301	8.93	336	3.49	371	0.85	406	0.17	441	0.03
302	8.79	337	3.37	372	0.82	407	0.17	442	0.03
303	8.65	338	3.25	373	0.78	408	0.16	443	0.03
304	8.50	339	3.14	374	0.75	409	0.15	444	0.02
305	8.34	340	3.03	375	0.71	410	0.15	445	0.02
306	8.18	341	2.92	376	0.68	411	0.14	446	0.02
307	8.01	342	2.82	377	0.65	412	0.13	447	0.02
308	7.84	343	2.71	378	0.62	413	0.12	448	0.02
309	7.68	344	2.61	379	0.59	414	0.12	449	0.02
310	7.51	345	2.50	380	0.57	415	0.11	450	0.02
311	7.34	346	2.41	381	0.55	416	0.10	451	0.01
312	7.17	347	2.40	382	0.52	417	0.10	452	0.01
313	6.99	348	2.24	383	0.50	418	0.10	453	0.01
314	6.81	349	2.16	384	0.48	419	0.09	454	0.01
315	6.64	350	2.08	385	0.46	420	0.08	455	0.01
316	6.49	351	2.00	386	0.43	421	0.07	456	0.01
317	6.32	352	1.92	387	0.41	422	0.07	457	0.01
318	6.15	353	1.84	388	0.39	423	0.07	458	0.01
319	5.98	354	1.76	389	0.38	424	0.07	459	0.01
320	5.82	355	1.68	390	0.36	425	0.07	460	0.01
321	5.67	356	1.62	391	0.35	426	0.06		
322	5.51	357	1.55	392	0.33	427	0.06		
323	5.35	358	1.49	393	0.31	428	0.06		
324	5.18	359	1.43	394	0.30	429	0.05		

Note: Units on σ are cm^2 $molecule^{-1}$.

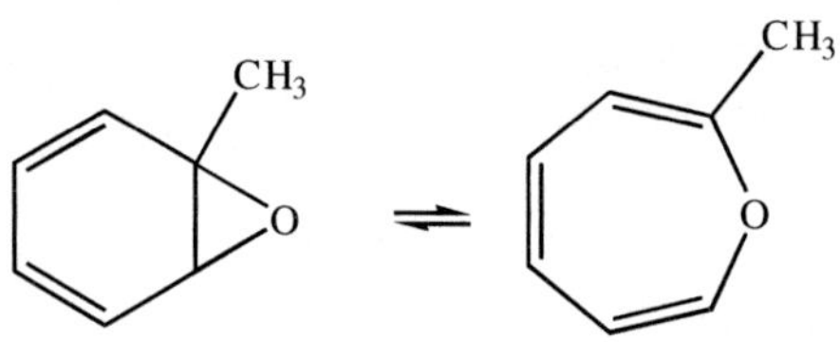

(1) (2)

These compounds absorb ultraviolet radiation that is present within the troposphere (see figure VII-C-26), and their photochemistry may be important. The primary process (I) has been suggested to explain the *o*-cresol product observed

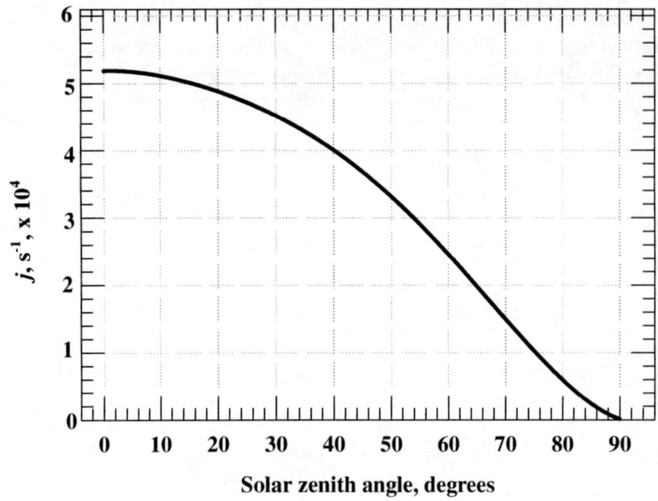

Figure VII-C-27. Estimated rate coefficients for the photodecomposition (disappearance) of benzene oxide/oxepin at 0.5 km altitude as a function of solar zenith angle using measured cross sections and an assumed quantum yield for photodecomposition of 0.12 to provide a match with the measured value of $j(C_6H_6O)/j(NO_2)$.

(Klotz et al., 1998a, 1998b). Process (II) has been suggested to occur from the 2-methyloxepin form (Vogel and Günther, 1967; Kaubisch et al., 1972):

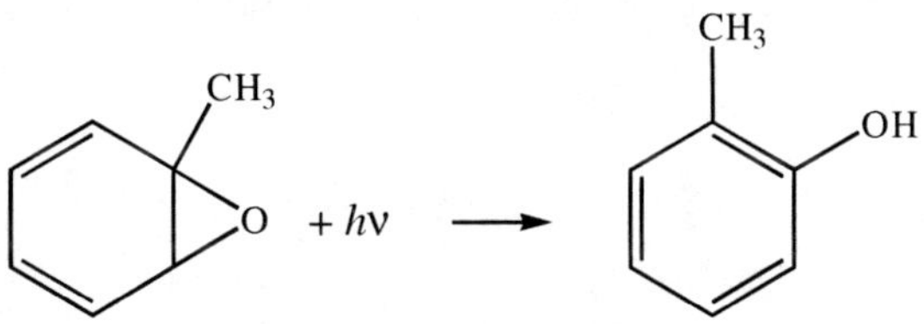

(I)

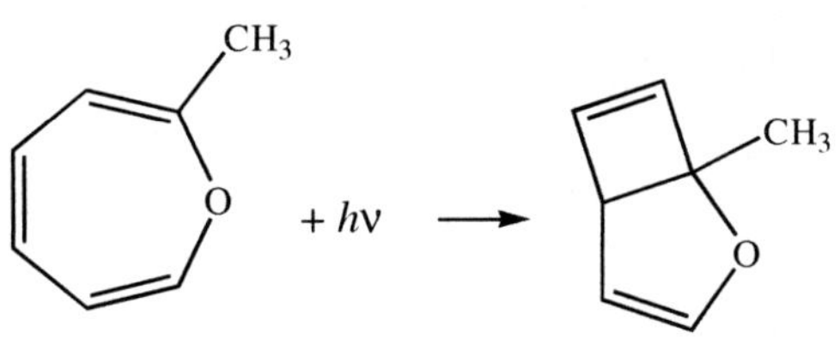

(II)

VII-C-16a. Cross Sections

The absorption cross sections for the toluene oxide/2-methyloxepin are plotted in Figure VII-C-26 and given in tabular form in table VII-C-17, where the high-resolution data of Klotz et al. (1998a) have been averaged in bins of 1 nm (centered on the wavelengths shown).

Table VII-C-17. Absorption cross sections for gas-phase toluene oxide/2-methyloxepin. The high-resolution data of Klotz et al. (1998a) have been averaged over 1-nm bins, with the center of the bins at the wavelengths shown

λ (nm)	$\sigma \times 10^{19}$	λ (nm)	$\sigma \times 10^{19}$	λ (nm)	$\sigma \times 10^{19}$	λ (nm)	$\sigma \times 10^{19}$	λ (nm)	$\sigma \times 10^{19}$
290	12.90	325	7.15	360	1.83	395	0.30	430	0.05
291	12.90	326	6.93	361	1.67	396	0.29	431	0.05
292	12.90	327	6.71	362	1.59	397	0.28	432	0.05
293	12.90	328	6.50	363	1.52	398	0.27	433	0.04
294	12.85	329	6.29	364	1.45	399	0.26	434	0.04
295	12.80	330	6.08	365	1.38	400	0.24	435	0.04
296	12.71	331	5.87	366	1.31	401	0.23	436	0.04
297	12.63	332	5.76	367	1.24	402	0.22	437	0.04
298	12.54	333	5.47	368	1.18	403	0.21	438	0.03
299	12.44	334	5.27	369	1.12	404	0.20	439	0.03
300	12.32	335	5.09	370	1.06	405	0.19	440	0.03
301	12.18	336	4.92	371	1.01	406	0.18	441	0.02
302	12.05	337	4.73	372	0.96	407	0.17	442	0.03
303	11.89	338	4.55	373	0.91	408	0.16	443	0.02
304	11.72	339	4.37	374	0.87	409	0.15	444	0.02
305	11.55	340	4.20	375	0.83	410	0.15	445	0.02
306	11.37	341	4.04	376	0.78	411	0.14	446	0.02
307	11.18	342	3.88	377	0.75	412	0.13	447	0.02
308	10.98	343	3.72	378	0.71	413	0.13	448	0.02
309	10.78	344	3.57	379	0.68	414	0.12	449	0.02
310	10.57	345	3.43	380	0.65	415	0.11	450	0.02
311	10.36	346	3.27	381	0.62	416	0.11	451	0.02
312	10.13	347	3.29	382	0.59	417	0.10	452	0.01
313	9.90	348	3.02	383	0.56	418	0.10	453	0.01
314	9.68	349	3.29	384	0.53	419	0.09	454	0.01
315	9.46	350	2.78	385	0.51	420	0.08	455	0.01
316	9.22	351	2.64	386	0.48	421	0.08	456	0.01
317	8.99	352	2.53	387	0.46	422	0.08	457	0.01
318	8.76	353	2.42	388	0.43	423	0.07	458	0.01
319	8.53	354	2.31	389	0.42	424	0.07	459	0.01
320	8.29	355	2.21	390	0.39	425	0.07	460	0.01
321	8.07	356	2.11	391	0.38	426	0.06		
322	7.83	357	2.01	392	0.36	427	0.06		
323	7.60	358	1.92	393	0.34	428	0.06		
324	7.38	359	1.83	394	0.32	429	0.05		

Note: Units on σ are cm^2 $molecule^{-1}$.

VII-C-16b. Quantum yields

No quantum yield data have been reported for processes (I) or (II), but the apparent *j*-values for toluene oxide/2-methyloxepin disappearance have been determined in the European outdoor smog chamber (EUPHORE) in Valencia, Spain (Klotz et al., 1998a). The ratio $j(CH_3C_6H_5O)/j(NO_2)$ was measured to be $(3.99 \pm 0.48) \times 10^{-2}$. Hence, for conditions with $j(NO_2) = 8.0 \times 10^{-3}$, $j(CH_3C_6H_5O) \sim 3.2 \times 10^{-4}\ s^{-1}$. Because of the similarity in the absorption spectra of the benzene and toluene oxides and oxepins, this value is probably nearly independent of solar zenith angle (or radiation intensity), although this hypothesis could not be tested with the available data. In this case, however, only 0.027 ± 0.022 moles of the expected product, *o*-cresol, were formed for each toluene oxide/2-methyloxepin reacted. No other products were detected. The small extent of this product has been rationalized by Klotz et al. (1998a); it is known that methylation of the carbon atoms adjacent to the ring oxygen in arene oxides results in a shift of the equilibrium toward the oxepin form (Vogel and Günther, 1967). While the arene oxide form is expected to photolyze to give phenol (benzene oxide) or cresols (toluene oxide), the oxepin isomer is assumed to be responsible for the formation of an observed cyclization product, 2-oxabicyclo(3.2.0]hepta-3,6-diene, or, in the case of 2-methyloxepin, the methylated derivative (Vogel and Günther, 1967; Kaubisch et al., 1972). Presumably, in this case the photolysis observed in the outdoor smog chamber may result largely in the formation of this latter product through process (II), with process (I) being much less important.

VII-C-16c. Tropospheric Photolysis

Photolysis rate coefficients for the photolysis of toluene oxide/2-methyloxepin representative of the lower atmosphere were calculated with the TUV model (as described in the introduction to this section) using the measured cross sections and an assumed effective quantum yield of 0.082 for formation of the unknown products. See figure VII-C-28.

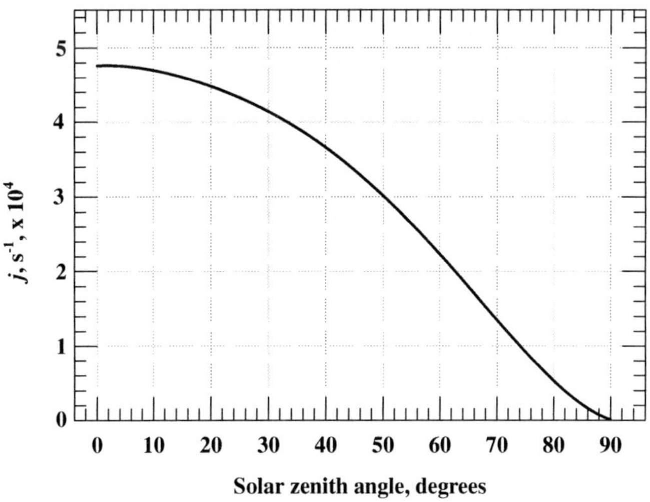

Figure VII-C-28. Toluene oxide/2-methyloxepin photodissociation coefficients at 0.5 km altitude as a function of solar zenith angle estimated using measured cross sections and assuming a quantum yield of 0.082 for photodecomposition into unknown products to provide a match with the $j(CH_3C_6H_5O)/j(NO_2)$ ratio observed experimentally.

VII-D. Photochemistry of Some Common Aromatic Ring-Retaining Products of Atmospheric Oxidation of Aromatic Hydrocarbons

In the tropospheric oxidation of many of the aromatic hydrocarbons, important products are formed in which the aromatic ring system is retained, and new functional groups are introduced that extend the absorption of the compounds into the region of the spectrum available within the troposphere. Hence, the photochemistry of these species can be an important consideration among the many routes of reaction within the atmosphere. For example, the oxidation of alkyl side chains in toluene leads to benzaldehyde or acetophenone in the case of ethylbenzene. The introduction of a carbonyl group attached to the aromatic ring leads to a significant shift in the absorption toward longer wavelengths. See figure VII-D-1. In toluene the absorption corresponds to the excitation of an electron in a π-orbital to an antibonding π^* orbital, while the new longer wavelength absorption band seen in its oxidation product, benzaldehyde, relates to an $n \rightarrow \pi^*$ transition in which a non-bonding electron on the carbonyl oxygen is promoted to an antibonding π-orbital in the aromatic system, a much lower energy transition. A large change in the spectrum of toluene occurs as a nitro-group is attached to the ring to form *o*-nitrotoluene. A similar but much smaller shift in the absorption spectra occurs as toluene is oxidized to *o*-cresol, by substitution of an OH-group for one of the H-atoms of the ring. In this case, the non-bonding electron on the O-atom requires significantly more energy to promote it to the antibonding π-system. The wavelength shift in the spectra for

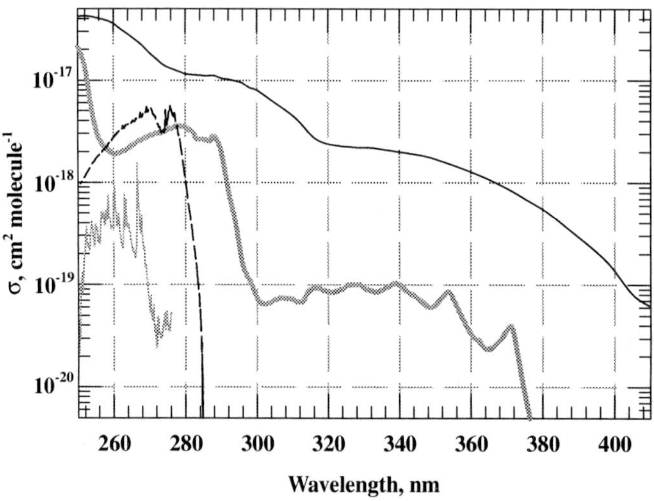

Figure VII-D-1. Comparison of the gas-phase absorption spectra of toluene (light gray curve) (Etzkorn et al., 1999), benzaldehyde solution in hexane (heavy gray curve) (Láng, 1961–1971), *o*-cresol gas (dashed black curve) (Etzkorn et al., 1999), and *o*-nitrotoluene solution in hexane (solid black curve) (Láng, 1961–1971).

this case is not sufficient to result in significant overlap with the wavelengths of sunlight present in troposphere. However, when the aromatic compounds that are transparent in the region of the sunlight available in the troposphere oxidize in the atmosphere to add $-CH{=}O$ and $-NO_2$ groups, they do become candidates for photochemical change within the troposphere. It is in this light that we discuss in this section the nature of the photochemistry of some of these products.

VII-D-1. The Photochemistry of Benzaldehyde

The gas-phase photochemistry of benzaldehyde has been studied by Berger et al. (1973), who irradiated benzaldehyde at low pressures with light at several selected wavelengths: 276, 284, 328, and 365 nm. The following photochemical reaction scheme was considered in explanation of the results:

$$C_6H_5CHO + h\nu \rightarrow C_6H_5CHO(S_1)$$

$$C_6H_5CHO(S_1) \rightarrow C_6H_5CHO(T_1)$$

$$C_6H_5CHO(S_1) \rightarrow C_6H_5CHO(T_2)$$

$$C_6H_5CHO(T_1) \rightarrow C_6H_5CHO$$

$$C_6H_5CHO(T_1 \text{ or } T_2) \rightarrow C_6H_6 + CO \qquad \text{(I)}$$

$$C_6H_5CHO(T_1 \text{ or } T_2) + M \rightarrow C_6H_5CHO + M \qquad \text{(II)}$$

$$C_6H_5CHO(T_1) \rightarrow C_6H_5CO + H \qquad \text{(III)}$$

$$C_6H_5CHO(T_1) \rightarrow C_6H_5 + CHO \qquad \text{(IV)}$$

$$C_6H_5CHO(T_1) + O_2 \rightarrow C_6H_5CHO + O_2(^1\Delta_g, \text{ other?}) \qquad \text{(V)}$$

The S_1 and T_1 refer to first excited singlet and triplet states, respectively; T_2 represents a ill-defined triplet-like upper state of short lifetime that is important in the experiments within the $\pi \rightarrow \pi^*$ region (276, 284 nm). Process (I), forming benzene and CO, is important in photolyses at 276 and 284 nm; in experiments at low pressure, ϕ_I is near unity. However, even at these short wavelengths, quenching in processes (II) or (V) is expected to be important for photolyses in air at atmospheric pressure. The rate coefficient k_V was estimated to be 4.8×10^{-12} $cm^3 \, molecule^{-1} \, s^{-1}$. The quantum yields of benzene and CO formed in process (I) following absorption within the $n \rightarrow \pi^*$ region (328, 365nm) were very small, even at low pressures, although quantum yields for benzaldehyde disappearance remained at about 0.4 for these conditions (as observed at 276 and 284 nm), and the yield was unaffected by the presence of triplet-state quenchers. It appears that process (I) can be neglected for photolyses within the troposphere. The relatively large quantum yield of disappearance of benzaldehyde remains unexplained. On prolonged and repeated photolysis, significant amounts of a pale yellow polymeric material built up on the walls of the photolysis vessel, an observation also made by previous workers (Almasy, 1933; DeMare et al., 1968). In experiments at 313 nm, Vanselow and Blacet (1957) reported finding the following products: H_2, benzil, benzophenone, and diphenyl, in addition to benzene and carbon monoxide. The most likely source of these compounds is reaction of radicals formed by bond rupture of the benzaldehyde in reaction (III), where $\Delta H^{\circ}_{298} \sim 364\,\text{kJ mol}^{-1}$, corresponding to a wavelength of absorbed light of 329 nm. Reaction (IV) appears to be less likely than reaction (III) for photolyses within the troposphere; $\Delta H^{\circ}_{298} \sim 409\,\text{kJ mol}^{-1}$, corresponding to a wavelength of absorbed light of 293 nm. Berger et al. (1973) found that the benzaldehyde-consuming process was not affected by the presence of a triplet scavenger, so they concluded that either this is initiated by free radicals originating from S_1 or possibly from a surface-initiated reaction. It has been shown that the efficiency of the $S_1 \rightarrow T_1$ conversion is probably very near 100% (Itoh (1988), and it is difficult to involve a significant fraction of the S_1 state formed in photolyses at the longer wavelengths in other reactions if this is the case. The phosphorescence lifetime for the gas-phase, vibrationally relaxed T_1 state at zero pressure was measured by Biron and Longin (1985) to be 7.1×10^{-4} s. If one couples this with the quenching rate coefficient for reaction with O_2, then 80% of excited triplets formed in 1 atm of air are quenched by O_2 in 1 μs. Obviously, if process (III) occurs rapidly from some vibrationally rich triplet state within a period of 1 μs, then these processes can compete with quenching. However, it remains uncertain as to the extent to which they occur in tropospheric photolyses.

Chamber irradiation of benzaldehyde–NO_x mixtures in synthetic air [$j(NO_2) \sim 6.7 \times 10^{-3}\,\text{s}^{-1}$] required 132 min to oxidize one-half of the NO to NO_2, a rather slow rate (Kuntz et al., 1973). The formation of benzoylperoxy nitrate, a highly eye-irritating compound, was observed as reported earlier by Heuss and Glasson (1968). When 2.5 ppmv of benzaldehyde was added to a synthetic air mixture containing 10 ppmv-C of reactive alkenes and 1 ppm NO_x, the rate of NO and alkene oxidation was suppressed. These data suggest that the rate of generation of free radicals from sunlight-irradiated benzaldehyde in air is low, and that it is a very inefficient chain carrier with peroxybenzoyl nitrate acting as a temporary sink for NO_x.

In a smog chamber study [$j(NO_2) = 6.6 \times 10^{-3}\,\text{s}^{-1}$] of the photooxidation of toluene and the cresols, Atkinson et al. (1980) found that an assumed j-value of $3 \times 10^{-5}\,\text{s}^{-1}$ for benzaldehyde photodecomposition was required to account for the observed benzaldehyde loss rates. The product data were best fit by assuming that benzaldehyde photolysis results in the formation of non-radical products, such as process (I), although the benzene yields observed were less than 1% of those expected if process (I) were the major photodecomposition mode.

VII-D-1a. Cross Sections

The gas-phase ultraviolet absorption spectrum of benzaldehyde (figure VII-D-2) shows some fine structure in the 265–290 nm wavelength range (Berger et al., 1973); the digitized data of Etzkorn et al. (1999) are very similar to those of Berger et al. Gas-phase absorption spectra for the region of tropospheric interest (290–360 nm) are difficult to determine at 25°C with the low vapor pressure of benzaldehyde, and no such spectral data are available at this writing. The solution-phase absorption data for this band, derived from the data of Láng (1961–1971), are given in figure VII-D-3. These data appear to be very similar

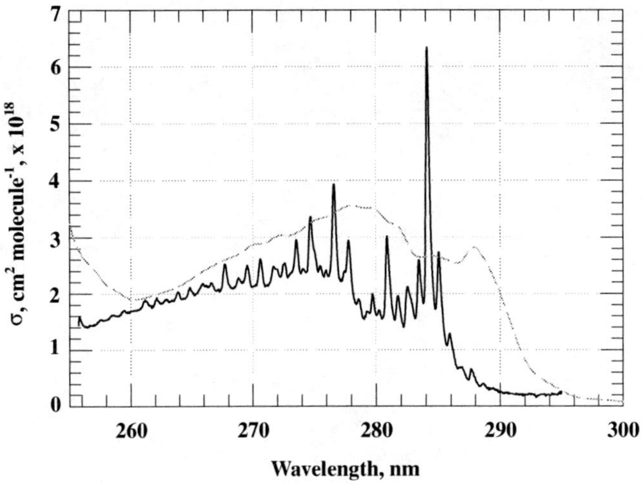

Figure VII-D-2. Plot of the cross sections for benzaldehyde (gas phase; black curve) versus wavelength for the 250–300 nm region; from measurements of Etzkorn et al. (1999) compared with the solution-phase spectrum (in hexane; gray dashed curve) given by Láng (1961–1971).

to the solution-phase spectrum given in figure 2 of Berger et al. (1973). The values for the cross sections for each wavelength (nm) are summarized in table VII-D-1.

VII-D-2b. Quantum Yields

The quantum yield data that are available at this writing were obtained in experiments at relatively low pressures that are not directly applicable to tropospheric

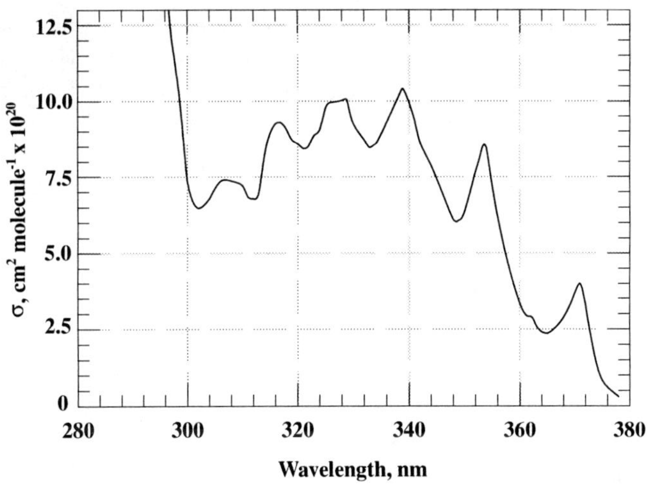

Figure VII-D-3. Absorption spectrum for benzaldehyde in the 290–360 nm region as determined in hexane solution by Láng (1961–1971).

Table VII-D-1. Benzaldehyde absorption cross sections in the 290–380 nm range as measured in hexane solution by Láng (1961–1971)

λ (nm)	$\sigma \times 10^{19}$	λ (nm)	$\sigma \times 10^{19}$	λ (nm)	$\sigma \times 10^{19}$	λ (nm)	$\sigma \times 10^{19}$
280	34.91	305	0.71	330	0.93	355	0.73
281	32.67	306	0.73	331	0.90	356	0.62
282	31.31	307	0.74	332	0.87	357	0.55
283	27.05	308	0.74	333	0.85	358	0.48
284	26.56	309	0.73	334	0.86	359	0.41
285	26.43	310	0.72	335	0.89	360	0.33
286	25.69	311	0.68	336	0.93	361	0.31
287	25.99	312	0.68	337	0.97	362	0.29
288	28.10	313	0.70	338	1.01	363	0.26
289	24.00	314	0.81	339	1.04	364	0.24
290	19.98	315	0.89	340	1.00	365	0.24
291	14.10	316	0.92	341	0.94	366	0.25
292	8.25	317	0.93	342	0.86	367	0.27
293	6.12	318	0.91	243	0.83	368	0.29
294	3.99	319	0.87	344	0.79	369	0.33
295	2.78	320	0.86	345	0.74	370	0.36
296	1.57	321	0.84	346	0.70	371	0.40
297	1.34	322	0.85	437	0.65	372	0.33
298	1.10	323	0.89	348	0.61	373	0.23
299	0.92	324	0.91	349	0.61	374	0.14
300	0.74	325	0.97	350	0.63	375	0.099
301	0.69	326	1.00	351	0.70	376	0.061
302	0.65	327	1.00	352	0.76	377	0.046
303	0.66	328	1.00	353	0.83	378	0.030
304	0.68	329	1.00	354	0.84		

Note: Units on σ are cm^2 molecule^{-1}.

conditions. The quantum yields of disappearance of benzaldehyde as determined by Berger et al. (1973) are independent of added triplet quenchers and wavelength over the limited pressure and wavelength ranges employed. If one assumes that this trend continues to tropospheric pressures and that process (III) is the dominant mode of decay here, then the assumption of $\phi_{III} = 0.39 \pm 0.03$ is appropriate. This procedure will likely lead to maximum estimates of the photodecomposition of benzaldehyde. However, at this stage of our knowledge, the alternative assumption that ϕ for decomposition by all processes is near zero can be equally well justified.

VII-D-1c. Tropospheric Photolysis

Photolysis rate coefficients representative of the lower atmosphere were calculated with the TUV model (as described in the introduction to this section) using the measured solution-phase cross sections and using alternative values of assumed quantum yield of decomposition of 0.39 and 0.039 at all wavelengths of absorbed light. The choice of the smaller values is more consistent with the smog chamber results of Atkinson et al. (1980), but the spectral distribution of the solar simu-

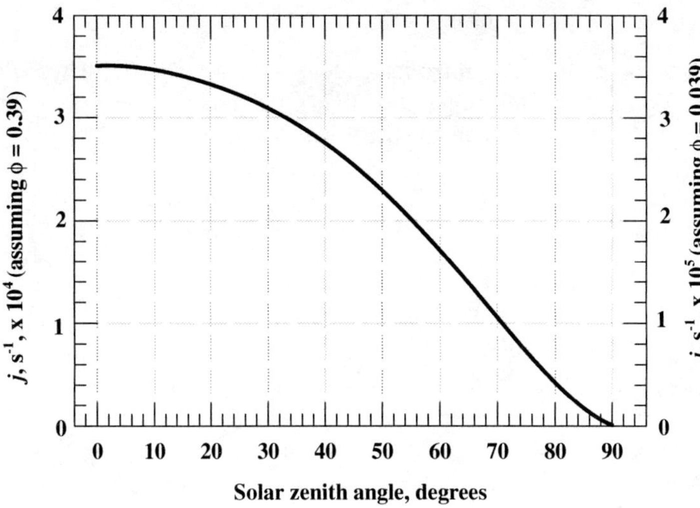

Figure VII-D-4. Estimated benzaldehyde photodissociation coefficients at 0.5 km altitude as a function of solar zenith angle calculated from the measured solution-phase cross sections and an assumed ϕ for photodecomposition of either 0.39 (left-axis scale) or 0.039 (right-axis scale).

lator used in the chamber may not reflect well the short-wavelength region of the solar spectrum. See figure VII-D-4.

VII-D-2. The Photochemistry of 2-Nitrobenzaldehyde

Nitrobenzaldehydes have not been reported as products of the atmospheric oxidation of toluene, but these compounds could conceivably be formed through the oxidation of toluene in an NO_x-rich atmosphere. *o*-Nitrobenzaldehyde has been used successfully as an actinometer in photochemical studies (Pitts et al., 1964), and for this reason a brief discussion of its photochemistry is given here. The photochemistry of 2-nitrobenzaldehyde has been studied in acetone and ligroin solutions (Leighton and Lucy, 1934a, 1934b), in the pure solid (Bowen et al., 1924), and in KBr matrices (Pitts et al., 1964). The only identified primary process involves the isomerization to *o*-nitrosobenzoic acid, process (I):

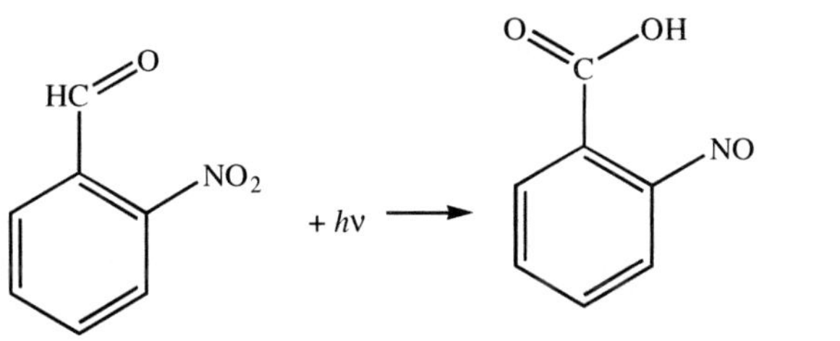

(I)

This interesting rearrangement occurs with high efficiency ($\phi_I = 0.5$) at all of the wavelengths that have been studied and in both the liquid and solid phases.

VII-D-2a. Absorption Spectrum

The absorption spectrum of *o*-nitrobenzaldehyde (2-nitrobenzaldehyde) as measured in a non-polar solvent [Láng (1961–1971] is given in figure VII-D-5, where it can be compared with that of unsubstituted benzaldehyde and 2-hydroxybenzaldehyde. 2-Nitrobenzaldehyde is a much stronger absorber and its absorption region is shifted into the visible spectral region. The spectral data obtained from its solutions in non-polar solvents are given in table VII-D-2.

VII-D-2b. Quantum Yields

Quantum yields appear to be relatively insensitive to the wavelength of absorbed light as well as the phase in which the compound occurs. In solution: $\phi_I = 0.46$ at 404 nm; 0.51 at 366 nm; in pure solid: $\phi_I = 0.51$ at 313 nm; in KBr matrix: $\phi_I = 0.5$. Because of its highly polar nature and relatively low vapor pressure (0.02 Torr at 298 K) this compound may end up in aerosols, and it is reasonable to assume that $\phi_I = 0.5$ for all $\lambda < 420$ nm available in sunlight within the troposphere.

VII-D-2c. Tropospheric Photochemistry

Photolysis rate coefficients representative of the lower atmosphere were calculated with the TUV model (as described in the introduction to this section) using the measured solution-phase cross sections and a quantum yield of photoisomerization of 0.5 at all wavelengths of absorbed light. See figure VII-D-6.

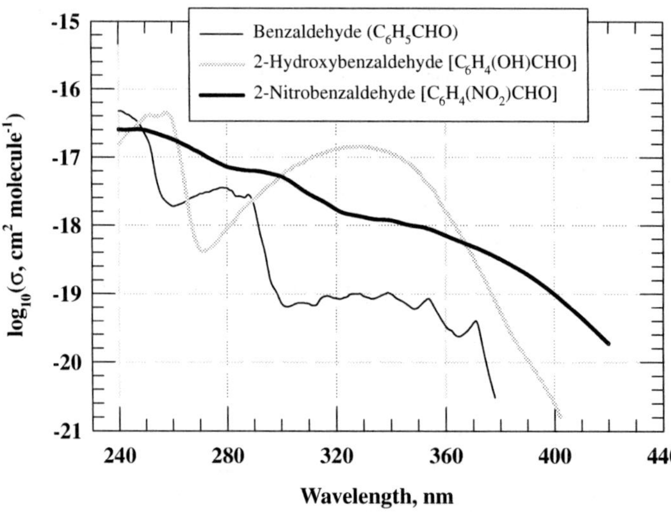

Figure VII-D-5. Plot of the absorption cross section (σ) for a series of substituted benzaldehydes; from measurements in non-polar solvents (Láng, 1961–1971).

Table VII-D-2. Cross sections for 2-nitrobenzaldehyde at wavelengths of 290–420 nm; from measurements in non-polar solvents

λ (nm)	$\sigma \times 10^{19}$	λ (nm)	$\sigma \times 10^{19}$	λ (nm)	$\sigma \times 10^{19}$	λ (nm)	$\sigma \times 10^{19}$
290	63.9	323	15.3	356	8.21	389	2.05
291	63.3	324	15.0	357	7.94	390	1.92
292	62.7	325	14.7	358	7.66	391	1.83
293	61.7	326	14.4	359	7.40	392	1.73
294	60.7	327	14.1	360	7.14	393	1.64
295	59.5	328	13.8	361	6.88	394	1.55
296	58.3	329	13.6	362	6.61	395	1.46
297	57.0	330	13.3	363	6.35	396	1.36
298	55.6	331	13.1	364	6.09	397	1.27
299	54.2	332	12.8	365	5.90	398	1.18
300	52.8	333	12.6	366	5.70	399	1.08
301	50.3	334	12.3	367	5.51	400	0.99
302	47.7	335	12.2	368	5.32	401	0.94
303	45.1	336	12.1	369	5.12	402	0.88
304	42.6	337	12.1	370	4.93	403	0.83
305	40.4	338	12.0	371	4.73	404	0.77
306	38.1	339	11.9	372	4.54	405	0.72
307	35.9	340	11.8	373	4.37	406	0.66
308	33.6	341	11.5	374	4.21	407	0.61
309	31.4	342	11.3	375	4.04	408	0.55
310	29.1	343	11.0	376	3.87	409	0.50
311	27.9	344	10.7	377	3.70	410	0.44
312	26.6	345	10.5	378	3.54	411	0.42
313	25.4	346	10.2	379	3.37	412	0.39
314	24.2	347	10.1	380	3.20	413	0.37
315	22.9	348	9.90	381	3.07	414	0.34
316	21.7	349	9.76	382	2.94	415	0.32
317	20.7	350	9.61	383	2.82	416	0.29
318	19.7	351	9.46	384	2.69	417	0.27
319	18.7	352	9.31	385	2.56	418	0.24
320	17.6	353	9.04	386	2.43	419	0.22
321	16.6	354	8.76	387	2.30	420	0.19
322	15.6	355	8.48	388	2.18		

Source: Láng (1961–1971).
Note: Units on σ are cm^2 $molecule^{-1}$.

VII-D-3. The Photochemistry of Nitrobenzene

The primary processes in the simplest of the nitro-substituted hydrocarbons, nitrobenzene, have received some attention, but the experimental conditions employed in most of the laboratory studies do not match those encountered in the troposphere; for example, irradiations used $\lambda < 290$ nm (Hastings and Matsen, 1948; Hurley and Testa, 1966; Yip et al., 1984; Galloway et al., 1993, 1994; Hwang et al., 1996; Takezaki et al., 1997a, 1997b). There are four primary processes that have been proposed to explain the results of these studies:

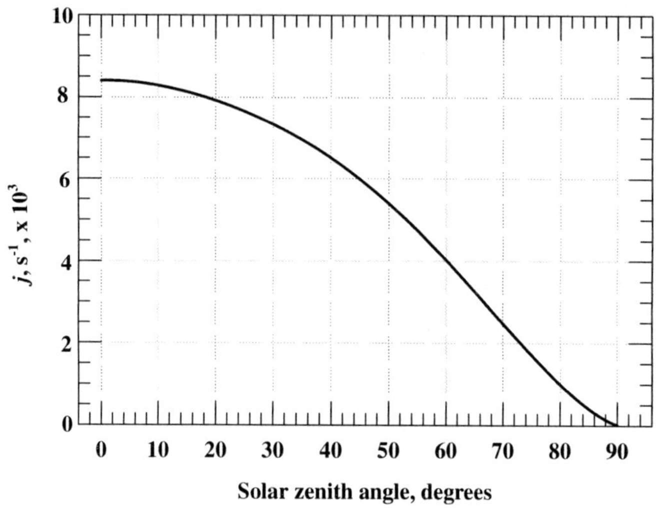

Figure VII-D-6. *o*-Nitrobenzaldehyde photodissociation coefficients at 0.5 km altitude as a function of solar zenith angle calculated using the measured solution-phase cross sections, and a quantum yield of photoisomerization of 0.5.

$$C_6H_5NO_2 + h\nu \rightarrow C_6H_5NO + O(^3P) \quad \text{(I)}$$
$$\rightarrow C_6H_5 + NO_2 \quad \text{(II)}$$
$$\rightarrow [C_6H_5ONO] \rightarrow C_6H_5O + NO \quad \text{(III)}$$
$$\rightarrow C_5H_5 + CO + NO \quad \text{(IV)}$$

In the full-arc irradiation of nitrobenzene vapors (in N_2), Hastings and Matsen (1948) identified nitrosobenzene and *p*-nitrophenol but found no evidence for NO_2 formation. The occurrence of primary process (I) alone was suggested to explain the results. The reactions of electronically excited nitrobenzene are believed to occur from the first excited n, π^*-triplet (243 kJ mol^{-1} above the ground state) (Yip et al., 1984). Vacuum UV photoionization–molecular beam mass spectrometry was used by Galloway et al. (1993) to identify the different fragments formed in the primary process in photolyses at wavelengths in the range 220–320 nm. The occurrence of process (I) was confirmed by photoionization–mass spectrometric observation of the nitrosobenzene molecule, but the $O(^3P)$ atom, with its very high ionization potential, is not detected by the technique employed. The photoionization cross section for nitrosobenzene is unknown, so estimates of the relative extent of the photolysis that proceeded by channel (I) could not be made. However, the relative yield of C_6H_5NO increased significantly from the onset at about 280 nm to the shorter wavelengths. The ratios of $(\phi_{III} + \phi_{IV})/\phi_I$ observed for photolyses at 280, 240, and 222 nm are 1.3 ± 0.3, 3.7 ± 0.2, and 5.9 ± 0.3, respectively. It is estimated that the minimum energies required for processes (I) and (II) are 389 and 297 kJ mol^{-1}, corresponding to possible onset wavelengths of 307 and 403 nm, respectively. In photolyses at the shorter wavelengths (280 nm), the NO_2 product may be electronically excited. The detection of the C_5H_5 fragment (presumably cyclopenta-1,3-dien-5-yl radical) from decomposition in process (IV) by Galloway et al. (1993) is consistent with the spectroscopic observation of cyclopenta-1,3-diene in the flash photolysis of nitrobenzene (Porter and Ward, 1968). Galloway et al. (1994) found that the NO molecule fragment formed in processes (III) and (IV) is very rich in rotational energy (but with little vibrational energy); the rotational temperatures of NO formed in 226-nm photolyses was 3700 ± 350 K, while that formed at 280 nm was 2400 ± 200 K. They rationalize the energy distribution in the products of processes (III) and/or (IV) by a mechanism of isomerization of nitrobenzene to C_6H_5ONO (benzene nitrite) followed by a rupture of the C_6H_5O-NO bond to form the various products. In solution-phase studies of Takezaki et al. (1997a), the lifetimes of the S_1 and T_1 states of nitrobenzene were found to be very short: < 10 ps and ~ 480 ps, respectively. They assumed that photodecomposition was unimportant for nitrobenzene and estimated the ϕ for intersystem crossing to be ≥ 0.8. In the reaction of the triplet nitrobenzene in H-atom–containing solvents, H-atom abstraction from solvent molecules occurs, but it is inefficient since the lifetime of the triplet is very short (Hurley and Testa, 1968; Yip et al., 1984). Takezaki et al. (1997b) have studied theoretically the geometries and energies of the excited states of nitrobenzene, and they report that the O-atoms of the normally in-plane nitro-group are largely bent out of the phenyl plane in the excited states.

VII-D-3a. Absorption Spectra

The solution-phase absorption spectrum of nitrobenzene (250–380 nm), shows both the strong $\pi \rightarrow \pi^*$ and the weaker $n \rightarrow \pi^*$ bands; the latter band alone is important in the troposphere, and this is given (for a hexane solution) in figure VII-D-7 (Láng, 1961–1971). The cross sections for nitrobenzene (in hexane solution) as a function of wavelength are summarized in table VII-D-3.

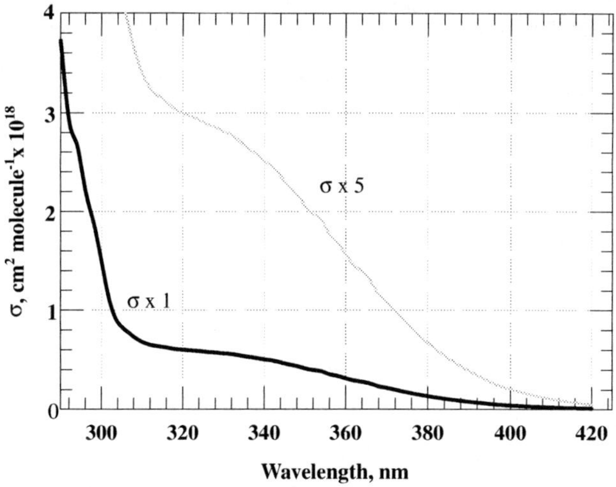

Figure VII-D-7. Absorption cross sections for the long-wavelength band in nitrobenzene as measured in ethanol solution (Láng, 1961–1971).

Table VII-D-3. Absorption cross sections for nitrobenzene as measured by Láng (1961–1971) in ethanol solution

λ (nm)	$\sigma \times 10^{19}$	λ (nm)	$\sigma \times 10^{19}$	λ (nm)	$\sigma \times 10^{19}$	λ (nm)	$\sigma \times 10^{19}$
290	37.25	323	5.89	356	3.54	389	0.82
291	33.24	324	5.86	357	3.45	390	0.77
292	29.22	325	5.82	358	3.36	391	0.73
293	27.98	326	5.77	359	3.24	392	0.69
294	26.74	327	5.74	360	3.12	393	0.64
295	24.45	328	5.71	361	3.02	394	0.60
296	22.16	329	5.66	362	2.91	395	0.56
297	20.54	330	5.62	363	2.84	396	0.53
298	18.91	331	5.59	364	2.77	397	0.50
299	16.91	332	5.56	365	2.68	398	0.46
300	14.90	333	5.49	366	2.59	399	0.43
301	12.90	334	5.42	367	2.45	400	0.40
302	10.89	335	5.36	368	2.31	401	0.38
303	9.84	336	5.29	369	2.24	402	0.36
304	8.79	337	5.23	370	2.16	403	0.34
305	8.37	338	5.16	371	2.07	404	0.32
306	7.95	339	5.09	372	1.98	405	0.31
307	7.62	340	5.02	373	1.89	406	0.29
308	7.30	341	4.96	374	1.80	407	0.27
309	7.05	342	4.91	375	1.71	408	0.25
310	6.80	343	4.81	376	1.64	409	0.23
311	6.65	344	4.71	377	1.56	410	0.21
312	6.49	345	4.62	378	1.49	411	0.20
313	6.42	346	4.52	379	1.41	412	0.19
314	6.34	347	4.44	380	1.34	413	0.18
315	6.29	348	4.35	381	1.28	414	0.17
316	6.23	349	4.24	382	1.21	415	0.16
317	6.15	350	4.12	383	1.15	416	0.15
318	6.07	351	4.04	384	1.08	417	0.14
319	6.03	352	3.95	385	1.02	418	0.13
320	6.00	353	3.89	386	0.97	419	0.12
321	5.96	354	3.83	387	0.92	420	0.10
322	5.93	355	3.69	388	0.87		

Note: Units on σ are cm^2 $molecule^{-1}$.

VII-D-3b. Quantum Yields

For the wavelengths available in sunlight within the troposphere, it is likely that process (I) is unimportant because of the energy limitation. The enthalpy changes for processes (II), (III), and (IV) are approximately 297, 71, and 155 kcal mol^{-1}, respectively, but since extensive rearrangements accompany processes (III) and (IV), significant unknown barriers to these reactions may exist. There are no direct measurements of the extent and wavelength dependence of the photodecomposition modes within the long-wavelength band, so we recommend the assumption that decomposition occurs by processes (II) and (III) with an efficiency similar to that observed in sunlight photolyses of other aromatic nitrocompounds. See section VII-D-4 on 1- and 2-nitronaphthalene and section VII-D-5 on 2-methyl-1-nitronaphthalene.

VII-D-3c. Tropospheric Photochemistry

Photolysis rate coefficients representative of the lower atmosphere were calculated using the TUV model (as described in the introduction to this section) with the measured solution-phase cross sections and quantum yield data representative of the nitronapthalenes ($\phi \sim 3.5 \times 10^{-3}$). See figure VII-D-8.

VII-D-4. Photochemistry of 1-Nitronaphthalene and 2-Nitronaphthalene.

Photochemical studies of the nitronaphthalenes have been made by Atkinson et al. (1989b) and Fielberg et al. (1999). Irradiations in both indoor and outdoor chambers have been carried out and rates of disappearance of the nitronaphthalenes have been determined. A product observed in 1-nitronaphthalene studies was 1,4-naphthoquinone (Atkinson et al., 1989b), which suggests that the primary processes (I) and (II) may occur in the photodecomposition at the wavelengths present within the troposphere:

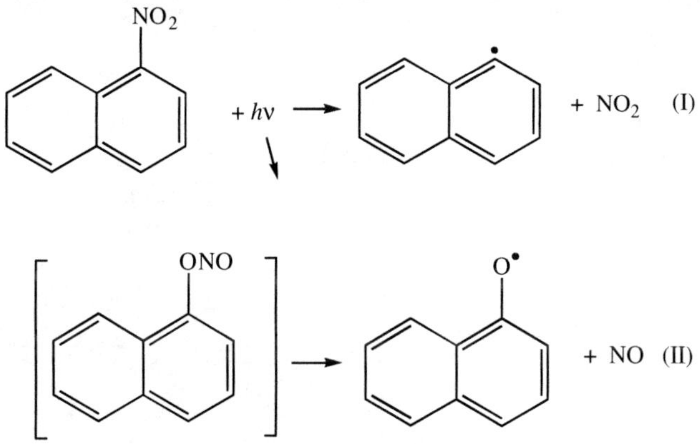

The similar rates of photolysis for 2-nitronaphthalene exposed to sunlight suggest that, perhaps, similar photodecomposition modes may occur in this compound, although no final reaction products were reported for 2-nitronaphthalene.

Somewhat different results were obtain in the chamber studies of Fielberg et al. (1999), where a more rapid photochemical decay of the 1-nitronaphthalene was observed.

VII-D-4a. Absorption Spectra

Gas-phase absorption spectra for 1- and 2-nitronaphthalene are not available at this writing, but as a reasonable surrogate we may use that observed for 2-methyl-1-nitronaphthalene in ethanol solution (Láng, 1961–1971). This is shown in figure VII-D-9. The cross sections at wavelengths within the troposphere are summarized in table VII-D-4.

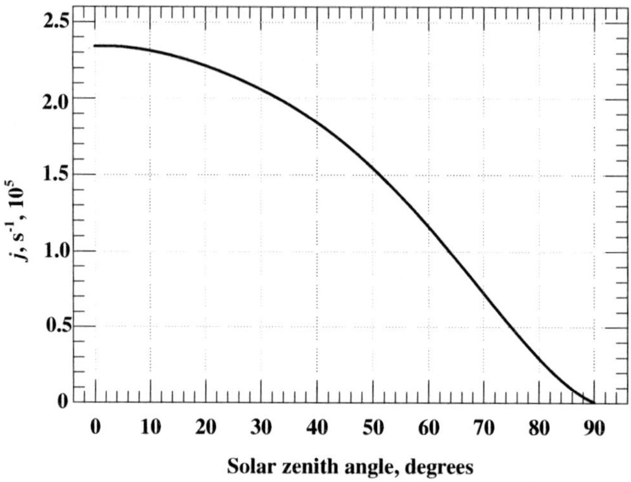

Figure VII-D-8. Nitrobenzene photodissociation coefficients at 0.5 km altitude as a function of solar zenith angle using the solution-phase absorption spectrum and quantum yield of decomposition typical of the nitronaphthalenes ($\phi \sim 3.5 \times 10^{-3}$).

VII-D-4b. Quantum Yields for 1- and 2-Nitronaphthalene

Atkinson et al. (1989b) carried out three experiments in an outdoor chamber in sunlight-irradiated 1-nitronaphthalene in mixtures of synthetic air with measured first-order rates of loss (min^{-1}) of 0.0130 ± 0.0022 (September 7, 14:18–16:42 Pacific daylight time, PDT), 0.0115 ± 0.0041 (September 8, 15:07–17:18 PDT), and 0.0082 ± 0.0006 (October 31, 10:36–15:25 Pacific standard time, PST). Similar results with 2-nitronaphthalene gave first-order rates of loss of

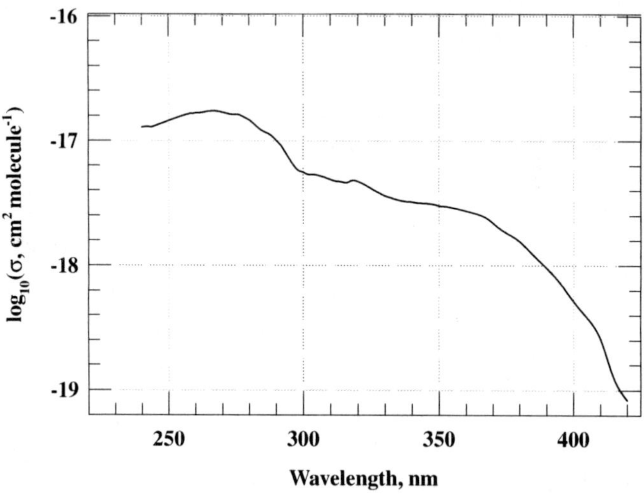

Figure VII-D-9. Cross sections for 2-methyl-1-nitronaphthalene as measured in ethanol solution by Láng (1961–1971).

Table VII-D-4. Cross sections at various wavelengths for 2-methyl-1-nitronaphthalene as measured in ethanol solutions by Láng (1961–1971)

λ (nm)	$\sigma \times 10^{19}$	λ (nm)	$\sigma \times 10^{19}$	λ (nm)	$\sigma \times 10^{19}$	λ (nm)	$\sigma \times 10^{19}$
290	101.2	323	43.7	356	28.6	389	10.1
291	96.0	324	42.5	357	28.3	390	9.57
292	90.8	325	41.2	358	28.0	391	9.11
293	83.9	326	40.0	359	27.6	392	8.66
294	77.0	327	38.9	360	27.3	393	8.20
295	71.4	328	37.9	361	26.9	394	7.74
296	65.8	329	37.0	362	26.5	395	7.29
297	61.9	330	36.0	363	26.2	396	6.87
298	58.1	331	35.4	364	25.8	397	6.45
299	56.9	332	34.9	365	25.3	398	6.04
300	55.7	333	34.3	366	24.7	399	5.62
301	54.5	334	33.7	367	24.2	400	5.20
302	53.3	335	33.4	368	23.6	401	4.93
303	53.3	336	33.0	369	22.7	402	4.66
304	53.3	337	32.7	370	21.9	403	4.39
305	52.7	338	32.4	371	21.0	404	4.12
306	52.1	339	32.3	372	20.1	405	3.85
307	51.4	340	32.2	373	19.5	406	3.60
308	50.6	341	31.9	374	18.9	407	3.35
309	49.7	342	31.6	375	18.3	408	3.10
310	48.7	343	31.5	376	17.7	409	2.85
311	47.9	344	31.4	377	17.2	410	2.60
312	47.1	345	31.3	378	16.6	411	2.33
313	46.9	346	31.2	379	16.1	412	2.06
314	46.6	347	31.0	380	15.5	413	1.79
315	46.3	348	30.8	381	14.8	414	1.52
316	45.9	349	30.4	382	14.2	415	1.25
317	46.8	350	30.0	383	13.5	416	1.17
318	47.8	351	29.8	384	12.9	417	1.08
319	47.4	352	29.7	385	12.2	418	1.00
320	47.0	353	29.4	386	11.7	419	0.92
321	46.0	354	29.1	387	11.1	420	0.83
322	45.0	355	28.9	388	10.6		

Note: Units on σ are cm^2 $molecule^{-1}$.

0.0101 ± 0.0030, 0.0034 ± 0.0040, and 0.0064 ± 0.0005 min^{-1}. Measurements of $j(NO_2)$ made at approximately the location of the chamber on October 10 were 4.16×10^{-3} s^{-1} at 11:15 PDT and 4.33×10^{-3} s^{-1} at 11:00 PST on October 31. Hence, for these conditions, Atkinson et al. (1989b) estimate j(decomposition) for 1- and 2-nitronaphthalenes are $(1.37 \pm 0.10) \times 10^{-4}$ s^{-1} and $(1.06 \pm 0.08) \times 10^{-4}$ s^{-1}, respectively.

Somewhat more rapid photolysis of 1-nitronaphthalene was observed in the outdoor chamber experiments of Fielberg et al. (1999). These experiments involved the sunlight irradiation of mixtures of the naphthalene, propylene, NO_x, and diesel soot particles and the modeling of the rates of formation and decay of the nitronaphthalene products. The experiments were designed primarily to test the understanding of the partitioning of the nitro-PAH on particles, and the composition of the starting mixtures was considerably more complex than that employed by Atkinson et al. (1989b). Fielberg et al. (1999) suggest that the gas-phase photolysis is the major degradation pathway for 1-nitronaphthalene in this system, with *j*(photodecomposition) $\sim 0.07 \times j(NO_2)$, a result about a factor of 2 higher than that of Atkinson et al. (1989b). Rates of photolysis for the 2-nitronaphthalene observed by Fielberg et al. (1999), *j*(photodecompostion) $\sim 0.005 \times j(NO_2)$, are about a factor of 5 lower than those observed by Atkinson et al. (1989b). While further study is needed, it is tentatively recommended that the photolysis rate data of Atkinson et al. (1989b) be used.

VII-D-5. Photochemistry of 2-Methyl-1-nitronaphthalene

The photochemistry of 2-methyl-1-nitronaphthalene in air was studied by Arey et al. (1990) in a sunlight-irradiated chamber. 2-Methyl-1,4-naphthoquinone was observed to be a major product, and photodecomposition modes analogous to those given for 1- and 2-nitronaphthalene are suggested:

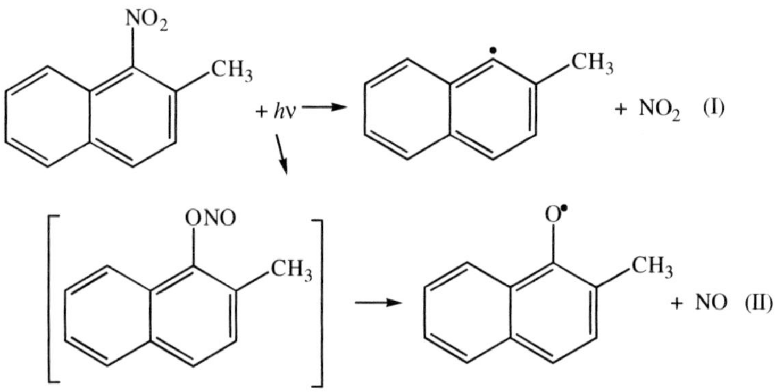

VII-D-5a. Absorption Spectra

The spectrum of 2-methyl-1-nitronaphthalene is given in figure VII-D-9, as determined in ethanol solution by Láng (1961–1971).

VII-D-5b. Quantum Yields

The rate of disappearance in the sunlight photolysis of the 2-methyl-1-nitronaphthalene was determined in the studies of Arey et al. (1990), from which an estimate of *j*(decomposition) $= (1.1 \pm 0.3) \times 10^{-4}\ s^{-1}$ was determined for irradiation at 10:54–15:31 PST on November 4 in Riverside, CA. This is similar to the values observed by Atkinson et al. (1989b) and is qualitatively in agreement with those reported by Fielberg et al. (1999) for the 1- and 2-nitronaphthalenes.

VII-D-5c: Tropospheric Photolyses

Photolysis rates representative of the lower atmosphere were calculated with the TUV model (as described in the introduction to this section) using the solution-phase cross sections and an assumed effective quantum yield of 3.5×10^{-3} to

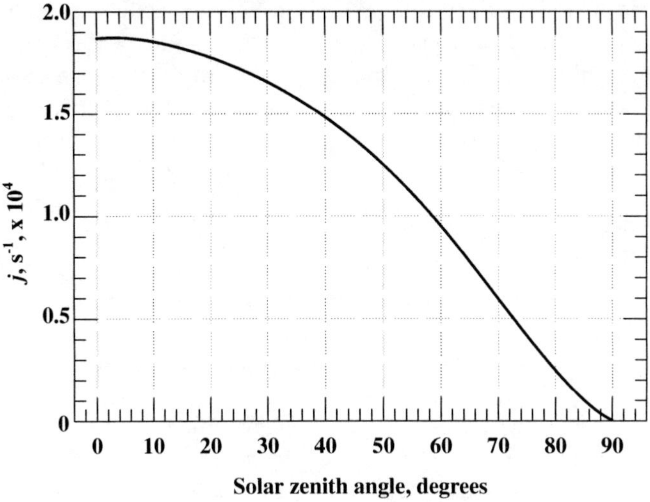

Figure VII-D-10. 2-Methyl-1-nitronaphthalene photodissociation coefficients at 0.5 km altitude as a function of solar zenith angle, using measured solution-phase cross sections and an assumed quantum yield of decomposition of 3.5×10^{-3} in order to obtain a match between the calculations and measurements.

make a rough match with the *j*-values observed experimentally by Atkinson et al. (1989b). See figure VII-D-10.

VII-D-6. Phenols, Cresols, and Other OH-Substituted Aromatic Compounds

There is little information concerning the photochemistry of the HO-substituted polycyclic aromatic species. However, the several studies of the photochemistry of the simple substituted phenols allow a significant insight into the primary processes present in the higher-molecular-weight, OH-substituted aromatic hydrocarbons, and for this reason a brief review of this work is given here.

Process (I), the rupture of the RO–H bond, appears to be the only significant primary process in the photodecomposition of phenol, alkyl-radical–substituted phenols, and, by analogy, the higher-molecular-weight, OH-substituted aromatic hydrocarbons; this is shown here for the substituted phenols (R = an H-atom or any alkyl group):

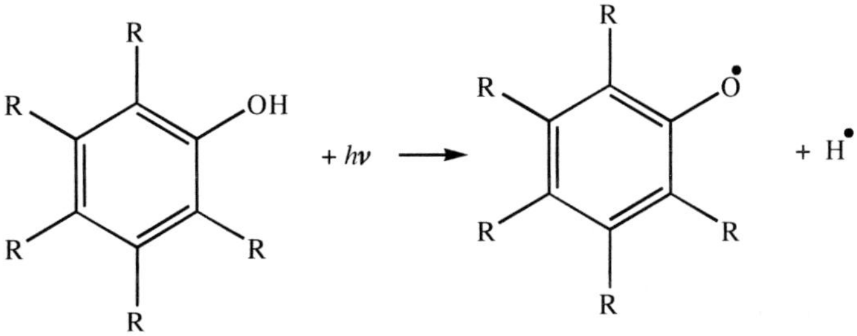

(I)

Land and Porter (1963) studied the flash photolysis of 18 different alkyl-radical–substituted phenols with R = CH_3, C_2H_5, *tert*-C_4H_9, and/or C_6H_5 in various nonpolar and polar solvents, and the absorption spectra of the corresponding phenoxyl radical was observed in all cases. Grabner et al. (1990) made an extensive study of the photochemistry of the methylated phenols in hydrocarbon solvents at 293 K and with $\lambda_{excit} = 275$ nm; they determined the quantum yields of fluorescence, triplet molecule formation, and photodecomposition. The quantum yields of photodecomposition of phenol and its 2-methyl-, 3-methyl-, 4-methyl-, 2,6-dimethyl-, 3,5-dimethyl-, 2,4,6-trimethyl-, and 2,3,5-trimethyl-derivatives, measured through the absorption of the phenoxyl radicals formed, were measured as 0.13, 0.18, 0.14, 0.08, 0.21, 0.14, 0.20, and 0.18, respectively. Evidence suggested that this photodecomposition occurred from a vibrationally excited triplet state.

Lipczynska-Kochany and Botton (1992) have studied the photolysis of phenol in undegassed aqueous solution and observed *p*-benzoquinone as a major product, with small amounts of hydroquinone. Presumably, these products can arise following photodissociation in reaction (I).

Grosjean (1984) has studied the photooxidation of *o*-cresol in NO_x mixtures in sunlight-irradiated air and found a first-order loss of *o*-cresol ($k \sim 1.5 \times 10^{-4}\ s^{-1}$), a slow generation of ozone, significant loss of NO_x, and formation of nitrocresols, pyruvic acid, acetaldehyde, formaldehyde, peroxyacetyl nitrate, and various particulate nitro-aromatic products. However, it is not clear to what extent the photolysis of the *o*-cresol was responsible for this observed chemistry, since very little absorption of tropospheric sunlight is expected for the *o*-cresol, and NO_3-radical chemistry may have been dominant. See the following section.

VII-D-6a. Absorption Spectra

The region of light absorption for the simple phenols can be observed in the data shown in figure VII-D-11 (Etzkorn et al., 1999). The onset of absorption shifts toward the longer wavelengths as methyl groups are added to the aromatic ring of phenol, but even with the highly substituted phenols, absorption does not overlap significantly with the actinic flux present in the lower troposphere.

Light absorption by the cresols, $CH_3C_6H_4{-}OH$, is shifted somewhat to the longer wavelengths than that observed in phenol, but there is still little overlap of this with the solar region available in the lower troposphere; see figure VII-D-12.

With the hydroxy-substituted aromatics, absorption of the light available in the lower troposphere increases as the number of fused aromatic rings increases; see figure VII-D-13.

VII-D-6b. Quantum Yields

We are not aware of any studies of the photochemistry of the OH-substituted polycyclic aromatic hydrocarbons, but by analogy with the photochemistry of the simple substituted phenols discussed in this section, we would expect photo-

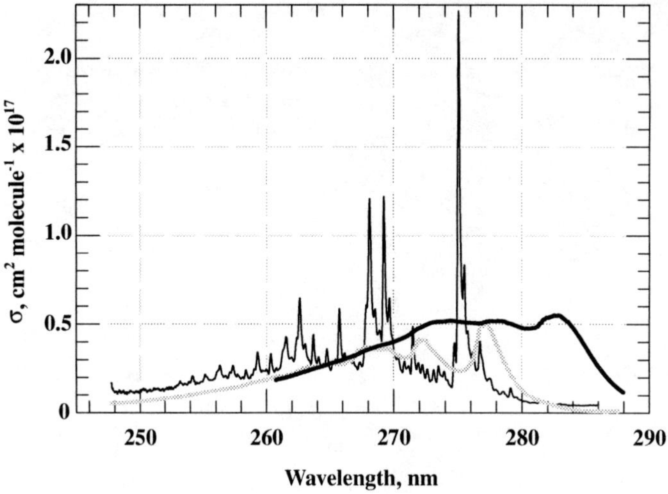

Figure VII-D-11. Cross sections as a function of wavelength for phenol vapor (thin black curve), 2,3-dimethylphenol (gray curve), and 2,4,6-trimethylphenol (heavy black curve); all data from Etzkorn et al. (1999).

dissociation by process (I) to be significant at the wavelengths of sunlight available within the troposphere. The minimum energy required for the occurrence of process (I) is provided at $\lambda \leq 330$ nm. The *j*(excitation) for 2-hydroxynaphthalene and 3-hydroxyphenanthrene, the maximum possible photodissociation rates in summer sunlight at 40° N latitude, shown in figure VII-D-14, suggest that their photodecomposition may be significant.

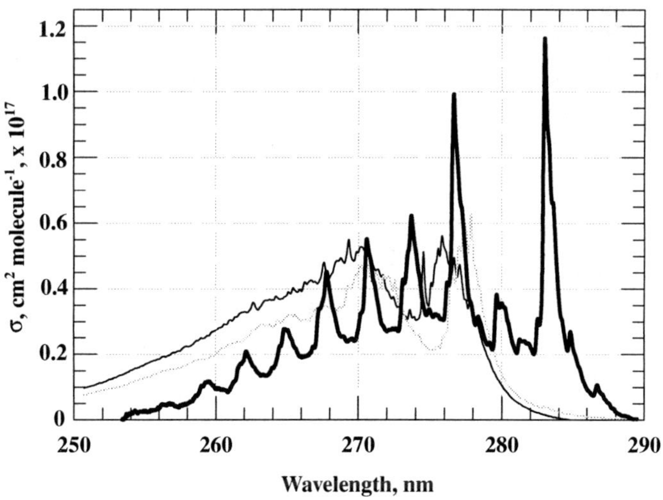

Figure VII-D-12. Cross sections as a function of wavelength for vapors of *o*-cresol (thin black curve), *m*-cresol (gray curve), and *p*-cresol (heavy black curve); all data from Etzkorn et al. (1999).

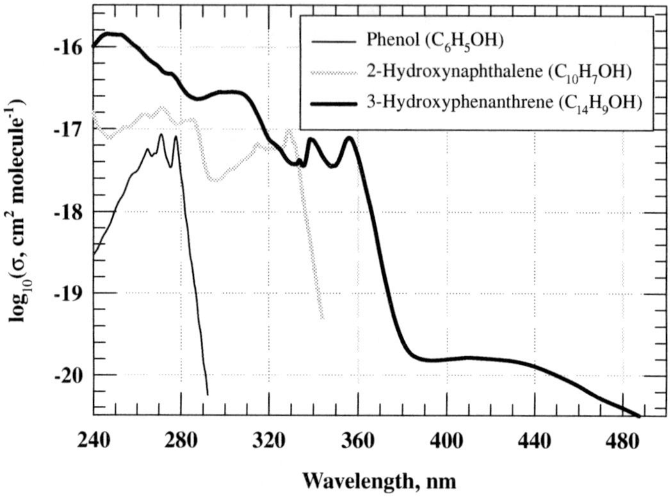

Figure VII-D-13. Absorption cross sections for a series of HO-substituted aromatic hydrocarbons, phenol, 2-hydroxynaphthalene, and 3-hydroxyphenanthrene, as measured in non-polar solvents; data from Láng (1961–1971).

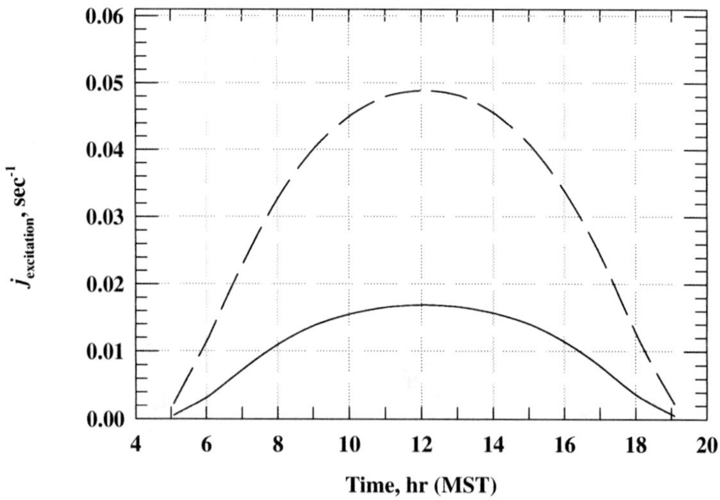

Figure VII-D-14. Approximate *j*(excitation), excited singlet molecule formation, estimated for 2-hydroxynaphthalene (solid curve) and 3-hydroxyphenanthrene (dashed curve); cross sections used in the calculation are from Láng (1961–1971) as measured in ethanol solutions; actinic flux data are for a cloudless summer day (June 9, 1998) in Boulder, CO.

VII-D-7. Nitrophenols and Other Nitro- and OH-Substituted Organic Compounds

The nitrophenols have been measured among the many other compounds present in motor-vehicle exhaust (Nojima et al., 1983; Tremp et al., 1993). Also, these compounds can be formed by the nitration of the phenols and other OH-substituted aromatic hydrocarbons present in the atmosphere. This can occur, especially at night—when reactions with NO_3 can take place rapidly; see figure I-C-2 in chapter I and rate data in chapter III. These compounds are highly soluble in cloud-water and rainwater, and they often appear in these aqueous media (Lüttke and Levsen, 1997; Lüttke et al., 1997). Information concerning the photochemistry of the nitrophenols is very limited, but in view of the rather large cross sections related to these compounds, some speculation here on the potential pathways for photodecomposition is deemed important. See figure VII-D-15.

In view of the minimum energy necessary to dissociate phenol and nitrobenzene, and the studies of their photochemistry, it is reasonable to assume that primary process (I) will dominate in the gas-phase photodecomposition of the nitrophenols:

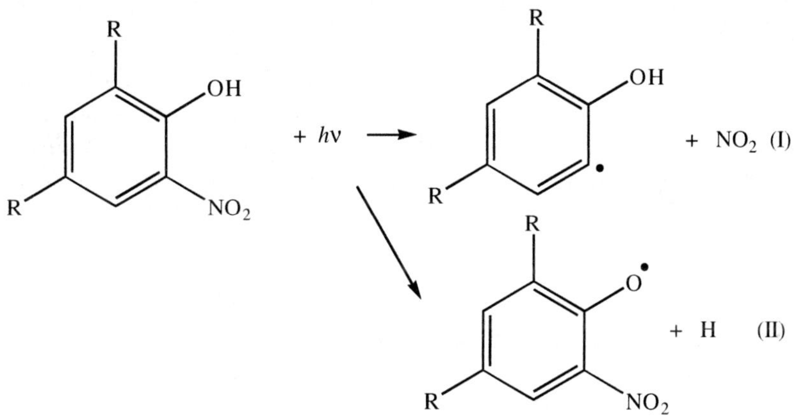

where R is either an NO_2-group or an H-atom. The minimum energy required for processes (I) and (II) is expected to be approximately 297 and 362 kJ mol^{-1}, respectively, corresponding to wavelengths for onset of approximately 403 and 330 nm, respectively.

Palm et al. (1999a) have determined that the direct photolysis of nitrophenols in water solution in sunlight is a negligible sink. Photolytic half-lives of 1,4-dinitrophenol, 4-methyl-2-nitrophenol, and 2,6-dinitro-*p*-cresol at 50° N (summer–winter) were found to be 5–35, 32–180, and 55–320 days, respectively. In daylight, the major sink was suggested to be the solution-phase reaction of these species with the OH radical, where it was estimated that half-lives were less than 1 h. Similar results were reported by Palm et al. (1999b); they estimated quantum yields of decay in aqueous solution for 2,4-dinitrophenol, 3-methyl-2-nitrophenol, and 3-methyl-2,6-dinitrophenol to be: 1.1×10^{-5} (pH = 2.6); 3.3×10^{-6} (pH = 6.0); and 0.4×10^{-6} (pH = 6.0), respectively.

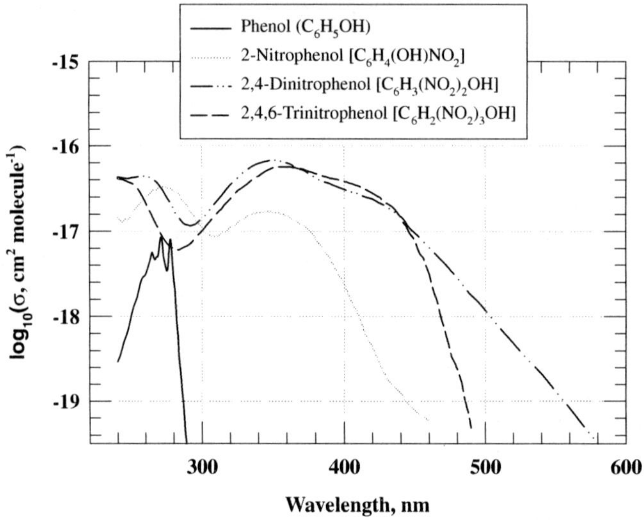

Figure VII-D-15. Plot of the cross sections for some nitro-substituted phenols; measurements were made in ethanol solutions. Data of Láng (1961–1971).

It is evident from the discussions of sections VII-C and VII-D that there is a great need to improve our quantitative understanding of the photochemistry of the many products of the atmospheric oxidation of the aromatic hydrocarbons. Not only is it important to enhance our knowledge of the many products discussed in this chapter, but also we require quantitative studies of the many newly discovered polycarbonyl products (Jang and Kamens, 2001), many of which promise to be highly photochemically active species; see section VIII-C.

VIII

Aerosol Generation in Atmospheric Oxidation of Aromatic Hydrocarbons

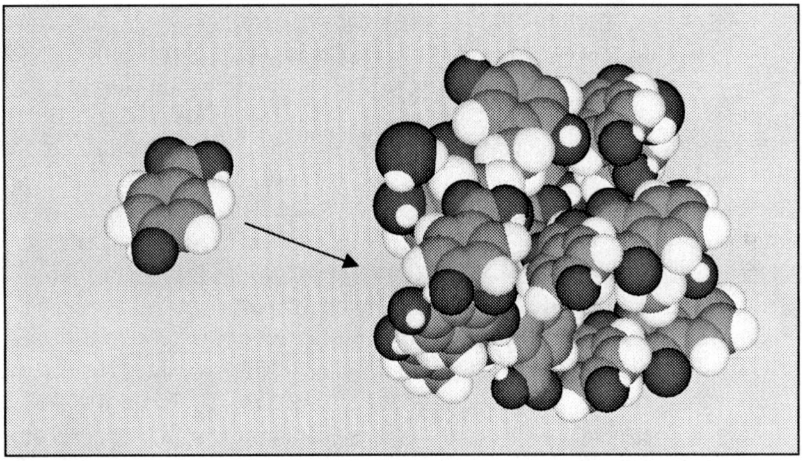

VIII-A. Organic Composition of Atmospheric Aerosols

Atmospheric aerosol in virtually all areas of the troposphere, but especially in urban areas, contains a significant fraction of carbonaceous material. Atmospheric particulate carbonaceous material is composed of a complex mixture of substances, frequently classified in two main fractions: elemental carbon (EC) and organic carbon (OC). Elemental, or black carbon is a primary pollutant, emitted directly as a result of the combustion of fossil and biomass fuels. Particulate organic carbon, a complex mixture of hydrocarbons and oxygenated hydrocarbons, has both primary and secondary origins. Primary particulate OC is emitted during combustion of fuels, cooking of meat, and so on, whereas secondary particulate OC results from gas-to-particle conversion of volatile organic compounds in the atmosphere. This secondary particulate OC is usually referred to as secondary organic aerosol (SOA). To develop strategies for controlling ambient levels of carbonaceous particulate matter, it is necessary to quantify both the primary and secondary contributions to carbonaceous aerosols; reduction of direct particulate emissions affects only the primary carbonaceous component.

Separation and quantification of the primary and secondary components of particulate OC has been challenging to accomplish. No direct analytical technique exists to determine readily which fraction of particulate OC is of primary versus secondary origin. As a result, a number of indirect approaches have been applied. Determination of the aerosol carbon isotopic composition ($^{14}C/^{12}C$) provides an indication of the fraction of OC from fossil sources as opposed to contemporary carbon sources (Hildemann et al., 1994; Kaplan and Gordon, 1994). Measurement of the OC/EC ratio under conditions of low atmospheric photochemical activity, when formation of secondary organic aerosol should be more or less non-existent, provides an estimate of the primary contribution to OC in the region of interest (Larson et al., 1989; Kadowaki, 1990; Turpin and Huntzicker, 1991, 1995; Turpin et al., 1991; Lee and Huang, 1993; Castro et al., 1999). Tracer, or fingerprint, compounds have been used to identify primary and secondary OC (Chu and Macias, 1981; Wolff et al., 1983; Gray et al., 1986; Huntzicker et al., 1986; Turpin et al., 1991). Atmospheric models that describe transport and transformation of emitted gases and particles have also been used to predict the primary and secondary OC fractions (Gray et al., 1986; Pilinis and Seinfeld, 1988; Larson et al., 1989; Pandis et al., 1992; Hildemann et al., 1993; Meng et al., 1997).

Quantitative information on the amount of secondary OC in different areas of the United States and the remainder of the world is scarce. By far the largest collection of information exists for the South Coast Air Basin of California (the Los Angeles Air Basin). Recent studies also address SOA levels in the San Joaquin Valley of California (Strader et al., 1999) and in Europe (Castro et al., 1999). Under intense photochemical smog episodes in the South Coast Air Basin of California from 40% to 80% of the particulate organic carbon has been estimated to be secondary in origin (Turpin and Huntzicker, 1991, 1995). Lee and Huang (1993) estimated SOA to be as much as 73% of OC in Taipei. Castro et al. (1999) analyzed aerosol samples from urban, rural, and coastal areas in Europe for carbonaceous content by a thermal–optical transmission method. The secondary fraction of particulate OC was estimated from the minimum OC/EC ratio, which occurred during periods of low photochemical activity. Values calculated in this way ranged from a minimum of 17%, in Birmingham, England, during winter, to a maximum of 78% at a rural coastal area in Portugal, where air masses were transported directly from the ocean. A clear seasonal dependence was observed at all sites, with minimum secondary organic carbon during winter. In analyzing data from the San Joaquin Valley of California during winter 1995–1996, Strader et al. (1999) found that under suitable conditions (clear skies, low winds, low mixing heights) as much as 15–20 μg C m^{-3} of SOA can be produced, mainly due to oxidation of aromatic compounds.

In estimating the amount of particulate secondary organic carbon from the minimum observed value of the OC/EC ratio, if, in all samples, the OC/EC ratio remains constant, the concentration of SOA can be calculated by

$$OC_{sec} = OC_{total} - \left(\frac{OC}{EC}\right)_{min} \times EC$$

This procedure will produce a correct estimate of SOA only if (1) the amount of SOA is negligible in samples used to calculate $(OC/EC)_{min}$; (2) the contribution of semivolatile organic compounds in comparison with non-volatile organic compounds is small; (3) the composition of primary carbonaceous aerosol sources and the relative contribution of each source to the aerosol concentration are spatially and temporally constant; and (4) the contribution of non-combustion primary particulate OC is small, or constant (Castro et al., 1999). If the fraction of semivolatile organic compounds in the aerosol is large, the concentration of particulate OC depends significantly upon temperature. In such a case, changes in ambient temperature—from winter to summer, for example—would result in a decrease of $(OC/EC)_{min}$ as a result of the evaporation of semivolatile primary organic compounds in summer. For point (3) above, if the contribution of sources that have different primary OC/EC ratios to the total carbonaceous aerosol varies strongly—say, from winter to summer, as might result from higher wood combustion in winter—the use of a constant OC/EC ratio is questionable. In their analysis of San Joaquin Valley data, Strader et al. (1999) discuss the usefulness of the OC/EC ratio approach further and show several alternate ways to estimate OC_{sec}. In short, there is still no definitive way to estimate the secondary contribution to particulate OC, although techniques based on atmospheric modeling are improving rapidly. These will be discussed later in this chapter.

VIII-B. Mechanisms of Aerosol Generation from Aromatic Hydrocarbons

VIII-B-1. General Nature of Secondary Organic Aerosol (SOA) Formation

Like ozone, secondary organic aerosol results from the atmospheric oxidation of primary organic compounds, but whereas the oxidation of most primary organic compounds results in ozone formation, SOA is generally formed only from the oxidation of organic molecules sufficiently large to lead to products that have vapor pressures low enough to enable them to condense into the aerosol phase.

One of the important factors that determines whether an oxidation product appears in the gas phase or in the aerosol phase in the atmosphere is the magnitude of its vapor pressure. The lower the vapor pressure of a compound, the greater is the chance that it will reach its saturation concentration in the atmosphere and condense to build secondary aerosols onto existing particles. Of course, other factors such as the solubility in aqueous solutions or organic liquids play an important role as well. In this regard, it is interesting to note the relationship of vapor pressure to the changes in molecular structure as oxidation proceeds in a series of related compounds. Products similar to those of aromatic ring rupture are illustrated by series of compounds shown in figure VIII-B-1. Here, the vapor pressures of a series of structurally related C_4-compounds are compared. As one progresses from the simple *n*-butane molecule, labeled 1 in the figure, to the highly oxidized tartaric acid [HOOCCH(OH)CH(OH)COOH], labeled 16 in the figure, the vapor pressure decreases by about 15 orders of

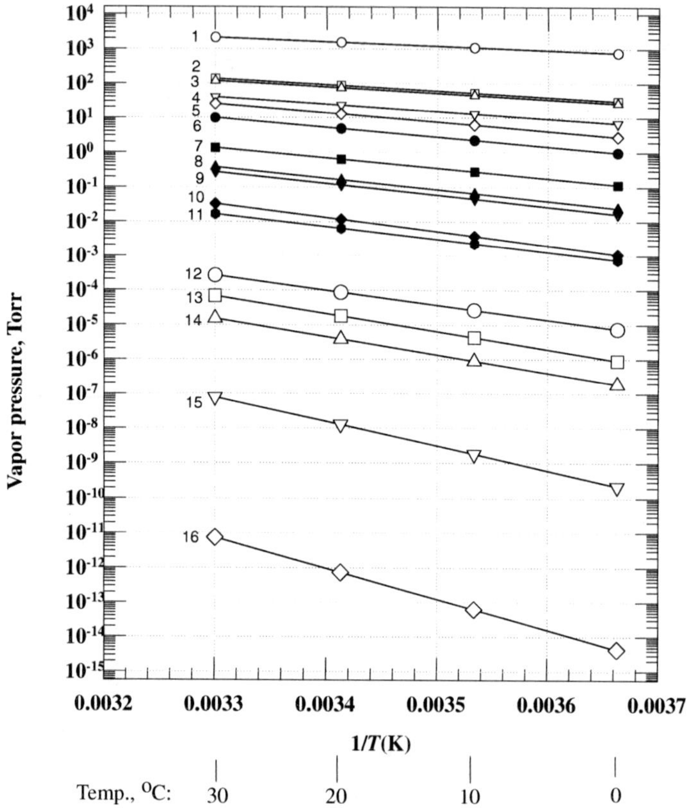

Figure VIII-B-1. Plot of vapor pressure of some C_4-compounds with increasing extent of oxidation: (1) *n*-C_4H_{10}; (2) *n*-$C_3H_7C(=O)H$; (3) $CH_3C(=O)C_2H_5$; (4) *trans*-$CH_3CH=CHC(=O)H$; (5) *sec*-C_4H_9OH; (6) *n*-C_4H_9OH; (7) *n*-$C_3H_7(C=O)OH$; (8) *trans*-$CH_3CH=CHC(=O)OH$; (9) $CH_3CH(OH)CH(OH)CH_3$; (10) $HOCH_2CH_2CH(OH)CH_3$; (11) $HOCH_2CH_2CH_2CH_2OH$; (12) *trans*-$HO(O=)CCH=CHC(=O)OH$; (13) *cis*-$HO(O=)CCH=CHC(=O)OH$; (14) $HO(O=)CCH_2CH_2C(=O)OH$; (15) $HO(O=)CCH(OH)$-$CH_2C(=O)OH$; (16) $HO(O=)CCH(OH)CH(OH)C(=O)OH$. Vapor pressure data were calculated from the equations for the compounds in the liquid state as given by Yaws et al. (1999); the data for the compounds of lowest vapor pressure are very approximate since they were obtained by extrapolation of measurements made at much higher temperatures.

magnitude. The data shown for the lowest vapor pressure compounds are very approximate since the existing data were extrapolated to 0°C, 10°C, 20°C, and 30°C from much higher temperatures at which measurements were made. However, the general trends of vapor pressure with molecular structure observed here are qualitatively useful. Note that as an increasing number of oxygenated groups, such as $-OH$, $-C(=O)H$, and $-C(=O)OH$, appear in the four-carbon chain of butane, the vapor pressure continues to fall. The increasingly negative slope of the logarithm of the vapor pressure versus $1/T$ curves seen as one proceeds to the compounds of lower vapor pressure reflects the greater enthalpies of vaporization that are characteristic of the more polar nature of the highly oxidized products [$d(\ln p)/d(1/T) = -\Delta H_{vap}/R$]. Vapor pressures of 10^{-3}, 10^{-4}, 10^{-5}, 10^{-6}, 10^{-7}, 10^{-8}, 10^{-9}, 10^{-10}, 10^{-11}, 10^{-12}, 10^{-13}, and 10^{-14} Torr correspond to vapor-phase concentrations at saturation in air at 25°C of 1.3 ppmv, 0.13 ppmv, 0.013 ppmv, 1.3 ppbv, 0.13 ppbv, 0.013 ppbv, 1.3 pptv, 0.13 pptv, 0.013 pptv, 0.0013 pptv, 0.00013 pptv, and 0.000013 pptv, respectively. Compounds with vapor pressures below about 10^{-8} Torr are often found in the aerosol phase in the atmosphere.

In figure VIII-B-2, a similar comparison is made of the vapor pressures for toluene and a series of its oxidation products that retain the aromatic ring structure. Note that there are large decreases in vapor pressure that occur as –OH and –NO_2 groups are added to the toluene ring and as the CH_3-group on toluene is oxidized to a –CH_2OH, (C(=O)H or a –C(=O)OH group. As a general rule, those oxidation products that contain multiple groups such as –OH, –NO_2, –CH_2OH, –C(=O)H, and –C(=O)OH have the lower vapor pressures and are more likely to appear in the aerosol phase in the atmosphere.

The chemical process of SOA formation can be depicted in general terms. Consider the gas-phase oxidation of an aromatic parent molecule A by the OH radical that leads to a set of stable semivolatile product molecules, S_1, S_2, . . .,

$$A + OH \rightarrow \alpha_1 S_1 + \alpha_2 S_2 + \cdots$$

where α_1, α_2 are stoichiometric product coefficients. The products S_1, S_2, . . ., may themselves undergo gas-phase reaction, creating second-generation, semivolatile products:

$$S_1 + OH \rightarrow \alpha_{11} V_1 + \alpha_{12} V_2 + \cdots$$
$$S_2 + OH \rightarrow \alpha_{21} W_1 + \alpha_{22} W_2 + \cdots$$

Secondary organic aerosol yields have been measured for individual parent organic species since 1990 (Izumi and Fukuyama, 1990; Pandis et al., 1991; Wang et al., 1992a, 1992b; Zhang et al., 1992; Odum et al., 1996, 1997a, 1997b; Hoffmann et al., 1997; Griffin et al., 1999; Kleindienst et al., 1999). Initially, it was believed that each parent organic should exhibit a unique value of its SOA yield, characterized by the values of the stoichiometric coefficients, such as α_1, α_2, α_{11}, α_{12}, and so on, perhaps depending on the prevailing NO_x level. Measured yields for an individual parent organic exhibited a degree of variation that could not be reconciled in terms of a single, unique SOA yield or of the expected dependence on NO_x level during photooxidation.

Following Pankow (1994a, 1994b), Odum et al. (1996) formulated a framework for explaining observed SOA yield data. It was recognized that semivolatile oxidation products, such as S_1, S_2, V_1, V_2, and so on, will partition themselves between the gas and aerosol phases at a concentration below their saturation concentration, analogous to the partitioning that occurs between the gas and aqueous phases of a water-soluble atmospheric constituent.

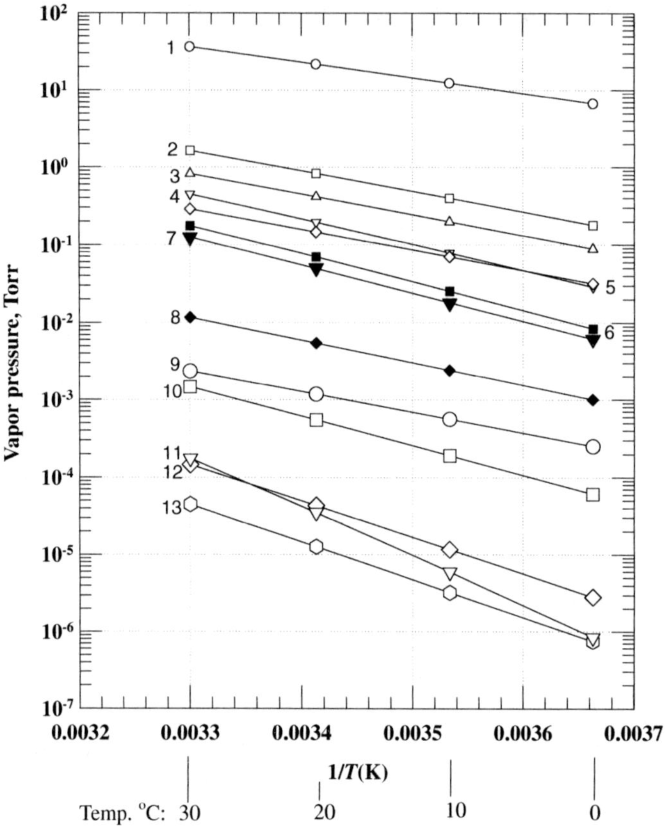

Figure VIII-B-2. Plot of vapor presssure of toluene and some of its ring-retaining oxidation products: (1) toluene; (2) benzaldehyde; (3) *o*-hydroxybenzaldehyde; (4) *o*-cresol; (5) *m*-nitrotoluene; (6) *m*-cresol; (7) benzyl alcohol; (8) benzoic acid; (9) 2,4,6-trinitrotoluene; (10) *p*-hydroxybenzaldehyde; (11) *o*-hydroxybenzoic acid; (12) 2,5-dinitrotoluene; (13) 3,4-dinitrotoluene. Vapor pressure data were calculated from the equations for the compounds in the liquid state as given by Yaws et al. (1999); the data for the compounds of lowest vapor pressure are very approximate since they were obtained by extrapolation of measurements made at much higher temperatures.

VIII-B-2. Gas/Particle Partitioning

The equilibrium gas/particle partitioning of a semivolatile organic species i between the gas phase and a particulate organic phase can be described by the vapor pressure relation,

$$p_i = x_i \gamma_i p_i^o \tag{1}$$

where p_i (Torr) is the gas-phase partial pressure of species i, x_i is the mole fraction of species i in the organic aerosol phase, γ_i is the activity coefficient of species i in the aerosol-phase organic mixture, and p_i^o is the vapor pressure of species i as a pure liquid (subcooled, if necessary). The gas-phase partial pressure, p_i, can be converted to gas-phase mass concentration, G_i, by the ideal gas relation,

$$G_i = \frac{p_i M_{wi}}{RT} 10^6 \tag{2}$$

where M_{wi} (g mol^{-1}) is the molecular weight of species i, R (6.2×10^{-2} Torr m^3 mol^{-1} K^{-1} = 8.314 J mol^{-1} K^{-1}) is the ideal gas constant, and T (K) is temperature. The factor 10^6 accomplishes the appropriate unit conversions.

Let A_i be the aerosol-phase mass concentration of species i (μg m^{-3}), where the volume (m^{-3}) referred to is volume of air. The mean molecular weight of the organic aerosol phase is

$$\overline{M}_w = \sum_{i=1}^{n} x_i M_{wi} \tag{3}$$

The total aerosol mass concentration is

$$M_o = \sum_{i=1}^{n} A_i \tag{4}$$

If the organic species have similar molecular weights, then the mole fractions in the aerosol phase are given by

$$x_i = \frac{A_i}{M_o} \tag{5}$$

Equations (2) and (5) can be substituted into equation (1) to yield

$$\frac{A_i}{G_i M_o} = \frac{RT}{\overline{M}_w \gamma_i p_i^o} 10^{-6} \tag{6}$$

The left-hand side of equation (6) is the ratio of the aerosol-to-gas-phase mass concentrations of species i, including the total mass of the organic particulate phase. The right-hand side can be defined as the equilibrium gas/particle partitioning coefficient of species i,

$$K_i = \frac{RT}{\overline{M}_w \gamma_i p_i^o} 10^{-6} \tag{7}$$

Note that $\gamma_i = \gamma_i(x_1, x_2, \ldots, x_n; T)$ and $p_i^o = p_i^o(T)$; K_i has units m^3 μg^{-1}. Since, at a particular temperature, K_i is essentially a constant for a given mixture of organic compounds, an important implication of equation (7) is that a greater fraction of each product will partition to the organic phase as the total organic aerosol concentration increases. Additional discussion of the temperature dependence of K_i is given in chapter IX.

VIII-B-3. Experimental Determination of SOA Yields

Secondary organic aerosol yields have been measured in chamber irradiations. The SOA yield for a given oxidation product i, Y_i, is the amount of aerosol mass of species i that is produced from the oxidation of the parent reactive organic gas (ROG) and is defined by

$$Y_i = \frac{A_i}{\Delta \text{ROG}} \tag{8}$$

when ΔROG (μg m^{-3}) is the reacted amount of the parent ROG. The total aerosol yield from that parent molecule is just the sum of the individual product yields,

$$Y = \sum_{i=1}^{n} Y_i \tag{9}$$

The total SOA yield from an ROG that generates n semivolatile products is computed as follows. The total concentration of product i, C_i, is related to the total amount of parent ROG that is consumed, ΔROG, by the stoichiometric coefficient for that product,

$$\alpha_i \Delta \text{ROG} = C_i \tag{10}$$

where α_i is now considered to be the mass-based stoichiometric coefficient. At any time, C_i is also equal to the sum of the gas- and aerosol-phase concentrations of species i,

$$C_i = A_i + G_i \tag{11}$$

Combining these relations with the definition of the gas/particle equilibrium partitioning coefficient, K_i, gives the expression for the total aerosol yield in terms of the individual product mass-based stoichiometric coefficients, α_i, and partitioning coefficients, K_i, and the total organic aerosol mass concentration, M_o,

$$Y = M_o \sum_{i=1}^{n} \frac{\alpha_i K_i}{1 + K_i M_o} \tag{12}$$

Note that in the limit of small organic aerosol mass concentration or high volatility of the products,

$$Y \sim M_o \sum_{i=1}^{n} \alpha_i K_i \tag{13}$$

so the SOA yield is directly proportional to M_o. In the limit of large organic aerosol mass concentration or low volatility of the products,

$$Y \sim \sum_{i=1}^{n} \alpha_i \tag{14}$$

and the total yield is independent of M_o and is just the sum of the mass-based stoichiometric coefficient of the products.

It is important to distinguish between aerosol yield, Y, and stoichiometric coefficients, α_i. Stoichiometric coefficients depend on the gas-phase chemical mechanism and represent the total amount of semivolatile product formed, in both gas and aerosol phases, per amount of parent organic compound reacted. The yield, on the other hand, which measures only the semivolatile products that have partitioned into the aerosol phase, depends on the gas-phase reaction mechanism as well as the amount of organic mass available as an absorption medium. Stoichiometric coefficients, by themselves, are therefore not sufficient to predict the amount of aerosol formation [except in the limits discussed previously, as in equation (14)]. In general, partitioning coefficients, stoichiometric coefficients, and organic aerosol mass are required to determine the SOA yield.

In the study of secondary organic aerosol formation, typically a smog chamber is initially filled with a mixture of NO_x, inorganic seed particles, and an aerosol-producing hydrocarbon. The chamber is then exposed to sunlight, or another UV source, that initiates photooxidation. (Dark experiments can also be conducted when O_3 or NO_3 is the oxidizing species.) As the hydrocarbon reacts, it forms semivolatile products that condense on the seed particles. If mass transport to the available particles cannot keep up with the rate of product formation or when a seed aerosol is not initially present, the semivolatile products accumulate in the gas phase until supersaturation is reached and nucleation occurs. Studies show that the amount of aerosol produced for a given amount of reacted ROG is independent of whether a seed aerosol is present or not (Odum et al., 1996, 1997a, 1997b). Reactions are ordinarily run until the entire initial amount of ROG is consumed. Typically, the volume of the initial seed aerosol is small compared with the organic aerosol volume generated.

In principle, the aerosol-forming potential of a parent organic could be calculated based on an atmospheric oxidation mechanism that includes all significant semivolatile product species. This ab initio approach represents a goal that is not yet attainable because of incomplete knowledge of the semivolatile oxidation products of the important aerosol-forming compounds. Thus, it is necessary to rely on experimentally measured aerosols yields. Aerosol yields Y are expressed theoretically as a function of the available M_o through equation (12). Observed aerosol yields as a function of measured M_o can be fit to equation (12) by determining the values of the stoichiometric coefficients, α_i, and gas/particle partitioning coefficients, K_i, that produce the best fit of observed and predicted yields.

Since the mid-1990s, SOA yields for over 30 aromatic and biogenic organics have been measured in the California Institute of Technology outdoor smog chamber (table VIII-B-1). In these experiments, M_o is generated completely by photooxidation of the parent. To fit the observed yields to equation (12), the unknown mix of semivolatile oxidation products for each parent compound has been represented by two empirical products, characterized by parameters α_1, K_1, α_2, and K_2. It has been determined that observed yields generally cannot be fit by assuming only a single product and that use of three products is superfluous (Odum et al., 1996). (An exception is those compounds with extremely large yields, which can be represented by a single product.) Roughly speaking, one of the empirical semivolatile products tends to represent a relatively lower vapor pressure compound and the other a relatively higher vapor pressure compound. In table VIII-B-1, low-yield aromatics (those with multiple methyl substituents) represent those species that fall on the lower of two curves of Y versus M_o described by Odum et al. (1996, 1997a). Correspondingly, high-yield aromatics (those with only one methyl substituent) represent the species falling on the higher of the two curves. These measurements were conducted in an outdoor smog

Table VIII-B-1. Secondary organic aerosol formation parameters α_i and $K_i(\mathrm{m}^3\ \mu\mathrm{g}^{-1})$ values obtained in the Caltech smog chamber

Hydrocarbon	α_1	K_1	α_2	K_2
Aromatic compounds				
Low-yield aromatics	0.038	0.042	0.167	0.0014
High-yield aromatics	0.071	0.053	0.138	0.0019
m-Xylene	0.030	0.032	0.167	0.0019
Diethylbenzene	0.083	0.093	0.220	0.0010
Methylpropylbenzene	0.050	0.054	0.136	0.0023
Biogenic compounds				
Δ^3-Carene	0.054	0.043	0.517	0.0042
β-Caryophyllene	1.000	0.0416	—	—
α-Humulene	1.000	0.0501	—	—
Limonene	0.239	0.055	0.363	0.0053
Linalool	0.073	0.049	0.053	0.0210
Ocimene	0.045	0.174	0.149	0.0041
α-Pinene	0.038	0.171	0.326	0.0040
β-Pinene	0.130	0.044	0.406	0.0049
Sabinene	0.067	0.258	0.399	0.0038
α- and *γ*-Terpinene	0.091	0.081	0.367	0.0046
Terpinene-4-ol	0.049	0.159	0.063	0.0045
Terpinolene	0.046	0.185	0.034	0.0024

Sources: Odum et al. 1996, 1997a, 1997b; Hoffmann et al., 1997; Griffin et al., 1999.

chamber, for which strict temperature control is not possible. The experiments were carried out from mid-morning to mid-afternoon, when temperatures were more or less constant over the duration of the experiment. Temperatures were generally in the range of 30–40°C.

With a purely inorganic seed aerosol in chamber experiments, the oxidation products first adsorb onto the seed particles. It can be shown that after a threshold amount of the parent compound reacts, $\Delta\mathrm{ROG}^*$, products condense onto the seed aerosol to form an organic layer that can then act as an absorptive medium. Because of interactions among the organic compounds, the products can condense onto this organic layer at concentrations lower than the saturation concentration of the compound (Seinfeld and Pandis, 1998).

For the purpose of illustration, assume that two semivolatile products are formed from the parent organic ROG_j:

$$\mathrm{ROG}_j + \text{oxidant} \rightarrow \alpha_{1j}S_1 + \alpha_{2j}S_2 \tag{15}$$

where α_{ij} represents the mass-based stoichiometric factors described earlier. Therefore, by a mass balance,

$$G_1 + A_1 = \alpha_{1j}\,\Delta\mathrm{ROG}_j \tag{16a}$$
$$G_2 + A_2 = \alpha_{2j}\,\Delta\mathrm{ROG}_j \tag{16b}$$

where G_i and A_i represent the gas- and aerosol-phase mass concentrations of product *i*, respectively, and $\Delta\mathrm{ROG}_j$ represents the amount of ROG_j consumed. As shown earlier, their gas-phase concentrations would be defined by

$$G_1 = \frac{x_1 \gamma_1 p_1^o M_{w1}}{RT} 10^6 \tag{17a}$$

$$G_2 = \frac{(1 - x_1) \gamma_2 p_1^o M_{w2}}{RT} 10^6 \tag{17b}$$

where x_1 represents the mole fraction of the first compound in the binary solution. If it is assumed that M_{w1} is approximately equal to M_{w2}, this mole fraction is given by

$$x_1 = \frac{A_1}{A_1 + A_2} \tag{18}$$

Combining equations (16)–(18), replacing vapor pressure with the expression for the equilibrium partitioning coefficient given in equation (7), and setting each γ_i equal to 1, since an ideal solution is assumed, results in a quadratic expression for x_1 in terms of α_{1j}, α_{2j}, K_{1j}, K_{2j}, and $\Delta\mathrm{ROG}_j$. This expression has real and positive roots. When solved subject to the constraints that A_1 and A_2 are greater than or equal to zero, it can then be shown that the threshold value of the parent organic that must be oxidized to form aerosol, $\Delta\mathrm{ROG}_j^*$ (μg m^{-3}) is (Seinfeld and Pandis, 1998)

$$\frac{1}{\Delta\mathrm{ROG}_j^*} = \sum_{i=1}^{2} \alpha_{i,j} K_{i,j} \tag{19}$$

Odum and colleagues (Odum et al., 1996, 1997a) showed that aerosol formation from the photooxidation of a mixture of parent hydrocarbons can be predicted simply from SOA yields for the individual parent compounds. This suggests that, at least for the case of a pure organic absorbing phase, oxidation products of different parent hydrocarbons are as soluble in a mixed organic product phase as in an organic phase consisting exclusively of their own oxidation products. Thus, the amount of organic aerosol produced from a mixture of parent hydrocarbons can be predicted as long as each $\Delta\mathrm{ROG}_j$ and the appropriate $\alpha_{i,j}$ and $K_{i,j}$ values are known. For a given M_o, the total yield for a parent organic molecule j can be calculated from equation (12):

$$Y_j = M_o \sum_i \left(\frac{\alpha_{i,j} K_{i,j}}{1 + M_o K_{i,j}} \right) \tag{20}$$

The amount of aerosol formed specifically from that parent of interest can then be calculated using this yield:

$$M_{oj} = Y_j \, \Delta\mathrm{ROG}_j \tag{21}$$

Summing the amount of aerosol generated by each parent in the mixture and adding any initial organic aerosol, M_{init}, gives the total amount of organic aerosol or

$$M_o = \sum_j M_{oj} + M_{\mathrm{init}} \tag{22}$$

Finally, combining equations (20)–(22) results in

$$1 = \sum_j \Delta\text{ROG}_j \sum_i \left(\frac{\alpha_{i,j}K_{i,j}}{1 + K_{i,j}M_o}\right) + \frac{M_{\text{init}}}{M_o} \tag{23}$$

in which M_o can be found iteratively if M_{init} is known. In this expression, ΔROG_j is actually equal to $\chi_j[\text{ROG}_j]_o$, where χ_j is the fraction of the parent hydrocarbon that reacts and $[\text{ROG}_j]_o$ is the initial concentration of ROG_j.

VIII-C. Observed Molecular Composition of Aromatic Secondary Organic Aerosols

Aromatics have been identified as aerosol precursors in a variety of laboratory chamber experiments (Stern et al., 1987; Izumi and Fukuyama, 1990; Wang et al., 1992a, 1992b; Odum et al., 1996, 1997a, 1997b; Kleindienst et al., 1999). In the studies cited, aerosol yields were measured, but generally the molecular composition of either the gas- or aerosol-phase products of aromatic oxidation was not measured. First, identification of the major gas-phase products of aromatic photooxidation has proved to be extremely challenging in its own right, and, second, since the aerosol produced generally constitutes less than, say, 10% by mass of the original aromatic, identification of the molecular composition of the aerosol phase has been even more difficult.

As noted earlier, the distribution of SOA products reflects the results of two distinct processes: the gas-phase photooxidation of the aromatic and the subsequent partitioning of the reaction products into the particle phase. First, there is a gas-phase yield associated with each first, second, third, and so on, generation product according to the oxidation chemistry resulting in its formation. As an example, consider benzaldehyde. Toluene reacts with OH radicals, with a benzaldehyde yield of approximately 10%. That is, for every 10 molecules of toluene that react with OH in the presence of sufficient NO_x, one molecule of benzaldehyde forms. Following that, there is another yield, or more precisely a gas/particle partitioning, associated with the benzaldehyde molecule, which depends on physicochemical parameters, such as the compound's vapor pressure, the amount of organic aerosol present, and the temperature. The overall measured contribution of a product like benzaldehyde to the measured aerosol mass therefore reflects both its gas-phase reaction yield and its gas/particle partitioning coefficient.

Up to the year 2000, there were only two studies in which molecular composition of aromatic oxidation products was identified in conjunction with aerosol formation. Forstner et al. (1997) reported the composition of aerosol generated in the photooxidation of seven individual aromatics, subsequently extracted from quartz filters and measured by GC/MS. In all cases, the products identified represented only 15–30% of the mass that was extracted from the filters and eluted through the gas chromatograph. This degree of identification is typical of molecular composition studies of atmospheric organic aerosol (Rogge et al., 1991, 1993a, 1993b, 1993c, 1993d). Kleindienst et al. (1999) reported gas-phase products detected but not quantified in the photooxidation of toluene, *p*-xylene, and 1,3,5-trimethylbenzene, in which aerosol formation was also measured. Specific molecular composition of aerosol products was not measured. Table VIII-C-1 summarizes products reported in these two studies.

A number of the products identified by Forstner et al. (1997) have sufficiently high vapor pressures that appreciable concentrations would not be expected to be found in the aerosol phase. The vapor pressures at 298 K for those compounds that were detected in the largest concentrations, for the parent aromatic, are estimated to be

	Torr
Benzaldehyde	1.04
Benzoic acid	0.01
3-Methyl-2,5-furandione	0.06
4-Methyl-2-nitrophenol	0.035
m-Toluic acid	0.005
4-Hydroxy-3-methylbenzaldehyde	0.011

The presence of such compounds in the aerosol phase could be a result of sampling artifacts (McDow and Huntzicker, 1990; Turpin et al., 1994) or of actual partitioning into the organic aerosol smog chamber mixture at levels well below their saturation concentration. Based on analysis of vapor pressure of many of the species identified by Forstner et al. (1997), their presence on the filter samples could have been a result of vapor adsorption onto the filter, since a denuder system was not employed in sampling during those experiments.

Little is known about any aerosol-phase reactions that might be occurring. Until very recently, it has been presumed that compounds produced in the gas phase subsequently partition into the aerosol phase, within which no additional reaction occurs. The existence of aerosol-phase chemistry involving SOA constituents is unknown and may be an important subject for future investigation. However, an outdoor chamber study by Jang and Kamens (2001) has given some new information on this subject. They have determined the nature of the gaseous and aerosol products formed in the sunlight-initiated oxidation of toluene in a propylene–NO_x–air mixture. Products in both the gas and aerosol phases were collected with a filter–filter denuder sampling system (see chapter IX for sampling artifact discussion), separated by gas chromatography, and detected with an electron-impact and chemical-impact ion trap mass spectrometry. In addition, secondary aerosols were collected directly by impaction onto an ungreased zinc selenide (ZnSe) FTIR disk (25 mm diameter). These samples were extracted and also analyzed directly by Fourier transform infrared spectrometry to obtain additional functional group information for products in the aerosol phase. Compounds analyzed from the FTIR disks represent aerosol-associated products and are not prone to the possible sampling artifacts often ascribed to filter sampling systems (see chapter IX).

Carbonyl functional groups in solvent extracts were determined by derivatizing with *O*-(2,3,4,5,6-pentafluorobenzyl) hydroxylamine hydrochloride (PFBHA) (Yu et al., 1997), and either carboxylic acids or phenolic hydroxyl groups, with pentafluorobenzyl bromide (PFBBr) (Chien et al., 1998). To confirm carboxylic acids and hydroxyl functional groups, the PFBHA-derivatives of products were re-

Table VIII-C-1. Compounds identified in secondary organic aerosol phase or gas phase from aromatic hydrocarbon photooxidation

Parent Aromatic Compound	Aerosol Species Identified	Reference
1. Toluene	2,5-Furandione	a
	2-Furaldehyde	a
	2-Hydroxy-5-nitrobenzaldehyde	a
	2-Methyl-1,4-benzoquinone	a
	2-Methyl-4,6-dinitrophenol	a
	3-Hydroxybenzaldehyde	a
	3-Methyl-2(5*H*)-furanone	a
	3-Methyl-2,5-furandione	a
	3-Methyl-4-nitrophenol	a
	3-Nitrotoluene	a
	4-Methyl-2-nitrophenol	a
	2-Methyl-4-nitrophenol	a
	5-Methyl-2(3*H*)-furanone	a
	5-Methyl-2-furancarboxaldehyde	a
	Benzaldehyde	a
	Benzoic acid	a,b
	Benzyl alcohol	a
	Dihydro-2,5-furandione	a
	m-Cresol	a
	o-Cresol	a,b
	p-Cresol	a,b
	Phenol	b
	2-Methylfuran	a
	o-, *m*-, and *p*-Nitrotoluene	b
	3-Nitrocresol isomers	b
	4-Oxo-2-pentanal or	b
	2-Methyl-2-butenedial	b
	Oxalic acid	b
	Maleic anhydride	b
	Furfural	b
	Methyl maleic anhydride	b
2. *m*-Xylene	2,4-Dimethylphenol	a
	2,5-Furandione	a
	2,6-Dimethyl-1,4-benzoquinone	a
	2,6-Dimethyl-4-nitrophenol	a
	2,6-Dimethylphenol	a
	2-Acetyl-5-methylfuran	a
	2-Methyl-4,6-dinitrophenol	a
	3,4-Dimethylfurandione	a
	3,5-Dimethyl-2(3*H*)-furanone	a
	3,5-Dimethyl-2*H*-pyran-2-one	a
	3-Methyl-2(5*H*)-furanone	a
	3-Methyl-2,5-furandione	a
	3-Methyl-4-nitrophenol	a
	3-Methylbenzyl alcohol	a
	4-Hydroxy-3-methylbenzaldehyde	a
	5-Methyl-2(3*H*)-furanone	a
	5-Methylfurfural	a
	Benzaldehyde	a
	m-Tolualdehyde	a
	m-Toluic acid	a

Table VIII-C-1. *(continued)*

Parent Aromatic Compound	Aerosol Species Identified	Reference
3. *p*-Xylene	2,5-Dimethyl-1,4-benzoquinone	a
	2,5-Dimethylphenol	a,b
	2,5-Furandione	a
	2,5-Hexanedione	a
	2-Acetyl-5-methylfuran	a
	2-Furaldehyde	a
	3,4-Dimethylfurandione	a
	3-Acetyl-2,5-dimethylfuran	a
	3-Methyl-2(5*H*)-furanone	a
	3-Methyl-2,5-furandione	a
	3-Nitrotoluene	a
	4-Methyl-2-nitrophenol	a
	4-Methylbenzyl alcohol	a
	5-Methyl-2-furancarboxaldehyde	a
	Benzaldehyde	a
	Dihydro-2,5-furandione	a
	p-Cresol	a
	p-Tolualdehyde	a,b
	p-Toluic acid	a
	2-Nitro-*p*-xylene	b
	cis- and *trans*-3-Hexene-2,5-dione	b
	4-Oxo-2-pentanal or	b
	2-Methyl-2-butenedial	b
	2,5-Dimethyl-2,5-cyclo-hexadiene-1,4-dione	b
	2,5-Dimethylfuran	b
	Methyl maleic anhydride	b
	Oxalic acid	b
4. Ethylbenzene	2,5-Furandione	a
	2-Acetyl-5-methylfuran	a
	2-Ethyl-1,4-benzoquinone	a
	3,4-Dimethylfurandione	a
	3-Ethyl-2,5-furandione	a
	3-Methyl-2,5-furandione	a
	3-Methyl-5-ethyl-2(3*H*)-furanone	a
	3′-Nitroacetophenone	a
	4′-Hydroxy-3′-nitroacetophenone	a
	4-Ethylbenzoic acid	a
	5-Ethyl-2(3*H*)-furanone	a
	5-Ethyl-2-furaldehyde	a
	5-Methyl-2(3*H*)-furanone	a
	5-Methyl-2-furancarboxaldehyde	a
	Acetophenone	a
	Benzaldehyde	a
	Dihydro-2,5-furandione	a
	Dihydro-5-methyl-2(3*H*)-furanone	a
	Ethyl-nitrophenol	a
	Phenol	a
	sec-Phenethyl alcohol	a

(continued)

Table VIII-C-1. (*continued*)

Parent Aromatic Compound	Aerosol Species Identified	Reference
5. *m*-Ethyltoluene	2′-Hydroxy-5′-methylacetophenone	a
	2,5-Furandione	a
	2,5-Hexanedione	a
	2-Acetyl-5-methylfuran	a
	3-Acetyl-2,5-dimethylfuran	a
	3-Ethyl-2(5*H*)-furanone	a
	3-Ethyl-2,5-furandione	a
	3-Ethyl-5-methyl-2(3*H*)-furanone	a
	3-Ethylbenzaldehyde	a
	3-Ethylbenzoic acid	a
	3′-Methylacetophenone	a
	3-Methyl-2,5-furandione	a
	3-Methyl-5-ethyl-2(3*H*)-furanone	a
	4′-Hydroxyacetophenone	a
	4-Ethylnitrobenzene	a
	5-Ethyl-2(3*H*)-furanone	a
	5-Ethyl-2-furaldehyde	a
	5-Methyl-2(3*H*)-furanone	a
	Acetophenone	a
	α-Methylbenzene methanol	a
	Benzaldehyde	a
	Benzoic acid	a
	Dihydro-2,5-furandione	a
	Dihydro-5-methyl-2(3*H*)-furanone	a
	m-Tolualdehyde	a
	m-Toluic acid	a
	Phenol	a
6. *p*-Ethyltoluene	2,5-Furandione	a
	2,5-Heptadione	a
	2,5-Hexanedione	a
	2-Acetyl-5-methylfuran	a
	3,4-Dimethylphenol	a
	3-Ethyl-2(5*H*)-furanone	a
	3-Ethyl-2,5-furandione	a
	3-Methyl-2(5*H*)-furanone	a
	3-Methyl-2,5-furandione	a
	4′-Hydroxy-3′-nitroacetophenone	a
	4′-Hydroxyacetophenone	a
	4-Ethylbenzaldehyde	a
	4-Ethylbenzoic acid	a
	4′-Methylacetophenone	a
	4-Methyl-2-nitrophenol	a
	5-Methyl-2(3*H*)-furanone	a
	5-Methyl-2-furancarboxaldehyde	a
	Acetophenone	a
	Benzaldehyde	a
	p-Cresol	a
	p-Tolualdehyde	a
	p-Toluic acid	a
	Phenol	a
	Terephthaldicarboxaldehyde	a

Table VIII-C-1. (*continued*)

Parent Aromatic Compound	Aerosol Species Identified	Reference
7. 1,2,4-Trimethylbenzene	2,3,5-Trimethyl-1,4-benzoquinone	a
	2,4-Dimethylbenzaldehyde	a
	2,4-Dimethylphenol	a
	2.5-Dimethylbenzaldehyde	a
	2,5-Dimethylphenol	a
	2,5-Furandione	a
	2-Acetyl-5-methylfuran	a
	3,4,5-Trimethyl-2(3*H*)-furanone	a
	3,4-Dimethylbenzaldehyde	a
	3,4-Dimethylbenzoic acid	a
	3,4-Dimethylfurandione	a
	3,4-/4,5-Dimethyl-2(3*H*)-furanone	a
	3-Acetyl-2,5-dimethylfuran	a
	3-Methyl-2,5-furandione	a
	3-Methyl-2,5-hexanedione	a
	4-Methylphthalic acid	a
	5-Methyl-2(3*H*)-furanone	a
	5-Methyl-2-furancarboxaldehyde	a
	Benzaldehyde	a
8. 1,3,5-Trimethylbenzene	3,5-Dimethylbenzaldehyde	b
	2,4,6-Trimethylphenol	b
	2-Methyl-4-oxo-2-pentanal	b
	3-Methyl-5-methylidene-(5*H*)-2-furanone	b
	Methyl maleic anhydride	b
	Oxalic acid	b

[a]Forstner et al. (1997).
[b]Kleindienst et al. (1999).

derivatized by silylation using *N*,*O*-bis(trimethylsilyl)-trifluoro-acetamide (BSTFA) (Yu et al., 1999).

As indicated in previous sections of this book, the photooxidation reaction of toluene in the gas phase leads to substituted aromatics, multifunctional non-aromatic ring-retaining and ring-opening products. In the Jang and Kamens (2001) study, the following ring-cleaved multifunctional oxy-products retaining a carboxylic acid, a carbonyl, and a hydroxyl group were newly identified: 4,5-dioxo-2-hexenoic acid [$CH_3C(=O)C(=O)CH{=}CHC(=O)OH$], 2-methyl-4,5-dioxo-2-pentenoic acid [$HC(=O)C(=O)CH{=}C(CH_3)C(=O)OH$], 6-oxo-2,4-heptadienoic acid [$CH_3C(=O)CH{=}CHCH{=}CHC(=O)OH$], 5-methyl-6-oxo-2,4-hexadienoic acid [$HC(=O)C(CH_3){=}CH{-}CH{=}CHC(=O)OH$], 4,5-dioxo-2-pentenoic acid [$HC(=O)C(=O)CH{=}CHC(=O)OH$], 2-methyl-4-oxo-2-butenoic acid [$HC(=O)CH{=}C(CH_3)C(=O)OH$], 4-oxo-2-butenoic acid [$HC(=O)CH{=}CHC(=O)OH$], methylglyoxylic acid [$CH_3C(=O)C(=O)OH$], glyoxylic acid [$HC(=O)C(=O)OH$], 2-methyl-2-hydroxy-3-oxo-1,4-butandial [$HC(=O)C(=O)C(OH)(CH_3)C(=O)H$], and hydroxy-$C_7$-tricarbonyls. The newly found ring-retaining products were methylcyclohexene tricarbonyls; for example:

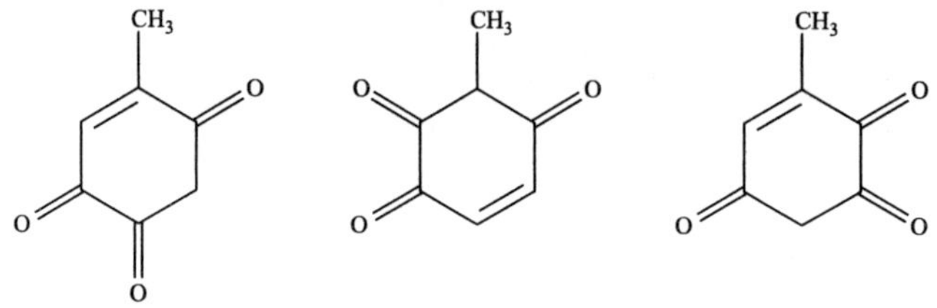

and hydroxy-methyl-cyclohexene tricarbonyls; for example:

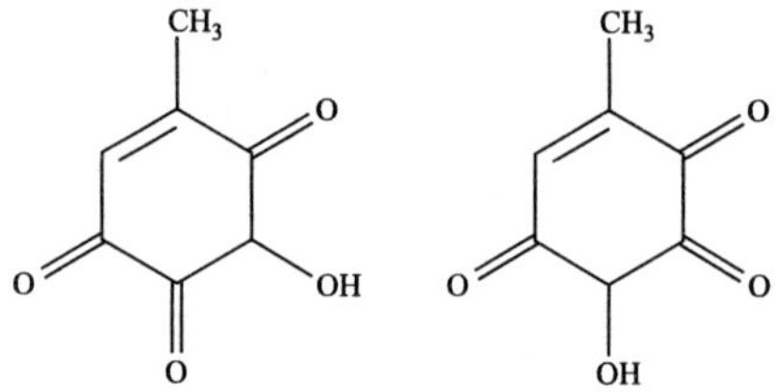

These highly oxidized products of gas-phase photooxidation reactions are semi-volatile; they can migrate from the gas to the particle phase, and they contribute to some of the major components in the particle phase. The carboxylic acids, carbonyls, and hydroxyl groups of reaction products may then be involved in other process in the particle phase.

The major aromatic products found in the particle phase under the high NO_x conditions of the experiment of Jang and Kamens (2001) included 2-methly-4-nitrophenols, 4-methyl-2-nitrophenol, and methyldinitrophenol isomers. The principal non-aromatic ring-retaining compounds were 2-methyl-1,4-benzoquinone,

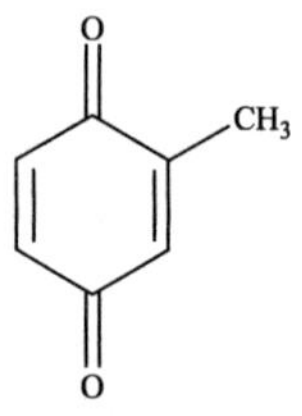

and methylcyclohexene tricarbonyls as illustrated previously. The major ring-opening products in the particle phase were 4-oxo-2-butenoic acid, 4,6-dioxo-2-hexenoic acid, 2-hydroxy-3-penten-1,5-dial [HC(=O)CH=CHCH(OH)C(=O)H], 2-hydroxy-5-oxo-3-hexenal [CH_3C(=O)CH=CHCH(OH)C(=O)H], 5-hydroxy-4,6-dioxo-2-heptenal [CH_3C(=O)CH(OH)C(=O)CH=CHC(=O)H], and their structural isomers.

The partitioning behavior of major secondary aerosol products was also explored by comparing experimentally observed and predicted ${}^{i}K_p$ values of products. Compounds that had both aldehydic and OH functional groups exhibited much higher experimental partitioning coefficients (more in the particle phase) than would be predicted from vapor pressure and activity coefficient relationships (Pankow, 1994a, 1994b; Jang et al., 1997; see chapter IX). The Jang and Kamens study (2001) suggests that chemical reactions in the particle phase may dramatically influence observed partitioning behavior. Possible particle reactions include hemiacetal or acetal formation as a "zipping" reaction. This can be illustrated for the case of 2-hydroxy-propanedial, because its predicted liquid vapor pressure is too high (~ 0.2 Torr) to permit it to appear to an appreciable extent in the particle phase. However, reacting it with formaldehyde, methylgloxal, or another 2-hydroxy-propanedial molecule in the particle phase will produce corresponding hemiacetal products with the vapor pressures that are respectively 1.3, 2, and 6 orders of magnitude lower than the original 2-hydroxy propanedial. These lower vapor pressures permit hemiacetals to exist to a greater extent in the particle phase than their parent compounds. Last, and most important, hemiacetals or acetals possibly can be "unzipped" to their original aldehydes and alcohols *via* backward reactions that involve hydrolysis (during sample workup, or instrumental analysis procedures that involve heat). Hence, these types of compounds can go undetected when conventional methods are used, and they have only recently been observed on secondary aerosols by Paul Ziemann's research group when they used direct "cold" particle collection and subsequent direct thermal desorption into the ion source of a mass spectrometer (Tobias and Ziemann, 2000).

VIII-D. Atmospheric Oxidation Mechanisms Inferred from Aromatic Aerosol Molecular Composition

Identification of aerosol-phase products provides a limited picture of the full spectrum of gas-phase oxidation products. This is because only those products with sufficiently low vapor pressure will be found in the aerosol phase, and such products generally constitute only a modest fraction of the total set of oxidation products. Nonetheless, SOA molecular speciation does provide important information that, together with gas-phase product identification, allows evaluation of gas-phase photooxidation mechanisms.

The gas-phase reaction paths that lead to those aerosol products that have been identified are generally not well established. Forstner et al. (1997) postulated mechanisms by which a number of the more prevalent SOA compounds found in their study could have been formed; however, almost every mechanism contains steps that are not well defined. We will review several of the possible mechanisms that lead to products identified in that study. This discussion is based on that in section II-E. We remind the reader that it is likely that the presence of these volatile products in the filter samples reported by Forstner et al. (1997) resulted from adsorption from the gas phase onto the filter surface rather than as a result of these products constituting the aromatic aerosol. The actual aerosol appears to have been largely contained in the 70% or so of the filter material that was not identified.

VIII-D-1. Toluene

The toluene–OH reaction results in about 10% H-atom abstraction from the methyl group and 90% OH addition to the ring (see section II-B-2). The

H-atom abstraction pathway of the toluene–OH reaction in the presence of NO_x leads to benzaldehyde, which was identified as a product by both Forstner et al. (1997) and Kleindienst et al. (1999) (table VIII-C-1). In addition, Forstner et al. (1997) identified benzoic acid as an aerosol product. It is presumed that benzoic acid is a product of photooxidation of benzaldehyde. A possible mechanism for this is given in figure VIII-D-1.

The principal photooxidation products of toluene identified by Forstner et al. (1997) in filter samples, in their order of abundance, were 3-methyl-2,5-furandione, dihydro-2,5-furandione, 2-methyl-4-nitrophenol, 2,5-furandione, 3-methyl-4-nitrophenol, benzoic acid, 2-methyl-4,6-dinitrophenol. A possible path-

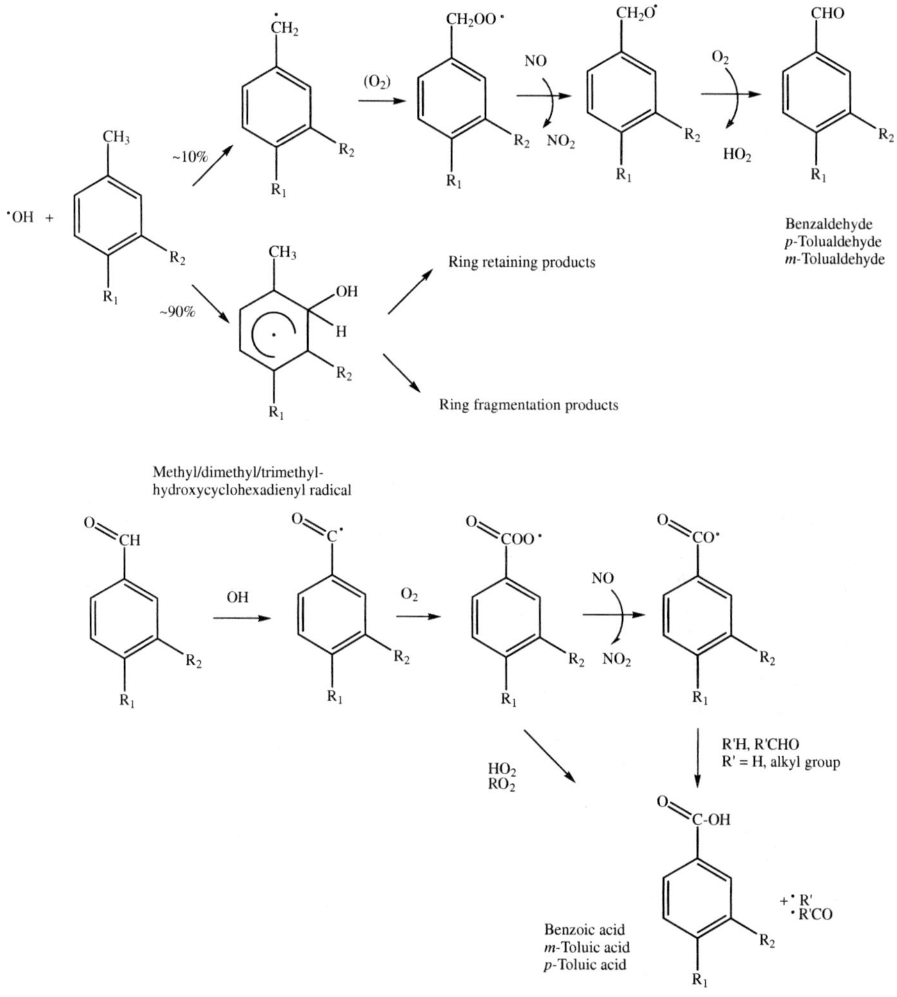

Figure VIII-D-1. Possible mechanism for the reaction of OH radical with toluene, *m*-xylene, and *p*-xylene showing the initial reaction split and details of the abstraction pathway; R_1, $R_2 = H$ for toluene; $R_1 = CH_3$, $R_2 = H$ for *p*-xylene; $R_1 = H$, $R_2 = CH_3$ for *m*-xylene.

way to form 2,5-furandione, suggested by Forstner et al. (1997), is shown in figure VIII-D-2.

Of the fraction of toluene SOA identified by Forstner et al. (1997), 3-methyl-2,5-furandione was the predominant species. Based on the photoisomerization and oxidation mechanism for unsaturated dicarbonyls proposed by Bierbach et al. (1994), the formation of 3-methyl-2,5-furandione is consistent with that from 2-methyl-2-butenedial. Figure VIII-D-3, panel (A), shows the detailed mechanism to form this anhydride from the unsaturated dicarbonyl, in addition to the rearrangement mechanism to form 3-methyl-2(5*H*)-furanone.

Significant quantities of dihydro-2,5-furandione (succinic anhydride) were detected in the toluene aerosol by Forstner et al. (1997). Previous studies have

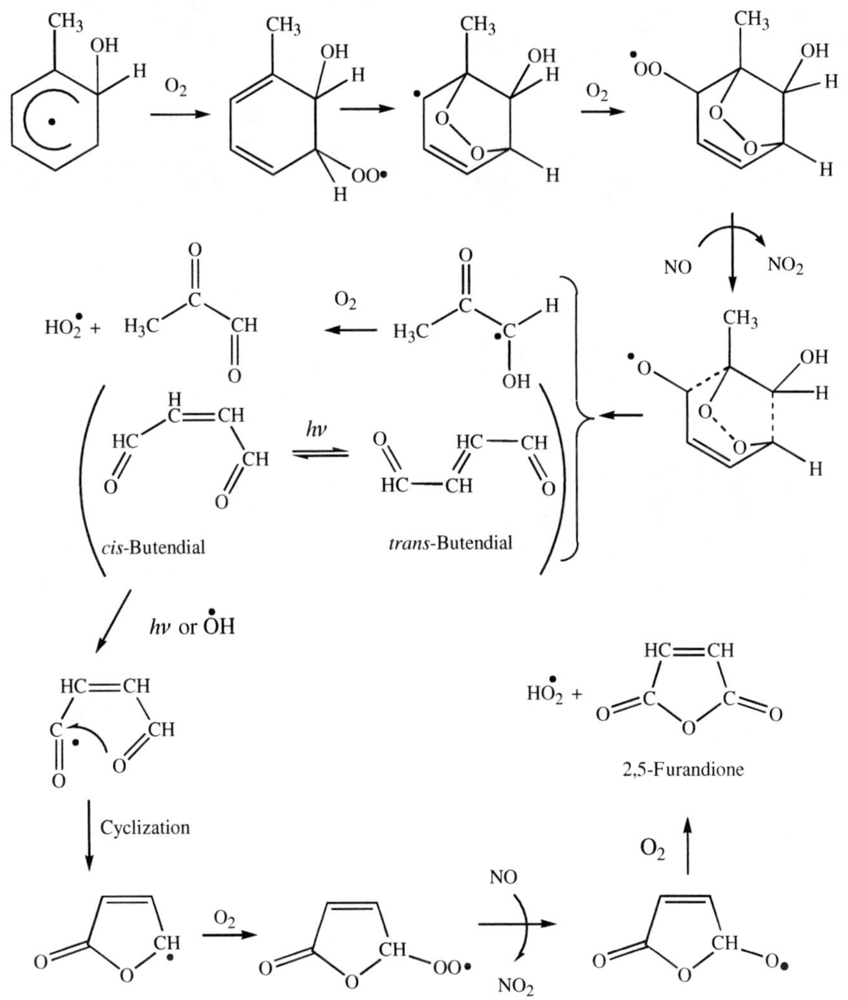

Figure VIII-D-2. Suggested reaction pathway for the OH-radical reaction with toluene that leads to 2,5-furandione; figure based on that given by Forstner et al. (1997).

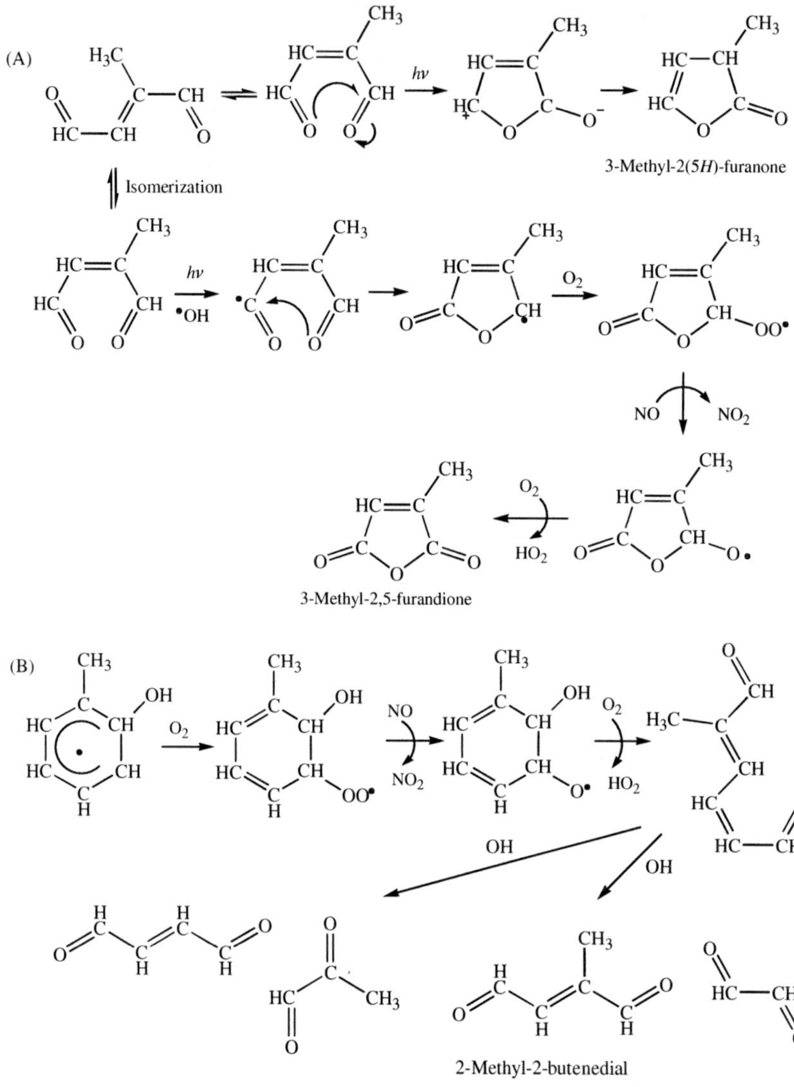

Figure VIII-D-3. Suggested mechanism of the OH-radical reaction with toluene that leads to 3-methyl-2,5-furandione and 3-methyl-2(5H)-furanone [panel (A)]; pathway to 2-methyl-2-butenedial from the substitution of the OH radical in the *ortho*-position [panel (B)].

not reported saturated compounds from aromatic photooxidation. Photoreduction of ketones in the solution phase is well established, and there is evidence to support intermolecular hydrogen abstraction of cyclic alkenes after photoexcitation (Turro, 1978). Given the lack of prior evidence of succinic anhydride as a product in gas-phase toluene photooxidation, it is possible that succinic anhydride forms from the intermolecular hydrogen abstraction of 2,5-furandione in the particle phase. Water, nitrophenols, and benzoic acid in the particle phase

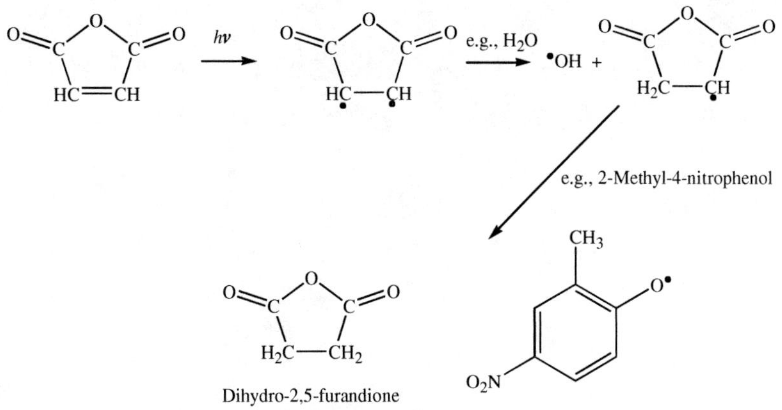

Figure VIII-D-4. A suggested possible photolytically induced transformation of 2,5-furandione to dihydro-2,5-furandione; figure based on that given by Forstner et al. (1997).

can contribute to a slightly acidic environment in the particle, allowing facile proton (H^+) exchange. Figure VIII-D-4 outlines a possible mechanism to produce dihydro-2,5-furandione.

Figure VIII-D-5 gives a possible mechanism for formation of observed nitrogen-containing aerosol products. These pathways could also involve NO_3-radical reactions, since NO_2 levels in these smog chamber experiments were about an order of magnitude higher than ambient levels.

VIII-D-2. *m*- and *p*-Xylene

Three compounds (3-methyl-2,5-furandione, *m*-toluic acid, and 2,5-furandione) comprise 75% of the *m*-xylene SOA identified by Forstner et al. (1997). The mechanism leading to *m*-toluic acid is readily extended from the corresponding toluene–OH mechanism that results in benzaldehyde and benzoic acid. Bierbach et al. (1994) postulate a mechanism by which 4-oxo-2-pentenal can form 2,5-furandione. This mechanism involves the initial loss of the hydrogen adjacent to the carbonyl group by photolysis or OH abstraction. Subsequent addition of oxygen to the alkyl radical, formation of an alkoxy radical, and cyclization lead to 2,5-furandione (figure VIII-D-6). The mechanism to yield 3-methyl-2,5-furandione is outlined in figure VIII-D-3 from the photoisomerization and oxidation of 2-methyl-2-butendial. Since 3-methyl-2,5-furandione contributed 60% by mass to the SOA measured by Forstner et al. (1997), another pathway leading to this compound may exist, in addition to the ring-fragmentation that leads to 2-methyl-2-butenedial. Extending the mechanism in figure VIII-D-6 to 2-methyl-4-oxo-pentenal, another pathway to 3-methyl-2,5-furandione is proposed.

Similar to toluene and *m*-xylene, OH can either abstract a hydrogen from one of the methyl groups on *p*-xylene or it can add to the ring. The hydrogen abstraction pathway leads to *p*-tolualdehyde, which is identified as a significant constituent of *p*-xylene SOA. Further oxidation of *p*-tolualdehyde leads to the

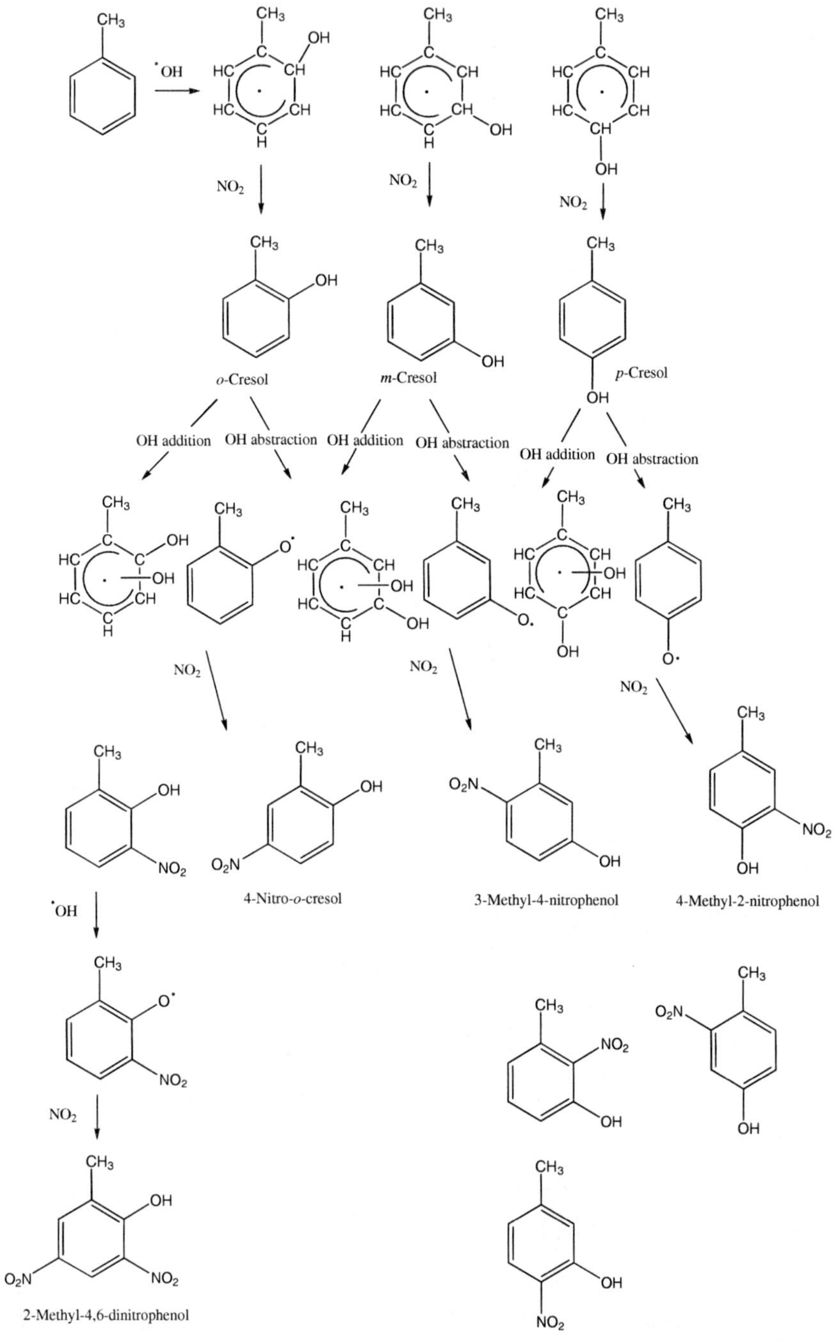

Figure VIII-D-5. A possible mechanism for formation of observed aerosol products of the reactions of NO_2 with the methyl hydroxycyclohexadienyl radical formed in the OH-radical reaction with toluene; figure is based on that given by Forstner et al. (1997).

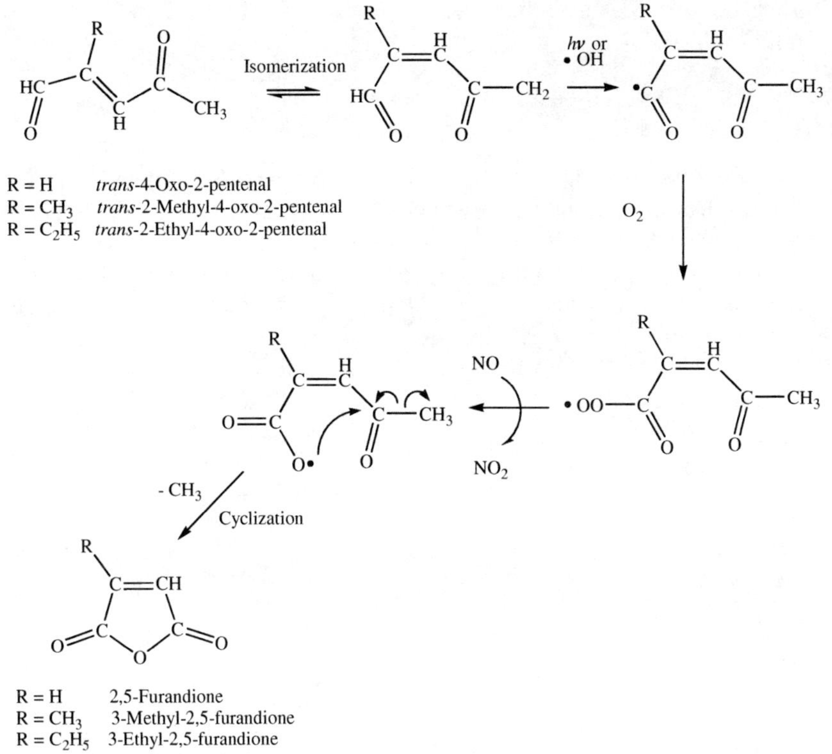

Figure VIII-D-6. Suggested mechanism for the formation of 2,5-furandione, 3-methyl-2,5-furandione, and 3-ethyl-2,5-furandione from *trans*-4-oxo-2-pentenal, *trans*-2-methyl-4-oxo-2-pentenal, and *trans*-2-ethyl-4-oxo-2-pentenal; figure is based on that given by Forstner et al. (1997).

formation of *p*-toluic acid. Andino et al. (1996) also studied ring-fragmentation products from the bicyclic oxy radicals from *p*-xylene–OH addition. Because of *p*-xylene's symmetry, only 2-methyl-2-butenedial and 3-hexene-2,5-dione result. Again, as described in figure VIII-D-3 and in the previous sections, 3-methyl-2,5-furandione is formed from the photoisomerization and oxidation of 2-methyl-2-butenedial. As suggested in the toluene mechanism, dihydro-2,5-furandione may arise from a photolytic hydrogenation process of 2,5-furandione in the particle phase. In an analogous fashion, formation of 2,5-hexanedione can be postulated. The 3-hexene-2,5-dione can partition into the particle phase and may be photolytically excited to break one bond of the double bond to produce an alkyl biradical. Water, *p*-toluic acid, and nitrophenols have pK_a values of 14.9, 4.38, and approximately 8, respectively (Streitwieser and Heathcock, 1976). Hence, upon interaction with this alkyl biradical, these compounds can yield a hydrogen, resulting in an alkyl radical and a hydroxyl or phenoxy radical. The alkyl radical will abstract another hydrogen from another water molecule or nitrophenol molecule to yield 2,5-hexanedione. The other saturated organic that comprises a significant portion of *p*-xylene SOA identified by Forstner et al. (1997) is dihydro-2,5-furandione. Its origin from 2,5-furandione has been outlined in figure VIII-D-4.

VIII-D-3. Formation of α,ω-Dicarboxylic Acids

One class of compounds that has been identified in aerosol samples, both in the laboratory and in the ambient atmosphere, is the dicarboxylic acids. Such compounds, moreover, have extremely low vapor pressures and, if formed in the gas phase, will partition preferentially into the organic aerosol phase. Indeed, it is likely that dicarboxylic acids are a substantial component of SOA from many classes of hydrocarbon parent molecules.

Carboxylic acids arise from the reactions of stabilized, monoalkyl-substituted Criegee biradicals with water:

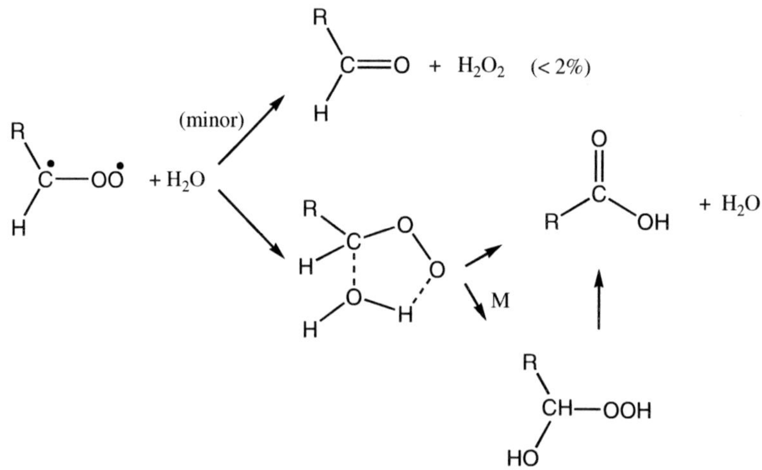

Under atmospheric conditions, the reaction of stabilized biradicals with water vapor will dominate over reactions with other species such as aldehydes, SO_2, CO, and NO_2. The yields of the stabilized biradicals depend on the alkene precursor. Generally, the stabilized biradicals are produced at significant yields ranging from 0.10 to 0.50 (Calvert et al., 2000). Calvert and Madronich (1987) estimated that for tropospheric air masses typical of those in the eastern United States, organic acids generated through the Criegee biradical pathway ranged from ~20% of the total carboxylic acids for very polluted air to ~5–10% for somewhat cleaner masses.

A second route for production of carboxylic acids is the reaction of peroxyacyl radicals and HO_2:

$$CH_3C(=O)OO^\bullet + HO_2 \rightarrow CH_3C(=O)OOH + O_2 \quad \text{(a)}$$
$$CH_3C(=O)OO^\bullet + HO_2 \rightarrow CH_3C(=O)OH + O_3 \quad \text{(b)}$$

For reaction of $CH_3C(=O)OO$ the ratio $k_b/(k_a + k_b)$ was determined to be 0.25 ± 0.05 by Niki et al. (1985b), 0.33 ± 0.07 by Moortgat et al. (1989), and 0.12 ± 0.04 by Crawford et al. (1999). Formation of O_3 along with acetic acid is observed. A similar reaction sequence starting with propanal was also observed with a similar yield of O_3 (Lightfoot et al., 1992). It is, therefore, reasonable to expect this reaction sequence to be operative for other aldehydes. Under polluted conditions where NO_x levels are high, this formation pathway is not expected to be important; for continental conditions, this pathway may represent a potential source of carboxylic acids.

In the aqueous phase, the OH radical can oxidize hydrated formaldehyde into formic acid:

$$H_2C(OH)_2 + OH\ (+\ O_2) \rightarrow HO_2 + HC(=O)OH + H_2O$$

McElroy and Waygood (1991) confirmed that the above reaction produces formic acid with unit yield. Similarly, higher carbonyl compounds may also be precursors of organic acids in the atmospheric aqueous phase (Chebbi and Carlier, 1996). In addition, a laboratory study of Vallalta et al. (1996) shows irreversible loss of gas-phase $CH_3C(O)OO$ to pure liquid water surfaces due to rapid hydrolysis:

$$CH_3C(=O)OO + H_2O \rightarrow CH_3C(=O)OH + HO_2$$

with a rate coefficient $> 7 \times 10^5\ M^{-1}\ s^{-1}$. The acid formed in the aqueous solution can be volatilized under suitable pH conditions.

Dicarboxylic acids usually occur in the α,ω-form $[HO(O{=}C){-}(CH_2)_n{-}C({=}O)OH]$. The immediate precursor of α,ω-dicarboxylic acids is likely the corresponding α,ω-dialdehyde or α,ω-oxo-acids. Photooxidation of a number of hydrocarbons, such as aromatics and cycloalkenes, leads to formation of α,ω-dialdehydes. For example, aerosol products found in the cyclohexene–O_3 system include α,ω-dialdehydes, α,ω-oxo-acids, and α,ω-diacids (table VIII-D-1).

At present, we do not understand the mechanism of formation of α,ω-dicarboxylic acids. For example, Hatakeyama et al. (1985) suggested that the sequential oxidation of C_5 and C_6 difunctional compounds takes place mainly on the wall (or particles) after deposition. The following questions need to be addressed to understand the source of the dicarboxylic acids:

Table VIII-D-1. Cyclohexene–O_3 Aerosol Products

Product	Structure	Yield	Reference
Succinic acid	$HO(O{=})C(CH_2)_2C({=}O)OH$	0.011	a
Glutaraldehyde	$HC({=}O)(CH_2)_3C({=}O)H$	0.02	a
5-Oxo-pentanoic acid	$HC({=}O)(CH_2)_3C({=}O)OH$	0.048	a
Glutaric acid	$HO(O{=})C(CH_2)_3C({=}O)OH$	0.029	a
Adipaldehyde	$HC({=}O)(CH_2)_4C({=}O)H$	0.011	a
6-Oxo-hexanoic acid	$H(C{=}O)(CH_2)_4C({=}O)OH$	0.047	a
Adipic acid	$HO(O{=})C(CH_2)_4C({=}O)OH$	0.042	a
Total yield		0.208	a
		0.183 ± 0.0036	b

[a]Hatakeyama et al. (1985).
[b]Izumi et al. (1988).

1. What is the mechanism of oxidation of α,ω-dialdehydes to α,ω-oxo-acids, and further to α,ω-diacids?
2. Where does the oxidation take place: on particle surfaces or in the gas phase where the process is initiated by oxidants such as OH and HO_2?
3. Are all α,ω-dialdehydes converted to α,ω-oxo-acids, provided that sufficient time and oxidizing surface or oxidants are available?

VIII-E. Possible Influence of Aromatic Aerosols on the Chemistry of Urban Ozone Formation

Formation of organic aerosols can potentially influence ozone chemistry in several ways: (1) First-generation gas-phase oxidation products that might, if they remain in the gas phase, continue to be oxidized by OH radicals in a way that promotes ozone formation, can partition into the aerosol phase and be removed from further gas-phase photooxidation chemistry. (2) Aerosols can serve as sites for condensation of other gas-phase species involved in photooxidation chemistry, thus removing these species from further participation in ozone formation. (3) Aerosols scatter and absorb solar radiation and, in so doing, alter the local actinic flux that drives photochemical reactions and ozone formation.

Aromatic oxidation products resulting from both ring-retaining and ring-cleavage reactions may react with the OH radical, themselves contributing to ozone formation. Prominent examples are glyoxal and methylglyoxal. Most of these products are sufficiently small molecules that they continue to reside in the gas phase even in the presence of an absorbing organic aerosol phase. As ring-retaining products acquire more functional groups, their volatility decreases, and they become more susceptible to condensation into the aerosol phase. At the same time, the ozone-forming potential of some of these molecules, as measured by kinetic and mechanistic reactivity, tends to decrease relative to those of the parent aromatic hydrocarbon. Both of these factors tend to limit the role of these products in further ozone formation.

VIII-E-1. Condensation of Gas-Phase Species on Aerosols

Organic aerosols, or for that matter any aerosols, can serve as sites for condensation of gas-phase species. A prominent example is the condensation of NH_4NO_3 on aerosols, which removes HNO_3 from the gas phase (Meng et al., 1997). Since HNO_3 is a product of photooxidation that is not recycled back into active ozone formation chemistry, whether it remains in the gas phase or is deposited on aerosol in the form of nitrate will not affect ozone formation chemistry. Perhaps the most prominent species involved in ozone formation chemistry that has been considered as susceptible to removal by aerosols is the HO_2 radical (Mozurkewich et al., 1987; Hanson et al., 1992; Cooper and Abbatt, 1996). The fraction of HO_2 radicals removed on collision with various aerosols and surfaces characteristic of the atmospheric aerosols can be large, but it is highly dependent on the composition of the aerosol. Mozurkewich et al. (1987) concluded from their study that reactions of HO_2 with Cu(I) and Cu(II) ions in submicrometer aerosols may have a significant role in converting atmospheric odd hydrogen radicals into H_2O_2; they found the mass accommodation coefficient for HO_2 on aqueous particles to be greater than 0.2. Hanson et al. (1992) found that the lower limits for the accommodation coefficients of HO_2 on 28% w/w sulfuric acid and water (at 275 K), respectively, were 0.05 and 0.01. Cooper and Abbatt (1996) reported that the mass accommodation coefficient was > 0.2 for HO_2 on supercooled sulfuric acid solutions (55 wt%; doped with 0.1 M $CuSO_4$; 223 K). During the year-long field studies made at Mauna Loa, Hawaii during 1991–1992, the measured $[HO_2] + [RO_2]$ were in reasonable accord with those calculated from the extensive list of measured precursors and j-values. However, at certain times, the measured values were well below those expected theoretically (Cantrell et al., 1996). Aerosol-surface-area measurements made during a few days correlated reasonably well with simultaneously measured condensation nuclei (CN) measurements. The CN measurements were made continuously during the studies, and these were used as a surrogate for aerosol surface area during the study. For several days of highest CN content (used to estimate aerosol surface area), the amount of decrease appeared to be roughly proportional to the increased CN count (and, presumably, the surface area) measured at that time. Whether this reflects removal of the HO_2 and RO_2 radicals by aerosol, one suggested explanation, is uncertain, but this remains a viable hypothesis. Obviously, additional atmospheric observations are needed to investigate the potential role of heterogeneous removal of peroxy radicals by aerosols.

VIII-E-2. Effect of Aerosols on Atmospheric Photolysis Reactions

Atmospheric ozone formation chemistry is driven by photolysis reactions. As discussed in chapter VII, the first-order rate coefficient for photolysis (s^{-1}) of an atmospheric species i is given by

$$j_i = \int_{\lambda_1}^{\lambda_2} \sigma_i(\lambda, T)\phi_i(\lambda, T)F(\lambda, z)d\lambda \tag{24}$$

where $\sigma_i(\lambda, T)$ (cm^2) is the absorption cross section of species i, $\phi_i(\lambda, T)$ is the quantum yield for photodecomposition, and $F(\lambda, z)$ (photons cm^{-2} nm^{-1} s^{-1}) is the solar actinic flux at altitude z in the atmosphere. The limits λ_1 and λ_2 represent the extremes of the solar spectral range for the troposphere, about $\lambda_1 = 290$ nm to $\lambda_2 = 700$ nm. Whereas σ_i and ϕ_i depend only on the properties of the molecule in question and the wavelength, the actinic flux depends on the presence in the atmosphere of gases, aerosols, and clouds (as well as latitude, solar zenith angle, etc.). See also section VII-C-1.

Aerosols both scatter and absorb solar radiation. The presence of scattering aerosols tends to enhance the actinic flux, whereas absorbing particles tend to reduce the actinic flux by removing photons in the air volume. Studies of the effects of aerosols on UV radiation have been carried out by Liu et al. (1991), Forster (1995), Reuder et al. (1996), Ma and Guicherit (1997), Erlick and Frederick (1998), Erlick et al. (1998), Papayannis et al. (1998), Repapis et al.

(1998), Jacobson (1999), and Reuder and Schwander (1999). Ma and Guicherit (1997) studied the effects of air pollutants on surface UV-B radiation by using a multilayer radiative transfer model, and they found that the increase of tropospheric pollution in non-urban regions over the last 50–100 years has reduced the surface UV-B, offsetting the effect of the increase due to stratospheric ozone depletion. More recently, Reuder and Schwander (1999) showed that potential day-to-day variability in non-urban atmospheric aerosols produces changes of spectrally integrated actinic flux by 10–50%. Jacobson (1999) demonstrated that nitrated and aromatic aerosols may be important in reducing UV radiation. Measurements in Athens, Greece, showed that urban aerosols can reduce clear-sky UV-B radiation by as much as 40% (Papayannis et al., 1998; Repapis et al., 1998). Studies of the effects of tropospheric aerosols on photolysis rates have been carried out by Demerjian et al. (1980), Ruggaber et al. (1994), Lantz et al. (1996), Castro et al. (1997), Dickerson et al. (1997), Jacobson (1998), Landgraf and Crutzen (1998), and Liao et al. (1999). Scattering aerosols in the boundary layer accelerate photochemical reactions within and above the aerosol layer (Ruggaber et al., 1994; Dickerson et al., 1997; Jacobson, 1998; Landgraf and Crutzen, 1998), while absorbing aerosols decrease photolysis rates (Ruggaber et al., 1994; Dickerson et al., 1997; Jacobson, 1998). Effects of aerosols on photolysis rates are sensitive to the aerosol optical depth, single-scattering albedo, and vertical profile, as well as relative humidity and solar zenith angle (Ruggaber et al., 1994).

Aerosols resulting from aromatic oxidation will both scatter and absorb radiation (Jacobson, 1998), although optical properties of such compounds are still somewhat uncertain. Liao et al. (1999) studied the effect of aerosols on tropospheric photolysis rates for the idealized cases of pure $(NH_4)_2SO_4$, pure soot, or mixtures of the two. Optical properties for these particles are better known than for organic aerosols from aromatics, and a calculation such as that presented by Liao et al. (1999) can serve to define the limits of behavior from purely scattering to strongly absorbing particles, as well as combinations in between.

When $(NH_4)_2SO_4$, soot, and mixtures thereof are present, Liao et al. (1999) assumed that an aerosol layer extends from the Earth's surface to 5 km altitude, with the mass concentration of each species decreasing linearly from its value at the surface to 0.1 of its surface value at 3 km altitude and remaining constant from 3 to 5 km. Column burdens of 12.5 mg m^{-2} for $(NH_4)_2SO_4$ and 2.5 mg m^{-2} for soot are representative of continental conditions. For urban conditions, column burdens of 125 mg m^{-2} for $(NH_4)_2SO_4$ and 25 mg m^{-2} for soot were assumed. Whereas an $(NH_4)_2SO_4$ column burden of 125 mg m^{-2} is somewhat high for sulfate alone, this value can account for the presence of other essentially non-absorbing components, such as nitrate. When a cloud was present, a stratus cloud layer centered at 950 m altitude was assumed, with liquid water content increasing linearly with height in the cloud and with a vertically averaged value of 0.2 g m^{-3} with an effective cloud drop radius of 10 μm.

Liao et al. (1999) performed detailed calculations of photolysis as a function of altitude, aerosol, and cloud state for the following 14 photochemical reactions:

$$O_3 + h\nu \rightarrow O(^3P) + O_2$$
$$O_3 + h\nu \rightarrow O(^1D) + O_2$$
$$NO_2 + h\nu \rightarrow NO + O$$
$$NO_3 + h\nu \rightarrow NO + O_2$$
$$NO_3 + h\nu \rightarrow NO_2 + O$$
$$N_2O_5 + h\nu \rightarrow NO_3 + NO_2$$
$$H_2O_2 + h\nu \rightarrow OH + OH$$
$$HONO + h\nu \rightarrow OH + NO$$
$$HO_2NO_2 + h\nu \rightarrow HO_2 + NO_2$$
$$HO_2NO_2 + h\nu \rightarrow OH + NO_3$$
$$HCHO + h\nu \rightarrow HCO + H$$
$$HCHO + h\nu \rightarrow CO + H_2$$
$$CH_3OOH + h\nu \rightarrow CH_3O + OH$$
$$CH_3COCH_3 + h\nu \rightarrow CH_3CO + CH_3$$

When $(NH_4)_2SO_4$ aerosol is added to a clear atmosphere, $j[O_3 \rightarrow O(^1D)]$ increases everywhere except at the surface, because the increased diffuse actinic flux offsets the loss in direct actinic flux at the surface. Soot aerosol in a cloud-free atmosphere reduces photolysis rates at all altitudes, with the maximum reduction occurring near the surface where the soot mass concentration is highest. In a cloud atmosphere, soot aerosol absorbs both incident radiation and radiation reflected by the high-albedo cloud layer; hence, it significantly reduces j-values above the cloud layer, with the magnitude of reduction being several times higher than that under clear-sky conditions.

Atmospheric aerosols exist as mixtures. For the two species $(NH_4)_2SO_4$ and soot, in an internal mixture, every particle contains both species, whereas an external mixture consists of pure $(NH_4)_2SO_4$ particles and pure soot particles. With every particle exhibiting some absorption, an internal mixture has a lower single-scattering albedo than the corresponding external mixture. Aerosol mixtures may either increase or decrease j- values under clear-sky conditions, depending on the mixture single-scattering albedo, solar zenith angle, and altitude. Regardless of whether aerosol mixtures enhance or reduce j-values in a clear atmosphere, they always lead to a reduction in j-values in the presence of a cloud.

Tropospheric averaged photolysis rates are presented by Liao et al. (1999) at a cloud thickness of 500 m and a solar zenith angle of 0°. Under clear-sky conditions, sulfate aerosol at the urban level increases all j-values by 10–18%, while soot aerosol reduces all j-values by 6–11%. When the 500-m cloud layer is present, sulfate aerosol increases all j-values by about 5%, and soot aerosol reduces j-values by amounts between 9% and 19%. The tropospheric average photolysis ratios of rates show that a cloud layer reduces the effect of sulfate aerosol but enhances that of soot aerosol. Thus, urban aerosols produce a change in photolysis rates of the order of 10–20% when averaged through the troposphere. This value can be larger locally.

In summary, in the absence of clouds, soot aerosol reduces photolysis rates at all altitudes, whereas sulfate aerosol generally increases photolysis rates above and in the upper part of the aerosol layer but reduces photolysis rates in the lower part of the aerosol layer and at the surface. Aerosol mixtures may reduce or increase photolysis rates from those under clear-sky conditions, depending on the single-scattering albedo, solar zenith angle, and altitude. The results obtained by Liao et al. (1999) for clear-sky conditions are consistent with those presented by Dickerson et al. (1997) and Jacobson (1998).

The results of Jacobson (1999) apply specifically to the effect of aerosols generated from aromatics on atmospheric UV absorption. With optical properties for such aerosols determined by Jacobson, it is possible to estimate effects of these aerosols at urban levels on photolysis rates, following the work of Liao et al. (1999). The results of Liao et al. (1999), which consider mixtures of sulfate and soot aerosols as models of scattering and absorbing aerosols, do provide indications of both the qualitative and quantitative effects of pollution aerosols on photolysis rates. Because the organic aerosols exhibit both scattering and absorption behavior, their effect on atmospheric actinic flux is complex, depending on the vertical distribution of aerosol and the presence or absence of clouds. It is not possible to say whether such aerosols will enhance or retard photolysis rates without detailed knowledge of such factors.

Some indication of the effect of aerosol composition is also found in the modeling study of He and Carmichael (1999). They studied the loading strength and radiative characteristics of four different types of aerosols (urban, rural, desert, and maritime) and the effects of relative humidity, aerosol vertical loading profile, and NO_x emissions. They found that ozone production in the upper troposphere can be either enhanced or weakened, depending on the scattering and absorbing ability of aerosol particles and the availability of NO_x. They concluded that in the lower troposphere, NO_2 photolysis and ozone production rates are most sensitive to urban aerosol, followed by rural, then desert, and, finally, maritime aerosol.

VIII-F. Fate and Lifetime of Atmospheric Organic Aerosols

Particles are eventually removed from the atmosphere by two mechanisms: deposition at the Earth's surface, so-called dry deposition, and scavenging by droplets, so-called wet deposition (Seinfeld and Pandis, 1998). Because wet and dry deposition lead to relatively short residence times in the troposphere, and because the geographical distribution of particle sources is highly non-uniform, tropospheric aerosols vary widely in concentration and composition over the Earth. Whereas atmospheric trace gases have lifetimes ranging from less than a second to a century or more, the residence times of particles in the troposphere vary only from a few days to a few weeks.

The dry deposition flux of particles to the surface, F_d, is assumed to be proportional to the particle concentration at a reference height, C; that is, $F_d = v_d C$, where the proportionally constant v_d, the deposition velocity, depends on the meteorological state of the atmosphere and the size of the particles. Three processes serve to deliver particles to the Earth's surface: gravitational settling, turbulent transport, and Brownian diffusion. Although virtually any atmospheric flow is turbulent, a very thin laminar sublayer exists immediately adjacent to the surface. Turbulence brings particles down to the laminar sublayer, through which Brownian diffusion and settling govern transport. Small particles have a relatively large Brownian diffusivity so they move efficiently through the sublayer, whereas larger particles transfer primarily via inertia or settling. Those in between, in the size range of 0.1–1 μm diameter, are deposited about an order of magnitude slower than those at either the small or large extremes, because none of the mechanisms is relatively effective in this intermediate size range.

Wet deposition refers to the scavenging of airborne particles by droplets, followed by subsequent removal by precipitation. Scavenging is necessary, but not sufficient, for wet deposition to occur since cloud-drops or raindrops can evaporate before reaching the Earth's surface, and, if this occurs, the scavenged particle is returned to the air mass. As opposed to dry deposition, which operates only at the Earth's surface, wet deposition serves to remove particles from the entire air mass. Most particles are removed when they become activated to form a cloud droplet. Some are collected by falling drops. The rate of particle collection by falling drops is proportional to the number of drops, their settling velocity, their cross-sectional area, and a collection efficiency. As with dry deposition, there is a minimum in the total collection efficiency in the 0.1–1 μm size range—small particles diffuse to the drop surface, larger ones collide with it, while in between neither process is very efficient.

Particles that become activated to grow to fog or cloud droplets are termed cloud condensation nuclei (CCN). At a given mass of water-soluble material in the particle, there is a critical value of the ambient water supersaturation, above which the particle undergoes an unstable process of spontaneous water accretion, leading to a cloud droplet (Seinfeld and Pandis, 1998). The number of particles from any given aerosol population that can act as CCN thus depends on the water supersaturation achieved in the cloud. For marine stratiform clouds, for example, maximum supersaturations achieved are in the range of 0.1–0.5%, which corresponds to a minimum CCN particle diameter in the range of 0.05–0.14 μm. The CCN number concentrations vary from fewer than 100 cm^{-3} in remote marine regions to a few thousand per cubic centimeter in polluted urban areas.

Atmospheric aerosol lifetimes depend primarily on the size of the particle and the height in the atmosphere at which the particle resides. The residence time τ can be viewed as an exponential half-life, the time required for a population of particles of a given size to decay to $1/e$ of its initial concentration.

An empirical expression for atmospheric particle residence time as a function of particle diameter D_p and altitude (Jaenicke, 1988) that is useful for estimates is

$$\frac{1}{\tau} = \frac{1}{\beta}\left[\left(\frac{D_p}{D_{\max}}\right)^2 + \left(\frac{D_p}{D_{\max}}\right)^{-2}\right] + \frac{1}{\tau_{\text{wet}}} \tag{25}$$

where $\beta = 1.28 \times 10^8$ s (constant), $D_{\max} = 0.6$ μm (the diameter of particle with maximum residence time), and τ_{wet} is the lifetime for removal of particles by wet deposition. The first term on the right-hand side of equation (25) represents dry removal at the Earth's surface. τ_{wet} depends mainly on the altitude in the atmosphere. Roughly, three altitude regions can be distinguished by the frequency with which precipitation scavenging occurs:

Height $\leq$ 1.5 km (lower troposphere):	$\tau_{wet} \cong 6.9 \times 10^5$ s (8 days)
Middle troposphere to tropopause:	$\tau_{wet} \cong 1.8 \times 10^6$ s (3 weeks)
Tropopause and above:	$\tau_{wet} \cong 1.7 \times 10^7$ s (200 days)

Organic aerosols derived from aromatic oxidation are not expected to behave differently from aerosols of other composition with respect to dry deposition, since it is only the physical size of the particle that is the essential property governing removal by this route. Wet removal, however, depends not only on particle size but also on the water solubility of its components. Water solubility properties, while well established for typical inorganic atmospheric aerosols, are less well known for organic aerosols, and it is the water solubility that is the crucial factor in determining whether a particle will activate to form a cloud droplet. Since activation as a CCN is the principal wet removal mechanism in the atmosphere, determining water solubility properties is paramount.

Available data indicate that from 20% to 60% of the carbonaceous mass present in submicrometer atmospheric aerosol can be classified as water-soluble organic compounds (WSOCs) (Cadle and Groblicki, 1982; Mueller et al., 1982; Sempere and Kawamura, 1994). Little of the molecular composition of this WSOC mass has been identified. Saxena and Hildemann (1996) concluded that the unidentified fraction could contain hundreds of different species; for example, highly polar C_2–C_6 multifunctional compounds. Moreover, much of SOA is suspected to fall into the WSOC fraction because of the polar nature of many of the SOA constituents that have been identified.

Saxena and Hildemann (1996) estimated Henry's law coefficients for a number of atmospheric organic species. They estimated that Henry's law coefficient values H for multifunctional species can be of the order of 10^{10} M atm^{-1} or larger. At this level of H, most of the compound can be expected to reside in the aqueous phase of atmospheric aerosols. For example, for $H = 10^{10}$ M atm^{-1}, at a liquid water fraction of 10^{-10} cm^3 of liquid water per cubic centimeter of air, typical for aerosol, the equilibrium ratio between the aqueous phase and the gas phase is 100.

While there are a number of studies in which the hygroscopic and/or CCN behavior of organic aerosols have been measured, very few focus on the behavior of secondary organic aerosol and none is directed specifically at aerosol from aromatic precursors. Cruz and Pandis (1997) studied experimentally the CCN activity of two model SOA compounds—glutaric acid and adipic acid—and dioctylphthalate at 0.3 and 1.0% supersaturation. Adipic acid and glutaric acid acted as CCN at both supersaturations; dioctylphthalate aerosols did not activate even at supersaturations as high as 1.2%. These results indicate that the CCN activity of hygroscopic organic aerosols may be comparable to that of some inorganic aerosols.

In summary, available data indicate that secondary organic aerosols are hygroscopic to an extent, with behavior lying between that of a relatively non-hygroscopic aerosol such as soot and a very hygroscopic aerosol like ammonium sulfate. As a result, secondary organic aerosols will act as CCN if there is an insufficient amount of more hygroscopic aerosols present to compete effectively for the available water vapor. The atmospheric lifetime of these organic aerosols will depend on their history with respect to cloud activation, which is difficult to predict. For representation in atmospheric models, the scavenging efficiency of secondary organic aerosol material can be considered to lie intermediate between those of a fully hygroscopic salt and a non-hygroscopic substance like elemental carbon.

IX

Atmospheric Chemistry of Polycyclic Aromatic Hydrocarbons Sorbed on Particles

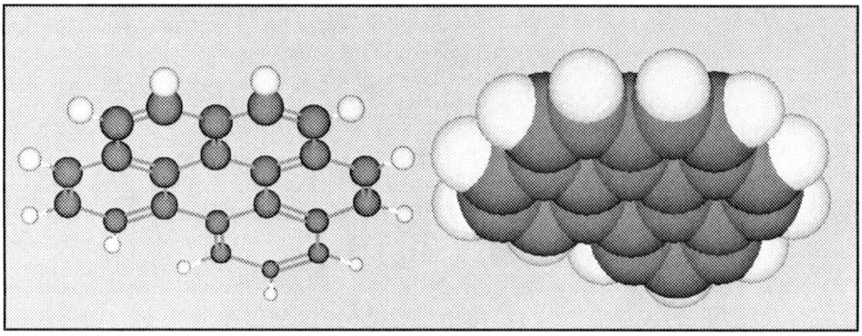

IX-A. The Polycyclic Aromatic Hydrocarbons

Polycyclic aromatic hydrocarbons (PAHs) arise in the environment primarily as direct atmospheric emissions or as combustion wastes that are either buried in, or applied to, the surface of dump sites (Muller et al., 1989; Luthy et al., 1994; Finlayson-Pitts and Pitts, 1999). As a class of compounds, PAHs contain two or more fused benzene rings. This class also includes aromatics with fused aliphatic and olefinic rings, as well as aliphatic substituent groups on the aromatic rings. In 1962, Badger proposed a simple flame mechanism in which acetylene radicals combine as building blocks to form PAHs (Badger, 1962; Badger et al., 1965). Current mechanisms involve hydrogen abstraction and acetylene addition to the growing PAH structure (Wang and Frenklach, 1997). Recent work suggests that gas-phase collisions may lead to PAH dimers, and trimers, and so on, which ultimately lead to a soot seed particle. It is also possible that growth to soot particles can take place via the formation of aromatic–aliphatic-linked hydrocarbons (Minutolo et al., 1988; Ciajolo et al., 1994; D'Anna et al., 1994; Dobbins and Subramaniasiavam, 1994; Hepp et al., 1995).

At room temperature, PAHs are solids and they have melting points that range from ~ 80°C to 400°C, and boiling points that range from ~ 200°C to 570°C, and they have subcooled liquid vapor pressures that range from 10^{-4} to 10^{-11} atm. Depending on their volatility, they can exist as free gases or sorbed onto particles (Yamasaki et al., 1982a; Pankow, 1987). Some properties of typical PAHs (boiling point, melting point, and vapor pressure) are given in table IX-A-1. The structures of some common PAHs are shown in figure IX-A-1 and in the appendix I-B, and the common abbreviations for the PAH names that are used throughout this book are summarized in table IX-A-2.

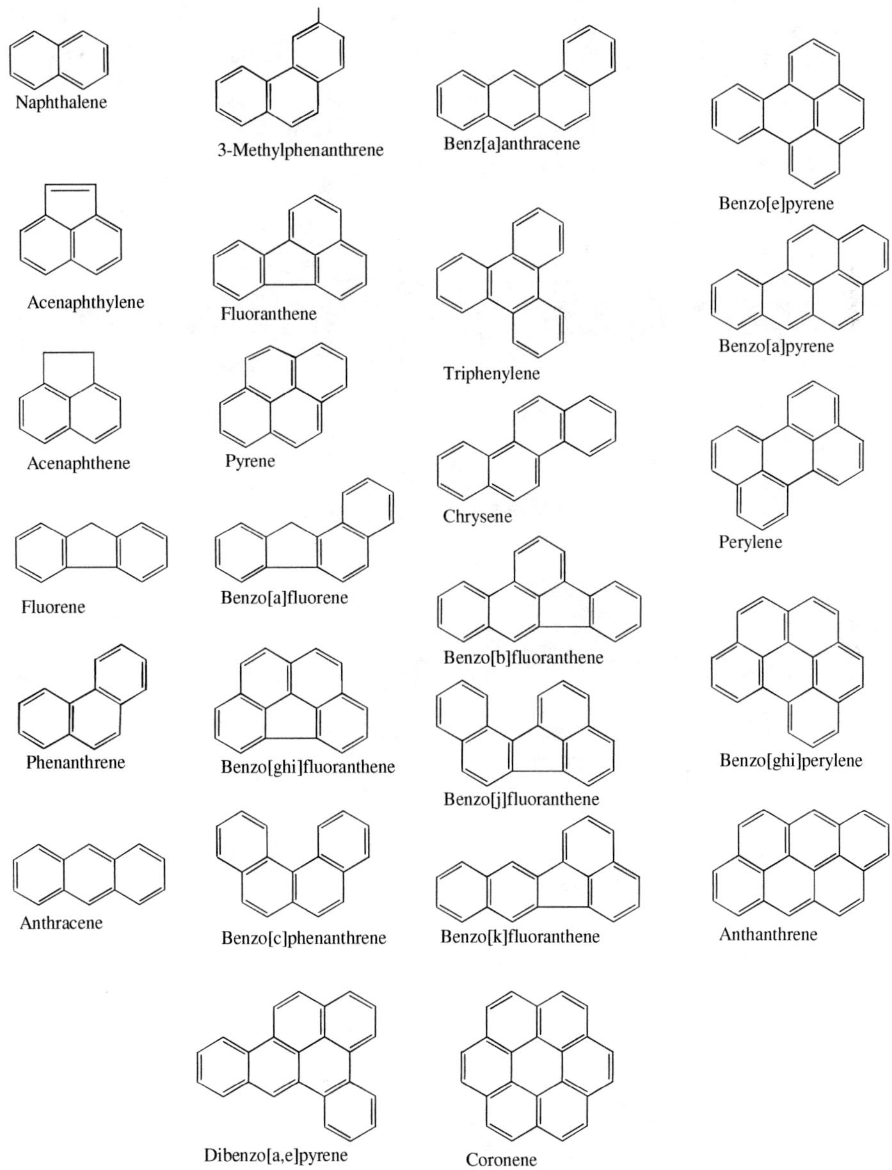

Figure IX-A-1. Some typical PAH structures.

Table IX-A-1. PAH physical and chemical data

	Name of PAH	Formula	MW[a]	mp (°C)[a,h]	bp (K)[a,h]	bp (K)[b]	P_L^o (atm)[c]	P_L^o (atm)[d]	P_S^o (atm)[e]	P_L^o (atm)[g]	C_{sat}^w (mg L^{-1})[e]	K_H (atm L mol^{-1})[e]	Log K_{ow}[e]	IP (eV)[f]
1	Naphthalene	$C_{10}H_8$	128.2	81	491	473.86	2.98×10^{-4}	6.76×10^{-4}	1.12×10^{-4}	4.01×10^{-4}	3.10×10^{1}	4.830×10^{-1}	3.30	8.14
2	Azulene	$C_{10}H_8$	128.2	100	543	473.86	2.27×10^{-5}	6.76×10^{-4}	6.43×10^{-4}	3.56×10^{-3}	5.02×10^{1}	$3.930 \times 10^{+1}$	3.20	7.41
3	Acenaphthylene	$C_{12}H_8$	152.2	93	543	535.88	2.27×10^{-5}	3.25×10^{-5}	1.20×10^{-6}	5.66×10^{-6}	1.61×10^{1}	1.140×10^{-1}	3.94	8.22
4	Acenaphthene	$C_{12}H_{10}$	154.2	96	552	536.72	1.43×10^{-5}	3.11×10^{-5}	3.29×10^{-6}	1.66×10^{-5}	3.90	1.550×10^{-1}	3.92	7.68
5	Fluorene	$C_{13}H_{10}$	166.2	117	567	563.03	6.63×10^{-6}	8.14×10^{-6}	1.66×10^{-6}	1.35×10^{-5}	1.34	1.670×10^{-1}	4.02	7.89
6	Phenanthrene	$C_{14}H_{10}$	178.2	101	611	589.34	6.58×10^{-7}	2.07×10^{-6}	1.47×10^{-7}	8.35×10^{-7}	1.15	3.600×10^{-2}	4.46	7.86
7	Anthracene	$C_{14}H_{10}$	178.2	216	613	589.34	5.91×10^{-7}	2.07×10^{-6}	3.51×10^{-9}	2.74×10^{-7}	4.34×10^{-2}	7.200×10^{-1}	4.45	7.45
8	3-Methylphenanthrene	$C_{15}H_{12}$	192.3	65	625	617.2	3.11×10^{-7}	4.72×10^{-7}	6.59×10^{-8}	1.64×10^{-7}	2.69×10^{-1}	5.670×10^{-2}	5.08	7.68
9	Fluoranthene	$C_{16}H_{10}$	202.3	111	656	651.36	5.77×10^{-8}	7.44×10^{-8}	6.91×10^{-8}	4.92×10^{-7}	2.60×10^{-1}	1.360×10^{-1}	5.16	7.95
10	Pyrene	$C_{16}H_{10}$	202.3	156	666	651.36	3.33×10^{-8}	7.44×10^{-8}	3.22×10^{-9}	6.41×10^{-8}	1.35×10^{-1}	1.100×10^{-2}	4.88	7.41
11	Benzo[*a*]fluorene	$C_{17}H_{12}$	216.3	190	680	678.51	1.53×10^{-8}	1.67×10^{-8}	6.16×10^{-9}	2.66×10^{-7}	4.54×10^{-2}	1.630×10^{-2}	5.19	
12	Benzo[*g,h,i*]fluoranthene	$C_{18}H_{10}$	226.3		705	713.38	3.79×10^{-9}	2.36×10^{-9}						
13	Benzo[*c*]phenanthrene	$C_{18}H_{12}$	228.3	68		704.82		3.83×10^{-9}	8.76×10^{-10}	2.34×10^{-9}	3.45×10^{-3}	5.010×10^{-3}	5.52	7.6
14	Benz[*a*]anthracene	$C_{18}H_{12}$	228.3	162	708	704.82	3.20×10^{-9}	3.83×10^{-9}	2.14×10^{-10}	4.88×10^{-9}	1.00×10^{-2}		5.61	7.43
15	Triphenylene	$C_{18}H_{12}$	228.3	199	712	704.82	2.56×10^{-9}	3.83×10^{-9}	2.24×10^{-9}	1.19×10^{-7}	4.11×10^{-2}	1.240×10^{-2}	5.49	7.84
16	Chrysene	$C_{18}H_{12}$	228.3	256	714	704.82	2.28×10^{-9}	3.83×10^{-9}	8.20×10^{-12}	1.60×10^{-9}	2.00×10^{-3}	9.460×10^{-2}	5.50	7.59
17	Benzo[*b*]fluoranthene	$C_{20}H_{12}$	252.3	168	754	766.84	2.33×10^{-10}	1.11×10^{-10}	6.58×10^{-10}	1.72×10^{-8}	1.50×10^{-3}	1.110×10^{-2}	5.78	
18	Benzo[*j*]fluoranthene	$C_{20}H_{12}$	252.3	166	753	766.84	2.47×10^{-10}	1.11×10^{-10}	3.16×10^{-11}	7.88×10^{-10}	2.50×10^{-3}	8.100×10^{-4}	6.11	
19	Benzo[*k*]fluoranthene	$C_{20}H_{12}$	252.3	217	754	766.84	2.33×10^{-10}	1.11×10^{-10}	1.27×10^{-12}	1.01×10^{-10}	8.00×10^{-4}	4.000×10^{-4}	6.11	
20	Benzo[*e*]pyrene	$C_{20}H_{12}$	252.3	179	766	766.84	1.17×10^{-10}	1.11×10^{-10}	7.50×10^{-12}	2.52×10^{-10}	6.30×10^{-3}	1.070×10^{-3}	6.44	7.41
21	Benzo[*a*]pyrene	$C_{20}H_{12}$	252.3	177	769	766.84	9.80×10^{-11}	1.11×10^{-10}	7.22×10^{-12}	2.32×10^{-10}	1.62×10^{-3}	2.450×10^{-3}	5.97	7.12
22	Perylene	$C_{20}H_{12}$	252.3	273	773	766.84	7.77×10^{-11}	1.11×10^{-10}	5.79×10^{-12}	1.66×10^{-9}	4.00×10^{-4}	3.650×10^{-3}	6.25	6.9
23	3-Methylcholanthrene	$C_{21}H_{16}$	268.4	178		795.54		2.09×10^{-11}	4.34×10^{-11}	1.43×10^{-9}	2.90×10^{-3}		6.42	
24	Indino[1,2,3-*c,d*]pyrene	$C_{22}H_{12}$	276.3	164	770	797.85	9.25×10^{-11}	1.82×10^{-11}	1.32×10^{-13}	3.11×10^{-12}	2.20×10^{-5}	1.600×10^{-3}	6.70	
25	Dibenz[*a,h*]anthracene	$C_{22}H_{14}$	278.4	270	797	820.3	1.92×10^{-11}	4.86×10^{-12}	1.32×10^{-13}	3.52×10^{-11}	2.49×10^{-6}		6.75	7.38
26	Benzo[*b*]chrysene	$C_{22}H_{14}$	278.4	294		820.3		4.86×10^{-12}	4.91×10^{-12}	2.27×10^{-9}	1.08×10^{-3}	4.890×10^{-4}	7.11	7.14
27	Benzo[*g,h,i*]perylene	$C_{22}H_{12}$	276.3	278	798	828.86	1.81×10^{-11}	2.93×10^{-12}	1.32×10^{-13}	4.23×10^{-11}	2.60×10^{-4}	1.600×10^{-3}	6.63	7.15
28	Anthanthrene	$C_{22}H_{12}$	276.3			828.86		2.93×10^{-12}						6.92
29	Coronene	$C_{24}H_{12}$	300.4	437	798	890.88	1.81×10^{-11}	7.10×10^{-14}	4.33×10^{-14}	5.28×10^{-10}	1.40×10^{-4}	2.120×10^{-5}	7.64	7.29
30	Dibenzo[*a,e*]pyrene	$C_{24}H_{14}$	302.4	233		882.32		1.19×10^{-13}	9.25×10^{-14}	1.07×10^{-11}	8.02×10^{-5}	1.410×10^{-5}	7.28	

[a]mp = melting point from Bjorseth (1983).
[b]Calculated boiling point (T_p) using group contribution method (Joback and Reid, 1985), where $T_b = 198 + \Sigma(T_{b,i})$ and $T_{b,i}$ is a weighted contribution that is assigned to each functional group.
[c]Calculated liquid vapor pressures (P_L^o) in atmospheres (atm) at 25°C with boiling points from Björseth (1983) according to the method of Mackay et al. (1982), or using equations (4)–(18) of Schwarzenbach et al. (1993) with estimated entropies of vaporization from the method of Kistiakowsky as illustrated in Schwarzenbach et al. (1993) equations (4)–(11).
[d]Calculated liquid vapor pressures P_L^o (atm.) with Joback estimate T_b (K) at 25°C using equations (4)–(18) of Schwarzenbach et al. (1993).
[e]From Howard and Meylan (1997).
[f]IP = Ionization potential from Lias et al. (1988).
[g]Calculated liquid vapor pressures (P_L^o) in atm at 25°C from solid-state vapor pressures (P_S^o) (Mackay et al., 1982; Schwarzenbach et al., 1993).
[h]Donald et al. (1992).

Table IX-A-2. Names of the common PAHs discussed in this chapter and the abbreviations for their names used in this book

Name of PAH	Abbreviation of Name (Used Here)
Acenaphthalene	Ace
Acenaphthene	Can
Aenaphthalene	Aen
Anthanthrene	Ananth
Anthracene	Ant
Benzo[*e*]acenaphthylene	BeA
Benz[*a*]anthracene	BaA
Benzo[*b*]chrysene	Bchry
Benzo[*b*]fluoranthene	BbFla
Benzo[*g,h,i*]fluoranthene	BghiFla
Benzo[*j*]fluoranthrene	BjFla
Benzo[*a*]fluorene	BaF
Benzo[*j*]fluoranthene	BjFla
Benzo[*k*]fluoranthene	BkFla
Benzo[*g,h,i*]perylene	BghiP
Benzo[*e*]phenanthrene	BePhe
Benzo[*a*]pyrene	BaP
Benzo[*e*]pyrene	BeP
Chrysene	Chry
Coronene	Cor
Cyclopenteno[*c,d*]pyrene	CpC
Dibenz[*a,c*]anthracene	DiBacAnt
Dibenz[*a,h*]anthracene	DiBahAnt
Dibenzo[*a,e*]pyrene	DiBaeP
Dimethylanthracenes	diMeAnt
Dimethylphenanthrenes	diMePhe
Fluoranthene	Fla
Fluorene	Flu
Indino[1,2,3-*c,d*]fluoranthene	Ind-Fla
Indino[1,2,3-*c,d*]pyrene	Ind
Methylanthracenes	MeAnt
3-Methylcholanthrene	3-MeCAnt
2-Methylnaphthylene	2-MeNap
Methylphenanthrenes	MePhe
3-Methylphenanthrene	3-MePhe
Naphthalene	Nap
Perylene	Pery
Phenanthrene	Phe
Pyrene	Py
Triphenylene	Tri

Interest in PAHs has been fueled by the discovery that some PAHs are carcinogens in animal test models (Schoental, 1984), and that benzo[a]pyrene (BaP) was the first PAH carcinogen isolated from coal-tar (Kennaway, 1955). According to the International Agency for Research on Cancer (IARC, 1983), many PAHs with more than three rings are believed to be human carcinogens. The IARC also indicated that complex environmental mixtures containing

PAHs from sources such as aluminum production, coal gasification, coke production, iron and steel founding, tobacco smoke, and other soots are thought to be carcinogenic to humans. Although PAHs do not account for a large part of the bacterial mutagenicity in reactions, they can be very potent direct-acting bacterial mutagens (Rosenkranz and Mermelstein, 1983), and they can contribute substantially to the overall bacterial mutagenicity in ambient particle extracts (Wang et al., 1980; Talcott and Harger, 1981). Other potential atmospheric products such as benzo[a]pyrene,4-5-epoxide (Pitts et al., 1980) are weak carcinogens when applied to mouse skin.

It was estimated during the early 1970s that over a lifetime, urban human beings accumulate about 2 g of particulate matter in their lungs (Lewis and Counghlin, 1973). It is thought that most human carcinomas are epithelium in origin, initiating in the skin, the lining of the digestive system, or the bronchial linings of the lung (Karahalil et al., 1999). Work by Grimmer and his colleagues (Grimmer, 1983, 1993; Grimmer et al., 1987) highlighted the carcinogenic potency of the PAHs by fractionating combustion emissions from different sources, by applying the different fractions to mouse skin, and by giving interlung injections of the different fractions to rats. As illustrated in figure IX-A-2, Grimmer observed that most of the carcinogenic potency of the extracts was expressed in the fraction that contained PAHs with more than three rings. This was observed for organic extracts from diesel and gasoline combustion exhaust, coal-fired power plant flue condensates, side-stream cigarette smoke, and used lubricating oil. The actual risk associated with PAH exposures is extremely difficult to estimate. Nielsen and coworkers (Nielsen et al., 1996) compared a number of approaches and suggested that the annual risk of increased lung cancer cases from continuous exposure to PAHs on a busy street in Copenhagen ranged from 0.33 to 17 per million population. Over a lifetime of exposure, this translated into 23 to 1180 excess lung cancers per million.

During the 1972–1999 period, there have been many excellent reviews on PAH atmospheric chemistry, and it is recommended that the reader consult these for additional background information (NAS, 1972; Nielsen et al., 1983; Pitts, 1983; Finlayson-Pitts and Pitts, 1986, 1999; Lane, 1989; Atkinson and Arey, 1994). The intent of this chapter is to highlight the important features of the atmospheric chemistry of PAHs and their rates of reaction in the atmosphere. In doing so, it will be necessary to discuss some of their physical and chemical attributes, their sources, their ambient concentrations, their distribution between the gas and the particle phases, and then their gas- and particle-phase reactions. An effort will be made to focus on research since about 1985. Methods of PAH analysis are not covered and the reader is referred to a recent publication on these topics (Neilson, 1998).

IX-B. Sources of PAH Emissions to the Atmosphere

In 1993, based on domestic and commercial coal use, wood and biomass burning, and a small fraction from diesel fuel use, Penner et al. (1993) estimated that 12

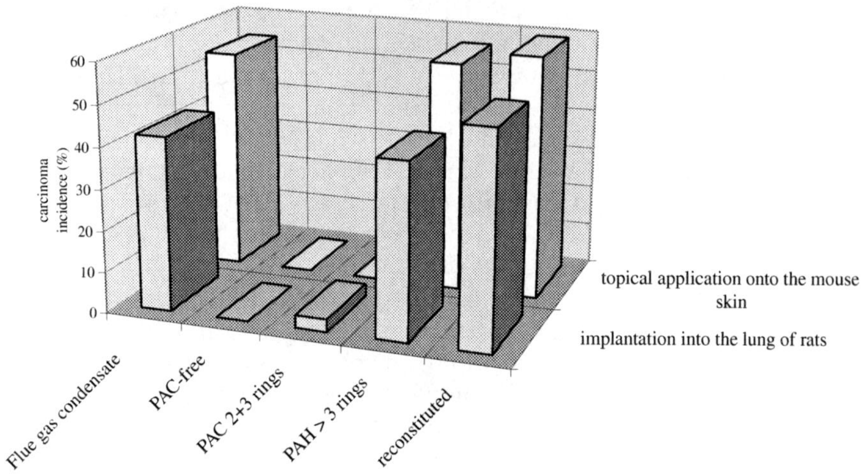

Figure IX-A-2. Comparison of the carcinoma incidence of flue gas condensate from coal combustion, its fractions, and the reconstituted coal fuel gas condensate (Grimmer, 1993).

teragrams (Tg) of black carbon is emitted per year into the global atmosphere. Using regional black carbon to sulfur emissions, a value of ~24 Tg C yr^{-1} was estimated. The contribution of different countries and regions to the total global black carbon emission rate is summarized in table IX-B-1. With these black carbon emissions come significant emissions of PAHs (Penner et al., 1993). If one assumes, as a very crude estimate, that 0.01–1% of the black carbon emissions are PAHs (Ramdahl and Moller, 1983; Freeman and Cattell, 1990) (see table IX-B-2), the total global emissions of PAHs are on the order of 0.001–0.2 Tg C yr^{-1}. It is estimated that more than one-half of these emissions come from Asia and the former Soviet Union.

Table IX-B-1. Regional black carbon emissions estimates

	Emissions ($Tg\,C\,yr^{-1}$)	%
Former USSR	5.60	23.49
Rest of Asia	4.96	20.81
China[a]	3.66	15.35
Western Europe	2.46	10.32
Africa	1.56	6.54
Eastern Europe	1.34	5.62
USA	1.1	4.61
Mexico	1.03	4.32
Brazil	0.44	1.85
Other	1.69	7.09
Total	23.84	

[a]Since these black carbon estimates were made, however, there has been a shift—for example, in China—to more centralized coal-fired power plants, natural gas, and industrial coal-fired boilers. Hence, their use here is for illustrative purposes only.
Source: Penner et al. (1993).

Table IX-B-2. Percent PAH emissions in particulate matter

	% (PAH/particle mass)
Barley cereal straw[a]	0.46
Barley cereal straw[b]	12.70
Wheat straw[b]	0.80
Rice straw[b]	5.40
Spruce, wood stove, not banked[c]	0.77
Douglas fir, high and low flaming[b]	7.60
Pine slash, high and low flaming[b]	3.30
Leaves, open fire[d]	0.01
Open wood fire[d]	0.04
Large-scale brush fire[d]	0.08
Coal-fired power plants[e]	7×10^{-4} to 0.5
Oil-fired power plants[f]	1×10^{-4} to 1×10^{-6}
Charcoal[c]	0.01
Sintering furnace[g]	0.02
Diesel car[g]	0.002
Heavy diesel truck (1987)[h]	0.05
Non-catalyst car[i] (1965–1976)[h]	2.37
Catalyst car (1977–1983)[h]	0.29

[a]See Ramdahl and Moller (1983); 75-kW continuous-feed stoker furnace; 127 kg h^{-1}; total PAHs.
[b]Jenkins et al. (1996); burn performed in a wind tunnel; cereals burned in spreading fires; wood burned in piles; particle PAHs.
[c]Ramdahl et al. (1982); small air tight wood stove; 4–5 kW; unbanked fire; total PAHs.
[d]Freeman and Cattell (1990); particle PAHs.
[e]Coal-fired power plants in Germany: Funcke et al. (1988); fluidized-bed coal-fired: Hunt et al. (1982).
[f]Bennett et al. (1979).
[g]Oda et al. (1998); particle PAHs.
[h]Rogge et al. (1993a); particle PAHs.
[i]Car without a catalytic converter.

On an urban scale, it is not surprising that automotive and other internal combustion engine emissions can be a significant source of PAHs in urban atmospheres, although they are not the exclusive source. Nielsen (1996), using ratios of benzo[*g*,*h*,*i*]perylene to benzo[*e*]pyrene, showed that mobile sources were the primary contributors to aerosol-bound PAHs in central Copenhagen during the winters of 1992 and 1993. Aceves and Grimalt (1993) investigated the 1989 sources of PAHs at both an urban and an industrial site in Barcelona by using a chemical mass balance model with PAH source signatures (Li and Kamens, 1993) and adjusted for possible PAH photochemical reactions on particles (Kamens et al., 1988). Their predictions indicated contributions of ~30–40% gasoline engine internal combustion exhaust and 60% diesel engine internal combustion exhaust at an urban site and almost 100% diesel exhaust at an industrial site. Freeman and Cattell (1990), using PAH source profiles, showed that brush fire burning ~50 km away from Sydney could contribute ~30% of the PAHs to the Sydney atmosphere.

A recent source reconciliation study of the 1982 Los Angeles atmosphere by Schauer et al. (1996) shows that diesel exhaust, followed by paved-road dust, gasoline exhaust, and wood smoke were the most important contributors to ambient fine particle concentrations. In this same study, however, non-catalytically controlled gasoline vehicles cars were the most significant source (70–90%) of cyclopenteno[*c*,*d*]pyrene, benzo[*e*]pyrene, benzo[*a*]pyrene, indino[1,2,3-*c*,*d*]pyrene, benzo[*g*,*h*,*i*]perylene, and coronene. On the other hand, natural gas combustion was the most significant source of fluoranthene, pyrene, benzo[*g*,*h*,*i*]fluoranthene, chrysene, and triphenylene. Catalytically controlled vehicles and wood burning each contributed on average ~5% to each of these PAHs.

In marine atmospheres, the ocean can be a source of PAHs. Marty et al. (1984) collected atmospheric samples in the tropical and equatorial east Atlantic (west of Africa) in May 1974 and May–June, 1977, respectively. Phenanthrene dominated the gas-phase samples, and it was proposed that it originated from dissolved phenanthrene in the sea-surface microlayer and from land-based continental emissions. The distributions of PAHs in the particle phase also resembled the dissolved higher-molecular-weight compounds in the sea-surface water, and suggest aerosolized sea spray as a source. Continental emissions, however, could not be ruled out. This, of course, invokes the notion of long-range transport of PAHs to remote environments, and this will be discussed in the following sections.

Emissions of PAHs vary widely between different sources and even within the same source. The difference between sources is illustrated in figure IX-B-1 (Li and Kamens, 1993) by the PAH chromatograms of wood soot, diesel soot, and gasoline soot particle extracts. Notice the prominence of the coronene (Cor) and benzo[*g*,*h*,*i*]perylene (BghiP) peaks in the gasoline soot extract compared to wood soot and diesel. In table IX-B-3, the PAH 1982 emissions from uncontrolled (non-catalyst) gasoline cars are compared with catalytically controlled cars and heavy-duty diesel trucks (Rogge et al., 1993a). Catalytically controlled cars of this vintage emit 10–20 μg less PAH per kilometer traveled than uncontrolled cars, and they emit different amounts during the summer and winter. In table IX-B-4, the emissions from different biomass combustion sources are shown.

IX-C. PAHs in the Atmosphere

IX-C-1. Urban Concentrations

Polycyclic aromatic hydrocarbon concentrations have been reported by many investigators for a number of different urban and rural locations throughout the world (Katz et al., 1978; Björseth et al., 1979; Faoro and Manning, 1981; Masclet et al., 1988; Gardner et al., 1995; Brown et al., 1996; Nielsen et al., 1996; Caricchia et al., 1999; Menichini et al., 1999; Odabasi et al., 1999b). Table IX-C-1 summarizes some recent gas/particle PAH data from four different cities. As can be seen, the most volatile PAHs, naphthalene, acenaphthene, fluorene, phenanthrene, fluoranthene, and pyrene, are present in the highest concentrations. These often exceed the least volatile compounds (benzo[*k*]fluorenes, benzo[*a*]pyrene, indino[1,2,3-*c*,*d*]pyrene, and benzo[*g*,*h*,*i*]pyrene) by a factor of 50–1000. From a concentration perspective, naphthalene is by far the most prominent PAH.

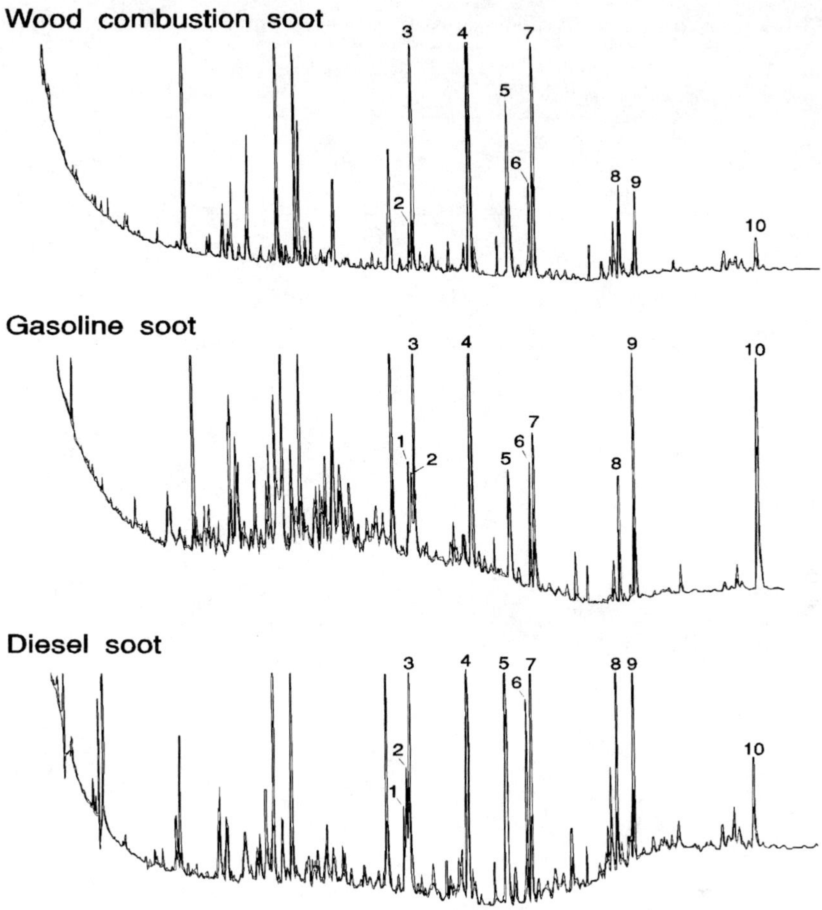

Figure IX-B-1. Illustration of the differences in PAH patterns from wood smoke, non-cat-gasoline soot and diesel soot filter extracts. Samples taken from the University of North Carolina's 25 m^3 chambers ~ 5–10 minutes after addition to chambers. Analysis was on a Carlo Erba 4130 gas chromatograph with flame ionization detection. A 30-m DB5 0.32-mm i.d. fused silica column was used with ultrapure hydrogen carrier gas. The initial oven temperature was 40°C; after injection, the oven temperature was rapidly increased to 100°C and then programmed at 8°C/min^{-1} to 300°C. Peak identification numbers are: (1) Cpc; (2) BaA; (3) Chry and Tri; (4) internal standard β,β-binaphthyl; (5) benzo[*b*]-, [*j*]-, and [*k*]fluoranthene; (6) BeP; (7) BaP; (8) Ind; (9) BghiP; (10) Cor. Figure taken from Li and Kamens (1993) with the permission of *Atmospheric Environment*.

On a seasonal basis, the most volatile PAHs (Nap, Phe, Fla, Py) tend to be higher in concentration during the summer months than the winter months (see Boston and Houston in table IX-C-1). The least volatile compounds, which are primarily associated with particulate matter (BaA through BghiP), are present at higher concentrations during winter than summer. Data reported by Brown et al. (1996) support this behavior for London air. This strongly suggests that source strengths with respect to one another are changing with the seasons and that the

Table IX-B-3. Comparison of PAH emissions[a] from different automotive sources

	Non-Catalyst Cars [c]		Catalyst Cars		Heavy-Duty Diesel
PAH[b]	Rogge et al. (1993a)	Rogge et al. (1993a)[d,e]	Cadle et al. (1999)[f] Summer	Cadle et al. (1999)[f] Winter	Rogge et al. (1993a)
Phe	17.3	0.9			12.2
Ant	5.1	0.1	13.7	77	1.6
MePhe + MeAnt	42.2	1.9			31.3
DiMePhe + diMeAnt	75.5	3.0			56.9
Fla	48.3	2.0	30.1	135	13.0
Py	31.0	2.5	34.5	159	22.6
Bghi-Fla	24.7	1.3			6.9
CpC	45.9	1.6			1.4
BaA	73.8	1.9	1.0	1.0	3.6
Chry + Tri	55.8	3.8	0.8	1.0	9.9
Me-228s[g]	143.5	3.4			2.7
DiMe202s[h]	161	4.1			14.6
BbFla + BkFla + BjFla	84.6	5.4	0.04	0.0	17.3
BeP	45.8	2.0			2.6
BaP	43.5	1.9	0.3	0.4	1.3
Pery	14.0	0.60			1.0
Ind	6.4	0.47	0.2	0.0	Nd[i]
Ind-Fla	32.6	1.7			Nd
BghiP	145.4	4.7	0.6	0.1	Nd
Anthanthrene	9.4	0.08			Nd
Cor	104.6	1.1			Nd

[a]Emission rates in $\mu g\,km^{-1}$ for fine particles.
[b]See table IX-A-2 for compound names corresponding to the abbreviations shown here.
[c]One-half of cars were 10 years old in 1982.
[d]Average mileage of catalyst cars was 27,186 miles.
[e]Cold-start U.S. Federal Test Procedure (FTP).
[f]Low-emission catalyst vehicles reported in 1999 for a range of vehicles called low emitters in Cadle et al. (1999); cold-start U.S. Federal Test Procedure (FTP) cycle.
[g]Me-228s includes MeBaA, MeChry, and MeTri.
[h]DiMe202s includes dimethyl pyrenes and fluoranthenes.
[i]Nd = not determined.

individual sources as discussed above have different PAH emission patterns. Similar seasonal trends were also observed for particle samples collected in New Jersey by Harkov and Greenberg (1985) in 1982–1983. It was shown that BaP was a factor of 5–10 higher in concentration during the winter months (1–1.6 $ng\,m^{-3}$) than in the summer months (0.1–0.2 $ng\,m^{-3}$). Energy-use patterns and specifically home heating with wood was related to the high winter emissions and ambient concentrations. Data collected in Rome, Italy, by Menichini et al. (1999) for the years 1993–1998 show that BaP levels were on the order of 4 $ng\,m^{-3}$ in the winter months and 0.1–0.2 $ng\,m^{-3}$ during the summer. These authors also suggest that BaA and BaP photodegraded on particles during the summer months; this is consistent with the observations and interpretations of Nielsen (1988) and Greenberg (1989).

Table IX-B-4. Comparison of PAH emissions[a] from different sources

PAH[b]	Birch, Normal Conditions[c]	Barley Straw[d]	Rice Straw[e]	Charcoal, Normal Conditions[c]
Nap			20,000	
2-MeNap			18,100	
Ace		5275	3100	
Acn		559	2900	
Flu		463	500	
Phe	5821	17,550	11,100	45.6
Ant	822	1180	2200	3.9
Flu	2843	8640	11,900	16
BeA	688	1240		1.2
Pry	2690	8470	14,500	20.1
CpC	381	665		
BaA	476	1003	26,100	
Chry + Tri	615	2070	29,500	1.6
Ind	415	623	14,500	0.2
DiBacAnt and/or DiBabAnt		87		
BghiP	544	891	8100	0.1
Ananth		70		
Cor		295		

[a]Emission rates in $\mu g\,kg^{-1}$ dry fuel.
[b]See table IX-A-2 for compound names corresponding to the abbreviations shown here.
[c]Total PAHs (gas + particle emissions); the wood stove was not banked and emissions are expressed as $\mu g\ kg^{-1}$ dry fuel burned (Ramdahl et al., 1982).
[d]Total PAHs (gas + particle) expressed as $\mu g\ kg^{-1}$ dry fuel burned; from a continuous-feed stoker cereal burner (Ramdahl and Moller, 1983).
[e]PAH concentrations in particulate matter ($\mu g\ kg^{-1}$); wind tunnel (Jenkins et al., 1996).

It is interesting to compare BaP concentrations compiled by Grimmer (table IX-C-2) for the 1950s and 1960s (Grimmer, 1983) with the concentrations observed in the 1990s (table IX-C-1). Although these data might suggest that concentrations are lower now in certain western cities than in the 1950s and 1960s, new data from some of the emerging nations and eastern European nations suggest high atmospheric concentrations of PAHs. Ando et al. (1996) reported an outdoor average winter concentration of $33.7\,ng\,m^{-3}$ for residential areas in Beijing, while these same authors observed a concentration of $2.3\,ng\,m^{-3}$ for residential areas in Tokyo. In Poland, eight or nine different sampling stations during the winters of 1988 and 1989 had average BaP concentrations of $26\,ng\,m^{-3}$ (Bodzek et al., 1993; see Note Added in Proof, p. 397).

IX-C-2. Rural PAHs

In rural environments, BaP concentrations range from 0.008 (Muller et al., 1998), 0.03 (Zheng et al., 1997) to $1.2\,ng\,mg^{-1}$ (Nielsen et al. 1996). The high value of $1.2\,ng\,mg^{-1}$ is reported for Denmark, and illustrates that rural areas can be heavily impacted with PAHs from nearby towns and urban centers. Björseth et al. (1979) reported on an association of PAH concentrations observed in southern

Table IX-C-1. Total PAHs (gas + particle) in different atmospheres (ng m^{-3})

PAH[a]	Boston[b] 1991–1992 Summer[c]	Boston[b] 1991–1992 Winter[c]	Houston[b] 1991–1992 Summer[d]	Houston[b] 1991–1992 Winter[d]	Chicago[e] 1995[c]	London[g] 1991[h]
Nap	606	452	515	671		
Ace	24.1	5.7	22.8	12.8	77.9	4.7
Flu	36.2	10.0	24.2	18.0	74.8	31.6
Phe	101	30.2	50.1	37.9	200.3	82.0
Ant	3.0	1.6	1.4	2.2	14.1	6.7
Fla	16.7	9.4	10.2	6.7	44.1	13.3
Py	8.4	8.3	7.9	6.4	24.6	12.3
CpC	0.3	1.1	0.2	0.2		
BaA	0.9	2.0	0.6	0.4	2.1	1.8
Chry	1.0	2.1	0.6	0.7	3.6	3.1
Bfs	1.1	2.9	0.5	1.0	4.2	3.4
BeP	0.4	1.0	0.2	0.4		
BaP	0.4	1.5	0.2	0.4	1.6	1.1
Ind	0.5	2.0	0.2	0.8	1.2	
BghiP	0.6	3.2	0.3	1.4	1.1	5.3
Cor	0.3	2.7	0.9	0.2		

[a]See table IX-A-2 for compound names corresponding to the abbreviations shown here.
[b]Lewis and Coutant (1999).
[c]USEPA Toxic Air Monitoring Site from August 1990 to August 1991; 24-h samples taken ~2 times per month. The Boston site is located downtown and is heavily impacted by automobile traffic and general urban sources; mean summer temperature = 24°C, winter = 1°C.
[d]The Houston site was 20 km east of downtown Houston, in Dear Park; site is in a semirural area, downwind of large refineries and chemical plants; average summer temperature = 28°C, winter = 12°C.
[e]Odabasi et al. (1999b).
[f]Sampling site located on the roof of a four-story building on the campus of the Illinois Institute of Technology, 5.6 miles south of Chicago's city center; samples taken from June 14 to October 23, 1995; average of 24-h samples; average temperature ~24°C, %RH = ~55%.
[g]Halsall et al. (1994).
[h]Sampled near the city center on top of a ~25 m building; sampled every other week for 30 min each hour of the sampling week.

Norway and western Sweden during the winter months, with air masses that originated in England or elsewhere in Europe. Their data strongly suggests the occurrence of long-range transport of PAHs on particulate matter. Specifically, PAH concentrations tended to be much higher at their sampling sites when air mass came from the southwest (England) and southeast (Europe) than from the north. Masclet et al. (1988) also associated PAH concentrations at a remote sampling site on the island of Corsica with wind patterns. When clean air masses arrived from the Mediterranean, particle PAH concentrations tended to be lower than when they arrived from southern France or Spain. In the arctic climate, similar observations were made by Halsall et al. (1997a). Air-mass back trajectories were computed for air arriving at three Arctic sites. The origins of these air masses, which were either relatively "polluted" or "clean," were consistent with observed concentrations at the different sites. Concentrations of PAHs for two rural/remote locations are shown in table IX-C-3.

Table IX-C-2. Selected BaP concentrations in ng m^{-3}

Country	Year	Winter	Summer
Australia	1962–1963	8	0.8
Belgium	1958–1962	110	15
Prague	1964	122 (high)	19 (low)
Copenhagen	1956	17	5
Budapest	1968	1000[a,b]	32[c]
Teheran	1971	6	0.6
Belfast	1961–1962	51	9
Genoa	1958–1960	212[a]	6[c]
Osaka	1965	50[c]	
Amsterdam	1969	18	2
Paris	1958	300–500	
Lancaster, USA	1953	20[d]	
Oslo	1956	15	1
Poland, average for 10 large cities	1966–1967	130	30
Poland, 8 or 9 stations[d]	1988–1989	24.1	13.2
USA, 100 urban communities[e]	1958–1959	6.6[d]	
	1962	5[d]	
USA, 32 large urban stations[e]	1966–1967	3.3[d]	
	1969	2.9[d]	
	1970	2[d]	
Beijing[f]	1991	33.7	
Tokyo[f]	1991	2.3	

Source: Reported by Grimmer (1983).
[a] High value.
[b] Air pollution episode.
[c] Low value.
[d] Not seasonal.
[e] Data for the United States between the 1950s and 1960s reported by Perera (1981).
[f] Ando et al. (1996).

IX-C-3. PAH Fluxes

That rural PAH concentrations tend to be significantly lower than urban concentrations leads to greater PAH fluxes at the Earth's surface in urban environments than in rural environments. In rural–semirural air, bulk fluxes (wet plus dry deposition) for a sum of PAH (Σ PAH compounds from Flu to BghiP) tend to be on the order of 0.2–1 μg $m^{-2}d^{-1}$ (Brorström-Lundén et al., 1994; Leister and Baker, 1994). In urban areas, for a sum of PAH (Σ PAH Ace to Cor), variations from 1 to 5 μg m^{-2} d^{-1} have been reported (Halsall et al., 1997b; Brorström-Lundén and Lövblad, 1991). Odabasi et al. (1999a) have summarized the individual fluxes of PAHs from a number of different studies and show orders of magnitude differences in PAH fluxes, measurements, and estimates. For the Chicago area, they measured fluxes for a ΣPAH of 14 compounds of ~360 μg m^{-2} d^{-1}, with volatile PAHs (Ace to Py) contributing more than 50% of the overall fluxes via gas-phase dry deposition.

Winter fluxes of PAHs in Manchester, England (Halsall et al., 1997b) tended to be a factor of 2 lower in summer than winter. However, this was attributed to a decline of the most volatile PAHs during the summer. For the heavier PAHs, such

Table IX-C-3. Rural/Remote PAH concentrations

	Gas Phase	Particle Phase	Gas + Particle Phases	Gas + Particle Phases
PAH[a]	Corsica[b] (ng mg^{-1})	Corsica (ng mg^{-1})	Birmingham[c] (ng mg^{-1})	Tagish[d] (pg m^{-3})
Nap	21.71	—	0.34	
2MeNap	10.40	—		
Can			0.66	
Ace			0.81	15.4
Flu	2.77	—	2.02	94.7
Phe	6.08	—	0.6	78.6
Ant	0.19	0.01	0.17	4.56
Fla	0.52	0.33	0.41	28.5
Py	0.23	0.11	1.00	22.8
Retene				62.1
BaA	0.00	0.02	0.08	3.99
Chry	0.00	0.03	0.15	7.79
Bfs	0.01	0.04	0.21	13.54
BeP	0.00	0.07		8.16
BaP	0.00	0.01	0.06	3.18
Ind	0.00	0.05	0.11	3.37
BghiP	0.00	0.10	0.21	8.93
Cor	0.00	0.14	0.06	

[a]See table IX-A-2 for compound names corresponding to the abbreviations shown here.
[b]Island, south of France; 24-h sample; particle concentration 0.9 ng m^{-3}; trajectory from the south over the Mediterranean sea; data from Masclet et al. (1988).
[c]Roof of a house in farmland ~2 km beyond the southernmost border of Birmingham, England; data from Smith and Harrison (1996).
[d]Tagish is in the western Yukon in the Canadian Arctic; samples from October–April, 1993–1994; data from Halsall et al. (1997a).

as BaP, it was evident that a greater deposition takes place in the winter than the summer. This may be due to increased precipitation/particle scavenging during the winter versus summer months. Overall mass transfer velocities (ν_{total}) were calculated from average PAH concentrations and flux measurements (where flux = $\nu_{\text{total}} \times [\text{PAH}]_{\text{gas+particle}}$). At two urban centers in England, Halsall et al. (1997b) calculated ν_{total} values for Flu, Phe, and Py of 0.015, 0.025, 0.075 cm s^{-1}, respectively. For less volatile PAHs such as Chry, BaA, BbFla, BaP, BghiP, and Cor, the following deposition velocities were calculated: 0.145, 0.155, 0.195, 0.22, 0.135, 0.15 cm s^{-1}, respectively. Higher deposition velocities (>0.2 cm s^{-1}) for low-volatility PAHs have been reported by McVeety and Hites (1988), and these were related to particle size. Since particulate matter tends to have higher dry deposition velocities than gases, it follows that less volatile PAHs would also have higher deposition velocities than more volatile PAHs.

Halsall et al. (2001) modeled the major decay processes of gas- and particle-bound PAHs during long-range transport from the United Kingdom to the Arctic. Wet scavenging by snow during the winter months was a very important loss mechanism for particle-phase PAHs. Halsall and coworkers also suggest that OH-radical attack was the main removal process for gas-phase PAHs under temperate conditions.

IX-C-4. PAH Distributions on Different Particle Sizes

Polycyclic aromatic hydrocarbons are primarily emitted from combustion sources, and combustion aerosols are typically below 1 µm in mass mean diameters (Miguel and Friedlander, 1978; Kamens et al., 1984, 1990; Benner et al., 1989; Hildemann et al., 1991; Rogge et al., 1993a; de Valk and van der Hage, 1994). Miguel and Friedlander (1978) published particle size distributions for benzo[*a*]pyrene (BaP) and coronene (Cor) in Pasadena, California. They observed that 75% of the particle BaP and 85% of the particle Cor were associated with particles less than 0.26 µm in diameter, and that almost half the mass of these two PAHs was in a 0.075–0.12 µm particle size range. About 10% of the BaP mass was associated with particles >2 µm. In Massachusetts in 1994, Allen et al. (1997) collected urban and rural particles and observed that a significant fraction of PAHs was associated with urban coarse particles (1.9 µm $< d_p <$ 19 µm). For example, phenanthrene and anthracene on coarse particles constituted ~55% of the total PAHs on all particles. Polycyclic aromatic hydrocarbons with molecular weights of 252 g mol^{-1} contributed only ~10% to the total PAHs. Some of these data are shown in table IX-C-4 and are directionally consistent with a studies by Venkataraman et al. (1999) in India and in Los Angeles, California (Venkataraman et al., 1994). For rural particles, however, Allen et al. (1997) observed that ~55% of the mass of low-volatility PAHs (MW 228–278) appeared in the coarse-particle fraction (1.9 µm $< d_p <$ 19 µm). This observation is also supported by the observations of others (Pierce and Katz, 1975; Venkataraman et al., 1994). This may be due to fine-particle scavenging by coarser aerosols or mechanical reentrainment of coarse particles.

Table IX-C-4. Fraction of individual PAHs on different-size aerosols at an urban sampling site in Boston

Compound[a]	Aerosol Concentration (µm m^{-3})	d_p[b] < 0.14 µm	0.14 < d_p < 1.9 µm	1.9 < d_p < 19 µm
Phe	13.23	0.03	0.35	0.61
Ant	1.28	0.06	0.43	0.51
Fla	14.7	0.06	0.48	0.46
Py	8.07	0.07	0.50	0.43
BaA	1.66	0.11	0.72	0.17
Chry	1.78	0.10	0.71	0.19
Bfs	3.22	0.16	0.74	0.10
BeP	1.34	0.19	0.71	0.10
BaP	1.17	0.18	0.74	0.08
Pery	0.22	0.14	0.79	0.07
Ind	1.03	0.25	0.68	0.07
BghiP	0.82	0.36	0.58	0.06
Cor	0.11	0.52	0.44	0.05

Source: Allen et al. (1997).
[a]See table IX-A-2 for compound names corresponding to the abbreviations shown here.
[b]d_p = aerodynamic particle diameter.

Seasonal PAH–particle size data have been presented by Aceves and Grimalt (1993) for Barcelona urban aerosols. During the month of March for a total of 16 PAHs, 77% of the PAH particle-associated mass was on particles below 0.5 μm, 90% was on those below 0.96 μm, and 97% was on those below 1.5 μm. During the month of December, particle PAH shifted very slightly to larger sizes, with 73% on particles below 0.05 μm, 87% on those below 0.5 μm, and 94% on those below 1.5 μm. Kaupp and McLachlan (1999) reported on samples taken in northern Bavaria, and directionally similar results were observed. In their study, however, there was a higher percentage of PAH on particles >1.35 μm than on particles <0.45 μm during the winter than during the summer. This difference is probably due to differences in temperature.

One would surmise that after dilution from the source, low-molecular-weight PAHs, such as fluorene and phenanthrene, quickly redistribute primarily into the gas phase (Kamens et al., 1995). This is generally observed for these compounds in ambient measurements (see table IX-C-1). These low-molecular-weight gas-phase PAHs can then migrate to the "PAH-naked surfaces" on coarse particles. In addition, the presence of coarse-particle solid surfaces provides a chemical potential difference between the fine and the coarse particles, resulting in a net migration of low-molecular-weight PAHs from fine to coarse particles in the gas phase. This volatilization of low-molecular-weight PAHs from very fine (0.01–0.1 μm) particles may also be somewhat enhanced by the increased vapor pressure/Kelvin effect that PAHs experience on very fine particles (Venkataraman et al., 1999). Given that PAHs of 5 rings or greater are primarily associated with the particle phase, and do not have a large companion gas-phase concentration, much less migration or mass transfer to the coarse-particle fraction is observed. What becomes clear from all of the studies cited is that the majority of the PAHs above molecular weight 228 will be in a fine-aerosol fraction. If the current USEPA standard of 2.5 μm is used, then perhaps as much as 90–95% of these compounds will be collected with samplers designed to collect particles equal to or below 2.5 μm in diameter.

IX-C-5. PAH Gas/Particle Partitioning

From the above discussion, it is apparent that semivolatile PAHs can be distributed between both the gas and the particle phase. As will be discussed in this section, PAHs can have very different atmospheric chemical fates depending on their phase. In addition, models that estimate atmospheric fluxes and deposition rates (Brorström-Lundén et al., 1994; Leister and Baker, 1994; Halsall et al., 1997b) and washout ratios (Ligocki et al., 1985a, 1985b) use PAH gas- and particle-phase inputs. Finally, the phase in which a semivolatile exists can dramatically influence the extent to which it is sorbed by the inner linings of the lungs (Pankow et al., 1997). During the 1970s and 1980s, much of the research on PAH gas/particle distributions was based on monolayer organic coverage or Langmuirian gas–solid equilibrium theory (Langmuir, 1918; Junge, 1977; Yamasaki et al., 1982; Pankow, 1987, 1991). Perhaps the first important relationship was reported by Junge (1977). He was able to relate the fraction of a semivolatile compound (ϕ) in the particle phases to the saturated vapor pressure (p° in

Torr), the concentration of aerosol surface area (θ_j in $cm^2\,cm^{-3}$ air), and a constant c_j:

$$\phi = \theta_j c_j/(p^\circ + \theta_j c_j) \tag{1}$$

Junge started with incomplete particle coverage of the adsorbing semivolatile compound. Under these conditions, the Langmuirian isotherm becomes a special case of the more general Brunauer–Emmett–Teller (BET) isotherm (Brunauer et al., 1938). The constant c_j can be evaluated (Junge, 1977; Pankow, 1987) such that $c_j = 760 b_B RTN_s$, where $b_B =$ the BET constant (Brunauer et al., 1938) or approximately, $e^{(Qd-Qv)/RT}$, where R is the gas constant in kcal $(\deg\ mol)^{-1}$ in the BET term and in cm^3 atm per (deg mol) in the preexponential term, T is the temperature (K), Q_d is the heat of desorption from a surface in kcal mol^{-1}, Q_v is the heat of vaporization of the pure substance in kcal mol^{-1}, and N_s is the moles of sorption sites per square centimeter of aerosol. Pankow gives a value for N of $\sim 4 \times 10^{-10}$ moles of sites per square centimeter for a number of PAHs (Pankow, 1987).

In table IX-C-5, the relationship of the impact of saturated vapor pressure and available particle surface area for three monodisperse aerosols on gas/particle (G/P) partitioning is illustrated. It is apparent that particle surface area can have an important impact on semivolatile compounds like pyrene, which have vapor pressures in the range of 10^{-5} Torr. At very low particle concentrations, these compounds are almost entirely in the gas phase, while at very high particle concentrations ($500\ \mu g\ m^{-3}$) they appear mostly on particles.

Yamasaki et al. (1982) presented a Langmuirian adsorption/desorption model to describe the gas/particle distribution of urban PAHs as a function of temperature and aerosol concentration (total suspended particulate, TSP).

$$\log K_y = \log[G/(A/\mathrm{TSP})] = -a(1/T) + b \tag{2}$$

Here, Ky is the equilibrium constant relating the gas and particle distribution, and the term $[G/(A/\mathrm{TSP})]$ is a partitioning or equilibrium constant, where A is the particle-phase concentration and G is the gas-phase concentration. The derivation of the relationship required that TSP be directly related to the Langmuirian number of adsorption sites on the aerosol surface area. Another implicit assumption is that the ratio of the aerosol surface and volume particle size distributions remains relatively constant over the study period. Lastly, this and the Junge relationship require that instantaneous G/P equilibrium is established and main-

Table IX-C-5. Fraction (ϕ) in the aerosol phase at different aerosol concentrations at 25°C

Aerosol Concentration (Size μm)	ϕ Phenanthrene (5.3×10^{-4} Torr)[a]	ϕ Pyrene (3.3×10^{-5} Torr)[a]	ϕ BaP (9×10^{-8} Torr)[a]
$10\ \mu g\ m^{-3}$ (0.5)	0.002	0.03	0.93
$100\ \mu g\ m^{-3}$ (0.35)	0.03	0.32	0.99
$500\ \mu g\ m^{-3}$ (0.25)	0.17	0.77	1.00

Source: Values taken from Kamens et al. (1995).

[a]Subcooled, saturated vapor pressure at 25°C computed from Hinckley et al. (1990); particle density = 1.25.

Table IX-C-6. Regression coefficients for the gas/particle partitioning relationship of Yamasaki et al. (1982)

PAH	a	b	r^2
Phe + Ant	4117	21.45	0.78
Flu	4421	21.52	0.73
Py	4183	20.55	0.80
BaA + Chry + Tri	5693	23.24	0.89
BeP + BaP	4864	19.99	0.70

tained. Individual PAH gas- and particle-phase data (A and G in ng m^{-3} and TSP in ng m^{-3}) were reported for Osaka, Japan, for November 1977 through 1978. These data were fitted with the above relationship over a temperature range of ~277–303 K. The PAH regression values for a and b from Yamasaki et al. (1982) are presented in table IX-C-6. Baek et al. (1991), using the same approach, provided another similar set of regression values for the London atmosphere. Pankow (1991) later suggested that a chemical class of partitioning organics should have the same y-intercept b in equation (2) above, and developed a set of regression equations (constant y-intercept regression, CYIR) with a constant y-intercept from the Yamasaki data set. Pankow (1992) then compared measured gas- and particle-phase PAH concentrations (G/A) in the Baltimore tunnel and in Chicago in 1986 (Benner et al., 1989) with predictions from the Yamasaki et al. and CYIR regressions developed from the Osaka atmosphere. These are summarized in table IX-C-7. In general, predictions agree to within a factor of 3 with either model. This is rather remarkable given the implicit assumptions in Langmuirian-based models.

In the mid-1980s, You and Bidleman and their coworkers (You and Bidleman, 1984; Bidleman et al., 1986) observed that ambient PAH partitioning (log of K_y values) varied directly with the log of subcooled liquid PAH vapor pressures. Foreman and Bidleman (1997) then reported that much better fits were obtained

Table IX-C-7. Average ratios of predicted C_{gas}/C_{part} to measured C_{gas}/C_{part} in the Baltimore tunnel and at an urban sampling site in Chicago[a]

	Chicago	Chicago	Baltimore	Baltimore
PAH[b]	Yamasaki et al. (1982)[a]	CYIR	Yamasaki et al. (1982)[a]	CYIR
Phe	2.39 (1.76)	2.41 (1.77)	1.65 (0.59)	1.67 (0.60)
Ant	1.84 (0.99)	1.85 (1.00)	2.29 (1.06)	2.32 (1.07)
Fla	1.42 (0.88)	1.40 (0.87)	3.02 (1.55)	2.39 (1.24)
Py	1.77 (1.45)	1.54 (1.24)	1.30 (0.54)	1.28 (0.53)
BaA	0.26 (0.14)	0.44 (0.23)		
Chry	0.46 (0.41)	0.73 (0.61)		

[a]For this presentation, extreme data points (21.19 and 45.2 and 38.0) were deleted from Pankow analysis (Pankow, 1991); $n = 10$ samples in Chicago and in Baltimore for most compounds; numbers in parentheses are standard deviations.

[b]See table IX-A-2 for compound names corresponding to the abbreviations shown here.

when subcooled vapor pressures ($p_{\mathrm{L}}^{\mathrm{o}}$) were used as opposed to PAH solid vapors ($p_{\mathrm{S}}^{\mathrm{o}}$). Starting with the derivation of the Yamasaki relationship (Yamasaki et al., 1982), the gas/particle kinetic representations in the Langmuirian isotherm (Pankow, 1987; Adamson and Gast, 1997), and the convergence of the Langmuirian and BET isotherms for a monolayer coverage, Pankow (1987) was able to show theoretically a log (K_y) versus log p relationship. This takes the form:

$$\log(K_p) = \log\{A/(G \times \mathrm{TSP})\} = \boldsymbol{m}\log(p_{\mathrm{L}}^{\mathrm{o}}) + c \tag{3}$$

where the partitioning coefficient K_p ($K_p = K_y^{-1}$) is also an equilibrium constant and the inverse of the K_y. In this form, the slope $\boldsymbol{m}$ at equilibrium, as predicted by Pankow and coworkers, takes on a value of −1 (Pankow, 1987, 1991; Pankow and Bidleman, 1992), and the intercept c equals $\log\{N_s A_{\mathrm{TSP}} T\, \mathrm{e}^{(Qd-Qv)/RT}\}$. Parameters are defined in the previous discussion of the Junge adsorption model and A_{TSP}, the specific particle surface area, is in $\mathrm{cm}^2\,\mu\mathrm{g}^{-1}$. A plot of typical Yamazaki data for five different PAHs as treated by Pankow (1991) is given in figure IX-C-1. Slopes close to −1 have been observed for PAHs and alkanes in the Portland area (Pankow and Bidleman, 1991, 1992) and for dilute diesel exhaust in an outdoor chamber study (Kamens et al., 1995). This is not exclusively the case, and values significantly less than 1 have been reported (Finizio et al., 1997; Simcik et al., 1998).

There has been mounting evidence, with respect to gas/particle partitioning, that many types of urban particles may behave as though sorption from the gas phase is taking place into an organic liquid. The first indication of this was that subcooled liquid vapor pressures gave better regression fits to the partitioning relationship in equation (3). Then, experiments by McDow et al. (1994) showed that fresh wood soot particles, when accelerated through an impactor and col-

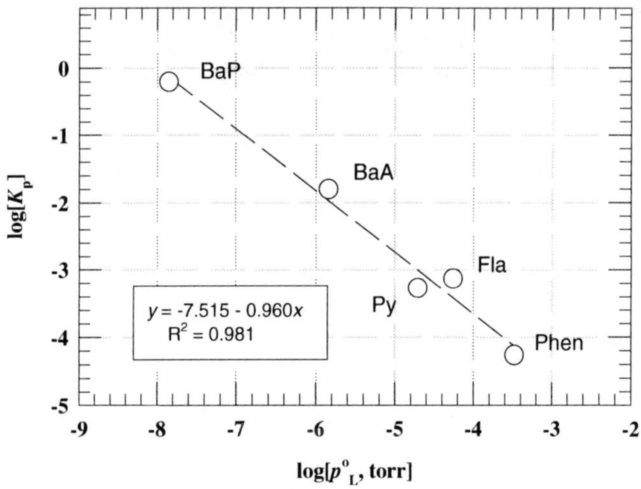

Figure IX-C-1. Log K_p versus log $p^{\mathrm{o}}{}_{\mathrm{L}}$ for one sampling period from the Yamasaki et al. (1982) data set; replotted from data in Pankow and Bidleman (1991).

lected on an indium foil, formed an amorphous liquid-like mass and lost all particle integrity. Diesel soot behaved similarly, but some particle integrity was maintained after impacting onto the foil. This is consistent with diesel soot having much more elemental carbon than wood soot. Additional evidence comes from estimates of the amount of organic coverage on ambient aerosols (McDow et al., 1996). A typical urban aerosol has from 1.5 to 2 times as much organic carbon as compared with elemental carbon. Diesel exhaust particles have a ratio of $\sim$0.4:0.25 organic to elemental carbon (Scheutzle et al., 1981; Japar et al., 1984; Rogge et al., 1993a). If one uses an estimate that only $\sim$4% organic carbon is needed to give a monolayer of coverage (Risby and Sehnert, 1988), diesel soot may contain many molecular layers of organic molecules coating an elemental carbon core. For ambient air it is even greater, and for wood smoke, which is almost entirely organic (Kamens et al., 1988), the layer could be hundreds of molecules thick. Finally, Liang et al. (1997) observed that the G/P partitioning to urban aerosols exhibits similar behavior to aerosols with a known liquid layer.

To treat G/P absorptive partitioning, Pankow (1994b) derived a relationship that has the same K_p functional relationship to p_L^o as equation (3), but also includes an activity coefficient, γ, to represent the non-ideality of a given semi-volatile absorbing into a given organic fraction of a liquid aerosol:

$$^iK_p = (760RT f_{om})/(10^6\,\mathrm{MW}_{om}\,{}^i\gamma_{om}\,{}^ip_L^o) \tag{4}$$

where ^iK_p has units of $m^3\,\mu g^{-1}$; T is the temperature (K); f_{om} is the mass fraction of organic material in particulate matter; MW_{om} is the average molecular weight of a given liquid medium ($g\,mol^{-1}$); $^i\gamma_{om}$ is the activity coefficient of a given organic compound, i, in a given organic mixture; $^ip_L^o$ is the saturated subcooled liquid vapor pressue of compound i (mmHg), and R is the gas constant (8.02 $m^3\,mol^{-1}\,K^{-1}$). Although one might assume that activity coefficients within a class of compounds on a given organic aerosol are relatively constant, it is not necessarily true for all organic molecules. Jang and coworkers computed activity coefficients for a number of PAHs as well as other non-polar and polar compounds on wood soot, diesel soot, and α-pinene/O_3 secondary aerosols (Jang et al., 1997; Jang and Kamens, 1998). A vast improvement in fits was observed when an activity-coefficient-adjusted $^iK_p'$ ($^iK_p\,{}^i\gamma_{om}$) was related to $^ip_L^o$ (log $^iK_p\,{}^i\gamma_{om}$ plotted vs. $^ip_L^o$). The PAHs tended to have activity coefficients ranging from 1.7 to 5.8 on wood soot, diesel soot, and α-pinene secondary aerosols, with most values falling between 2 and 4.

Another approach that has recently been reported by Harner and Mackay (1995) and Finizio et al. (1997) uses the relationship between G/P partitioning and an octanol/air partitioning coefficient. The octanol/air partitioning coefficient (K_{oa}) is defined as the octanol/water partitioning coefficient K_{ow} divided by the dimensionless Henry's law constant K_H, where K_H' equals the Henry's law value, in atm $L\,mol^{-1}$ divided by the gas constant R (in this case, 0.082 $L\,atm\,mol^{-1}\,K^{-1}$) and the temperature. The partitioning expression takes the form:

$$\log(K_p) = \boldsymbol{m}\log(K_{oa}) + b \tag{5}$$

The intercept includes the activity coefficients of a partitioning semivolatile in the octanol and organic particle phases. This approach permitted reasonable fitting of a number of different studies that did not necessarily exhibit a slope close to -1 in log ${}^{i}K_p$ versus log ${}^{i}p_L^o$ plots (Finizio et al., 1997). Since K_H and K_{ow} values are available for many PAHs (see table IX-A-1) and can be calculated for different temperatures, this approach eliminates the need to estimate vapor pressures.

In all of the above discussion, an implicit assumption is that the gas and particle phases rapidly come to equilibrium and rapidly redistribute as changes in one phase occur or the system undergoes further dilution of concentration. Kamens and Coe (1997) were able to demonstrate experimentally for fresh diesel exhaust that low-volatility PAHs on particles respond on time-scales of less than 1 s to rapid changes in gas-phase concentrations. Other experiments in outdoor chambers indicated that under moderate temperatures ($\sim 20^\circ$C), fresh PAHs seem to be in G/P equilibrium. However, under cooler conditions ($\sim 0^\circ$C), if the system gets out of equilibrium, tens of hours may be required to reapproach 90% of equilibrium (Kamens et al., 1995). When the system cannot maintain conditions close to equilibrium, it is necessary to use radial diffusion models to estimate G/P partitioning (Rounds et al., 1993; Strommen and Kamens, 1997, 1999).

In addition to equilibrium, the above discussion also implies that simultaneous measurement of gas- and particle-phase PAHs is possible given current and previous sampling technology. In actuality, most of the reported measurements were subject to sampling artifacts. Cadle and Groblicki (1982), Van Vaeck et al. (1984), McDow and Huntzicker (1990), and Turpin et al. (1994) show that the glass fiber filter material has an adsorptive capacity and a significant positive artifact can result for organic carbon with aerosol concentrations in the low $\mu g\, m^{-3}$ range. This results because gas-phase semivolatiles are retained on filters and are reported as particle phase. This phenomenon decreases with increasing face velocity and pressure drop across the filter (McDow and Huntzicker, 1990) and can cause a +50% positive artifact of filter-phase organic carbon when the total ambient particle concentration is in the range of $\sim 5\, \mu g\, m^{-3}$. The artifact decreases to approximately +20% when $\sim 20\, \mu g\, m^{-3}$ were sampled. A backup glass fiber filter that just captures adsorbed gas-phase PAHs behind an upstream Teflon filter may be used to correct this problem (Turpin et al., 1994).

Coutant et al. (1988), Eatough et al. (1989), and Tang et al. (1994), however, have reported negative sampling artifacts when sampling ambient aerosols. The negative artifact is an observed loss of semivolatiles from a filter sampler, sampling directly from an atmosphere, due to volatilization ("blow off") as air passes over the filter. If one assumes that the gas phase and particle phase are at equilibrium in the sample of air, and the PAH concentration in both phases does not change, a negative artifact is difficult to explain. This is because as particulate matter accumulates on the filter, both the PAH particle concentration in grams of PAH per volume of air, and the TSP in grams of particulate matter per volume of air, increase at the same rate. Hence, in the equilibrium expression $K_p = A/(G \times \text{TSP})$, with a constant K_p at a given temperature and a constant ratio of A/TSP, G remains constant, and there is no chemical potential imbalance between the gas phase and particle phase. McDow and Huntziker (1990) do, however, show that with increasing face velocity or pressure drop across their filters, the positive artifact decreases. One might surmise that with a signifiant pressure drop, a negative sampling artifact might result.

Another confounding issue results because multihour filter samplers often experience ambient changes in gas- and particle-phase semivolatiles. During the initial part of the sampling period, there is probably a tendency for positive artifacts to result because of adsorption of gas-phase semivolatile compounds on the filter. As particulate mass accumulates, and if the sample air mass becomes appreciably lower in gas- and particle-phase semivolatiles, then semivolatile "off-gassing" will most probably occur. If the air mass becomes substantially "dirtier" than the previously sampled air, positive filter artifacts may occur. Changes in ambient temperature during the sampling period could have a similar effect, with higher-to-lower temperature changes causing a positive artifact and lower-to-higher temperature changes resulting in a negative artifact. Hence, a major problem with direct filter sampling of the atmosphere is that these systems exhibit artifacts that may be dramatically influenced by the sampling history or changes in the concentrations of the semivolatiles in the air with time.

Given the above-mentioned problems associated with direct filter sampling of the atmosphere, a number of investigators first remove gas-phase semivolatiles and then collect the particle semivolatiles (Coutant et al., 1988; Lane et al., 1992; Gundel et al., 1995; Kamens et al., 1995). In a typical sampling arrangement, air first enters an upstream semivolatile gas-phase denuder. Sample air then passes through a filter or series of filters and a downstream adsorbent. The sum of the semivolatiles on filter and downstream adsorbent represents the particle-associated semivolatiles. Ideally, the gas phase is removed with very high efficiency and the residence time in the denuder is short enough that significant "off-gassing" does not occur as particles traverse the denuder. In this regard, the reader is referred to a book edited by Lane (1999) on the use of denuders in sampling. Kamens and Coe (1997) have reported that for a 0.3-s residence time in a 40-cm five-channel annular denuder, 10–15% of the particle-phase fluorene (Flu) and phenanthrene (Phe) in fresh diesel soot is lost due to particle off-gassing. Longer denuder residence times lead to substantially higher particle-phase losses. Fluoranthrene and pyrene are less volatile than Phe and Flu and were lost at lower rates than Phe and Flu. Almost no benz[*a*]anthracene and chrysene were lost for residence times up to 8 s. However, in recent experiments (unpublished) by the Kamens group, almost 100% of a pure hexadecane aerosol may have been lost in the above gas-phase denuder system due to particle off-gassing. Unfortunately, the denuder below the filter was not analyzed for hexadecane and particle evaporation from the filter may have occurred. These results, however, simply suggest that sampling systems that use an upstream gas-phase denuder should be designed with extremely short sampling times, and that, at the present time, both filter–adsorbent or denuder–filter–adsorbent sampling systems can potentially exhibit significant artifact sampling problems for the most volatile of the semivolatile PAHs.

IX-D. Reactions, Mechanisms, Products, and Reaction Rates of PAHs Sorbed on Particles

As noted in the previous sections of this chapter, PAH partition between the gas phase and particle phase, and different mechanisms, dominate the reactions of PAHs in these two phases. In the gas phase, reactions with the OH radical are very important during the daytime (Atkinson et al., 1987a, 1989a, 1990a; Atkinson and Arey, 1994). In the particle phase, depending on the nature of the particles, PAH reactions are associated with a photoinduced or O_3-mediated process. Photoinduced reactions can involve direct photolysis in an organic liquid layer of particles or photosensitized reactions (Fox and Olive, 1979; Kamens et al., 1990; McDow et al., 1994).

IX-D-1. PAH Reactions on Particles in Light

While the reaction of gas-phase PAHs are somewhat understood and documented, interpretation of experiments conducted to elucidate the atmospheric reactions of PAHs on particles are less straightforward and strongly depend on the nature of the particles onto which the PAHs are sorbed. For example, PAHs associated with fly ash and carbon black tend not to react upon exposure to light (Korfmacher et al., 1980; Behymer and Hites, 1985, 1988; Dunstan et al., 1989) while they can degrade very quickly when sorbed on silica or fresh composition emissions (Kamens et al., 1985, 1986, 1988; Behymer and Hites, 1988; Dunstan et al., 1989).

Falk et al. (1956) exposed spark ignition and diesel exhaust particles collected on filters to clean air and sunlight. They also carried out the same experiment with pure PAHs on filters. The authors observed a 20% decay of benzo[*a*]pyrene (BaP) exposed on filters for 48 h of daylight compared with a 10% decay on soot particles. They suggested that the soot particles seemed to impart photostability to PAHs. Benzo[*a*]pyrene did not decay in the dark on filters. Pure lower-molecular-weight PAHs (phenanthrene through pyrene) on filters decayed at the same rate in the light and the dark (the intensity of the light is not described). This was probably due to volatilization from the filters, given the tendency of these compounds to partition to the gas phase (Yamasaki et al., 1982). Hence it is reasonable to assume that the losses would be similar under light and darkness if this is the main loss process.

Cimberle et al. (1983) also exposed to sunlight BaP sorbed onto glass fiber filters from a benzene solution and found a direct relationship between the rate coefficient of BaP decay and the total solar intensity (TSR) in calories cm^{-2} min^{-1}:

$$k_{rate} = 0.0015 + 0.0042\,\mathrm{TSR}$$

If one assumes an average $j(NO_2)$ value of $5.2 \times 10^{-3}\ s^{-1}$, an equivalent TSR of ~ 0.69 calories $cm^{-2}\ min^{-1}$ would be estimated; this gives an estimated lifetime of about 3.8 h. One explanation for the difference between the results of this study and the similar study of Falk et al. (1956) is the difference in intensity and wavelength distribution of the light sources employed.

In 1978, Fox and Olive (1979) investigated the ambient particles on glass fiber filters collected at two locations in Texas. Anthracene was sublimed onto the particle-laden filters and exposed to bright Texas sunlight for 4 days. After the 4-day exposure, only 10% of the anthracene and 15% of the indigenous BaP remained. As shown in figure IX-D-1, approximately a third of the reacted anthracene was accounted for as oxygenated products.

Fataidi (1967) exposed solvent-extracted garden soil particles coated with pyrene in a borosilicate flask to 366-nm light (8000 $\mu W\,cm^{-2}\,s^{-1}$) for 240 h. About 8–10% of the reacted pyrene appeared as oxygenated products. Tebbens et al. (1966) reported on the irradiation of propane combustion particles in a 15.2-cm cylindrical Pyrex flow reactor, of which 3.35 m was irradiated with fluorescent lights. From *o*-nitrobenzaldehyde actinometry it was estimated that the light flux was about one-quarter of the maximum solar noon intensity in July in northern California (Thomas et al., 1968). The combustion conditions used in this study must have been very fuel-rich, given the extremely high reported BaP particle loadings of $> 30{,}000\,ng\,mg^{-1}$ of soot particles. On a particle mass basis, 35–65% of the BaP reacted between the inlet and outlet in 40 min. These experiments were repeated by Thomas et al. (1968) with a 58% photolytic loss of BaP. If one assumed that an average of 50% of the BaP is lost at the solar intensity of the experimental conditions, then at a solar intensity equivalent to $j(NO_2) = 5.2 \times 10^{-3}\,s^{-1}$, the estimated lifetime for BaP would be $\sim$25 min on these particles.

A number of studies have been conducted with PAHs doped onto carbon black, silica gel, alumina, and coal fly ash (Korfmacher et al., 1980; Daisey et al., 1982; Behymer and Hites, 1985, 1988; Dunstan et al., 1989). Both vapor

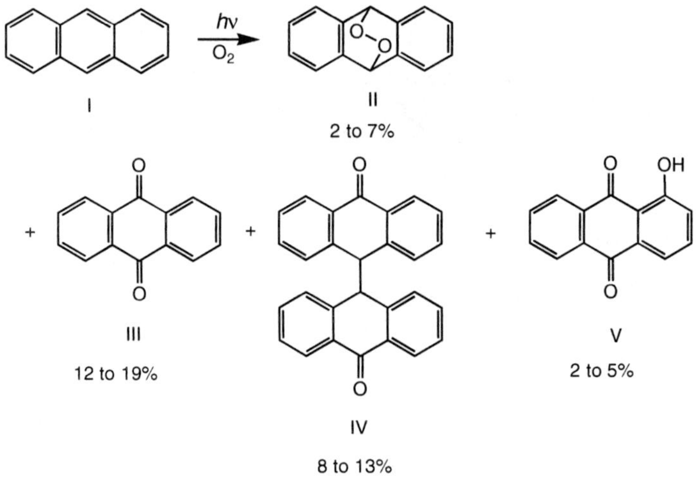

Figure IX-D-1. Products observed from sunlight irradiation of anthracene on filter-collected ambient particles in sunlight.

doping (Miguel et al., 1979; Korfmacher et al., 1980) and liquid evaporation (Behymer and Hites, 1985) of PAHs onto the substrate surface have been employed. Fluidized-bed reactors and rotary evaporators (Korfmacher et al., 1980; Daisey et al., 1982; Behymer and Hites, 1985) were used for exposure of PAHs. The light sources were 200-W (Daisey et al., 1982) and 450-W (Behymer and Hites, 1988) medium-pressure mercury vapor lamps with irradiances of $14\,W\,m^{-2}$ (Behymer and Hites, 1985, 1988) to $35\,W\,m^{-2}$ (Daisey et al., 1982), which translate into light intensities of 0.2 and 0.5 calories $cm^{-2}\,min^{-1}$, respectively. These studies reported that PAHs are very resistant to photodegradation on carbon black, with lifetimes of 400 to >1400 h (Korfmacher et al., 1980). On silica and alumina (Behymer and Hites, 1985, 1988), many PAHs react quickly, with lifetimes that range from 0.3 to 30 h. On fly ash, most PAHs tend to have lifetimes of ~ 60 h (Behymer and Hites, 1985, 1988) with very little difference in reactivity among the different PAHs.

Korfmacher et al. (1980) also exposed fly ash vapor doped with PAHs to a variety of different light sources including sunlamps, xenon lamps, Quartzline lamps and a high-pressure mercury arc lamp. Though their findings support the observation that PAHs are not very photoreactive on fly ash, light fluxes at the point of sample exposure were not given. Behymer and Hites (1988) also reported on results for 18 different substrates that included carbon black, cyclone dusts, different fly ashes, silica, and alumina. They observed that the carbon content of the substrate had the greatest influence on the photostability of 19 different PAHs. The PAHs were stabilized on fly ash if the carbon content was greater than 5%. These authors did not see a relationship between PAH reactivity and substrate surface area, porosity, particle size, surface, pH, or iron content. Dunstan et al. (1989) also observed that the carbon fraction of fly ash has a tendency to photostabilize PAHs. The mineral (silica–alumina) subfraction of fly ash tended to adsorb pyrene more weakly than the carbon fraction, although pyrene was much more photoactive on this fraction than the carbon subfraction. This is consistent with the smaller half-lives observed for PAHs on silica and alumina (Behymer and Hites, 1985) than on carbon black. Lastly, the magnetic subfraction of fly ash also weakly adsorbs pyrene, but appears to make pyrene resistant to photodegradation.

Between 1986 and 1994, a number of outdoor chamber studies addressed PAH photostability on fresh wood smoke and diesel and gasoline combustion exhaust (Kamens et al., 1985, 1986, 1988, 1994). Under relatively constant midday sunlight, PAH decay could be fit with first-order kinetics. In these studies, fresh combustion emissions were directly added to outdoor chambers and the diluted gas and particles were permitted to age in the presence of sunlight, or in darkness in the presence of O_3 and NO_2. Typically, particle concentrations ranged from 100 to 5000 $\mu g\,m^{-3}$. Under darkness (O_3 and NO_2 <0.07 ppm), it was observed that particle benz[*a*]anthracene, chrysene, benzofluorenes, benzo[*a*]pyrene, benzo[*e*]-pyrene, and benzo[*g,h,i*]pyrene tended to be stable on a mass PAH/particle mass basis (Kamens et al., 1985, 1986, 1988, 1994). In sunlight, PAHs tended to react quickly. This is illustrated in figure IX-D-2 for PAHs associated with gasoline exhaust and wood soot.

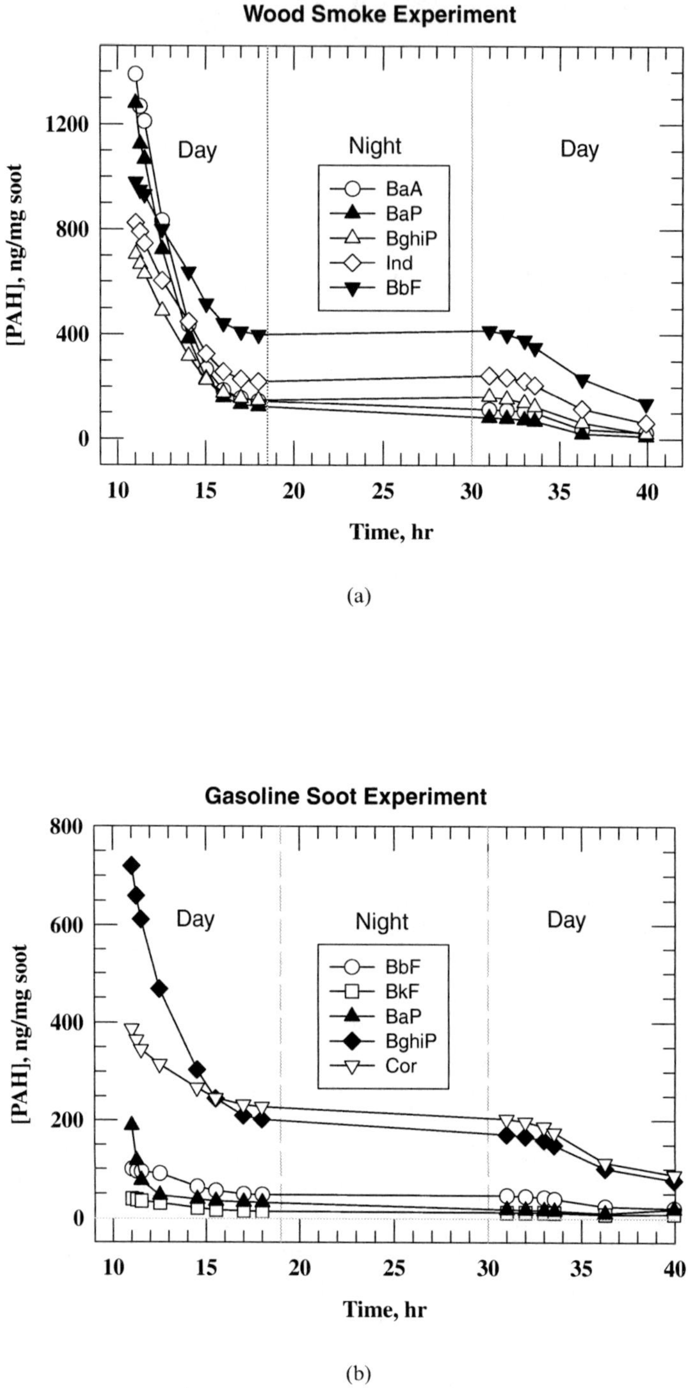

Figure IX-D-2. Decay of PAHs on soot particles at 15°C under clear North Carolina April skies; see table IX-A-2 for compound names corresponding to the abbreviations shown here.

Kamens and coworkers attempted to characterize photodecay rates for low-volatility PAHs on wood soot particles (Kamens et al., 1988). The PAHs were measured in outdoor chamber experiments over time in the presence of different solar intensities, humidities, and temperatures (Kamens et al., 1988). Generally, NO_x was below 0.07 ppm. Initial PAH decay rates (60–80% of the reaction) tended to be rapid and could be fit with a first-order rate law (Kamens et al., 1986). This was followed by a slower decay rate. When individual PAH particle loadings ranged from 30 to 350 ng mg^{-1} of soot, the individual PAH decay rates were found to fit statistically the following expression during the first few hours of irradiation:

$$\ln(k_{\text{phot}}) = a_1(1/T) + a_2 \ln(I) + a_3 \ln[\text{H}_2\text{O}] + \text{constant} \tag{6}$$

where k_{phot} is the first-order rate coefficient (min^{-1}), T is the temperature (K), I is the solar intensity (calories cm^{-2} min^{-1}), and $[H_2O]$ is the water vapor concentration (g m^{-3}). Most of the variability in the data can be accounted for by the light and water terms, and hence a simpler expression can be used:

$$\ln(k_{\text{phot}}) = a_2 \ln(I) + a_3 \ln[\text{H}_2\text{O}] + \text{constant} \tag{7}$$

The regression coefficients for equation (7) are given in table IX-D-1. To estimate the dew-point water vapor concentrations in g m^{-3}, the saturated water vapor concentration (Sat_{H2O}) in g m^{-3} was related to the absolute temperature, T, by fitting a simple exponential form to saturated water concentration in g m^{-3} versus dew-point temperature data:

$$\text{Sat}_{\text{H2O}} = \exp(-5928/T + 20.95); \qquad r^2 = 0.9996 \tag{8}$$

If equation (8) is multiplied by the relative humidity (RH) in fractional units, the water vapor concentration in g m^{-3} at a given relative humidity can be computed and used in equation (9):

$$[\text{H}_2\text{O}] = \text{RH} \times \exp(-5298/T + 20.95) \tag{9}$$

Lifetimes of PAHs at different solar intensities, relative humidities, and temperatures (to compute %RH) are given in table IX-D-2. Note the dramatic

Table IX-D-1. Equation (7) regression parameters; data used to compute PAH first-order photodecay coefficients, k_{phot}, on wood soot

PAH[a]	Constant	a_2	a_3	Adjusted r^2
CpC	−4.695	0.836	0.700	0.914
BaA	−6.298	0.948	1.159	0.830
Chry	−7.783	0.884	1.345	0.767
BbF	−6.312	1.051	0.699	0.679
BkF	−6.347	1.216	0.897	0.738
BaP	−6.155	0.958	1.042	0.871
Ind	−8.416	1.069	1.839	0.749
BghiP	−6.405	1.382	1.035	0.754

Source: Kamens et al. (1988).

[a]See table IX-A-2 for compound names corresponding to the abbreviations shown here.

Table IX-D-2. PAH lifetimes in hours using rate coefficients calculated from equation (7) on wood soot particles

PAH	25°C, 60% RH TSR[b] = 0.693 cal cm^{-2} min^{-1}	15°C, 50% RH TSR = 0.8 cal cm^{-2} min^{-1}	−10°C, 40% RH TSR = 0.3 cal cm^{-2} min^{-1}
CpC	0.3	1.2	0.2
BaA	0.4	2.6	2.6
Chry	1.1	7.9	11
BbF	1.4	6.3	3
BkF	0.9	4.7	4
BaP	0.5	2.8	2
Ind	0.6	6.1	27
BghiP	0.6	4.0	5

Source: Kamens et al. (1988).
[a]See table IX-A-2 for compound names corresponding to the abbreviations shown here.
[b]$j(NO_2) \sim 5.2 \times 10^{-3}$ s^{-1} or ~ 0.69 cal cm^{-2} min^{-1}.

effect that low %RH can have on PAH lifetimes. For PAH particle loadings greater than 1000 ng mg^{-1} of wood soot, Kamens et al. (1988) observed that PAHs tend to have sunlight decay rates about a factor of 2 lower than wood soot particles, which have lower PAH loadings (30–350 ng mg^{-1}). Additional unpublished data from chamber experiments by the Kamens' group with other combustion particles (which included emissions from non-catalyst gasoline and diesel cars, and China coal combustion soot particles) were compared with predicted wood soot PAH decay coefficients. In general, the wood soot equations in table IX-D-3 fit the observed decay of PAHs on other particles to within a factor of 2 or 3 (figure IX-D-3).

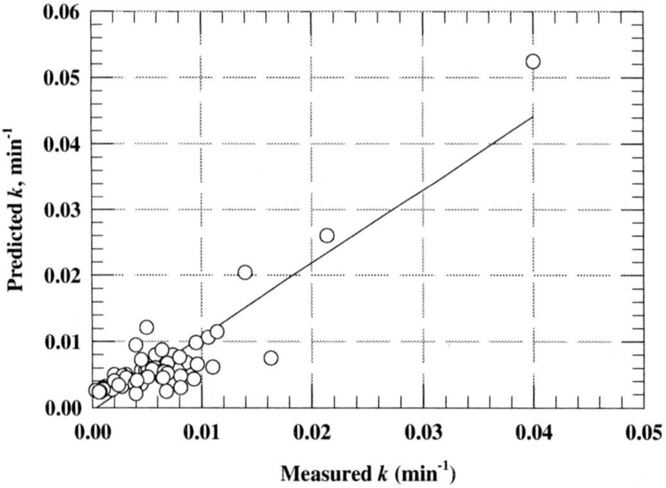

Figure IX-D-3. Measured versus predicted photolytic rate coefficients for benzo[*b*]fluorene, benzo[*k*]fluorene, benzo[*a*]pyrene, and benzo[*g,h,i*]pyrene on gasoline, diesel, and China coal soot using wood soot regression coefficents.

Similar outdoor experiments were conducted by combusting polyurethane foam (PUF) (Birla and Kamens, 1994) and wood chips plus polyvinyl chloride (PVC) shavings laced with pentachlorophenol (Pennise and Kamens, 1996). A combustion vessel with a propane burner was used to combust rapidly the PUF or wood chips, and the emissions were ducted into outdoor chambers (Birla and Kamens, 1994; Pennise and Kamens, 1996). The burner could be controlled so that the wood chips and PUF could be combusted at low temperatures (~350°C) and high temperatures (600–800°C). The PAHs on particles generated at high temperatures tend to be much more stable in the outdoor chamber atmosphere in direct sunlight than PAHs on particles generated at lower temperatures (Pennise and Kamens, 1996). This is illustrated in figure IX-D-4 for PAHs reacting in sunlight on the same day on particles combusted at 800°C and 350°C. Extraction of the particles with methylene chloride showed that the low-temperature combustion particles had more extractable mass than the high-temperature particles. It was surmised (Birla and Kamens, 1994) that the low-temperature particles contained more liquid mass and that liquid-based reactions such as the photosensitized process could occur more readily on the low-temperature particles than on the high-temperature particles.

It is important to note that the data in tables IX-D-1 and IX-D-2 were generated from "cool" combustion sources. The gasoline and diesel engines were uncontrolled and "high emitters." Since automobile catalytic converters operate at high temperatures, it is expected the PAHs on these particles will be more stable when compared with non-catalytic engines than is suggested by tables IX-D-1 and IX-D-2. The same would be true for PAHs on particles emitted from modern high-temperature incinerators and power plants. The high-versus cool-temperature experiments also help to interpret the differences in atmospheric lifetimes that have been observed on fly ash, carbon black, and fresh combustion particles. The organic non-elemental carbon portion of carbon black and fly ash is very low, and therefore one would expect these to lack an organic liquid layer. As will be discussed later in this section, organic liquid films on particles appear to promote photoinduced reactions (McDow et al., 1994, 1996).

IX-D-2. Mechanisms of PAH Photolysis on Particles

To explain the observed products from anthracene exposure to sunlight on ambient particles, Fox and Olive (1979) suggested photosensitized singlet-oxygen formation and reaction as a possible mechanism. This was supported by their observation of the formation of 9,10-anthracene endoperoxide, known to form from a singlet oxygen addition to a PAHs (NAS, 1972). As outlined in detail in chapter VII, a ground-state PAH(S_0) absorbs light and is raised to its first excited singlet state. The PAH excited singlet [PAH(S_1)] then undergoes intersystem crossing to form a first excited triplet state PAH(T_1), which then transfers its energy to a ground-state oxygen [$O_2(^3\Sigma_g^-)$] to form the first excited singlet state of oxygen [$O_2(^1\Delta_g)$]. It is also possible for the first excited singlet of the PAH to react with a ground-state O_2 to form a first excited triplet PAH [PAH(T_1)], and $O_2(^1\Delta_g)$. Singlet oxygen then proceeds to react with a proximal PAH molecule to form PAH products:

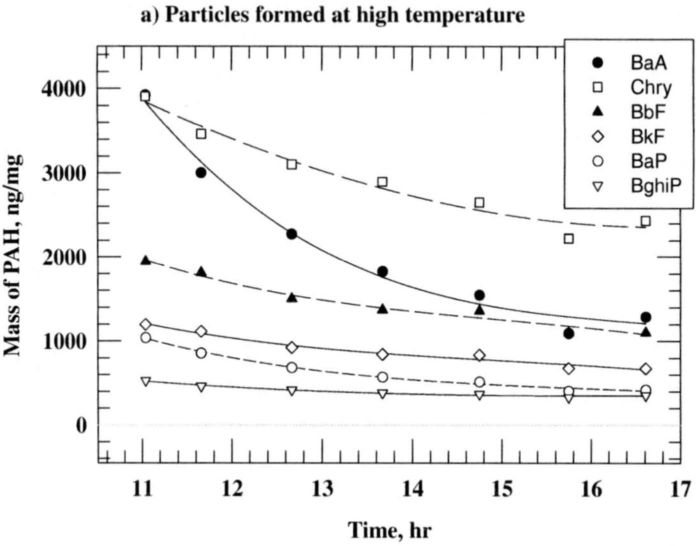

(a)

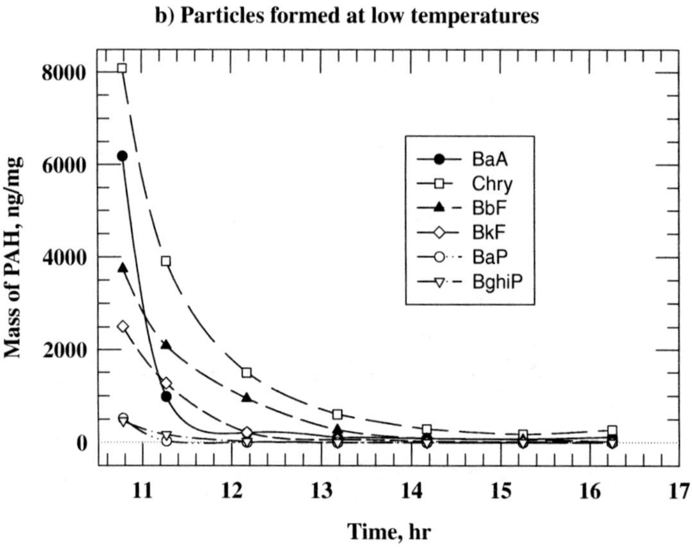

(b)

Figure IX-D-4. Reaction of PAHs on (a) high and (b) low-temperature particles in 25-m^3 outdoor Teflon film chambers on June 14, 1994, under relatively clear sunlight; graphs taken from Pennise and Kamens (1996). See table IX-A-2 for compound names corresponding to the abbreviations shown here.

$$PAH(S_0) + h\nu \rightarrow PAH(S_1)$$

$$PAH(S_1) \rightarrow PAH(T_1)$$

$$PAH(T_1) + O_2(^3\Sigma_g^-) \rightarrow PAH(S_0) + O_2(^1\Delta_g)$$

$$PAH(S_1) + O_2 \rightarrow PAH(T_1) + O_2(^1\Delta_g)$$

$$PAH + O_2(^1\Delta_g) \rightarrow PAH_{products}$$

That PAHs can photosensitize the formation of singlet oxygen was demonstrated by Eisenberg and coworkers by using a flow cell coated with PAHs and a singlet oxygen trap (Eisenberg et al., 1984). In the case of anthracene, $O_2(^1\Delta_g)$ adds across the 9,10-carbons to form an *endo*- (or *meso*-) peroxide. This is consistent with the observed anthracene–sunlight exposure experiments of Fox and Olive (1979). See figure IX-D-5.

Barbas et al. (1996) proposed a similar singlet oxygen mechanism for irradiated phenanthrene adsorbed on silica (SiO_2, N_2 BET surface area of $274\,m^2\,g^{-1}$). Exposures were conducted with 245-nm light at 3.4×10^{16} photons $cm^{-2}\,s^{-1}$ or 350-nm light at 1.23×10^{16} photons $cm^{-2}\,sec^{-1}$ for 3–4 h. About 6–15% of the phenanthrene had reacted and ~90% of the reacted phenanthrene was recovered in nine products with ~70% of the product mass being almost equally distributed between products 1 and 2 (figure IX-D-6). Product number 4 is a dibenzopyranone, and, as mentioned in the gas-phase discussion, its nitro-derivatives have been measured from the gas-phase reaction of phenanthrene and NO_3 radicals (Kwok et al., 1994c).

More recently, the photooxidation of pyrene coated on unactivated and activated silica powder was studied in a rotary reactor (Reyes et al., 2000). An Oriel 1000-W xenon–mercury light source and associated filters permitted exposures over 320–420 nm at an irradiance of $66\,W\,m^{-2}$. The major products were 1-hydroxypyrene, 1,6-pyrenedione, and 1,8-pyrenedione. At a surface coverage of $2 \times 10^{-5}\,mol\,g^{-1}$, small amounts of a 1,1′-bipyrene dimer were observed. These authors suggest that light absorption by pyrene results in a radical cation, which then reacts with physically sorbed water at the surface to form 1-hydroxypyrene.

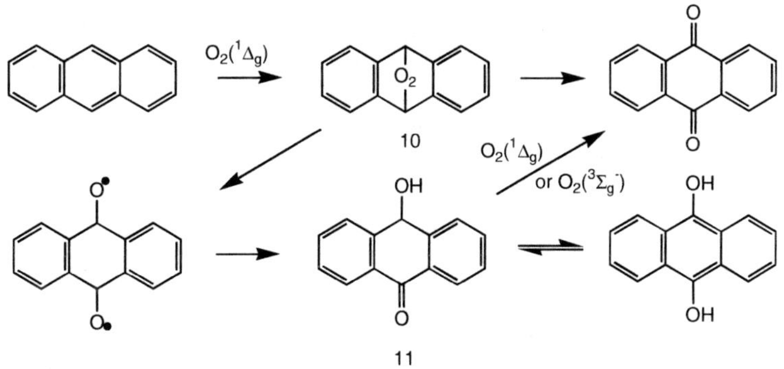

Figure IX-D-5. Photosensitization of singlet oxygen by anthracene.

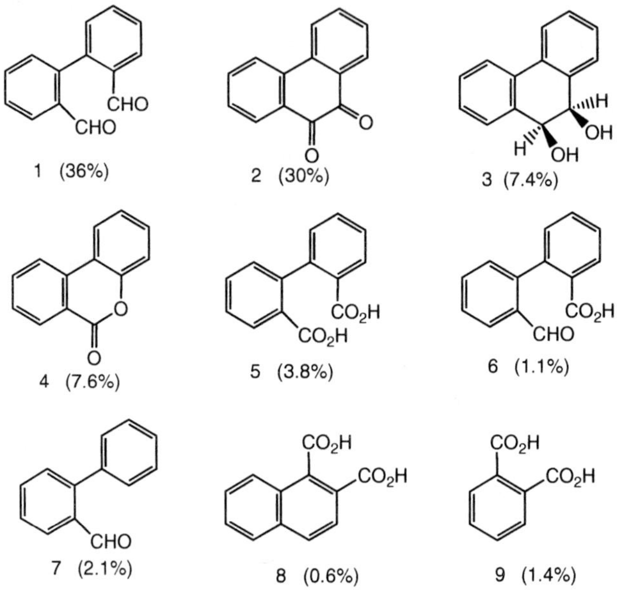

Figure IX-D-6. Products from phenanthrene irradiation on silica. Numbers in parentheses are yields from reacted phenanthrene (Weldon and Ogilby, 1998).

It was further suggested that subsequent photolysis of 1-hydroxypyrene on silica leads to dihydroxypyrene, and that this undergoes rapid oxidation in air to the observed dione products. When methylene blue, a singlet oxygen sensitizer (at ($\lambda \sim 650$ nm) was coadsorbed onto the silica with pyrene, no evidence for pyrene reaction was observed under the irradiation with 650-nm light. This suggest that singlet oxygen was not the operative mechanism. When similar experiments at 650-nm light were performed by Barbas et al. (1996) in the previously noted phenanthrene study, the major products 1, 2, and 3 shown in figure IX-D-6 were observed, suggesting that a singlet oxygen mechanism was indeed operating. This interpretation was further supported by the addition of 2,5-dimethylfuran, a singlet oxygen trap, to the phenanthrene–silica system. Under these conditions, no phenanthrene reaction was observed (Barbas et al., 1996). Since both of these studies were performed by the same group, it is reasonable to assume that, depending on the system and the nature of the PAHs employed, more than one mechanism of PAH oxidation is possible.

Lee-Ruff et al. (1986) exposed aerated benzene solutions containing 0.001 M benzo[*a*]pyrene (BaP) to a 450-W medium-pressure Hg lamp with a Pyrex filter and 0.001 M BaP in $CHCl_3$ to a 500-W visible light source. The UV irradiation of BaP produced only diones II–IV, while irradiation with visible light generated diones (II–IV) and a major product identified as VI, a 6-seco-BaP derivative (figure IX-D-7). Irradiation was also carried out with either methylene blue or tetraphenylporphrin, also a known singlet oxygen sensitizer. Appropriate filters with a transparent window between 405 and 520 nm were used to significantly

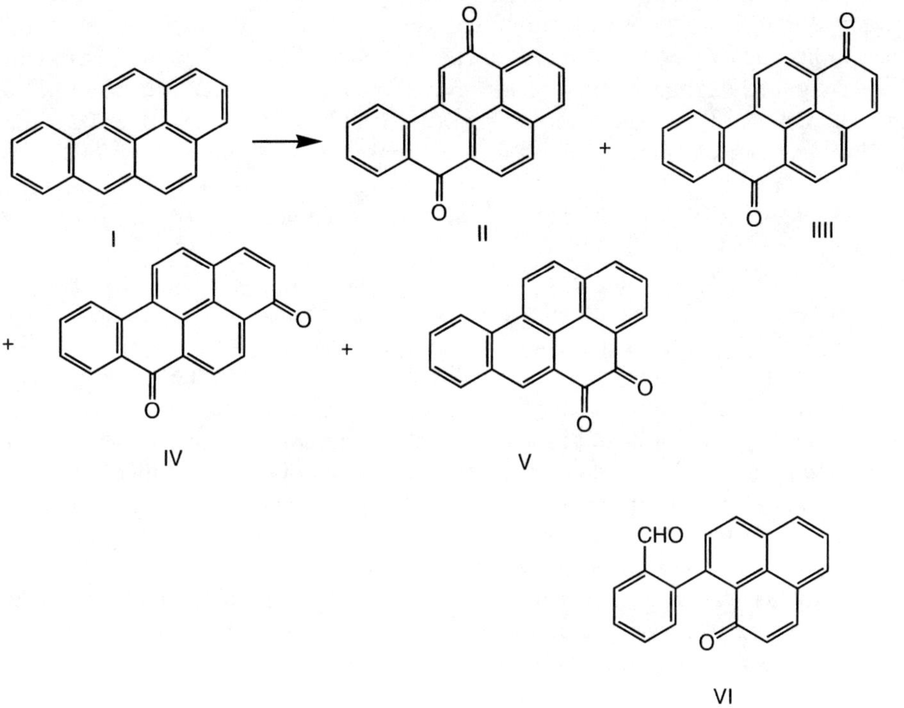

Figure IX-D-7. Products of BaP photolysis (Lee-Ruff et al., 1986).

reduce BaP light absorption and yet permit irradiation of the sensitizers. As with Barbas et al. (1996), the same products were observed for the sensitizer experiments, including the seco-BaP product, as for the aerated BaP experiments in visible light. This suggests that sensitized singlet oxygen formation was the operative mechanism promoting BaP reaction. Further evidence was obtained by adding the singlet oxygen trap, diphenylanthracene to the reaction mixture. It quenches the $PAH + O_2(^1\Delta_g)$ reaction by reacting with the singlet oxygen to form the endoperoxide of diphenylanthracene.

In the previous discussion of the gas/particle partitioning, arguments were presented to support the notion that the organic portion of organic aerosols tends to behave as a liquid (McDow et al., 1994, 1996). Given that the composition of the organic portion of the particle will differ from combustion source to combustion source (Schuetzle et al., 1981; Rogge et al., 1991, 1993a, 1993d, 1994, 1997, 1998; McDow et al., 1994; Jang and McDow, 1995), Jang and McDow (1997) investigated the importance of chemical composition on PAH photoinduced decay. They used a turntable reactor submerged in water at 16.5°C, a medium-pressure Hg lamp, and a borosilicate immersion well. The light intensity inside the borosilicate reaction tubes, based on ferrioxalate chemical actinometry, was $\sim 1.8 \times 10^{17}$ photons s^{-1} $sample^{-1}$. Methoxyphenols were investigated because wood soot has a high methoxyphenol content (Hawthorne et al., 1989, 1992), while diesel soot has high alkane content (Rogge et al., 1993a). It was observed that PAHs photoreacted much faster in dilute methoxyphenol mixtures than in hexane (McDow et al., 1994). The reactivity of PAHs also increased with increasing methoxyphenol concentrations (0.01 to 50% by mass) with toluene as a diluent. On an individual methoxyphenol basis (1% by mass in toluene), benz[*a*]anthracene (BaA) reactivity was greatest in isoeugenol, followed by 2-methylcatechol, syringol, guaiacol and then toluene. Irradiated methoxyphenols decayed in the same order as BaA. The lifetimes of BaA in isoeugenol, methylcatechol, and syringol ranged from 5 to 30 min.

Since diesel, wood, and gasoline soot particles are somewhat hydroscopic (Vartiainen et al., 1996), the effects of water on the rate of PAH degradation were also investigated in the same photoreactor, using a methoxyphenolic mix composed of 12 different compounds (McDow et al., 1996). The water concentration was varied from 0% to 15%. This corresponds to the amount of water that wood soot particles can absorb over a relative humidity range of 40–90% at room temperature (Vartiainen et al., 1996). A threefold increase in the rate of photodegradation was observed for benzo[*k*]fluoranthene as the water percentage increased from 0% to 10%. For water above 10%, the rate fell off slightly. For BaA, there was a twofold increase from 0% to 5% water content, but the rate leveled off above this amount. For benzo[*a*]pyrene, only a slight increase was observed, from 0% to 1% water content. This is directionally consistent with the influence of water on photoinduced rate coefficients derived from wood soot. The chamber studies suggested, however, a linear relationship with atmospheric water vapor concentrations for most of the PAHs and over a wide range of humidities (Kamens et al., 1988). One possible explanation for the PAH reactivity/water enhancement is the photolysis of HONO (from sorbed gas-phase NO_2) on the wood soot particles, resulting in a direct OH-particle source.

The major product observed by Jang and McDow (1997) from the irradiation of BaA dissolved into solutions with benzene or toluene was 7,12-benzanthracenedione. Other observed products in this study are shown in figure IX-D-8. Also illustrated in this figure is the formation of ring-cleavage products, consistent with studies of other PAHs (Mariconi and Taranko, 1963; Lee-Ruff et al., 1986; Barbas et al., 1996). Note that different BaA product distributions resulted when benzene or toluene were used, or when the toluene or benzene solutions contained 9,10-anthraquinone, or 9-xanathone, or vanillin as cosolvents. A singlet oxygen mechanism was suggested for the formation of the ring-cleavage products. Compounds XI, XII, and XIII (figure IX-D-8) form in a toluene solvent system and most probably result from a mechanism involving abstraction of a methyl hydrogen from toluene. Jang and McDow (1997) proposed the mechanism as shown in figure IX-D-9. Here, anthraquinone forms an excited state that can abstract a methyl hydrogen from toluene. Certain quinones can also photosensitize the formation of triplet PAH. In another study, Jang and McDow (1995) observed that anthraquinone, phenanthraquinone, and furaldehyde cosolvents all significantly enhanced the rate of BaA photolysis. For this to occur, the energy gap between the first excited singlet and excited triplet states of the sensitizer must be smaller than the same transition of the PAH, and the energy gap between the ground-state singlet and the first excited triplet of the PAH must be less than that of the sensitizer.

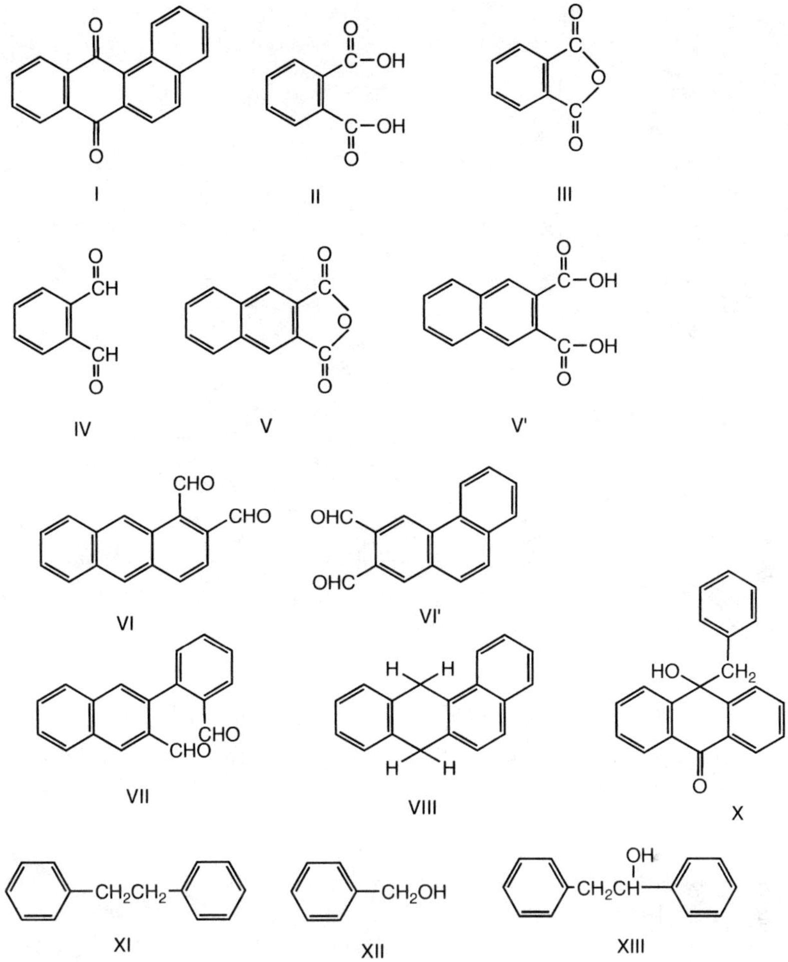

Figure IX-D-8. Products observed from the photolysis of BaA in the presence of different solvents (Jang and McDow, 1997).

Odum et al. (1994) also used the turntable irradiation system described previously with a quartz immersion well and a 366-nm filter to investigate possible reaction mechanisms. Logarithms of BaA photodegradation rate coefficients in different *para*-substituted guaicols (x-2-methoxyphenols) were found to be directly related to linear free-energy Hammet σ values and to resonance stabilization energies (Bordwell and Cheng, 1991). Based on these results, Odum et al. (1994) suggested that BaA triplets resulting from BaA light absorption could abstract a methoxyphenolic hydrogen. The ability to do this was related to the stability of the phenoxy radical formed by loss of a phenolic hydrogen. A singlet oxygen mechanism was also explored, in which $O_2(^1\Delta_g)$ mediated the loss of a phenolic hydrogen, and the resulting phenoxy radical reacted with a PAH mole-

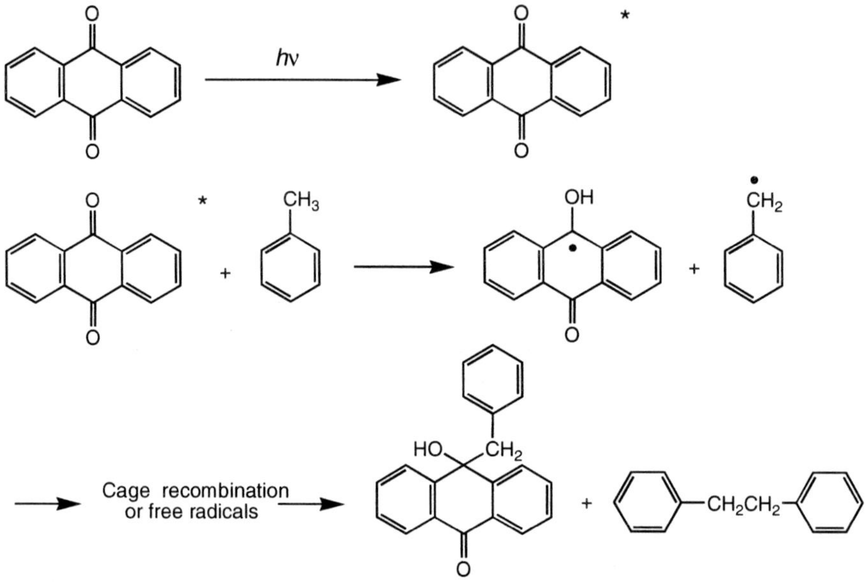

Figure IX-D-9. Proposed mechanism for anthraquinone reaction with solvent toluene (Jang and McDow, 1995).

cule. Odum and coworkers concluded, however, that more unsubstantiated assumptions were required to make this mechanism compatible with observations than the PAH triplet phenolic hydrogen abstraction mechanism (figure IX-D-10).

In summary, a number of studies indicate that the products formed from the photooxidation of PAHs on surfaces or in the liquid particle phase are consistent with a PAH photosensitization–singlet oxygen mechanism (Fox and Olive, 1979; Eisenberg et al., 1984; Lee-Ruff et al., 1986; Jang and McDow, 1995, 1997; Barbas et al., 1996). In addition, when a singlet oxygen trap was used, the apparent PAH $+ O_2(^1\Delta_g)$ reaction was quenched (Lee-Ruff et al., 1986). Triplets of PAHs can result from direct PAH light absorption forming a singlet PAH, which degrades to a triple state. It is also possible for quinones and carbonyls to photosensitize the formation of a triplet PAH (Jang and McDow, 1997). The evidence, however, for singlet oxygen oxidation is not conclusive. It may be that other reaction initiators bring about single-oxygen-type PAH products. Another possible mechanism involves PAH triple abstraction of a phenolic hydrogen (Odum et al., 1994). Hence, it is clear at this time that this entire area requires much further investigation.

IX-D-3. Reactions of Particle PAH with Gas-Phase O_3

Lane and Katz (1977) exposed benzo[*a*]pyrene- (BaP)-coated dishes to 0.19 ppm of O_3 and observed a rapid decay of BaP with a lifetime of ~ 50 min. Benzo[*k*]fluoranthene reacted much more slowly. Pitts et al. (1980) exposed BaP on glass fiber filters to 0.2 ppm of O_3 and observed a BaP lifetime of ~ 70 min. The

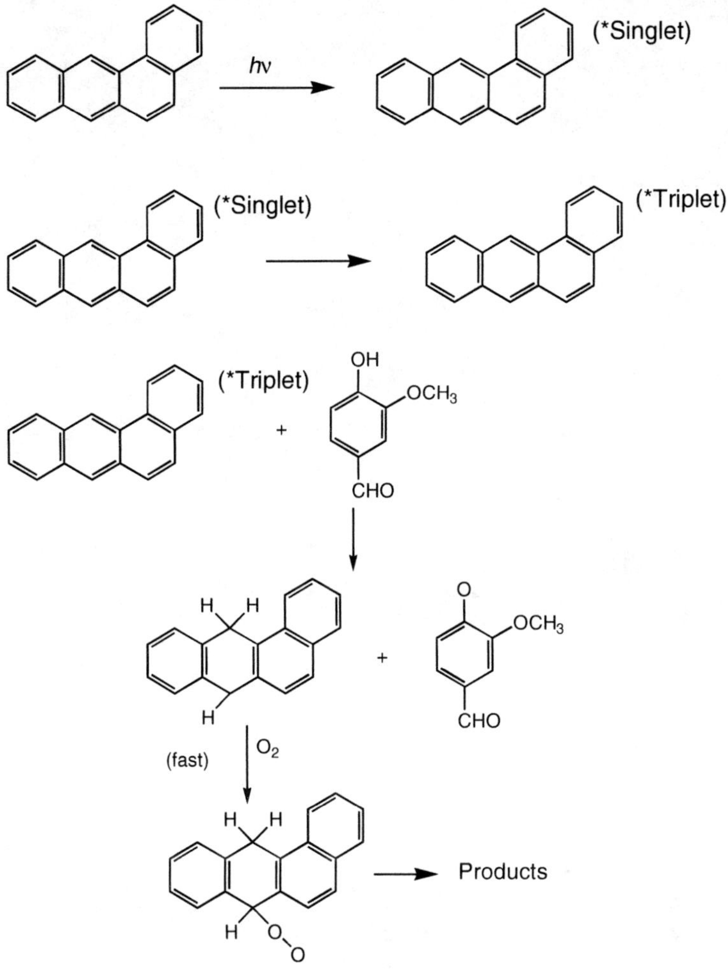

Figure IX-D-10. Triplet BaA mechanism for abstraction of a phenolic hydrogen.

formation of benzo[*a*]pyrene,4-5-epoxide, which has weak carcinogenic properties on mouse skin, was observed along with BaP quinones (Pitts et al., 1980). Products that have been identified from ozonolysis of BaP-coated filters by Van Cauwenberghe et al. (1984) are shown in figure (IX-D-11).

Limited data exist for the reaction of particle-bound PAH with O_3. Kamens et al. (1985) reported on the dark exposure of airborne wood smoke particles to gas-phase ozone in an outdoor chamber experiment. The average O_3 concentration was 0.52 ppm, NO_2 was 0.05 ppm, and the outdoor temperature was 5.8°C. About 60–70% of the PAHs was lost over a period of 2.5 h. Based on unpublished data from the Kamens' group, there is little temperature effect associated with this reaction. To be consistent with gas-phase rate coefficients, second-order PAH–O_3 rate coefficients were derived and are presented in table IX-D-3. These rates were not corrected for possible reaction losses on filters (discussed later) and represent

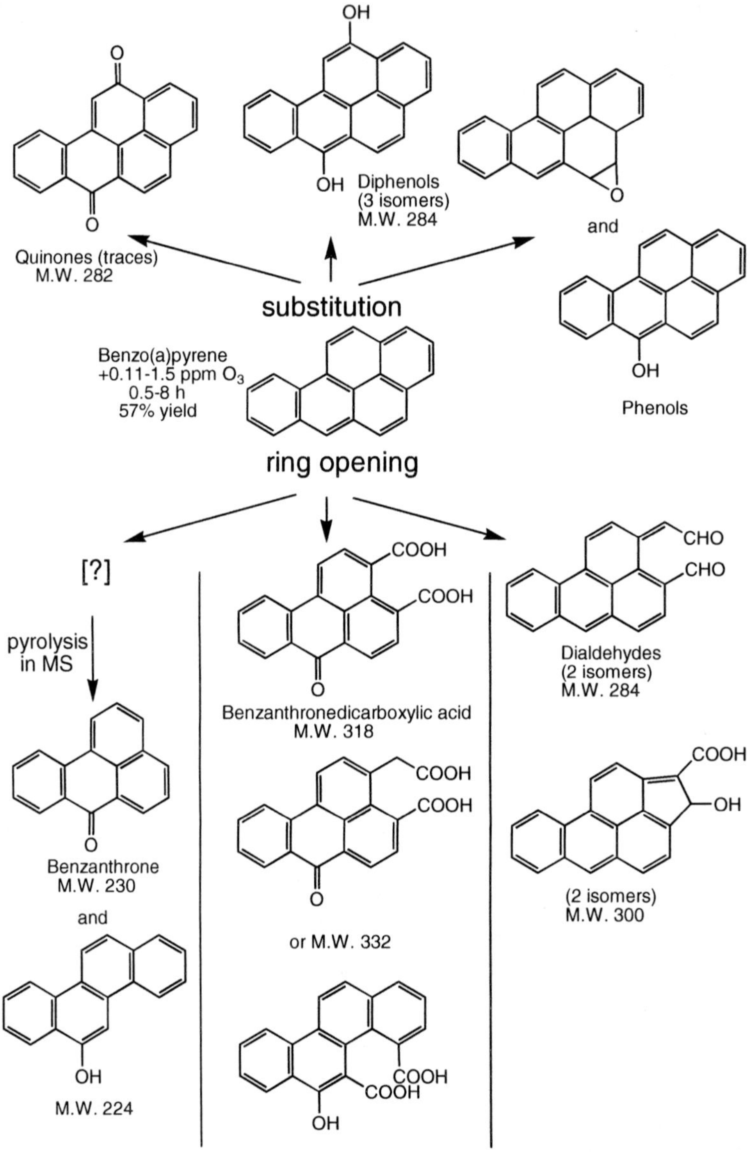

Figure IX-D-11. Products from the reaction of BaP on filters with O_3; redrawn from Van Cauwenberghe et al. (1984).

upper limits. The BaP–O_3 decay rates on soot particles appear to be a factor of 5 slower than for BaP on filters reacting with O_3.

When ~2.5 mg of precollected fresh wood soot filter samples (47-mm Teflon-impregnated glass fiber filters) were exposed to 0.5 ppm of O_3 by pulling the O_3 across the filters at a flow of 70 L min^{-1} for 5 min, ~9% of the BaP mass reacted on the filters (Kamens et al., 1985). Van Vaeck and Van

Table IX-D-3. Rate coefficients for selected particle-bound PAHs (cm^3 $molecule^{-1}$ s^{-1})

PAH[a]	O_3 [b,c] Wood	NO_2 [d] Wood	NO_2 [e] Wood	HNO_3 [f,g] Wood	N_2O_5 [g,h] Wood	N_2O_5 [g,i] Diesel
BaA	1.3×10^{-17}	3.7×10^{-18}				
Chry/Tri	7.2×10^{-18}	3.7×10^{-18}				
BkF		3.7×10^{-18}			1.8×10^{-18}	9.3×10^{-19}
BbF			2.6×10^{-19}	2.7×10^{-19}	1.7×10^{-18}	5.3×10^{-18}
BaP	1.0×10^{-17}	3.7×10^{-18}	6.2×10^{-19}	3.4×10^{-19}	4.7×10^{-18}	2.9×10^{-17}
BghiP	4.1×10^{-18}[j]				1.6×10^{-18}	1.9×10^{-17}

[a]See table IX-A-2 for compound names corresponding to the abbreviations shown here.
[b]Low PAH particle loading (< 350 ng individual PAH per milligram of particles); outdoor chamber experiment; average $[O_3] = 0.52$ ppm (1.37×10^{13} molecules cm^{-3}) at 5.8°C and a dew point of 4.1°C.
[c]Unpublished data for wood smoke by the Kamens' research group; temperature 21°C; average $[O_3] = 0.29$ ppm gave the following rate coefficients: $k(\text{BaP}) = 1.6 \times 10^{-17}$ cm^3 $molecule^{-1}$ s^{-1}, $k(\text{BaA}) = 1.3 \times 10^{-17}$ cm^3 $molecule^{-1}$ s^{-1}; $k(\text{Chry}) = 4.6 \times 10^{-18}$ cm^3 $molecule^{-1}$ s^{-1}.
[d]Low PAH particle loading; outdoor chamber; $[NO_2] = 0.54$ ppm (1.41×10^{13} molecules cm^{-3}) at 7.8°C and a dew point of 7.2°C.
[e]High PAH particle loading (> 1000 ng individual PAH per milligram of particles); CSTR experiment; average $[NO_2] = 6.54 \times 10^{14}$ molecules cm^{-3} at 24°C and a dew point of 15°C.
[f]Intermediate–low PAH; CSTR loadings; $[HNO_3] = 44.8$ ppm (1.1×10^{15} molecules cm^{-3}) at 24°C and a dew point of 5.0°C.
[g]Given the uncertainty in HNO_3 and N_2O_5 concentrations, rate coefficients are assumed to be within a factor of 3 at best.
[h]Low PAH; CSTR loadings; $[N_2O_5] = 12$ ppm (3.1×10^{14} molecules cm^{-3}) at 15.2°C.
[i]Low PAH; CSTR loadings; $[N_2O_5] = 10$ ppm (2.5×10^{14} molecules cm^{-3}) at 16°C.
[j]Unpublished data by Kamens' research group; temperature 21°C; average $[O_3] = 0.29$ ppm.

Cauwenberghe (1984) exposed fresh diesel soot on filters to ~9 times the O_3 exposure/particle mass compared with the previous study (Kamens et al., 1985) and observed a 62% loss in BaP, a 25% loss of benzofluoranthene, and 38% loss in benzo[*g*,*h*,*i*]perylene. These results could have important implications for samplers that use long sampling periods and expose filter-collected PAHs to appreciable ambient levels of O_3.

IX-D-4. Reactions with NO_2, HNO_3 and N_2O_5/NO_3 with Sorbed PAHs

Van Cauwenberghe et al. (1984) exposed BaP on filters to 1 ppm of NO_2 in nitrogen, and observed a 56% loss of BaP in 8 h. Three mono-nitro-BaP derivatives were observed: 6-nitro-, 3-nitro-, and 1-nitro-BaP. The formation ratio of the 6-nitro- to the 1-nitro- plus 3-nitro-BaP derivatives was 4:1. When 0.25 ppm NO_2 was used, ~20% of the BaP was lost in 8 h. Trace amounts of HNO_3 were measured in the NO_2 and this apparently catalyzed the reactions, since in a similar experiment, but with the removal of HNO_3 from the gaseous NO_2 exposure stream with nylon wool, little BaP degradation was observed. When 25 ppb of HNO_3 was added back to the NO_2 stream, the reaction of BaP with NO_2 returned. Just HNO_3 in the flow stream over the filters did not produce any nitro-aromatic products. As with O_3, outdoor chamber data were developed for an experiment with 0.54 ppm NO_2 at 7.8°C, and rate coefficients for this experiment are presented in table IX-D-3. At the time of this writing, we do not know the extent to which this reaction is temperature sensitive.

Nielsen (1984) exposed PAHs in the liquid phase to 0.16 M fuming HNO_3 and 0.16 M $NaNO_3$. The half-lives ($t_{1/2}$) of 25 different PAHs ranged from 2.2 min for anthanthrene, 200 min for BaP, 5.5 days for BghiP, 34 days for BaA, to greater than 300 days for Bkf and BbF. Nielsen was able to demonstrate that there was a reasonable correlation between the logarithm of his observed $t_{1/2}$ and PAH ionization potentials, first π-band wavelengths in Ångström, and Clar's degree of aromaticity index (Clar, 1964). Clar's index is calculated from $a \times 100/n_C n_H$, where a is the maximum number of aromatic sextets for a given PAH structure, n_C is the number of carbon atoms, and n_H is the number of hydrogen atoms.

$$t_{1/2} = 12.4 \times \text{ionization potential} - 83, \quad r^2 = 0.83, \quad \text{number of PAHs} = 16$$
$$t_{1/2} = -0.0014 \times \lambda - 48.7, \quad r^2 = 0.85, \quad \text{number of PAHs} = 19$$
$$t_{1/2} = 13.2 \times (a \times 100/n_C n_H) + 5.11, \quad r^2 = 0.85, \quad \text{number of PAHs} = 15$$

Pitts et al. (1985d) reported that the reaction of N_2O_5 with fluoranthene-coated filters generates the 1-, 3-, 7-, and 8-isomers of nitrofluoranthene. Reaction of fluoranthene in the gas phase with N_2O_5, however, yields primarily 2-nitrofluoranthene (Sweetman et al., 1986). Mechanisms for the reaction of PAHs in the liquid phase with N_2O_5 are discussed by Zielinska et al. (1986) and Pitts et al. (1985b). In polar solvents, N_2O_5 is primarily in an ionic state ($NO_2^+NO_3^-$) and electrophilic substitution involving the NO_2^+ ion results. In aprotic solvents, N_2O_5 exists in a covalent form. Wood smoke particles have a high methoxyphenolic content (Hawthorne et al., 1992) and are considered to be polar. Diesel soot is composed largely of alkanes and alkanoic acids (Rogge et al., 1993a). Although the organic phase of diesel soot is thought to be non-polar, the presence of alkanoic acids may permit N_2O_5 to exist, to some extent, as the ion pair $NO_2^+NO_3^-$. Nitration of PAHs under these conditions may result directly from the electrophylic attack by the nitronium ion, NO_2^+. Using carbon tetrachloride (CCl_4) solvent, Zielinska et al. (1986) proposed a radical mechanism to illustrate nitration in the gas-phase and non-polar solvents (figure IX-D-12). Products 2, 3, 11, and 12 were identified. Product 2 (2-nitrofluoranthene) is indicative of radical attack, while the 1-, 3-, 7-, and 8-isomers are formed by electrophylic nitration of fluoranthene (Lee and Schwartz, 1981).

In experiments with soot particles, an outdoor chamber was charged with wood soot or diesel soot particles, and used as a reservoir to feed a 200 L continuously stirred tank reactor (CSTR) at a constant flow of 10–11 L min^{-1}. Then, N_2O_5, NO_2, or HNO_3 (Kamens et al., 1990; Guo and Kamens, 1991) were added to the CSTR and rate coefficients were determined as a function of inlet and outlet concentrations (Guo and Kamens 1991). Rate coefficients and lifetimes for these experiments are also presented in tables IX-D-3 and IX-D-4. As with photolytic processes, rates of NO_2 decay appear to be lower when PAH particle loadings are high rather than low. According to Lee and Schwartz (1981), NO_2 on particulate matter may redistribute to HONO and $H^+NO_3^-$. If this is the case, then one would expect the lifetimes of BaP and the benzofluoranthenes to be very different (with BaP exhibiting much greater reactivity; assuming a similar nitration mechanism) for the NO_2 and HNO_3 PAH experiments. Although directionally this was

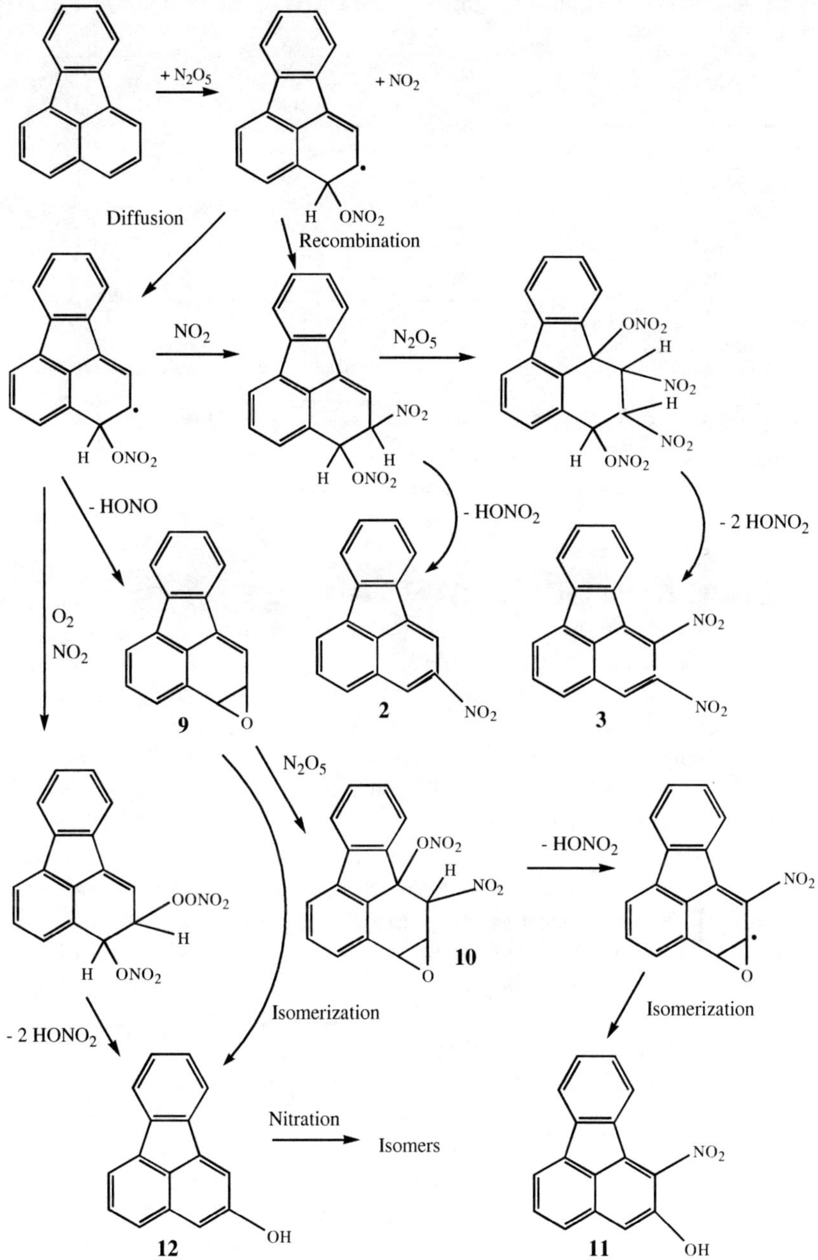

Figure IX-D-12. Postulated N_2O_5/NO_3–fluoranthene reaction in non-polar solvents and the gas phase (after Zielinska et al., 1986).

Table IX-D-4. Lifetimes for selected particle-bound PAHs on either wood soot or diesel soot

PAH[a]	O_3 [b] Wood	NO_2 [c] Wood	NO_2 [d] Wood	HNO_3 [e] Wood	N_2O_5 [f] Wood	N_2O_5 Diesel
BaA	31 h	13 d				
Chry/tri	55 h	13 d				
BkF		16 d			1.8 yr	3.4 yr
BbF			183 d	2.3 yr	1.9 yr	0.6 yr
BaP	39 h	13 d	77 d	1.8 yr	0.7 yr	25 d
BghiP					2.0 yr	72 d

[a]See table IX-A-2 for compound names corresponding to the abbreviations shown here.
[b]Average O_3 concentration = 7×10^{11} molecules cm^{-3} (Logan, 1985) or 0.029 ppm at 298 K.
[c]Low particle loadings; average NO_2 concentration = 2.4×10^{11} molecules cm^{-3} (Atkinson and Arey, 1994) or 9.9 ppb.
[d]High particle loadings; average NO_2 concentration = 2.4×10^{11} molecules cm^{-3} (Atkinson and Arey, 1994) or 9.9 ppb.
[e]Average HNO_3 concentration = 5×10^{10} molecules cm^{-3} or 2 ppb.
[f]Average $[N_2O_5] = 2 \times 10^{10}$ molecules cm^{-3} (Atkinson et al., 1986) or 0.8 ppb.

observed, the rate coefficients for NO_2 + wood smoke and HNO_3 + wood smoke do not exhibit the large differences that would be predicted by the Nielsen nitric acid experiments described previously (Nielsen, 1984). A reasonable correlation, however, does exist in PAH reactivity between the N_2O_5–diesel PAH reactivity and the Nielsen nitric acid experiments, suggesting that both may have a common reaction mechanism.

That NO_2 may redistribute into different oxidation species once sorbed onto atmospheric particles may help explain the reported influence of ambient water vapor concentrations on particle PAH sunlight reactivity as reported by Kamens et al. (1988). If sorbed gas-phase NO_2 equilibrates in water on aerosols to form HONO and $H^+NO_3^-$, HONO could photolyze to form particle-associated OH radicals. These OH radicals could then react directly with particle PAH. Wood soot and gasoline exhaust particles have an affinity for water, being capable of absorbing ~6–12% of their mass in water over a 40–90% relative humidity change at room temperature (McDow et al., 1995; Vartiainen et al., 1996). As the water content of particles increases, so would the amount of sorbed NO_2 from the gas phase, which would result in more HONO.

From the rate coefficients in table IX-D-3, lifetimes in the atmosphere were estimated for selected PAHs (table IX-D-4) based on reactions with average levels of O_3 (Logan, 1985), NO_2 (Atkinson and Arey, 1994), HNO_3, and N_2O_5 (Atkinson et al., 1986). Comparison of the lifetimes in tables IX-D-2 and IX-D-4 suggests that sunlight has the greatest potential to degrade PAHs associated with fresh combustion emissions. In average sunlight, particle PAH can have lifetimes on the order of hours. This is followed by O_3, NO_2, N_2O_5, and HNO_3. Losses of PAHs from particles in sunlight are of the same order as gas-phase PAH losses via OH attack. Lifetimes of PAHs on particles due to reaction with gas-phase O_3 are in the range of tens of hours, while those for O_3 attack on gas PAH are generally a factor of 10–20 longer. Removal of PAHs in either the gas phase or the particle phase from reaction with HNO_3 or N_2O_5/NO_3 is not an important process. However, PAH reaction with N_2O_5/NO_3 does lead to the formation of nitro-PAH in both the gas phase and particle phase (Arey et al., 1989b; Kamens et al., 1990; Atkinson and Arey, 1994; Kwok et al., 1994c), and this combined with OH-initiated reaction may account for much of the nitro-PAHs in the atmosphere.

IX-D-5. Reactions of Nitro-PAHs (NPAHs) and Oxygenated PAHs (oxy-PAHs) Sorbed on Particles

A paucity of data exists for the reaction of PAH products on atmospheric particles. Fan et al. (1995, 1996a) vaporized selected deuterated nitropyrenes and nitrofluoranthenes into an outdoor chamber that contained preexisting diesel or wood soot particles. In these studies, the temperature ranged from −19 to 38° C, natural sunlight irradiated the particles, and the apparent first-order rate coefficients were computed with respect to NO_2 photolysis rate coefficients. This was done because, like NO_2, these nitro-PAHs absorb in the UV, and visible regions of the spectrum. Temperature was not an important parameter, but photolysis rates were approximately twice as fast on wood smoke as on diesel soot. Rate coefficients and lifetimes for some selected nitropyrenes and fluoranthenes are tabulated in table IX-D-5. These are comparable to the photolysis lifetimes of gas-phase nitronaphthalenes.

Compounds such as 9,10-anthraquinone, cyclopenta[*d,e,f*]phenanthrone, benzanthrone, and so on (here called oxy-PAHs), are combustion products and are coemitted with PAHs (in the case of wood smoke) in approximately the same particle-phase concentrations (Kamens et al., 1989). At night, on wood soot particles, they appear to be stable. Benzanthrone is stable on these particles in the dark and in the presence of sub-ppm levels of O_3 for several hours. Cope and Kalkwarf (1987) observed a significant loss of pyrenequinones, hydroxy-9-fluorenone, and methylanthraldehyde, coated on a glass surface,

Table IX-D-5. Rate coefficients and lifetimes for selected nitropyrenes and nitrofluoranthenes

Compound	$k_{NPAH} = A \times j(NO_2)$	Lifetime (h)
1-Nitropyrene $(d)^3$	$0.03 \times j(NO_2)$	2.1
2-Nitropyrene $(d)^2$	$0.033 \times j(NO_2)$ $(\pm 0.0025)^a$	1.9
2-Nitrofluoranthene $(d)^2$	$0.023 \times j(NO_2)$ $(\pm 0.0015)^a$	2.8
d_9-3-Nitrofluoranthene $(d)^3$	$0.025 \times j(NO_2)$ $(\pm 0.008)^b$	2.6
d_9-8-Nitrofluoranthene $(d)^3$	$0.025 \times j(NO_2)$ $(\pm 0.008)^b$	2.6
d_9-1-Nitropyrene (w)	$0.05 \times j(NO_2)$	1.3
2-Nitrofluoranthene (w)	$0.045 \times j(NO_2)$	1.4
d_9-3-Nitrofluoranthene (w)	$0.05 \times j(NO_2)$	1.3
d_9-8-Nitrofluoranthene (w)	$0.05 \times j(NO_2)$	1.3

Note: symbols used are $(d)^2$, $(d)^3$ = diesel exhaust particles with two or three experiments, respectively; w = wood soot with only one experiment; d_9 = experiment using nine deuterium atoms per molecule of reactant.

[a] Range.

[b] Two standard deviations, $j(NO_2) = 5.2 \times 10^{-3}\ s^{-1}$.

when exposed to 0.16 ppm O_3. Grosjean et al. (1983) were unable to detect a loss of anthraquinone coated on Teflon, cellulose, or silica gel in the presence of "high" O_3. In outdoor chamber experiments, oxy-PAHs in dilute wood smoke atmospheres in clear North Carolina sunlight tend to be stable, while their companion PAH compounds degrade (Benner et al., 1989). The exception is naphthalic anhydride, which tends to increase in concentration on particles with time. Attack by OH on the double bond of gas-phase acenaphthylene or OH abstraction of one of the CH_2 hydrogen atoms of acenaphthene could potentially lead to the formation of this product. In the presence of sunlight, however, an active photochemical smog system generated 0.22 ppm of O_3, and oxy-PAH decay with lifetimes of ~2–5 h. Cope and Kalkwarf (1987) also report that pyrenequinones are stable in simulated sunlight (1.37 kW m^{-2}), but react quickly in simulated sunlight when O_3 is present. To explain these results, the mechanism shown in figure IX-D-13 was proposed; it begins with light absorption by pyrenequinone followed by O_3 attack (Cope and Kalkwarf, 1987).

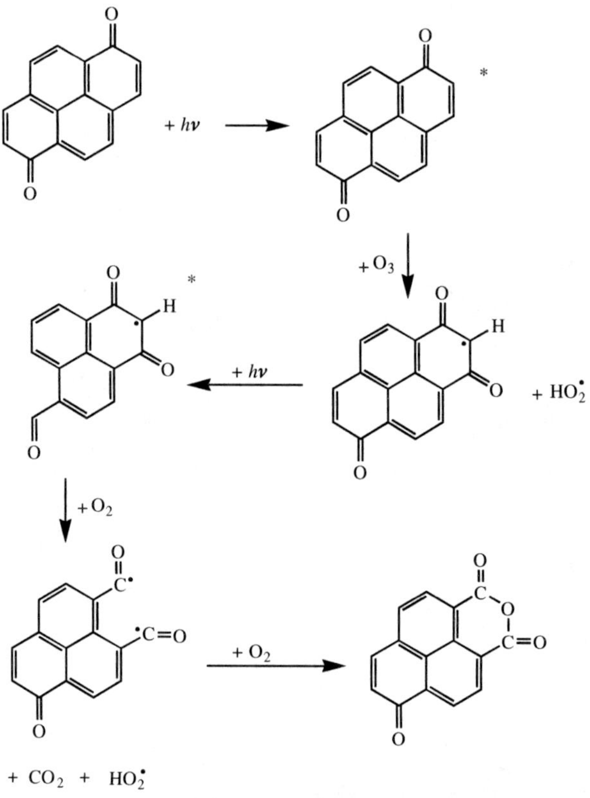

Figure IX-D-13. Reaction of pyrenequinone in sunlight in the presence of ozone (Cope and Kalkwarf, 1987); the asterisk designates an electronically excited molecule.

IX-D-6. Evidence for PAH Reaction in the Gas Phase and Particle Phase

Evidence for the reaction of gas-phase PAHs in the atmosphere was provided by Arey et al. (1989a). The ratio of evening to daytime concentrations of selected PAHs was linearly correlated (figure IX-D-14) with the OH-reaction rate coefficient for individual PAHs. Although there is an implicit assumption that emissions are relatively similar for both day- and nighttime periods and that O_3 and NO_3 reactions are relatively unimportant compared with OH, the apparent relationship between the k(OH) and $(PAH]_{nighttime}/[PAH]_{daytime}$ is rather striking. That PAHs react in the gas phase is further supported by the observation of nitro-PAH photochemical reaction products. For example, OH attack on gas-phase fluoranthene or pyrene will lead to the formation of 2-nitrofluoranthene (2-NF) and 2-nitropyrene (2-NP). These products are not emitted directly into the atmosphere (as illustrated by figure IX-D-15), and must be formed by OH attack and subsequent NO_2 addition (Arey et al., 1987) or NO_3 attack. 1-Nitropyrene and 8-nitrofluoranthene, on the other hand, can be formed shortly after the combustion process by electrophylic PAH nitration. Both 2-NF and 2-NP have been observed in both urban (Pitts et al., 1985c; Arey et al., 1987, 1990; Ciccioli et al., 1993) and rural (Nielsen et al., 1984) environments. Lastly, evidence for the ambient formation of methylnitronaphthalenes at night from possible reaction with NO_3 was provided by Gupta et al. (1996).

Since little ambient PAH product data exists for particle-phase reactions of PAHs, ratios of selected relatively "unreactive" PAHs versus reactive PAHs are sometimes used as indicators of reactivity. Nielsen was able to demonstrate via

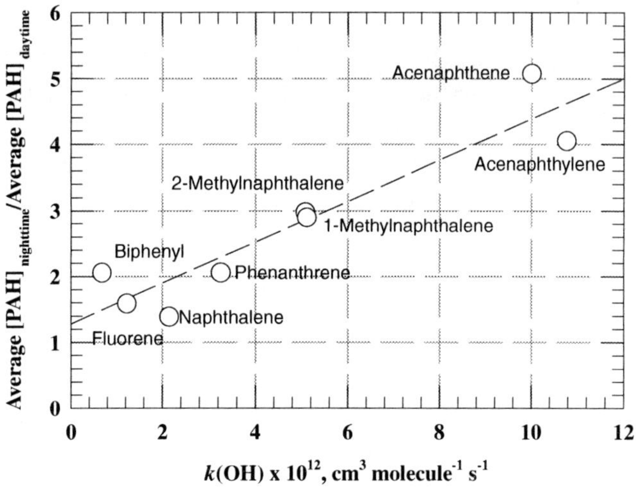

Figure IX-D-14. Ratio of nighttime to daytime gas-phase PAH concentrations versus PAH–OH rate constants. Data collected from Glendora, California during August 1986. Figure from Arey et al. (1989a) with the permission of *Environmental Science and Technology*.

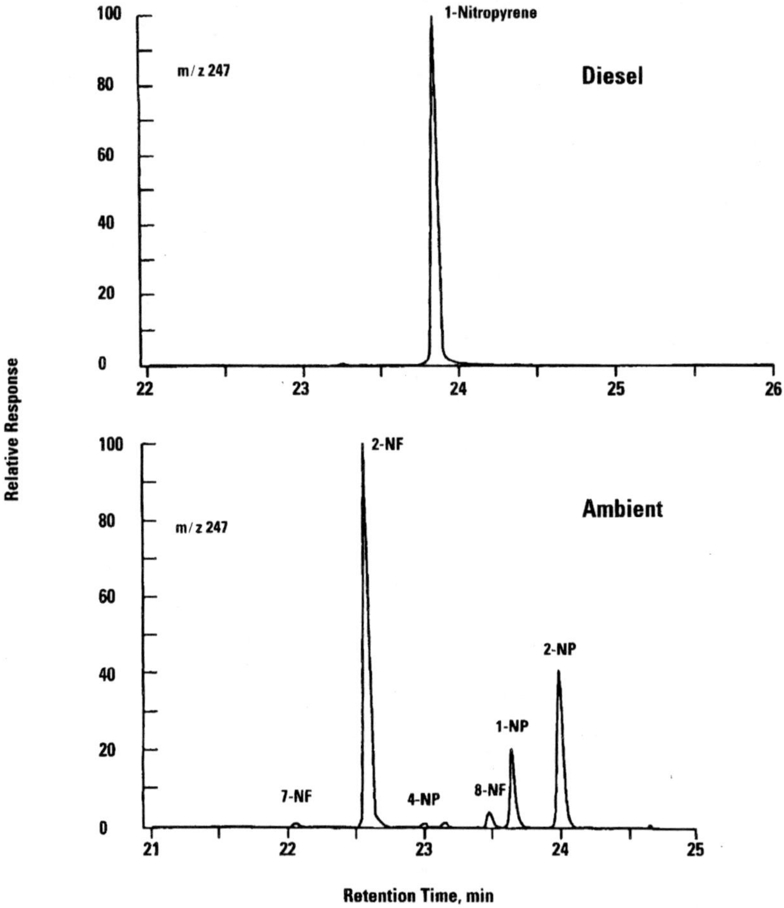

Figure IX-D-15. Nitro-PAHs emitted directly from diesel exhaust (top) and observed in ambient air (bottom). Figure is reproduced from Arey et al. (1987).

concentration ratios of Cpc/(Chry + Tri) and BaP/BeP and wind directions that significant Cpc and BaP degradation occurred during transport under winter conditions (Nielsen, 1988). When air was transported relatively long distances compared with air that had come from nearby Copenhagen or the town of Roskilde, ratios were much lower, suggesting reaction relative to (Chry + Tri) or BeP. Pistikopoulos et al. (1990) attempted to estimate the reactivity of PAHs during long transport from Paris to the island of Corsica. The ratio of BaP/BeP for air originating in Paris was 2.54, and this dropped to 0.11 in Corsica. Ratios of BbF/BeP and BbF/BkF decreased by ~10%. This appears to be a reasonable occurrence since the compounds have sunlight reactivities similar to those in the outdoor chamber studies of Kamens et al. (1986). Greenberg (1989) evaluated summer/winter PAH data from three urban locations in New Jersey during the early 1980s. Average ratios of Py, Ba, CpC, BaA, BkF, and Ind were determined with respect to BbF, which was assumed to be non-reactive. Ratios of BaP/BbF, CpC/BbF, BaA/BbF, and Py/BbF for the summer were significantly lower than for the winter. It is possible that partitioning to the gas phase for Py and BaA is greater during the summer than winter, and this partially explains the summer/winter low ratios for these compounds. It is clear from these data, however, that BaP and CpC exhibit considerable reactivity; BkF and Ind have slightly more sunlight reactivity than BbF in chamber experiments (Kamens et al., 1988) but did not exhibit any reactivity compared with BbF in ambient New Jersey data. Freeman and Cattell (1990) measured PAH concentrations over the course of a sunny summer day in Sydney and normalized these with respect to Ind. No change was observed in the ratios over time, and this was interpreted as a lack of PAH atmospheric reactivity. It is also possible that this sampling site was heavily influenced by local automotive sources throughout the day and that these dominated any atmospheric losses.

In a study in Czechoslovakia, reported in 1967, BaP particle loadings decreased dramatically with distance (0–900 m) from a coking plant (Masek, 1967). The decrease was much greater in summer than winter, suggesting the possibility that sunlight modified the particle-bound BaP. Tebbens et al. (1971) compiled BaP per extractable particle mass for a number of U.S. cities during the summer and winter months. It was observed that BaP particle extract loadings were on average 50% lower during the summer than the winter. Similar trends could be observed in the Sydney study mentioned above (Freeman and Cattell, 1990) for reported monthly total PAH concentrations normalized to TSP. It is important to note, however, that although these decreases are consistent with more light being available during the summer, one cannot rule out the possibility that emissions of PAHs per particle mass or particle extract are lower during the warmer months than cooler months due to changes in energy usage.

From measurements in downtown Los Angeles in 1982, Schauer et al. (1996) plotted the ratio of different ambient organic concentrations to their estimated emissions. The ratios of particle-bound PAHs tended to be at or below the minimum for conserved species such as aluminum, silicon, or elemental carbon. When compared with elemental carbon, which could be considered a specific unreactive tracer for PAHs, very significant atmospheric PAH reactivity is suggested. Unreactive compounds such as alkanes like tricosane, heptacosane, and so on, were very close to the conserved range for elemental carbon, and did not suggest any atmospheric reactivity. Organic compounds, however, that were formed in the atmosphere, such as benzenedicarboxylic acids, were far above the conserved ratio range.

Of additional interest is the relative photochemical reactivity that is imparted to the atmosphere by PAHs. As illustrated in the previous section, where ambient concentrations were discussed, naphthalene is the most ubiquitous PAH and can be present in the gas phase at concentrations of 500–1000 ng m^{-3}. For naphthalene, this translates into approximately 1–2 ppbC. For the PAHs of molecular weight higher than naphthalene, ambient gas PAHs decline rapidly in concentration, and their impact on O_3 formation is probably not very great in urban atmospheres that have 500–1000 total ppbC.

IX-D-7. A Gas–Particle-PAH Kinetic Mechanism

Given the available rate coefficients for PAHs reacting in both the gas phase and particle phase, kinetic modeling mechanisms have been developed for nitro-PAH from naphthalene (Fielberg et al., 1999), pyrene and fluoranthene (Kamens et al., 1994; Fan et al., 1995, 1996b). The model outputs were compared with outdoor chamber experiments in which propylene, NO_x, and dilute diesel exhaust were reacted under natural sunlight. Two possible reaction sequences have been used to represent the initial OH attack and NO_2 addition on fluoranthene (Fla) and pyrene (Py) (Kamens et al., 1994; Fan et al., 1995, 1996b; Fielberg et al., 1999). In the first sequence, OH attacks the PAH and forms an OH-aromatic adduct. Then, NO_2 adds to aromatic carbons of the OH–adduct at sites of high electron density to produce nitro-PAHs (Kamens et al., 1994; Fielberg et al., 1999); O_2 also reacts with the OH–aromatic adduct and competes with NO_2. In the second sequence, the reactions with OH and NO_2 addition to the aromatic ring to form a nitro-PAH are combined into one reaction. This is illustrated for Fla and Py in table IX-D-6. Although both of these reaction schemes adequately model nitro-PAH formation in relatively high concentrations in smog chamber experiments, the first, which is similar to simple aromatic reactions (see chapters II and III), grossly underpredicts the observed formation of nitro-PAHs at ambient concentrations. Given the current rate coefficient for O_2 addition to the OH-aromatic adduct (Knispel et al., 1990), this problem results because at ambient concentrations O_2 completely outcompetes NO_2 for the OH–aromatic adduct. Although Carter (1999) suggests that other alternatives for the relative rates of the NO_2 OH–adduct and O_2 OH-adduct are possible (see chapter X), in his 1999 aromatic mechanism he still recommends a much higher rate for the OH–aromatic adduct with O_2 than with NO_2. This suggests that our understanding of PAH reactions with OH and NO_2 is far from complete and these reactions may be different for PAHs than for single-ring aromatics. See also section II-G-1.

The reactions used to model Py and Fla and nitro-PAHs by the Kamens' group are listed in table IX-D-6. These were appended to the Carbon Bond VI mechanism as published by Gery et al. (1989). Dark losses of PAHs to the chamber walls were determined in separate nighttime experiments. As it can be seen from the smog reactants in figure IX-D-16, it is possible to fit the reaction of gas- and particle-phase fluoranthene and, in figure IX-D-17, the appearance of 2-nitrofluoranthene in the particle phase from OH/NO_2 addition on gas-phase fluoranthene. Experimental data show that for the reaction of Fla and Py, all of the product nitro-PAH was found to appear almost immediately in the particle phase (Kamens et al., 1994; Fan et al., 1995, 1996a). When semivolatile products form, as in the case of naphthalene atmospheric reactions (Fielberg et al., 1999), it is necessary to add semivolatile gas/particle partitioning "reactions" to the mechanism summarized in table IX-D-6.

If current PAH kinetic mechanisms (Kamens et al., 1994; Fan et al., 1995, 1996b; Fielberg et al., 1999) are to be used in larger urban grid models, the semivolatile portion of models must be able to respond to temperature and particle concentration changes since this will influence the extent of partitioning to the gas/particle phases. For particles with a significant organic layer, partitioning

Table IX-D-6. PAH and NPAH reactions in modeling

Reaction		k (min^{-1} or $ppm^{-1}\ min^{-1}$)	Reference(s)
Gas-phase reactions:[a]			
$PY_{(g)}$[b] + OH	$= 0.995 \times PY - OH_{(g)} + 0.005 \times 2NP_{(p)}$[b]	7.4×10^4	Atkinson et al. (1990a)
$PY_{(g)} - OH + O_2$	= Products[c]	0.296	Knispel et al. (1990)
$PY_{(g)} + N_2O_5$	= $1NP_{(p)}$[b]	0.081	Atkinson et al. (1990a)
$FL_{(g)}$[b] + OH	$= 0.9 \times FL_{(g)} - OH + (0.1 \pm 0.05) \times 2NF_{(p)}$[b]	7.4×10^4	Atkinson et al. (1990a)
$FL_{(g)} - OH + O_2$	= Products[c]	0.296	Knispel et al. (1990)
$FL_{(g)} + N_2O_5$	$= 0.5 \times 2NF_{(p)} + 0.5 \times$ products	0.029	Atkinson et al. (1990a)
Particle-phase reactions:[a]			
$PY_{(p)} + N_2O_5$[d]	= $1NP_{(p)}$	0.01	Kamens et al. (1990)
$FL_{(p)} + N_2O_5$[d]	= $3NF_{(p)}$[b]	0.01	Kamens et al. (1990)
$PY_{(p)}\ \{+h\nu\}$	= Products	$(0.04 \pm 0.01) \times j(NO_2)$	Kamens et al. (1994); Fan et al. (1995)
$FL_{(p)}\ \{+h\nu\}$	= Products	$(0.02 \pm 0.01) \times j(NO_2)$	Kamens et al. (1994); Fan et al. (1995)
$FL_{(p)} + O_3$	= Products	0.004 ± 0.002	Kamens et al. (1994); Kamens et al. (1985)
$PY_{(p)} + O_3$	= Products	0.004 ± 0.002	Kamens et al. (1994); Kamens et al. (1985)
$1NP_{(p)}\ \{+h\nu\}$	= Products	$(0.04 \pm 0.01) \times j(NO_2)$	Kamens et al. (1994); Fan et al. (1995)
$2NP_{(p)}\ \{+h\nu\}$	= Products	$(0.04 \pm 0.01) \times j(NO_2)$	Fan et al. (1996b)
$2NF_{(p)}\ \{+h\nu\}$	= Products	$(0.025 \pm 0.005) \times j(NO_2)$	Fan et al. (1996b)
$1NP_{(p)} + O_3$	= Products	0.0025 ± 0.001	Fan et al. (1996b)
$2NF_{(p)} + O_3$	= Products	0.0025 ± 0.001	Fan et al. (1996b)
$1NP_{(p)} + N_2O_5$[d]	= Products	< 0.001	Fan et al. (1996b)

[a] All gas-phase PAHs were lost to the walls at a rate of 0.0021 min^{-1}. Particle PAH and NPAH wall losses (p) follow the total particulate wall loss.
[b] Here, PY = pyrene; FL = fluorene; 1NP = nitropyrene; 2NP = 2-nitropyrene; 2NF = 2-nitrofluorene; 3NF = 3-nitrofluorene.
[c] Unknown products.
[d] N_2O_5 is used as a proxy for NO_3 as per chapter II.

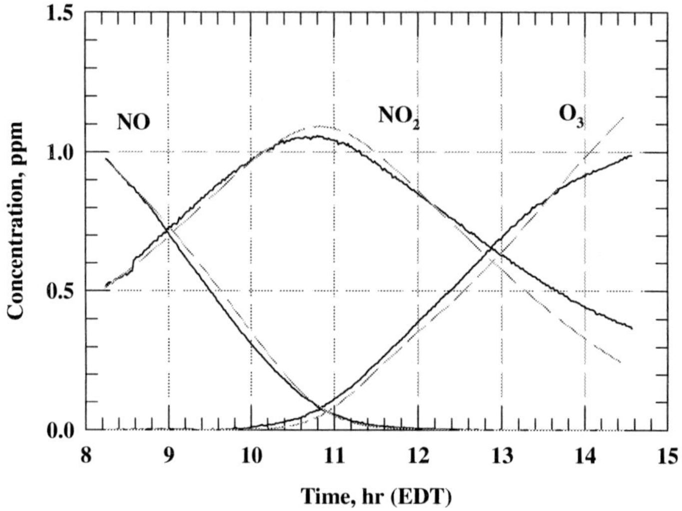

Figure IX-D-16. Model (dashed gray lines) and experimental data (solid lines) for the concentration of NO, NO_2, and O_3 as a function of time in a propylene–NO_x + dilute diesel exhaust (720 μg m^{-3}) experiment on June 16, 1994 under clear mid-North Carolina skies.

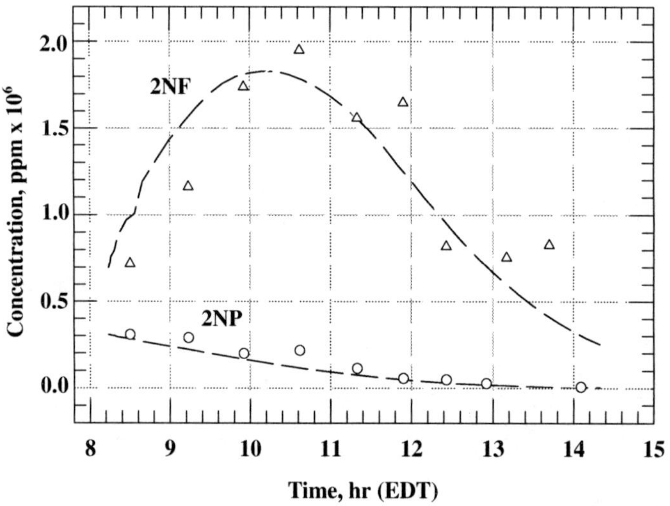

Figure IX-D-17. Model (dashed lines) and experimental data (symbols) for the formation and decay of 2-nitrofluorene (2NF, triangles) and 2-nitropyrene (2NP; circles) in the smog atmosphere with dilute diesel exhaust; results are from the same University of North Carolina outdoor chamber experiment for which [NO], [NO_2], and [O_3] versus time results are shown in figure IX-D-16.

coefficients as illustrated in section IX-C-5 can be estimated, and then a rate coefficient can be computed that characterizes the rate at which PAHs or their products leave the particle phase, ${}^{i}k_{\text{off}}$. This gives the rate coefficient ${}^{i}k_{\text{on}}$ for sorption back to the particle phase:

$${}^{i}K_p = {}^{i}k_{\text{on}}/{}^{i}k_{\text{off}}$$

It can be shown that ${}^{i}k_{\text{off}}$ is a function of the energy barrier that a single molecule must overcome to evaporate from the surface (Glasstone et al., 1941; Matsumoto et al., 1994), or an activation energy, E_a in kJ mol^{-1}. Matsumoto et al. (1994) suggest that the surface rate coefficient in the flux expression for evaporation can be approximated by:

$${}^{i}k_{\text{surf}} = \frac{k_b T}{h} \exp \frac{-E_a}{RT}$$

where k_b is Boltzmann's constant (1.381×10^{-23} J K^{-1}); h is Planck's constant (6.626×10^{-34} J s), T is the temperature (K), and R is the gas constant. Using the particle radius as the diffusion length, Kamens et al. (1999) show that ${}^{i}k_{\text{off}} = {}^{i}k_{\text{surf}}$. Estimates of E_a values for fluorene, phenanthrene, and pyrene of 73.9, 74.6, and 80.7 kJ mol^{-1}, respectively, have recently been reported by Kamens et al. (1999) with the use of a large gas-phase stripping device (Kamens and Coe, 1997). It is anticipated that the gas-phase uptake by the particles will not be very sensitive to temperature changes since the rate coefficient, k_{on}, for the gas-phase uptake by particles varies with $T^{0.5}$:

$${}^{i}k_{\text{on}} \propto [RT/(2\pi \text{MW}_{\text{I}})]^{1/2}$$

Since particle mass and surface are constantly changing in the atmosphere, the absorption of a gas-phase PAH to a particle is represented as

$$\text{PAH}_{\text{gas-phase}} + \text{particle} \rightarrow \text{PAH}_{\text{particle-phase}} + \text{particle} \qquad {}^{i}k_{\text{on}}$$

and loss of a particle-associated PAH to the gas-phase from a particle as

$$\text{PAH}_{\text{particle-phase}} \rightarrow \text{PAH}_{\text{gas-phase}} \qquad {}^{i}k_{\text{off}}$$

As new products from PAH reactions are identified, reaction rate coefficients in the presence of sunlight, OH, NO_2, NO_3, and so on, will be required to explain them properly. These, along with estimates of their vapor pressures and E_a values, will permit estimates of ${}^{i}k_{\text{off}}$ and ${}^{i}k_{\text{on}}$, and wet and dry deposition parameters will permit the prediction of their behavior in dynamic models of the atmosphere.

NOTE ADDED IN PROOF

Bodzek, D., K. Luks-Betlej, and L. Warzecha (1993), Determination of particle-associated polycyclic aromatic hydrocarbons in ambient samples from the upper Silesia region of Poland, *Atmos. Environ.*, *27A*, 759–764.

X

Application of Findings to Modeling the Atmospheric Chemistry of Aromatic Hydrocarbons

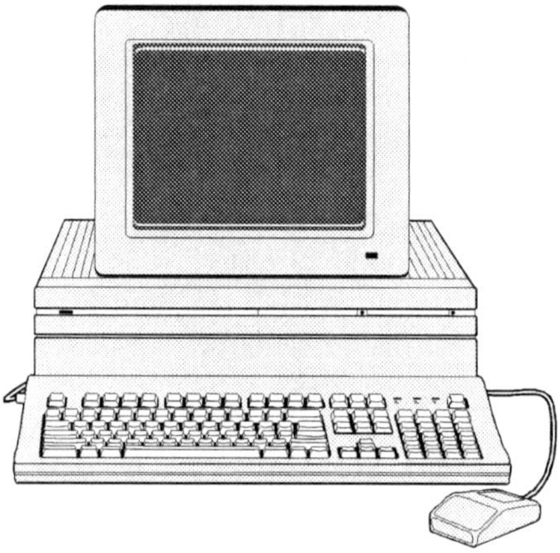

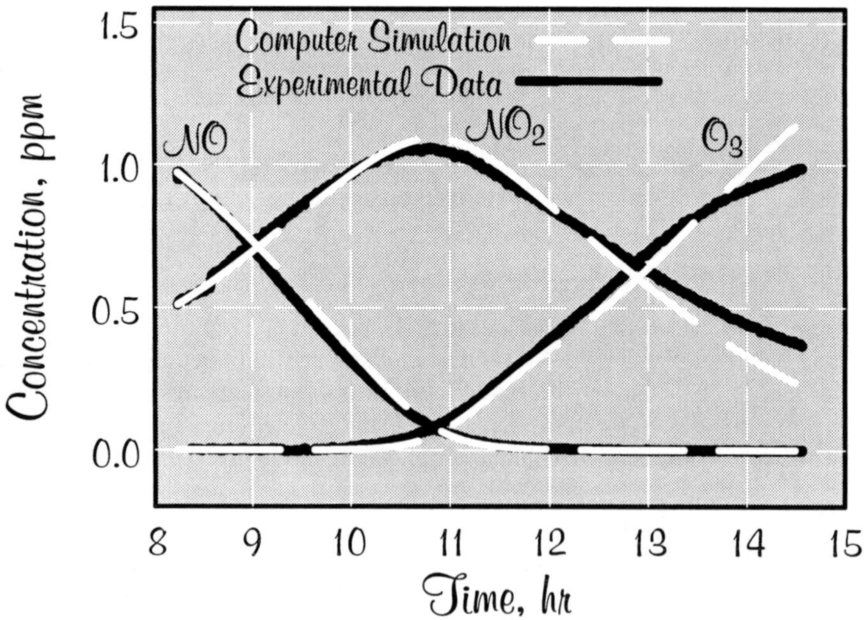

X-A. Modeling Urban Air Chemistry

The modeling of the atmospheric chemistry of aromatic hydrocarbons and the influence of these hydrocarbons on ozone generation within the urban atmosphere is an especially challenging task, since many of the details of the complex reactions that occur are unknown at present. One must resort to parameterized mechanisms, with the yields of surrogate model species representing reactive, uncharacterized products that are adjusted to fit the experimentally measured O_3, NO, and NO_2, and possibly other measured products that are observed in smog chamber experiments.

Some features of the atmospheric oxidation of the aromatic hydrocarbons are reasonably well established and can be treated in atmospheric chemistry mechanisms with a high degree of confidence. Thus, it is clear that reactions of the OH radical with the aromatic hydrocarbons are the primary, rate-determining steps in the removal of the aromatic hydrocarbons in atmospheric oxidation. The OH–aromatic hydrocarbon rate coefficients in our reaction mechanisms can be formulated accurately (commonly to within (±20%). However, to evaluate the rate of loss of the *i*th aromatic hydrocarbon in the system, and ultimately the subsequent rates of reaction of its reaction products and all the reactions that lead to ozone generation, we must start with the simple rate expression:

$$-\mathrm{d}[\text{Aromatic}]_i/\mathrm{d}t = [\text{Aromatic}]_i[\text{OH}]k_i$$

It is apparent that in addition to the OH–aromatic rate coefficient (k_i) and the measured hydrocarbon concentration at a given time, the OH-radical concentration is required. Although current techniques allow the direct measurement of the [OH], this has rarely been attempted in smog chamber experiments. Commonly, [OH] is estimated by one of two methods: (1) through the measured rate of loss of a species that reacts only with the OH radical, together with the known rate coefficients; or (2) theoretically through the solution of coupled differential equations that involve NO_x, hydrocarbons, and the oxidation products that take part in its many formation and destruction steps. It is the uncertainty in the nature of these subsequent reactions of aromatic hydrocarbon oxidation that is the major problem in developing models of urban ozone generation.

In the discussion of chapter II, it has been recognized that in the reactions of OH radical with toluene, the xylenes, and other alkyl-substituted monocyclic aromatics, only a small fraction of the reaction (4–12%) occurs by H-atom abstraction from the alkyl groups. This fraction of the reaction can be described reasonably well from present knowledge. The immediate product of the reaction is a benzyl-type radical. For monosubstituted benzenes, $C_6H_5C(\cdot)H{-}R$ type radicals are formed, where R is an H-atom or alkyl group; these subsequently react with O_2 to form benzylperoxy-like radicals [$C_6H_5C(OO\cdot)H{-}R$]. Various alternative pathways of reaction of the benzylperoxy-type radicals can be written that allow (1) NO to be oxidized to NO_2; or (2) formation of aromatic nitrates [$C_6H_5C(ONO_2)H{-}R$]; and (3) reaction with HO_2 and RO_2 species. The fraction that oxidizes NO to NO_2 forms a benzyloxy-type radical [$C_6H_5C(O\cdot)H{-}R$] that will generate HO_2 radicals and an aromatic aldehyde or ketone by reaction with

O_2; thus, if R = H, $C_6H_5C(=O)H$ forms; if R is an alkyl group, then an aromatic ketone results, $C_6H_5C(=O)R$.

The largest fraction of the OH-radical–aromatic hydrocarbon reaction proceeds through the addition of the OH radical to the aromatic ring: at 298 K with toluene, ~93%; *o*-xylene, ~90%; *m*-xylene, ~96%; *p*-xylene, ~92%. Unfortunately, it is the mechanism involving this larger fraction of the OH–aromatic hydrocarbon reaction that is least well understood, and the subsequent reactions of the OH–aromatic hydrocarbon intermediate adduct species remain speculative.

There have been several representations of the chemistry of the aromatic hydrocarbons suggested for use in urban and regional atmospheric chemical modules. For examples, see Gery et al. (1989), Carter (1990), Jenkin et al. (1997), and Stockwell et al. (1997). In the chemistry modules of urban air-shed models, only a very limited number of the many aromatic molecules that actually appear in the atmosphere are included; this is a consequence of the rapid increase in the time required for the computer solution as the number of reactant molecules added to the mechanism is increased. Thus, in the extended formulation of Carter (1990), 16 different aromatic hydrocarbons are represented. In the model of Stockwell et al. (1990, 1997), the aromatic hydrocarbons are separated into groups with similar reactivity toward OH: TOL (toluene and the less reactive aromatic compounds), XYL (xylene and the more reactive aromatic compounds), and CSL (cresol and other hydroxy-substituted aromatic compounds). In the Gery et al. (1989) carbon bond model (CBM-EX) used in many urban air-shed models, toluene (TOL) and *m*-xylene (XYL) are the surrogate aromatic species, and other aromatics are represented by surrogate molecular components as required in the carbon bond formulations; for example, ethylbenzene is represented by 1 TOL and 1 PAR (1 CH2 unit).

In their chemical mechanism, Jenkin et al. (1997) have included most of the reactions that have been suggested or postulated as involving aromatic hydrocarbons. Their basic module is the so-called "Master Chemical Mechanism" (MCM), which is an explicit gas-phase reaction scheme designed for the description of chemical processes in the troposphere. It includes the degradation of 123 VOCs and a set of reactions describing the inorganic chemistry. In principle, any and all aromatic compounds deemed to be significant can be included in the reaction mechanism as possible reacting species. The current version 2.0 consists of more than 10,000 reactions and more than 3400 species (including aromatic hydrocarbons, alkenes, alkanes, and their various oxidation products). The MCM is regularly updated using the results of new laboratory studies; the version 2.0 of April, 2001 was available at this writing; see

http://cast.nerc.ac.uk/LIBRARY/MCM2/

At present, there is a real uncertainty as to whether the use of MCM or other highly detailed reaction schemes in the simulation of urban ozone generation will lead to a significant increase in accuracy over that obtained using the highly abbreviated reaction schemes. In the years ahead, as we continue to increase both our knowledge of the mechanisms of atmospheric oxidation of the aromatic hydrocarbons and the capacity and efficiency of computers, it is probable that interest of atmospheric scientists in the use of the extensive explicit mechanisms such as the MCM in the simulation of urban ozone generation will continue to grow. Certainly, the use of such detailed mechanisms will be most useful in predicting the influence of a specific chemical species on the generation of urban ozone or other potentially harmful products of photochemical smog.

Carter's extensive model (Carter, 1999) of the atmospheric chemistry of the aromatic hydrocarbons is used at the time of this writing in estimating maximum incremental reactivities (MIR values; see sections I-B and I-C) of selected organic molecules. It is instructive to look at the methods that he has employed to model the aromatic hydrocarbon oxidation in the atmosphere. He considers the following generalized reaction mechanisms:

(a) H-atom abstraction by OH

OH + aromatic → H_2O + benzyl-type radical (e.g., $C_6H_5CH_2\cdot$) (1)
Benzyl-type radical + O_2 → benzylperoxy-type radical (e.g., $C_6H_5CH_2OO\cdot$) (2)
Benzylperoxy-type radical + NO → aromatic nitrate (e.g., $C_6H_5CONO_2$) (3)
Benzylperoxy-type radical + NO → NO_2 + benzyloxy-type radical (e.g., $C_6H_5CH_2O\cdot$) (4)
Benzylperoxy-type radical + O_2 → HO_2 + aromatic aldehyde or ketone (e.g., $C_6H_5C(H){=}O$) (5)

(b) OH addition to the aromatic ring

OH + aromatic → OH–aromatic adduct (6)
OH–aromatic adduct + O_2 → HO_2 + phenol or cresol (7)
OH–aromatic adduct + O_2 → OH–aromatic–O_2 adduct (8)
OH–aromatic–O_2 adduct + NO → uncharacterized nitrate products (9)
OH–aromatic–O_2 adduct + NO → NO_2 + uncharacterized radical intermediates (10)
Uncharacterized radical intermediates → → HO_2 + α-dicarbonyls and other fragmentation products (11)

Carter points out that alternative mechanism formulations can be used. For example, one can (1) assume that the OH–aromatic adduct reacts with NO_2 at a rate competing with or exceeding its reaction with O_2; (2) assume radical intermediates react with NO_2 to form stable products; or (3) assume that additional NO to NO_2 conversions are involved in the formation of α-dicarbonyls or other fragmentation products. However, except for the naphthalenes and tetralin, Carter's experience has shown that parameterizations based upon these alternative mechanisms do not fit the available smog chamber data as well as those based on the general reactions (6) through (11) outlined in this section (Carter, 1999).

In the Carter (1999) treatment of the aromatic hydrocarbon oxidation, selected surrogates are used to represent the known and uncharacterized aromatic ring-fragmentation products. The three experimentally identified α-dicarbonyl products from the methylbenzenes—glyoxal (GLY), methylglyoxal (MGLY), and biacetyl (BACL)—are represented explicitly, and three model species are used to represent the non-photoreactive (DCB1) and the two types of photoreactive, uncharacterized ring-fragmentation products (DCB2 and DCB3). The specific mechanisms for the DCB species are based roughly on those estimated experimentally for unsaturated dicarbonyls. Carter's mechanism includes several types of radicals: HO_2 (HO2), and RO_2 radicals (RO2-R), which can oxidize NO to NO_2 and promote O_3 generation; RO_2 radicals, which can lead to organic nitrate formation (RO2-N); and a fraction of the RO_2 radical that is a peroxyacyl nitrate precursor (RCO-O2). Specific products of ring-retaining reactions, such as phenol-like molecules (PHEN), cresol-like molecules (CRES), benzaldehyde-like molecules (BALD), and a representative aromatic ketone (PROD2), are considered. In terms of the model species used in Carter's mechanism (Carter, 1999), the overall reactions of the aromatic hydrocarbons are represented as follows:

$$\begin{aligned} \mathrm{OH} + \text{aromatic} \rightarrow\ & y_{\mathrm{RH}}\,\mathrm{HO2} + y_{\mathrm{RR}}\,\mathrm{RO2\text{-}R} + y_{\mathrm{NR}}\,\mathrm{RO2\text{-}N} + y_{\mathrm{Q2}}\,\mathrm{RCO\text{-}O2} \\ & + y_{\mathrm{PH}}\,\mathrm{PHEN} + y_{\mathrm{CR}}\,\mathrm{CRES} + y_{\mathrm{BL}}\,\mathrm{BALD} + y_{\mathrm{k6}}\,\mathrm{PROD2} + y_{\mathrm{GL}}\,\mathrm{GLY} \\ & + y_{\mathrm{MG}}\,\mathrm{MGLY} + y_{\mathrm{BA}}\,\mathrm{BACL} + y_{\mathrm{D1}}\,\mathrm{DCB1} + y_{\mathrm{D2}}\,\mathrm{DCB2} + y_{\mathrm{D3}}\,\mathrm{DCB3} \end{aligned}$$

where the $y_{\mathrm{RH}} \ldots y_{\mathrm{D3}}$ are stoichiometric parameters that must be specified to define the mechanism; they are optimized to fit smog chamber data (and are different) for each aromatic hydrocarbon.

As expected with such a large number of adjustable parameters, the surrogate mechanisms for the atmospheric oxidation of the aromatic hydrocarbons can be fit reasonably well to the limited smog chamber results for ozone generation. Of course, a reasonably good fit does not prove the correctness of the details in the assumed mechanism.

In terms of our limited knowledge of the chemistry of the atmospheric oxidation of the aromatic hydrocarbons, the highly simplified chemical modules (Gery et al., 1989; Stockwell et al., 1997), as well as the more detailed modules (e.g., Jenkin et al., 1997; Carter, 1999), appear to provide reasonable simulations of chamber measurements of the ozone and nitrogen oxide profiles. However, the detailed chemistry of the many products formed cannot be treated in the simplified mechanisms, and it will be necessary to resort to a more detailed treatment as our knowledge of the mechanism details grows and profiles of other important products become available. In this chapter, we attempt to identify certain findings in this study related to the atmospheric chemistry of the aromatic hydrocarbons that we recommend be utilized in the development of future models of the urban atmosphere.

X-B. Modeling the Atmospheric Chemistry of the Aromatic Hydrocarbons

Care should be taken by urban air-shed modelers to represent, as accurately as our present knowledge allows, the chemistry of those aromatic hydrocarbons that are observed in the urban atmosphere to be most reactive toward the OH radical and to have the highest MIRs. These are toluene, the xylenes, and the trimethyl benzenes. Among the aromatic compounds, these species have the greatest influence on the extent of ozone generation in the urban atmosphere. However, the task is not a simple one since, as noted in the previous discussions, many of the details of these reactions remain uncertain, and product mass balances are far from complete. One must recognize the high degree of uncertainty associated with the application of these surrogate mechanisms in predicting ozone formation, as well as other important products of the oxidation of the aromatic hydrocarbons that occur in the polluted urban atmosphere. Realistic estimation of the range of uncertainties in the results from such applications is an obligation of the modeling community.

Although one cannot gain great confidence in a surrogate mechanism from the fact that a given mechanism can be made to reproduce the extent of ozone generation in a smog chamber experiment, simulations of the time dependence of the measured reactant and products concentration profiles still remain as the major test of aromatic reaction mechanisms at the time of this writing. Modelers should continue to apply this procedure in the development of surrogate mechanisms.

Protocols for evaluating oxidant mechanisms in the urban and regional models have been discussed actively for many years, and a special philosophy related to the problem of photochemical "smog" mechanism construction and testing has developed. It is instructive to consider an example that relates directly to the view of the problem as seen in 1991 by U.S. Environmental Protection Agency scientists and its consultants and formulated by Jeffries et al. (1992). We feel that many of its features are important to all who are involved in atmospheric modeling. These authors state:

> "Mechanism evaluation is more than just establishing the agreement of the model with observations. The notion of acceptance is very complex and perhaps ambiguous; there is no single criterion of merit. In part this is because knowledge is incomplete and somewhat uncertain and therefore one must accept a particular explanation as the best one *now* available. As with other scientific theories, acceptable chemical reaction mechanisms have several characteristics* (this list is not exhaustive, but the items are most important and certainly show the type of elements that enter into such judgments):
>
> (i) *Consistency*—a mechanism's formulation must be consistent, not only within itself, but with currently accepted theories and facts applicable to these theories; the most obvious example of this characteristic is that acceptable mechanisms must incorporate consensus kinetics values such as rate coefficients and reaction products;
>
> (ii) *Accuracy*—the mechanism's consequences of predictions should be in demonstrated agreement with the results of existing experiments and observations; such agreement must not arise from compensating errors among a mechanism's inner workings;
>
> (iii) *Simplicity*—the mechanism should not be burdened with a large amount of mathematical apparatus required to explain the basic features (i.e., be simple and elegant in symbolic expression), yet should contain all the terms that are needed in explanation, so one does not have to substitute a large number of external facts: it should have the "feel" of a good solution—there should be a certain sense of congruity or recognition. Thus, we prefer explanations of events in terms of necessary causal forces to explanations of events in terms of other empirical generalizations;

*These are adapted from Thomas Kuhn's "Objectivity, Value Judgment, and Theory Choice" in *The Essential Tension*, University of Chicago Press, 1977.

(iv) *Scope* —the mechanism's predictions should extend beyond the particular observations it was initially designed to explain; for example, a mechanism's predictions should show agreement with chamber data that were not used in its development, and, in another form, a mechanism initially intended for the prediction of ozone, should also be capable of predicting other associated intermediate or final products such as formaldehyde or hydrogen peroxide;

(v) *Fertility*—the mechanism is expected to predict novel phenomena, that is, phenomena that were not parts of the set to be explained. The display of such fertility reduces the likelihood that his/her mechanism is just an *ad hoc* one—that is, invented just for the original occasion but of limited use. Or, where an anomaly is encountered, a good model should itself serve to suggest possible modifications or extensions.

Individually these characteristics as criteria are imprecise: individuals may legitimately differ about their application to concrete cases. In addition, when used together, they conflict with each other; i.e., a simple mechanism may be limited in scope. Nevertheless, these characteristics must be used to warrant the arguments that mechanism developers and evaluators must make in building a case that their mechanism should be used by EPA in ambient air modeling.

X-C. Improvements in Tests of Aromatic Hydrocarbon Atmospheric Oxidation Mechanisms and Experimental Product Data

As described in section X-B, many of the current mechanisms of atmospheric oxidation of the aromatic hydrocarbons that are employed in urban air-shed models are highly simplified. A major rationale for this has been the fact that so little about the detailed mechanisms has been established in aromatic hydrocarbon oxidation. Most modelers believe that the use of a reasonably good parameterization of the processes using hypothetical surrogate species is the more appropriate pathway to follow, rather than the alternative design of a detailed mechanism with many reaction steps expected to occur in theory and with assigned rate coefficients based necessarily on empirical thermochemical kinetic data or more sophisticated theoretical calculations (e.g., determination of potential energy surfaces for the specific reaction channels). The parameterization of the various steps is usually formulated to match smog chamber experiments in which only the aromatic hydrocarbon, ozone, and oxides of nitrogen are commonly measured. A reasonable match or "fit" to chamber data is often considered "verification" of the mechanism. The modeler who follows this protocol may find useful the archives of databases for photochemical smog chamber experiments available on the internet; for example,

http://www.cert.ucr.edu/~carter/bycarter.htm
http://airchem.sph.unc.edu/Research/TVAData/
http://airchem.sph.unc.edu/Research/UNCData/

However, one must retain a high degree of skepticism as to the reality of the product of this procedure, since some reactions in the smog chamber can be unique to the chamber and do not occur in the real atmosphere. For example, in the smog chambers, wall reactions can occur and are a definite problem, both in the heterogeneous generation of free radicals unrelated to the homogeneous gas-phase fundamental chemistry that we wish to study, and in loss of important products at the wall of the chamber. The rates of radical formation in wall reactions necessitate correction (often 10–20% of the homogeneous radical generation rates), and these corrections must be made in order to analyze the results and extrapolate them to the conditions of the real atmosphere where walls equivalent to those of the chamber are rarely encountered by the reacting atmospheric gases.

In addition, using VOC mixtures (containing alkanes, alkenes, and aromatic hydrocarbons) for "verification" experiments often artificially minimizes the uncertainties in the model. In these cases, only a small portion of the OH radical is consumed by the aromatic VOC, and the errors of the models are damped. In contrast, the results from "verification" experiments using single aromatic compounds are more sensitive to changes in the details of the mechanism and are more appropriate for use in mechanism testing.

There is another significant complication in a strong dependence on the match of the existing smog chamber studies to surrogate mechanisms as proof of their validity and usefulness. This relates to the concentrations that are employed in such studies; these are usually more than 10 times those encountered in the polluted atmosphere, a limitation forced largely by the limited sensitivity of the analytical techniques commonly employed. Because of this, the concentrations of products formed are necessarily much higher than those present in the atmosphere. In the case of the aromatic hydrocarbons, especially, this can lead to a potentially serious problem. Many of the products of oxidation of the aromatic hydrocarbons are highly polar in nature and have low vapor pressures, and the loss of this material to the chamber walls, as well as through aerosol generation, can remove species that, at their much lower concentrations in the real atmosphere, will remain largely in the gas phase. These species under atmospheric conditions would potentially continue to react in the gas phase and lead to further ozone generation as well as other products.

The development of reliable methods of simulating atmospheric chemistry, other than the use of conventional smog chamber data, should be encouraged, although, at the time of this writing, this latter practice will necessarily continue. Studies based upon chamber experiments at much lower concentrations, as present in the urban atmosphere, would be preferred to test the development of surrogate mechanisms. We feel that the lofty goal of alternative method development is worth seeking, but the pathways to it are not well defined at this writing.

The development of atmospheric pressure flow reactors is one possible route to the desired goal (e.g., see Donahue et al., 1998). At least, in principle, in such a system, the critical transient species could be measured directly as a function of time or distance of the flow (e.g., [OH] as well as O_3, NO_x, etc.). With modern analytical techniques that measure the reactive components in the atmosphere, it should be possible to use reacting mixtures of the aromatic hydrocarbons, NO_x, and so on, at low concentrations typical of those encountered in the atmosphere. Problems with wall reactions are more easily minimized in the atmospheric flow systems than in the current smog chamber studies where turbulent mixing and contact of reactants and products with the walls of the chamber are inadvertently encouraged through the use of circulating fans within the chamber.

X-D. Modeling of Aerosol-Phase Atmospheric Chemistry of Aromatic Hydrocarbons

In chapters VIII and IX, some discussion of aerosol formation and the modeling of the heterogeneous chemistry of the aromatic hydrocarbons has been given and reference to the details given there is suggested. Obviously, further development of more realistic chemical modules to handle the aerosol-phase physics and atmospheric chemistry should be encouraged as our knowledge of the products and mechanisms of the atmospheric chemistry of the aromatic hydrocarbons improves.

XI

Important Areas for Further Laboratory Studies

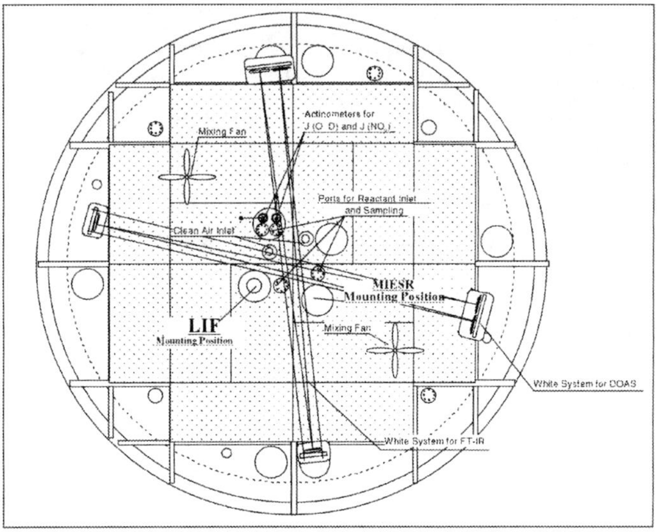

(European Photoreactor, EUPHORE, Valencia, Spain)

Although a great deal of quantitative information related to the atmospheric oxidation of the aromatic hydrocarbons has been obtained in recent years, there is a large uncertainty related to many aspects of our knowledge. In this chapter we attempt to identify some of these critical areas to provide direction for new and definitive laboratory and theoretical work. The discussion in the first two sections is separated into work related to (a) measurement of rate coefficients for the reactions of the important oxidizing species in the atmosphere (section XI-A); and (b) the products and mechanisms of the reactions (section XI-B). The areas of knowledge that require further study are considered in the order they first appear in this book. In section XI-C, the needed studies of the photochemistry of the products of the aromatic hydrocarbon oxidation are discussed. In sections XI-D, suggested further studies are described related to aerosol generation in the oxidation of the aromatic hydrocarbons; and in section XI-E, studies of reactions of PAHs sorbed on aerosols are considered. In section XI-F, an attempt is made to prioritize the areas of needed study.

XI-A. Kinetic Studies of Gas-Phase Atmospheric Oxidation of Aromatic Hydrocarbons

XI-A-1. The Rate Coefficients for the OH-Radical Reactions with the Monocyclic Aromatic Compounds

This is the area of greatest strength of our knowledge at the time of this writing. However, there are certain aspects of the area that we identify for further studies. First, rate coefficients of additional aromatic hydrocarbons of relevance to atmospheric chemistry should be studied. Of the 23 different monocyclic aromatic hydrocarbons that have been studied to date, five have been studied by only one laboratory and confirmation of these results should be obtained: *p*-cymene, tetralin, α-methylstyrene, β-methylstyrene, β,β-dimethylstyrene. Seven of the 18 different aromatic hydrocarbons studied by more than one laboratory have estimated overall uncertainties greater than 30%, and further studies and a resulting lowering of the uncertainty in the estimate are needed: *iso*-propylbenzene, the *o*-, *m*-, and *p*-ethyltoluenes, *tert*-butylbenzene, indan, and indene.

Rate coefficients for only 20 of the many aromatic products of the oxidation of the aromatic hydrocarbons have been determined at the time of this writing, and 14 of these are from single studies from one laboratory: the 2,4-, 2,5-, and 3,4-dimethylbenzaldehydes, *trans*-cinnamaldehyde, naphthaquinone, the 2,3-, 2,4-, 2,5-, 2,6-, 3,4-, and 3,5-dimethylphenols, the 2,3,5-, and 2,3,6-trimethylphenols, and *m*-nitrotoluene. The database for these molecules should be enhanced by further work.

While the method of estimation of rate coefficients for the reactions of OH radicals with monocyclic aromatic species, involving the use of σ^+ values for substituent groups, appears to provide reasonably good results, whenever possible rate coefficients should be measured experimentally rather than relying on estimation methods. However, comparison of estimates using this method with the measured rate coefficient data suggests that further studies are required to understand the reasons for the differences (greater than a factor of 2) seen with *o*-xylene, *p*-xylene, indan, tetralin, 2,5-dimethylbenzaldehyde, 3,5-dimethylphenol, 2,3,5-trimethylphenol, and *m*-nitrotoluene.

XI-A-2. The Rate Coefficients for the OH-Radical Reactions with the Polycyclic Aromatic Compounds

A number of the polycyclic aromatic hydrocarbons and other compounds have sufficient vapor pressure to remain largely in the gas phase. Rate coefficients for reaction with OH radical have been measured for only 16 of these species and additional PAHs should be studied. Many of the presently available measurements (eight) come from the work of a single laboratory and should be confirmed: 1- and 2-methylnaphthalenes, 2,3-dimethylnaphthalene, acenaphthalene, fluoranthene, 1- and 2-nitronaphthalenes, and 2-methyl-1-nitronaphthalene. Existing rate coefficient measurements for six of these PAHs have estimated uncertainties (given in parentheses) $\geq$ 30% and should be redetermined in additional studies: biphenyl (30%), fluorene (factor of 1.5), acenaphthene (factor of 2), phenanthrene (factor of 2); dibenzo-*p*-dioxin (factor of 1.5), and dibenzofuran (30%). The rate coefficient for the seventh PAH, anthracene, one of the most abundant and important PAHs in the atmosphere, is highly uncertain, with a range of (18–289) $\times 10^{-12}$ cm^3 $molecule^{-1}$, and should be remeasured and the precision of the rate data improved. The rate coefficients for PAHs that have not yet been studied and that have vapor pressures greater than about 10^{-5} Torr (298 K) should be determined, as the gas-phase reaction with OH may be an important removal mechanism for these species; for example, 3-methylphenanthrene, pyrene, and benzo[*a*]fluorene.

The existing methods for estimation of rate coefficients for the PAHs (see section II-H) require significantly more development. Presently, it is recommended that experimental measurements of rate coefficients be made. When kinetic data do not exist, the use of the Hückel hardness factor, rather than the ionization potential, appears to offer the best correlation with the measured rate coefficients for the five PAHs for which data are available. It is clear that a more reliable estimate of the rate coefficient for anthracene is a key to the testing of this and other models for prediction. Many additional PAH rate coefficients should be measured to help development of estimation methods. The low volatility of the PAHs has limited experimental studies in the past. Measurement at elevated temperatures, with extrapolation to determine rate coefficients at 298 K, can be employed; however, a high precision of the measurements is necessary to obtain data that when extrapolated to 298 K will have relatively low uncertainty.

XI-A-3. Rate Coefficients for NO_3-Radical Reaction with Aromatic Compounds

The reactions of the NO_3 radical with the monocyclic aromatic compounds in the atmosphere appear to be unimportant unless there is an alkene-like substituent (e.g., indene and styrene) or an $-$OH group attached to the aromatic ring (e.g., phenols, cresols). Reactions with NO_3 radical can be an important removal mechanism for these species. Of this class of compounds, only phenol, *o*-, *m*-, and *p*-cresols, and *m*-nitrophenol have been studied, and these rate coefficients have uncertainty factors that are rather large (>1.5 in most cases). Studies are needed of the many other phenolic compounds that are oxidation products of the aromatic hydrocarbons.

XI-A-4. Rate Coefficients for Reactions of Aromatic Compounds with Ozone

Reactions of the monocyclic aromatic hydrocarbons with O_3 are very slow and are unimportant in the atmosphere unless there is an alkene-like substituent attached to the aromatic ring (e.g., indene, *trans*-cinnamaldehyde, and acenaphthylene). It appears that the cresols also react, although slowly, by some ill-defined mechanism and with a high uncertainty in the measured rate coefficients (factor of 2). In view of the relative unimportance of the O_3 reactions with phenols, kinetic studies of these reaction is of low priority at this time.

XI-A-5. The Rate Coefficients for Reactions of $O(^3P)$-Atoms with Aromatic Compounds

Rate coefficient data for 18 different aromatic compounds that have been measured at this writing correlate well with the matching measurements for reactions with OH radical. Hence, one can use this correlation and the measured OH–aromatic rate coefficient data to estimate a reasonably reliable rate coefficient for those $O(^3P)$–aromatic reactions that have not been measured and for which correction may be required in the extrapolation from smog chamber studies to atmospheric mechanisms. Since $O(^3P)$-atom reactions with the aromatics are unimportant in the urban atmosphere, further studies on this system appear to be unnecessary for use in urban air chemistry at the time of this writing.

XI-A-6. Rate Coefficients for Non-Aromatic Products of Oxidation of Aromatic Compounds

Of the 25 different non-aromatic products of the oxidation of the aromatic compounds for which rate coefficients for reaction with OH have been reported, 19 were determined by only one laboratory, and four have been determined by two or more laboratories; all four are estimated to be uncertain by 30% or more. Rate coefficients for reactions of 22 of the 25 compounds are relatively large ($k > 10^{-11}$ cm^3 molecule^{-1} s^{-1}). It is clear that additional kinetic studies of the rates of OH-radical–initiated oxidation of the non-aromatic products of the aromatic hydrocarbon oxidation are required.

XI-A-7. Reactions with the NO_3 Radical

The rate coefficients for the reactions of the NO_3 radical with the non-aromatic, first-generation products of the oxidation of the aromatic hydrocarbons are even less extensive than those for OH radical. Of the 12 different species studied at the time of this writing, 10 are from a single laboratory and two have uncertainties of a factor of ≥ 1.5. Since the NO_3 reactions with these species are probably much less important than those of OH radical in ozone generation within the urban atmosphere, the further study of these reactions is of lower priority.

XI-A-8. Reactions with Ozone

The rate coefficients for the reactions of the non-aromatic, first-generation products of the oxidation of the aromatic hydrocarbons can be important for those species that contain $>C{=}C<$ bonds (e.g., 2,4-hexadiendial). Of the nine different compounds that have been studied at the time of this writing, five of the rate coefficient measurements represent upper limits, and four were determined by a single laboratory. Additional measurements for the various observed unsaturated diones should be made to build confidence in the value of rate coefficients for these apparently important product species.

XI-B. Product and Mechanism Studies of Gas-Phase Atmospheric Oxidation of Aromatic Compounds

XI-B-1. Products and Mechanisms of the Aromatic Compound Reactions with OH Radical

In view of the discussions in section II-D, it is apparent that there is no aromatic hydrocarbon or aromatic compound for which a reasonably good carbon mass balance has been obtained from the product analysis. In the best of cases, the mass balance for carbon is less than 70% and typically it is less than 50%. There are three main reasons for this. First, the analytical methods employed by most laboratories are not able to identify and quantify many of the possible classes of products. Second, many of the products are highly polar or have relatively low vapor pressures and can be lost at the walls of the reaction vessel or form aerosols and thereby escape detection by the analytical techniques commonly employed. Third, many of the primary products of aromatic ring-opening are more reactive than the original aromatic hydrocarbon (dicarbonyls, etc.) and these can be removed rapidly by reaction (photolysis and reaction with OH radical), although this complication should not in itself cause a loss of carbon.

In some studies, specialized techniques involving derivatization of carbonyls have been used to identify the many carbonyl products, but standards for these compounds are often not available, and the amounts of product formed remain very uncertain. It is recommended that additional (well established, as well as entirely new) analytical methods be applied to the identification and quantification of reaction products of oxidation of aromatic hydrocarbons. This is particularly important for studies of the products of the ubiquitous, reactive aromatic compounds such as toluene, the xylenes, and trimethylbenzenes, so that a reasonable product mass balance can be obtained, and our preliminary judgments about mechanisms can be tested more quantitatively. In particular, the influence of the available NO_x in the system on the product yields should be studied more carefully. These data should allow a realistic extrapolation to the ambient air case. It is of crucial importance that we establish the extent of NO_x requirement in the ring-opening step that leads to highly reactive products, as this bears directly on the significance of the contribution of the aromatic hydrocarbons to ozone formation during their dispersion within the atmosphere.

It is quite possible that some as yet unidentified reaction pathways exist in the aromatic oxidation mechanism of aromatic compounds. Until such time as product studies are able to achieve carbon balances approaching 100% and we are able to rationalize the observed formation of products, it will not be possible to ensure that the proposed reaction mechanisms are accurate. Control strategies for urban smog abatement that are not based upon an accurate description of the atmospheric oxidation mechanism of aromatic hydrocarbons may be flawed.

In view of the importance of the phenolic compounds among the products of the atmospheric oxidation of the aromatic compounds, more detailed studies of the their atmospheric behavior are required.

XI-B-2. The Mechanisms and Products of Arene Oxides/ Oxepin Reactions with OH Radical

The possible importance of these interesting reactions in the atmospheric oxidation of the aromatic hydrocarbons is unknown at the time of this writing, although they have been proposed as intermediates in explanation of some ill-understood aspects of aromatic hydrocarbon oxidation. The general features of reaction mechanisms have been defined (section VI-B), but their significance to atmospheric chemistry should be tested further.

XI-B-3. The Mechanism and Products of the Reactions of Unsaturated Dicarbonyls with Atmospheric Oxidizing Species

There are three classes of dicarbonyls formed in the oxidation of the aromatic hydrocarbons: 1,2-dicarbonyls, unsaturated 1,4-dicarbonyls, and 1,6-dicarbonyls. There is no question that each of these is formed among the many products, regardless of which ring-opening mechanism is operative. However, the nature and the yields of many of the products remain uncertain. The information on the reactions of the unsaturated 1,6-dicarbonyls is most limited; the product yields are often unknown, and hence their importance remains highly uncertain. Further work on the mechanisms and products of these reactions is suggested to be an important link to our understanding of the mechanisms of atmospheric oxidation of the aromatic hydrocarbons.

XI-B-4. Summary of Current Knowledge of the Mechanism of Aromatic Compound Oxidation

Although our knowledge has increased significantly in recent years, our current understanding of the overall atmospheric oxidation mechanisms of aromatic compounds remains rather limited. We have a reasonably clear picture of the first two steps of the mechanism, which, in general, are reaction with OH radicals to give an adduct, followed by reaction of the OH–aromatic adduct with O_2. However, the kinetics and mechanisms of the reactions that occur subsequent to reaction of the OH–aromatic adduct with O_2 are not understood. There are many possible transient radical species formed during the OH-initiated oxidation of aromatic compounds (see figures II-E-1 and II-E-2). The relative importance of the numerous possible radical species and reaction pathways in the oxidation mechanism is unclear. There are few convenient, clean laboratory sources of these radicals, and consequently there have been few direct laboratory studies of their chemistry. Convenient sources of these radicals need to be developed and their chemical behavior needs to be determined. High-level ab initio quantum mechanical calculations have provided valuable insight since 1990 and, with ever-increasing computational resources, are likely to play an important role in future advances in this area. It should be noted that there are essentially no available data concerning the dependence of product yields on temperature, total pressure, and partial oxygen pressure. Such data are essential to construct accurate mechanisms for the atmospheric oxidation of aromatic compounds.

XI-C. Gas-Phase Photochemistry of Aromatic Hydrocarbons and Their Oxidation Products

XI-C-1. Ring-Fragmentation Products

Quantitative information on the photochemistry of many of the oxidation products of the aromatic hydrocarbons is very limited, although many of these species absorb sunlight, and qualitative data suggest that they undergo photodecomposition that leads to the generation of radicals. To calculate a meaningful apparent first-order reaction rate coefficient for a given photochemical processes (*j*-value) occurring in the atmosphere, both the absorption cross sections of the absorbing molecule and the quantum yields of the specific process must be known as a function of wavelength of the absorbed light and the pressure. For many of the simple carbonyl molecules, both of these quantities have been studied—for example, CH_2O, CH_3CHO, CH_3CH_2CHO, CHOCHO, CH_3COCHO, CH_3COCH_3—and seemingly reliable *j*-values have been estimated for a variety of conditions (e.g., see Calvert et al., 2000). This is not the case for many of the other photochemically reactive carbonyl products of aromatic hydrocarbon oxidation.

XI-C.1a. Absorption Cross-Section Measurements

Often, because of the low vapor pressure of the compounds, the absorption spectra of products of the oxidation of the aromatic hydrocarbons have been measured in aqueous solutions or solutions of organic solvents. Such spectra may be significantly different from those measured in the gas phase; for example, compare those reported for benzaldehyde in the gas phase and in hexane solution in figure VII-D-2. Many of the unsaturated dicarbonyl products appear to be very photochemically active, and absorption cross sections that we now must employ are only determined in solution or arise from highly uncertain gas-phase measurements. Thus, it is important that reliable gas-phase spectra be determined for *trans*-butenedial, 4-oxo-2-pentanal, 3-hexene-2,5-dione, and the 2,4-hexadienedials.

XI-C-1b. Quantum Yield Measurements

Many of the studies of the photochemistry of the ring-fragmentation products of aromatic hydrocarbon oxidation provide only qualitative evidence of the magnitude of the quantum yield of the products formed, and often these are effective quantum yields averaged over a broad spectral range or determined at one or two wavelengths within the absorption region. The *j*-values obtained from the use of such qualitative information are of rather limited value in atmospheric modeling. Of course, the wavelength distribution of the solar radiation varies with the location of the observer, the time of day (solar zenith angle), the elevation in the atmosphere, and other factors such as surface albedo and extent and type of cloud cover. The use of "effective" quantum yields determined for a given wavelength region cannot be used to calculate accurate *j*-values for light conditions other than those provided by the particular lamps or sunlight employed in the studies. For many of the oxidation products (e.g., the unsaturated carbonyls), the *j*-value estimates are particularly sensitive to the quantum yields for dissociation within the longer wavelength region (400–500 nm) of the absorption where the solar flux is highest. Commonly, quantum yields of photodecomposition are strongly dependent on the wavelength of absorbed light, and tend to fall off rapidly to zero at the longer wavelengths where the energy available in the quantum of light approaches that of the minimum necessary for dissociation. However, the solar flux (F) increases rapidly in this region of fall-off of the quantum yield values, so even relatively small quantum yields (ϕ) and cross sections (σ) may contribute significantly to the *j*-values [given by the integral of the product, $\phi(\lambda)\,\sigma(\lambda)\,F(\lambda)\,d\lambda$, over all wavelengths]. Thus, a quantitative measure of the wavelength dependence of the quantum yields, as well as the absorption cross sections, is a necessary prerequisite to the calculation of meaningful *j*-values for tropospheric conditions.

The needed measurements of quantum yields versus wavelength for many of the complex molecular products of aromatic hydrocarbon oxidation are very difficult and time-consuming to make. Often, several photochemical channels are operative within the photochemically active region of the molecule, and it is difficult to sort out the nature of the primary processes and secondary reactions of the primary products from which these products are ultimately formed. Thus, in the simple acrolein molecule [$CH_2{=}CHC({=}O)H$], it appears that four primary modes of photodissociation occur, and each of these is wavelength- and pressure-dependent (see section VII-C-4). Although the data for this molecule are very limited (mostly determined in experiments using only 313-nm radiation), they do suggest that the photodissociation is probably a minor process for removal of acrolein in the lower atmosphere compared with the reaction with OH radical. By analogy, one might predict that the photodecomposition modes of methacrolein [$CH_2{=}CH(CH_3)CH({=}O)$] are similar to those of acrolein and are probably relatively unimportant in the urban atmosphere compared with the reactions with OH radical. Data for the methyl vinyl ketone quantum yields are also lacking, and we have assumed that this molecule behaves like acetone, which has a similar CH_3–COR bond. Quantitative studies of the quantum yields as a function of wavelength and pressure will be necessary to test these conclusions for methacrolein and methyl vinyl ketone.

Probably of greater importance to our understanding of the generation of urban ozone is the determination of the quantum yields as a function of wavelength for the following photochemically active molecules: butenedial, 4-oxo-pentenal, 3-hexene-2,5-dione, *E,E*-2,4-hexadienedial, and *E,E*-2-methyl-2,4-hexadenedial. In some of these cases, we have qualitative information on experimentally measured *j*-values for at least one set of atmospheric conditions. Future quantitative studies of the photochemistry of the many newly discovered polycarbonyl products of aromatic hydrocarbon oxidation (Jang and Kamens, 2001) are especially important in view of their potential for strong sunlight absorption and high photochemical activity within the troposphere; see section VIII-C.

XI-C-2. The Photochemistry of Some Common Aromatic Ring-Retaining Products of the Oxidation of the Aromatic Hydrocarbons

Photochemical data for many of the different aromatic ring-retaining oxidation products are very limited. Again, as with the ring-fragmentation products, at the time of this writing, in the calculation of *j*-values one must often resort to the use of solution-phase spectra and qualitative evidence of the quantum yields of photodecomposition as a function of wavelength.

XI-C-2a. Absorption Cross-Section Data

Although gas-phase spectra of some ring-retaining products have been measured (e.g., benzaldehyde), for many of these compounds only the solution-phase spectra are available at this writing. Gas-phase spectra of the nitrobenzaldehydes, nitrobenzene, nitronaphthalenes, and their methyl-substituted derivatives should be determined. From the solution-phase spectra of the phenols and cresols, it appears that the photochemistry of these compounds will be relatively unimportant as removal mechanisms within the troposphere (see section VII-D-6). Solution-phase spectra and the minimum energy required for photodissociation of the OH-substituted PAHs (e.g., hydroxynaphthalenes, hydroxyphenanthrenes) and the nitrophenolic compounds (e.g., the nitro- and polynitrophenols) suggest the possible significance of their photochemical processes within the troposphere. Gas-phase absorption spectra for these rather low vapor pressure species would be an important input to the test of this hypothesis.

XI-C-2b. Quantum Yield Data

Quantum yields as a function of wavelength have not been measured for any of the gas-phase, ring-retaining products of the aromatic hydrocarbons. The nitro-substituted phenols, the nitro-substituted PAHs, nitro-substituted benzaldehyde and its methyl-substituted derivatives, and the quinones appear to have a high potential for removal by gas-phase photochemistry within the troposphere. Measurements of quantum yields versus wavelength will be required to evaluate the potential importance of photolysis of these compounds in the atmosphere. If the benzene oxide/oxepin and the higher analogues (toluene oxide/2-methyloxepin) are identified unambiguously as intermediates in the oxidation of the aromatic hydrocarbons, then determination of the wavelength dependence of the quantum yields of their reactions should be made to confirm the very limited experimentally measured *j*-values (see sections VII-C-15 and VII-C-16).

XI-D. Aerosol Generation in Oxidation of Aromatic Hydrocarbons

XI-D-1. The Mechanism of Secondary Aerosol Generation in the Oxidation of the Aromatic Hydrocarbons

It has been estimated that a large fraction (~ 40–80%) of the carbonaceous aerosol present in some urban environments (e.g., in the Los Angeles area) is composed of secondary organic aerosols, which are generated in the photochemical oxidation of the organic compounds present in the air mass. It is not clear what fraction of this material results from the oxidation of the aromatic hydrocarbons, but it is highly probable that the aromatic compounds are important in the generation of aerosols. The chemical nature of only a small fraction of the secondary aerosol mass has been identified. The application of current theories of secondary aerosol formation and the development and testing of more detailed theories in the future requires a much more detailed and complete analysis of the aerosol composition. Of course, the compounds that are present in the aerosol are those of low vapor pressure, and the identification of all components offers a significant experimental challenge. As for the gas-phase measurements, additional (possibly in situ) analytical methods should be developed so that a reasonably good accounting (mass balance) can be made of the aerosol chemical composition. The development of a detailed and accurate mechanism of oxidation of a given aromatic hydrocarbon depends on our knowledge of the product species formed, and those present in the aerosol portion of these products are an important part of the total input to this development (noting that they may be second- or later-generation products).

Current theoretical treatment of aerosol generation in smog chamber experiments is necessarily highly simplified. Aerosol-formation data are commonly fit empirically by assigning two types of semivolatile products of oxidation with assumed aerosol formation parameters (α_i and K_i) that allow the best fit of the data; see section VIII-B-3. Usually, one empirical compound is considered to be of relatively low vapor pressure and the other of relatively higher vapor pressure. Using this technique, significant progress has been made in understanding the general features of secondary organic aerosol growth, even in mixtures of various organic compounds that form aerosols. As our knowledge of the aerosol product compositions grows and actual measurements of appropriate values for stoichiometric product coefficients (α_i) and gas/particle partitioning coefficients (K_i) can be made, more rigorous and realistic tests of current theories will be developed, and appropriate improvements in theory should result.

Organic acids appear to be important constituents of the secondary organic aerosols generated in the laboratory and in the atmosphere, although the nature and extent of their formation pathways are not well understood. To understand the mechanisms of dicarboxylic acid generation, we need to learn how α,ω-dialdehydes are converted to α,ω-oxo-acids and hence to α,ω-diacids. Experiments should be designed to answer such questions as: Are these transformations ones that occur in gas-phase reactions or in the aerosols? What is the nature of the oxidizing species responsible (OH, O_3, HO_2, H_2O_2, etc.)?

Experiments should be designed to test the significance of various mechanisms that have suggested the possible involvement of atmospheric aerosols on the extent of ozone generation in the urban atmosphere. For example, we should answer the question: How important is the removal of OH, HO_2, and RO_2 radicals, the most important chain-carrying intermediates involved in ozone generation in the troposphere, by interaction with the aerosols of different composition?

Our knowledge of the effects of the secondary organic aerosols (SOA) on the magnitude and wavelength distribution of solar radiation present in the urban atmosphere should be enhanced by new experimental and theoretical studies. Appropriate consideration of these effects can be used in newly developed urban air-shed models and to assess the contribution, if any, of SOA to global climate change.

XI-E. Atmospheric Chemistry of Polycyclic Aromatic Hydrocarbons Sorbed on Particles

It is well demonstrated that PAHs sorbed on particles (other than those high in black carbon or on fly ash) are oxidized relatively rapidly when exposed to sunlight (see section IX-C). There are limited rate data for particle-bound PAHs on wood soot (benz[*a*]anthracene, chrysene, benzofluorenes, benzo[*a*]pyrene, benzo-[*g,h,i*]pyrene) and much less data for other types of PAH reactions on other types of particles. Some mechanism development work has taken place to explain PAH photoinduced reactions. Although many different oxidizing species may be involved in this process, the nature of the oxidation process(es) is still in debate at the time of this writing, and additional studies are needed. It is clear that the generation of OH radicals, $O_2(^1\Delta_g)$, O_3, and possibly other oxidizing agents can lead to the oxidation of the PAHs in the complex mixture of liquid and solid compounds that constitutes the aerosol. The marked influence of relative humidity and the water content of the aerosol on the oxidation rate is an interesting aspect of the reactions and may be an important factor in explaining the detailed mechanisms.

New experiments are required to shed more light on the relative importance of possible pathways that are operative in the atmosphere. Quantitative photochemical laboratory experiments involving aerosols into which PAHs are sorbed are very difficult to perform and interpret. Perhaps use can be made of relatively simple but representative surrogates of the aerosol substrates present in the real world onto which single non-volatile PAHs are sorbed. The determination of the rate of oxidation of the PAHs as stimulated by a fixed number of quanta s^{-1} at selected individual wavelengths within the absorption band of the PAH (HONO and/or other light-absorbing compounds) and at varied intensities would be useful. Whether the photochemical action spectrum reflects the particular PAH absorption cross sections or possibly other light absorbers present in the aerosol (e.g., absorption-enhanced PAH ground-state singlet to PAH triplet absorption) will be an important factor in evaluation of the mechanism.

The nature of the products formed in the photodegradation of PAHs on aerosols has been used in the past and clearly can provide further evidence of the reaction pathways involved. Many of these products are aromatic ring-cleavage products, quinones, and hydroxy compounds. A possible influence of the organic composition of the particles on the rate of PAH photodecomposition is an interesting aspect of the reactions and should be exploited to provide further clues as to the mechanism. Further product studies should be carried out in concert with experiments that target the effects of different chemical and physical particle compositions.

XI-F. Qualitative Prioritization of Research Efforts in Areas Suggested for Further Studies

Effort should be made eventually to resolve all of the recognized uncertainties that remain in the mechanisms of the atmospheric oxidation of the aromatic hydrocarbons, and which have been outlined briefly in the preceding sections of this chapter. However, from our current knowledge of the relative importance of the reactions of the various aromatic hydrocarbons with the several transient species, some guidance can be given to researchers for ordering the relative importance of possible further work.

Probably the most uncertain area related to the atmospheric oxidation of the aromatic compounds is the nature and the yields of the products formed. We suggest that a major effort be made to obtain a complete analysis of the gaseous and aerosol products of the most important aromatic hydrocarbons including toluene, the xylenes, and the trimethyl-substituted benzenes. We must enhance significantly our knowledge in this area to arrive at realistic atmospheric reaction mechanisms and realistic modules for urban chemistry models.

Modelers should incorporate into urban chemical models, modules that account for sunlight attenuation and enhancement caused by the presence of aerosols, as existing evidence suggests that these effects can be very important at the relatively high aerosol concentrations encountered in most urban areas.

The mechanisms of the atmospheric oxidation of the important polycyclic aromatic hydrocarbons sorbed on particles is an important area for further study. Definitive proof of the nature and extent of the oxidizing species

[$O_2(^1\Delta_g)$, OH, O_3, etc.] is a prerequisite to our understanding of the mechanism and incorporation of this into future modules of our urban chemistry models in order to predict more reliably the fate and lifetime of these important PAH species.

Further determination should be made of important rate coefficients for OH-radical reactions with selected aromatic hydrocarbons and PAHs, including (but not limited to) those that are now based on measurements from a single laboratory study and those with very high uncertainties (e.g., anthracene).

Further studies of the photochemical mechanisms are required, including the measurement of gas-phase cross sections and quantum yields of the various primary processes as a function of wavelength. These data are needed particularly for the highly photochemically active dicarbonyl products formed from aromatic-ring cleavage. In addition to laboratory measurements of the necessary cross sections and quantum yields as a function of wavelength, we suggest that the use of outdoor chamber experiments be extended to determine j-values for seemingly important light-absorbing products of the aromatic hydrocarbons. Such measurements should be made with suitable OH-radical scavengers present so that the observed rate of loss of the compound by OH attack and other non-photochemical loss processes can be minimized. The simultaneous measurement of $j(NO_2)$ values should be made as well. Such data provide an important test of the reliability of j-values calculated from the quantum yield and cross-section data.

Experimental tests of the current theories of aerosol generation that results from the atmospheric oxidation of the aromatic hydrocarbons and their products will be important in refining and expanding the current theories.

Although not specific to aromatic hydrocarbon atmospheric chemistry, definitive tests are desired to determine the extent of the alteration of the concentration fields of HO, HO_2, and RO_2 radicals present in the urban atmosphere through their removal (or generation) by aerosols. We assume in current modeling of the chemistry of the urban atmosphere that these species are created and destroyed only through gas-phase processes. Yet the relatively long-lived peroxy-radical species (HO_2 and RO_2), often with lifetimes of the order of seconds, can undergo many collisions with aerosols during their lifetimes. If only a small fraction of these collisions results in removal from the gas phase, then, in theory, the efficiency of ozone generation can be altered significantly by the presence of aerosols.

Further laboratory work on selected reactions under different atmospheric conditions, including temperature variation, simulation experiments on more complex chemical systems in smog chambers with well-characterized radical sources, and measurements of radicals and concomitant species in the field, all combined with chemical modeling, are required to reveal quantitatively the oxidation mechanisms of aromatic hydrocarbons. Undoubtedly, the aromatic hydrocarbons contribute significantly to the ozone and particle formation in polluted urban atmospheres and, therefore, require improved chemical treatment in models.

XII

Summary

$$\sum_{i=\mathrm{I}}^{i=\mathrm{XI}} (\mathbf{Chapter})_i$$

XII-A. Chapter I: Importance of Aromatic Hydrocarbons in the Chemistry of Ozone Generation in the Urban Atmosphere

Aromatic hydrocarbons are released into the atmosphere of an urban community from sources involving the combustion of fossil fuels (e.g., exhaust of internal combustion engines), solvent use, or other sources. During the daylight hours they react rapidly, often within hours, with OH radicals. When these reactions occur by attack on the aromatic ring, the ultimate result is, in large part, rupture of the aromatic ring and generation of a variety of multifunctional carbonyl compounds. These are reactive both photochemically and toward OH radicals, and they form a variety of peroxy radicals (RO_2). The attack of the OH radical on an alkyl side chain attached to the aromatic ring may also occur, with the generation of benzylperoxy-like radicals (e.g., $C_6H_5CH_2O_2$). The dominant fate of the peroxy radicals is reaction with NO, which proceeds in two channels; the main channel gives RO radicals and NO_2, while the minor channel gives organic nitrates $RONO_2$: RO_2 (HO_2) + NO → RO (OH) + NO_2; RO_2 + NO → $RONO_2$. The RO radicals formed in these NO → NO_2 oxidation steps regenerate other RO_2 and HO_2 radicals in a multitude of chain-carrying reactions that result in the transformation of additional NO to NO_2. The subsequent photodissociation of this continually regenerated NO_2, as well as that added from emissions, provides in the sunlight the major source of O_3 generation within the urban atmosphere: $NO_2 + h\nu \rightarrow O(^3P) + NO$; $O(^3P) + O_2\ (+\ N_2/O_2) \rightarrow O_3\ (+\ N_2/O_2)$. A major interest in the chemistry of the aromatic hydrocarbons in the urban atmosphere is the ability of the aromatic hydrocarbons to generate ozone through their reactions, largely initiated through the reactions with OH radical.

Although the aromatic hydrocarbons constituted only 19% of the carbon atoms present in the multitude of non-methane hydrocarbons found in the morning hours in Los Angeles in 1995, and about 18% of the species with which OH reacted, they represented the largest share (30%) of the maximum incremental reactivity (MIR) among the various classes of hydrocarbons present. The MIR is a theoretical measure of the ozone formed per specific hydrocarbon reacted when an incremental increase in the particular hydrocarbon is effected in simulated urban air of typical compositions. Similar importance is assigned to the presence of aromatic hydrocarbons in the urban atmosphere of Boston in the morning hours in the Fall of 1995: the aromatic hydrocarbons made up about 30% of the carbon atoms in the non-methane hydrocarbons present, but they represented about 39% of the species that react with OH, and about 42% of the MIR of the intitial hydrocarbon mixture.

From the MIR values calculated for the hydrocarbons present in the 1995 measurements of the morning air composition in Los Angeles, the biggest contribution to the MIRs came from the xylenes (*m*-xylene + *p*-xylene, 9.4%; *o*-xylene, 3.0%); 1,2,4-trimethylbenzene (7.1%); toluene (6.1%); and 1,3,5-trimethylbenzene (2.3%). Thus, these six aromatic hydrocarbons alone can account for about 93% of the total MIR estimated for all of the aromatic compounds measured in the early morning in the urban atmosphere of Los Angeles. [(9.4 + 3.0 + 7.1 + 6.1 + 2.3)%/29.9% (total MIR for all the aromatics; from table I-A-2) = 93%.]

In the Boston atmosphere in Fall of 1995, the largest contributions to the total MIR of all the hydrocarbons were from 1,2,4-trimethylbenzene (15.4%); the xylenes: *m*-xylene + *p*-xylene (5.0%); *o*-xylene (1.9%); toluene (4.5%); ethylbenzene (3.1%); *m*-ethyltoluene (2.2%); and 1,3,5-trimethylbenzene (2.0%). Here, about 82% of the MIR contributed by all of the aromatic compounds arises from these eight aromatic hydrocarbons. [(15.4 + 5.0 + 1.9 + 4.5 + 3.1 + 2.2 + 2.0)%/41.5% (total MIR for all the aromatics; from table I-A-3) = 82%.]

The concentration of ambient VOCs in the atmosphere of several European sites provides a perspective of the differences within Europe and the United States. The percentages of aromatic hydrocarbons found among the VOCs in European ambient air are generally higher than those observed in the United States, and this reflects the higher limits on the aromatic content of fuels set by the European Commission.

In the United Kingdom the largest contribution of the POCP (photochemical ozone creation potential) weighted emission of 40% for a single group of hydrocarbons comes from the aromatic hydrocarbons (aromatics: 40%; alkanes: 30%, alkenes: 15%). In the city of Wuppertal, Germany, in summer 1998, an average of 44% of total carbon of the anthropogenic NMVOC mix in ambient air has been identified as aromatic compounds, similar to 47% of carbon measured in a nearby traffic tunnel. In both cases, it is the largest contribution from one class of hydrocarbons. However, the largest fraction of the OH reactivity is due to the alkenes, 65% in the city and 62% in the traffic tunnel, with 24% and 25%, respectively, attributed to the aromatic hydrocarbons. In the city of Munich, 34–36% of total

NMVOC carbon has been attributed to aromatic hydrocarbons, with a contribution of 31% to the total OH reactivity and 51% to the reactivity as calculated by the MIR factors. In this case, again, the carbon contribution and the MIR reactivities are highest for the group of aromatic hydrocarbons. The highest contribution to the OH reactivity of 54% comes from the group of the alkenes.

XII-B. Chapter II: Reactions of Aromatic Compounds with OH Radicals

From the viewpoint of ozone generation, the aromatic hydrocarbons are the single most important class of compounds in current urban atmospheres. During daylight hours, the major atmospheric reactant of the aromatic hydrocarbons is the OH radical. Chapter II provides a review and evaluation of this important chemistry.

In section II-B, the rate coefficients for the OH-radical reaction with 23 different monocyclic aromatic hydrocarbons and 20 different monocyclic aromatic products of their oxidation are evaluated. The kinetic information is given for each rate coefficient reported in the literature for the primary OH–aromatic compounds reaction, the experimental methods employed, and the estimated uncertainties. When sufficient temperature-dependent rate data are available, preexponential factors and activation energies for the reactions are also evaluated and reported. Summaries of these rate coefficients are given in tables II-B-44 and II-B-45.

The primary step in the reaction mechanism of OH radical with the aromatic compounds involves a number of possible reaction channels. The most important of these involves addition of OH radical to the aromatic ring with the formation of the OH–aromatic adduct ($> 90\%$). Also, a somewhat less important channel involves the OH-radical abstraction of an H-atom from alkyl groups attached to the aromatic ring with the generation of benzyl-like radicals ($< 10\%$). When a double bond $>C{=}C<$ is attached to the ring, then an additional channel exists that is common to all alkene–OH reactions: OH-radical addition to this double bond. If –OH or –C(=O)H groups are attached to the aromatic ring, then abstraction of the hydrogen atom may occur as well.

Methods of estimation of rate coefficients for reactions of OH radicals with monocyclic aromatic compounds are reviewed in section II-C. The rate coefficient for the addition of OH to the aromatic ring in the hydrocarbons is a function of the number and type of substituents that are attached to the ring. The logarithm of the rate coefficient for the addition reaction is roughly proportional to the summation of the σ^+ values of the substituents on the ring, and the addition occurs at the position yielding the most negative value of $\Sigma\sigma^+$. The rate coefficients recommended in this study for 36 different monocyclic aromatic compounds are in reasonable agreement with those calculated using the $\Sigma\sigma^+$ relationship. The calculated values agree within a factor of 2 for 28 of the compounds, and for all but one compound, the agreement is within a factor of 3.

The major products of the OH-radical reaction with the aromatic hydrocarbons are described in section II-D. An important problem encountered in all existing product studies of the OH-radical reaction with the aromatic hydrocarbons is the deficient carbon balance. Generally, only about 50% of the carbon can be accounted for in the products observed. The high reactivity of the primary products, usually much greater than that of the parent hydrocarbon, as well as the low volatility of the further products, contributes to this problem. The primary products of the reactions of OH radical with benzene, toluene, the xylenes, and the trimethylbenzenes are reviewed in this section. These products consist largely of aromatic aldehydes, arising from H-atom abstraction pathway ($< 10\%$) from the alkyl groups attached to the ring, and phenolic compounds and nitro-aromatics produced by reactions following the OH addition to the aromatic ring. A variety of reactive carbonyl compounds are produced as ring-fragments by the reaction of O_2 with the OH–aromatic compound adduct. Product studies of the OH-radical reaction with phenol, the cresols, the methyl-substituted styrenes, the aromatic aldehydes, and 2,3-benzofuran are also reviewed.

In section II-E, mechanisms of reaction of OH radicals with aromatic hydrocarbons are outlined and evaluated in terms of the kinetic and product data that are available as of this writing. It still remains an open question as to what extent the yields of the ring-cleavage products are influenced by NO_x. Several recent observations support the conclusion that the ring cleavage is controlled largely by the reactions of O_2 with the OH–adduct under atmospheric conditions.

Section II-F examines the existing rate coefficient data for the first step in the reactions of the OH radical with the polycyclic hydrocarbons (PAHs). As with the monocyclic aromatic hydrocarbons, the dominant pathway for the reactions of the PAHs involves the initial addition of OH radical to the aromatic ring system. Reviewed and evaluated here are the gas-phase rate data for 11 of the more volatile PAHs and five other polycyclic compounds that have been studied at this writing. A summary of these rate coefficients is given in table II-F-17.

The products and complex mechanisms of reaction are discussed in section II-G for several PAHs: naphthalene, fluorene, phenanthrene, fluoranthene, pyrene, and biphenyl. The extent of nitroarene product formation is reviewed for reactions of 12 of the PAHs with OH radical in the presence of NO_2. In section II-H, a discussion is given of some possible ways of estimating the rate coefficients for the OH-radical reactions with the PAHs.

XII-C. Chapter III: Reactions of Monocyclic and Polycyclic Aromatic Compounds with NO_3 Radicals

The reaction of NO_3 radical with the aromatic hydrocarbons is only important in the atmosphere for those aromatic hydrocarbons that contain a $>C=C<$ side chain attached to the aromatic ring (e.g., styrene and indene), for phenolic compounds (e.g., phenol and cresols), and some PAHs and PACs. For methyl-substituted benzenes, the mechanism appears to involve largely the abstraction of an H-atom from the methyl groups attached to the ring. However, with naphthalene and other multiring PAHs, addition of NO_3 to the aromatic ring is the major pathway of the reaction. With the phenolic compounds, NO_3-radical addition to the aromatic ring may be the dominant first step in the reaction mechanism. For aromatic compounds that contain $>C=C<$ groups attached to the aromatic ring, addition of NO_3 to this group appears to be the favored pathway for the initial reaction.

In section III-B, the existing rate data for the NO_3-radical reaction with the aromatic hydrocarbons are reviewed and evaluated for 14 monocyclic aromatic hydrocarbons, eight other monocyclic aromatic compounds, and 16 different polycyclic aromatic compounds. Summaries of these rate coefficients are given in tables III-B-41, III-B-42, and III-B-43.

A review of the present empirical methods of estimation of rate coefficients for the NO_3 reactions with the hydrocarbons is given in section III-C. The existing methods appear to be unreliable for application to the aromatic compounds.

Products studies of the reactions of NO_3 and N_2O_5 with the monocyclic aromatic hydrocarbons are limited. In section III-D, the existing data are reviewed, and the mechanism for reactions of NO_3 with phenolic compounds is discussed. In section III-E, the mechanisms of the reaction of NO_3 with the PAHs are reviewed.

XII-D. Chapter IV: Reactions of Ozone with Aromatic Compounds

The gas-phase reactions of the aromatic hydrocarbons with ozone are only significant for those aromatic hydrocarbons that contain $>C=C<$ bonds. Given in section IV-B are the rate coefficients for the reactions of ozone with the 12 monocyclic aromatic hydrocarbons that have been studied. For 10 of these compounds, the rate coefficients measured represent upper limits. Rate coefficients for five other monocyclic aromatic compounds are also presented, one of which is an upper limit. Rate data for the reactions of 15 different PAHs are given, 14 of which are upper limits. Summaries of these rate coefficients are given in tables IV-B-33, IV-B-34, and IV-B-35.

In section III-C the methods of estimation of rate coefficients for aromatic compounds are reviewed, and in section III-D the very limited product and mechanism studies of the ozone–aromatic hydrocarbon reactions are described.

XII-E. Chapter V: Reactions of $O(^3P)$ Atoms with Aromatic Compounds

Reactions of $O(^3P)$ atoms with aromatic hydrocarbons are of little importance in atmospheric chemistry. However, they can be of significance in laboratory studies involving the irradiation of NO_x–aromatic hydrocarbon–air mixtures and possibly in stack plumes containing high concentrations of NO_2. The $O(^3P)$ atom reaction rate data are often necessary to account for, and to eliminate, the contribution from these reactions in the extrapolation of the environmental chamber data to atmospheric conditions. As with OH-radical reactions, the primary step in the reaction of $O(^3P)$ atoms with aromatic hydrocarbons is O-atom addition to the aromatic ring. There is a reasonably good correlation between $O(^3P)$ and OH-rate coefficients for the 15 different aromatic hydrocarbons for which data for both reactants have been determined.

Rate coefficients for 16 different aromatic hydrocarbons and two other aromatic compounds are reviewed and evaluated in section V-B. The products and mechanisms for the $O(^3P)$ atom reactions with the aromatic hydrocarbons are reviewed in section V-C.

XII-F. Chapter VI: Reactions of Non-Aromatic Products of Aromatic Compound Oxidation

The available rate coefficient data for the reactions of OH, NO_3, and O_3 with the products of ring-opening in aromatic hydrocarbon oxidation are reviewed and discussed in chapter VI. In section VI-A, rate coefficients are given for reactions of OH with 25 different oxidation products, 12 different oxidation products reacting with NO_3, and nine different oxidation products reacting with O_3. Summaries of these rate coefficients are given in tables VI-A-29, VI-A-30, and VI-A-31.

In section V-B, the products and mechanisms of the reaction of OH with some oxidation products of aromatic hydrocarbons are given (arene oxides, unsaturated dicarbonyls, 1,2-dicarbonyls, and maleic anhydride).

XII-G. Chapter VII: Primary Photochemical Processes of the Aromatic Hydrocarbons and Some of Their Common Oxidation Products

The monocyclic aromatic hydrocarbons, benzene and the methyl-substituted benzenes, absorb very little of the sunlight present in the lower troposphere, and as a consequence their gas-phase photochemistry is unimportant in the urban atmosphere. However, as each additional aromatic ring fragment is fused to benzene to form the various larger PAH species (naphthalene, anthracene, etc.), the conjugated π-system grows, the energy to promote an electron to the first excited state decreases, the light absorption increases in magnitude, and the absorption band extends further into the longer-wavelength region of light available within the troposphere. In fact, the very large cross sections typical of the PAHs result in very large rates of sunlight absorption. For example, the *j*-value for excitation of naphthacene (tetracene) is $>0.05\ s^{-1}$ at high solar zenith angles. This is 5 times the first-order rate coefficient for photodissociation of NO_2, $j(NO_2)$, the important process responsible for ozone generation within the troposphere. These large rates of excitation of the PAHs within the troposphere and the potential for generation of $O_2(^1\Delta_g)$ through electronically excited PAH quenching by atmospheric O_2, coupled with the recent observation of large concentrations of a singlet oxygen-like oxidant (SOLO) in the Kyoto atmosphere, and the possible importance of $O_2(^1\Delta_g)$ in atmospheric, gas-phase chemistry is revisited in section VII-B. Largely as a consequence of its rapid rate of physical quenching in the atmosphere, current evidence suggests that this oxidant is not important in the gas-phase chemistry of the urban atmosphere, although $O_2(^1\Delta_g)$ is possibly an oxidant in reactions in the aqueous and organic solutions of which many atmospheric aerosols are composed. The large concentrations of SOLO observed in Kyoto do not seem to be attributable to the presence of $O_2(^1\Delta_g)$. The identity of the SOLO species observed in Kyoto remains unexplained.

In section VII-C, the photochemistry of several common ring-fragmentation products of the atmospheric oxidation of the aromatic hydrocarbons is reviewed and evaluated. Our previous recommendations of photochemical data for formaldehyde, acetaldehyde, propionaldehyde, glyoxal, methylglyoxal, acetone, methylhydroperoxide, and methyl nitrate have not changed (Calvert et al., 2000) and the reader is referred to those data for use in describing these compounds. In this section, a review and evaluation is made of the existing photochemical data for several other light-absorbing products of atmospheric oxidation of the hydrocarbons. The data are very limited in many cases. Cross sections and quantum yields for several additional compounds are reviewed and evaluated, and *j*-values are calculated for typical tropospheric conditions at different solar zenith angles. Recommended photochemical data are presented for glycolaldehyde, hydroxyacetone, acrolein, glycidaldehyde, pyruvic acid, butenedial, methacrolein, methyl vinyl ketone, 4-oxo-2-pentenal, 3-hexene-2,5-dione, *E,Z*-2,4-hexadienedial, *E,E*-2,4-hexadienedial, and *E,E*-2-methyl-2,4-hexadienedial.

In section VII-D, the photochemistry of some aromatic ring-retaining products of the atmospheric oxidation of the aromatic hydrocarbons is reviewed and evaluated. Recommended cross sections, quantum yields, and representative *j*-values are estimated for several compounds: benzaldehyde, 2-nitrobenzaldehyde, nitrobenzene, 1-nitronaphthalene, 2-nitronaphthalene, 2-methyl-1-nitronaphthalene, benzene oxide/oxepin, and toluene oxide/2-methyloxepin. A very general discussion of the photochemistry of some compounds with very limited data is also given for the phenols, cresols, 2-hydroxynaphthalene, and the nitrophenols.

XII-H. Chapter VIII: Aerosol Generation in Atmospheric Oxidation of Aromatic Hydrocarbons

Atmospheric aerosol is ubiquitous in virtually all areas of the troposphere, but especially in urban areas. It contains a significant fraction of carbonaceous material. Evidence is strong that a large portion of this material accumulates on the aerosol from the sorption of secondary oxygenated compounds that are relatively non-volatile products of the oxidation of the hydrocarbons, especially the aromatic hydrocarbons. The mechanisms of aerosol generation from the aromatic hydrocarbons are discussed in section VIII-B. Described here are gas/particle partitioning theory, experimental methods of determination of the yields of secondary organic aerosol, and practical mathematical methods of simulating these phenomena.

In section VIII-C, discussion is given of the available (and limited) data concerning the molecular composition of secondary organic aerosols formed in the oxidation of the aromatic hydrocarbons, toluene, *m*- and *p*-xylenes, ethylbenzene, *m*- and *p*-ethyltoluene, and 1,2,4- and 1,3,5-trimethylbenzenes. In section VIII-D, a discussion is given of the mechanisms of atmospheric oxidation of the several aromatic hydrocarbons that can be inferred from the composition of the aerosol formed. Mechanisms for the formation of aerosol products of toluene, and *m*- and *p*-xylenes are presented.

Important ingredients of many secondary organic aerosols are the organic acids, particularly α,ω-dicarboxylic acids; in section VIII-D-3, the possible mechanisms of the acid formation are considered.

There are several possible influences on atmospheric chemistry that the aerosols, including those from the oxidation of the aromatic hydrocarbons, can have. These are discussed in section VIII-E: (1) The aerosols can incorporate into their mass, and at least temporarily essentially remove from further chemical change, the first-generation (or later) gas-phase products that normally would oxidize further to aid in ozone generation. (2) Aerosols can serve as sites for the condensation of the highly reactive, transient gas-phase species involved in photooxidation chemistry (e.g., OH, HO_2, RO_2) and hence remove these from further participation in ozone generation. (3) Aerosols scatter and absorb solar radiation and, in so doing, alter the local actinic flux that drives photochemical reactions and ozone formation.

The fates and lifetimes for typical organic aerosols are described in section VIII-F.

XII-I. Chapter IX: Atmospheric Chemistry of Polycyclic Aromatic Hydrocarbons Sorbed on Particles

Polycyclic aromatic hydrocarbons are created primarily during the combustion of carbon-based fuels. They contain two or more fused benzene rings. Depending on their volatility, they can exist as free gases or sorbed onto particles. Some general properties of the PAHs are presented in section IX-A. The atmospheric chemistry and photochemistry of the polycyclic aromatic hydrocarbons (PAHs) sorbed on particles can be a major sink for removing and altering the composition of these species. The experimental evidence for these processes is reviewed and discussed in this chapter. Interest in the PAHs in the atmosphere has arisen because some of them are carcinogens in animal test models.

In section IX-B, the various sources of PAH emissions in various parts of the world are reviewed. Emissions of PAHs are ubiquitous throughout the world, but significant amounts are emitted over Asia and the former USSR. At an urban level, automotive emissions are significant contributors to urban atmospheres, but, as in Los Angeles, other energy sources like natural gas combustion are also important contributors, especially to semivolatile PAHs such as fluoranthene and pyrene. Rural PAH concentrations tend to be a factors of 0.2–0.05 lower than urban concentrations. In urban environments, higher-molecular-weight PAHs such as benzo[*a*]pyrene or benzo[*g,h,i*]pyrene are associated with particles below 1 μm in aerometric diameter. More volatile PAHs such as fluorene, phenanthrene and fluoranthene can migrate to larger particles, and a significant fraction of these compounds are found on particles greater than 2 μm. Semivolatile PAHs such as naphthalene, phenanthrene, pyrene, and benz[*a*]anthracene can simultaneously exist in the gas and particle phases. Currently, there are equilibrium models that permit estimates of the phase distributions of these compounds, but there is, at best, a factor of 3 uncertainty associated with estimates from current models.

In section IX-C, we consider the concentrations of the PAHs that have been measured in urban centers and in rural areas in several diverse regions of the world. A review of the estimated fluxes of PAHs is given in section IX-C-3, and the published data on the distribution of PAHs on aerosols of different sizes is reviewed in section IX-C-4. For example, in Pasadena, California, in 1978, it was observed that 75% of the benz[*a*]anthracene and 85% of the coronene were associated with particles less than 0.26 μm diameter, and that almost half of the mass of these two PAHs was in the size range 0.075–0.12 μm.

In section IX-C-5, the observations and theory related to the gas/particle partitioning of the PAHs are discussed. The fraction in each phase is dependent on the concentration of aerosol, the temperature, and the humidity. For relatively high-vapor-pressure PAHs such as phenanthrene, the fraction of the compound that appears on an aerosol (0.5 μm diameter) at an aerosol concentration of 10 $\mu g\ m^{-3}$ is calculated to be 0.002; while at an aerosol concentration of 500 $\mu g\ m^{-3}$ the fraction that is sorbed on aerosol increases to 0.17. For a relatively lower-vapor-pressure PAH such as benzo[*a*]pyrene, the calculated fraction sorbed onto the aerosol is 0.93 and 1.00, respectively, for the same two cases of aerosol concentration and size distribution.

In section IX-D, the rates of reactions, the products, and the mechanisms of PAHs sorbed on particles are reviewed. A number of authors have suggested that a photosensitized singlet oxygen mechanism is operative in the particle phase, although the extent of the oxidation that results from this and other oxidizing species remains highly uncertain. The most complete product data exists for naphthalene in the gas phase and for phenanthrene, anthracene, benz[*a*]anthracene, and benzo[*a*]pyrene in the sorbed phase. Many of the products from these photoinduced reactions are aromatic ring-cleavage products, quinones, and hydroxy compounds. The organic composition of particles can potentially influence the rate of PAH photodecomposition.

There are limited rate data for primarily particle-bound PAHs on wood soot (benz[*a*]anthracene, chrysene, benzofluorenes, benzo[*a*]pyrene, benzo[*g,h,i*]pyrene) and even less data for other types of PAH reactions on other types of particles. Some mechanism development work has taken place to explain PAH photoinduced reactions. Photosensitized singlet oxygen formation has been suggested as a possible mechanism for PAH reactions of surfaces and in liquid aerosols. The mechanisms are still in debate and much work is needed to clarify them.

In section IX-D-8, an example is given of computer modeling of a gas/particle kinetic mechanism of nitrofluoranthene and nitropyrene formation in a sunlight-irradiated mixture of propene, NO_x, and dilute diesel exhaust containing fluoranthene and pyrene (among many other species) in air.

XII-J. Chapter X: Application of Findings to Modeling the Atmospheric Chemistry of Aromatic Hydrocarbons

In the chapter X, several features of the modeling of the atmospheric chemistry of the aromatic hydrocarbons are discussed in view of current methods employed and the findings in this work. It is the hope of the authors that modelers of the urban atmospheric chemistry of the aromatic compounds will find the discussion and suggestions given here of value in further model development work.

XII-K. Chapter XI: Important Areas for Further Laboratory Studies

In view of the present review and evaluation, remaining areas of greatest uncertainty are identified in this chapter, along with recommendations of further laboratory research needed to improve our present understanding of the reaction kinetics and mechanisms that are operative in the atmospheric oxidation of the aromatic hydrocarbons. Efforts should be made eventually to resolve all of the recognized uncertainties that remain in the mechanisms of the atmospheric oxidation of the aromatic hydrocarbons and that have been outlined in chapter XI. However, from our current knowledge of the relative importance of the reactions of the various aromatic hydrocarbons with the several transient species, some guidance is given to researchers for ordering the relative importance of possible further work.

References

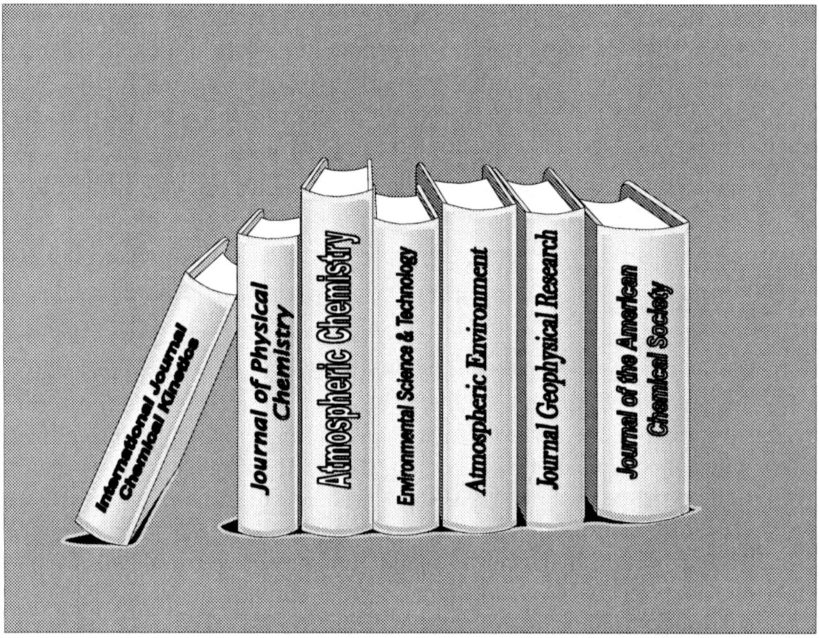

Aceves, M., and J.O. Grimalt (1993), Seasonally dependent size distributions of aliphatic and polycyclic aromatic hydrocarbons in urban aerosols from densely populated areas, *Environ. Sci. Technol.*, *27*, 2896–2908.

Adamson, A.W., and A.P. Gast (1997), *Physical Chemistry of Surfaces*, 6th ed., John Wiley & Sons, New York.

Allen, J.O., N.M. Dookeran, K. Taghizadeh, A.L. Lafleur, K.A. Smith, and A.F. Sarofim (1997), Measurement of oxygenated polycyclic aromatic hydrocarbons associated with a size-segregated urban aerosol, *Environ. Sci. Technol.*, *31*, 2064–2070.

Almasy, F. (1933), Polyatomic molecules. II. Structure and activation of benzaldehyde, *J. Chim. Phys.*, *30*, 634–645.

Alvarado, A., J. Arey, and R. Atkinson (1998), Kinetics of the gas-phase reactions of OH and NO_3 radicals and O_3 with the monoterpene reaction products pinonaldehyde, caronaldehyde, and sabinaketone, *J. Atmos. Chem.*, *31*, 281–297.

Alvarado, A., R. Atkinson, and J. Arey (1996), Kinetics of the gas-phase reactions of NO_3 radicals and O_3 with 3-methylfuran and the OH radical yield from the O_3 reaction, *Int. J. Chem. Kinet.*, *28*, 905–909.

Anderson, P.N., and R.A. Hites (1996a), OH radical reactions: the major removal pathway for polychlorinated biphenyls from the atmosphere, *Environ. Sci. Technol.*, *30*, 1756–1763.

Anderson, P.N., and R.A. Hites (1996b), System to measure relative rate constants of semivolatile organic compounds with hydroxyl radicals, *Environ. Sci. Technol.*, *30*, 301–306.

Andino, J.M., J.N. Smith, R.C. Flagan, W.A. Goddard, III, and J.H. Seinfeld (1996), Mechanism of atmospheric photooxidation of aromatics: a theoretical study, *J. Phys. Chem.*, *100*, 10967–10980.

Ando, M., K. Katagiri, K. Tamura, S. Yamamoto, and M. Matsumoto (1996), Indoor and outdoor air pollution in Tokyo and Beijing supercities, *Atmos. Environ.*, *30*, 695–702.

AQIRP (1995), Effects of gasoline T50, T90 and sulfur on exhaust emissions of current and future technology vehicles, Auto/Oil Air Quality Improvement Research Program, Technical Bulletin No. 18, available from the Coordinating Research Council, 3650 Mansell Road, Suite 140, Alpharetta, GA 30022–8246.

Arey, J., R. Atkinson, S.M. Aschmann, and D. Schuetzle (1990), Experimental investigation of the atmospheric chemistry of 2-methyl–1-nitronaphthalene and a comparison of predicted nitroarene concentrations with ambient air data, *Polycyclic Aromat. Compd.*, *1*, 33–50.

Arey, J., R. Atkinson, B. Zielinska, and P.A. McElroy (1989a), Diurnal concentrations of volatile polycyclic aromatic hydrocarbons and nitroarenes during a photochemical air pollution episode in Glendora, California, *Environ. Sci. Technol.*, *23*, 321–327.

Arey, J., B. Zielinska, R. Atkinson, and S.M. Aschmann (1989b), Nitroarene products from the gas-phase reactions of volatile polycyclic aromatic hydrocarbons with the OH radical and N_2O_5, *Int. J. Chem. Kinet.*, *21*, 775–799.

Arey, J., B. Zielinska, R. Atkinson, and A.M. Winer (1987), Polycyclic aromatic hydrocarbon and nitroarene concentrations in ambient air during a winter-time high NO_x episode in the Los Angeles basin, *Atmos. Environ.*, *21*, 1437–1444.

Arey, J., B. Zielinska, R. Atkinson, A.M. Winer, T. Ramdahl, and J.N. Pitts, Jr. (1986), The formation of nitro-PAH from the gas-phase reactions of fluoranthene and pyrene with the OH radical in the presence of NO_x, *Atmos. Environ.*, *20*, 2339–2345.

Aschmann, S.M., and R. Atkinson (1998), Kinetics of the gas-phase reactions of the OH radical with selected glycol ethers, glycols, and alcohols, *Int. J. Chem. Kinet.*, *30*, 533–540.

Ashton, L., G.V. Buxton, and C.R. Stuart (1995), Temperature dependence of the rate of reaction of OH with some aromatic compounds in aqueous solution, *J. Chem. Soc., Faraday Trans.*, *91*, 1631–1633.

Atkinson, R. (1986), Kinetics and mechanisms of the gas-phase reactions of the hydroxyl radical with organic compounds under atmospheric conditions, *Chem. Rev.*, *86*, 69–201.

Atkinson, R. (1987), A structure–activity relationship for the estimation of the rate constants of OH radicals with organic compounds, *Int. J. Chem. Kinet.*, *19*, 799–828.

Atkinson, R. (1989), Kinetics and mechanisms of the gas-phase reactions of the hydroxyl radical with organic compounds, *J. Phys. Chem. Ref. Data, Monograph 1*, 1–246.

Atkinson, R. (1991a), Atmospheric lifetimes of dibenzo-*p*-dioxins and dibenzofurans, *Sci. Total Environ.*, *104*, 17–33.

Atkinson, R. (1991b), Kinetics and mechanisms of the gas-phase reactions of the NO_3 radical with organic compounds, *J. Phys. Chem. Ref. Data*, *20*, 459–507.

Atkinson, R. (1994), Gas-phase tropospheric chemistry of organic compounds, *J. Phys. Chem. Ref. Data, Monograph No. 2*, 1–216.

Atkinson, R. (1997), Gas-phase tropospheric chemistry of volatile organic compounds: 1. Alkanes and alkenes, *J. Phys. Chem. Ref. Data*, *26*, 215–290.

Atkinson, R. (2000), Atmospheric Oxidation, in *Handbook of Property Estimation Methods for Chemicals: Environmental Health Sciences*, edited by D. Mackay and R.S. Boething, CRC Press, Boca Raton, FL.

Atkinson, R., and J. Arey (1994), Atmospheric chemistry of gas-phase polycyclic aromatic hydrocarbons: formation of atmospheric mutagens, *Environ. Health Perspect*, *102*, 117–124.

Atkinson, R., and S.M. Aschmann (1985), Rate constants for the gas-phase reaction of hydroxyl radicals with biphenyl and monochlorobiphenyls at 295 $\pm$ 1 K, *Environ. Sci. Technol.*, *19*, 462–464.

Atkinson, R., and S.M. Aschmann (1986), Kinetics of the reactions of naphthalene, 2-methylnaphthalene, and 2,3-dimethylnaphthalene with OH radicals and with O_3 at 295 $\pm$ 1 K, *Int. J. Chem. Kinet.*, *18*, 569–573.

Atkinson, R., and S.M. Aschmann (1987), Kinetics of the gas-phase reaction of alkyl-naphthalenes with O_3, N_2O_5, and OH radicals at 298 $\pm$ 2 K, *Atmos. Environ.*, *21*, 2323–2326.

Atkinson, R., and S.M. Aschmann (1988), Kinetics of the reactions of acenaphthene and acenaphthylene and structurally-related aromatic compounds with OH and NO_3 radicals, N_2O_5 and O_3 at 296 $\pm$ 2 K, *Int. J. Chem. Kinet.*, *20*, 513–539.

Atkinson, R., and S.M. Aschmann (1989), Rate constants for the gas-phase reactions of the OH radical with a series of aromatic hydrocarbons at 296 $\pm$ 2 K, *Int. J. Chem. Kinet.*, *21*, 355–365.

Atkinson, R., and S.M. Aschmann (1990), Rate constants for the gas-phase reactions of the OH radical with the cresols and dimethylphenols at 296 $\pm$ 2 K, *Int. J. Chem. Kinet.*, *22*, 59–67.

Atkinson, R., and S.M. Aschmann (1992), unpublished data referenced in Atkinson (1994).

Atkinson, R., and S.M. Aschmann (1994), Products of the gas-phase reactions of aromatic hydrocarbons: effect of NO_2 concentration, *Int. J. Chem. Kinet.*, *26*, 929–944.

Atkinson, R., and W.P.L. Carter (1984), Kinetics and mechanisms of the gas-phase reactions of ozone with organic compounds under atmospheric conditions, *Chem. Rev.*, *84*, 437–470.

Atkinson, R., and J.N. Pitts, Jr. (1974), Absolute rate constants for the reaction of $O(^3P)$ atoms with selected alkanes, alkenes, and aromatics as determined by a modulation technique, *J. Phys. Chem.*, *78*, 1780–1784.

Atkinson, R., and J.N. Pitts, Jr. (1975a), Absolute rate constants for the reaction of oxygen (^3P) atoms with methoxybenzene and *o*-cresol, *J. Phys. Chem.*, *79*, 541–542.

Atkinson, R., and J.N. Pitts, Jr. (1975b), Temperature dependence of the absolute rate constants for the reaction of $O(^3P)$ atoms with a series of aromatic hydrocarbons over the range 299–392 K, *J. Phys. Chem.*, *79*, 295–297.

Atkinson, R., and J.N. Pitts, Jr. (1979), Rate constants for the reactions of $O(^3P)$ atoms with benzene and toluene over the temperature range 299–440 K, *Chem. Phys. Lett.*, *63*, 485–489.

Atkinson, R., J. Arey, B. Zielinska, and S.M. Aschmann (1987a), Kinetics and products of the gas-phase reactions of OH radicals and N_2O_5 with naphthalene and biphenyl, *Environ. Sci. Technol.*, *21*, 1014–1022.

Atkinson, R., J. Arey, B. Zielinska, and S.M. Aschmann (1990a), Kinetics and nitro-products of the gas-phase OH and NO_3 radical-initiated reactions of naphthalene-d_8, fluoranthene-d_{10}, and pyrene, *Int. J. Chem. Kinet.*, *22*, 999–1014.

Atkinson, R., S.M. Aschmann, and J. Arey (1991), Formation of ring-retaining products from the OH radical-initiated reactions of *o*-, *m*-, and *p*-xylene, *Int. J. Chem. Kinet.*, *23*, 77–97.

Atkinson, R., S.M. Aschmann, and J. Arey (1992), Reactions of OH and NO_3 radicals with phenol, cresols, and 2-nitrophenol at 296 $\pm$ 2 K, *Environ. Sci. Technol.*, *26*, 1397–1403.

Atkinson, R., S.M. Aschmann, J. Arey, and W.P.L. Carter (1989a), Formation of ring-retaining products from the OH radical-initiated reactions of benzene and toluene, *Int. J. Chem. Kinet.*, *21*, 801–827.

Atkinson, R., S.M. Aschmann, J. Arey, B. Zielinska, and D. Schuetzle (1989b), Gas-phase atmospheric chemistry of 1- and 2-nitronaphthalene and 1,4-naphthoquinone, *Atmos. Environ.*, *23*, 2679–2690.

Atkinson, R., S.M. Aschmann, and W.P.L. Carter (1983a), Kinetics of the reactions of O_3 and OH radicals with furan and thophene at 298 ± 2 K, *Int. J. Chem. Kinet.*, *15*, 51–61.

Atkinson, R., S.M. Aschmann, and W.P.L. Carter (1983b), Rate constants for the gas-phase reactions of OH radicals with a series of bi- and tricycloalkanes at 299 ± 2 K: effects of ring strain, *Int. J. Chem. Kinet.*, *15*, 37–50.

Atkinson, R., S.M. Aschmann, D.R. Fitz, A.M. Winer, and J.N. Pitts, Jr. (1982), Rate constants for the gas-phase reactions of O_3 with selected organics at 296 K, *Int. J. Chem. Kinet.*, *14*, 13–18.

Atkinson, R., S.M. Aschmann, and J.N. Pitts, Jr. (1984a), Kinetics of the reactions of naphthalene and biphenyl with OH radicals and with O_3 at 294 ± 1 K, *Environ. Sci. Technol.*, *18*, 110–113.

Atkinson, R., S.M. Aschmann, E.C. Tuazon, J. Arey, and B. Zielinska (1989c), Formation of 3-methylfuran from the gas-phase reaction of OH radicals with isoprene and the rate constant for its reaction with OH radical, *Int. J. Chem. Kinet.*, *21*, 593–604.

Atkinson, R., S.M. Aschmann, and A.M. Winer (1987b), Kinetics of the reactions of NO_3 radicals with a series of aromatic compounds, *Environ. Sci. Technol.*, *21*, 1123–1126.

Atkinson, R., S.M. Aschmann, A.M. Winter, and W.P.L. Carter (1985), Rate constants for the gas-phase reactions of NO_3 radicals with furan, thiophene, and pyrrole at 295 ± 1 K and atmospheric pressure, *Environ. Sci. Technol.*, *19*, 87–90.

Atkinson, R., S.M. Aschmann, A.M. Winer, and J.N. Pitts, Jr. (1981), Rate constants for the gas-phase reactions of O_3 with a series of carbonyls at 296 K, *Int. J. Chem. Kinet.*, *13*, 1133–1142.

Atkinson, R., D.L. Baulch, R.A. Cox, R.F. Hampson, Jr., J.A. Kerr, M.J. Rossi, and J. Troe (1997a), Evaluated kinetic and photochemical data for atmospheric chemistry: Supplement VI, *J. Phys. Chem. Ref. Data*, *26*, 1329–1499.

Atkinson, R., D.L. Baulch, R.A. Cox, R.F. Hampson, Jr., J.A. Kerr, M.J. Rossi, and J. Troe (1997b), Evaluated kinetic, photochemical, and heterogeneous data for atmospheric chemistry. Supplement V, *J. Phys. Chem. Ref. Data*, *26*, 521–1011.

Atkinson, R., D.L. Baulch, R.A. Cox, R.F. Hampson, Jr., J.A. Kerr, M.J. Rossi, and J. Troe (1999), Evaluated kinetic and photochemical data for atmospheric chemistry: Supplement VII, Organic species, *J. Phys. Chem. Ref. Data*, *28*, 191–393.

Atkinson, R., W.P.L. Carter, K.R. Darnall, A.M. Winer, and J.N. Pitts, Jr. (1980), A smog chamber and modeling study of the gas phase NO_x–air photooxidation of toluene and the cresols, *Int. J. Chem. Kinet.*, *12*, 779–836.

Atkinson, R., W.P.L. Carter, C.N. Plum, A.M. Winer, and J.N. Pitts, Jr. (1984b), Kinetics of the gas-phase reactions of NO_3 radicals with a series of aromatics at 296 ± 2 K, *Int. J. Chem. Kinet.*, *16*, 887–898.

Atkinson, R., W.P.L. Carter, and A.M. Winer (1983c), Effects of pressure on product yields in the NO_x photooxidations of selected aromatic hydrocarbons, *J. Phys. Chem.*, *87*, 1605–1610.

Atkinson, R., K.R. Darnall, and J.N. Pitts, Jr. (1978), Rate constants for the reactions of OH radicals and ozone with cresols at 300 ± 1 K, *J. Phys. Chem.*, *82*, 2759–2761.

Atkinson, R., C.N. Plum, W.P.L. Carter, A.M. Winer, and J.N. Pitts, Jr. (1984c), Rate constants for the gas-phase reactions of nitrate radicals with a series of organics in air at 298 ± 1 K, *J. Phys. Chem.*, *88*, 1210–1215.

Atkinson, R., E.C. Tuazon, and J. Arey (1990b), Reactions of naphthalene in N_2O_5–NO_3–NO_2–air mixtures, *Int. J. Chem. Kinet.*, *22*, 1071–1082.

Atkinson, R., E.C. Tuazon, I. Bridier, and J. Arey (1994), Reactions of NO_3–naphthalene adducts with O_2 and NO_2, *Int. J. Chem. Kinet.*, *26*, 605–614.

Atkinson, R., E.C. Tuazon, T.J. Wallington, S.M. Aschmann, J. Arey, A.M. Winer, and J.N. Pitts, Jr. (1987c), Atmospheric chemistry of aniline, *N*,*N*-dimethylaniline, pyridine, 1,3,5-triazine, and nitrobenzene, *Environ. Sci. Technol.*, *21*, 64–72.

Atkinson, R., A.M. Winer, and J.N. Pitts, Jr. (1986), Estimation of night-time N_2O_5 concentrations from ambient NO_2 and NO_3 radical concentrations and the role of N_2O_5 in night-time chemistry, *Atmos. Environ.*, *20*, 331–339.

Bacher, C., G.S. Tyndall, and J.J. Orlando (2001), The atmospheric chemistry of glycoaldehyde, *J. Atmos. Chem.*, *39*, 171–189.

Badger, G.M. (1962), Mode of formation of carcinogenis in human environment, *National Cancer Institute Monograph, No. 9*, 1–16.

Badger, G.M., J.K. Donnelly, and T.M. Spottswood (1965), The formation of aromatic hydrocarbons at high temperatures, *Aust. J. Chem.*, *18*, 1249–1266.

Baek, S.O., M.E. Goldstone, R.W.W. Kirk, J.N. Lester, and R. Perry (1991), Phase distribution and particle size dependency of polycyclic aromatic hydrocarbons in the urban atmosphere, *Chemosphere*, *22*, 503–520.

Bandow, H., and N. Washida (1985a), Ring-cleavage reactions of aromatic hydrocarbons studied by FT-IR spectroscopy. II. Photooxidation of *o*-, *m*- and *p*-xylenes in the NO_x–air system, *Bull. Chem. Soc. Jpn.*, *58*, 2541–2548.

Bandow, H., and N. Washida (1985b), Ring-cleavage reactions of aromatic hydrocarbons studied by FT-IR spectroscopy. III. Photooxidation of 1,2,3-, 1,2,4-, and 1–3,5-trimethylbenzenes in the NO_x–air system, *Bull. Chem. Soc. Jpn.*, *58*, 2549–2555.

Bandow, H., N. Washida, and H. Akimoto (1985), Ring-cleavage reactions of aromatic hydrocarbons studied by FT-IR spectroscopy. I. Photooxidation of toluene and benzene in the NO_x–air system, *Bull. Chem. Soc. Jpn.*, *58*, 2531–2540.

Barbas, J.T., M.E. Sigman, and R. Dabestani (1996), Photochemical oxidation of phenanthrene sorbed on silica gel, *Environ. Sci. Technol.*, *30*, 1776–1780.

Barnes, I., V. Bastian, K.H. Becker, E.H. Fink, and F. Zabel (1982), Reactivity studies of organic substances towards hydroxyl radicals under atmospheric conditions, *Atmos. Environ.*, *16*, 545–550.

Barnes, I., K.H. Becker, A. Bierbach, and E. Wiesen (1991), Mechanism of the OH radical initiated oxidation of aromatics in the absence of NO_x, in *Laboratory Studies on Atmospheric Chemistry*, edited by R.A.Cox, pp. 183–189, Air Pollution Research Report 42, EUR 14756, European Commission, Brussels.

Baulch, D.L., I.M. Campbell, and S.M. Saunders (1988), The rate constants for the reaction of the hydroxyl radical with benzene, *J. Chem. Soc., Faraday Trans. 2*, *84*, 377–384.

Baulch, D.L., I.M. Campbell, S.M. Saunders, and P.K.K. Louie (1989), Rate constants for the reactions of the hydroxyl radical with indane, indene, and styrene, *J. Chem. Soc., Faraday Trans. 2*, *85*, 1819–1826.

Becker, K.H., and F. Graedler (1999), Photochemie ausgewählter Dicarbonylverbindungen unter atmosphärischen Bedingungen, Bericht Nr. 49, Physikalische Chemie (Fachbereich 9), Bergische Universität, GH-Wuppertal, December.

Becker, K.H., I. Barnes, A. Bierbach, K.J. Brockmann, F. Kirchner, B. Klotz, H.G. Libuda, A. Mayer-Figge, S. Mönninghoff, L. Ruppert, W. Thomas, E. Wiesen, K. Wirtz, and F. Zabel (1997), OH initiated oxidiation of VOC under variable NO_x conditions, in *Chemical Processes in Atmospheric Oxidation*, edited by G. Le Bras, pp. 79–90, Springer-Verlag, Berlin, Germany.

Becker, K.H., I. Barnes, A. Bierbach, F. Kirchner, L. Ruppert, E. Wiesen, and F. Zabel (1992), OH-initiated oxidation of NMVOC under variable NO_x conditions, in *EUROTRAC Annual Report 1991*, *Part 8, LACTOS*, edited by K.H. Becker, pp. 71–80, EUROTRAC International Scientific Secretariat, Garmisch-Partenkirchen, Germany.

Becker, K.H., I. Barnes, A. Bierbach, F. Kirchner, W. Thomas, E. Wiesen, and F. Zabel (1993), OH-initiated oxidation of NMVOC under variable NO_x conditions, in *EUROTRAC Annual Report 1992, Part 8, LACTOS*, edited by K.H. Becker, pp. 87–98, EUROTRAC International Scientific Secretariat, Garmisch-Partenkirchen, Germany.

Becker, K.H., and Th. Klein (1987), OH-initiated oxidation of *p*-xylene under atmospheric conditions, in *Physico-Chemical Behaviour of Atmospheric Pollutants*, edited by G.A. Restelli, pp. 320–326, D. Riedel, Dordrecht, The Netherlands.

Beeby, A., D.B.H. Mohammed, and J.R. Sodeau (1987), Photochemistry and photophysics of glycolaldehyde in solution, *J. Am. Chem. Soc.*, *109*, 857–861.

Behymer, T.D., and R.A. Hites (1985), Photolysis of polycyclic aromatic hydrocarbons adsorbed on simulated atmospheric particulates, *Environ. Sci. Technol.*, *19*, 1004–1006.

Behymer, T.D., and R.A. Hites (1988), Photolysis of polycyclic aromatic hydrocarbons adsorbed on fly ash, *Environ. Sci. Technol.*, *22*, 1311–1319.

Benner, B.A., G.E. Gorden, and S.A. Wise (1989), Mobile sources of atmospheric polycyclic aromatic hydrocarbons: a road way tunnel study, *Environ. Sci. Eng.*, *23*, 1269–1278.

Bennett, R.L., K.T. Knapp, P.W. Jones, J.E. Wilkerson, and P.E. Strup (1979), A measurement of polynuclear aromatic hydrocarbons and other hazardous organics in stack gases, in *Polynuclear Aromatic Hydrocarbons, 3rd International Symposium on Chemistry and Biology, Carcinogenesis and Mutagenesis*, editor, P.W. Jones, pp. 419–428, Ann Arbor Sciences, Ann Arbor, Michigan.

Benson, S.W. (1976), *Thermochemical Kinetics*, Wiley, New York.

Berger, M., I.L. Goldblatt, and C. Steel (1973), Photochemistry of benzaldehyde, *J. Am. Chem. Soc.*, *95*, 1717–1725.

Berho, F., and R. Lesclaux (1997), The phenoxy radical: UV spectrum and kinetics of gas-phase reactions with itself and with oxygen, *Chem. Phys. Lett.*, *21*, 289–296.

Berho, F., F. Caralp, M.-T. Rayez, R. Lesclaux, and E. Ratajczak (1998), Kinetics and thermochemistry of the reversible combination reaction of the phenoxy radical with NO, *J. Phys. Chem. A*, *102*, 1–8.

Berho, F., M.-T. Rayez, and R. Lesclaux (1999), UV absorption spectrum and self-reaction kinetics of the cyclohexadienyl radical, and stability of a series of cyclohexadienyl-type radicals, *J. Phys. Chem. A*, *103*, 5501–5509.

Berndt, T., O. Böge, and H. Herrmann (1999), On the formation of benzene oxide/oxepin in the gas-phase reaction of OH radicals with benzene, *Chem. Phys. Lett.*, *314*, 435–442.

Berndt, T., O. Böge, and W. Rolle (1997), Products of the gas-phase reactions of NO_3 radicals with furan and tetramethylfuran, *Environ. Sci. Technol.*, *31*, 1157–1162.

Besemer, A.C. (1982), Formation of chemical compounds from irradiated mixtures of aromatic hydrocarbons and nitrogen oxides, *Atmos. Environ.*, *16*, 1599–1602.

Bidleman, T.F., W.N. Billings, and W.T. Foreman (1986), Vapor-particle partitioning of semivolatile organic compounds: estimates from field collections, *Environ. Sci. Technol.*, *20*, 1038–1043.

Bierbach, A., I. Barnes, and K.H. Becker (1992), Rate coefficients for the gas-phase reactions of hydroxyl radicals with furan, 2-methylfuran, 2-ethylfuran and 2,5-dimethylfuran at 300 ± 2 K, *Atmos. Environ.*, *26A*, 813–817.

Bierbach, A., I. Barnes, and K.H. Becker (1995), Product and kinetic study of the OH-initiated gas-phase oxidation of furan, 2-methylfuran and furanaldehydes at ~300 K, *Atmos. Environ.*, *29*, 2651–2660.

Bierbach, A., I. Barnes, K.H. Becker, B. Klotz, and E. Wiesen (1993), OH-radical initiated degradation of aromatic hydrocarbons, in *Proceedings of the 6th European Symposium on Physico-Chemical Behaviour of Atmospheric Pollutants*, edited by G. Angeletti and G. Restelli, pp. 129–136, Report EUR15609/IEN, European Commission, Brussels.

Bierbach, A., I. Barnes, K.H. Becker, and E. Wiesen (1994), Atmospheric chemistry of unsaturated carbonyls: butenedial, 4-oxy–2-pentenal, 3-hexene–2,5-dione, maleic anhydride, 3*H*-furan–2-one, and 5-methyl–3*H*-furan–2-one, *Environ. Sci. Technol.*, *28*, 715–729.

Biermann, H.W., H. MacLeod, R. Atkinson, A.M. Winer, and J.N. Pitts, Jr. (1985), Kinetics of the gas-phase reactions of the hydroxyl radical with naphthalene, phenanthrene, and anthracene, *Environ. Sci. Technol.*, *19*, 244–248.

Bignozzi, C.A., A. Maldotti, C. Chiorboli, C. Bartocci, and V. Carassiti (1981), Kinetics and mechanism of reactions between aromatic olefins and hydroxyl radicals, *Int. J. Chem. Kinet.*, *13*, 1235–1242.

Birla, P., and R.M. Kamens (1994), Effects of combustion temperature on the atmospheric stability of polybrominated dibenzo-*p*-dioxins and dibenzofurans, *Environ. Sci. Technol.*, *28*, 1437–1443.

Biron, M., and P. Longin (1985), Triplet dynamics of gas-phase benzaldehyde after excitation of the forbidden $T_1(n, \pi^*) \rightarrow S_0$ transition, *Chem. Phys. Lett.*, *116*, 250–253.

Bjergbakke, E.A., A. Sillesen, and P. Pagsberg (1996), UV spectrum and kinetics of hydroxycyclohexadienyl radicals, *J. Phys. Chem.*, *100*, 5729–5736.

Björseth, A. (1983), *Handbook of Polycyclic Aromatic Hydrocarbons*, Marcel Deckker, New York.

Björseth, A., L. Gulbrand, and A. Linskog (1979), Long-range transport of polycyclic aromatic hydrocarbons, *Atmos. Environ.*, *13*, 45–53.

Bolzacchini, E., St. Caserini, and M. Maugeri (1998), Microinquinanti organici in atmosfera, in *Il benzene e altri comoosti aromatici: monitorraggio e rischi per l'uomo*, edited by B. Rindone, P. Beltrame, and A.L. De Cesaris, pp. 28–74, Fondazione Lombardia per l'Ambiente, Milano.

Bolzacchini, E., S. Meinardi, M. Orlandi, B. Rindone, J. Hjorth, and G. Restelli (1999), Nighttime tropospheric chemistry: kinetics and product studies in the reaction of 4-alkyl- and 4 alkoxytoluenes with NO_3 in gas phase, *Environ. Sci. Technol.*, *33*, 461–468.

Bonanno, R.A., P. Kim, J.-H. Lee, and R.B. Timmons (1972), Kinetics of the reaction of $O(^3P)$ atoms with benzene, *J. Chem. Phys.*, *57*, 1377–1380.

Boocock, G., and R.J. Cvetanovic (1961), Reaction of oxygen atoms with benzene, *Can. J. Chem.*, *39*, 2436–2443.

Borders, R.A., and J.W. Birks (1982), High-precision measurement of activation energies over small temperature intervals: curvature in the Arrhenius plot for the reaction $NO + O_3 \rightarrow NO_2 + O_2$, *J. Phys. Chem.*, *86*, 3295–3302.

Bordwell, F.G., and J.P. Cheng (1991), Substituent effects on the stabilities of phenoxy radicals and acidities of phenoxy radical cations, *J. Am. Chem. Soc.*, *113*, 1736–1743.

Bourmada, N., M. Carlier, J.-F. Pauwels, and P. Devolder (1988a), Mesure de las constante de vitesse de la réaction du radical OH avec le toluène par las technique du tube à écoulement rapide et décharge microonde, *J. Chim. Phys.*, *85*, 881–887.

Bourmada, N., P. Devolder, and L.-R. Sochet (1988b), Rate constant for the termolecular reaction of OH + toluene + helium in the fall-off range below 10 torr, *Chem. Phys. Lett.*, *149*, 339–342.

Bowen, E.J., H. Hartley, W.D. Scott, and H.G. Watts (1924), Rates of photochemical change in solids, *J. Chem. Soc.*, *125*, 1218–1221.

Brorström-Lundén, E, and A.G. Lövblad (1991), Deposition of soot related hydrocarbons during long-range transport of pollution to Sweden, *Atmos. Environ.*, *25A*, 2251–2257.

Brorström-Lundén, E., A. Lindskog, and J. Mowrer (1994), Concentrations and fluxes of organic compounds in the atmosphere of the Swedish west coast, *Atmos. Environ.*, *28*, 3605–3615.

Brown, H.C., and Y. Okamoto (1958), Electrophilic substituent constants, *J. Am. Chem. Soc.*, *80*, 4979–4987.

Brown, J.R., R.A. Field, M.E. Goldstone, J.N. Lester, and R. Perry (1996), Polycyclic aromatic hydrocarbons in central London air during 1991–1992, *Sci. Total Environ.*, *177*, 73–84.

Brubaker, W.W., Jr., and R.A. Hites (1998a), Gas-phase oxidation products of biphenyl and polychlorinated biphenyls, *Environ. Sci. Technol.*, *32*, 3913–3918.

Brubaker, W.W., Jr., and R.A. Hites (1998b), OH reaction kinetics of polycyclic aromatic hydrocarbons and polychlorinated dibenzo-*p*-dioxins and dibenzofurans, *J. Phys. Chem. A*, *102*, 915–921.

Brunauer, S., P.H. Emmett, and E.J. Teller (1938), Absorption of gases in multimolecular layers, *J. Am. Chem. Soc.*, *60*, 309–319.

Buchta, C., H. Ferichs, D.-V. Stucken, M. Tappe, and H.G. Wagner (1993), Kinetic and primary product studies of the reaction of $O(^3P)$ with styrene in the gas phase, *Ber. Bunsen-Ges. Phys. Chem.*, *97*, 658–662.

Bufalini, J.J., and A.P. Altshuller (1965), Kinetics of vapor-phase hydrocarbon-ozone reactions, *Can. J. Chem.*, *43*, 2243–2250.

Bunce, N.J., L. Liu , J. Zhu, and D.A. Lane (1997), Reaction of naphthalene and its derivatives with hydroxyl radicals in the gas phase, *Environ. Sci. Technol.*, *31*, 2252–2259.

Cadle, S.H., and P.J. Groblicki (1982), *An Evaluation of Methods for the Determination of Organic and Elemental Carbon in Particulate Samples*, pp. 89–109, Plenum Press, New York.

Cadle, S.H., P.A. Mulawa, E.C. Hunsanger, K. Nelson, R.A. Ragazzi, R. Barrett, G.L. Gallagher, D.R. Lawson, K.T. Knapp, and R. Snow (1999), Composition of light-duty motor vehicle exhaust particulate matter in the Denver, Colorado area, *Environ. Sci. Technol.*, *33*, 2328–2339.

Calvert, J.G., and S. Madronich (1987), Theoretical study of the initial products of the atmospheric oxidation of hydrocarbons, *J. Geophys. Res.*, *92*, 2211–2220.

Calvert, J.G., R. Atkinson, J.A. Kerr, S. Madronich, G.K. Moortgat, T.J. Wallington, and G. Yarwood (2000), *The Mechanisms of Atmospheric Oxidation of the Alkenes*, Oxford University Press, New York.

Cantrell, C.A., J.A. Davidson, A.H. McDaniel, R.E. Shetter, and J.G. Calvert (1990), Temperature-dependent formaldehyde cross sections in the near-ultraviolet spectral region, *J. Phys. Chem.*, *94*, 3902–3908.

Cantrell, C.A., J.A. Lind, R.E. Shetter, J.G. Calvert, P.D. Goldan, W. Custer, F.C. Fehsenfeld, S. Montzka, D.D. Parrish, E.J. Williams, M.P. Buhr, H.H. Westberg, G. Allwine, and R. Martin (1992), Peroxy radicals in the ROSE experiment: measurement and theory, *J. Geophys. Res.*, *97*, 20671–20686.

Cantrell, C.A., R.E. Shetter, J.G. Calvert, D.D. Parrish, F.C. Fehsenfeld, P.D. Goldan, W. Kuster, E.J. Williams, H.H. Westberg, G. Allwine, and R. Martin (1993), Peroxy radicals as measured in ROSE and estimated from photostationary state deviations, *J. Geophys. Res.*, *98*, 18355–18366.

Cantrell, C.A., R.A. Shetter, T.M. Gilpin, J.G. Calvert, F.L. Eisele, and D.I. Tanner (1996), Peroxy radical concentrations measured and calculated from trace gas measurements in Mauna Loa Photochemical Experiment–2, *J. Geophys. Res.*, *101*, 14653–14664.

Caralp, F., V. Foucher, R. Lesclaux, T.J. Wallington, and M.D. Hurley (1999), Atmospheric chemistry of benzaldehyde: UV absorption spectrum and reaction kinetics and mechanisms of the $C_6H_5C(O)O_2$ radical, *Phys. Chem. Chem. Phys.*, *1*, 3509–3517.

Caricchia, A.M., S. Chaiavarini, and P. Massimo (1999), Polycyclic aromatic hydrocarbons in the urban atmosphere particulate matter in the city Naples (Italy), *Atmos. Environ.*, *33*, 3731–3738.

Carter, W.P.L. (1990), A detailed mechanism for the gas-phase atmospheric reactions of organic compounds, *Atmos. Environ.*, *24A*, 481–518.

Carter, W.P.L. (1994), Development of reactivity scales for volatile organic compounds, *J. Air Waste Manage. Assoc.*, *44*, 881–899.

Carter, W.P.L. (1999), Documentation of the SAPRC–99 chemical mechanism for VOC reactivity assessment, Report to California Air Resources Board, Contracts 92–329 and 95–308, September 13.

Carter, W.P.L., and R. Atkinson (1987), An experimental study of incremental hydrocarbon reactivity, *Environ. Sci. Technol*, *21*, 670–679.

Carter, W.P.L., A.M. Winer, and J.N. Pitts, Jr. (1981), Major atmospheric sink for phenol and cresols. Reaction with the nitrate radical, *Environ. Sci. Technol.*, *15*, 829–831.

Castro, L.M., C.A. Pio, R.M. Harrison, and D.J.T. Smith (1999), Carbonaceous aerosol in urban and rural European atmospheres: estimation of secondary organic carbon concentrations, *Atmos. Environ.*, *33*, 2771–2781.

Castro, T., L.G. Ruiz-Suárez, J.C. Ruiz-Suárez, M.J. Molina, and M. Montero (1997), Sensitivity analysis of a UV radiation transfer model and experimental photolyis rates of NO_2 in the atmosphere of Mexico City, *Atmos. Environ.*, *31*, 609–620.

Chebbi, A., and P. Carlier (1996), Carboxylic acids in the troposphere, occurrence, sources, and sinks: a review, *Atmos. Environ.*, *30*, 4233–4249.

Chien, C., M.J. Charles, K.G. Sexton, and H.E. Jeffries (1998), Analysis of airborne carboxylic acids and phenols as their pentafluorobenzyl derivatives: gas chromatography/ion trap mass spectrometry with a novel chemical ionization reagent, PFBOH, *Environ. Sci. Technol.*, *32*, 299–309.

Chiodini, G., B. Rindone, F. Cariati, S. Polesello, G. Restelli, and J. Hjorth (1993), Comparison between the gas-phase and the solution reaction of the nitrate radical and methylarenes, *Environ. Sci. Technol.*, *27*, 1659–1664.

Chiorboli, C., C.A. Bignozzi, A. Maldotti, P.F. Giardini, A. Rossi, and V. Carassiti (1983), Rate constants for the gas-phase reactions of OH radicals with β-dimethylstyrene and acetone. Mechanism of β-dimethylstyrene NO_x–air photooxidation, *Int. J. Chem. Kinet.*, *15*, 579–586.

Christensen, L.K., J. Sehested, O.J. Nielsen, T.J. Wallington, A. Guschin, and M.D. Hurley (1998), Kinetics and mechanisms of the reaction of 2,3-butadiene with F and Cl atoms, U.V. absorption spectra of $CH_3C(O)C(O)CH_2(\cdot)$ and $CH_3C(O)C(O)CH_2O_2(\cdot)$ and atmospheric fate of $CH_3C(O)C(O)CH_2O(\cdot)$ radicals, *J. Phys. Chem. A*, *102*, 8913–8923.

Chu, W.P.L., and E.S. Macias (1981), Primary or secondary?, in *Atmospheric Aerosol: Source/Air Quality Relationships*, edited by E.S. Macias, and P.K. Hopke, pp. 251–268, American Chemical Society, Washington, DC.

Ciajolo, A., A. D'Anna, R. Barbella, and A. Tregrossi (1994), The formation of aromatic carbon in sooting ethylene flames, *25th Symposium (International) on Combustion*, pp. 679–685, The Combustion Institute, Pittsburgh, PA.

Ciccioli, P., E. Brancaleoni, and M. Frattoni (1999), Reactive hydrocarbons in the atmosphere at urban and regional scales, in *Reactive Hydrocarbons in the Atmosphere*, edited by C.N. Hewitt, pp. 159–207, Academic Press, San Diego.

Ciccioli, P., A. Cencinato, R. Cabella, and E. Brancaleoni (1993), The contribution of gas-phase reactions to the nitroarene fraction of molecular weight 247 present in carbon particles sampled in an urban area of northern Italy, *Atmos. Environ.*, *27A*, 1261–1270.

Cimberle, M.R., P. Bottino, and F. Valerio (1983), Decomposition of benzo(a)pyrene deposited on glass fiber filters and exposed to sunlight, *Chemosphere*, *12*, 317–324.

Clar, E. (1964), *Polycyclic Aromatic Hydrocarbons*, Academic Press, New York.

Colussi, A.J., D.L. Singleton, R.S. Irwin, and R.J. Cvetanovic (1975), Absolute rates of oxygen (O^3P) atoms reactions with benzene and toluene, *J. Phys. Chem.*, *79*, 1900–1903.

Coomber, J.W., and J.N. Pitts, Jr. (1969), Molecular structure and photochemical reactivity. XII. The vapor-phase chemistry of acrolein at 3130 Å, *J. Am. Chem. Soc.*, *91*, 547–550.

Cooper, P.L., and J.P.D. Abbatt (1996), Heterogeneous interactions of OH and HO_2 radicals with surfaces characteristic of atmospheric particulate matter, *J. Phys. Chem.*, *100*, 2249–2254.

Cope, W.C., and D.R. Kalkwarf (1987), Photooxidation of selected polycyclic aromatic hydrocarbons and pyrenequinones coated on glass surfaces, *Environ. Sci. Technol.*, *21*, 643–648.

Corchnoy, S.B., and R. Atkinson (1990), Kinetics of the gas-phase reactions of OH and NO_3 radicals with 2-carene, 1,8-cineole, *p*-cymene, and terpinolene, *Environ. Sci. Technol.*, *24*, 1497–1502.

Coutant, R.W., L. Brown, J.C. Chaung, R.M. Riggin, and R.G. Lewis (1988), Phase distribution and artifact formation in ambient air sampling for polynuclear aromatic hydrocarbons, *Atmos. Environ.*, *22*, 403–409.

Cox, R.A., R.G. Derwent, and M.R. Williams (1980), Atmospheric photooxidation reactions, rates, reactivity, and mechanism for reaction of organic compounds with hydroxyl radicals, *Environ. Sci. Technol.*, *14*, 57–61.

Crawford, M.A., T.J. Wallington, J.J. Szente, and M.M. Maricq (1999), Kinetics and mechanism of the acetylperoxy + HO_2 reaction, *J. Phys. Chem. A*, *103*, 365–378.

Criegee, R. (1957), The course of ozonization of unsaturated compounds, *Record Chem. Progress*, *18*, 111–120.

Cruz, C.N., and S.N. Pandis (1997), A study of the ability of pure secondary organic aerosol to act as cloud condensation nuclei, *Atmos. Environ.*, *31*, 2205–2214.

Dagaut, P., T.J. Wallington, R. Liu, and M.J. Kurylo (1988), A kinetics investigation of the gas-phase reactions of OH radicals with cyclic ketones and diones: mechanistic insights, *J. Phys. Chem.*, *92*, 4375–4377.

Daisey, J.M., C.G. Lewandowski, and M. Zorz (1982), A photoreactor for investigations of the degradation of particle-bound polycyclic aromatic hydrocarbons under simulated atmospheric conditions, *Environ. Sci. Technol.*, *16*, 857–861.

D'Anna, A., A. D'Alessio, and P. Minutolo (1994), Spectroscopic and chemical characterization of soot inception processes in premixed laminar flames at atmospheric pressure, in *Soot Formation in Combustion: Mechanisms and Models*, edited by H. Bockhorn, pp. 83–103, Springer-Verlag, Heidelberg.

Darnall, K.R., R. Atkinson, and J.N. Pitts, Jr. (1979), Observation of biacetyl from the reaction of OH radicals with *o*-xylene. Evidence for ring cleavage, *J. Phys. Chem.*, *83*, 1943–1946.

Davis, D.D., W. Bollinger, and S. Fischer (1975), A kinetics study of the reaction of the OH free radical with aromatic compounds. 1. Absolute rate constants for reaction with benzene and toluene at 300° K, *J. Phys. Chem.*, *79*, 293–294.

DeMare, G.R., M.C. Fontaine, and P. Goldfinger (1968), Heterogeneous photosensitization, *J. Org. Chem.*, *32*, 2528–2529.

Demerjian, K.L., J.A. Kerr, and J.G. Calvert (1972), The relative importance of the various intermediate species in olefin removal reactions in photochemical smog, *Environ. Lett.*, *3*, 137–149.

Demerjian, K.L., J.A. Kerr, and J.G. Calvert (1974), The mechanism of photochemical smog formation, *Adv. Environ. Sci. Technol.*, *4*, 1–262.

Demerjian, K.L., K.L. Schere, and J.T. Peterson (1980), Theoretical estimates of actinic (spherically integrated) flux and photolytic rate constants of atmospheric species in the lower troposphere, *Adv. Environ. Sci. Technol.*, *10*, 369–459.

DeMore, W.B., S.P. Sander, D.M. Golden, R.F. Hampson, M.J. Kurylo, C.J. Howard, A.R. Ravishankara, C.E. Kolb, and M.J. Molina (1997), Chemical kinetics and photochemical data for use in stratospheric modeling, NASA Evaluation No. 13, JPL Publication 97–4, Jet Propulsion Laboratory, Pasadena, CA.

Derwent, R.G., and M.E. Jenkin (1991), Hydrocarbons and the long-range transport of ozone and PAN across Europe, *Atmos. Environ.*, *25A*, 1661–1678.

Derwent, R.G., T.J. Davies, M. Delaney, G.J. Dollard, R.A. Field, P. Dumitrean, P.D. Nason, B.M.R. Jones, and S.A. Pepler (2000), Analysis and interpretation of the continuous hourly monitoring data for 26 C_2–C_8 hydrocarbons at 12 United Kingdom sites during 1996, *Atmos. Environ.*, *34*, 297–312.

Derwent, R.G., M.E. Jenkin, and S.M. Saunders (1996), Photochemical ozone creation potentials for a large number of reactive hydrocarbons under European conditions, *Atmos. Environ.*, *30*, 181–199.

Derwent, R.G., M.E. Jenkin, S.M. Saunders, and M.J. Pilling (1998), Photochemical ozone creation potential for organic compounds in Northwest Europe calculated with a master chemical mechanism, *Atmos. Environ.*, *32*, 2429–2441.

de Valk, J.P.J.M.M., and J.C.H. van der Hage (1994), A model for cloud chemistry processes suitable for use in long range transport models: a sensitivity study, *Atmos. Environ.*, *28*, 1653–1663.

Dewulf, J.P., H.R. van Langehove, and L.F. van der Auwera (1998), Air/water exchange dynamics of 13 volatile chlorinated C_1- and C_2-hydrocarbons and monocyclic aromatic hydrocarbons in the southern North Sea and the Scheldt estuary, *Environ. Sci. Technol.*, *32*, 903–911.

Dickerson, R.R., S. Kondragunta, G. Stenchikov, K.L. Civerolo, B.G. Doddridge, and B.N. Holben (1997), The impact of aerosols on solar UV radiation and photochemical smog, *Science*, *278*, 827–830.

Dietzel, B., and S.H. Kägler (1994), DGMK-Projekt 502, *Zusammensetzung von Ottokrafftstoffen aus deutschen Raffinerien*, edited by H. Jungen, DGMK-Fachbereich Verarbeitung und Anwendung, Hamburg.

Dobbins, R.A., H. Subramaniasiavam (1994), Soot precursor particles in flames, in *Soot Formation in Combustion: Mechanisms and Models*, edited by H. Bockhorn, pp. 290–301, Springer-Verlag, Heidelberg.

Donahue, M.N., J.H. Kroll, J.G. Anderson, and K.L. Demerjian (1998), Direct observations of OH production from the ozonolysis of olefins, *Geophys. Res. Lett.*, *25*, 59–62.

Donald, M., W.Y. Shiu, and K.C. Ma, (1992), *Illustrated Handbook of Physical Chemical Properties and Environmental Fate for Organic Chemicals*, Volume II, *Polynuclear Aromatic Hydrocarbons, Polychlorinated Dioxins, and Dibenzofurans*, Lewis Publishers, Chelsea, MI.

Dorfman, L.M., I.A. Taub, and R.E. Bühler (1962), Pulse radiolysis studies. I. Transient spectra and reaction-rate constants in irradiated aqueous solutions of benzene, *J. Chem. Phys.*, *36*, 3051–3061.

Doyle, G.J., A.C. Lloyd, K.R. Darnall, A.M. Winer, and J.N. Pitts, Jr. (1975), Gas phase kinetic study of relative rates of reaction of selected aromatic compounds with hydroxyl radicals in an environmental chamber, *Environ. Sci. Technol.*, *9*, 237–241.

Dumdei, B.E., and R.J. O'Brien (1984), Toluene degradation products in simulated atmospheric conditions, *Nature*, *311*, 248–250.

Dumdei, B.E., D.V. Kenny, P.B. Shepson, T.E. Kleindienst, C.M. Nero, L.T. Cupitt, and L.D. Claxton (1988), MS/MS analysis of the products of toluene photooxidation and measurement of their mutagenic activity, *Environ. Sci. Technol.*, *22*, 1493–1498.

Dunstan, T.D.J., R.F. Mauldin, Z. Jinxin, A.D. Hipps, E.L. Wehry, and G. Mamantov (1989), Adsorption of photodegradation of pyrene on magnetic, carbonaceous, and mineral subfractions of coal stack ash, *Environ. Sci. Technol.*, *23*, 303–308.

Eatough, D.J., B. Sedar, L. Lewis, L.D. Hansen, E.A. Lewis, and R. J. Farber (1989), Determination of semivolatile organic compounds in particles in the Grand Canyon area, *Aerosol Sci. Technol.*, *10*, 438–449.

Edney, E.O., T.E. Kleindienst, and E.W. Corse (1986), Room temperature rate constants for the reaction of OH with selected chlorinated and oxygenated hydrocarbons, *Int. J. Chem. Kinet.*, *18*, 1355–1371.

Ehhalt, D.H. (1999), Gas phase chemistry of the troposphere, in *Global Aspects of Atmospheric Chemistry*, edited by R. Zellner, pp. 21–109, Steinkopff-Verlag, Darmstadt, and Springer-Verlag, New York.

Eisenberg, W.C., K. Taylor, and R.W. Murray (1984), Production of singlet delta oxygen by atmospheric pollutants, *Carcinogenesis (London)*, *5*, 1095–1096.

Elterman, L. (1968), UV, visible and IR attenuation for altitudes to 50 km, AFCRL–68-0153, No. 285, Air Force Cambridge Research Laboratories, Bedford, MA.

EPA (1998), National air quality and emissions trends report, 1997. Office of Air Quality Planning and Standards, EPA 454/R–98-016, December.

Erlick, C., and J.E. Frederick (1998), Effects of aerosols on the wavelength dependence of atmospheric transmission in the ultraviolet and visible. 2. Continental and urban aerosol in clear skies, *J. Geophys. Res.*, *103*, 23275–23285.

Erlick, C., J.E. Frederick, V.K. Saxena, and B.N. Wenny (1998), Atmospheric transmission in the ultraviolet and visible: aerosols in cloudy atmospheres, *J. Geophys. Res.*, *103*, 31541–31556.

Etzkorn, T., B. Klotz, S. Sørensen, S. Patroescu, I. Barnes, K.H. Becker, and U. Platt (1999), Gas-phase absorption cross sections of 24 monocyclic aromatic hydrocarbons in the UV and IR spectral ranges, *Atmos. Environ.*, *33*, 525–540.

European Commission (1999), *Energy in Europe*, Directorate General for Energy, Brussels.

Fahr, A., W. Braun, and A.H. Laufer (1993), Photolysis of methyl vinyl ketone at 193.3 nm: quantum yield determinations of methyl and vinyl radicals, *J. Phys. Chem.*, *97*, 1502–1506.

Falk, H.L., I. Marloul, and P. Kotin (1956), Aromatic hydrocarbons (IV): their fate following emission into the atmosphere and experimental exposure, *A.M.A. Arch. Ind. Health*, *13*, 13–17.

Fan, Z., D. Chen, P. Birla, and R.M. Kamens (1995), Modeling of nitro-polycyclic aromatic hydrocarbon formation and decay in the atmosphere, *Atmos. Environ.*, *29*, 1171–1181.

Fan, Z., R.M. Kamens, J. Hu, J. Zhang, and S.R. McDow (1996a), Photostability of nitro-polycyclic aromatic hydrocarbons on combustion particles in sunlight, *Environ. Sci. Technol.*, *30*, 1358–1364.

Fan, Z., R.M. Kamens, J. Zhang, and J. Hu (1996b), Ozone–nitrogen dioxide–NPAH heterogeneous soot particle reactions and modeling NPAH in the atmosphere, *Environ. Sci. Technol.*, *30*, 2821–2827.

Faoro, R.B., and J.A. Manning (1981), Trends in benzo(a)pyrene 1996–97, *J. Air Pollut. Control Assoc.*, *42*, 162–164.

Fataidi, A. (1967), Effects of temperature and of ultraviolet radiation on pyrene adsorbed on garden soil, *Environ. Sci. Technol.*, *1*, 570–572.

Fenter, F.F., B. Noziére, F. Caralp, and R. Lesclaux (1994), Study of the kinetics and equilibrium of the benzyl-radical association reaction with molecular oxygen, *Int. J. Chem. Kinet.*, *26*, 171–189.

Fielberg, A., R.M. Kamens, M.R. Strommen, and T. Nielsen (1999), Modeling the formation, decay and partitioning of 1- and 2-nitronaphthalene in the atmosphere, *Atmos. Environ.*, *33*, 1231–1243.

Finizio, A., D. Mackay, T.F. Bidleman, and T. Harner (1997), Octanol–air partitioning coefficient as a predictor of partitioning of semi-volatile organic chemicals to aerosols, *Atmos. Environ.*, *31*, 2289–2296.

Finlayson-Pitts, B.J., and J.N. Pitts, Jr. (1986), *Atmospheric Chemistry*, John Wiley & Sons, New York.

Finlayson-Pitts, B.J., and J.N. Pitts, Jr. (1999), *Chemistry of the Upper and Lower Atmosphere*, Academic Press, San Diego.

Finlayson-Pitts, B.J., S.K. Hernandez, and H.N. Berko (1993), A new dark source of the gaseous hydroxyl radical for relative rate measurements, *J. Phys. Chem.*, *97*, 1172–1177.

Foreman, W.T., and T.F. Bidleman (1997), An experimental system for investigating vapor-pressure partitioning of trace organic pollutants, *Environ. Sci. Technol.*, *21*, 869–883.

Forster, P.M.D.F. (1995), Modeling ultraviolet radiation at the earth's surface. I. The sensitivity of ultraviolet irradiances to atmospheric changes, *J. Appl. Meteorol.*, *34*, 2412–2425.

Forstner, H.J.L., R.C. Flagan, and J.H. Seinfeld (1997), Secondary organic aerosol from the photooxidation of aromatic hydrocarbons: molecular composition, *Environ. Sci. Technol.*, *31*, 1346–1358.

Fox, M.A., and S. Olive (1979), Photooxidation of anthracene on atmospheric particulate matter, *Science*, *205*, 582–583.

Frankiewicz, T., and R.S. Berry (1972), Singlet O_2 production from photoexcited NO_2, *Environ. Sci. Technol.*, *6*, 365–366.

Freeman, D.J., and F.C.R. Cattell (1990), Woodburning as a source of atmospheric polycyclic aromatic hydrocarbons, *Environ. Sci. Technol.*, *24*, 1181–1185.

Frerichs, H., R. Koch, V. Schliephake, M. Tappe, and H.G. Wagner (1990a), Investigation of the reactions of phenol, anisole and cumene with $O(^3P)$ in the gas phase, *Z. Phys. Chem., Neue Folge*, *166*, 145–155.

Frerichs, H., R. Koch, M. Tappe, and H.G. Wagner (1989a), Investigation of the reaction of naphthalene with $O(^3P)$ in the gas phase, *Z. Naturforschung.*, *44a*, 1119–1121.

Frerichs, H., V. Schliephake, M. Tappe, and H.G. Wagner (1989b), Reactions of *o*-xylene *m*-xylene and *p*-xylene with atomic oxygen (O^3P) in the gas phase, *Z. Phys. Chem., Neue Folge*, *165*, 9–22.

Frerichs, H., D.-V. Stucken, M. Tappe, and H.G. Wagner (1991), Investigation of the reactions of benzonitrile, ethylbenzene and cumene with $O(^3P)$ in the gas phase, *Z. Phys. Chem. Neue Folge*, *174*, 1–9.

Frerichs, H., M. Tappe, and H.G. Wagner (1990b), Comparison of the reactions of mono- and polycyclic aromatic hydrocarbons with oxygen atoms, *Ber. Bunsen-Ges. Phys. Chem.*, *94*, 1404–1407.

Friedrich, R., and A. Obermeier (1999), Anthropogenic emissions of volatile organic compounds, in *Reactive Hydrocarbons in the Atmosphere*, edited by C.N. Hewitt, pp. 1–39, Academic Press, San Diego.

Fritz, B., V. Handwerk, M. Preidel, and R. Zellner (1985), Direct detection of hydroxy-cyclohexadienyl in the gas phase by cw-UV-laser absorption, *Ber. Bunsen-Ges. Phys. Chem.*, *89*, 343–344.

Fujimoto, G.T., M.E. Ulmstead, and M.C. Lin (1985), CO product energy distribution in the photodissociation of methylketene and acrolein at 193 nm, *J. Chem. Phys.*, *82*, 3042–3044.

Fujita, E.M., Z. Lu, L. Sheetz, G. Harshfeld, and B. Zielinska (1997), Determination of mobile source emission fraction using ambient field measurements, Report available from the Coordinating Research Council, 3650 Mansell Road, Suite 140, Alpharetta, GA, 30022–8246.

Funcke, W., J. Konig, and E. Balfanz (1988), Determination of polycyclic aromatic hydrocarbons in flue gases from coal-fired power plants, in *Polynuclear Aromatic Hydrocarbons, Physical and Biological Chemistry, A Decade of Progress, Tenth International Symposium,* edited by M. Cooke, and A.J. Dennis, pp. 277–284, Battelle Press, Columbus, Ohio.

Gaffney, J.S., R. Atkinson, and J.N. Pitts, Jr. (1976), Reaction of $O(^3P)$ atoms with toluene and 1-methylcylohexene, *J. Am. Chem. Soc.*, *98*, 1828–1832.

Galloway, D.B., J.A. Bartz, L.G. Huey, and F.F. Crim (1993), Pathways and kinetic energy disposal in the photodissociation of nitrobenzene, *J. Chem. Phys.*, *98*, 2107–2114.

Galloway, D.B., T. Glenewinkel-Meyer, J.A. Bartz, L.G. Huey, and F.F. Crim (1994), The kinetic and internal energy of NO from the photodissociation of nitrobenzene, *J. Chem. Phys.*, *100*, 1946–1952.

Gardner, B., C.N. Hewitt, and K.C. Jones (1995), PAHs in air adjacent to two inland water bodies, *Environ. Sci. Technol.*, *29*, 2405–2413.

Gardner, E.P., P.D. Sperry, and J.G. Calvert (1987), Photodecomposition of acrolein in O_2, N_2 mixtures, *J. Phys. Chem.*, *91*, 1922–1930.

Geron, C.D., A.B. Guenther, and T.E. Pierce (1994), An improved model for estimating emissions of biogenic volatile organic compounds from forests in the eastern United States, *J. Geophys. Res., 99*, 12773–12791.

Gery, M.W., D.L. Fox, H.E. Jeffries, L. Stockburger, and W.S. Weathers (1985), A continuous stirred tank reactor investigation of the gas-phase reaction of hydroxyl radicals and toluene, *Int. J. Chem. Kinet.*, *17*, 931–955.

Gery, M.W., D.L. Fox, R.M. Kamens, and L. Stockburger (1987), Investigation of hydroxyl radical reactions with *o*-xylene and *m*-xylene in a continuous stirred tank reactor, *Environ. Sci. Technol.*, *21*, 339–348.

Gery, M.W., G.Z. Whitten, J.P. Killus, and M.C. Dodge (1989), A photochemical kinetics mechanism for urban and regional scale computer modeling, *J. Geophys. Res.*, *94*, 12925–12956.

Ghigo, G., and G. Tonachini (1998), Benzene oxidation in the troposphere: theoretical investigation on the possible competition of three postulated reaction channels, *J. Am. Chem. Soc.*, *120*, 6753–6757.

Glasstone, S., K.J. Laidler, and H. Eyring (1941), *The Theory of Rate Processes*, McGraw-Hill, New York.

Goldan, P.D., W.C. Kuster, F.C. Fehsenfeld, and S.A. Montzka (1995), Hydrocarbon measurements in the southeast United States: the Rural Oxidants in the Southern Environment (ROSE) Program 1990, *J. Geophys. Res., 100*, 25945–25963.

Goumri, A., J.-F. Pauwels, and P. Devolder (1991), Rate of the OH + C_6H_6 + He reaction in the fall-off range by discharge flow and OH resonance fluorescence, *Can. J. Chem.*, *69*, 1057–1064.

Grabner, G., G. Köhler, G. Marconi, S. Monti, and E. Venuti (1990), Photophysical properties of methylated phenols in nonpolar solvents, *J. Phys. Chem.*, *94*, 3609–3613.

Graedel, T.E. (1978), *Chemical Compounds in the Atmosphere*, Academic Press, New York.

Gray, H.A., G.R. Cass, J.J. Huntzicker, E.K. Heyerdahl, and J.A. Rau (1986), Characteristics of atmospheric organic and elemental carbon particles concentrations in Los Angeles, *Environ. Sci. Technol.*, *20*, 580–589.

Green, M., G. Yarwood, and H. Niki (1990), FTIR study of the Cl-atom initiated oxidation of methylglyoxal, *Int. J. Chem. Kinet.*, *22*, 689–699.

Greenberg, A. (1989), Phenomenological study of benzo(a)pyrene and cyclopenteno(cd)-pyrene decay in ambient air using winter/summer comparisons, *Atmos. Environ.*, *23*, 2777–2799.

Griffin, R.J., D.R. Cocker, III, R.C. Flagan, and J.H. Seinfeld (1999), Organic aerosol formation from the oxidation of biogenic hydrocarbons, *J. Geophys. Res.*, *104*, 3555–3567.

Grimmer, G. (1983), Profile analysis of polycyclic aromatic hydrocarbons in air, in *Handbook of Polycyclic Aromatic Hydrocarbons*, edited by A. Bjorseth, pp. 149–182, Marcel Dekker, New York.

Grimmer, G. (1993), Relevance of polycyclic aromatic hydrocarbons as environmental carcinogens, in *Polycyclic Aromat. Compd.*, Vol. 3 supplement, pp. 31–41.

Grimmer, G., H. Brune, R. Deutsch-Wenzel, G. Dettbarn, J. Jacob, K.-W. Naujack, U. Morh, and H. Ernst (1987), Contribution of polycyclic aromatic hydrocarbons and nitroderivatives to the carcinogenic impact of diesel exhaust condensates evaluated by implantation into the lungs of rats, *Cancer Lett.*, *37*, 173–180.

Grosjean, D. (1983), Atmospheric reactions of pyruvic acid, *Atmos. Environ.*, *17*, 2379–2382.

Grosjean, D. (1984), Atmospheric reactions of *o*-cresol: gas phase and aerosol products, *Atmos. Environ.*, *18*, 1641–1652.

Grosjean, D. (1985), Reactions of *o*-cresol and nitrocresol with NO_x in sunlight and with ozone–nitrogen dioxide mixtures in the dark, *Environ. Sci. Technol.*, *19*, 968–974.

Grosjean, E., and D. Grosjean (1996), Carbonyl products of the gas-phase reaction of ozone with 1-alkenes, *Atmos. Environ.*, *30*, 4107–4113.

Grosjean, D., K. Fung, and J. Harrison (1983), Interactions of polycyclic aromatic hydrocarbons with atmospheric pollutants, *Environ. Sci. Technol.*, *17*, 673–679.

Grosjean, E., D. Grosjean, M.P. Fraser, and G.R. Cass (1996), Air quality model evaluations data for organics. 2. C_1–C_{14} carbonyls in Los Angeles air, *Environ. Sci. Technol.*, *30*, 2687–2703.

Grovenstein, E., Jr., and A.J. Mosher (1970), Reaction of atomic oxygen with aromatic hydrocarbons, *J. Am. Chem. Soc.*, *92*, 3810–3812.

Guenther, A., P. Zimmerman, and M. Wildermuth (1994), Natural volatile organic compound emission rate estimates for U.S. woodland landscapes, *Atmos. Environ.*, *28*, 1197–1210.

Gundel, L.A., V.C. Lee, R.R. Mahanama, R.K. Stevens, and J.M. Daisey (1995), Direct determination of the phase distributions of semi-volatile polycyclic aromatic hydrocarbons using annular denuders, *Atmos. Environ.*, *29*, 1719–1733.

Guo, Z., and R. Kamens (1991), An experimental technique for studing heterogeneous reactions of polyaromatic hydrocarbons on particle surfaces, *J. Atmos. Chem.*, *12*, 137–151.

Gupta, P., W.P. Harger, and J. Arey (1996), The contribution of nitro- and methylnitronaphthalenes to the vapor-phase mutagenicity of ambient air samples, *Atmos. Environ.*, *30*, 3157–3166.

Haas, B.-M., T.K. Minton, P. Felder, and J.R. Huber (1991), Photodissociation of acrolein and propynal at 193 nm in a molecular beam. Primary and secondary reactions, *J. Phys. Chem.*, *95*, 5149–5159.

Halsall, C.J., L.A. Barrie, P. Fellin, S.C.G. Muir, B.N. Billeck, L. Lockhard, F.Y. Rovinsky, E.Y. Kononov, and B. Pastukhov (1997a), Spatial and temporal variation of polycyclic aromatic hydrocarbons in the arctic atmosphere, *Environ. Sci. Technol.*, *31*, 3593–3599; (Correction: *Environ. Sci. Technol.*, *32*, 2480, 1998).

Halsall, C.J., P.J. Coleman, B.J. Davis, V. Burnett, K.S. Waterhouse, P. Harding-Jones, and K.C. Jones (1994), Polycyclic aromatic hydrocarbons in U.K. urban air, *Environ. Sci. Technol.*, *28*, 2380–2386.

Halsall, C.J., P.J. Coleman, and K.C. Jones (1997b), Atmospheric deposition of polychlorinated dibenzo-*p*-dioxins/dibenzofurans (PCDD/Fs) and polycyclic aromatic hydrocarbons (PAHs) in two UK cities, *Chemosphere*, *35*, 1919–1931.

Halsall, C.J., A.J. Sweetman, L.A. Barrie, and K.C. Jones (2001), Modelling the behaviour of PAH's during atmospheric transport from the UK to the Arctic, *Atmos. Environ.*, *35*, 255–267.

Hansen, D.A., R. Atkinson, and J.N. Pitts, Jr. (1975), Rate constants for the reaction of OH radicals with a series of aromatic hydrocarbons, *J. Phys. Chem.*, *79*, 1763–1766.

Hanson, D.R., J.B. Burkholder, C.J. Howard, and A.R. Ravishankara (1992), Measurement of OH and HO_2 radical uptake coefficients on water and sulfuric acid surfaces, *J. Phys. Chem.*, *96*, 4979–4985.

Harkov, R., and A. Greenberg (1985), Benzo(a)pyrene in New Jersey—results for a twenty-seven site study, *J. Air Pollut. Control Assoc.*, *35*, 238–243.

Harner, T., and D. Mackay (1995), Measurement of octanol–air partitioning coefficients for chlorobenzenes, PCBs and DDT, *Environ. Sci. Technol.*, *29*, 1599–1606.

Hastings, S.H., and F.A. Matsen (1948), The photodecomposition of nitrobenzene, *J. Am. Chem. Soc.*, *70*, 3514–3515.

Hatakeyama, S., T. Tanonaka, J. Weng, H. Bandow, H. Takagi, and H. Akimoto (1985), Ozone–cyclohexene reation in air: quantitative analysis of particulate products and reaction mechanism, *Environ. Sci. Technol.*, *19*, 935–942.

Hawthorne, S.B., M.S. Kreiger, D.J. Miller, and M.B. Mathiason (1989), Collection and quantitation of methoxylated phenol tracers for atmospheric pollution from residential wood stoves, *Environ. Sci. Technol.*, *23*, 470–475.

Hawthorne, S.B., D.J. Miller, J.J. Langenfeld, and M.S. Krieger (1992), PM–10 high-volume collection and quantitation of semi- and nonvolatile phenols, methoxyphenols, alkenes and polycyclic aromatic hydrocarbons from winter urban air and their relationship to wood smoke emissions, *Environ. Sci. Technol.*, *26*, 2251–2262.

He, S., and G.R. Carmichael (1999), Sensitivity of photolysis rates and ozone production in the troposphere to aerosol properties, *J. Geophys. Res.*, *104*, 26307–26324.

Heiden, A.C., K. Kobel, M. Komenda, R. Koppmann, M. Shao, and J. Wildt (1999), Toluene emissions from plants, *Geophys. Res. Lett.*, *26*, 1283–1286.

Helmig, D., J. Arey, R. Atkinson, W.P. Harger, and P.A. McElroy (1992a), Products of the OH radical-initiated gas-phase reaction of fluorene in the presence of NO_x, *Atmos. Environ.*, *26A*, 1735–1745.

Helmig, D., J. Arey, W.P. Harger, R. Atkinson, and J. López-Cancio (1992b), Formation of mutagenic nitrodibenzopyranones and their occurrence in ambient air, *Environ. Sci. Technol.*, *26*, 622–624.

Helmig, D., J. López-Cancio, J. Arey, W.P. Harger, and R. Atkinson (1992c), Quantification of ambient nitrodibenzopyranones: further evidence for atmospheric mutagen formation, *Environ. Sci. Technol.*, *26*, 2207–2213.

Hepp, H., K. Siegmann, and K. Sattler (1995), New aspects of growth mechanisms for polycyclic aromatic hydrocarbons in diffusion flames, *Chem. Phys. Lett.*, *233*, 16–22.

Heuss, J.M., and W.A. Glasson (1968), Hydrocarbon reactivity and eye irritation, *Environ. Sci. Technol.*, *2*, 1109–1116.

Hildemann, L.M., G.R. Cass, M.A. Mazurek, and B.R. Simoneit (1993), Mathematical modelling of urban organic aerosol: properties measured by high resolution gas chromatography, *Environ. Sci. Technol.*, *27*, 2045–2055.

Hildemann, L.M., D.B. Klinedinst, G.A. Klouda, L.A. Currie, and G.R. Cass (1994), Sources of urban contemporary carbon aerosol, *Environ. Sci. Technol.*, *28*, 1565–1576.

Hildemann, L.M., G.R. Markowski, R.G. Jones, and G.R. Cass (1991), Submicrometer aerosol mass distributions of emissions from boilers, fireplaces, automobiles, diesel trucks, and meat-cooking operations, *Aerosol Sci. Technol.*, *14*, 138–152.

Hinckley, D.A., T.F. Bidleman, and W.T. Foreman (1990), Determination of vapor pressures of nonpolar and semipolar organic compounds from gas chromatographic retention data, *J. Chem. Eng. Data*, *35*, 232–237.

Hoffmann, T., J.R. Odum, F. Bowman, D. Collins, D. Klockow, R.C. Flagan, and J.H. Seinfeld (1997), Formation of organic aerosols from the oxidation of biogenic hydrocarbons, *J. Atmos. Chem.*, *26*, 189–222.

Hoshino, M., H. Akimoto, and M. Okuda (1978), Photochemical oxidation of benzene, toluene, and ethylbenzene initiated by OH radicals in the gas phase, *Bull. Chem. Soc. Jpn.*, *51*, 718–724.

Howard, P.H., and W.M. Meylan (1997), *Handbook of Physical Properties of Organic Chemicals*, CRC, Lewis Publishers, New York.

Hufford, D.L., D.S. Tarbell, and T.R. Koszalka (1952), Maleic and fumaric dialdehydes, Δ^4-tetrahydrophthaldehyde and related compounds, *J. Am. Chem. Soc.*, *74*, 3014–3018.

Hunt, G.T., R.J. Kindya, R.R. Hall, R.F. Fennelly, and M. Hoyt (1982), The polycyclic aromatic environment of the fluidized bed coal combustion process—An Investigation of chemical and biological activity in polynuclear aromatic hydrocarbons, in *Physical and Biological Chemistry, 6th International Symposium,* edited by M. Cooke, A.J. Dennis, and G.L. Fisher, pp. 367–381, Battelle Press, Columbus, OH.

Huntzicker, J.J., E.K. Heyerdahl, J.A. Rau, W.H. Criest, and C.S. MacDougall (1986), Carbonaceous aerosol in the Ohio Valley, *J. Air Pollut. Control Assoc.*, *36*, 705–709.

Hurley, R., and A.C. Testa (1966), Photochemical $n \rightarrow \pi^*$ excitation of nitrobenzene, *J. Am. Chem. Soc.*, *88*, 4330–4332.

Hurley, R., and A.C. Testa (1968), Triplet-state yield of aromatic nitro-compounds, *J. Am. Chem. Soc.*, *90*, 1949–1952.

Hwang, W.G., M.S. Kim, and J.C. Choe (1996), Photodissociation dynamics of nitrobenzene molecular ion on a nanosecond time scale, *J. Phys. Chem.*, *100*, 9227–9234.

IARC (1983), *Polyaromatic Hydrocarbons. Part 1. Chemical, Environmental and Experimental Data*, International Agency for Research on Cancer, Lyon, France.

Itoh, T. (1988), The evidence showing that the intersystem crossing yield of benzaldehyde vapour is unity, *Chem. Phys. Lett.*, *151*, 166–168.

Izumi, K., and T. Fukuyama (1990), Photochemical aerosol formation from aromatic hydrocarbons in the presence of NO_x, *Atmos. Environ.*, *24A*, 1433–1441.

Izumi, K., K. Murano, M. Mizuochi, and T. Fukuyama (1988), Aerosol formation by the photooxidation of cyclohexene in the presence of nitrogen oxides, *Environ. Sci. Technol.*, *22*, 1207–1215.

Jacobson, M.Z. (1998), Studying the effects of aerosols on vertical photolysis rate coefficient and temperature profiles over an urban airshed, *J. Geophys. Res.*, *103*, 10593–10604.

Jacobson, M.Z. (1999), Isolating nitrated and aromatic aerosols and nitrated aromatic gases as sources of ultraviolet light absorption, *J. Geophys. Res.*, *104*, 3527–3542.

Jaenicke, R. (1988), Aerosol physics and chemistry, in *Meteorology, Landolt-Börnstein: Numerical Data and Functional Relationships in Science and Technology*, pp. 391–456, Springer-Verlag, New York.

Jaffé, H.H., and M. Orchin (1962), *Theory and Applications of Ultraviolet Spectroscopy*, John Wiley and Sons, Inc., New York.

Jang, M., and R.M. Kamens (1998), A thermodynamic approach for modeling partitioning of semivolatile organic compounds on atmospheric particulate matter: humidity effects, *Environ. Sci. Technol.*, *32*, 1237–1243.

Jang, M., and R.M. Kamens (2001), Characterization of secondary aerosols from the photooxidation of toluene in the presence of NO_x and propylene, *Environ. Sci. Technol.*, accepted for publication.

Jang, M., and S.R. McDow (1995), Benz(a)anthracene photodegradation in the presence of known organic constituents of atmospheric aerosols, *Environ. Sci. Technol.*, *29*, 2654–2660.

Jang, M., and S.R. McDow (1997), Products of benz[*a*]anthracene photodegradation in the presence of known organic constituents of atmospheric aerosols, *Environ. Sci. Technol.*, *31*, 1046–1053.

Jang, M., R.M. Kamens, B.K. Leach, and M.R. Strommen (1997), A thermodynamic approach using group contribution methods to model the partitioning of semi-volatile organic compounds on atmospheric particulate matter, *Environ. Sci. Technol.*, *31*, 2805–2811.

Japar, S.M., and H. Niki (1975), Gas-phase reactions of the nitrate radical with olefins, *J. Phys. Chem.*, *79*, 1629–1632.

Japar, S.M., A.C. Szkarlat, R.A. Gorse, E.K. Heyerdahl, R.L. Johnson, J.A. Rau, and J.J. Huntzicker (1984), Comparison of solvent extraction and thermal–optical carbon analysis methods: applications to diesel vehicle exhaust, *Environ. Sci. Technol.*, *18*, 231–234.

Jeffries, H.E., M.W. Gery, and W.P.L. Carter (1992), Protocols for evaluating oxidant mechanisms used in urban and regional models, Report based on findings of EPA workshops held in 1990 and 1991, Cooperative Agreement 815779, Project Officer, M. Dodge, Atmospheric Research and Exposure Assessment Laboratory, Office of Research and Development, U.S. Environmental Protection Agency, Research Triangle Park, NC 27711.

Jenkin, M.F., S.M. Saunders, and M.J. Pilling (1997), The tropospheric degradation of volatile organic compounds: a protocol for mechanism development, *Atmos. Environ.*, *31*, 81–104.

Jenkins, B.M., A.D. Jones, S.Q. Turn, and R.B. Williams (1996), Particle concentrations, gas–particle partitioning, and species intercorrelations for polycyclic aromatic hydrocarbons (PAH) emitted during biomass burning, *Atmos. Environ.*, *30*, 2825–2835.

Joback, K.G., and R.C. Reid (1985), Estimation of pure-component properties from group contributions, *Chem. Eng. Commun.*, *57*, 233–243.

Jones, G.R.H., and R.J. Cvetanovic (1961), Reaction of oxygen atoms with toluene, *Can. J. Chem.*, *39*, 2444–2451.

Jones, I.T.N., and K.D. Bayes (1971), Energy transfer from electronically excited NO_2, *Chem. Phys. Lett.*, *11*, 163–166.

Junge, C.E. (1977), Basic considerations about trace constituents in the atmosphere as related to the fate of global pollutants, in *Fate of Pollutants in the Air and Water Environment*, edited by J.H. Suffet, pp. 7–25, Wiley Interscience, New York.

Jüttner, F. (1988), Benzene in the anoxic hypolimnion of a freshwater lake, *Naturwissenschaften*, *75*, 151–153.

Jüttner, F., and J.J. Henatsch (1986), Anoxic hypolimnion is a significant source of biogenic toluene, *Nature*, *323*, 797–798.

Kadowaki, S. (1990), Characterisation of carbonaceous aerosols in the Nagoya urban area, 1. Elemental and organic carbon concentrations and the origin of organic aerosols, *Environ. Sci. Technol.*, *24*, 741–744.

Kamens, R.M., and D.L. Coe (1997), A large gas-phase stripping device to investigate rates of PAH evaporation from airborne diesel soot particles, *Environ. Sci. Technol.*, *31*, 1830–1833.

Kamens, R.M., Z.-H. Fan, and Y. Yao (1994), A methodology for modeling the formation and decay of nitro-PAH in the atmosphere, *Chemosphere*, *28*, 1623–1632.

Kamens, R.M., J.N. Fulcher, and Z. Guo (1986), Effects of temperature on wood soot PAH decay in atmospheres with sunlight and low NO_x, *Atmos. Environ.*, *20*, 1579–1587.

Kamens, R.M., J. Guo, Z. Guo, and S.R. McDow (1990), Polynuclear aromatic hydrocarbon degradation by heterogeneous reactions with N_2O_5 on atmospheric particles, *Atmos. Environ.*, *24A*, 1161–1173.

Kamens, R.M., Z. Guo, J.N. Fulcher, and D.A. Bell (1988), The influence of humidity and temperature on the daytime decay of polyaromatic hydrocarbons on atmospheric soot particles, *Environ. Sci. Technol.*, *22*, 103–108.

Kamens, R., M. Jang, C.-J. Chien, and K. Leach (1999), Aerosol formation from the reaction of α-pinene and ozone using a gas-phase kinetics-aerosol partitioning model, *Environ. Sci. Technol.*, *33*, 1430–1438.

Kamens, R.M., H. Karam, J. Guo, J.M. Perry, and L. Stockburger (1989), The behavior of oxygenated polycyclic aromatic hydrocarbons on atmospheric soot particles, *Environ. Sci. Technol.*, *23*, 801–806.

Kamens, R., J.R. Odum, and Z. Fan (1995), Observations on the times to equilibrium for semi-volatile polycyclic hydrocarbons, *Environ. Sci. Technol.*, *29*, 43–50.

Kamens, R.M., J.M. Perry, D.A. Saucy, D.A. Bell, D.L. Newton, and B. Brand (1985), Factors which influence PAH decomposition on airborne wood smoke particles, *Environ. Int.*, *11*, 131–136.

Kamens, R.M., G.D. Rives, J.M. Perry, D.A. Bell, R.F. Paylor, Jr., R.G. Goodman, and L.D. Claxton (1984), Mutagenic changes in dilute wood smoke as it ages and reacts with ozone and nitrogen dioxide: an outdoor chamber study, *Environ. Sci. Technol.*, *18*, 523–530.

Kaplan, I.R., and R.J. Gordon (1994), Non fossil fuel fine particle organic carbon aerosols in Southern California determined during the Los Angeles Aerosol Characterization and Source Apportionment Study, *Aerosol. Sci. Technol.*, *21*, 343–359.

Karahalil, B., A.E. Karakaya, and S. Burgaz (1999), The micronucleus assay in exfoliated buccal cels: application to occupational exposure to polycyclic aromatic hydrocarbons, *Mutat. Res.*, *442*, 29–35.

Katz, M., T. Sakuma, and A. Ho (1978), Chromatographic and spectral analysis of polynuclear aromatic hydrocarbons—quantitative distribution in air of Ontario cities, *Environ. Sci. Technol.*, *12*, 909–915.

Kaubisch, N., J.W. Daly, and D.M. Jerina (1972), Arene oxides as intermediates in the oxidative metabolism of aromatic compounds. Isomerization of methyl-substituted arene oxides, *Biochemistry*, *11*, 3080–3088.

Kaupp, H., and S. McLachlan (1999), Atmospheric particle size distributions of polychlorinated dibenzo-*p*-dioxins and dibenzofurans (PCDD/Fs) and polycyclic aromatic hydrocarbons (PAH) and their implication for wet and dry deposition, *Atmos. Environ.*, *33*, 85–95.

Kearns, D.R., A.U. Khan, C.K. Duncan, and A.H. Maki (1969), Detection of naphthalene-photosensitized generation of singlet ($^1\Delta_g$) oxygen by paramagnetic resonance spectroscopy, *J. Am. Chem. Soc.*, *91*, 1039–1040.

Kenley, R.A., and D.G. Hendry (1982), Generation of peroxy radicals from peroxynitrates ($ROONO_2$). Decomposition of peroxybenzoyl nitrate (PBzN), *J. Am. Chem. Soc.*, *104*, 220–224.

Kenley, R.A., J.E. Davenport, and D.G. Hendry (1978), Hydroxyl radical reactions in the gas phase. Products and pathways for the reaction of OH with toluene, *J. Phys. Chem.*, *82*, 1095–1096.

Kenley, R.A., J.E. Davenport, and D.G. Hendry (1981), Gas-phase hydroxyl radical reactions. Products and pathways for the reaction of OH with aromatic hydrocarbons, *J. Phys. Chem.*, *85*, 2740–2746.

Kennaway, E. (1955), The identification of a carcinogenic compound in coal-tar, *Br. Med. J.*, 749–752.

Kern, T., N. Metz, and D. Kley (1997), Untersuchungen von Verkehrsabgasemissionen insbesondere im Hinblick auf die Ozonbildung, Abschlußbericht Forschungszentrum Jülich, D–52425 Jülich und BMW AG, Energie und Umwelt, D–80788 München.

Kerr, J.A., and D.W. Sheppard (1981), Kinetics of the reactions of hydroxyl radicals with aldehydes studied under atmospheric conditions, *Environ. Sci. Technol.*, *15*, 960–963.

Khan, A.U., J.N. Pitts, Jr., and E.B. Smith (1967), Singlet oxygen in the environmental sciences: the role of singlet molecular oxygen in the production of photochemical air pollution, *Environ. Sci. Technol.*, *1*, 656–657.

Kind, I., T. Berndt, O. Boge, and W. Rolle (1996), Gas-phase rate constants for the reactions of NO_3 radicals with furan and methyl-substituted furans, *Chem. Phys. Lett.*, *256*, 679–683.

Kirchner, F., A. Mayer-Figge, F. Zabel, and K.H. Becker (1999), Thermal stability of peroxynitrates, *Int. J. Chem. Kinet.*, *31*, 127–144.

Kirchner, F., F. Zabel, and K.H. Becker (1992), Kinetic behaviour of benzoylperoxy radicals in the presence of NO and NO_2, *Chem. Phys. Lett.*, *191*, 169–174.

Kleindienst, T.E., G.W. Harris, and J.N. Pitts, Jr. (1982), Rates and temperatures dependences of the reaction of OH with isoprene, its oxidation products, and selected terpenes, *Environ. Sci. Technol.*, *16*, 844–846.

Kleindienst, T.E., D.F. Smith, W. Li, E.O. Edney, D.J. Driscoll, R.E. Speer, and W.S. Weathers (1999), Secondary organic aerosol formation from the oxidation of aromatic hydrocarbons in the presence of dry submicron ammonium sulfate aerosol, *Atmos. Environ.*, *33*, 3669–3681.

Klöpffer, W., R. Frank, E.-G. Köhl, and F. Haag (1986), Quantification of photochemical transformation processes in the troposphere, *Chemiker-Zeitung*, *110*, 57–61.

Klotz, B., I. Barnes, and K.H. Becker (1998a), New results on the atmospheric photooxidation of simple alkylbenzenes, *Chem. Phys.*, *231*, 289–301.

Klotz, B., I. Barnes, and K.H. Becker (1999), Kinetic study of the gas-phase photolysis and OH radical reaction of *E,Z*- and *E,E*–2,4-hexadienedial, *Int. J. Chem. Kinet.*, *31*, 689–697.

Klotz, B., I. Barnes, K.H. Becker, and B.T. Golding (1997), Atmospheric chemistry of benzene oxide/oxepin, *J. Chem. Soc., Faraday Trans.*, *93*, 1507–1516.

Klotz, B., I. Barnes, B.T. Golding, and K.H. Becker (2000), Atmospheric chemistry of toluene–1,2-oxide/2-methyloxepin, *Phys. Chem. Chem. Phys.*, *2*, 227–235.

Klotz, B.G., A. Bierbach, I. Barnes, and K.H. Becker (1995), Kinetic and mechanistic study of the atmospheric chemistry of muconaldehydes, *Environ. Sci. Technol.*, *29*, 2322–2332.

Klotz, B., S. Sørensen, I. Barnes, K.H. Becker, T. Etzkorn, R. Volkamer, U. Platt, K. Wirtz, and M. Martin-Reviejo (1998b), Atmospheric oxidation of toluene in a large-volume outdoor photoreactor: in situ determination of ring-retaining product yields, *J. Phys. Chem. A*, *102*, 10289–10299.

Knispel, R., R. Koch, M. Siese, and C. Zetzsch (1990), Adduct formation of OH radicals with benzene, toluene, and phenol and consecutive reactions of the adduct with NO_x and O_2, *Ber. Bunsen-Ges. Phys. Chem.*, *94*, 1375–1379.

Kobayashi, Y. (1979), Study on the role of toluene to photochemical smog, *Kanaganwaqken Taiki Osen Chosa Kenkyu Hokoka*, *21*, 87–94.

Koch, R. (1997), Comment on "UV spectrum and kinetics of hydroxycyclohexadienyl radicals," *J. Phys. Chem. B*, *101*, 293–294.

Kockarts, G. (1994), Penetration of solar radiation in the Schumann–Runge bands of molecular oxygen: a robust approximation, *Ann. Geophysicae*, *12*, 1207–1217.

Korfmacher, W.A., E.L. Wehry, G. Mamantov, and D.F.S. Natusch (1980), Resistance to photochemical decomposition of polycyclic aromatic hydrocarbons vapor-adsorbed on coal fly ash, *Environ. Sci. Technol.*, *14*, 1094–1099.

Kramp, F., and S.E. Paulson (1998), On the uncertainties in the rate coefficients for OH reactions with hydrocarbons, and the rate coefficients for 1,3,5-trimethylbenzene and *m*-xylene with OH radicals in the gas phase, *J. Phys. Chem. A*, *102*, 2685–2690.

Kummler, R.H., and M.H. Bortner (1969), Production of $O_2(^1\Delta_g)$ by energy transfer from excited benzaldehyde, *Environ. Sci. Technol.*, *3*, 944–946.

Kummler, R.H., M.H. Bortner, and T. Baurer (1969), The Hartley photolysis of ozone as a source of singlet oxygen in polluted atmospheres, *Environ. Sci. Technol.*, *3*, 248–250.

Kuntz, R.L., S.L. Kopczynski, and J.J. Bufalini (1973), Photochemical reactivity of benzaldehyde–NO_x and benzaldehyde–hydrocarbon–NO_x mixtures, *Environ. Sci. Technol.*, *13*, 1119–1123.

Kurtenbach, R. (1999), VOC-Measurements in rural air within the BERLOIOZ campaign (in German), in *Proceedings 2nd BERLIOZ Workshop, Munich*, edited by K.H. Becker, pp. 115–128, University of Wuppertal, Wuppertal, Germany.

Kurtenbach, R., R. Ackermann, K.J. Brockman, A. Geyer, J.A.G. Gomes, J.C. Lörzer, A. Niedojadlo, U. Platt, and K.H. Becker (1999), VOC-Measurements in a road traffic tunnel and in urban air of the city of Wuppertal (in German), in *TFS-LT3 Annual Report 1998*, edited by K.H. Becker, pp. 103–112, Universty of Wuppertal, Wuppertal, Germany.

Kurtenbach, R., R. Ackermann, K.J. Brockmann, A. Geyer, J.A.G. Gomes, J.C. Lörzer, A. Niedojadlo, U. Platt, and K.H. Becker (2000), VOC-Measurements in a road traffic tunnel and in urban air of the city of Wuppertal (in German), in *TFS-LT3 Annual Report 1999*, edited by K.H. Becker, pp. 207–214, University of Wuppertal, Wuppertal, Germany.

Kwok, E.S.C., and R. Atkinson (1995), Estimation of hydroxyl radical reaction rate constants for gas-phase organic compounds using a structure–reactivity relationship: an update, *Atmos. Environ., 29*, 1685–1695.

Kwok, E.S.C., J. Arey, and R. Atkinson (1994a), Gas–phase atmospheric chemistry of dibenzo-*p*-dioxin and dibenzofuran, *Environ. Sci. Technol., 28*, 528–533.

Kwok, E.S.C., S.M. Aschmann, R. Atkinson, and J. Arey (1997a), Products of the gas-phase reactions of *o*-, *m*-, and *p*-xylene with the OH radical in the presence and absence of NO_x, *J. Chem. Soc., Faraday Trans., 93*, 2847–2854.

Kwok, E.S.C., R. Atkinson, and J. Arey (1994b), Kinetics and mechanisms of the gas-phase reactions of the NO_3 radical with aromatic compounds, *Int. J. Chem. Kinet., 26*, 511–525.

Kwok, E.S.C., R. Atkinson, and J. Arey (1997b), Kinetics of the gas-phase reactions of indan, indene, fluorene, and 9,10-dihydroanthracene with OH radicals, NO_3 radicals, and O_3, *Int. J. Chem. Kinet., 29*, 299–309.

Kwok, E.S.C., W.P. Harger, J. Arey, and R. Atkinson (1994c), Reactions of gas-phase phenanthrene under simulated atmospheric conditions, *Environ. Sci. Technol., 28*, 521–527.

Land, E.J., and G. Porter (1963), Primary photochemical processes in aromatic molecules. Part 7. Spectra and kinetics of some phenoxyl derivatives, *Trans. Faraday Soc., 59*, 2016–2026.

Landgraf, J., and P.J. Crutzen (1998), An efficient method for online calculations of photolysis and heating rates, *J. Atmos. Sci., 55*, 863–878.

Lane, D.A. (1989), The fate of polycyclic aromatic compounds in the atmosphere and during sampling, in *Chemical Analysis of Polcyclic Aromatic Compounds*, edited by T. Vo-Dlinh, pp. 31–58, John Wiley, New York.

Lane, D.A., and M. Katz (1977), The photomodification of benz(a)pyrene, benzo(b)fluoranthene, and benzo(k)fluoranthene under simulated atmospheric conditions, in *Fate of Pollutants. I. The Air and Water Environments, Part 2, IA*, edited by I. Suffet, pp. 137–154, John Wiley–Interscience, New York.

Lane, D.A., N.D. Johnson, M.J.J. Henley, W.H. Schroeder, and D.T. Ord (1992), Gas and particle characterization of concentrations of α-hexachlorocyclohexane, γ-hexachlorocyclohexane, and hexachlorobenzene in Ontario air, *Environ. Sci. Technol., 22*, 941–947.

Lane, D.A.E. (1999), *Gas and Particle Phase Measurements of Atmospheric Organic Compounds*, Gordon and Breach Science Publishers, Amsterdam.

Láng, L. (1961–1971), *Absorption Spectra in the Ultraviolet and Visible Region*, Academic Press, New York.

Langmuir, I. (1918), The adsorption of gases on plane surfaces of glass, mica and platinum, *J. Am. Chem. Soc., 40*, 1361–1402.

Lantz, K.O., R.E. Shetter, C.A. Cantrell, S.J. Flocke, J.G. Calvert, and S. Madronich (1996), Theoretical, actinometric, and radiometric determination of the photolysis rate coefficient of NO_2 during the Mauna Loa Observatory Photochemistry Experiment 2, *J. Geophys. Res., 101*, 14613–14629.

Larson, S.M., G.R. Cass, and H.A. Gray (1989), Atmospheric carbon particles and the Los Angeles visibility problem, *Aerosol Sci. Technol.*, *10*, 118–130.

Lay, T.H., J.W. Bozzelli, and J.H. Seinfeld (1996), Atmospheric photochemical oxidation of benzene: benzene + OH and the benzene–OH adduct (hydroxyl–2,4-cyclohexadienyl) + O_2, *J. Phys. Chem.*, *100*, 6543–6554.

Lee, J.H., and I.N. Tang (1982), Absolute rate constants for the hydroxyl radical reactions with ethane, furan, and thiophene at room temperature, *J. Chem. Phys.*, *77*, 4459–4463.

Lee, W.M.G., and M.I. Huang (1993), Diurnal variation of OC in a severely polluted atmosphere, *J. Environ. Sci. Health*, *28*, 1565–1579.

Lee, Y.-N., and S.E. Schwartz (1981), Reaction kinetics of nitrogen dioxide with liquid water at low pressure, *J. Phys. Chem.*, *85*, 840–848.

Lee-Ruff, E., H. Kazarians-Moghaddam, and M. Katz (1986), Controlled oxidations of benzo(a)pyrene, *Can. J. Chem.*, *64*, 1297–1302.

Leighton, P.A., and F.A. Lucy (1934a), Photoisomerization of *o*-nitrobenzaldehydes. I. Photochemical results, *J. Chem. Phys.*, *2*, 756–759.

Leighton, P.A., and F.A. Lucy (1934b), Photoisomerization of *o*-nitrobenzaldehydes. II. Mathematical treatment, *J. Chem. Phys.*, *2*, 760–766.

Leister, D.L., and J.E. Baker (1994), Atmospheric deposition of organic contaminants to the Chesapeake Bay, *Atmos. Environ.*, *28*, 1499–1520.

Leone, J.A., R.C. Flagan, D. Grosjean, and J.H. Seinfeld (1985), An outdoor smog chamber and modeling study of toluene–NO_x photooxidation, *Int. J. Chem. Kinet.*, *17*, 177–216.

Leppard, W.R., J.D. Benson, V.R. Burns, R.A. Gorse, A.M. Hochhauser, J.C. Knapper, W.J. Koehl, L.A. Rapp, and R. M. Reuter (1993), How heavy hydrocarbons in the fuel affect exhaust mass emissions: correlation of fuel, engine-out, and tailpipe speciation—The Auto/Oil Air Quality Improvement Research Program, SAE paper 932725, Society of Automotive Engineers.

Leppard, W.R., W.J. Koehl, J.D. Benson, V.R. Burns, A.M. Hochhauser, J.C.Knapper, L.J Painter, L.A. Rapp, B.H. Rippon, R.M. Reuter, and J.A. Rutherford (1995), Effects of gasoline properties (T_{50}, T_{90}, and sulfur) on exhaust hydrocarbon emissions of current and future vehicles: speciation analysis—The Auto/Oil Air Quality Improvement Research Program, SAE paper 952505, Society of Automotive Engineers.

Lewis, G.P., and L. Counghlin (1973), Lung "soot" accumulation in man, *Atmos. Environ.*, *7*, 1249–1255.

Lewis, R.G., and R.W. Coutant (1999), Determination of phase distributed polycylic aromatic hydrocarbons in air by grease-coated denuders, in *Gas and Particle Phase Measurements of Atmospheric Organic Compounds*, edited by D.A. Lane, pp. 201–232, Gordon and Breach Science Publishers, Amsterdam.

Li, C.K., and R.M. Kamens (1993), The use of polycyclic aromatic hydrocarbons as source signatures in receptor modeling, *Atmos. Environ.*, *27A*, 523–532.

Liang, C., J.F. Pankow, J.R. Odum, and J.H. Seinfeld (1997), Gas/particle partitioning of semivolatile organic compounds to model inorganic, organic and ambient smog aerosols, *Environ. Sci. Technol.*, *31*, 3086–3092.

Liao, H., Y.L. Yung, and J.H. Seinfeld (1999), Effects of aerosols on tropospheric photolysis rates in clear and cloudy atmospheres, *J. Geophys. Res.*, 104, 23697–23707.

Lias, S.G., J.E. Bartmess, J.F. Liebman, J.L. Holmes, R.D. Levine, and W.G. Mallard (1988), Gas-phase ion and neutral thermochemistry, *J. Phys. Chem. Ref. Data, 17, Supplement 1*, 1–861.

Lide, D.R. (1999), *CRC Handbook of Chemistry and Physics*, 80th ed., CRC Press, Boca Raton, FL.

Lightfoot, P.D., R.A. Cox, J.N. Crowley, M. Destriau, G.D. Hayman, M.E. Jenkin, G.K. Moortgat, and F. Zabel (1992), Organic peroxy radicals: kinetics, spectroscopy and tropospheric chemistry, *Atmos. Environ.*, *26*, 1805–1961.

Ligocki, M.P., C. Leuenberger, and J.F. Pankow (1985a), Trace organic compounds in rain. II. Gas scavenging of neutral organic compounds, *Atmos. Environ.*, *19*, 1609–1618.

Ligocki, M.P., C. Leuenberger, and J.F. Pankow (1985b), Trace organic compounds in rain. III. Particle scavenging of neutral organic compounds, *Atmos. Environ.*, *19*, 1619–1626.

Lin, S.-C., T.-C. Kuo, and Y.-P. Lee (1994), Detailed rate coefficients and the enthalpy change of the equilibrium reaction OH + C_6H_6 (M) $\rightarrow$ HOC_6H_6 over the temperature range 345–385 K, *J. Chem. Phys.*, *101*, 2098–2105.

Lipczynska-Kochany, E., and J.R. Botton (1992), Flash photolysis/high-performance liquid chromatography method for studying the sequence of photochemical reactions: direct photoylsis of phenol, *Environ. Sci. Technol.*, *26*, 2524–2527.

Liu, S.C., S.A. McKeen, and S. Madronich (1991), Effect of anthropogenic aerosols on biologically active ultraviolet radiation, *Geophys. Res. Lett.*, *18*, 2265–2268.

Liu, X. (1999), Atmospheric photochemistry study of 1,4-unsaturated dicarbonyls and dienes, Ph.D. Thesis, University of North Carolina Chapel Hill, Chapel Hill, NC.

Liu, X., H.E. Jeffries, and K.G. Sexton (1999), Atmospheric photochemical degradation of 1,4-unsaturated dicarbonyls, *Environ. Sci. Technol.*, *33*, 4212–4220.

Lloyd, A.C., K.R. Darnall, A.M. Winer, and J.N. Pitts, Jr. (1976), Relative rate constants for reaction of the hydroxyl radical with a series of alkanes, alkenes and aromatic hydrocarbons, *J. Phys. Chem.*, *80*, 789–794.

Logan, J.A. (1985), Tropospheric ozone: season behavior, trends, and anthropogenic influence, *J. Geophys. Res.*, *90*, 10463–10482.

Lorenz, K., and R. Zellner (1983), Kinetics of the reactions of hydroxyl radicals with benzene, benzene-d_6 and naphthalene, *Ber. Bunsen-Ges. Phys. Chem.*, *87*, 629–636.

Luthy, R.G., D.A. Dzombak, C.A. Peters, S.M. Roy, A. Ramaswami, D.V. Nackles, and B.R. Noot (1994), Remediating tar-contaminated soils at manufactured gas plant sites, *Environ. Sci. Technol.*, *28*, 266–276.

Lüttke, J., and K. Levsen (1997), Phase partitioning of phenol and nitrophenols in clouds, *Atmos. Environ.*, *31*, 2649–2655.

Lüttke, J., V. Scheer, K. Levsen, G. Wünsch, J.N. Cape, K.J. Hargreaves, R.L. Storeton-West, K. Acker, W. Wieprecht, and B. Jones (1997), Occurrence and formation of nitrated phenols in and out of cloud, *Atmos. Environ.*, *31*, 2637–2648.

Ma, J., and R. Guicherit (1997), Effects of stratospheric ozone depletion and tropospheric pollution on UV-B radiation in the troposphere, *Photochem. Photobiol.*, *66*, 346–355.

Ma, S., I. Barnes, and K.H. Becker (1998), Atmospheric degradation of glycidaldehyde: photolysis and reaction with OH radicals, *Environ. Sci. Technol.*, *32*, 3515–3521.

Mackay, D., A. Bobra, D.W. Chan, and W.Y. Shiu (1982), Vapor pressure correction for low-volatility environmental chemicals, *Environ. Sci.. Technol.*, *16*, 16645–16649.

Madronich, S., and W. Felder (1985), Kinetics and mechanism of the reaction of OH with C_6H_6 over 790–1410 K, *J. Phys. Chem.*, *89*, 3556–3561.

Madronich, S., and S. J. Flocke (1998), The role of solar radiation in atmospheric chemistry, in *Handbook of Environmental Chemistry*, Vol. II, Part L., *Reactions and Processes*, edited by P. Boule, pp. 1–26, Springer-Verlag, Heidelberg.

Madronich, S., and G. Weller (1990), Numerical integration errors in calculated tropospheric photodissociation rate coefficients, *J. Atmos. Chem.*, *10*, 289–300.

Magneron, I., J. Tadic, F. Sauer, and G. K. Moortgat (1999), Photolytic studies on glyoxal, crotondaldehyde, acrolein, and glycolaldehyde, in *EC/EUROTRAC-2, Joint Workshop Chemical Mechanism Development (CMD)*, edited by R. Vogt and G. Axelsdottir, pp. 66–69, Ford Forschungszentrum, Aachen, Germany.

Mani, I., and M.C. Sauer, Jr. (1968), A pulsed-radiolysis study of the gas-phase reaction of oxygen atoms with benzene and related compounds: rate constants and transient spectra, *Adv. Chem. Ser.*, *82*, 142–152.

Mantaka, A., D.G. Marketos, and G. Stein (1971), Continuous γ and pulse radiolysis of aqueous benzene solutions: some reactions of the hydroxycyclohexadienyl radical, *J. Phys. Chem.*, *75*, 3886–3889.

Mariconi, E.J., and L.B. Taranko (1963), Ozonolysis of polycyclic aromatics. XI. 3-Methylcholanthrene, *J. Org. Chem.*, *28*, 2526–2529.

Markert, F., and P. Pagsberg (1993), UV spectra and kinetics of radicals produced in the gas phase reactions of Cl, F, and OH with toluene, *Chem. Phys. Lett.*, *209*, 445–454.

Martinez, R.D., A.A. Buitrago, N.W. Howell, C.H. Hearn, and J.A. Joens (1992), The near u.v. absorption spectra of several aliphatic aldehydes and ketones at 300 K, *Atmos. Environ.*, *26A*, 785–792.

Marty, J.C., M.J. Tussier, and A. Saliot (1984), Gaseous and particulate polycyclic aromatic hydrocarbons (PAH) from the marine atmosphere, *Atmos. Environ.*, *18*, 2183–2190.

Masclet, P., P. Pistikopoulous, S. Beyne, and G. Mouvier (1988), Long-range transport and gas/particle distributions of polycyclic aromatic hydrocarbons at a remote site in the Mediterranean Sea, *Atmos. Environ.*, *22*, 639–650.

Masek, V. (1967), Effect of solar radiation on 3,4-benzopyrene in stack emissions, *Chem. Prum.*, *17*, 99–103, 1967; *Chem. Abstr.*, *66*, 108014.

Matsumoto, M., K. Yasuoka, and Y. Kataoka (1994), Evaporation and condensation at a liquid surface: 1. Argon, *J. Chem. Phys.*, *101*, 7904–7911.

McDade, C.E., T.M. Lenhardt, and K.D. Bayes (1982), The rate of reaction of acetyl and benzoyl radicals with O_2, *J. Photochem.*, *20*, 1–7.

McDow, S.R., and J.J. Huntzicker (1990), Vapor adsorption artifact in the sampling of organic aerosol: face velocity effects, *Atmos. Environ.*, *24*, 2563–2571.

McDow, S.R., M. Jang, Y. Hong, and R.M. Kamens (1996), An approach to studying the effect of organic composition on atmospheric aerosol photochemistry, *J. Geophys. Res.*, *101*, 19593–19600.

McDow, S.R., Q.-R. Sun, M. Vartiainen, Y.-S. Hong, Y.-L. Yao, T. Fister, R.-Q. Yao, and R.M. Kamens (1994), Effect of composition and state of organic components on polycyclic aromatic hydrocarbon decay in atmospheric aerosols, *Environ. Sci. Technol.*, *28*, 2147–2153.

McDow, S.R., M. Vartiainen, Q. Sun, Y. Hong, Y. Yao, and R.M. Kamens (1995), Combustion aerosol water content and its effect on polycyclic aromatic hydrocarbon reactivity, *Atmos. Environ.*, *29*, 791–797.

McElroy, W.J., and S.J. Waygood (1991), Oxidation of formaldehyde by the hydroxyl radical in aqueous solution, *J. Chem. Soc., Faraday Trans.*, *87*, 1513–1521.

McKeen, S.A., T. Gierczak, J.B. Burkholder, P.O. Wennberg, T.F. Hanisco, E.R. Keim, R.-S. Gao, S. C. Liu, A.R. Ravishankara, and D.W. Fahey (1997), The photochemistry of acetone in the upper atmosphere: a source of odd-hydrogen radicals, *J. Geophys. Res.*, *24*, 3177–3180.

McMillan, G.R., J.G. Calvert, and J.N. Pitts, Jr. (1964), Detection and lifetime of enol-acetone in the photolysis of 2-pentanone vapor, *J. Am. Chem. Soc.*, 86, *3602–3605*.

McVeety, D.G., and R.A. Hites (1988), Atmospheric deposition of polycyclic aromatic hydrocarbons to water and surface: a mass balance approach, *Atmos. Environ.*, *22*, 511–536.

Meng, Z., D. Dabdub, and J.H. Seinfeld (1997), Chemical coupling between atmospheric ozone and particulate matter, *Science*, *277*, 116–119.

Menichini, E., F. Monfedini, and F. Merli (1999), The temporal variability of the profile of carcinogenic polycyclic aromatic hydrocarbons in urban air: a study in a medium traffic area in Rome, 1993–1998, *Atmos. Environ.*, *33*, 3739–3750.

Michael, J.V., J.E. Allen, Jr., and W.D. Brobst (1981), Temperature dependence of the $NO + O_3$ reaction rate from 195 to 369 K, *J. Phys. Chem.*, *85*, 4109–4117.

Miguel, A.H., and S.K. Friedlander (1978), Distribution of benzo(a)pyrene and coronene with respect to particle size in Pasadena aerosols in the submicron range, *Atmos. Environ.*, *12*, 2407–2413.

Miguel, A.H., W.A. Korfmacher, E.L. Wehry, G. Mamanatov, and D.F.S. Natusch (1979), Apparatus for vapor-phase adsorption of polycyclic organic matter onto particulate surfaces, *Environ. Sci. Technol.*, *13*, 1229–1232.

Minutolo, P., G. Gambi, A. D'Alessio, and S. Carlucci (1988), Spectroscopic characterization of carbonaceous nanoparticles in premixed flames, *Atmos. Environ.*, *22*, 2725–2732.

Molina, L.T., and M.J. Molina (1986), Absolute absorption cross sections of ozone in the 185–350 nm wavelength range, *J. Geophys. Res.*, *91*, 14501–14508.

Moortgat, G.K., B. Veyret, and R. Lesclaux (1989), Absorption spectrum and kinetics of the reactions of the acetylperoxy radical, *J. Phys. Chem.*, *93*, 2362–2368.

Moschonas, N., D. Danalytos, and S. Glavas (1999), The effect of O_2 and NO_2 on the ring retaining products of toluene with hydroxyl radicals, *Atmos. Environ.*, *33*, 111–116.

Mozurkewich, M., P.A. McMurry, A. Gupta, and J.G. Calvert (1987), The mass accommodation coefficient of HO_2 radicals on aqueous aerosols, *J. Geophys. Res.*, *92*, 4163–4170.

Mueller, P.K., K.K. Fung, S.L. Heisler, D. Grosjean, and G.M. Hidy (1982), Atmospheric particulate carbon observations in urban and rural areas of the United States, in *Particulate Carbon–Atmospheric Life Cycle*, edited by G.T. Wolff, and R.L. Klimisch, pp. 343–370, Plenum Press, New York.

Muller, J.G., P.J. Chapman, and P.H. Pritchard (1989), Creosote-contaminated sites, *Environ. Sci., Technol.*, *23*, 1197–1201.

Muller, R.F., D.W. Hawker, and D.W. Connell (1998), Polycyclic aromatic hydrocarbons in the atmosphere environment of Brisbane, Australia, *Chemosphere*, *37*, 1369–1383.

NAS (1972), *Particulate Polycyclic Organic Matter*, National Academy Press, Washington, DC.

Neckel, H., and D. Labs (1984), The solar radiation between 3300 and 12500 A, *Solar Phys.*, *90*, 205–258.

Neilson, A.H.E. (1998), *PAHs and Related Compounds*, Springer-Verlag, New York.

Nicol, C.H., and J.G. Calvert (1967), Relations between photodecomposition modes and molecular structure in the series of carbonyl compounds, n-C_3H_7COR, *J. Am. Chem. Soc.*, *89*, 1790–1798.

Nicovich, J.M., C.A. Gump, and A.R. Ravishankara (1982a), Rates of reaction of $O(^3P)$ with benzene and toluene, *J. Phys. Chem.*, *86*, 1684–1690.

Nicovich, J.M., C.A. Gump, and A.R. Ravishankara (1982b), Rates of reaction of $O(^3P)$ with xylenes, *J. Phys. Chem.*, *86*, 1690–1694.

Nicovich, J.M., R.L. Thompson, and A.R. Ravishankara (1981), Kinetics of the reactions of the hydroxyl radical with xylenes, *J. Phys. Chem.*, *85*, 2913–2916.

Nielsen, T. (1984), Reactivity of polycyclic aromatic hydrocarbons toward nitrating species, *Environ. Sci. Technol.*, *18*, 157–163.

Nielsen, T. (1988), The decay of benzo(a)pyrene and cyclopenteno(cd)pyrene in the atmosphere, *Atmos. Environ.*, *22*, 2249–2254.

Nielsen, T. (1996), Traffic contribution of polycyclic aromatic hydrocarbons in the center of a large city, *Atmos. Environ.*, *30*, 3481–3490.

Nielsen, T., H.E. Jorgensen, J.C. Larsen, and M. Poulsen (1996), City air pollution of polycyclic aromatic hydrocarbons and other mutagens, occurrence, sources and health effects, *Sci. Total Environ.*, *189/190*, 41–49.

Nielsen, T., T. Ramdahl, and A. Björseth (1983), The fate of airborne polycyclic organic matter, *Environ. Health Perspect.*, *47*, 103–114.

Nielsen, T., T. Seitz, and T. Ramdahl (1984), Occurrence of nitro-PAH in the atmosphere in a rural environment, *Atmos. Environ.*, *18*, 2159–2165.

Niki, H., P.D. Maker, C.M. Savage, and L.P. Breitenbach (1978), Relative rate constants for the reaction of hydroxyl radical with aldehydes, *J. Phys. Chem.*, *82*, 132–134.

Niki, H., P.D. Maker, C.M. Savage, and L.P. Breitenbach (1979), Fourier transform infrared (FTIR) studies of gaseous and particulate nitrogenous compounds, in *Nitrogenous Air Pollutants*, edited by D. Grosjean, pp. 1–16, Ann Arbor Science, Ann Arbor, MI.

Niki, H., P.D. Maker, C.M. Savage, and L.P. Breitenbach (1985a), An FTIR study of the Cl-atom-initiated reaction of glyoxal, *Int. J. Chem. Kinet.*, *17*, 547–558.

Niki, H., P.D. Maker, C.M. Savage, and L.P. Breitenbach (1985b), FTIR study of the kinetics and mechanism for Cl-atom initiated reactions of acetaldehyde, *J. Phys. Chem.*, *89*, 588–591.

Niu, Q.J., and G.D. Mendenhall (1992), Yields of singlet molecular oxygen from peroxyl radical termination, *J. Am. Chem. Soc.*, *114*, 165–172.

Nojima, K., and C. Isogami (1994), Studies on photochemical reactions of air pollutants. XII. Formation of nitrophenols by reactions of three toluene oxides with nitrogen dioxide in air, *Chem. Pharm. Bull.*, *42*, 2426–2429.

Nojima, K., K. Fukaya, S. Fukui, and S. Kanno (1974), The formation of glyoxals by the photochemical reaction of aromatic hydrocarbons in the presence of nitrogen monoxide, *Chemosphere*, *5*, 247–252.

Nojima, K., A. Kawaguchi, T. Ohya, S. Kanno, and M. Hirobe (1983), Studies on photochemical reactions of air pollutants. X. Identification of nitrophenols in suspended particulates, *Chem. Pharm. Bull.*, *31*, 1047–1051.

Nojima, K., T. Ohya, S. Kanno, and M. Hirobe (1982), Studies on photochemical reactions of air pollutants. IX. Formation of *o*-nitrophenol and *p*-nitrophenol by reaction of benzene oxide with NO_2 in air, *Chem. Pharm. Bull.*, *30*, 4566–4569.

Nozière, B., R. Lesclaux, M.D. Hurley, M.A. Dearth, and T.J. Wallington (1994), Kinetic and mechanistic study of the self-reaction and the reaction with HO_2 of the benzylperoxy radical, *J. Phys. Chem.*, *98*, 2864–2873.

O'Brien, R.J., B.E. Dumdei, S.V. Hummel, and R.A. Yost (1984), Determination of atmospheric degradation products of toluene by tandem mass spectrometry, *Anal. Chem.*, *56*, 1329–1335.

Oda, J., I. Maeda, T. Mori, A. Yasuhara, and Y. Saito (1998), The relative proportions of polycyclic aromatic hydrocarbons and oxygenated derivatives in accumulated organic particulates as affected by air pollution sources, *Environ. Tech.*, *19*, 961–976.

Odabasi, M., A. Sofuoglu, N. Vardar, T. Tasdemir, and T.M. Holsen (1999a), Measurement of dry deposition and air–water exchange of polycyclic aromatic hydrocarbons with the water surface sampler, *Environ. Sci. Technol.*, *33*, 426–434.

Odabasi, M., N. Vardar, A. Sofuoglu, Y. Tasdemir, and T.M. Holsen (1999b), Polycyclic aromatic hydrocarbons in Chicago air, *Sci. Total Environ.*, *277*, 57–67.

Odum, J.R., T. Hoffmann, F. Bowman, D. Collins, R.C. Flagan, and J.H. Seinfeld (1996), Gas/particle partitioning and secondary organic aerosol yields, *Environ. Sci. Technol.*, *30*, 2580–2585.

Odum, J.R., T.P.W. Jungkamp, R.J. Griffin, H.J.L. Forstner, R.C. Flagan, and J.H. Seinfeld (1997a), Aromatics, refomulated gasoline, and atmospheric organic aerosol formation, *Environ. Sci. Technol.*, *31*, 1890–1897.

Odum, J.R., T.P.W. Jungkamp, R.J. Griffin, R.C. Flagan, and J.H. Seinfeld (1997b), The atmospheric aerosol-forming potential of whole gasoline vapor, *Science*, *276*, 96–99.

Odum, J.R., S.R. McDow, and R.M. Kamens (1994), Mechanistic and kinetic studies of the photodegradation of benz[*a*]anthracene in the presence of methoxyphenols, *Environ. Sci. Technol.*, *26*, 1285–1290.

Ogawa, S., S. Fukui, Y. Hanasaki, K. Asano, H. Uegaki, S. Fujita, and R. Shimazaki (1991), Determination method of singlet oxygen in the atmosphere by use of α-terpinene, *Chemosphere*, *22*, 1211–1225.

Ogawa, S., R. Shimazaki, A. Soejina, E. Takamure, Y. Hanasaki, and S. Fukui (1996), Diurnal changes of singlet oxygen like oxidant concentrations in polluted ambient air, *Chemosphere*, *32*, 1823–1832.

Ohta, T., and I. Mizoguchi (1981), Thermal decomposition rate constant of peroxybenzoyl nitrate in the gas phase, *Environ. Sci. Technol.*, *15*, 1229–1232.

Ohta, T., and T. Ohyama (1985), A set of rate constants for the reactions of hydroxyl radicals with aromatic hydrocarbons, *Bull. Chem. Soc. Jpn.*, *58*, 3029–3030.

Olzmann, M., E. Kraka, D. Cremer, R. Gutbrod, and S. Andersson (1997), Energetics, kinetics, and product distribution of the reactions of ozone with ethene and 2,3-dimethyl-2-butene, *J. Phys. Chem. A*, *101*, 9421–9429.

Orlando, J.J., G.S. Tyndall, J.-M. Fracheboud, E.G. Estupinan, S. Haberkorn, and A. Zimmer (1999), The rate and mechanism of the gas-phase oxidation of hydroxyacetone, *Atmos. Environ.*, *33*, 1621–1629.

Pagsberg, P. (1997), Reply to Comment on "UV spectrum and kinetics of hydroxycyclohexadienyl radicals," *J. Phys. Chem. B*, *101*, 294.

Palm, W.-U., W. Behnke, and C. Zetzsch (1999a), Are light-induced reactions of nitrophenols an efficient sink in atmospheric droplets?, in *Chemical Mechanisms of Atmospheric Processes*, eds. K. H. Becker, and G. Angeletti, pp. 377–382, Air Pollution Research Report 67, EUR 18765, European Communities, Luxembourg.

Palm, W.-U., H.-U. Krüger, W. Behnke, and C. Zetsch (1999b), Uptake of NO_2 on aqueous solutions containing phenol and degradation of some nitroaromatics by photolysis and OH radicals, in *EC/EUROTRAC-2 Joint Workshop Chemical Mechanism Development (CMD)*, edited by R. Vogt and G. Axelsdottir, pp. 171–174, Ford Forschungszentrum, Aachen, Germany.

Pandis, S.N., R.A. Harley, G.R. Cass, and J.H. Seinfeld (1992), Secondary organic aerosol formation and transport, *Atmos. Environ.*, *26A*, 2269–2282.

Pandis, S.N., S.E. Paulson, J.H. Seinfeld, and R.C. Flagan (1991), Aerosol formation in the photooxidation of isoprene and α-pinene, *Atmos. Environ.*, *25A*, 997–1008.

Pankow, J.F. (1987), Review and comparative analysis of the theories on partitioning between the gas and aerosol particulate phases in the atmosphere, *Atmos. Environ.*, *21*, 2275–2283.

Pankow, J.F. (1991), Common y-intercepts and single compound regressions of gas–particle partitioning vs. $1/T$, *Atmos. Environ.*, *25A*, 2229–2239.

Pankow, J.F. (1992), Application of common y-intercept regression parameters for log K_p vs. $1/T$ for predicting gas–particle partitioning in the urban environment, *Atmos. Environ.*, *26*, 2489–2497.

Pankow, J.F. (1994a), An absorption model of gas/particle partitioning of organic compounds in the atmosphere, *Atmos. Environ.*, *28*, 185–188.

Pankow, J.F. (1994b), An absorption model of the gas/aerosol partitioning involved in the formation of secondary organic aerosol, *Atmos. Environ.*, *28*, 189–193.

Pankow, J.F., and T.F. Bidleman (1991), Effects of temperature, TSP, and percent nonexchangeable material in the determining the gas–particle partitioning of organic compounds, *Atmos. Environ.*, *25A*, 2241–2249.

Pankow, J.F., and T. Bidleman (1992), Interdependence of the slopes and intercepts from log–log correlations of measured gas–particle partitioning and vapor pressure—I. Theory and analysis of available data, *Atmos. Environ.*, *26A*, 1071–1080.

Pankow, J.F., B.T. Mader, L.M. Isabelle, W. Luo, A. Pavick, and C. Liang (1997), Conversion of nicotine in tobacco smoke to its volatile and available free-base form through the action of gaseous ammonia, *Environ. Sci. Technol.*, *31*, 2428–2433.

Papayannis, A., D. Balis, A. Bais, H. van der Bergh, B. Calpini, E. Durieux, L. Fiorani, L. Jaquet, I. Ziomas, and C.S. Zerefos (1998), Role of urban and suburban aerosols on solar UV radiation over Athens, Greece, *Atmos. Environ.*, *32*, 2193–2201.

Pate, C., R. Atkinson, and J.N. Pitts, Jr. (1976), The gas phase reaction of O_3 with a series of aromatic hydrocarbons, *J. Environ. Sci. Health*, *A11*, 1–10.

Penner, J.C., H. Eddleman, and T. Novakov (1993), Towards the development of a global inventory for black carbon emissions, *Atmos. Environ.*, *27A*, 1277–1295.

Pennise, D.M., and R.M. Kamens (1996), Atmospheric behavior of polychlorinated dibenzo-*p*-dioxins and dibenzofurans and the effect of combustion temperature, *Environ. Sci. Technol.*, *30*, 2832–2842.

Penzhorn, R.-D., H. Güsten, U. Schurath, and K. H. Becker (1974), Quenching of singlet molecular oxygen by some atmospheric pollutants, *Environ. Sci. Technol.*, *8*, 907–909.

Perera, R. (1981), Carcinogenicity of airborne fine particulate benzo[*a*]pyrene: an appraisal of the evidence and the need for control, *Environ. Health Perspect.*, *42*, 163–185.

Perry, R.A., R. Atkinson, and J.N. Pitts, Jr. (1977a), Kinetics and mechanism of the gas phase reaction of OH radicals with aromatic hydrocarbons over the temperature range 296–473 K, *J. Phys. Chem.*, *81*, 296–304.

Perry, R.A., R. Atkinson, and J.N. Pitts, Jr. (1977b), Kinetics and mechanism of the gas phase reaction of OH radicals with methoxybenzene and *o*-cresol over the temperature range 299–435 K, *J. Phys. Chem.*, *81*, 1607–1611.

Piccot, S.D., J.J. Watson, and J.W. Jones (1992), A global inventory of volatile organic compounds emissions from anthropogenic sources, *J. Geophys. Res.*, *97*, 9897–9912.

Pierce, R.C., and M. Katz (1975), Dependency of polynuclear aromatic hydrocarbon content on size distribution of atmospheric aerosols, *Environ. Sci. Technol.*, *9*, 347–353.

Pilinis, C., and J.H. Seinfeld (1988), Development and evaluation of an Eulerian photochemical gas–aerosol model, *Atmos. Environ.*, *22*, 1985–2001.

Pistikopoulos, P., H.M. Wortham, L. Gomes, S. Masclet-Beyne, E.B. Nguyen, P.A. Masclet, and G. Mouvier (1990), Mechanisms of formation of particulate polycyclic aromatic hydrocarbons in relation to the particle size distribution; effects on mesoscale transport, *Atmos. Environ.*, *24A*, 2573–2584.

Pitts, J.N., Jr. (1983), Formation and fate of gaseous and particulate mutagens and carcinogens in real and simulated atmospheres, *Environ. Health Perspect.*, *47*, 115–140.

Pitts, J.N., Jr., R. Atkinson, J.A. Sweetman, and B. Zielinska (1985a), The gas-phase reaction of naphthalene with N_2O_5 to form nitronaphthalenes, *Atmos. Environ.*, *19*, 701–705.

Pitts, J.N., Jr., A.U. Khan, E.B. Smith, and R.P. Wayne (1969), Singlet oxygen in the environmental sciences: singlet molecular oxygen and photochemical air pollution, *Environ. Sci. Technol.*, *3*, 241–247.

Pitts, J.N., Jr., D.M. Lokensgard, P.S. Ripley, K.A. van Cauwenberghe, L. van Vaeck, S.D. Shaffer, A.J. Thill, and W.L. Belser, Jr. (1980), "Atmospheric" epoxidation of benzo(a)pyrene by ozone: formation of the metabolite benzo(a)pyrene,4–5-epoxide, *Science*, *210*, 1347–1349.

Pitts, J.N., Jr., J.A. Sweetman, B. Zielinska, R. Atkinson, A.M. Winer, and W.P. Harger (1985b), Formation of nitroarenes from the reaction of polycyclic aromatic hydrocarbons with dinitrogenpentoxide, *Environ. Sci. Technol.*, *19*, 1115–1121.

Pitts, J.N., Jr., J.A. Sweetman, B. Zielinska, A.M. Winer, and R. Atkinson (1985c), Determination of 2-nitrofluoranthene and 2-nitropyrene in ambient particulate organic matter: evidence for atmospheric reactions, *Atmos. Environ.*, *19*, 1601–1608.

Pitts, J.N., Jr., J.K.S. Wan, and E. A. Schuck (1964), Photochemical studies in an alkali halide matrix. I. An *o*-nitrobenzaldehyde actinometer and its application to a kinetic study of the photoreduction of benzophenone by benzhydrol in a pressed potassium bromide disk, *J. Am. Chem. Soc.*, *86*, 3606–3610.

Pitts, J.N., Jr., B. Zielinska, J. Arey, T. Ramdahl, A.M. Winer, R. Atkinson, and W.P. Harger (1985d), Formation of nitroarenes from the reaction of polycyclic aromatic hydrocarbons with dinitrogen pentoxide, *Environ. Sci. Technol.*, *19*, 1115–1121.

Platt, U., A.M. Winer, H.W. Biermann, R. Atkinson, and J.N. Pitts, Jr. (1984), Measurement of nitrate radical concentrations in continental air, *Environ. Sci. Technol.*, *19*, 365–369.

Platz, J., O.J. Nielsen, T.J. Wallington, J.C. Ball, M.D. Hurley, A.M. Straccia, W.F. Schneider, and J. Sehested (1998), Atmospheric chemistry of the phenoxy radical, $C_6H_5O^{\cdot}$: UV spectrum and kinetics of reactions with NO, NO_2 and O_2, *J. Phys. Chem. A*, *102*, 7964–7974.

Plum, C.N., E. Sanhueza, R. Atkinson, W.P.L. Carter, and J.N. Pitts, Jr. (1983), OH radical rate constants and photolysis rates of α-dicarbonyls, *Environ. Sci. Technol.*, *17*, 479–484.

Porter, G., and B. Ward (1968), The photolytic preparation of cyclopentadienyl and phenyl nitrene from benzene derivatives, *Proc. R. Soc. London, Ser. A*, *303*, 139–156.

Raber, W.H., and G.K. Moortgat (1995), Photooxidation of selected carbonyl compounds in air: methyl ethyl ketone, methyl vinyl ketone, methacrolein and methylglyoxal, in *Progress and Problems in Atmospheric Chemistry*, edited by J.R. Barker, pp. 318–373, World Scientific, New York.

Ramdahl, T., I. Altheim, S. Rustad, and T. Olsen (1982), Chemical and biological characterization of emissions from small residential wood stoves burning wood and charcoal, *Chemosphere*, *11*, 601–611.

Ramdahl, T., and M. Moller (1983), Chemical and biological characterization of emissions from a cereal straw burning furnace, *Chemosphere*, *12*, 23–34.

Ravishankara, A.R., S. Wagner, S. Fischer, G. Smith, R. Schiff, R.T. Watson, G. Tesi, and D.D. Davis (1978), A kinetics study of the reactions of OH with several aromatic and olefinic compounds, *Int. J. Chem. Kinet.*, *10*, 783–804.

Repapis, C.C., H.T. Mantis, A.G. Paliatsos, C.M. Philandras, A.F. Bais, and C. Meleti (1998), Case study of UV-B modification during episodes of urban air pollution, *Atmos. Environ.*, *32*, 2203–2208.

Reuder, J., and H. Schwander (1999), Aerosol effects on UV radiation in nonurban regions, *J. Geophys. Res.*, *104*, 4065–4077.

Reuder, J., T. Gori, L. Kins, and R. Dlugi (1996), Determination of photolysis frequencies of ozone and nitrogen dioxide during SANA 2: the influence of tropospheric aerosol particles, *Meteorol. Z. N.F.*, *5*, 234–244.

Reyes, A.C., M. Medina, C. Crespo-Hernandez, M.Z. Cedeno, R. Arce, O. Rosario, D.M. Steffenson, I.N. Ivanov, M.E. Sigman, and R. Debestani (2000), Photochemistry of pyrene on unactivated and activated silica surfaces, *Environ. Sci. Technol.*, *34*, 415–421.

Rindone, B., G. Cariati, G. Restelli, and J. Hjorth (1991), Products and mechanism of the gas phase reaction between nitrate radical and arenes, *Fresenius, J. Anal. Chem.*, *339*, 673–675.

Rinke, M., and C. Zetzsch (1984), Rate constants for the reaction of OH radicals with aromatics: benzene, phenol, aniline, and 1,2,4-trichlorobenzene, *Ber. Bunsen-Ges. Phys. Chem.*, *88*, 55–62.

Risby, T.H., and S.S. Sehnert (1988), A model for the formation of airborne particulate matter based on the gas phase adsorption on amorphous carbon blacks, *Environ. Health Prespect.*, *77*, 131–140.

Rogge, W.F., L.M. Hildemann, M.A. Mazurek, and G.R. Cass (1991), Sources of fine organic aerosol. 1. Charbroilers and meat cooking operations, *Environ. Sci. Technol.*, *25*, 1112–1125.

Rogge, W.F., L.M. Hildermann, M.A. Mazurek, and G.R. Cass (1993a), Sources of fine organic aerosol. 2. Noncatalyst and catalyst-equipped automobiles and heavy-duty diesel trucks, *Environ. Sci. Technol.*, *27*, 636–651.

Rogge, W.F., L.M. Hildemann, M.A. Mazurek, and G.R. Cass (1993b), Sources of fine organic aerosol. 3. Road dust, tire debris, and organometallic brake lining dust: roads as sources and sinks, *Environ. Sci. Technol.*, *27*, 1892–1904.

Rogge, W.F., L.M. Hildemann, M.A. Mazurek, and G.R. Cass (1993c), Sources of fine organic aerosol. 4. Particulate abrasion products from leaf surfaces of urban plants, *Environ. Sci. Technol.*, *27*, 2700–2711.

Rogge, W.F., L.M. Hildermann, M.A. Mazurek, and G.R. Cass (1993d), Sources of fine organic aerosol. 5. Natural gas home appliances, *Environ. Sci. Technol.*, *27*, 2736–2744.

Rogge, W.F., L.M. Hildermann, M.A. Mazurek, and G.R. Cass (1994), Sources of fine organic aerosol. 6. Cigarette smoke in the urban atmosphere, *Environ. Sci. Technol.*, *28*, 1375–1388.

Rogge, W.F., L.M. Hildemann, M.A. Mazurek, G.R. Cass, and B.R.T. Simoneit (1997), Sources of fine organic aerosol. 8. Boilers burning No. 2 distillate fuel oil, *Environ. Sci. Technol.*, *31*, 2731–2737.

Rogge, W.F., L.M. Hildemann, M.A. Mazurek, G.R. Cass, and B.R.T. Simoneit (1998), Sources of fine organic aerosol. 9. Pine, oak, and synthetic log combustion in residential fireplaces, *Environ. Sci. Technol.*, *32*, 13–22.

Rosenkranz, H.S., and R. Mermelstein (1983), Mutagenicity and genotoxicity of nitroarenes: all nitro-containing chemical are not created equal, *Mutat. Res.*, 114, 217–267.

Röth, E.-P., R. Ruhnke, G. Moortgat, R. Meller, and W. Schneider (1997), UV/VIS-absorption cross sections and quantum yields for use in photochemistry and atmospheric modeling. Part 2: Organic substances, Institut für Chemie und Dynamik der Geosphäre 1: Stratosphärische Chemie, Forschungszentrum Jülich, Jülich, Germany.

Rounds, S.A., B.A. Tiffany, and J.F. Pankow (1993), Description of gas/particle sorption kinetics with an interparticle diffusion model: desorption experiments, *Environ. Sci. Technol.*, *27*, 366–377.

Ruggaber, A.R., R. Dlugi, and T. Nakajima (1994), Modeling radiation quantities and photolysis frequencies in the troposphere, *J. Atmos. Chem.*, *18*, 171–210.

Sasaki, J., S.M. Aschmann, E.S.C. Kwok, R. Atkinson, and J. Arey (1997), Products of the gas-phase OH and NO_3 radical-initiated reactions of naphthalene, *Environ. Sci. Technol.*, *31*, 3173–3179.

Saxena, P., and L.M. Hildemann (1996), Water-soluble organics in atmospheric particles: a critical review of the literature and application of thermodynamics to identify candidate compounds, *J. Atmos. Chem.*, *24*, 57–109.

Schauer, J.J., W.F. Rogge, L.M. Hildemann, M.A. Mazurek, G.R. Cass, and B.R.T. Simoneit (1996), Source apportionment of airborne particulate matter using organic compounds as tracers, *Atmos. Environ.*, *30*, 3837–3855.

Schliephake, V., K.-H. Mix, and H.G. Wagner (1986), Investigations of the kinetics of the reactions of tetrahydropyrane, cyhclopenanone and cyclohexanone with atomic oxygen, *Z. Phys. Chem. (Munich)*, *150*, 1–15.

Schoental, R. (1984), Carcinogenesis by polycyclic aromatic hydrocarbons and by certain other carcinogens, in *Polycyclic Aromatic Hydrocarbons*, edited by E. Clar, pp. 133–160, Academic Press, London.

Schuetzle, D., F.S. Lee, T.J. Prater, and S.B. Tajada (1981), The identification of polycyclic aromatic hydrocarbon derivatives in mutagenic fractions of diesel particulate extracts, *Int. J. Environ. Anal. Chem.*, *9*, 93–144.

Schwarzenbach, R.P., P.M. Gschwend, and D.M. Imboden (1993), *Environmental Organic Chemistry*, John Wiley & Sons, New York.

Seinfeld, J.H., and S.N. Pandis (1998), *Atmospheric Chemistry and Physics: From Air Pollution to Climate Change*, John Wiley, New York.

Semadeni, M., D.W. Stocker, and J.A. Kerr (1995), The temperature dependence of the OH radical reactions with some aromatic compounds under simulated tropospheric conditions, *Int. J. Chem. Kinet.*, *27*, 287–304.

Sempere, R., and K. Kawamura (1994), Comparative distributions of dicarboxylic acids and related polar compounds in snow, rain, and aerosols from urban atmosphere, *Atmos. Environ.*, *28*, 449–459.

Seuwen, R., and P. Warneck (1996), Oxidation of toluene in NO_x free air: product distribution and mechanism, *Int. J. Chem. Kinet.*, *28*, 315–332.

Shepson, P.B., E.O. Edney, and E.W. Corse (1984), Ring fragmentation reactions in the photooxidations of toluene and *o*-xylene, *J. Phys. Chem.*, *88*, 4122–4126.

Shinohara, H., and N. Nishi (1982), Laser photofragmentation dynamics of an acrolein supersonic molecular beam at 193 nm, *J. Chem. Phys.*, *77*, 234–245.

Sibener, S.J., R.J. Buss, P. Casavecchia, T. Hirooka, and Y.T. Lee (1980), A crossed molecular beams investigation of the reactions $O(^3P) + C_6H_6$, C_6D_6, *J. Chem. Phys.*, *72*, 4341–4349.

Simcik, M.F., T.P. Franz, H. Zhang, and S.J. Eisenreich (1998), Gas–particle partitioning of PCBs and PAHs in the Chicago urban and adjacent coastal atmosphere: states of equilibrium, *Environ. Sci. Technol.*, *32*, 251–257.

Sloane, T.M. (1977), Reaction product identification from $O(^3P)$ + benzene, toluene, and 1,3,5-trimethylbenzene collisions in crossed molecular beams, *J. Chem. Phys.*, *67*, 2267–2274.

Smith, A.M., E. Rigler, E.S.C. Kwok, and R. Atkinson (1996), Kinetics and products of the gas-phase reactions of 6-methyl-5-hepten-2-one and *trans*-cinnamaldehyde with OH and NO_3 radicals and O_3 at 296 ± 2 K, *Environ. Sci. Technol.*, *30*, 1781–1785.

Smith, D.F., T.E. Kleindienst, and C.D. McIver (1999), Primary product distributions from the reaction of OH with *m*-, *p*-xylene, 1,2,4- and 1,3,5-trimethylbenzene, *J. Atmos. Chem.*, *34*, 339–364.

Smith, D.F., C.D. McIver, and T.E. Kleindienst (1998), Primary product distribution from the reaction of hydroxyl radicals with toluene at ppb NO_x mixing ratios, *J. Atmos. Chem.*, *30*, 209–228.

Smith, D.J.T., and R.M. Harrison (1996), Concentrations, trends and vehicle source profile of polynuclear aromatic hydrocarbons in the UK atmosphere, *Atmos. Environ.*, *30*, 2513–2525.

Snelling, D.R. (1968), Production of singlet oxygen in the benzene oxygen photochemical system, *Chem. Phys. Lett.*, *2*, 346–348.

Sommerlade, R., H. Parlar, D. Wrobel, and P. Kochs (1993), Product analysis and kinetics of the gas-phase reactions of selected organosilicon compounds with OH radicals using a smog chamber–mass spectrometer system, *Environ. Sci. Technol.*, *27*, 2435–2440.

Sørensen, S., and I. Barnes (1998), Photolysis of unsaturated 1,4-dicarbonyls, in *EUPHORE Report 1997*, edited by I. Barnes and J. Wenger, pp. 149–164, University of Wuppertal, Wuppertal, Germany.

Stamnes, K.S., S.C. Tsay, W. Wiscombe, and K. Jayaweera (1988), Numerical Table algorithm for discrete-ordinate-method radiative transfer in multiple scattering and emitting layered media, *Appl. Opt.*, *27*, 2501–2509.

Stedman, D.H., and H. Niki (1973), Ozonolysis rates of some atmospheric gases, *Environ. Lett.*, *4*, 303–310.

Steer, R.P., J.L. Sprung, and J.N. Pitts, Jr. (1969), Singlet oxygen in the environmental sciences. Evidence for the production of $O_2(^1\Delta_g)$ by energy transfer in the gas phase, *Emviron. Sci. Technol.*, *3*, 946–947.

Stern, J.E., R.C. Flagan, D. Grosjean, and J.H. Seinfeld (1987), Aerosol formation and growth in atmospheric aromatic hydrocarbon photooxidation, *Environ. Sci. Technol.*, *21*, 1224–1231.

Stockwell, W.R., F. Kirchner, M. Kuhn, and S. Seefeld (1997), A new mechanism for regional atmospheric chemistry modeling, *J. Geophys. Res.*, *102*, 25847–25879.

Stockwell, W.R., P. Middleton, J.S. Chang, and X. Tang (1990), The second generation regional acid deposition model chemical mechanism for regional air quality modeling, *J. Geophys. Res.*, *95*, 16343–16367.

Strader, R., F. Lurmann, and S.N. Pandis (1999), Evaluation of secondary organic aerosol formation in winter, *Atmos. Environ.*, *33*, 4849–4863.

Streitwieser, A., and C.H. Heathcock (1976), *Introduction to Organic Chemistry*, MacMillan, New York.

Strommen, M.R., and R.M. Kamens (1997), Development and application of a dual-impedance radial diffusion model to simulate the partitioning of semivolatile organic compounds in combustion aerosols, *Environ. Sci. Technol.*, *31*, 2983–2990.

Strommen, M.R., and R.M. Kamens (1999), Simulation of semivolatile organic compound microtransport at different time scales in airborne diesel soot particles, *Environ. Sci. Technol.*, *33*, 1738–1746.

Sugimoto, H., and D.T. Sawyer (1988), Transient formation of hydrogen tetraoxide (HOOOOH) from the dimerization of perhydroxyl (HO_2): concerted homolytic dissociation to singlet oxygen ($^1O_2$) and hydrogen peroxide, *J. Am. Chem. Soc.*, *110*, 8707–8708.

Sweetman, J.A., B. Zielinska, R. Atkinson, T. Ramdahl, A.M. Winer, and J.N. Pitts, Jr. (1986), A possible formation pathway for the 2-nitrofluoranthene observed in ambient particulate matter, *Atmos. Environ.*, *20*, 235–238.

Takagi, H., N. Washida, H. Akimoto, K. Nagasawa, Y. Usui, and M. Okuda (1980), Photooxidation of *o*-xylene in the NO–H_2O–air system, *J. Phys. Chem.*, *84*, 478–483.

Takagi, H., N. Washida, H. Akimoto, and M. Okuda (1982), Observation of 3-hexene-2,5-dione in the photooxidation of 1,2,3-trimethylbenzene in the NO–H_2O–air system, *Spectrosc. Lett.*, *15*, 145–152.

Takezaki, M., N. Hirota, and M. Terazima (1997a), Nonradiative relaxation processes and electronically excited states of nitrobenzene studied by picosecond time-resolved transient grating method, *J. Phys. Chem. A*, *101*, 3443–3448.

Takezaki, M., N. Hirota, M. Terazima, H. Sato, T. Nakajima, and S. Kato (1997b), Geometries and energies of nitrobenzene studied by CAS-SCF calculations, *J. Phys. Chem. A*, *101*, 5190–5195.

Talcott, R.E., and W. Harger (1981), Chemical characterization of direct-acting airborne mutagens: the functional group, *Mutat. Res.*, *91*, 433–436.

Tang, H., E. A. Lewis, D.J. Eatough, R.M. Burton, and R.J. Farber (1994), Determination of the particle size distribution and chemical compounds in the atmospheric fine particles with a diffusion denuder sampling system, *Atmos. Environ.*, *28*, 939–947.

Tao, Z, and Z. Li (1999), A kinetics study on reactions of C_6H_5O with C_6H_5O and O_3 at 298 K, *Int. J. Chem. Kinet.*, *31*, 65–72.

Tappe, M., V. Schliephake, and H.G. Wagner (1989), Reaction of benzene, toluene and ethylbenzene with atomic oxygen (O^3P) in the gas phase, *Z. Phys. Chem., Neue Folge*, *162*, 129–145.

Tebbens, B.D., M. Mitsugi, and J.F. Thomas (1971), Fate of arenes incorporated with airborne soil. Effect of radiation, *Am. Ind. Hyg. Assoc. J.*, *32*, 365–372.

Tebbens, B.D., J.F. Thomas, and M. Mukai (1966), Fate of arenes incorporated with airborne soot, *Am. Ind. Hyg. Assoc. J.*, *27*, 415–422.

Thijsse, T.R., R.F. van Oss, and P. Lenschow (1999), Determination of source contributions to ambient volatile organic compound concentrations in Berlin, *J. Air Waste Manage. Assoc.*, *49*, 1394–1404.

Thomas, J.F., M. Mukai, and B.D. Tebbens (1968), Fate of airborne benzo[a]pyrene, *Environ. Sci. Technol.*, *2*, 33–39.

Thüner, L.P., T. Maurer, B. Donner, I. Barnes, K.H. Becker, J. Bea, M. Ponns, and K. Wirtz (1998), Exhaust gas experiment in the EUPHORE simulation chamber, in *Influence of Fuel Formulation on Reactivity of Exhaust Gases (INFORMATEX)*, edited by K.H. Becker, Chapter B1.3, EC-Project ENV$-CT95-0015, Final Report, University of Wuppertal, Germany.

Tobias, H.J., and P.J. Ziemann (2000), Thermal desorption mass spectrometric analysis of organic aerosol formed from reactions of 1-tetradecene and O_3 in the presence of alcohols and carboxylic acid, *Environ. Sci. Technol.*, *34*, 2105–2115.

Toby, S., L.J. Van de Burgt, and F.S. Toby (1985), Kinetics and chemiluminescence of ozone–aromatic reactions in the gas phase, *J. Phys. Chem.*, *89*, 1982–1986.

Tremp, J., P. Mattrel, S. Fingler, and W. Giger (1993), Phenol and nitrophenols as tropospheric pollutants: emissions from automobile exhausts and phase transfer in the atmosphere, *Water, Air, Soil Pollut.*, *68*, 113–123.

Tse, C.W., R.C. Flagan, and J.H. Seinfeld (1997), Rate constants for the gas-phase reaction of the hydroxyl radical with a series of dimethylbenzaldehydes and trimethylphenols at atmospheric pressure, *Int. J. Chem. Kinet.*, *29*, 523–525.

Tuazon, E.C., J. Arey, R. Atkinson, and S.M. Aschmann (1993), Gas-phase reactions of 2-vinylpyridine and styrene with OH and NO_3 radicals and O_3, *Environ. Sci. Technol.*, *27*, 1832–1841.

Tuazon, E.C., R. Atkinson, and W.P.L. Carter (1985), Atmospheric chemistry of *cis*- and *trans*–3-hexene–2,5-dione, *Environ. Sci. Technol.*, *19*, 265–269.

Tuazon, E.C., R. Atkinson, H. MacLeod, H.W. Biermann, A.M. Winer, W.P.L. Carter, and J.N. Pitts, Jr. (1984a), Yields of glyoxal and methylglyoxal from the NO_x–air photooxidations of toluene and *m*- and *p*-xylene, *Environ. Sci. Technol.*, *18*, 981–984.

Tuazon, E.C., R. Atkinson, A.M. Winer, and J.N. Pitts, Jr. (1984b), A study of the atmospheric reactions of 1,3-dichloropropene and other selected organochlorine compounds, *Arch. Environ. Contam. Toxicol.*, *13*, 691–700.

Tuazon, E.C., H. MacLeod, R. Atkinson, and W.P.L. Carter (1986), α-Dicarbonyl yields from the NO_x–air photooxidations of a series of aromatic hydrocarbons in air, *Environ. Sci. Technol.*, *20*, 383–387.

Tully, F.P., A.R. Ravishankara, R.L. Thompson, J.M. Nicovich, R.C. Shah, N.M. Kreutter, and P.H. Wine (1981), Kinetics of the reactions of hydroxyl radical with benzene and toluene, *J. Phys. Chem.*, *85*, 2262–2269.

Turpin, B.J., and J.J. Huntzicker (1991), Secondary formation of organic aerosol in the Los Angeles basin: a descriptive analysis of organic and elemental carbon concentrations, *Atmos. Environ.*, *25A*, 207–215.

Turpin, B.J., and J.J. Huntzicker (1995), Identification of secondary organic aerosol episodes and quantification of primary and secondary organic aerosol concentration during SCAQS, *Atmos. Environ.*, *29A*, 3327–3344.

Turpin, B.J., S.V. Hering, and J.J. Huntzicker (1994), Investigation of organic aerosol sampling artifacts in the Los Angeles Basin, *Atmos. Environ.*, *28*, 3062–3071.

Turpin, B.J., J.J. Huntzicker, S.M. Larson, and G.R. Cass (1991), Los Angeles summer midday particulate carbon: primary and secondary aerosol, *Environ. Sci. Technol.*, *25*, 1788–1793.

Turro, N.J. (1978), *Modern Molecular Photochemistry*, Benjamin/Cummings, Menlo Park, CA.

Tyndall, G.S., T.A. Staffelbach, J.J. Orlando, and J.G. Calvert (1995), Rate coefficients for the reactions of OH radicals with methylglyoxal and acetaldehyde, *Int. J. Chem. Kinet.*, *27*, 1009–1020.

UBA (1999), http://www.umweltbundesamt.de/uba-info-presse/pressemitteilungen/okap399.htm

Ulmstead, M.E., R.G. Shortridge, and M.C. Lin (1978), Energy partitioning in the photodissociation of C_3H_4O near 200 nm, *J. Phys. Chem.*, *82*, 1455–1460.

U.S. Standard Atmosphere (1976), National Oceanic and Atmospheric Administration, National Aeronautics and Space Administration, United States Air Force, Washington, DC, October.

Vallalta, P.W., E.R. Lovejoy, and D.R. Hanson (1996), Reaction probability of peroxyacetyl radical on aqueous surfaces, *Geophys. Res. Lett.*, *23*, 1765–1768.

Van Cauwenberghe, K.A., L. Van Vaeck, and J.N. Pitts, Jr. (1984), Physical and chemical transformations of organic pollutants during sampling, *Forensic Environ. Applic.*, 1499–1507.

Van Hoosier, M.E., J.-D.F. Bartoe, G.E. Brueckner, and K.D. Prinz (1988), Absolute solar spectral irradiance 120 nm–400 nm (Results from the Solar Ultraviolet Spectral Irradiance Monitor—SUSIM—Experiment on board Spacelab 2), *Astro. Lett. Commun.*, 27, 163–168.

Vanselow, R.D., and F.E. Blacet (1957), in *Abstracts of the 131th National Meeting of the American Chemical Society*, Miami Beach, FL.

Van Vaeck, L., and K. Van Cauwenberghe (1984), Conversion of polycyclic aromatic hydrocarbons on diesel particulate matter upon exposure to ppm levels of ozone, *Atmos. Environ.*, *18*, 323–328.

Van Vaeck, L., K. Van Cauwenberghe, and J. Janessens (1984), The gas–particle distribution of organic constituents: measurement of the volatilization artifact in hi-vol cascade impactor sampling, *Atmos. Environ.*, *18*, 417–430.

Vartiainen, M., S.R. McDow, and R.M. Kamens (1996), Water uptake by sunlight and ozone exposed diesel exhaust particles, *Chemosphere*, *32*, 1319–1325.

Venkataraman, C., J.M. Lyons, and S.K. Friedlander (1994), Size distributions of polycyclic aromatic hydrocarbons and elemental carbon. 1. Sampling, measurement methods, and source characterization, *Environ. Sci. Technol.*, *28*, 555–562.

Venkataraman, C., S. Thomas, and P. Kulkarni (1999), Size distributions of polycyclic aromatic hydrocarbons–gas/particle partitioning to urban aerosols, *J. Aerosol Sci.*, *30*, 759–770.

Vesly, G.F., and P.A. Leermakers (1964), The photochemistry of α-keto acids and α-keto esters. III. Photolysis of pyruvic acid in the vapor phase, *J. Phys. Chem.*, *68*, 2364–2366.

Vogel, E., and H. Günther (1967), Benzene oxide–oxepin valence tautomerism, *Angew. Chem., Int. Ed. Engl.*, *6*, 385–401.

Wahner, A., and C. Zetzsch (1983), Rate contants for the addition of OH to aromatics (benzene, *p*-chloroaniline, and *o*-, *m*- and *p*-dichlorobenzene) and the unimoecular decay of the adduct. Kinetics into a quasi-equilibrium. 1, *J. Phys. Chem.*, *87*, 4945–4951.

Wallington, T.J., P. Dagaut, and M.J. Kurylo (1992), Ultraviolet absorption cross sections and reaction kinetics and mechanisms for peroxy radicals in the gas phase, *Chem. Rev.*, *92*, 667–710.

Wallington, T.J., D.M. Neuman, and M.J. Kurylo (1987), Kinetics of the gas phase reaction of hydroxyl radicals with ethane, benzene, and a series of halogenated benzenes over the temperature range 234–438 K, *Int. J. Chem. Kinet.*, *19*, 725–739.

Wang, C.Y., M.-S. Lee, C.M. King, and O.P. Warner (1980), Evidence of nitroaromatics as direct acting mutagens of airborne particulate matter, *Chemosphere*, *9*, 83–87.

Wang, H., and M. Frenklach (1997), A detailed kinetic modeling study of aromatic formation in laminar premixed acetylene and ethylene flames, *Combust. Flame*, *110*, 173–221.

Wang, S.-C., R.C. Flagan, and J.H. Seinfeld (1992a), Aerosol formation and growth in atmospheric organic/NOx systems—II. Aerosol dynamics, *Atmos. Environ.*, *26A*, 421–434.

Wang, S.-C., S.E. Paulson, D. Grosjean, R.C. Flagan, and J.H. Seinfeld (1992b), Aerosol formation and growth in atmospheric organic/NOx systems—I. Outdoor smog chamber studies of C_7- and C_8-hydrocarbons, *Atmos. Environ.*, *26A*, 403–420.

Wängberg, I., T. Etzkorn, I. Barnes, U. Platt, and K.H. Becker (1997), Absolute determination of the temperature behavior of the $NO_2 + NO_3 + (M) \leftrightarrow N_2O_5 + (M)$ equilibrium, *J. Phys. Chem. A*, *101*, 9694–9698.

Warneck, P. (1988), *Chemistry of the Natural Atmosphere*, Academic Press, New York.

Wasserman, E., V.J. Kuck, W.M. Delavan, and W.A. Yager (1969), Electron paramagnetic resonance of $^1\Delta$ oxygen produced by the gas-phase photosensitization with naphthalene derivatives, *J. Am. Chem. Soc.*, *91*, 1040–1041.

Weldon, D., and P.R. Ogilby (1998), Time-resolved absorption spectrum of singlet oxygen in solution, *J. Am. Chem. Soc.*, *120*, 12978–12979.

Wiesen, E., I. Barnes, and K.H. Becker (1995), Study of the OH-initiated degradation of the aromatic photooxidation product 3,4-dihydroxy–3-hexene–2,5-dione, *Environ. Sci. Technol.*, *29*, 1380–1386.

Wine, P.H., and R.J. Thompson (1984), Kinetics of OH reactions with furan, thiophene, and tetrahydrothiophene, *Int. J. Chem. Kinet.*, *16*, 867–878.

Winer, A.M., J. Arey, R. Atkinson, S.M. Aschmann, W.D. Long, C.L. Morrison, and D.M. Olszyk (1992), Emission rates of organics from vegetation in California's central valley, *Atmos. Environ.*, *26*, 2647–2659.

Witte, F., E. Urbanik, and C. Zetzsch (1986), Temperature dependence of the rate constants for the addition of OH to benzene and to some monosubstituted aromatics (aniline, bromobenzene, and nitrobenzene) and the unimolecular decay of the adducts. Kinetics into a quasi-equilibrium. 2, *J. Phys. Chem.*, *90*, 3251–3259.

WMO (1986), Atmospheric Ozone 1985, Assessment of our understanding of the processes controlling its present distribution and change, World Meteorological Organization Global Ozone Research and Monitoring Project—Report No. 16.

Wolff, G.T., P.J. Groblicki, S.H. Cadle, and R.J. Countess (1983), Particulate carbon in various location in the United States, in *Particulate Carbon: Atmospheric Life Cycle*, edited by G.T. Wolff, and R.L. Klimisch, pp. 297–315, Plenum Press, New York.

Yamamoto, S., and R.A. Back (1985), The photolysis and thermal decomposition of pyruvic acid in the gas phase, *Can. J. Chem.*, *63*, 549–554.

Yamasaki, H., K. Kuwata, and H. Miyamoto (1982), Effects of ambient temperature on aspects of airborne polycyclic aromatic hydrocarbons, *Environ. Sci. Technol.*, *16*, 189–194.

Yaws, C.L., X. Lin, L.Bu, D.R. Balundgi, and S. Tripathi (1999), Vapor pressure, in *Chemical Properties Handbook: thermodynamic, environmental, transport, safety, and health related properties for organic and inorganic chemicals*, edited by C.L. Yaws, pp. 159–179, McGraw-Hill, New York.

Yip, R.W., D.K. Sharm, R. Giasson, and D. Gravel (1984), Picosecond excited-state absorption of alkyl nitrobenzenes in solution, *J. Phys. Chem.*, *88*, 5770–5772.

You, F., and T. Bidleman (1984), Influence of volatility on the collection of polycyclic aromatic hydrocarbon vapors with polyurethane foam, *Environ. Sci. Technol.*, *18*, 330–333.

Yu, J., and H.E. Jeffries (1997), Atmospheric photooxidation of alkylbenzenes—II. Evidence of formation of epoxide intermediates, *Atmos. Environ.*, *31*, 2281–2287.

Yu, J., D.R. Cocker, R.J. Griffin, R.C. Flagan, and J.H. Seinfeld (1999), Gas-phase ozone oxidation of monoterpenes gaseous and particulate products, *J. Atmos. Chem.*, *34*, 207–258.

Yu, J., H.E. Jeffries, and K.G. Sexton (1997), Atmospheric photooxidation of alkylbenzenes—I. Carbonyl product analyses, *Atmos. Environ.*, *31*, 2261–2280.

Yu, T., and M.C. Lin (1994), Kinetics of the $C_6H_5 + O_2$ reaction at low temperatures, *J. Am. Chem. Soc.*, *116*, 9571–9576.

Zellner, R., B. Fritz, and M. Preidel (1985), A cw-UV-laser absorption study of the reactions of the hydroxy-cyclohexadienyl radical with NO_2 and NO, *Chem. Phys. Lett.*, *121*, 412–416.

Zervas, E., X. Montagne, and J. Lahaye (1999), The influence of gasoline formulation on specific pollution emissions, *J. Air Waste Manage. Assoc.*, *49*, 1304–1314.

Zetzsch, C. (1982a), Predicting the rate of OH-addition to aromatics using σ^+-electrophilic substituent constants for mono- and polysubstituted benzene, in 15th Informal Conference on Photochemistry, Stanford, CA.

Zetzsch, C. (1982b), Report to Bundesminister für Forschung und Technologie, Projektträger für Umweltchemikalien; private communication to R. Atkinson, 1985.

Zetzsch, C., and B. Bohn (1999), Gas-phase reaction of the OH–benzene adduct with O_2; reversibility and secondary formation of HO_2, *Phys. Chem. Chem. Phys.*, *1*, 5097–5107.

Zetzsch, C., R. Koch, B. Bohn, R. Knispel, M. Siese, and F. Witte (1997), Adduct formation of OH with aromatics and unsaturated hydrocarbons and consecutive reactions with O_2 and NO_x to regenerate OH, in *Chemical Processes in Atmospheric Oxidation*, edited by G. LeBras, pp. 247–256, Springer-Verlag, Berlin.

Zetzsch, C., R. Koch, M. Siese, F. Witte, and P. Devolder (1989), Adduct formation of OH with benzene and toluene and reaction of the adducts with NO and NO_2, in *Proceedings of the 5th European Symposium on Physico-Chemical Behaviour of Atmospheric Pollutants*, edited by G. Restelli, and G. Angeletti, pp. 320–327, Reidel Publishing Co., Dordrecht, The Netherlands.

Zhang, J., W.E. Wilson, and P.J. Lioy (1994), Indoor air chemistry: formation of organic acids and aldehydes, *Environ. Sci. Technol.*, *28*, 1975–1982.

Zhang, S.-H., M. Shaw, J.H. Seinfeld, and R.C. Flagan (1992), Photochemical aerosol formation from α- and β-pinene, *J. Geophys. Res.*, *97*, 20717–20730.

Zheng, M., T.S.M. Wan, M. Fang, and F. Wang (1997), Characterization of the nonvolatile organic compounds in the aerosols of Hong Kong—identification, abundance and origin, *Atmos. Environ.*, *31*, 227–237.

Zhou, Z., and R. G. Parr (1989), New measures of aromaticity: absolute hardness and relative hardness, *J. Am. Chem. Soc.*, *111*, 7371–7379.

Zielinska, B., and E.M. Fujita (1994), The composition and concentration of hydrocarbons in the range C_7 to C_{18} in downtown Los Angeles, CA, *Res. Chem. Intermed.*, *20*, 321–334.

Zielinska, B., J. Arey, R. Atkinson, and P.A. McElroy (1988), Nitration of acephenanthrylene under simulated atmospheric conditions and in solution and the presence of nitroacephenanthrylene(s) in ambient particles, *Environ. Sci. Technol.*, *22*, 1044–1048.

Zielinska, B., J.A. Arey, R. Atkinson, T. Ramdahl, A.M. Winer, and J.N. Pitts, Jr. (1986), Reaction of dinitrogen pentoxide with fluoranthene, *J. Am. Chem. Soc.*, *108*, 4126–4132.

Appendix I

Nomenclature and Molecular Structure of Aromatic Hydrocarbons

A. Substituted Benzenes

Name of Compound	Molecular Formula	C-Skeletal Structure	3-D-Structural Formulas	
Benzene	C_6H_6		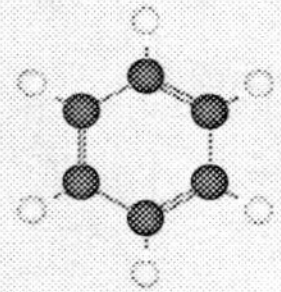	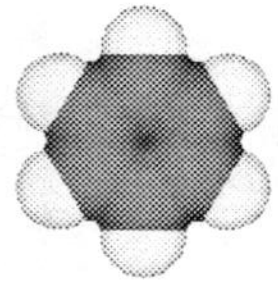
Toluene	C_7H_8	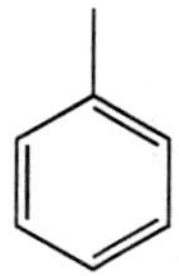	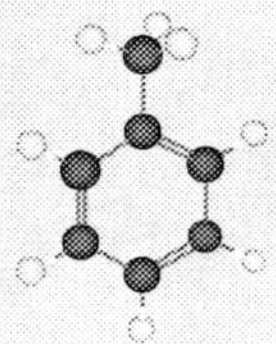	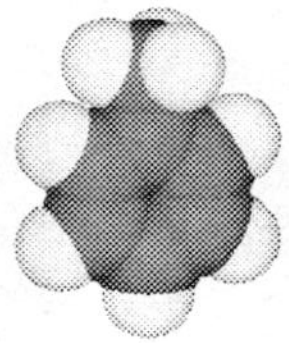
o-Xylene	C_8H_{10}	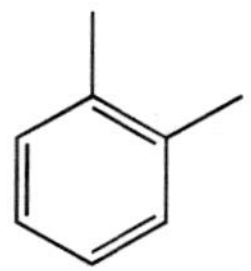	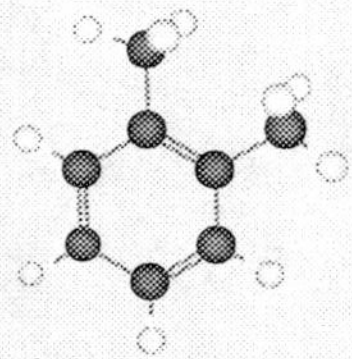	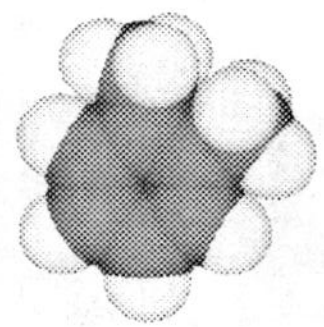

m-Xylene	C_8H_{10}	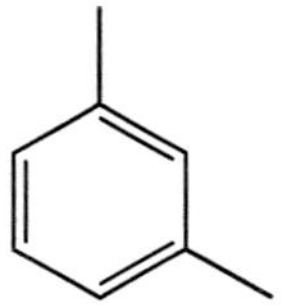	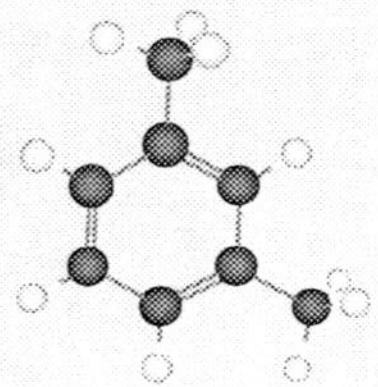	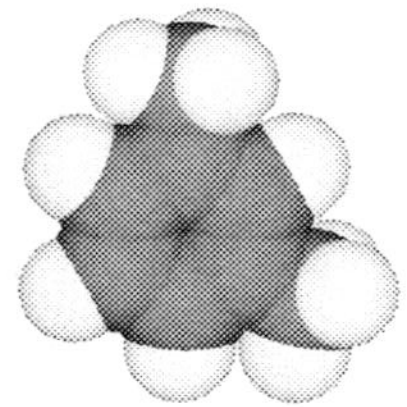
p-Xylene	C_8H_{10}		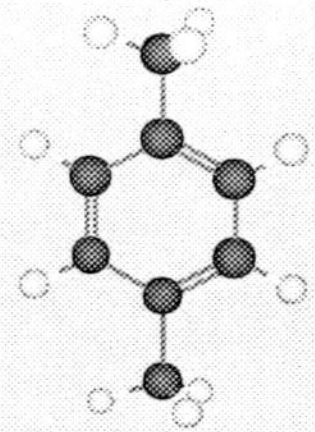	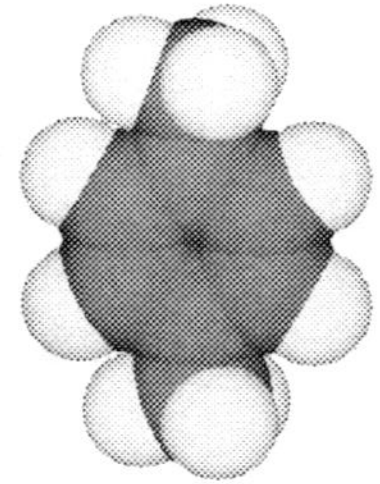
Styrene	C_8H_8	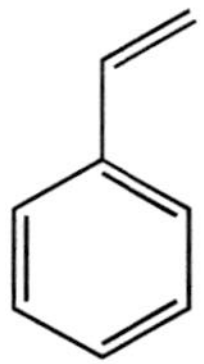	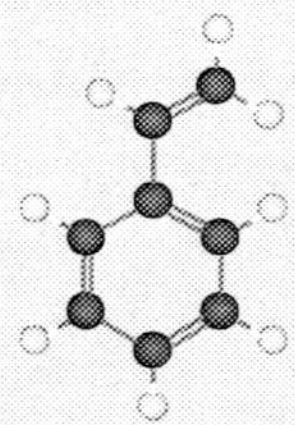	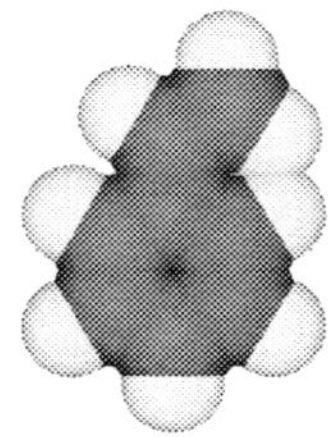

Ethylbenzene C_8H_{10}

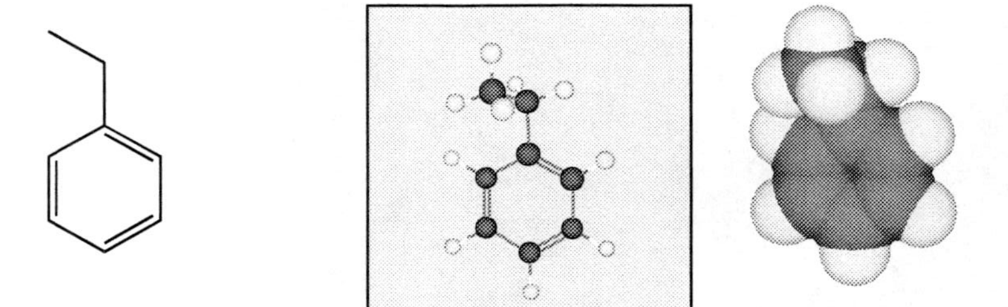

Indene C_9H_8

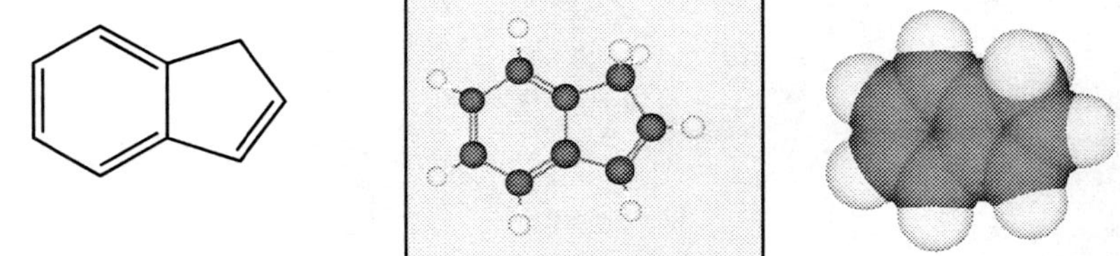

Indan C_9H_{10}

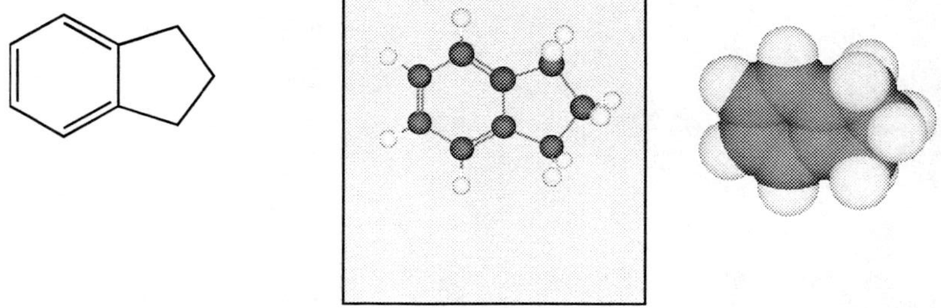

α-Methylstyrene
(1-methylethenylbenzene)

C_9H_{10}

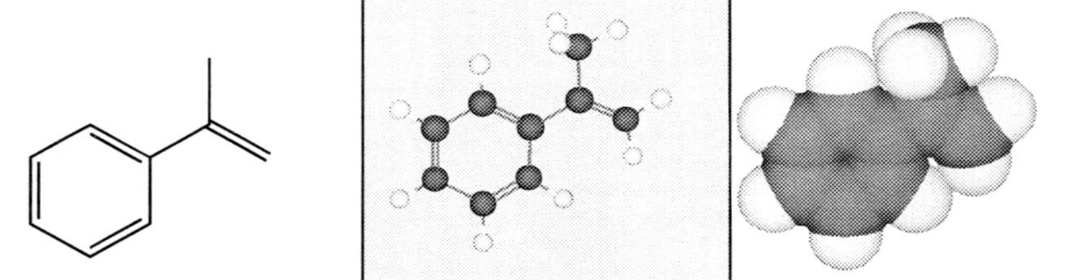

β-Methylstyrene)
(1-propenylbenzene)

C_9H_{10}

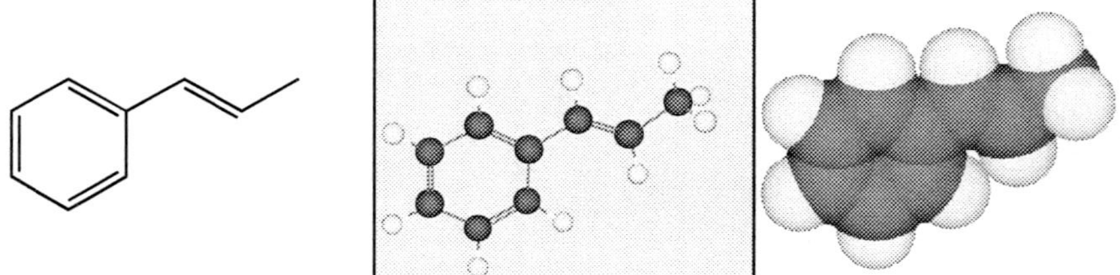

o-Ethyltoluene
(2-ethyl-1-methyl-benzene)

C_9H_{12}

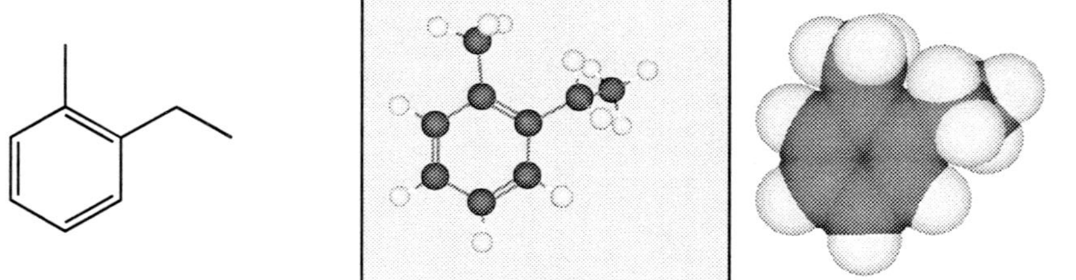

m-Ethyltoluene
(3-ethyl-1-methyl-benzene)

C_9H_{12}

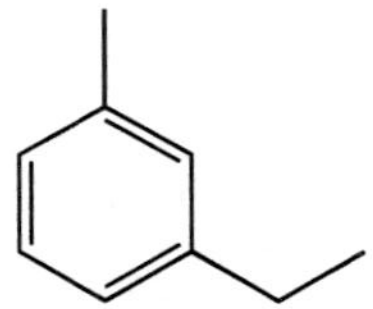

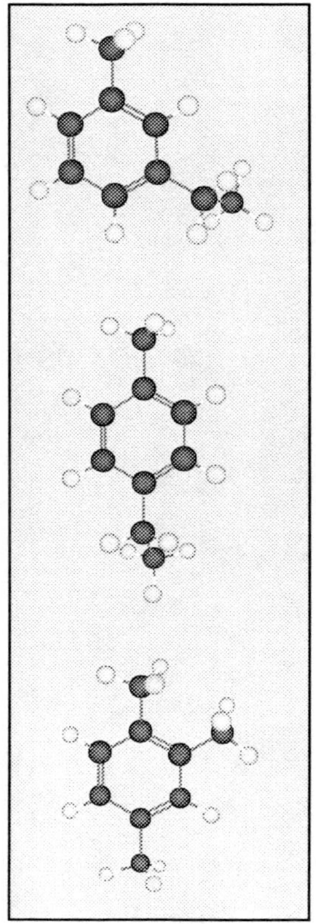

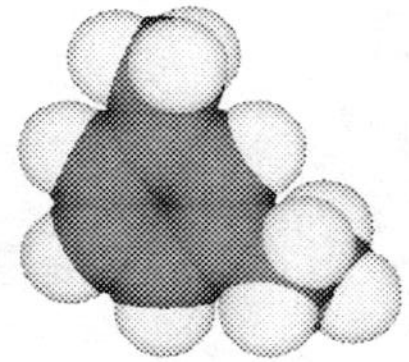

p-Ethyltoluene
(4-ethyl-1-methyl-benzene)

C_9H_{12}

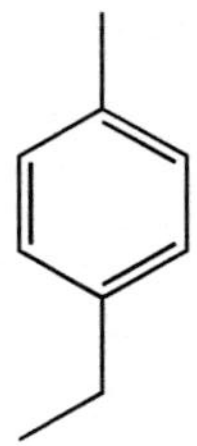

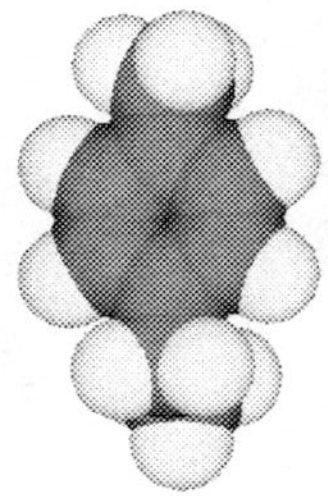

1,2,4-Trimethylbenzene

C_9H_{12}

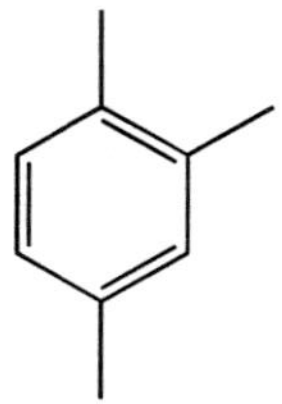

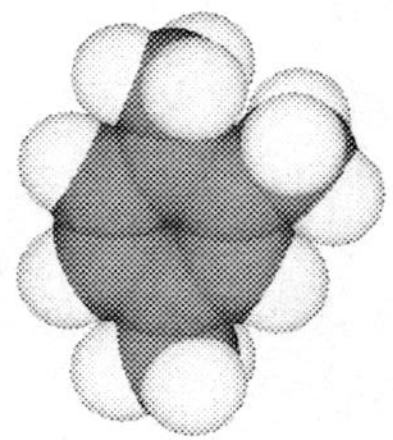

Name	Formula	Structure	Space-filling model
1,3,5-Trimethylbenzene (mesitylene)	C_9H_{12}	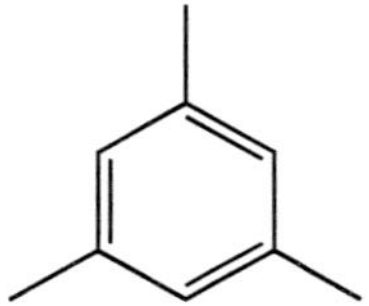	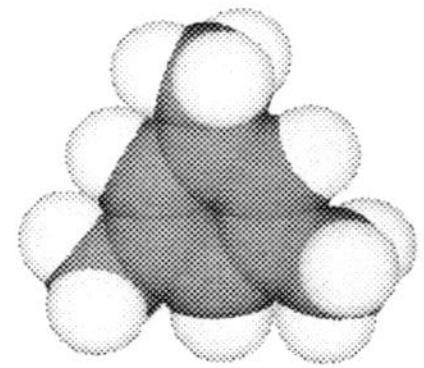
Isopropylbenzene	C_9H_{12}	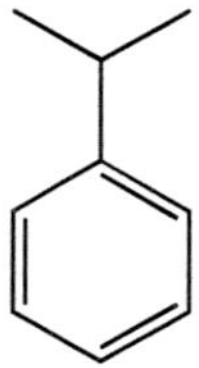	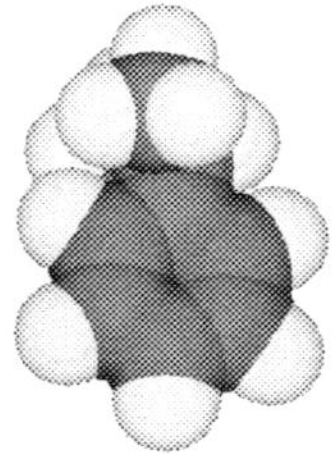
n-Propylbenzene	C_9H_{12}	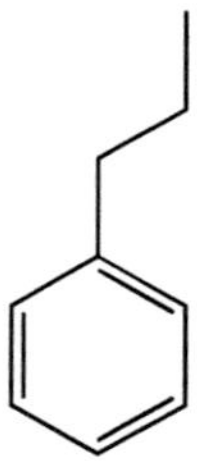	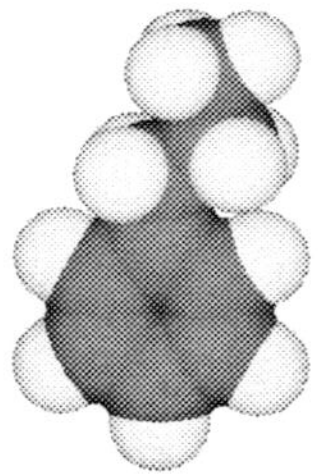

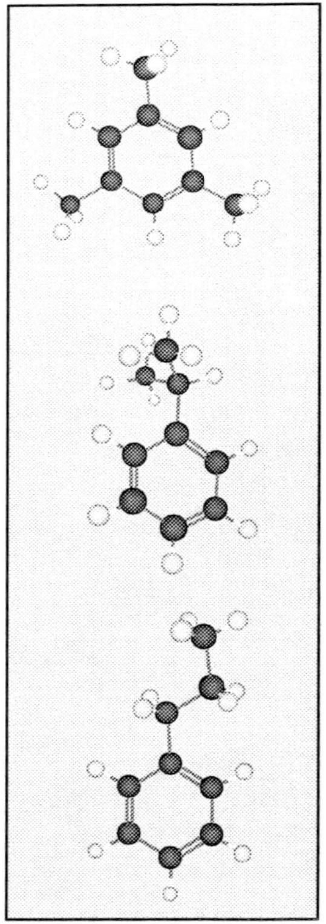

β, β-Dimethylstyrene $C_{10}H_{12}$

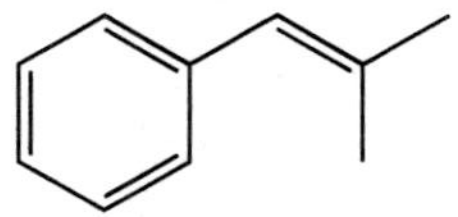
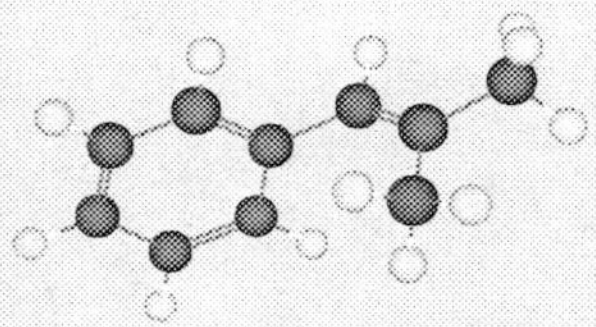
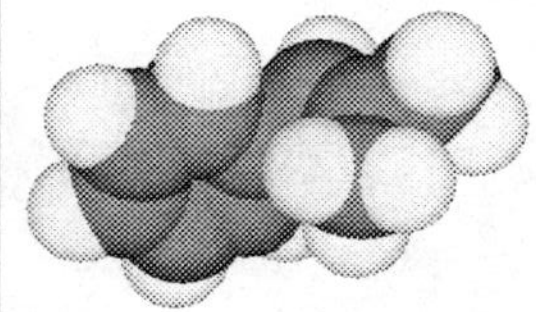

Tetralin $C_{10}H_{12}$

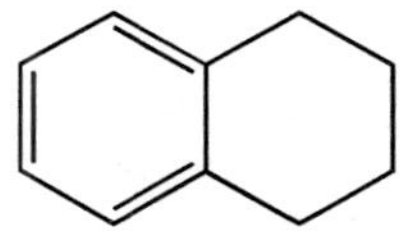
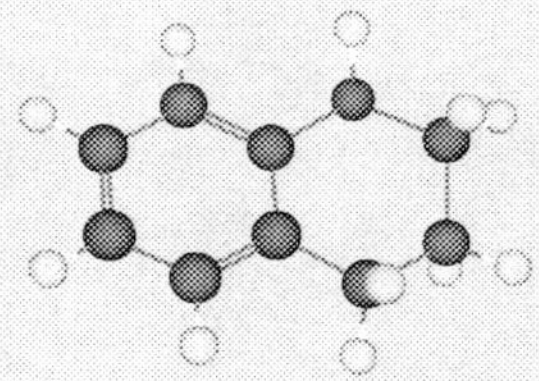
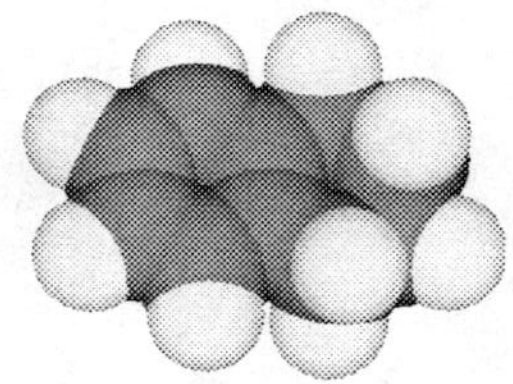

n-Butylbenzene $C_{10}H_{14}$

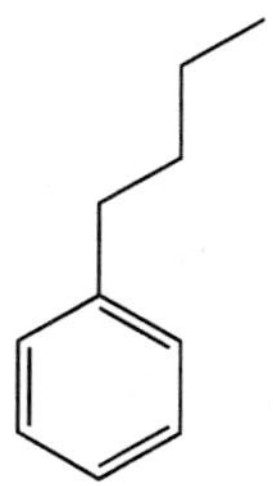
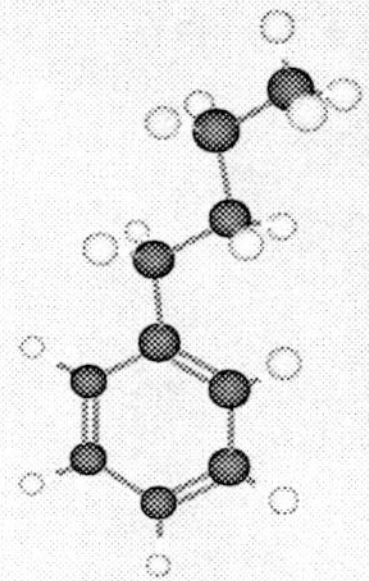
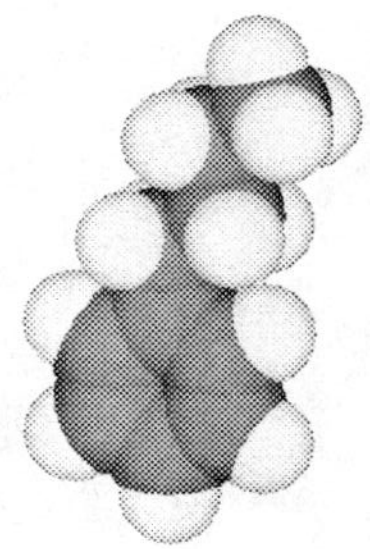

Isobutylbenzene $C_{10}H_{14}$

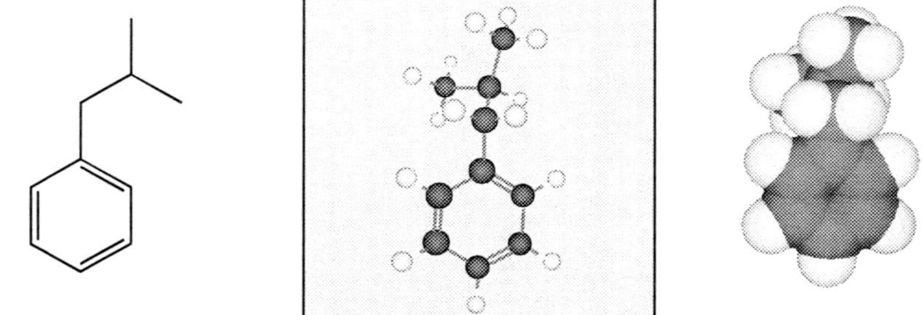

sec-Butylbenzene $C_{10}H_{14}$

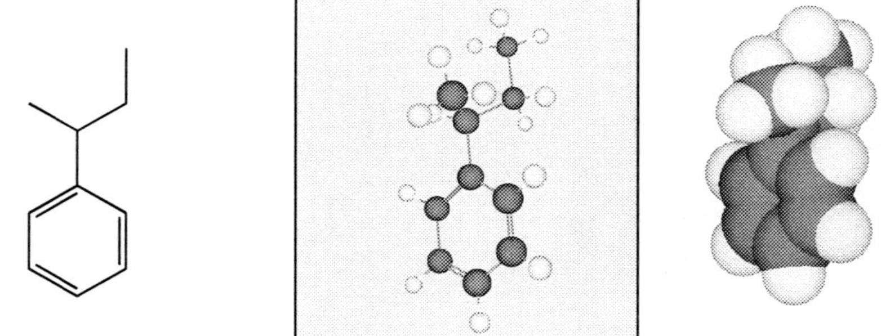

tert-Butylbenzene $C_{10}H_{14}$

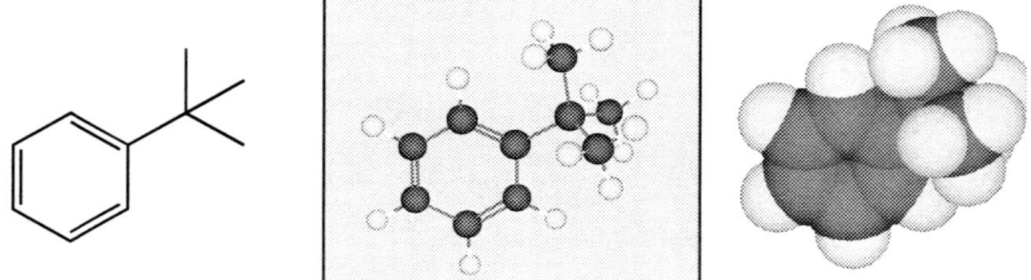

p-Cymene

$C_{10}H_{14}$

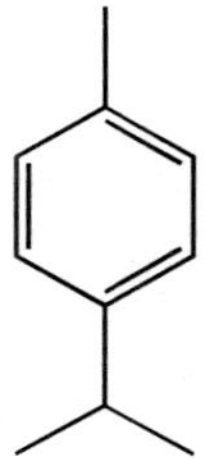

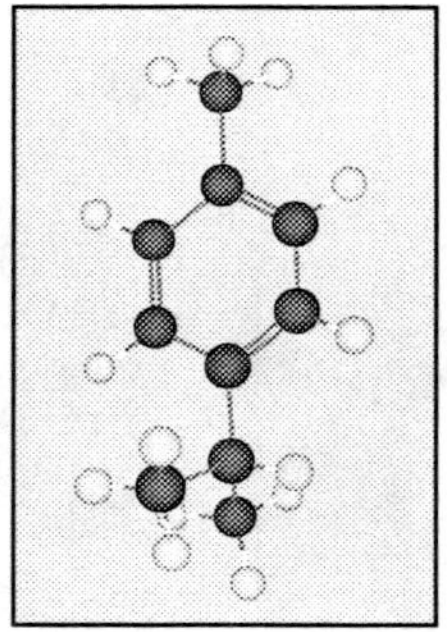

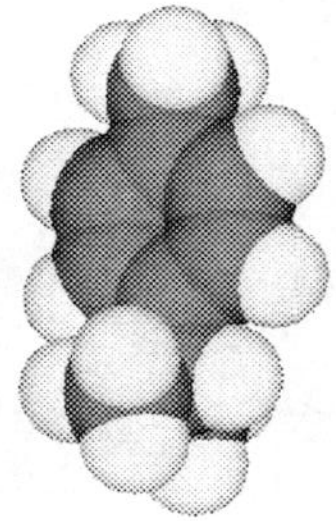

B. Polycyclic Aromatic Hydrocarbons

Name of Compound	Molecular Formula	C-Skeletal Structure	3D-Structural Formula
Naphthalene	$C_{10}H_8$	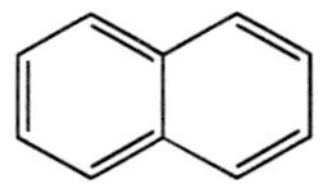	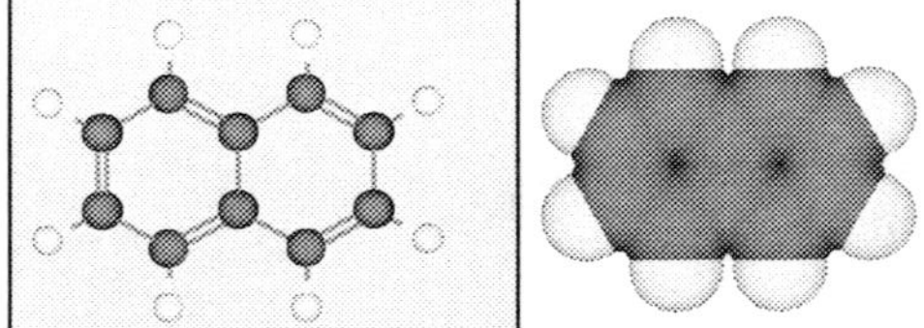
Azulene	$C_{10}H_8$	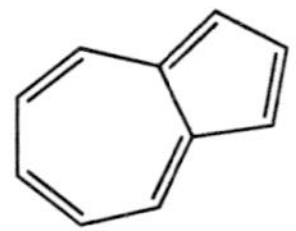	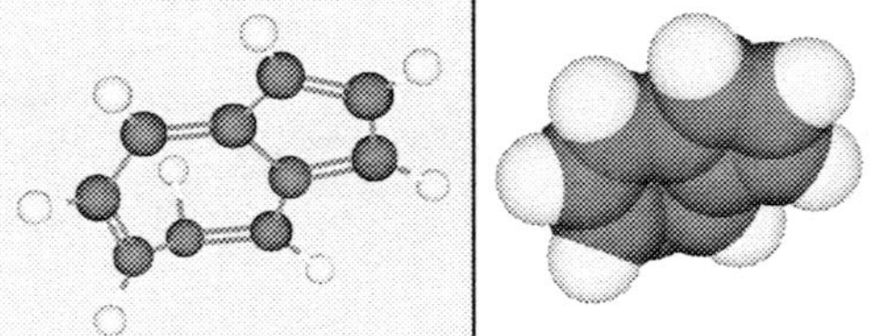
1-Methylnaphthalene	$C_{11}H_{10}$	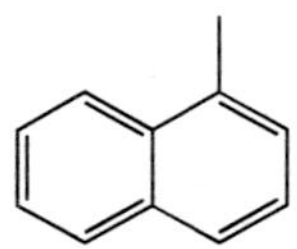	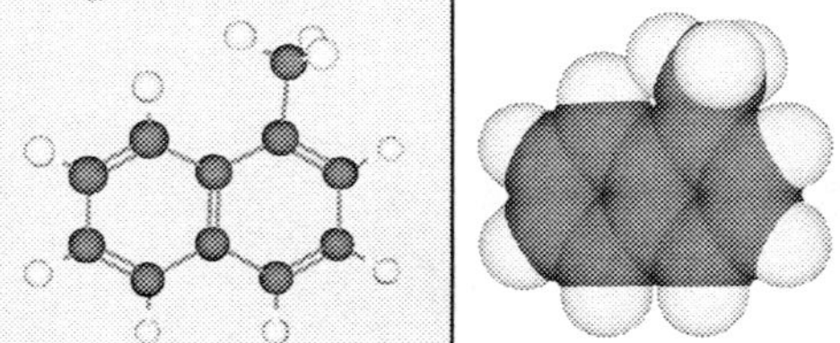

2-Methylnaphthalene $C_{11}H_{10}$

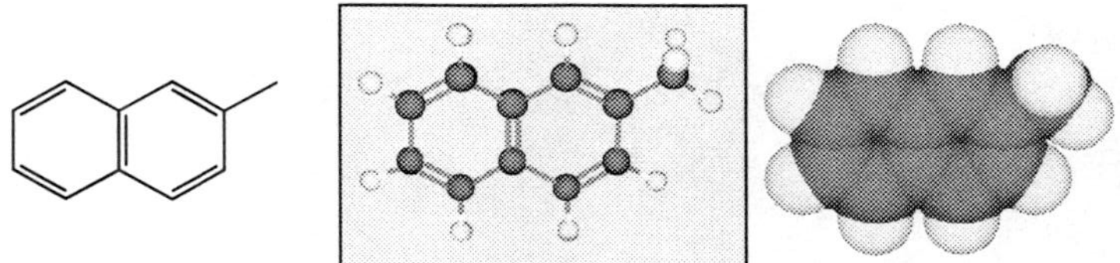

Acenaphthalene $C_{12}H_8$

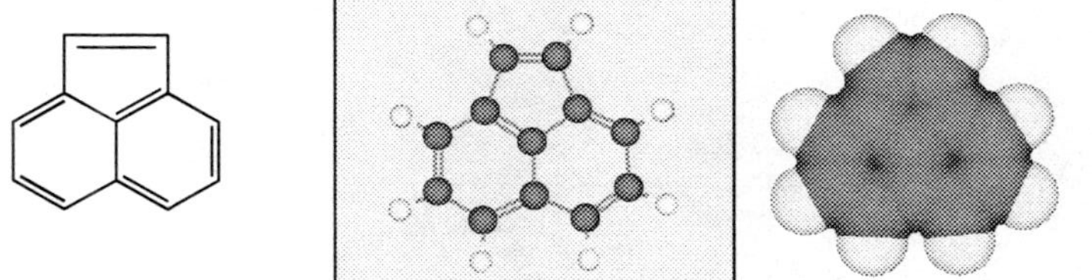

Acenaphthene $C_{12}H_{10}$

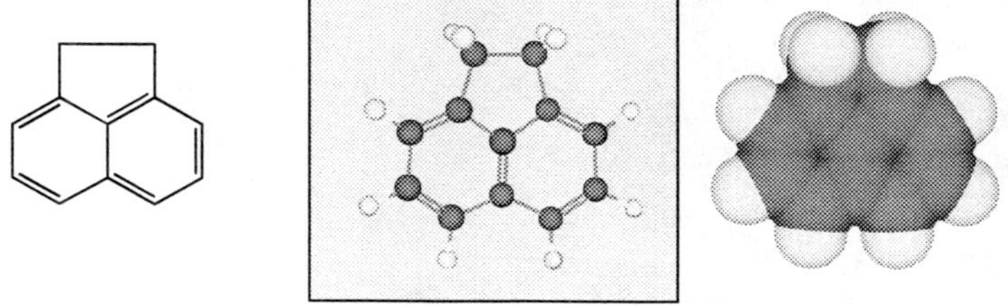

Fluorene $C_{13}H_{10}$

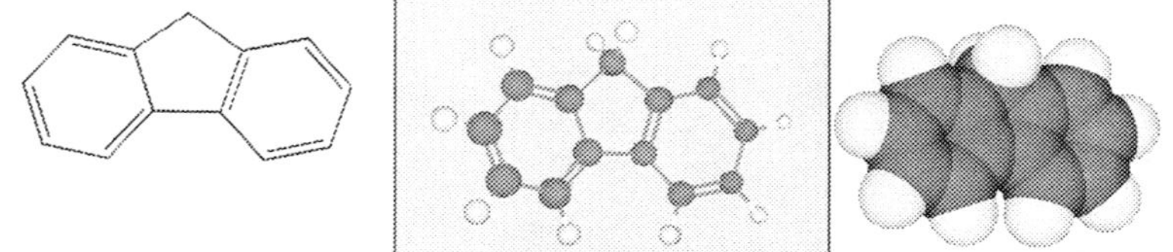

Anthracene $C_{14}H_{10}$

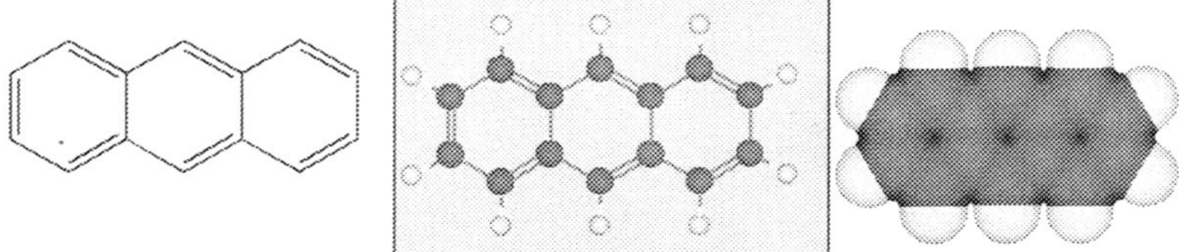

Phenanthrene $C_{14}H_{10}$

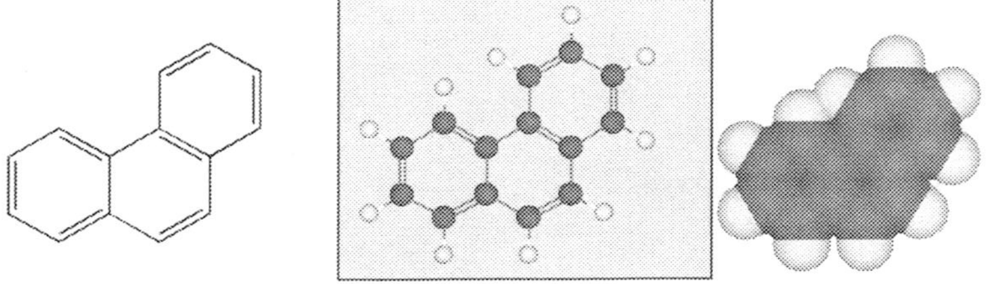

Benzo[*e*]acenaphthylene	$C_{16}H_{10}$	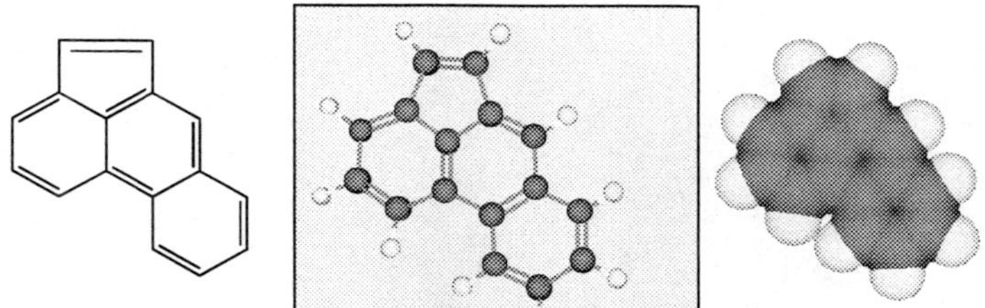
Fluoranthene	$C_{16}H_{10}$	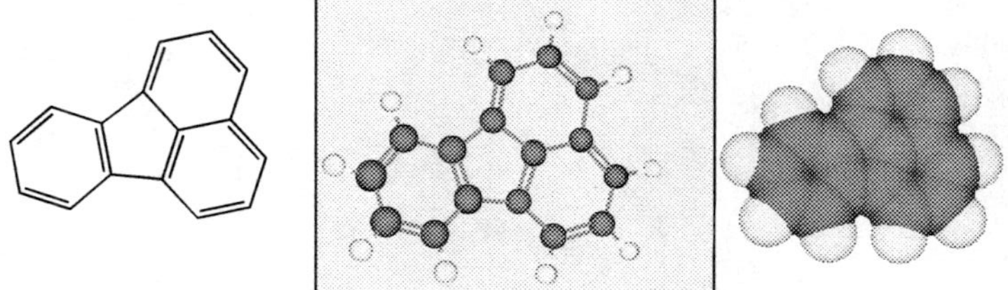
Pyrene	$C_{16}H_{10}$	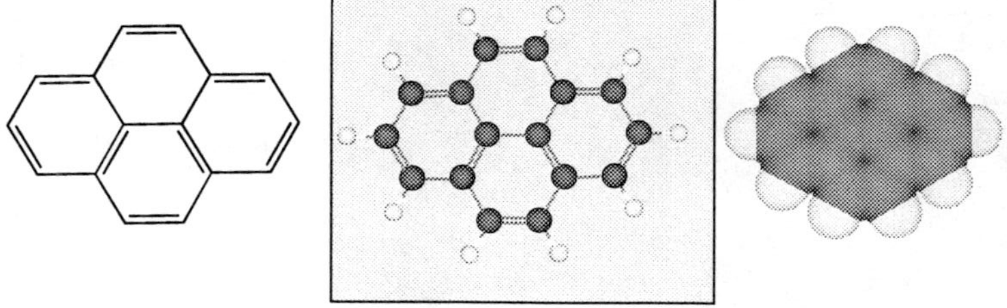

Benzo[*a*]fluorene	$C_{17}H_{12}$	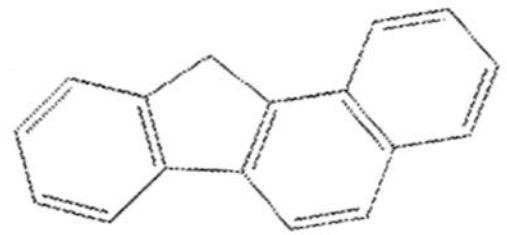	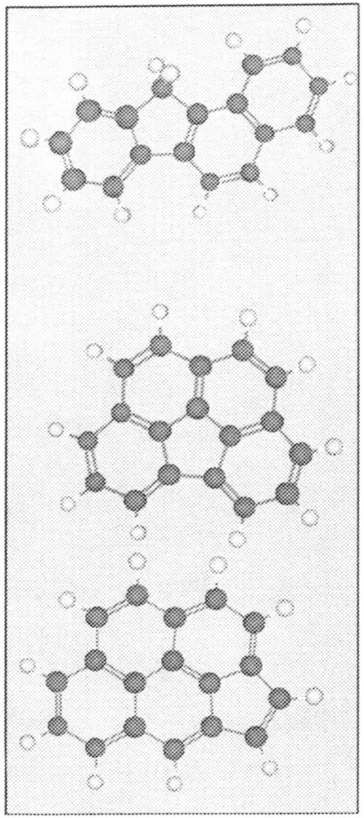	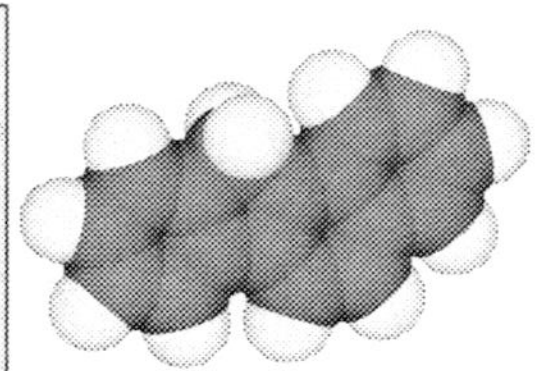
Benzo[*g,h,i*]fluoranthene	$C_{18}H_{10}$	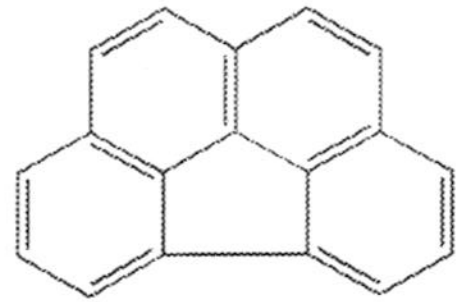		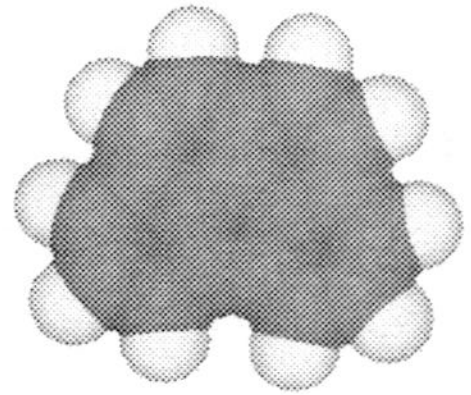
Cyclopenteno[*c,d*]pyrene	$C_{18}H_{10}$	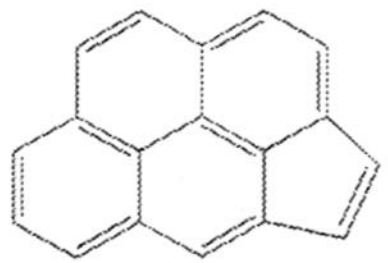		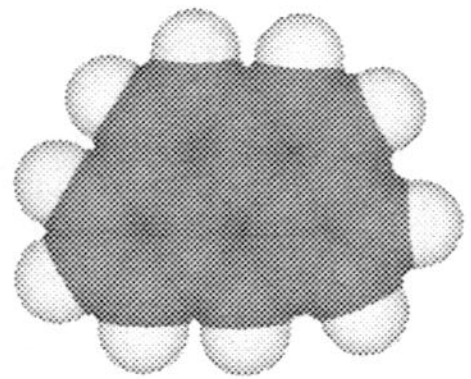

Benz[*a*]anthracene $C_{18}H_{12}$

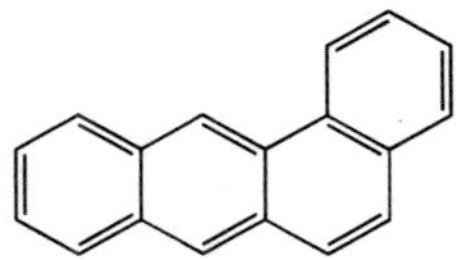
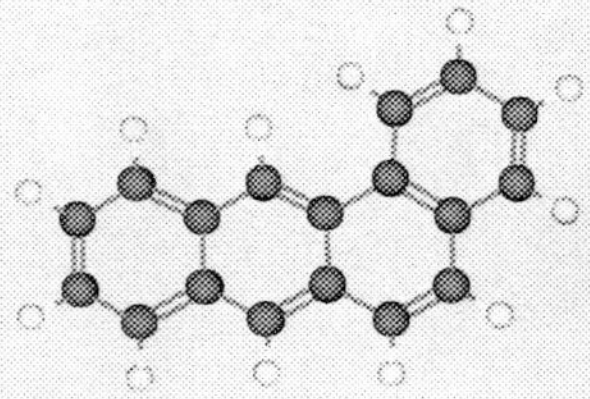
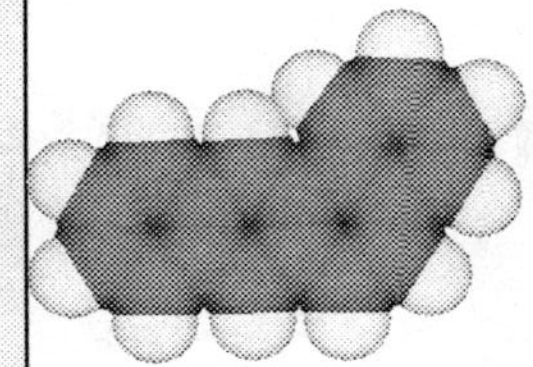

Benzo[*c*]phenanthrene $C_{18}H_{12}$

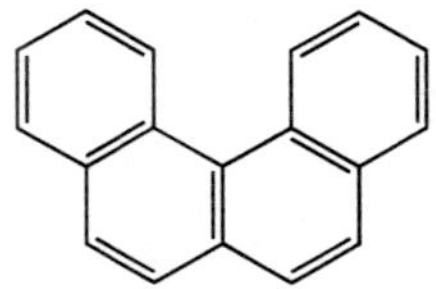
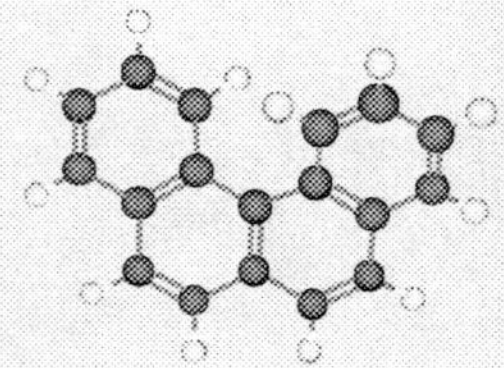
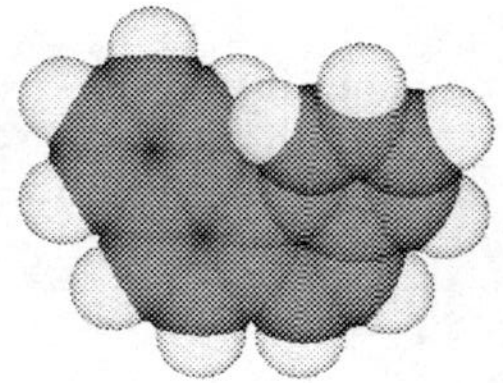

Chrysene $C_{18}H_{12}$

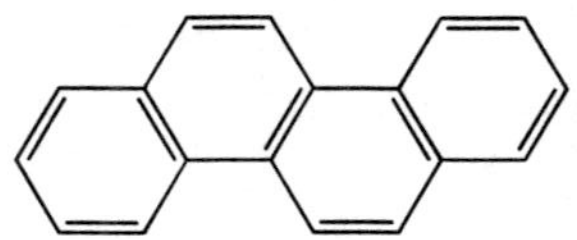
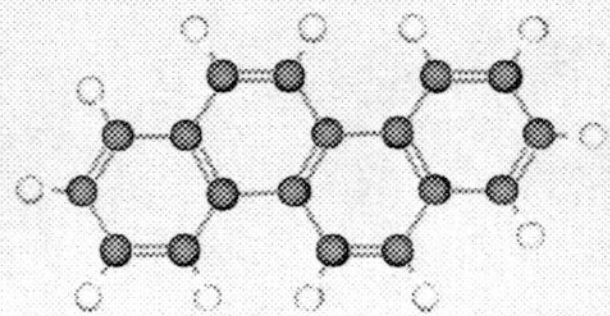
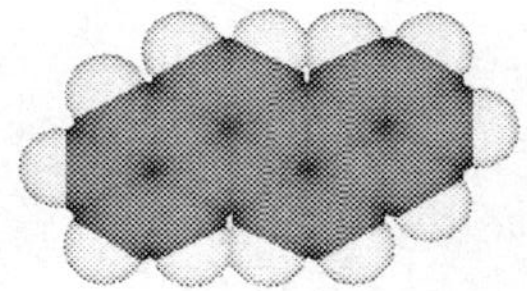

Naphthacene (tetracene) $C_{18}H_{12}$

Triphenylene $C_{18}H_{12}$

Benzo[*b*]fluoranthene $C_{20}H_{12}$

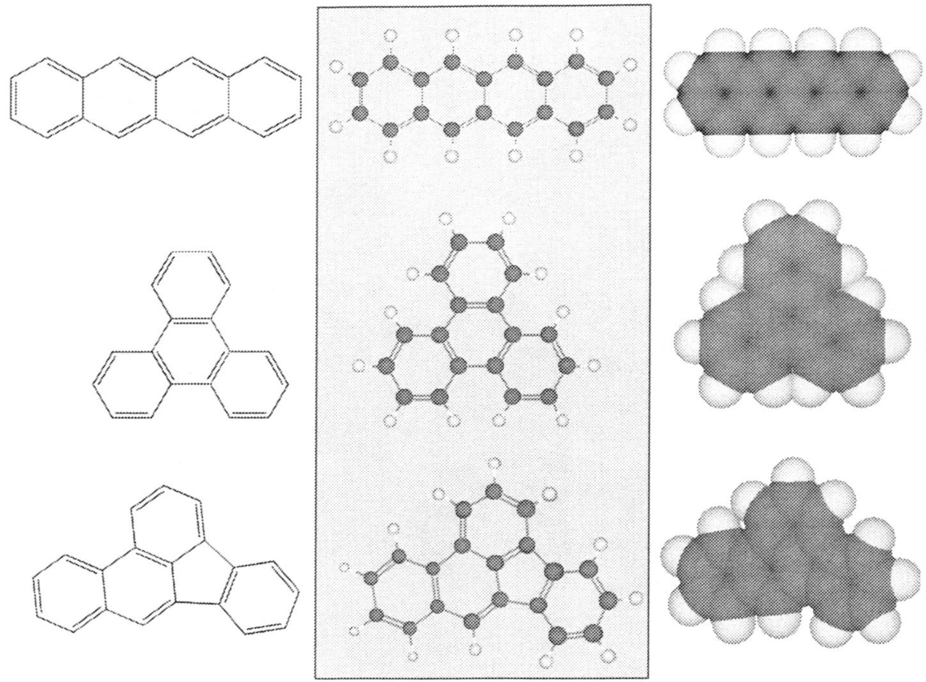

Benzo[*j*]fluoranthene $C_{20}H_{12}$

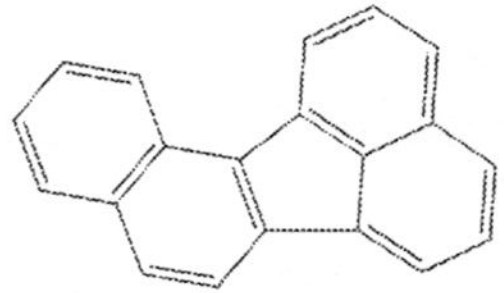
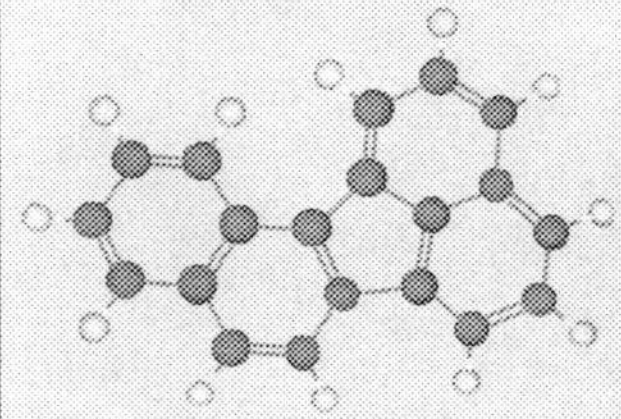
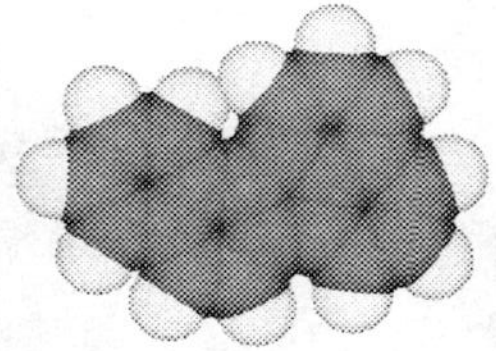

Benzo[*k*]fluoranthene $C_{20}H_{12}$

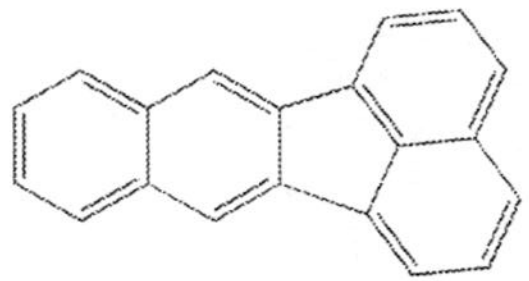
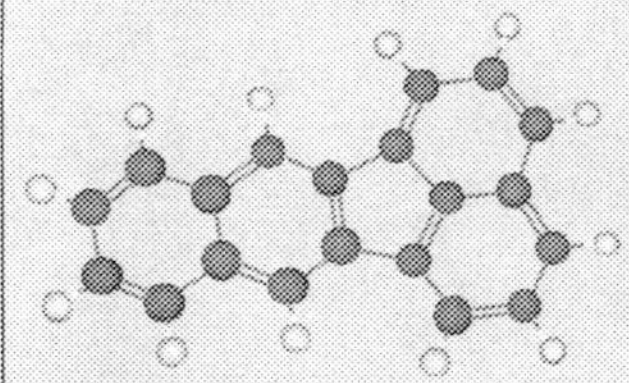
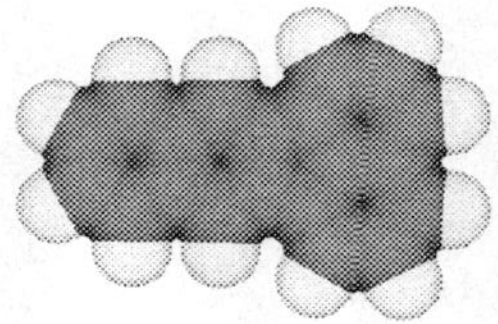

Benzo[*a*]pyrene $C_{20}H_{12}$

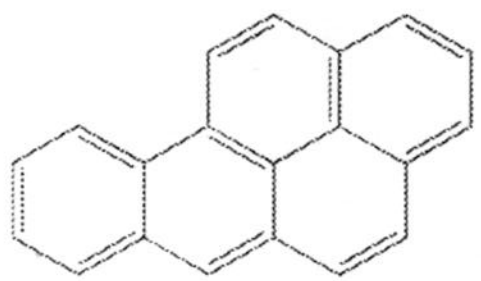
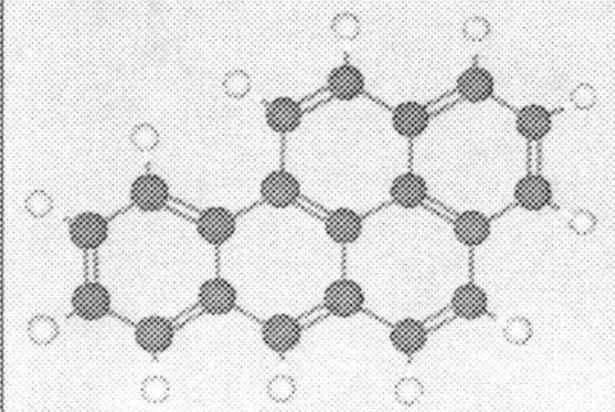
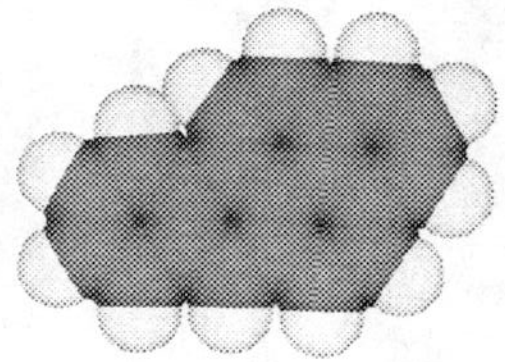

Benzo[e]pyrene $C_{20}H_{12}$

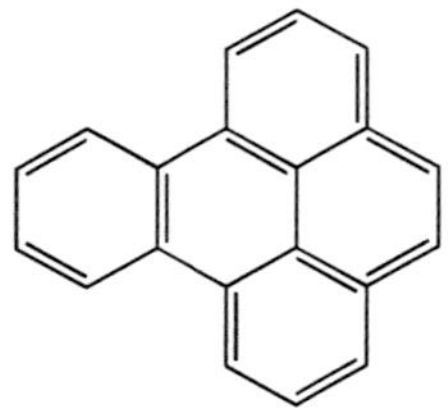

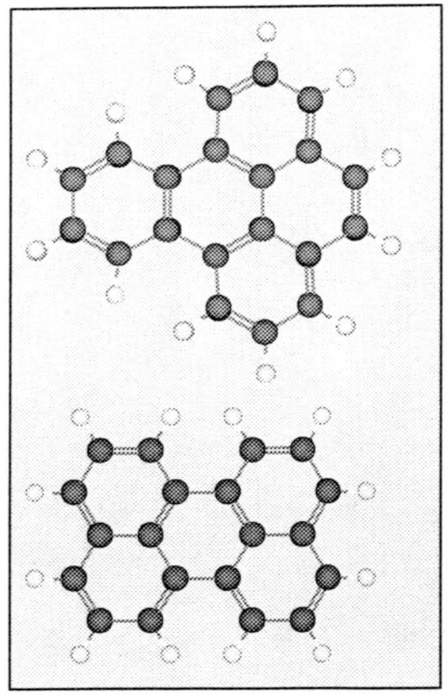

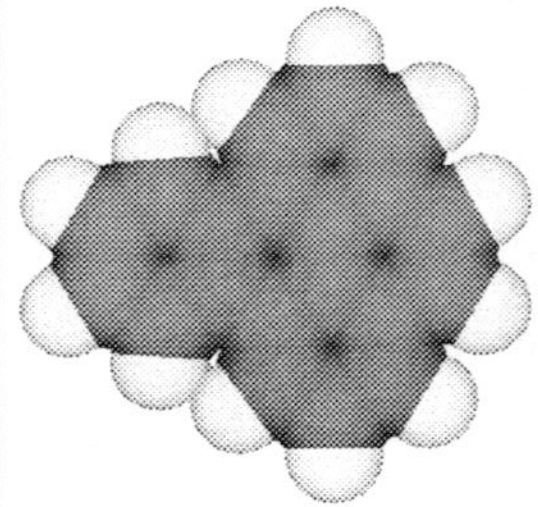

Perylene $C_{20}H_{12}$

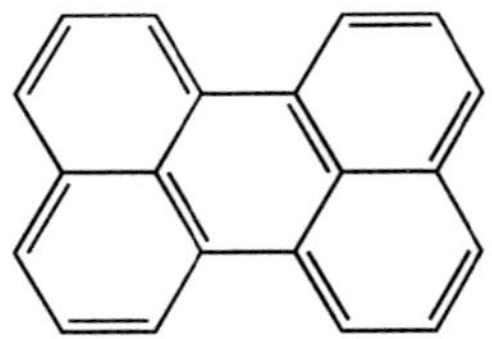

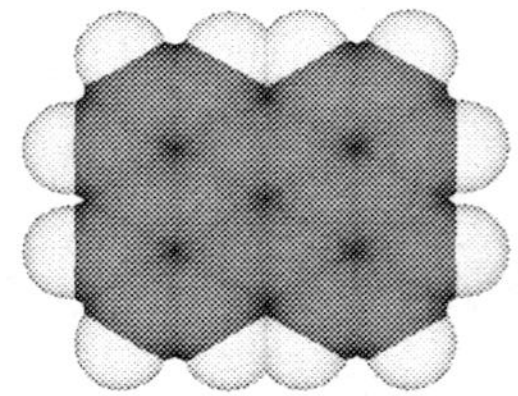

Cholanthrene

$C_{20}H_{14}$

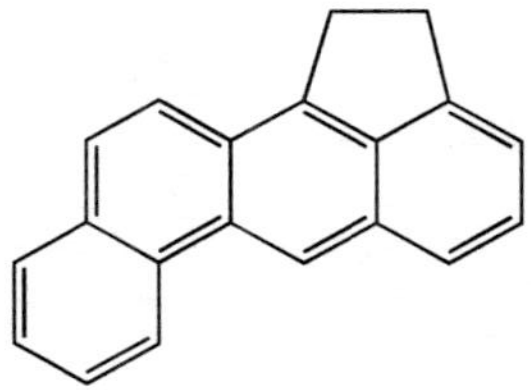

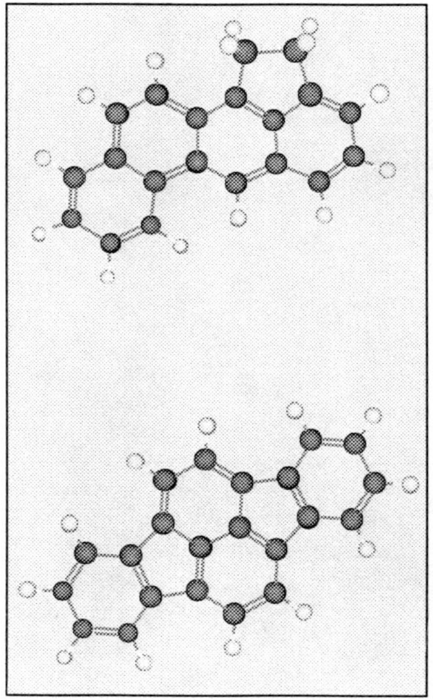

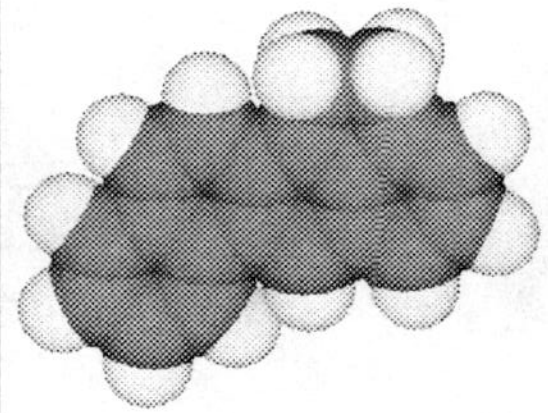

Indino[1,2,3-*c*,*d*]fluoranthene

$C_{22}H_{12}$

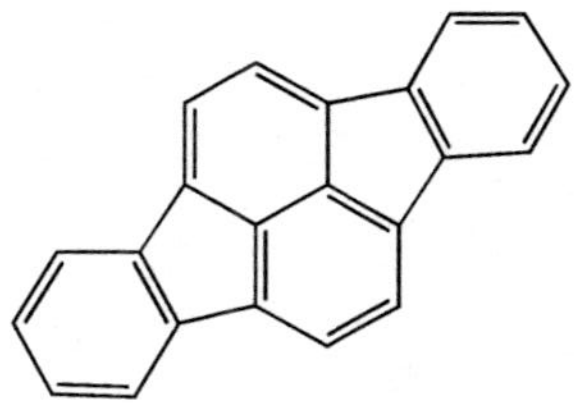

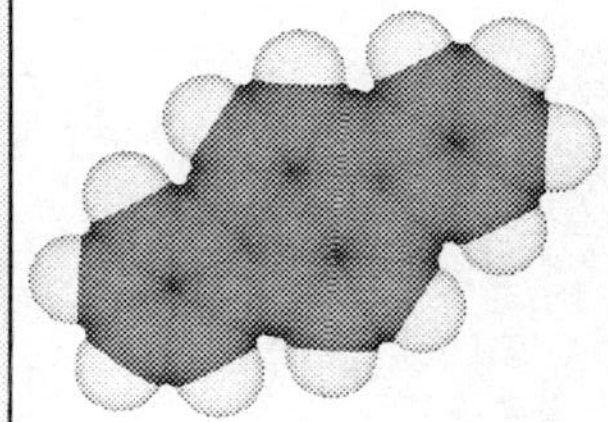

Indino[1,2,3-*c*,*d*]pyrene $C_{22}H_{12}$

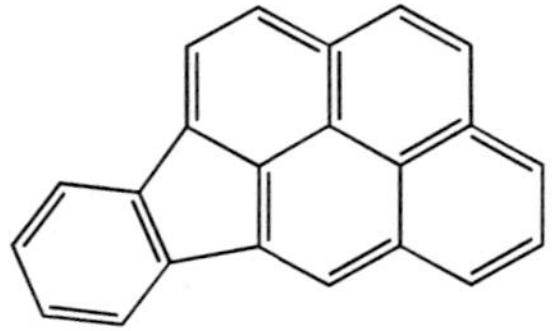

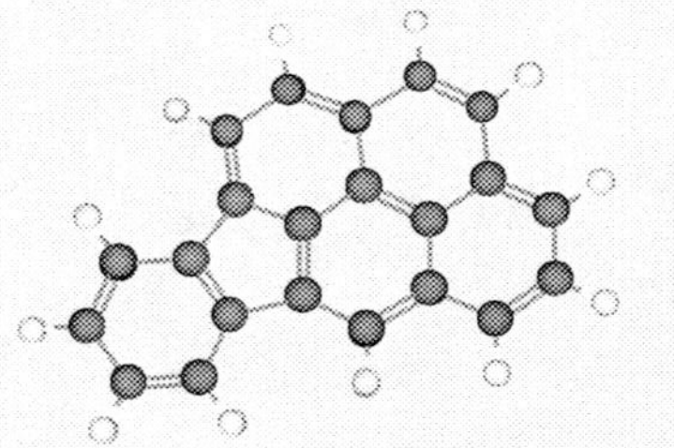

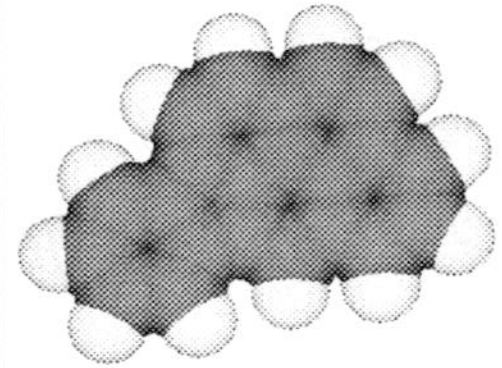

Benzo[*b*]chrysene $C_{22}H_{14}$

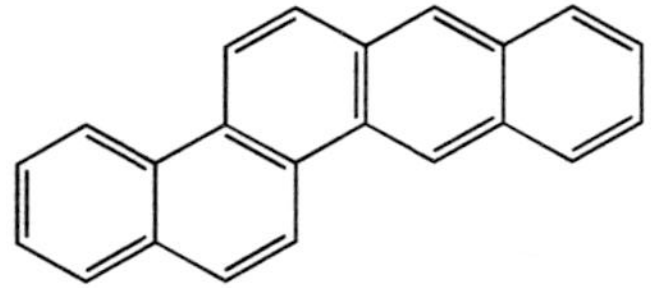

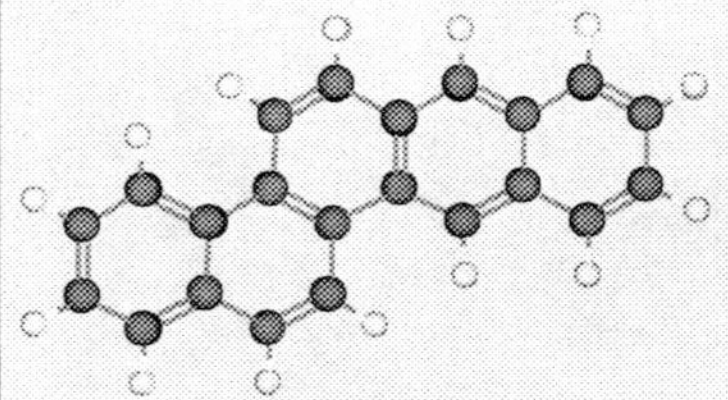

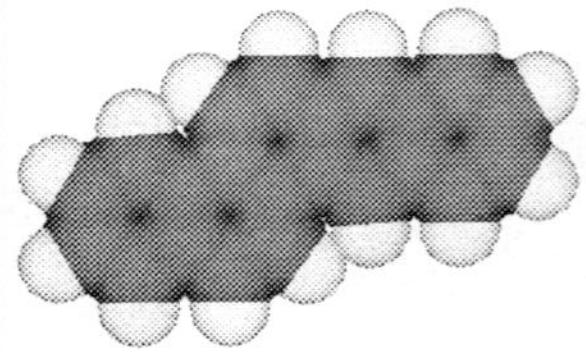

Dibenz[a,c]anthracene $C_{22}H_{14}$

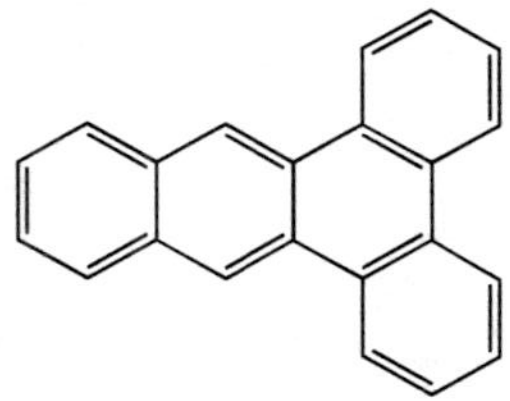

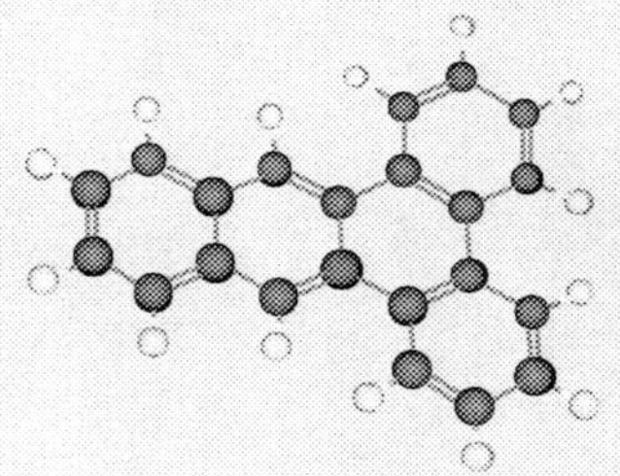

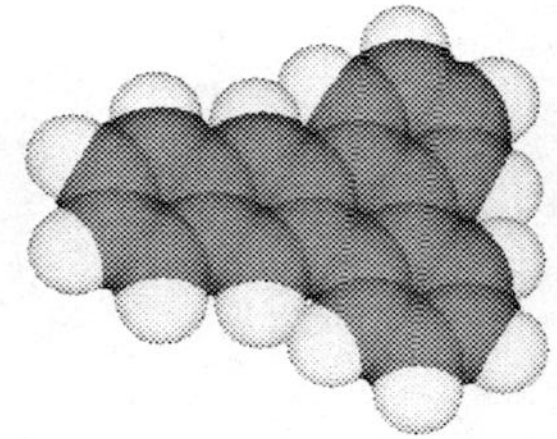

Dibenz[a,h]anthracene $C_{22}H_{14}$

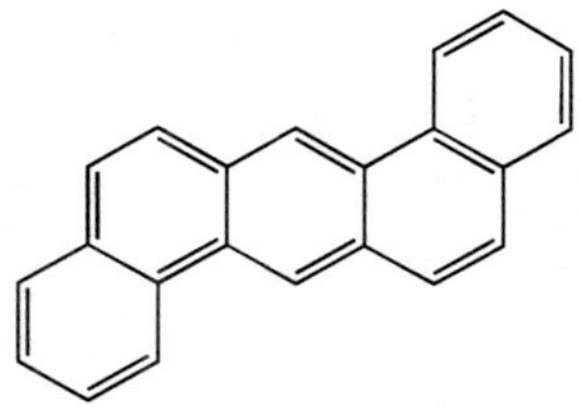

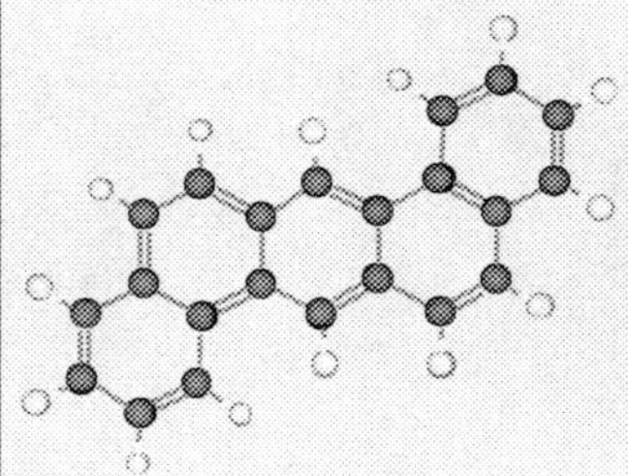

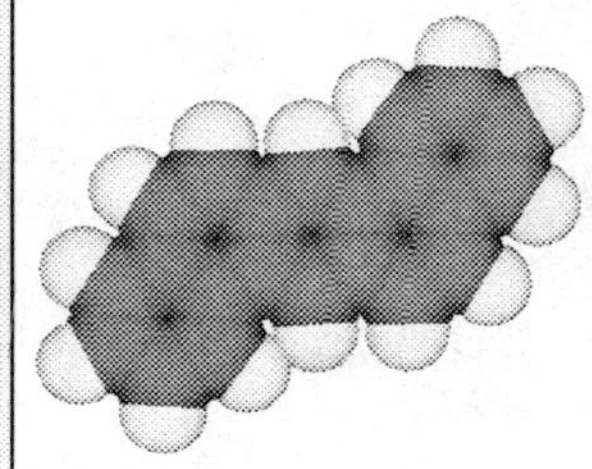

Pentacene $C_{22}H_{14}$ 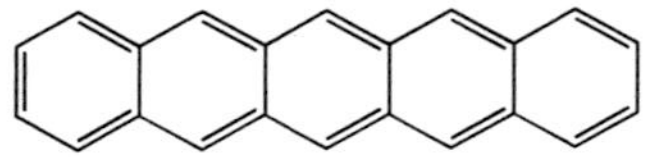

Coronene $C_{24}H_{12}$ 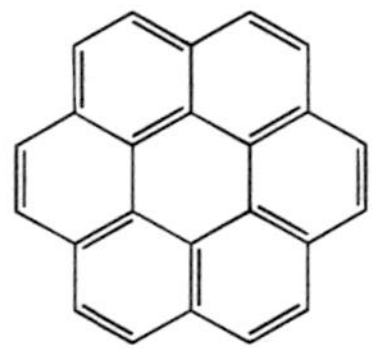

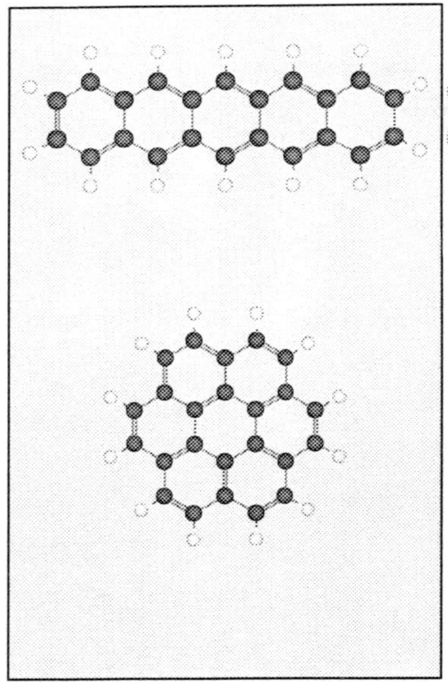

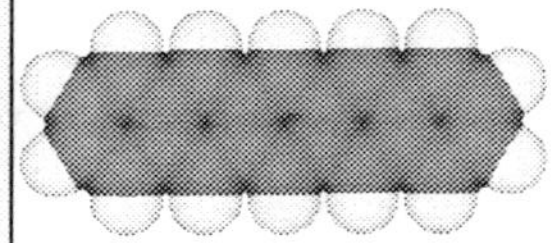

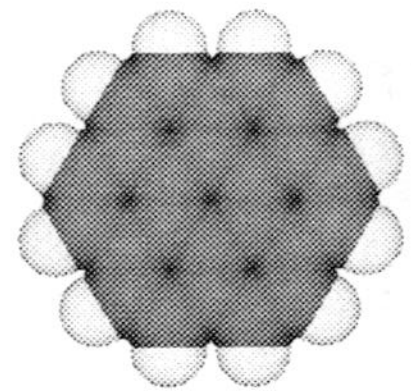

Dibenzo[*a*,*e*]pyrene	$C_{24}H_{14}$	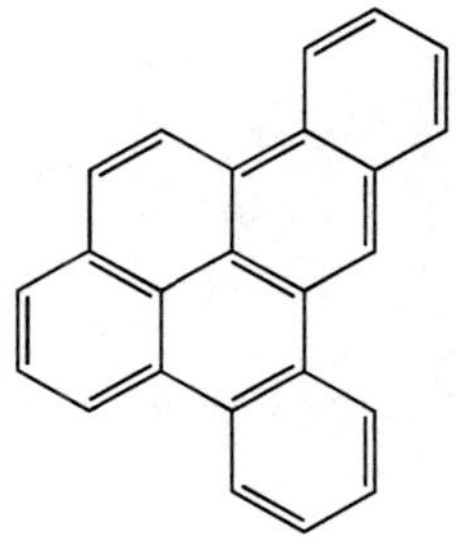	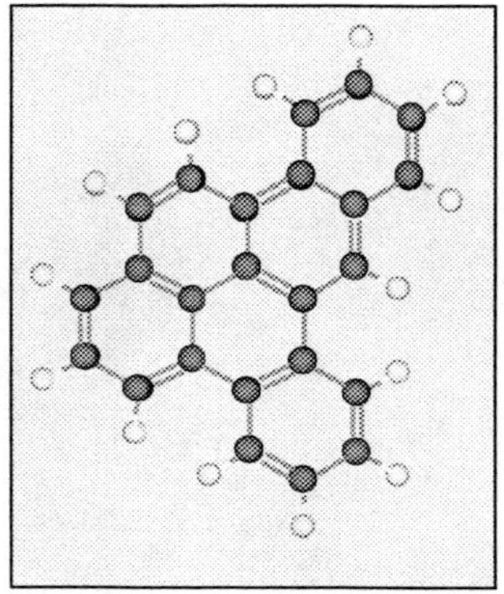	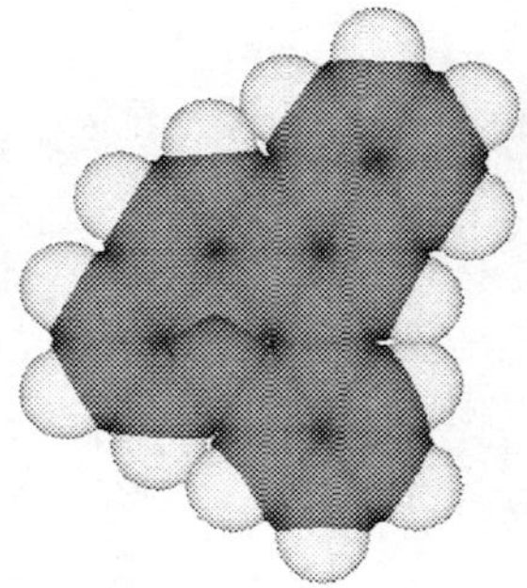

Appendix II

Nomenclature and Molecular Structure of Some Common Products of the Atmospheric Oxidation of the Aromatic Hydrocarbons

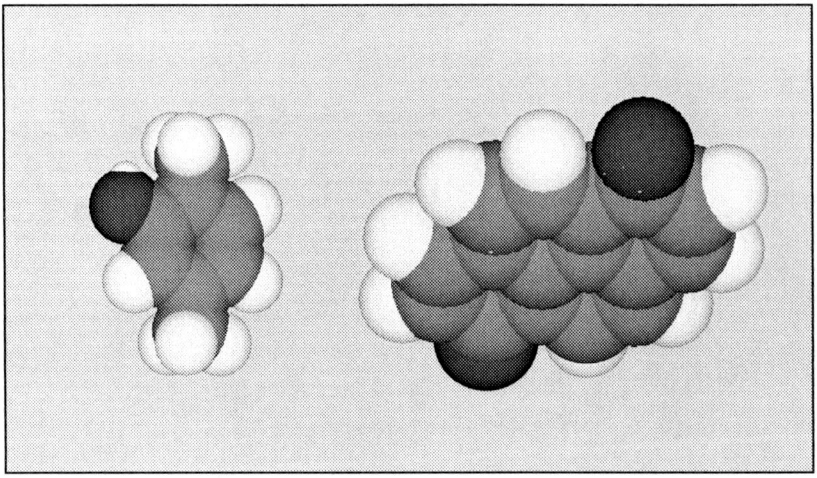

A. Aromatic Ring-Retaining Products

Name of Compound	Molecular Formula	C-Skeletal Structure	3D-Structural Formulas	
Phenol	C_6H_6O	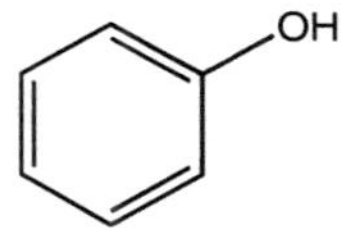	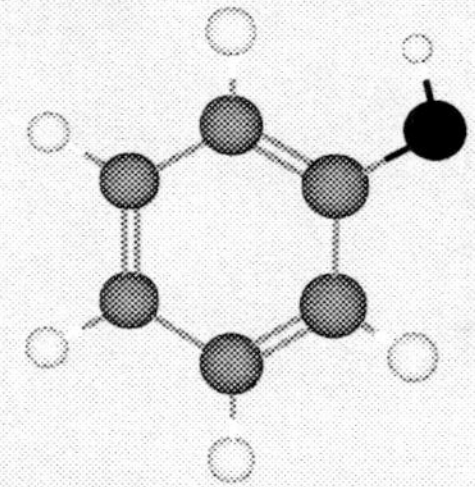	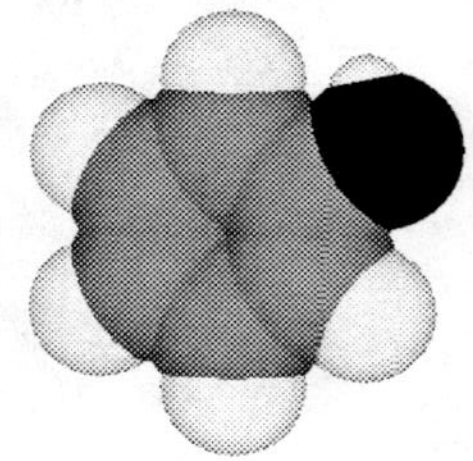
Nitrobenzene	$C_6H_5O_2N$	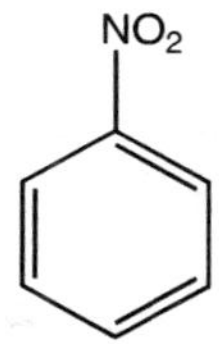	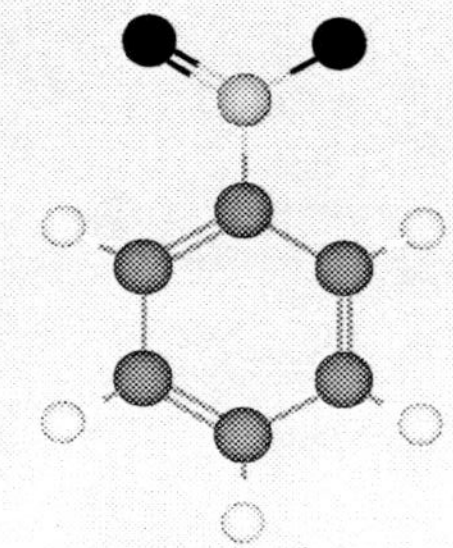	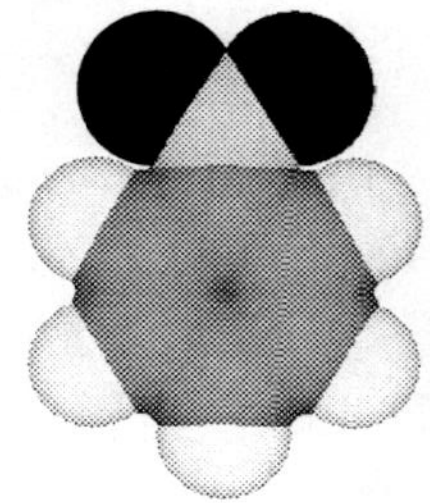

2-Nitrophenol $C_6H_5O_3N$

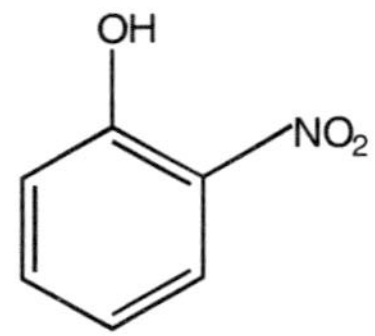

2,4-Dinitrophenol $C_6H_4O_5N_2$

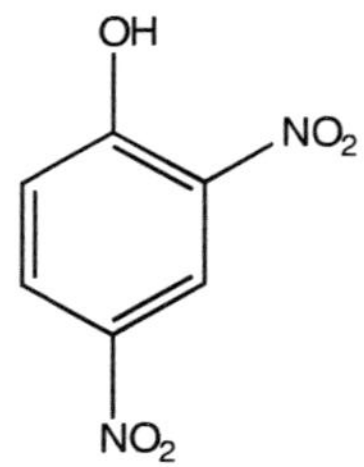

2,4,6-Trinitrophenol $C_6H_3O_7N_3$

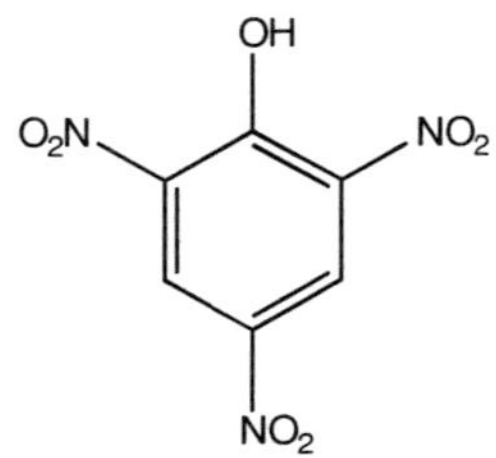

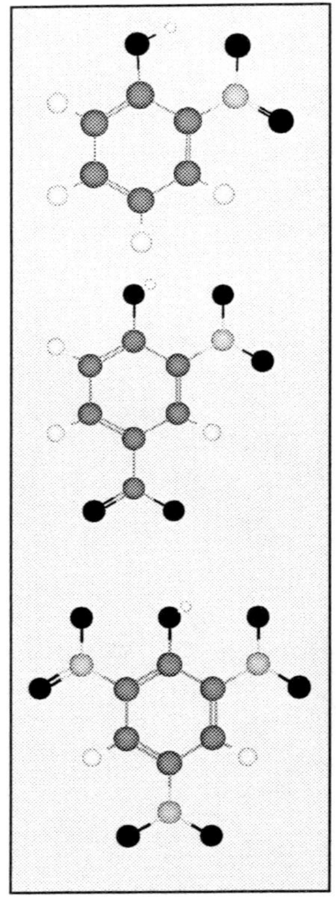

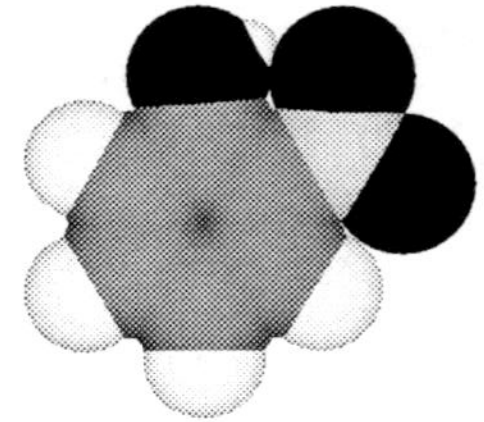

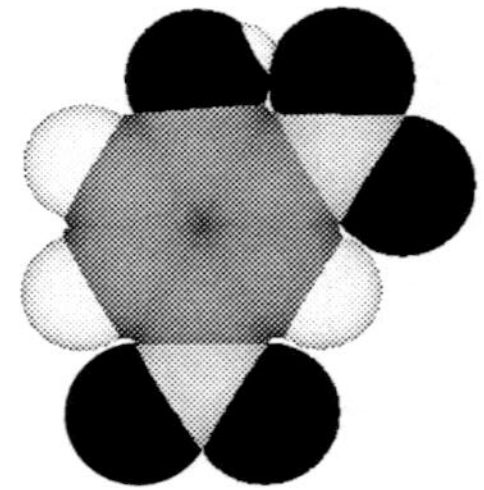

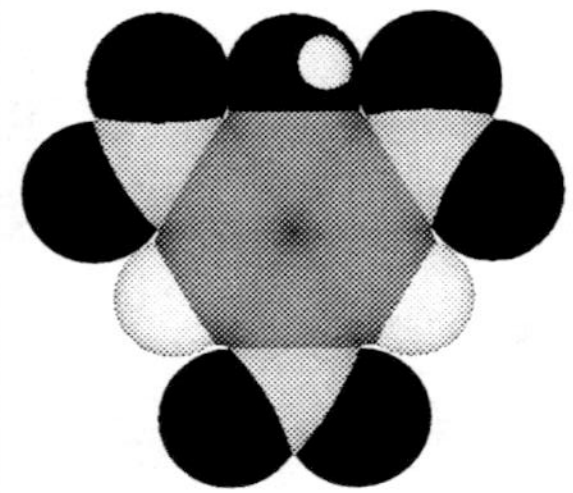

2-Nitrobenzoic acid $C_7H_5O_4N$

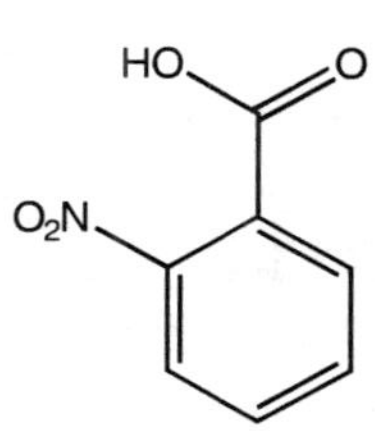

Benzaldehyde C_7H_6O

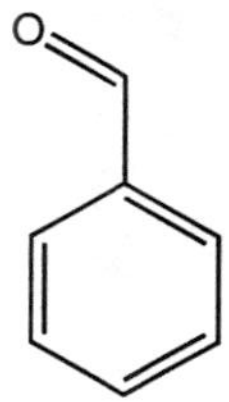

Benzoic acid $C_7H_6O_2$

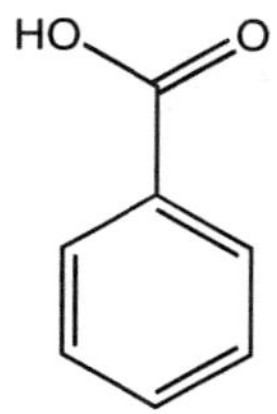

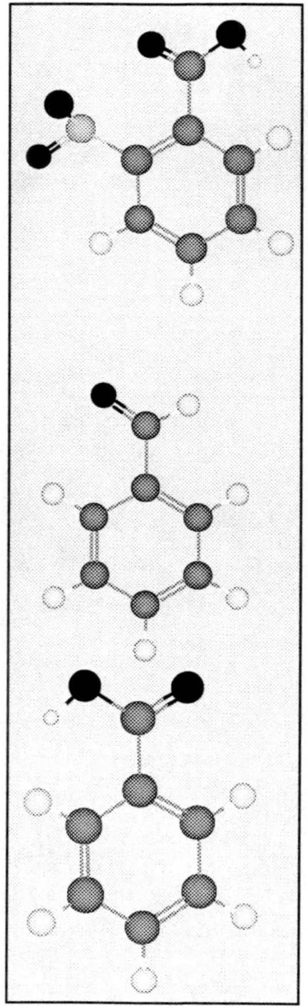

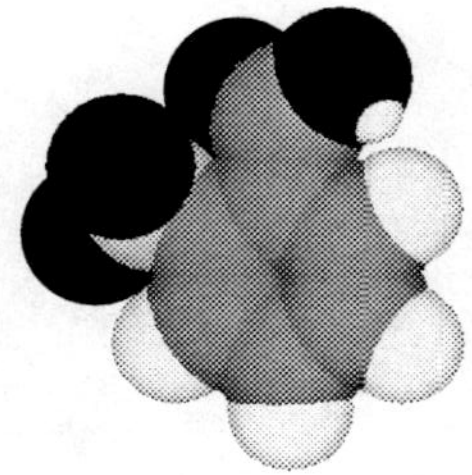

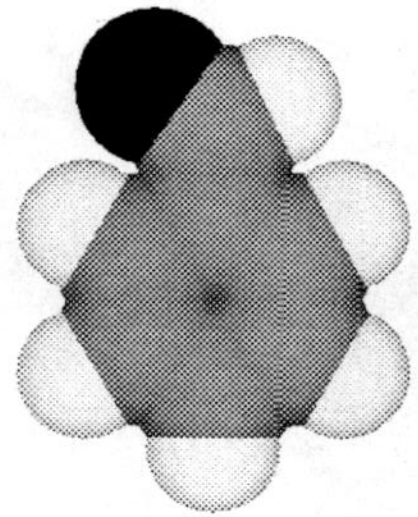

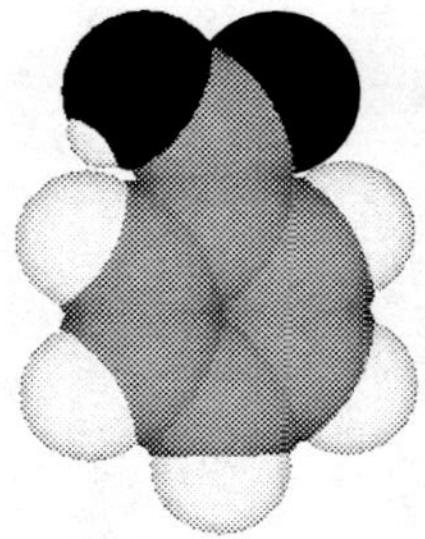

2-Hydroxybenzoic acid

$C_7H_6O_3$

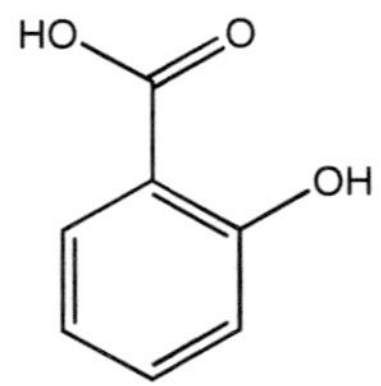

o-Nitrotoluene

$C_7H_7O_2N$

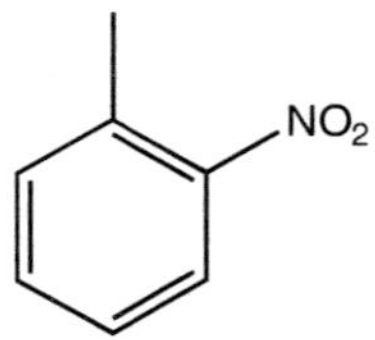

m-Nitrotoluene

$C_7H_7O_2N$

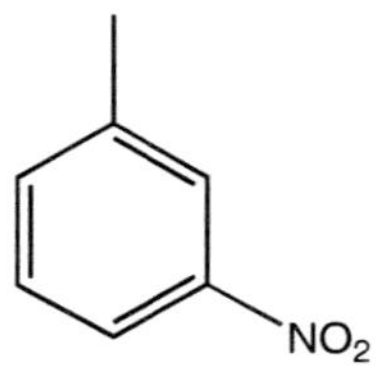

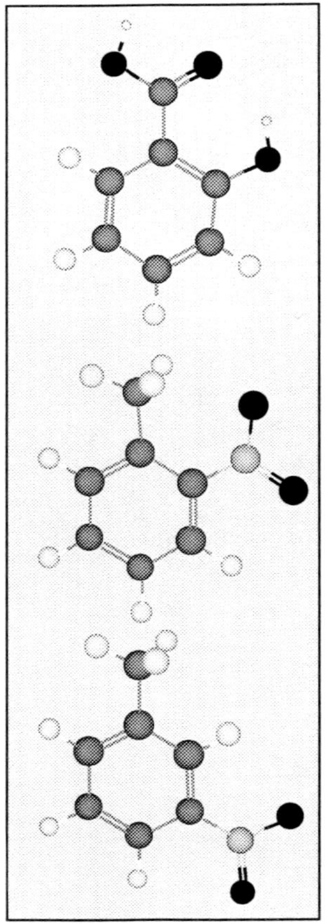

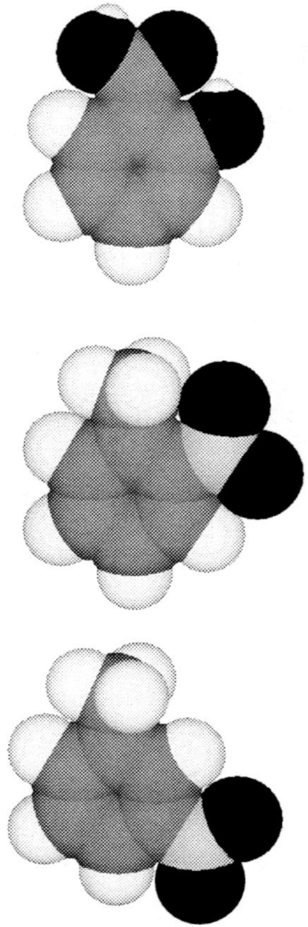

o-Cresol	C_7H_8O	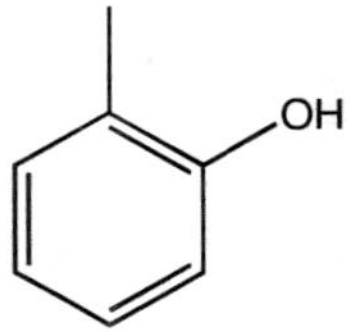	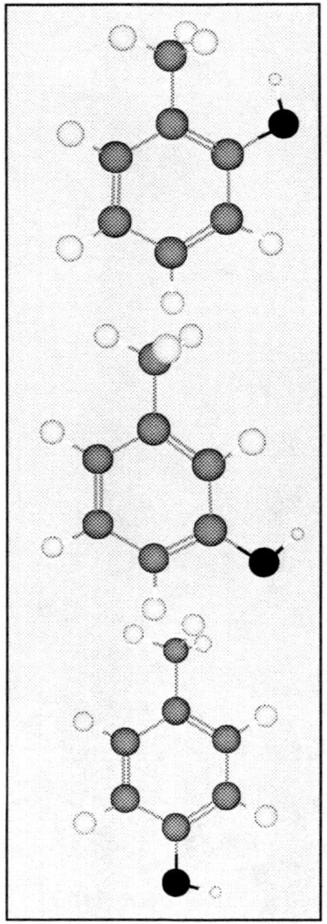	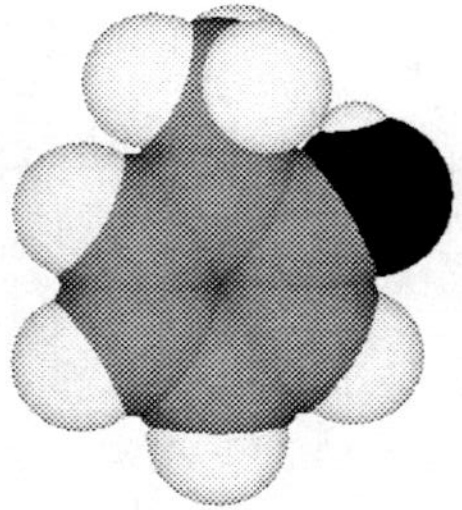
m-Cresol	C_7H_8O	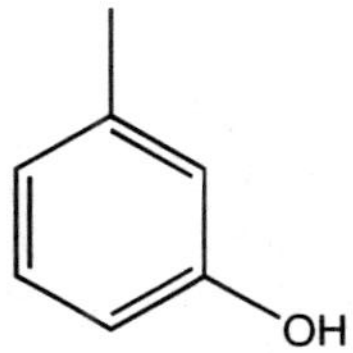		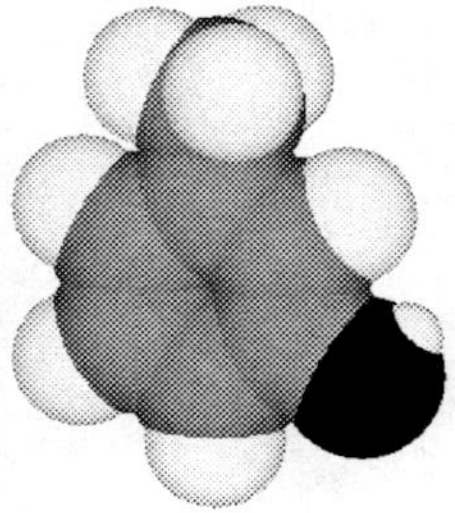
p-Cresol	C_7H_8O			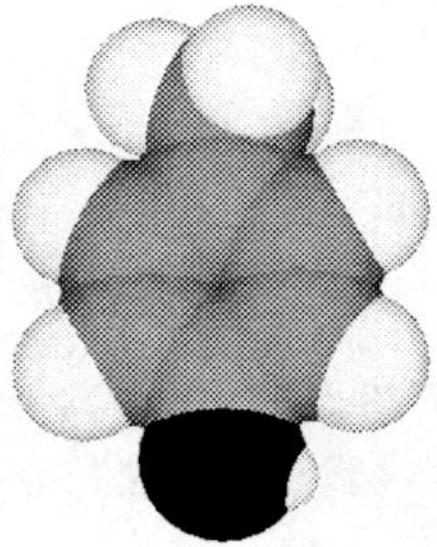

Phthalic acid
(1,2-benzene dicarboxylic acid) $C_8H_6O_4$

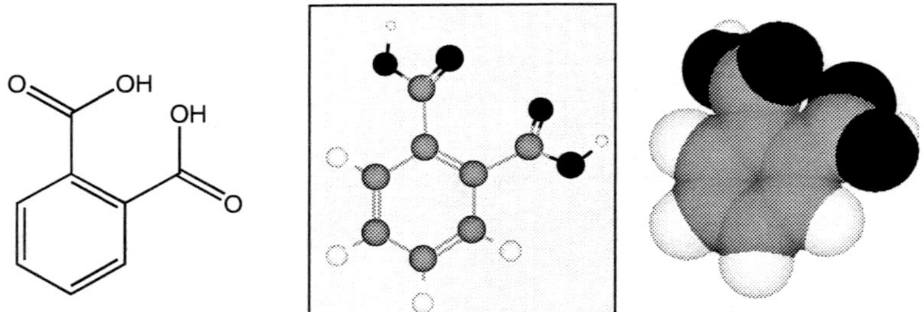

2-Methylbenzaldehyde C_8H_8O

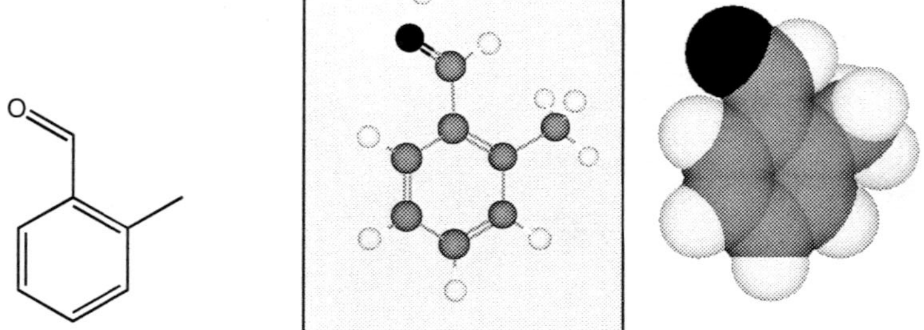

3-Methylbenzaldehyde C_8H_8O

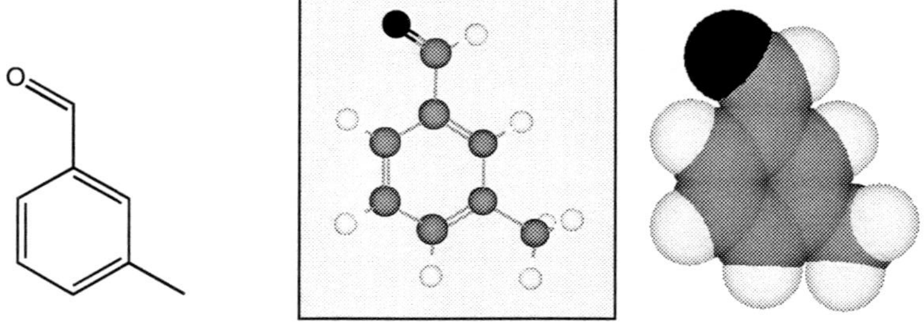

4-Methylbenzaldehyde C_8H_8O

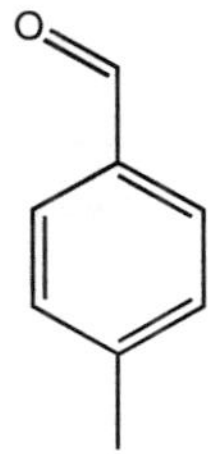

2-Methylbenzoic acid $C_8H_8O_2$

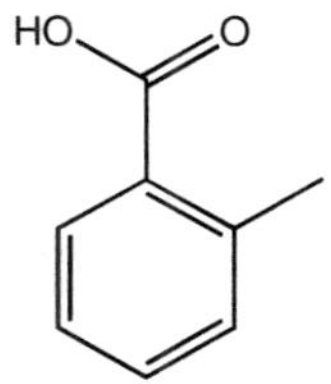

2,3-Dimethylphenol $C_8H_{10}O$

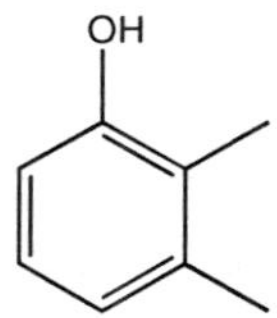

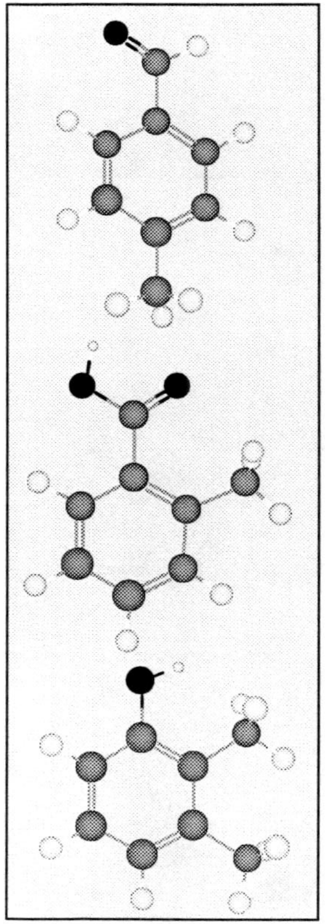

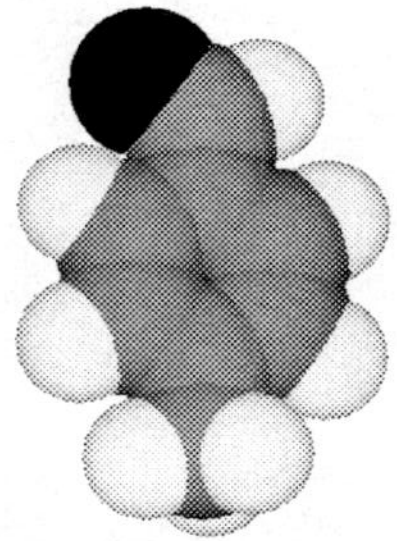

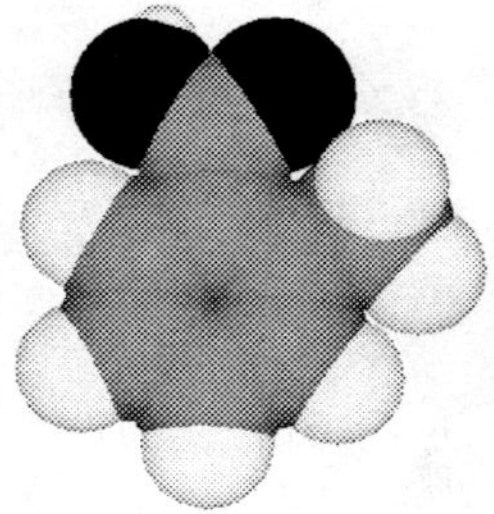

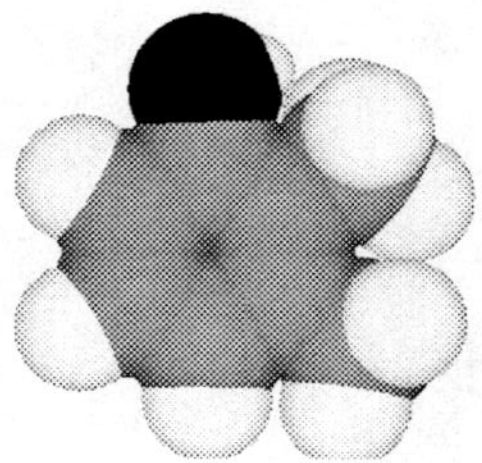

2,4-Dimethylphenol

$C_8H_{10}O$

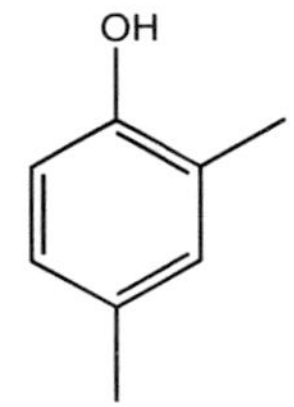

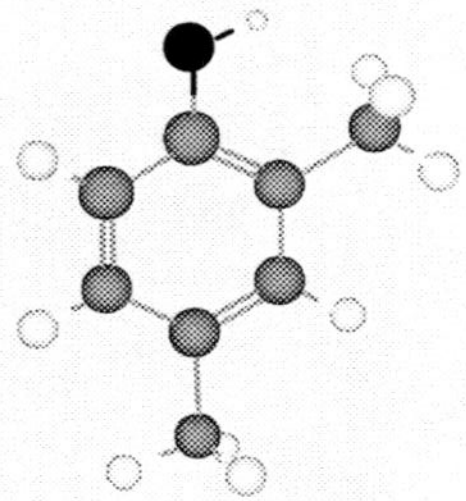

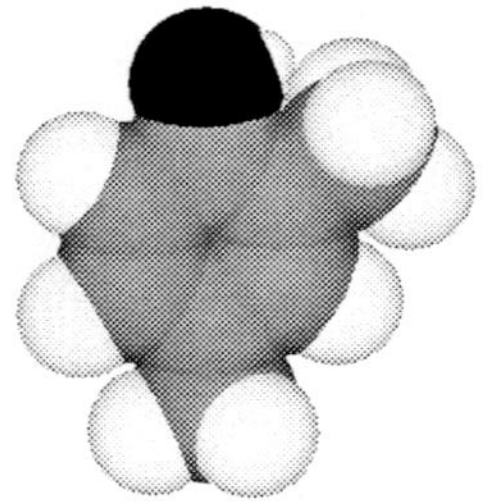

2,5-Dimethylphenol

$C_8H_{10}O$

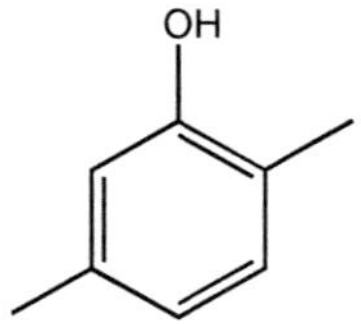

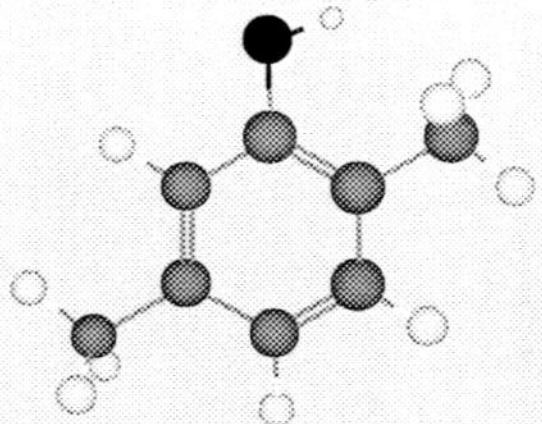

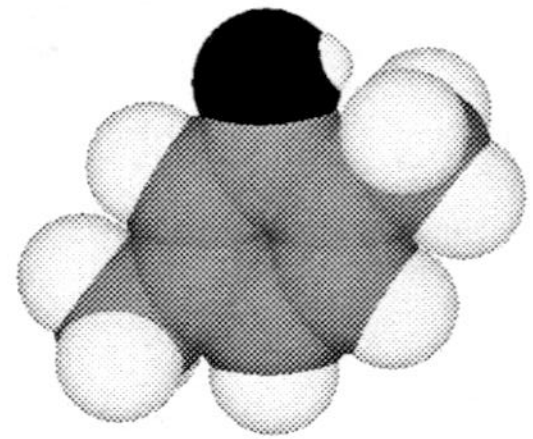

2,6-Dimethylphenol

$C_8H_{10}O$

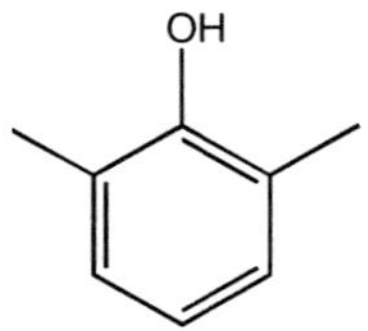

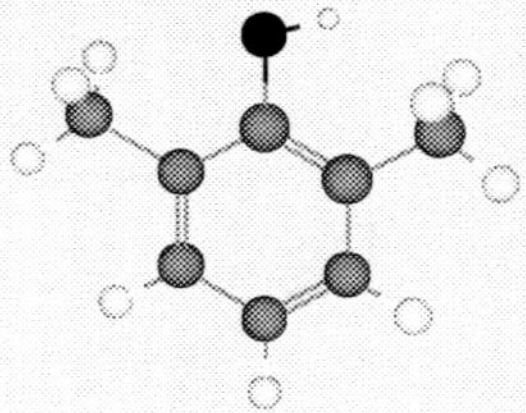

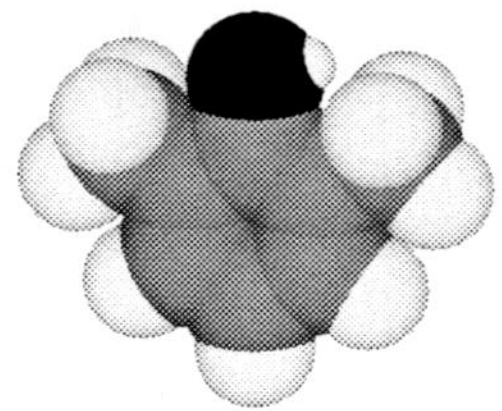

3,4-Dimethylphenol	$C_8H_{10}O$	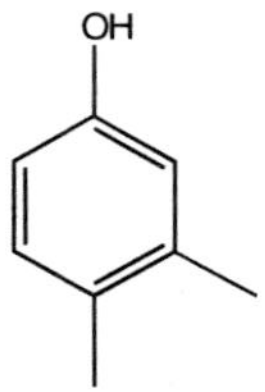	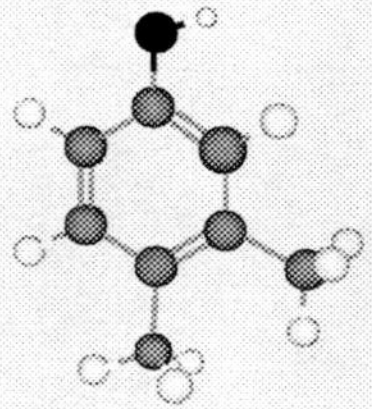	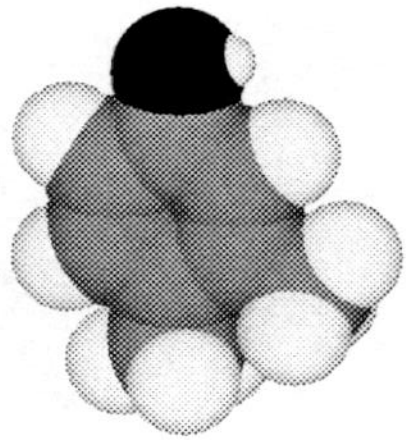
3,5-Dimethylphenol	$C_8H_{10}O$	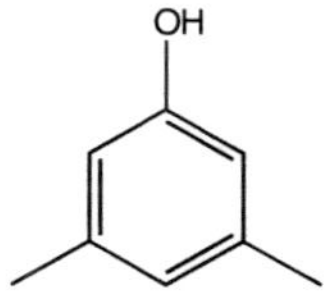	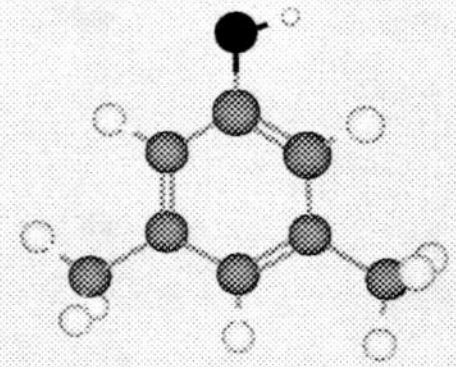	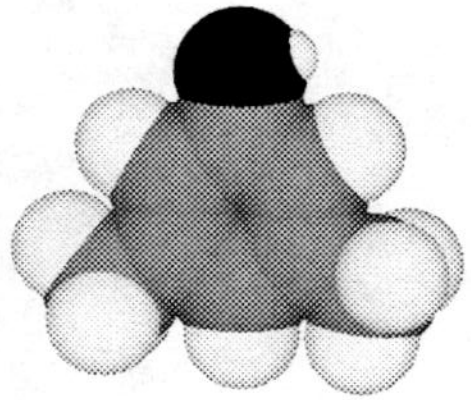
trans-Cinnamaldehyde	C_9H_8O	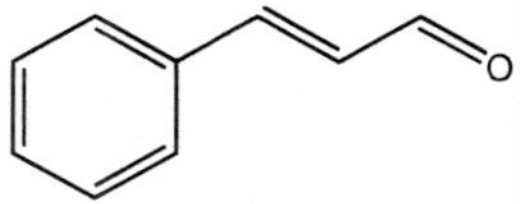	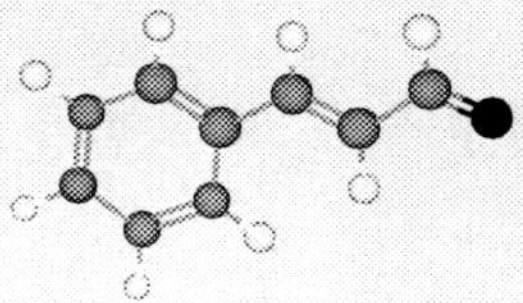	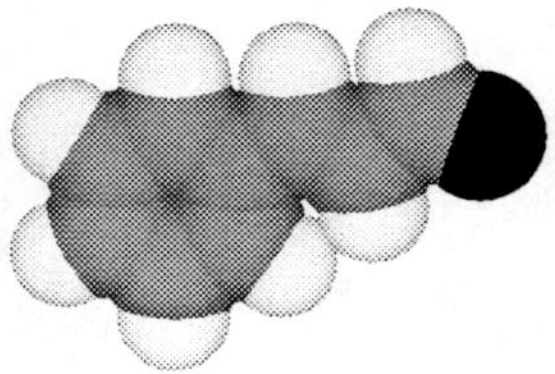

4-Acetylbenzoic acid $C_9H_8O_3$

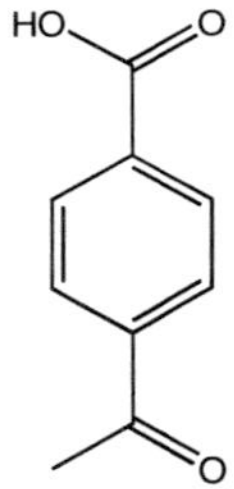

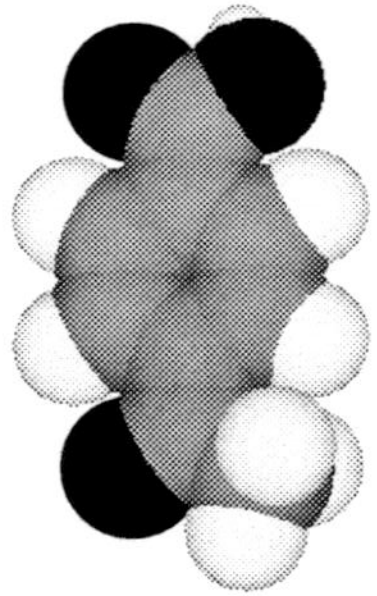

2,4-Dimethylbenzaldehyde $C_9H_{10}O$

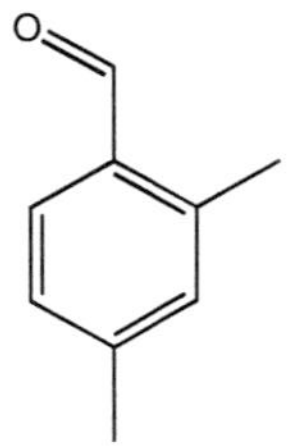

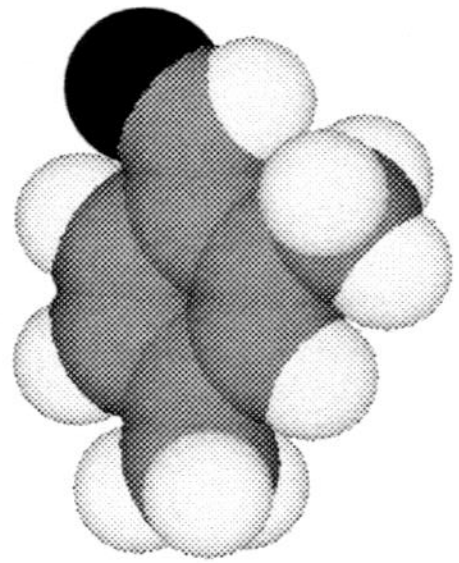

2,5-Dimethylbenzaldehyde $C_9H_{10}O$

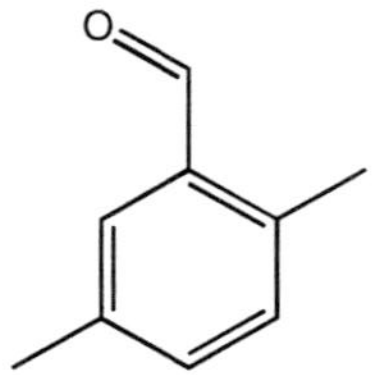

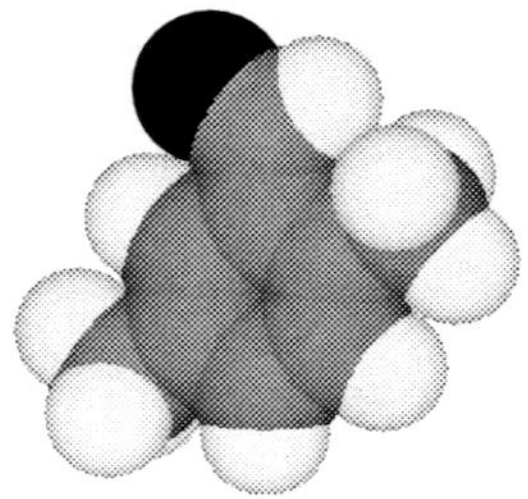

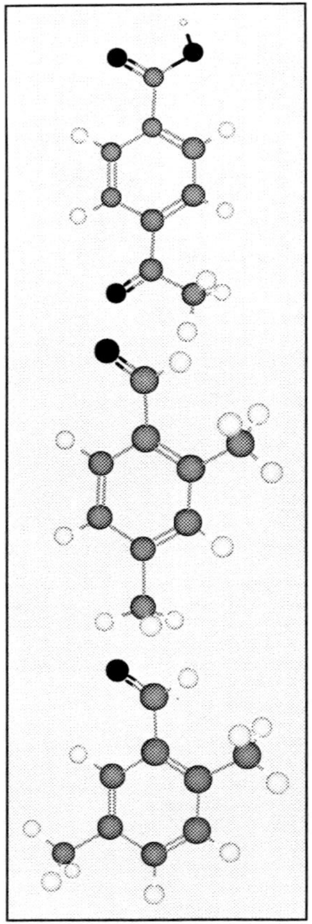

3,4-Dimethylbenzaldehyde $C_9H_{10}O$

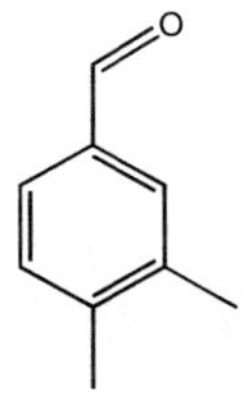

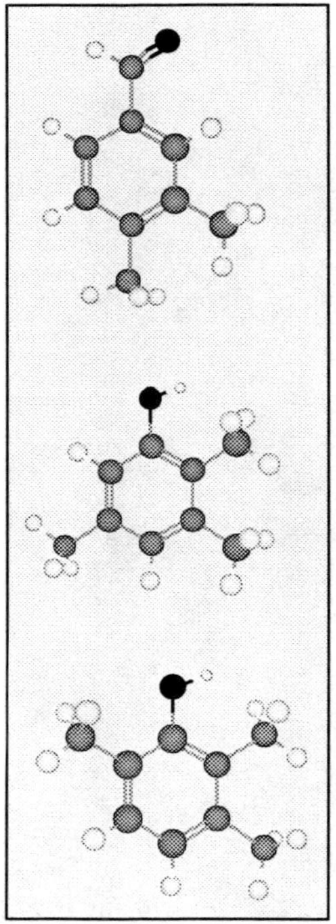

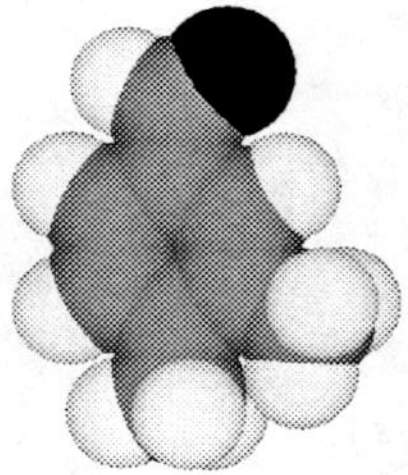

2,3,5-Trimethylphenol $C_9H_{12}O$

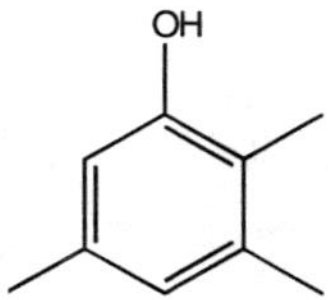

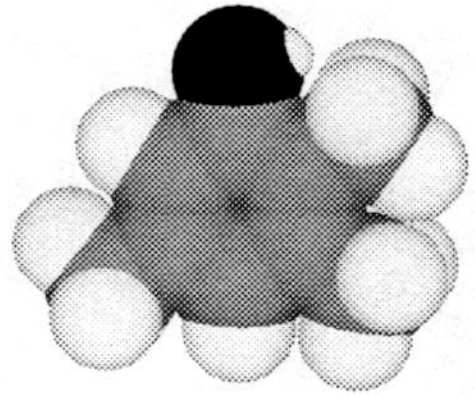

2,3,6-Trimethylphenol $C_9H_{12}O$

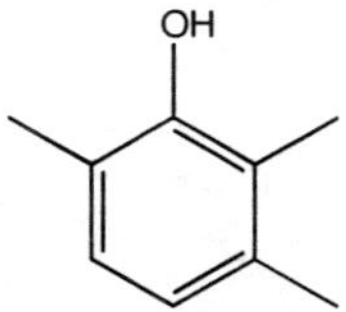

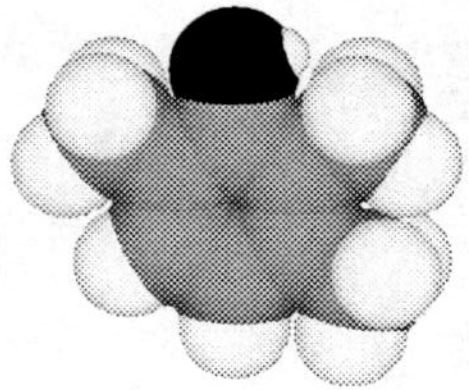

2,4,6-Trimethylphenol $C_9H_{12}O$

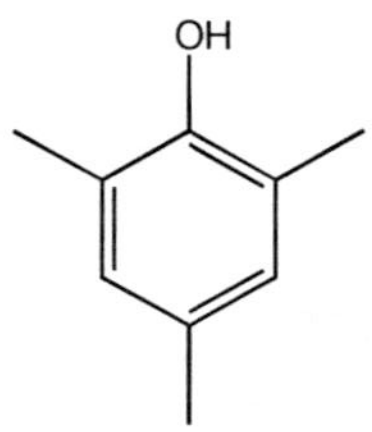

1,4-Naphthoquinone $C_{10}H_6O_2$

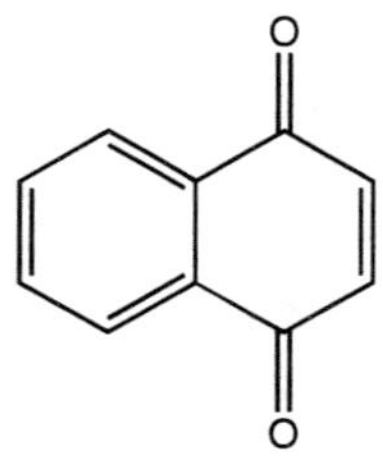

1-Nitronaphthalene $C_{10}H_7O_2N$

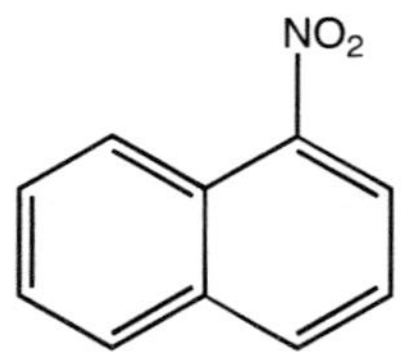

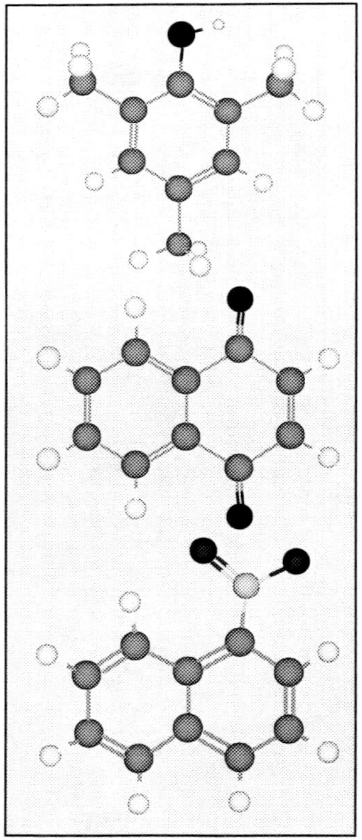

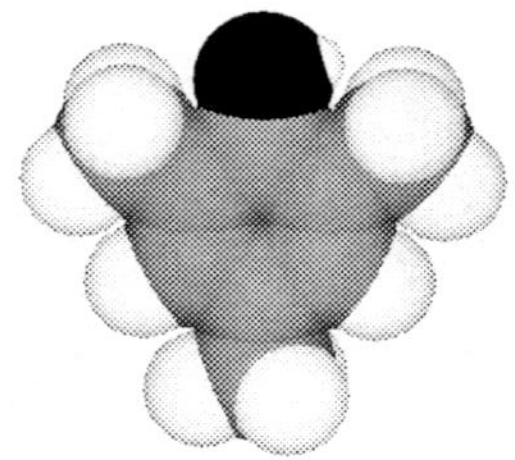

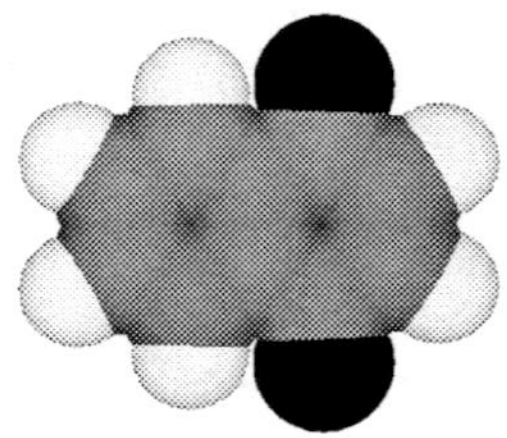

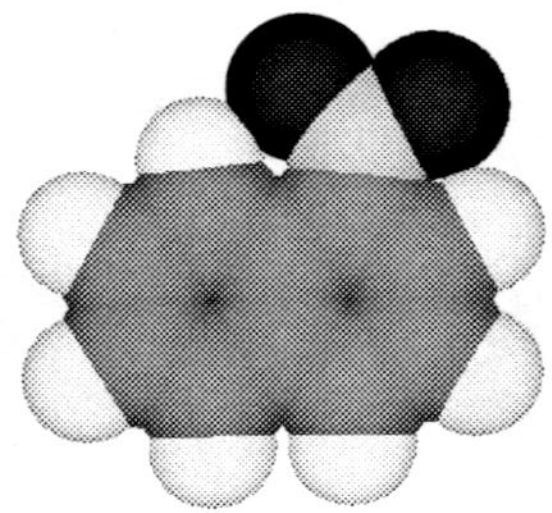

Name	Formula	Structure	Models	
2-Nitronaphthalene	$C_{10}H_7O_2N$	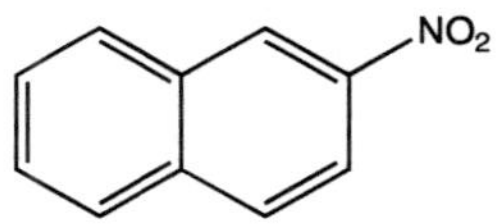	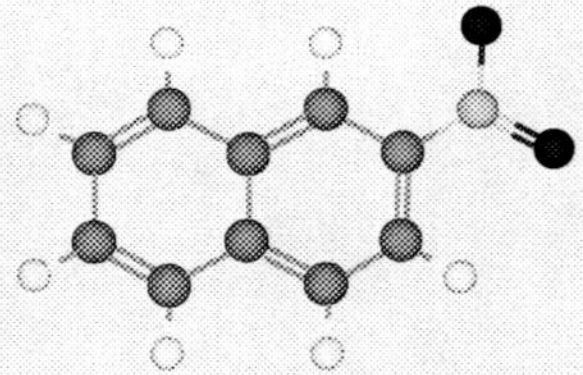	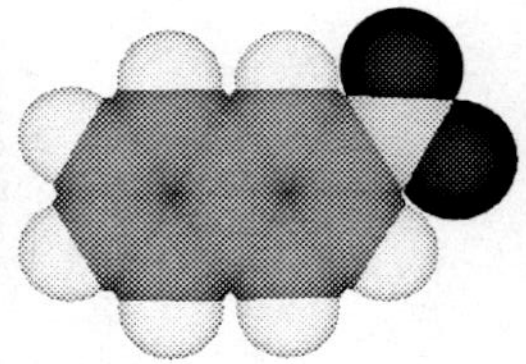
1-Hydroxynaphthalene	$C_{10}H_8O$	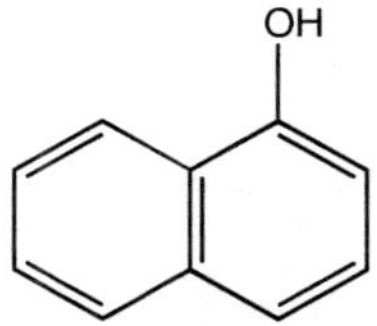	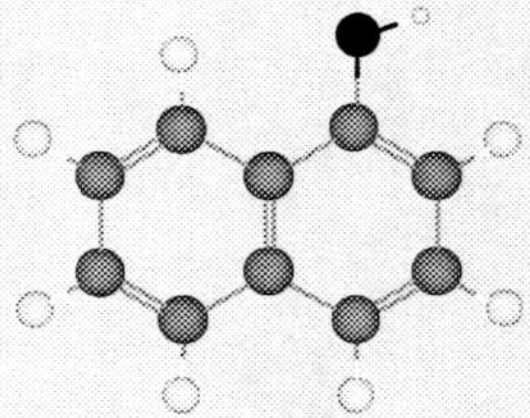	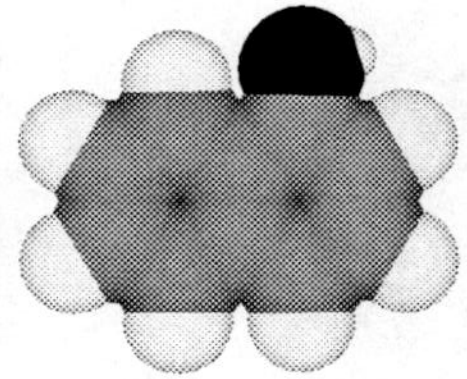
2-Hydroxynaphthalene	$C_{10}H_8O$	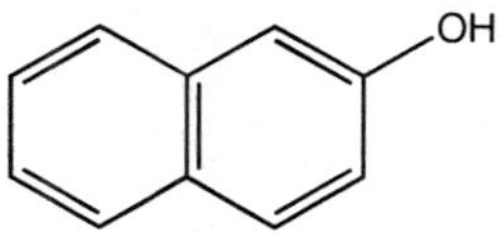	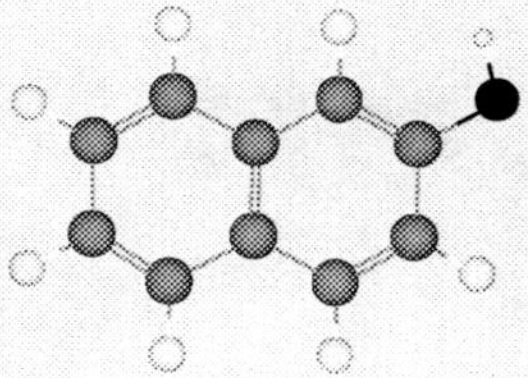	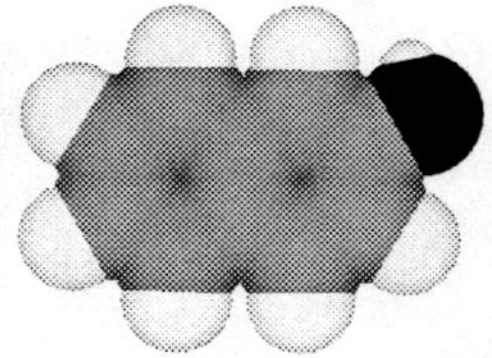

2-Formylcinnamaldehyde $C_{10}H_8O_2$

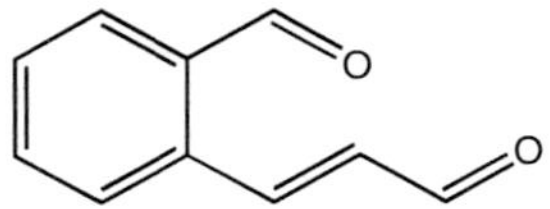

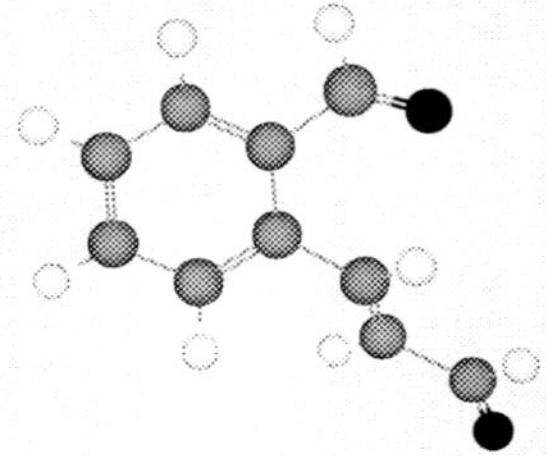

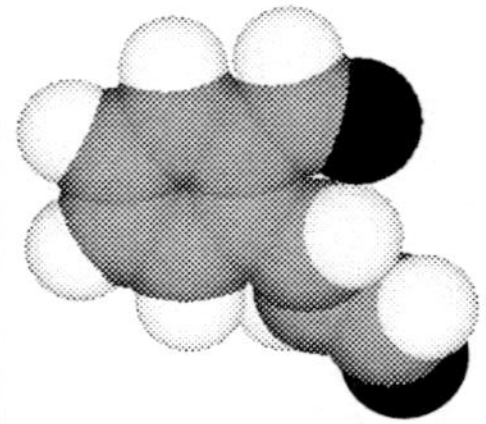

2-Methyl-1-nitronaphthalene $C_{11}H_9O_2N$

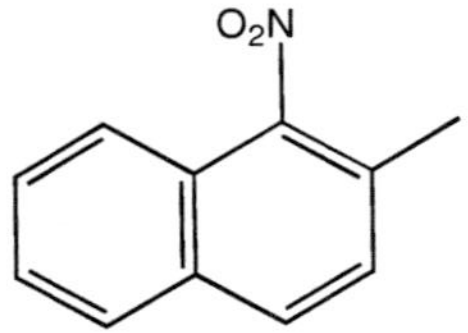

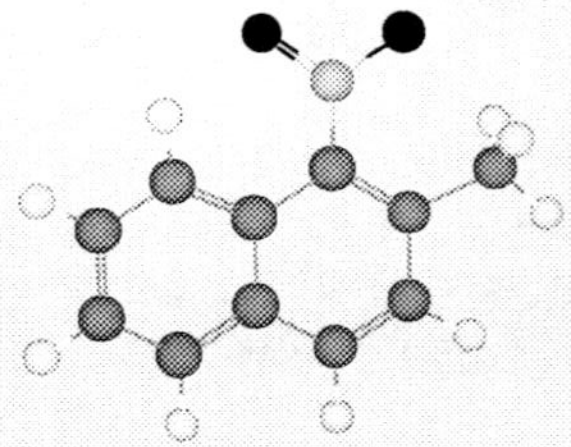

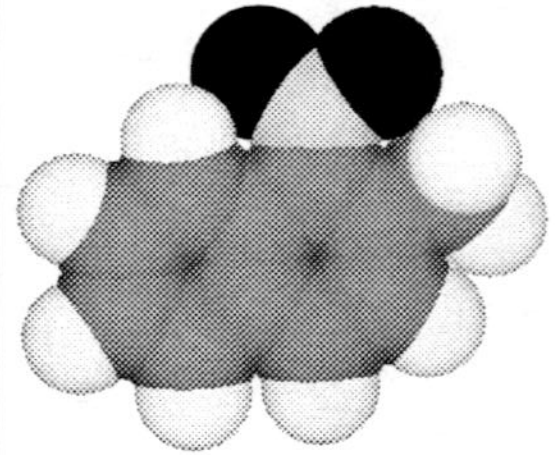

4-Nitroacenaphthylene $C_{12}H_7O_2N$

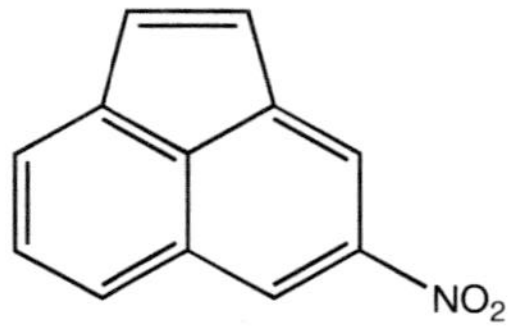

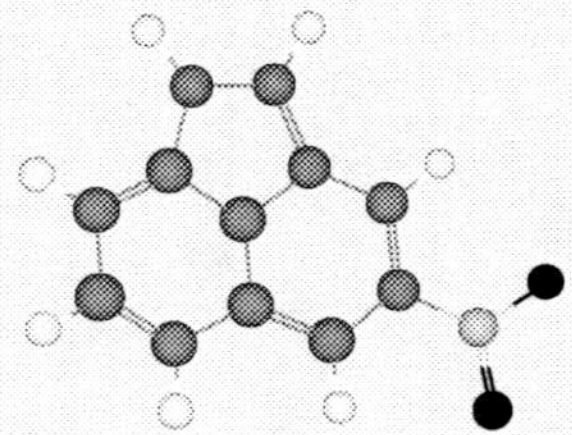

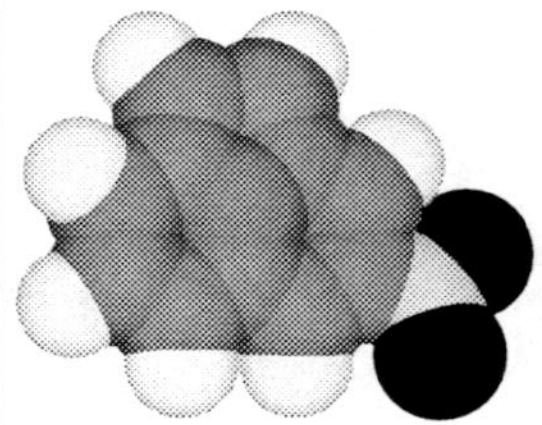

5-Nitroacenaphthene $C_{12}H_9O_2N$

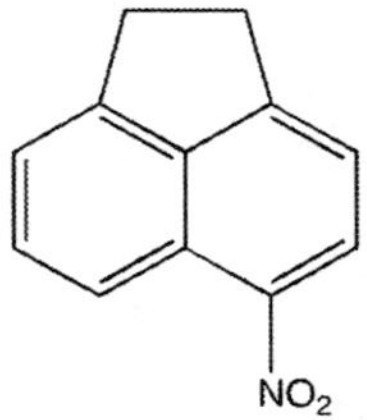

9-Fluorenone $C_{13}H_8O$

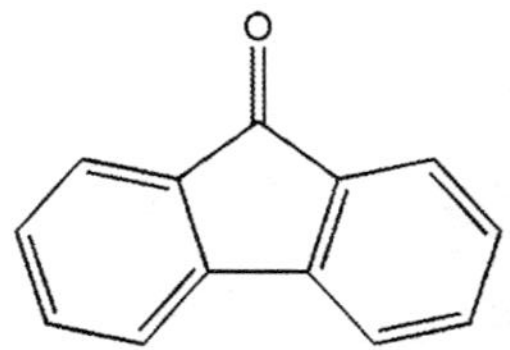

3-Nitrofluorene $C_{13}H_9O_2N$

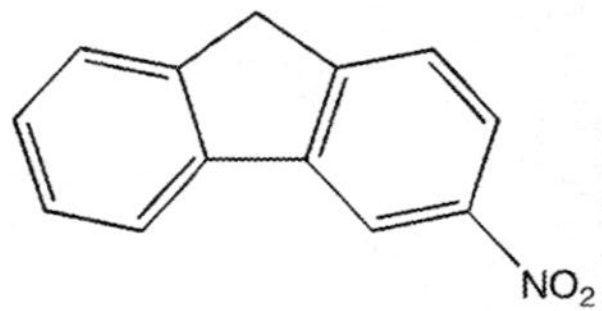

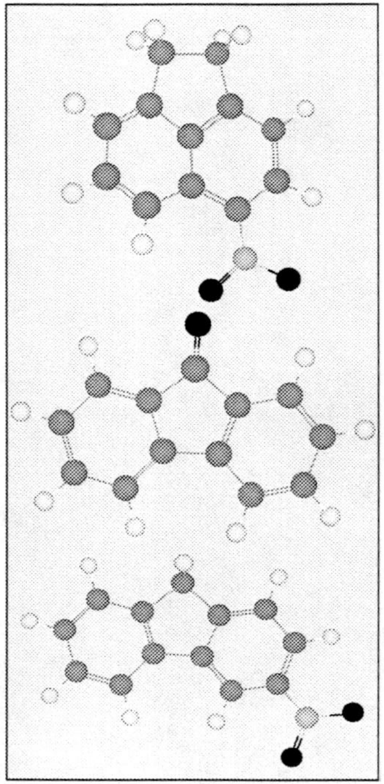

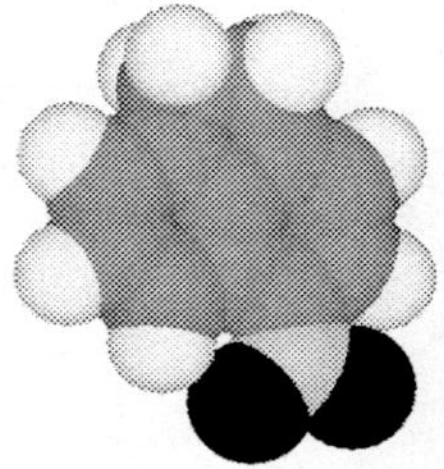

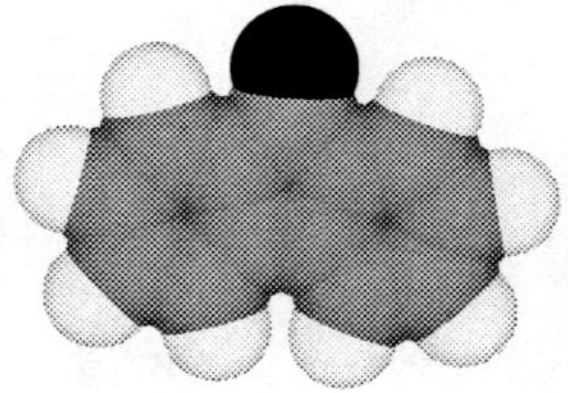

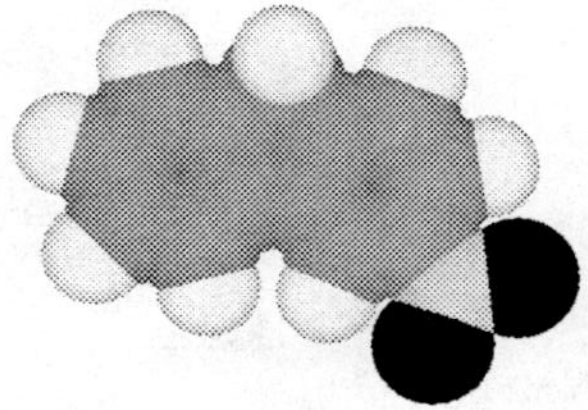

2-Nitro-6H-dibenzo[b,d]pyran-6-one $C_{13}H_7O_4N$

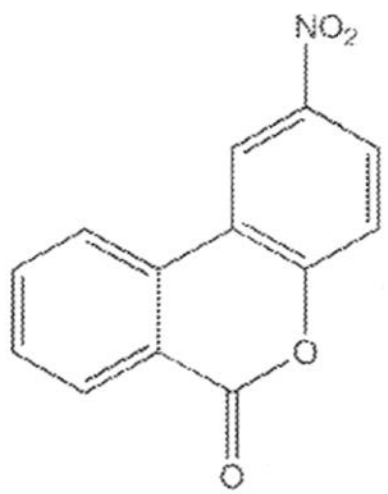

4-Nitro-6H-dibenzo[b,d]pyran-6-one $C_{13}H_7O_4N$

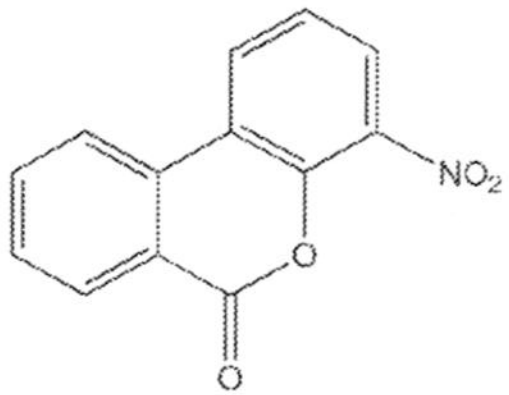

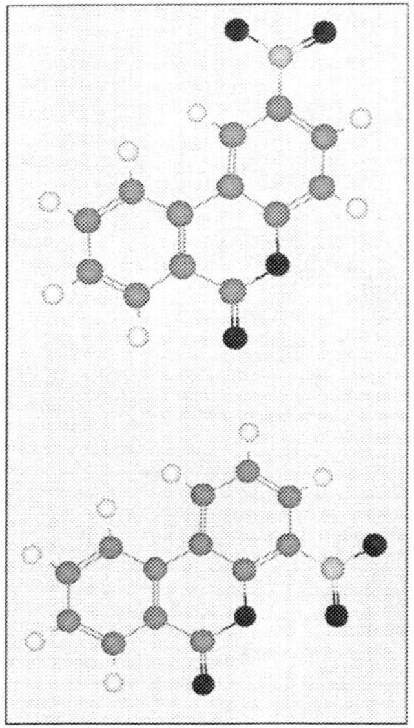

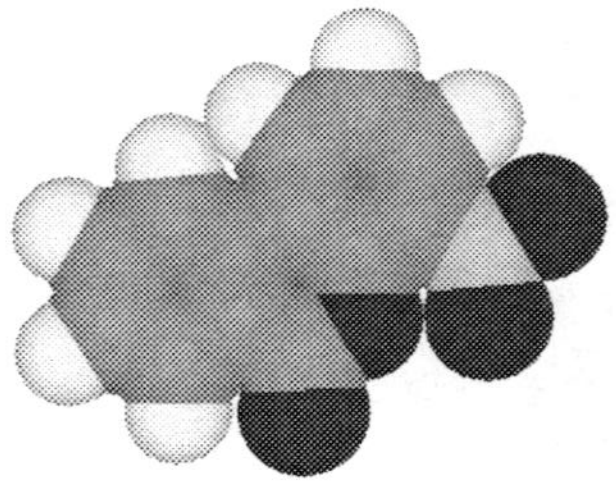

Anthraquinone $C_{14}H_8O_2$

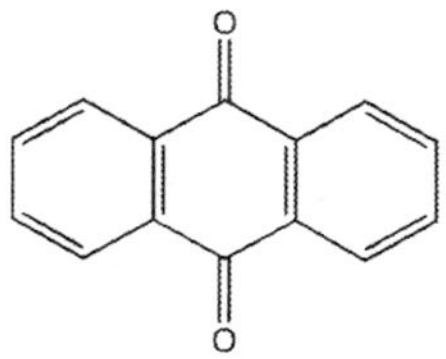

4-Nitrophenanthrene $C_{14}H_9O_2N$

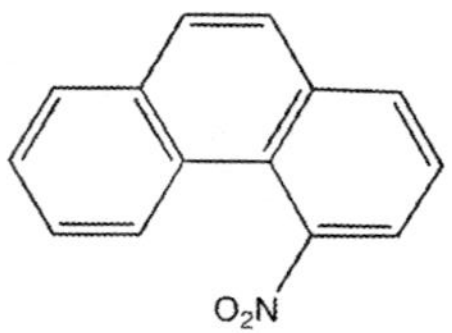

3-Hydroxyphenanthrene $C_{14}H_{10}O$

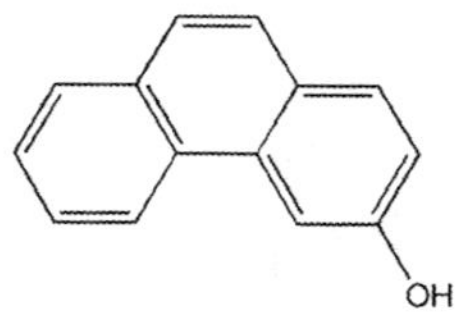

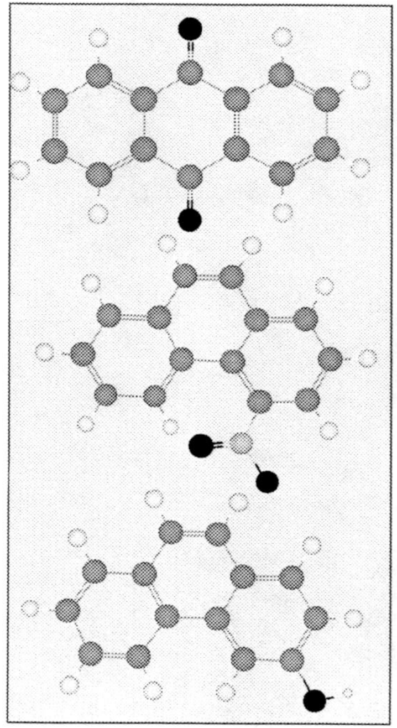

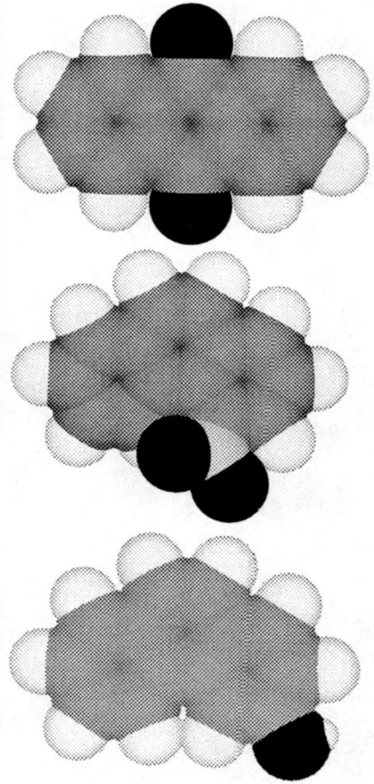

9,10-Dihydroxyanthracene $C_{14}H_{10}O_2$

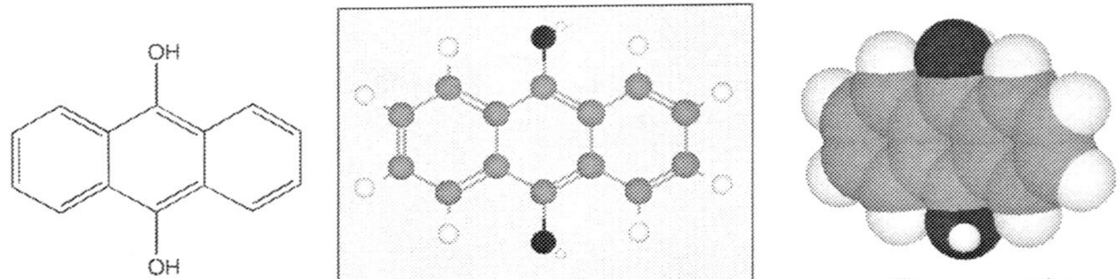

Anthracene-9,10-endoperoxide $C_{14}H_{10}O_2$

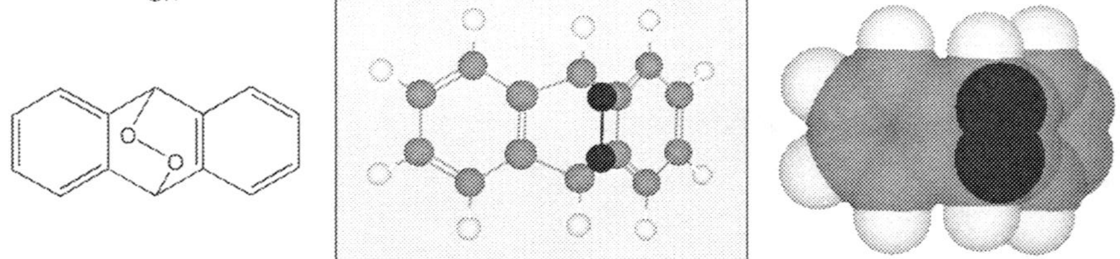

Pyrene-1-6-quinone $C_{16}H_8O_2$

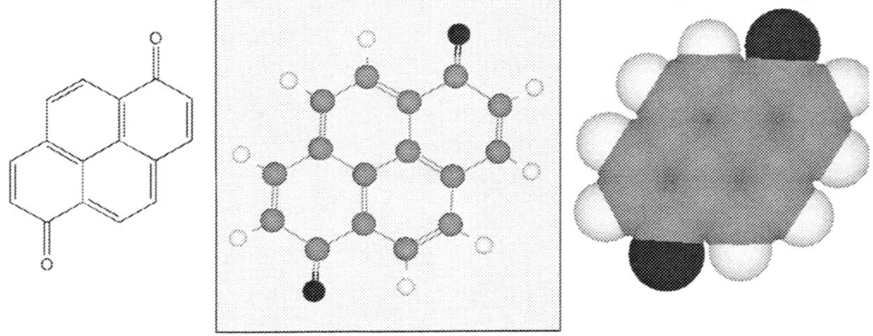

2-Nitrofluoranthene $C_{16}H_9O_2N$

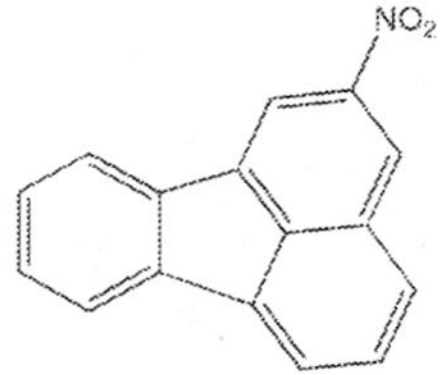

3-Nitrofluoranthene $C_{16}H_9O_2N$

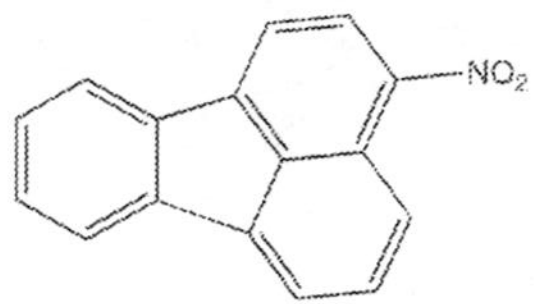

2-Nitropyrene $C_{16}H_9O_2N$

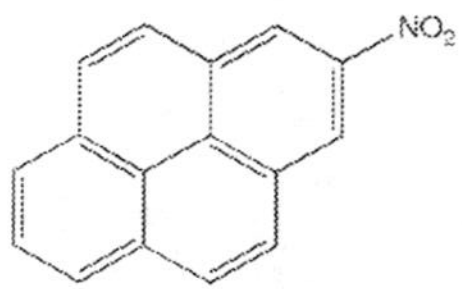

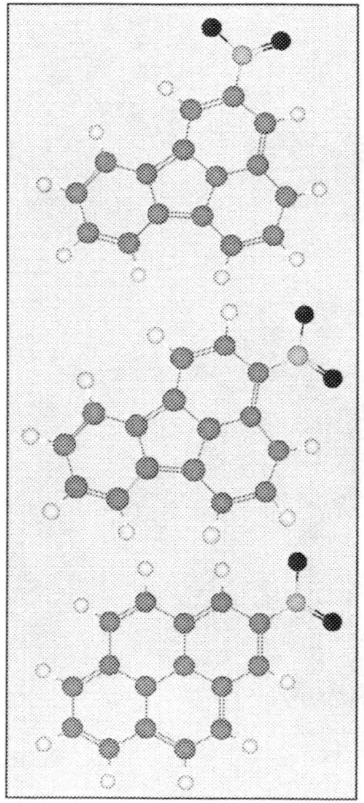

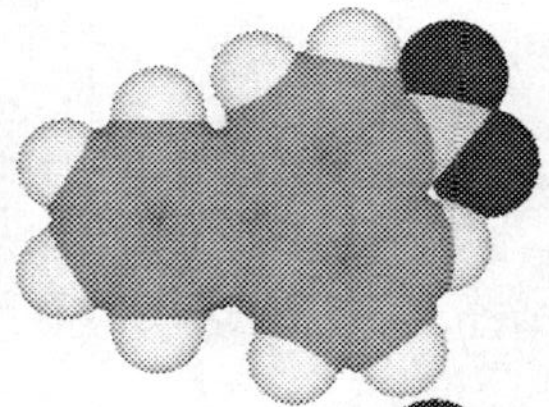

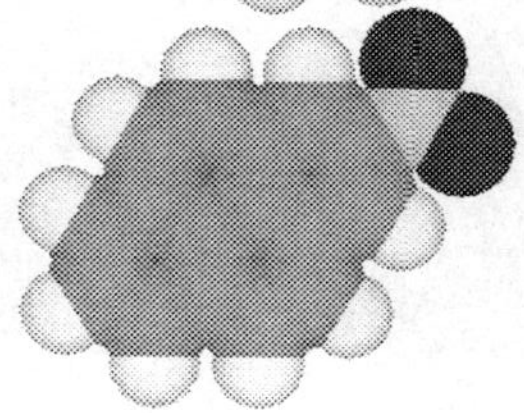

3-Hydroxyfluoranthene $C_{16}H_{10}O$

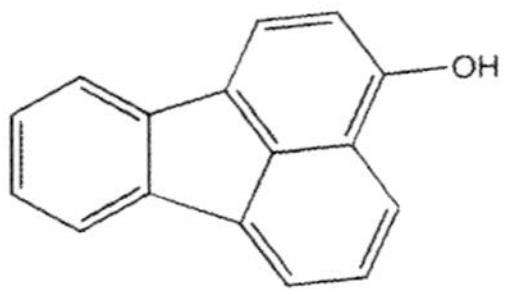

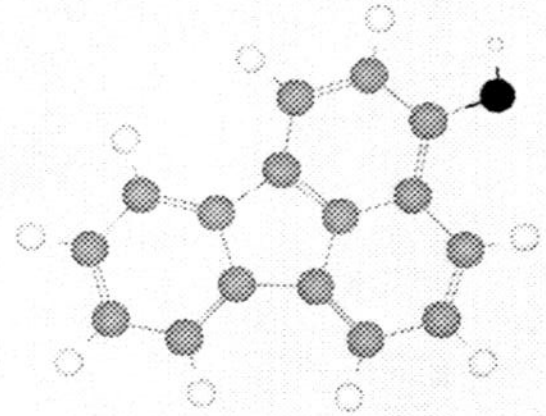

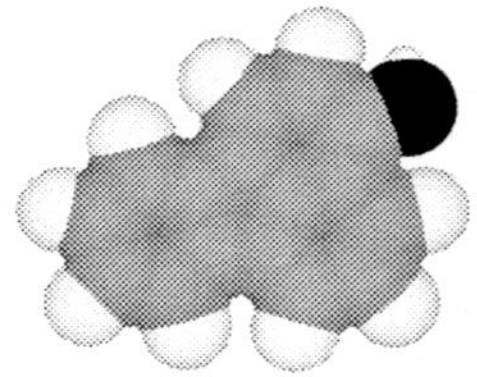

Benz[*a*]anthracene-7,12-quinone $C_{18}H_{10}O_2$

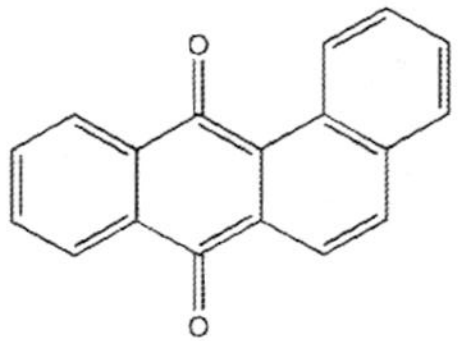

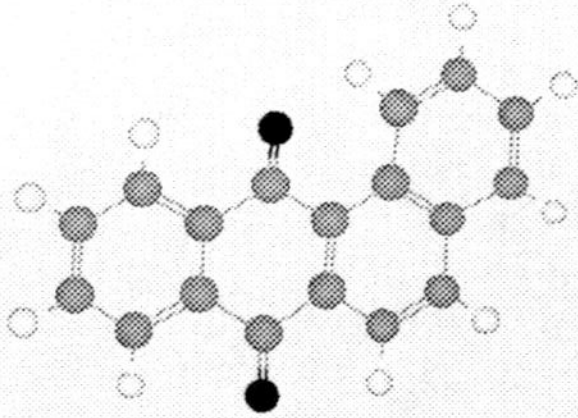

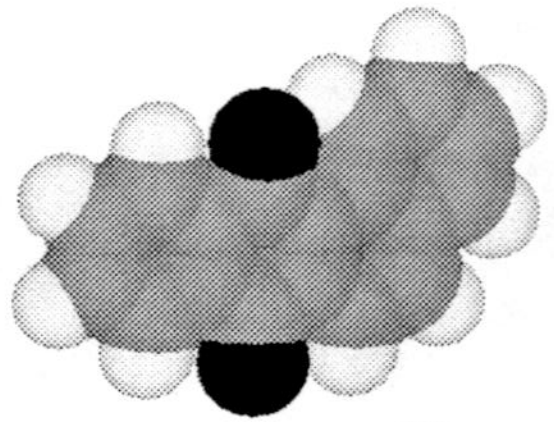

Benzo[*a*]pyrene-4,5-quinone $C_{18}H_{10}O_2$

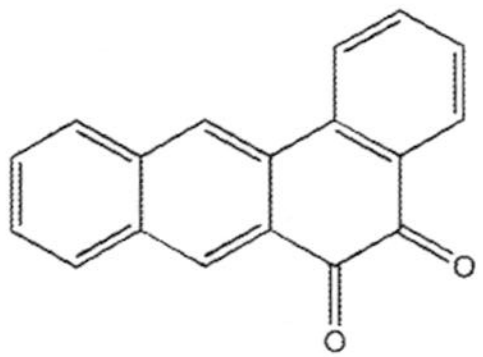

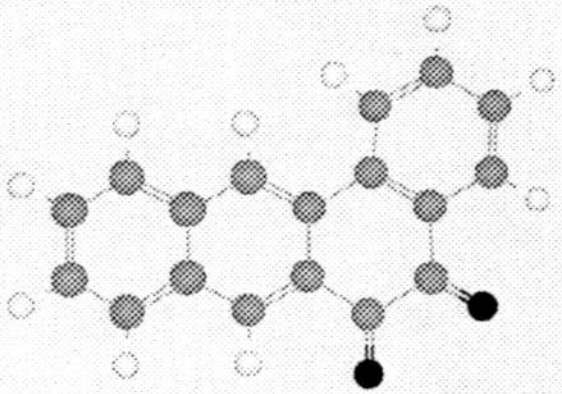

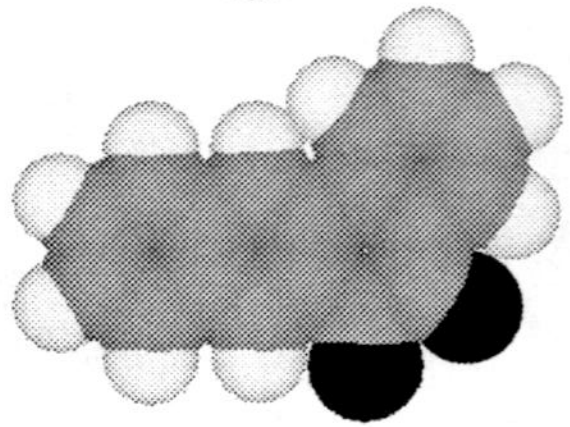

Naphthacene-5,12-quinone $C_{18}H_{10}O_2$

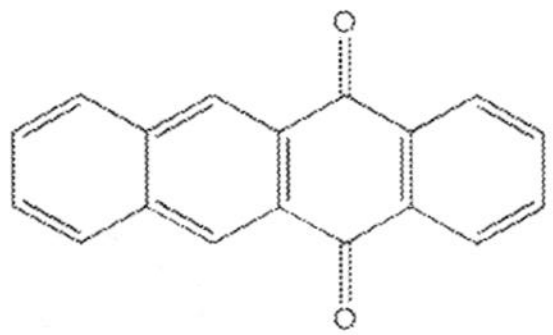

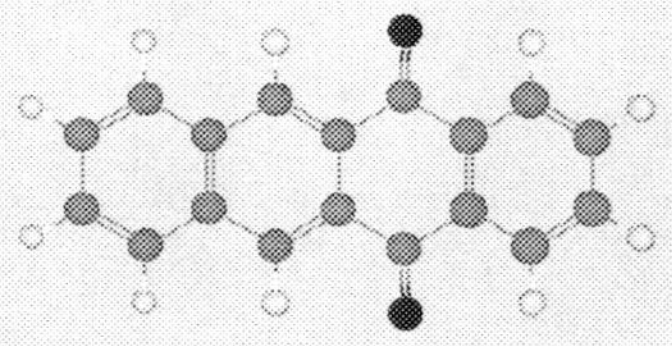

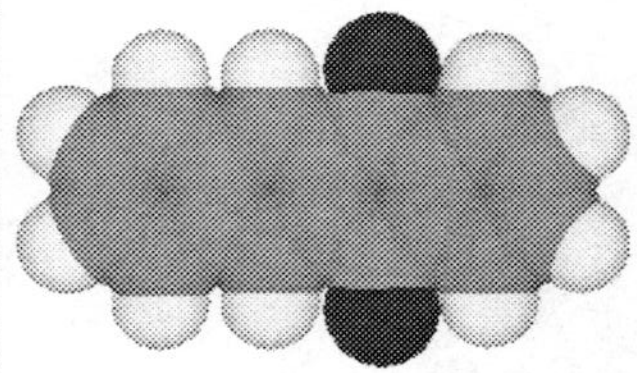

Benzo[*a*]pyrene-1,6-quinone $C_{20}H_{10}O_2$

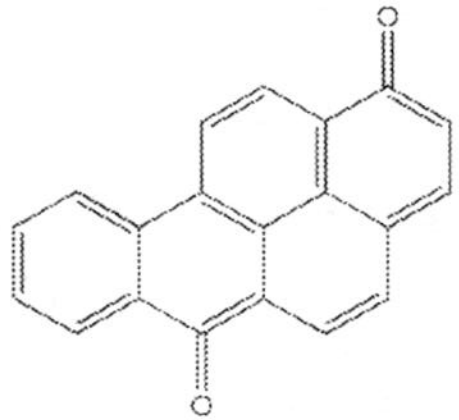

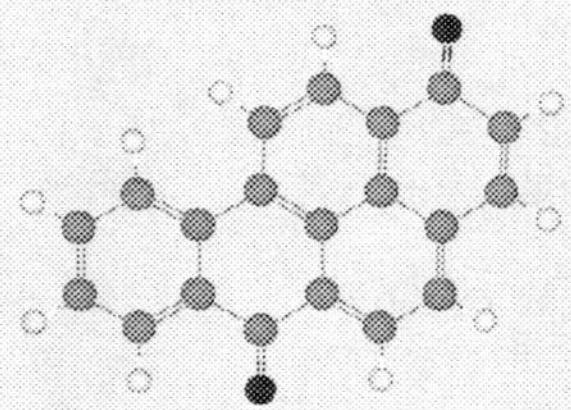

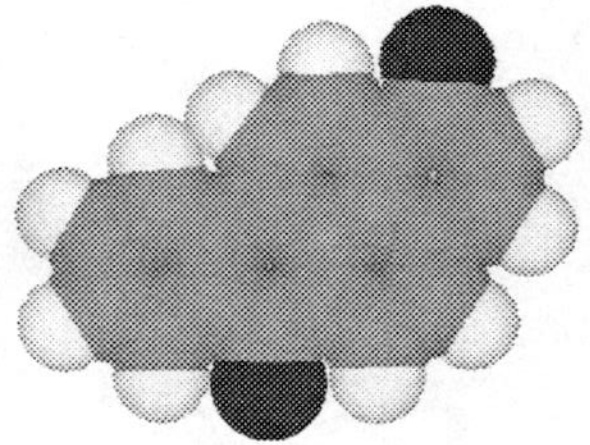

Benzo[*a*]pyrene-3,6-quinone $C_{20}H_{10}O_2$

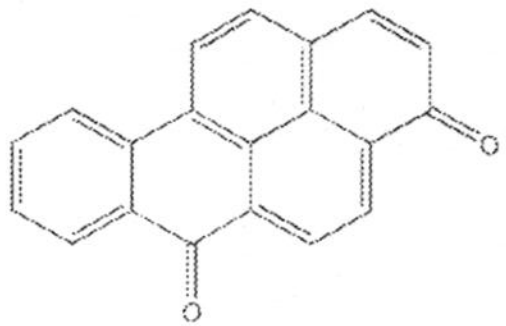

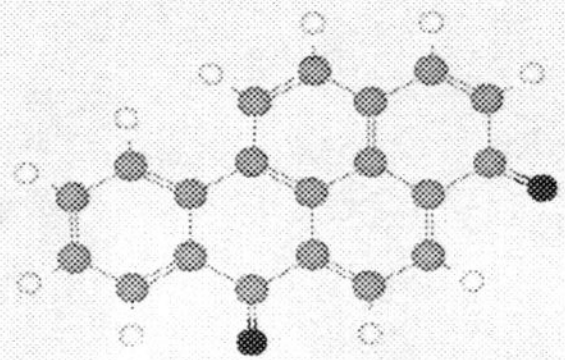

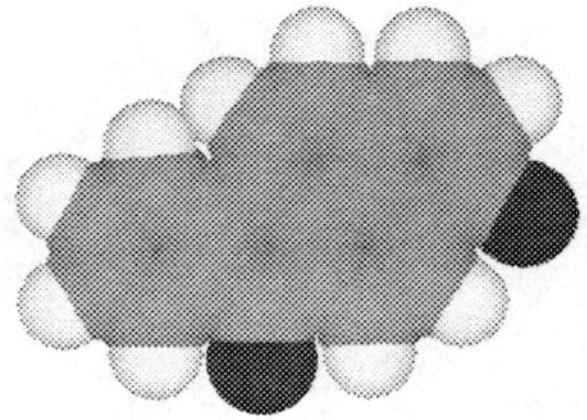

Compound	Formula	Structure	Ball-and-stick	Space-filling
Benzo[*a*]pyrene-6,12-quinone	$C_{20}H_{10}O_2$	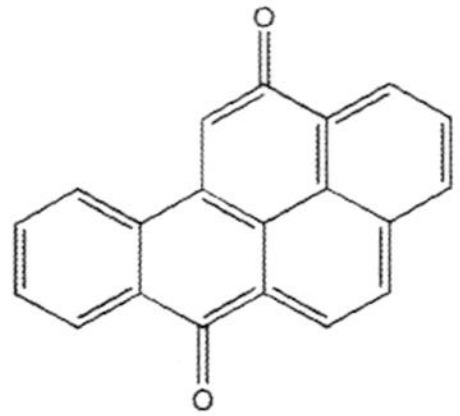	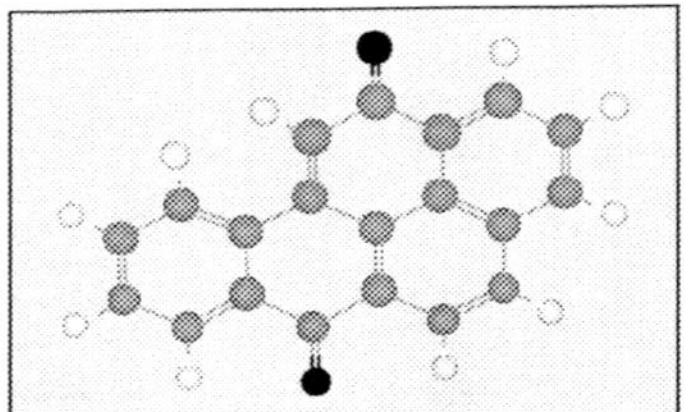	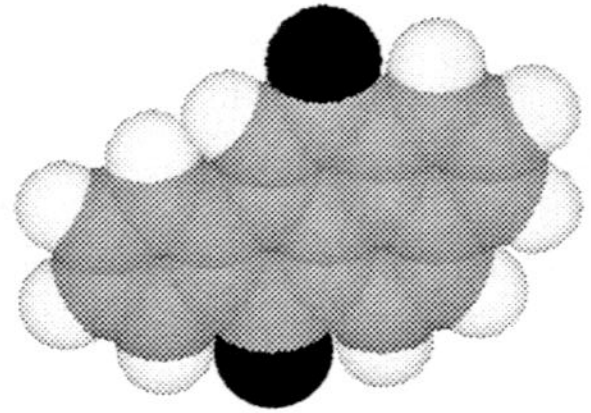
3-Nitroperylene	$C_{20}H_{11}O_2N$	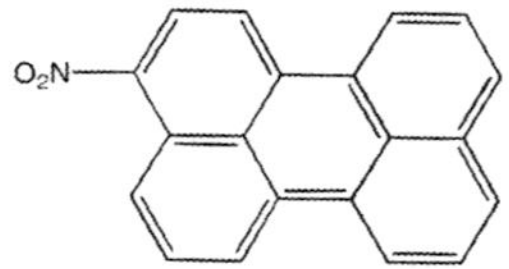	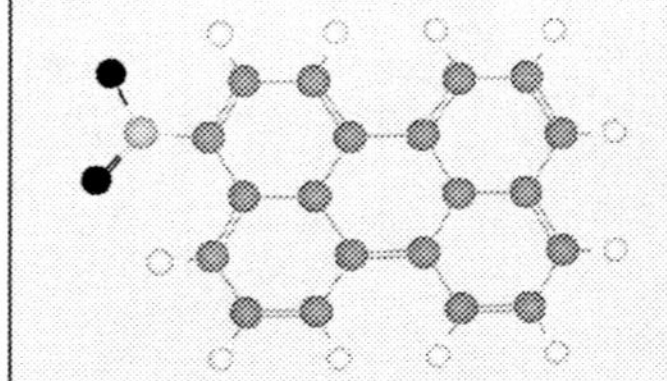	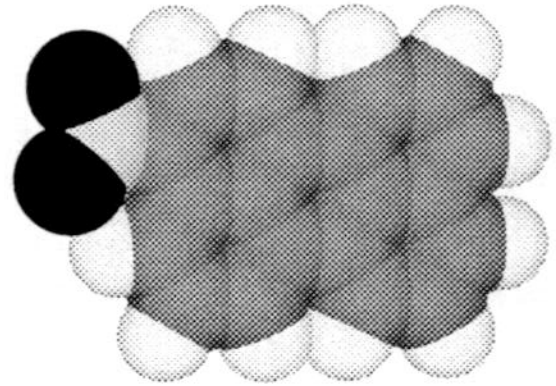
Pentacene-6,13-quinone	$C_{22}H_{12}O_2$	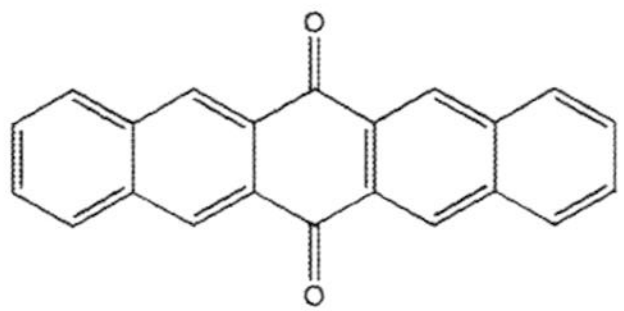	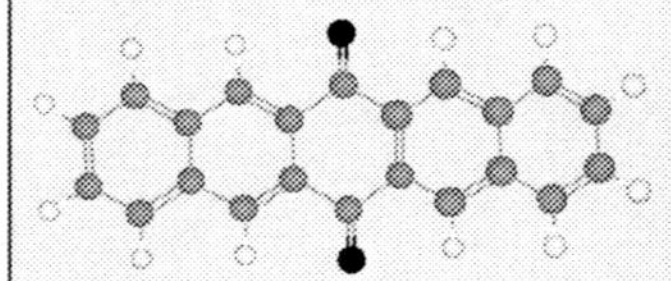	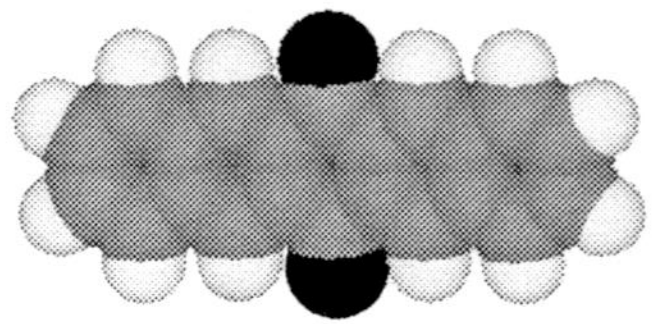

B. Aromatic Ring-Opening Products

Name of Compound	Molecular Formula	C-Skeletal Structure	3D-Structural Formulas	
Formaldehyde	CH_2O		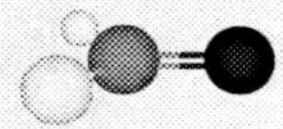	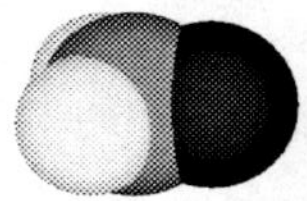
Glyoxal	$C_2H_2O_2$	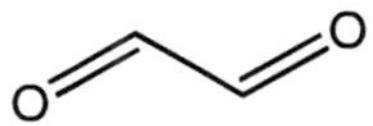	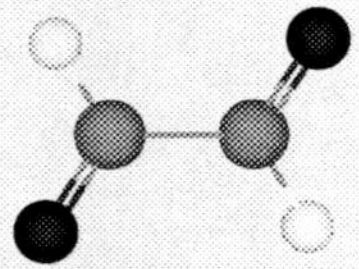	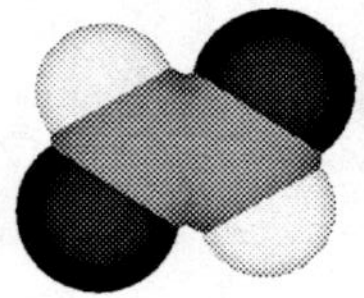
Acetaldehyde	C_2H_4O		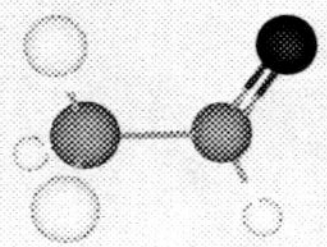	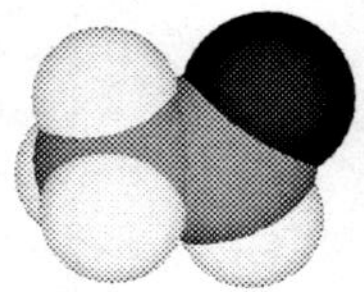
Hydroxyacetaldehyde (glycolaldehyde)	$C_2H_4O_2$	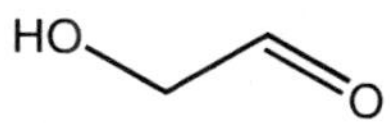	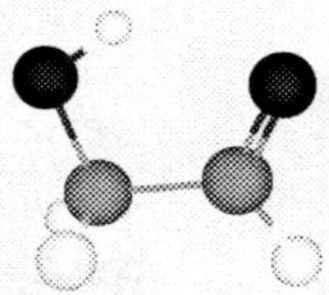	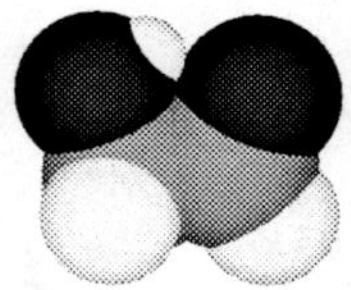

Acrolein (propenal) — C_3H_4O

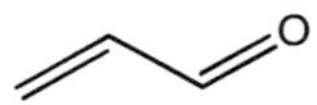

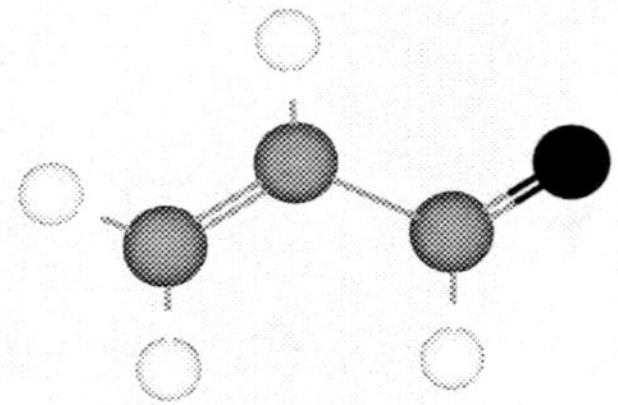

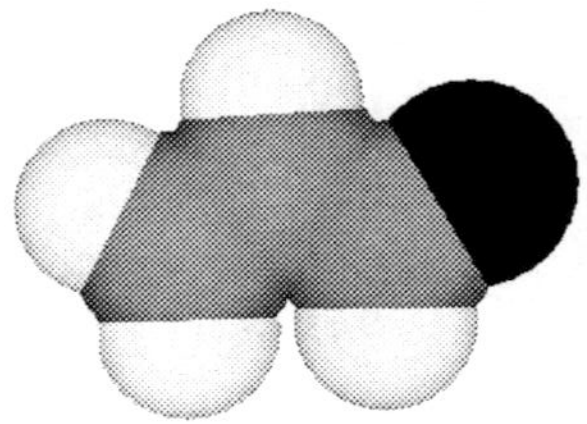

2,3-Epoxypropanal (glycidaldehyde) — $C_3H_4O_2$

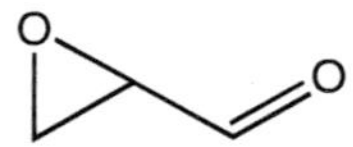

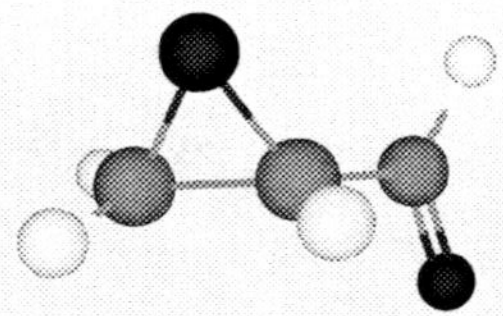

Malonaldehyde (propanedial) — $C_3H_4O_2$

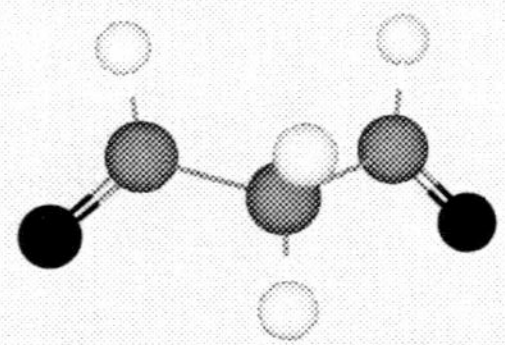

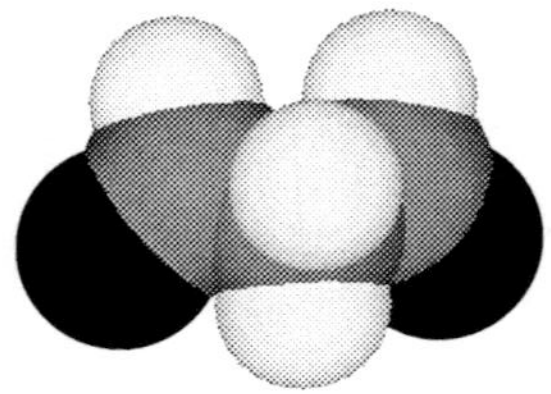

Methylglyoxal $C_3H_4O_2$

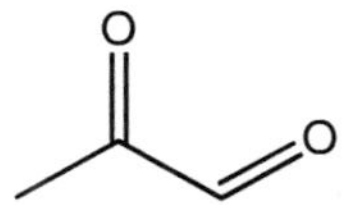

Pyruvic acid $C_3H_4O_3$

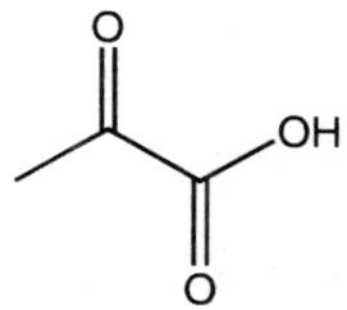

Acetone C_3H_6O

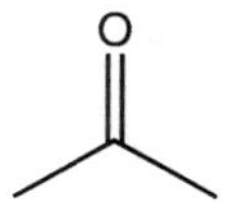

Hydroxyacetone $C_3H_6O_2$

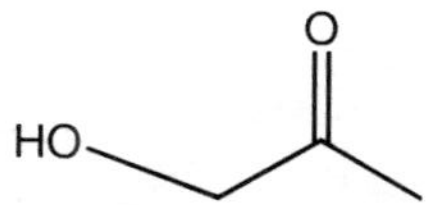

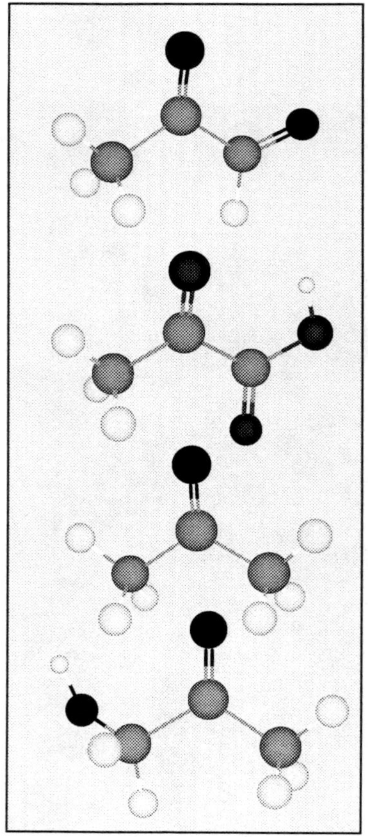

Furan C_4H_4O  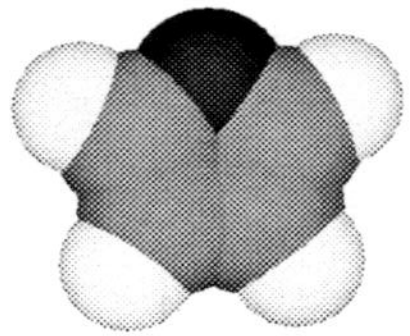

Maleic anhydride $C_4H_2O_3$  

3-*H*-Furan-2-one $C_4H_4O_2$ 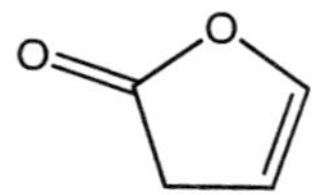 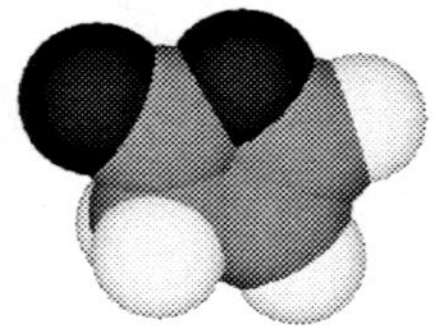

cis-Butenedial $C_4H_4O_2$  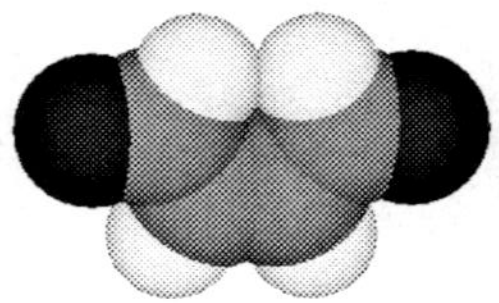

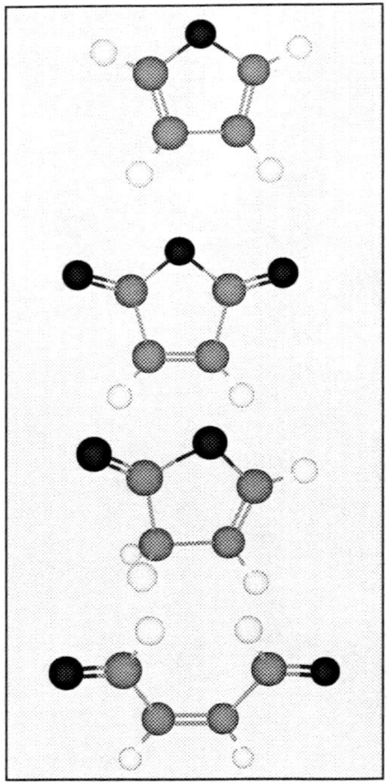

trans-Butenedial	$C_4H_4O_2$	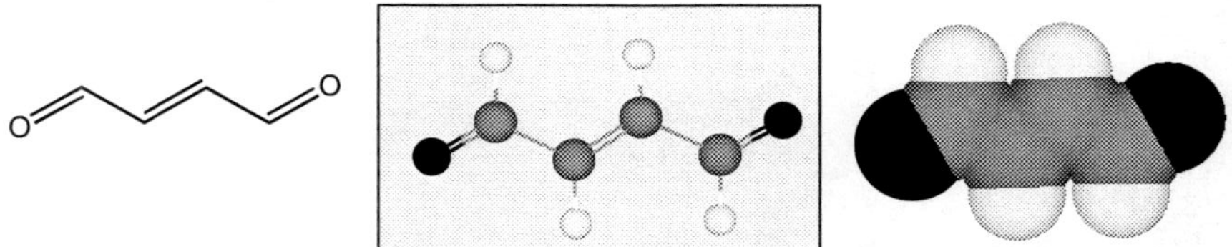
Methacrolein (2-methylpropenal)	C_4H_6O	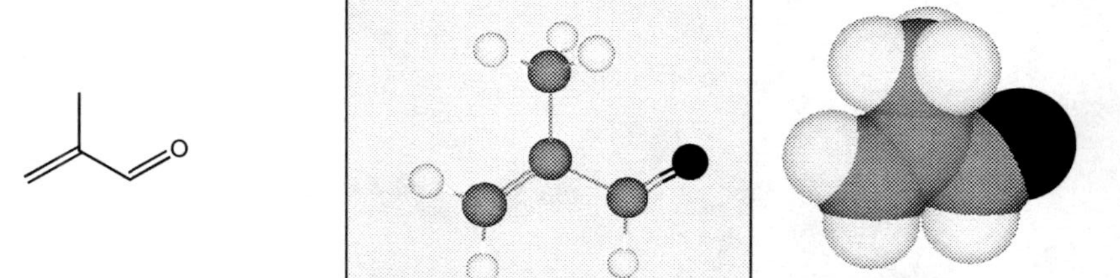
Methyl vinyl ketone (3-buten-2-one)	C_4H_6O	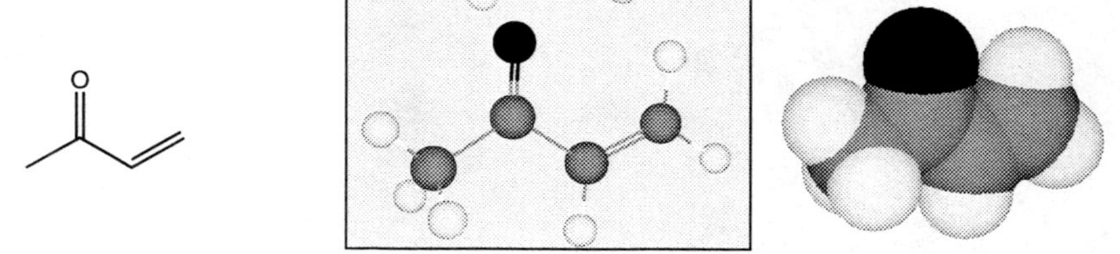

Biacetyl
(2,3-butandione)

$C_4H_6O_2$

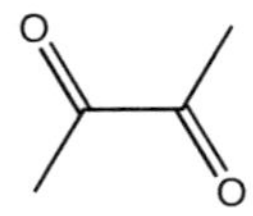

Furan-2-aldehyde
(furfural; 2-furancarboxaldehyde)

$C_5H_4O_2$

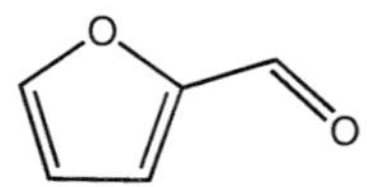

Furan-3-aldehyde
(3-furancarboxaldehyde)

$C_5H_4O_2$

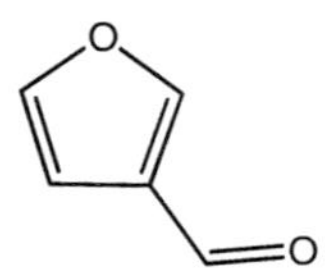

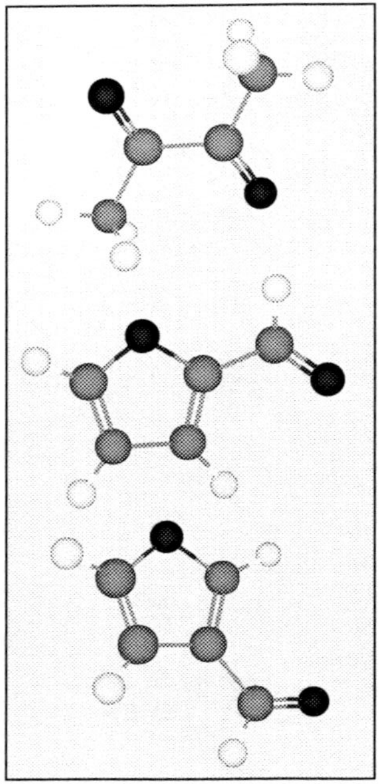

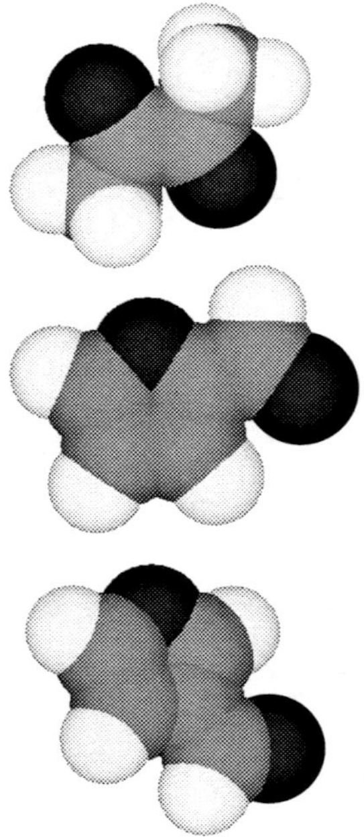

2-Methylfuran C_5H_6O

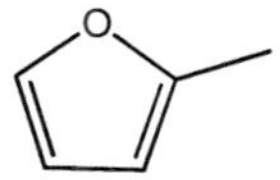

3-Methylfuran C_5H_6O

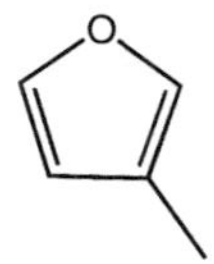

4-Oxo-2-pentenal $C_5H_6O_2$

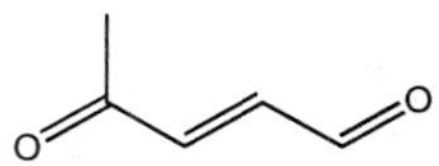

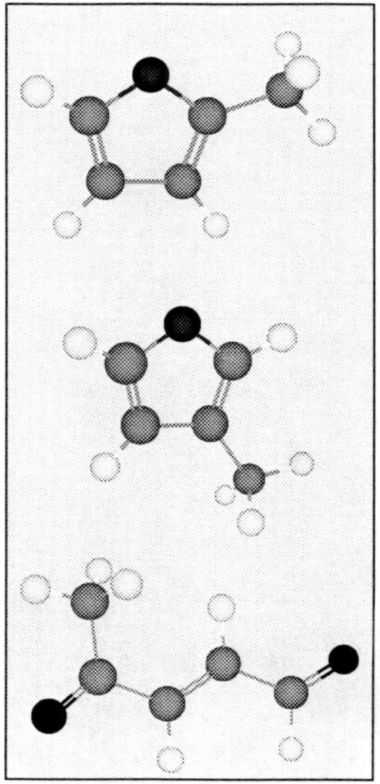

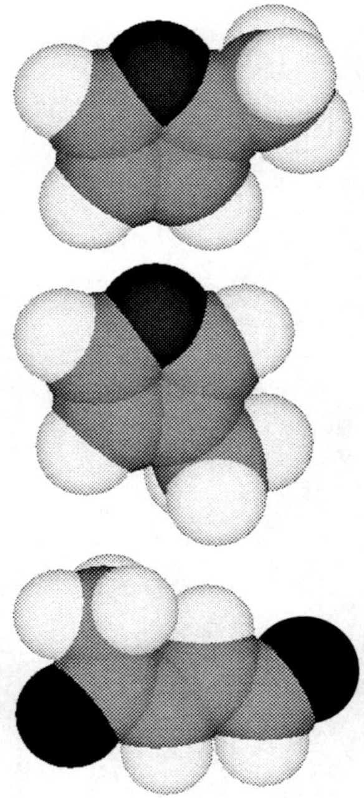

5-Methyl-3*H*-furan-2-one	$C_5H_6O_2$	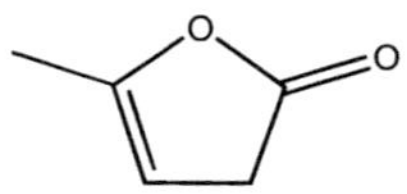	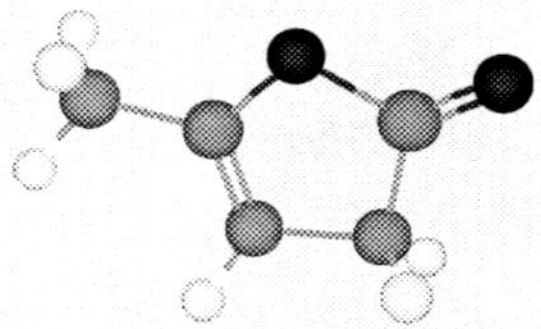	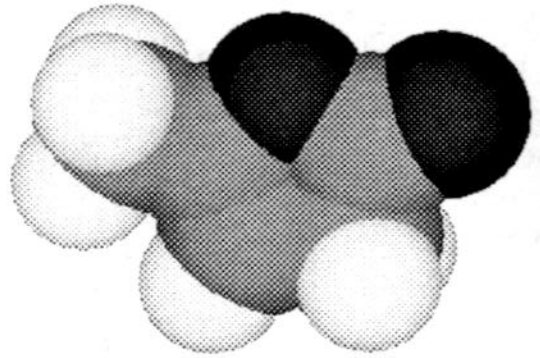
Benzene oxide	C_6H_6O	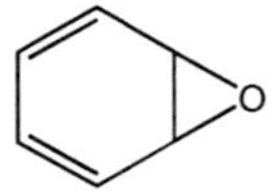	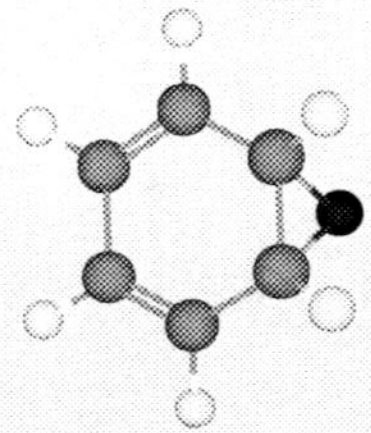	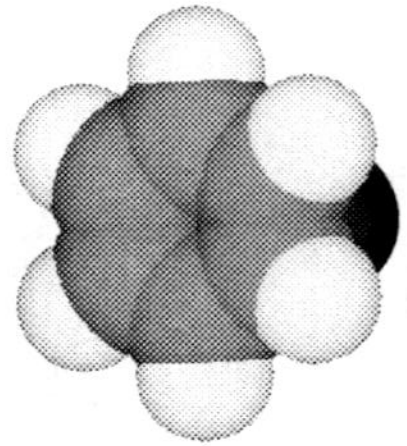
Oxepin	C_6H_6O	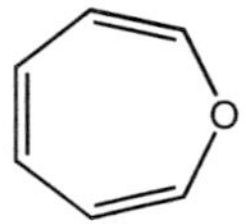	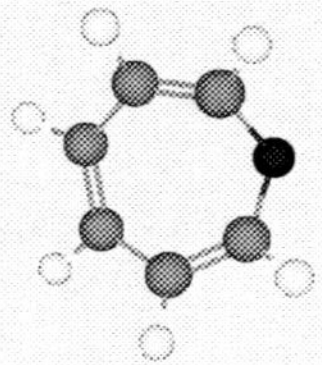	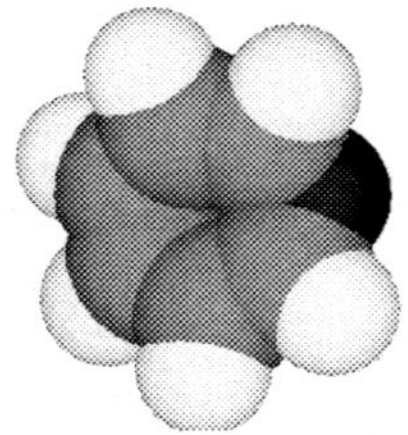

2-Butenal-4-yl-ketene $C_6H_6O_2$

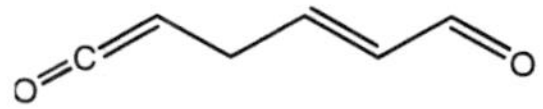

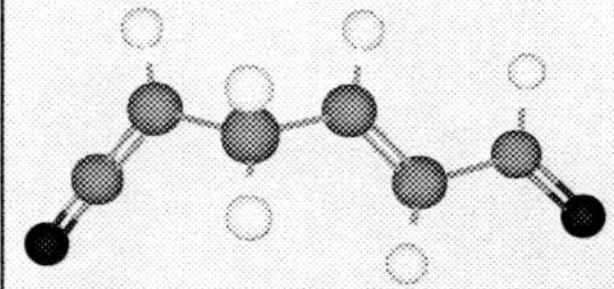

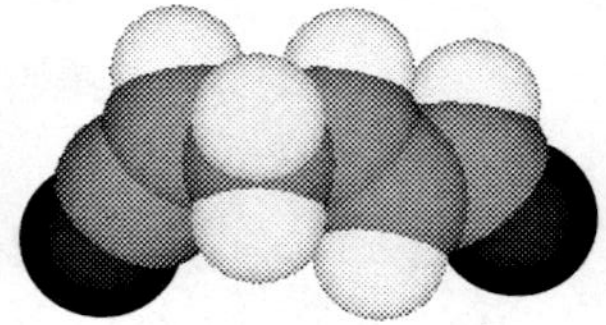

Z-3,4-Diformylcyclobutene $C_6H_6O_2$

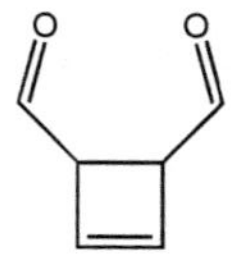

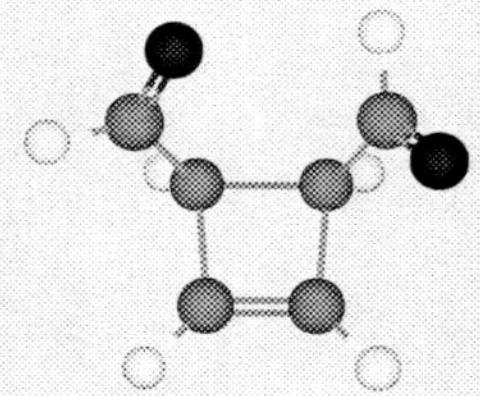

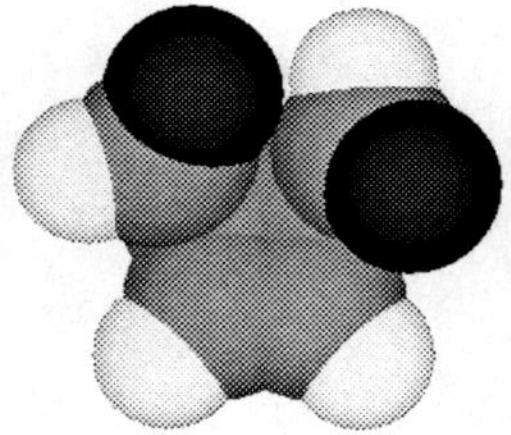

E,*E*-2,4-Hexadienedial (*trans*,*trans*-2,4-Hexadienedial) $C_6H_6O_2$

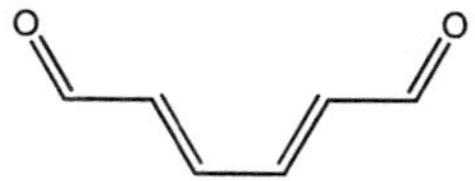

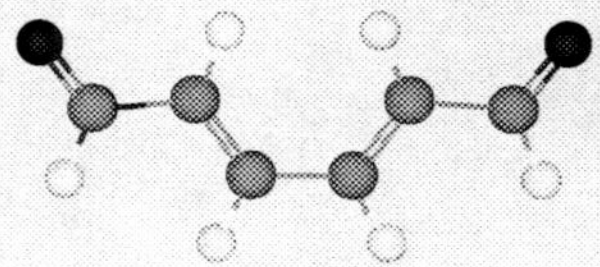

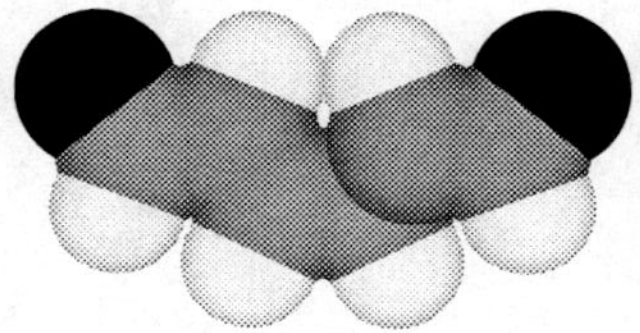

Name	Formula	Structure
E,Z,-2,4-Hexadienedial (*cis,trans*-2,4-Hexadienedial)	$C_6H_6O_2$	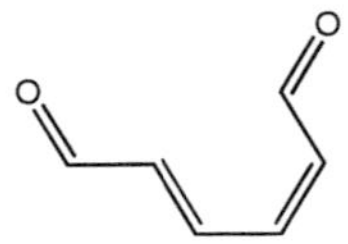
2,3-Dimethylfuran	C_6H_8O	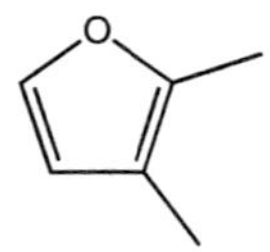
2,4-Dimethylfuran	C_6H_8O	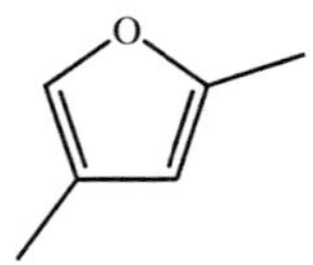

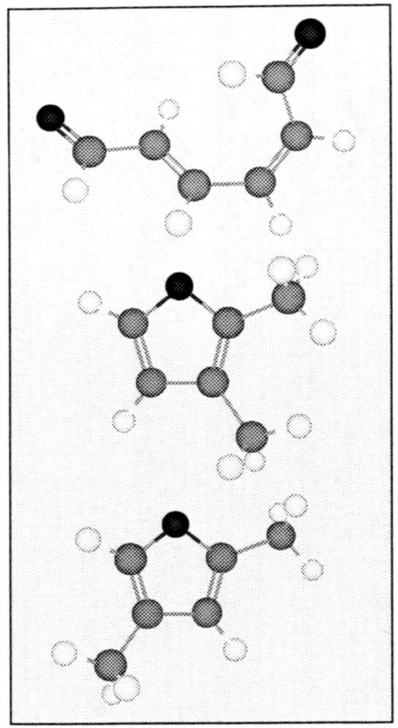

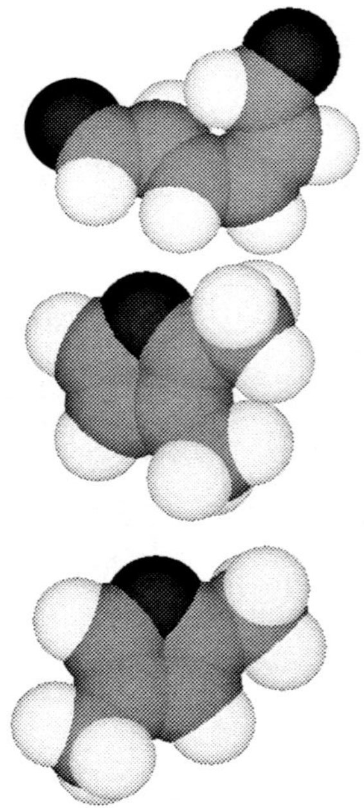

2,5-Dimethylfuran C_6H_8O

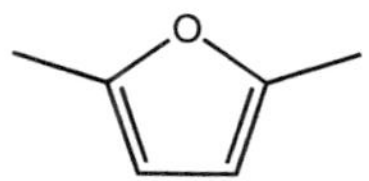

2-Ethylfuran C_6H_8O

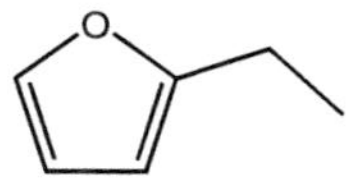

cis-3-Hexene-2,5-dione $C_6H_8O_2$

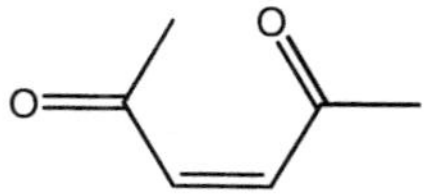

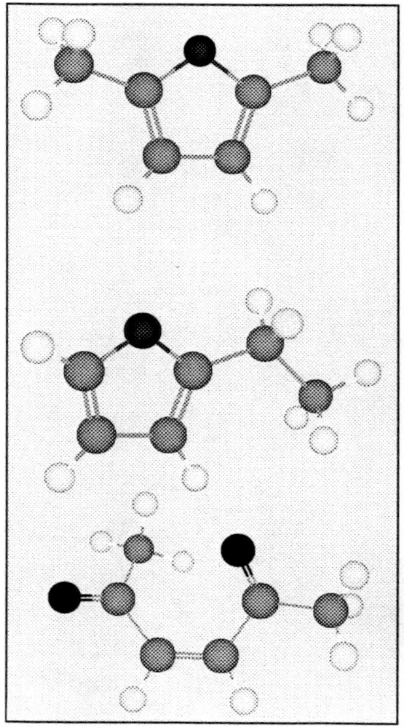

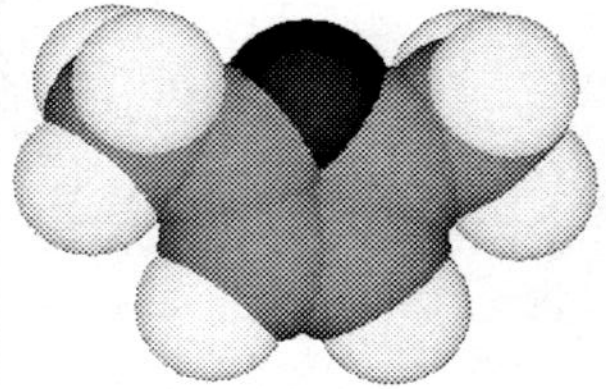

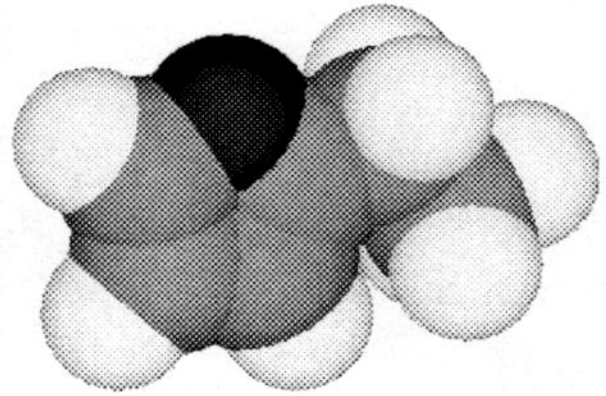

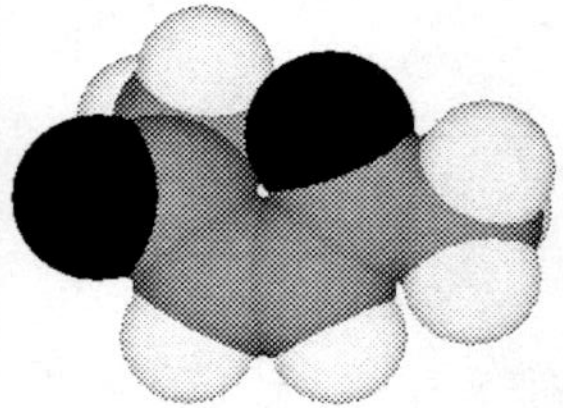

trans-3-Hexene-2,5-dione	$C_6H_8O_2$	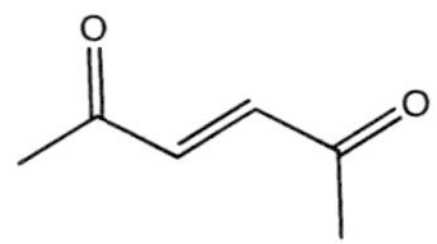	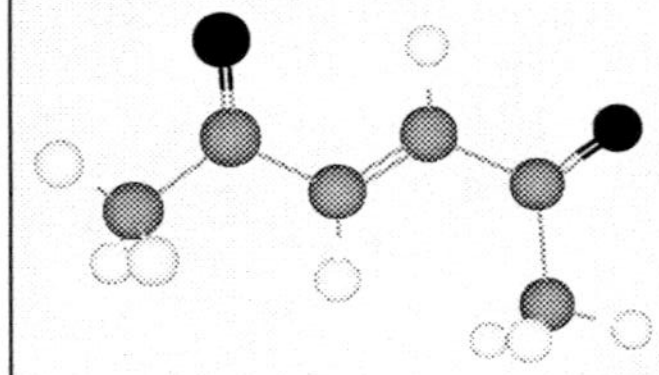	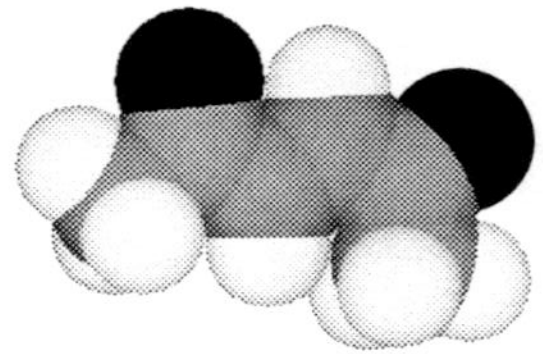
E,'*E*-2-Methyl-2,4-hexadienedial	$C_7H_8O_2$	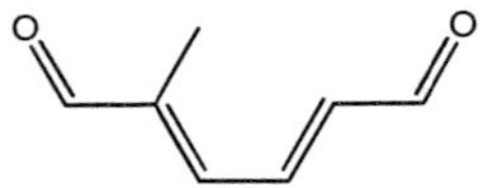	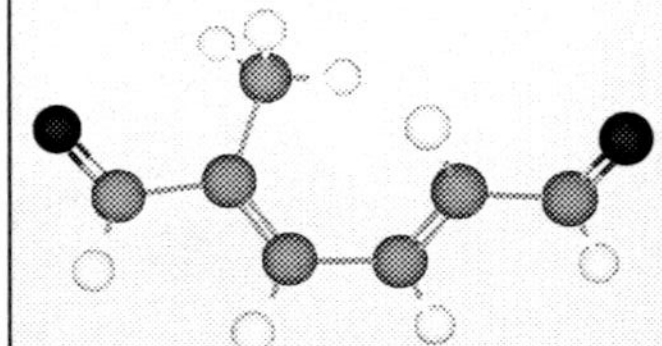	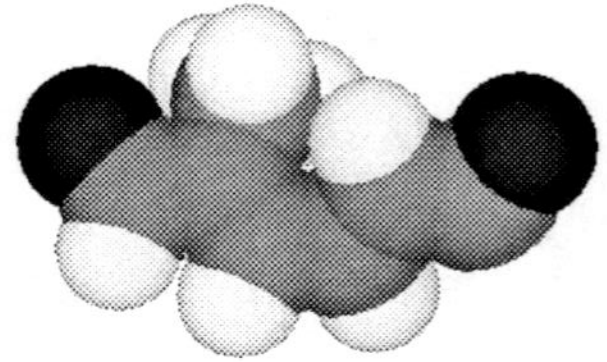
Tetramethylfuran	$C_8H_{12}O$	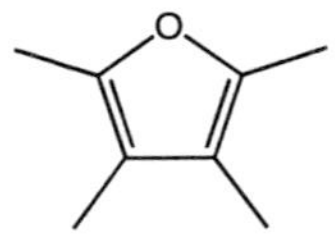	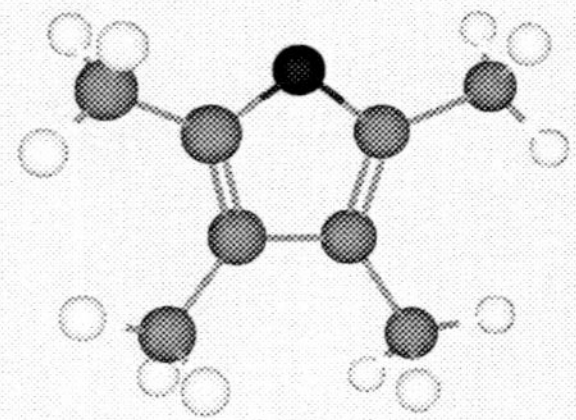	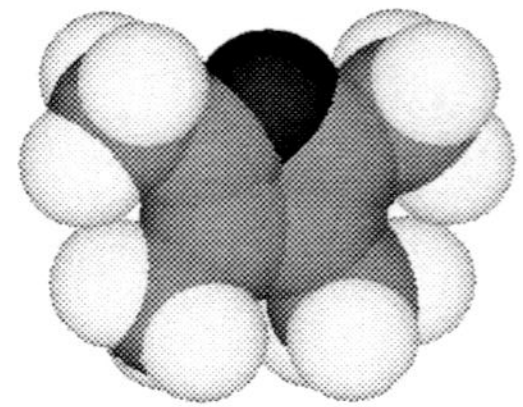

Author Index

Subject Index